StarGuides 2001

StarGuides 2001

A World-Wide Directory of Organizations in Astronomy, Related Space Sciences, and Other Related Fields

Compiled by

André Heck
Strasbourg Astronomical Observatory, France

SPRINGER-SCIENCE+BUSINESS MEDIA, B.V.

A C.I.P. Catalogue record for this book is available from the Library of Congress.

ISBN 978-94-010-5873-5   ISBN 978-94-011-4349-3 (eBook)
DOI 10.1007/978-94-011-4349-3

While every effort has been made to ensure the reliability of the information presented in this publication, neither Springer-Science+Business Media, B.V. nor the compiler guarantees the accuracy of the data contained herein. Neither Springer-Science+Business Media, B.V. nor the compiler accept payments for listing; and inclusion in the publication of any organization, agency, institution, publication, service, or individual does not imply endorsement by the compiler or publishers.
Neither publisher nor the compiler can accept responsibility, express or implied, with regard to the accuracy of the information contained in this publication and cannot accept any legal responsibility, or liability, for any errors or omissions that may be made, nor can they be held liable for any damages resulting in use of the material contained herein.

Note: This publication is seeded with detectors for copyright protection purposes.

Printed on acid-free paper

Table of Contents

Foreword

StarGuides 2001 is the enlarged and updated version of a long line of directories published since 1978 (see *e.g.* Heck 1995 & 1997) and gathering together all practical data available on associations, societies, scientific committees, agencies, companies, institutions, universities, etc., and more generally organizations, involved in astronomy and space sciences. Many other types of entries have also been included such as academies, advisory and expert committees, bibliographic services, consultants, data and documentation centres, dealers, distributors, funding agencies and organizations, journals, manufacturers, meteorological services, museums, norms and standards offices, planetariums, public observatories, publishers, research institutions in related fields, software producers and distributors, etc.

Besides astronomy and space sciences, related fields such as aeronautics, aeronomy, astronautics, atmospheric sciences, chemistry, communications, computer sciences, data processing, education, electronics, energetics, engineering, environment, geodesy, geophysics, information retrieval, management, mathematics, meteorology, optics, physics, remote sensing, and so on, are also covered when justified.

The basic philosophy of this directory (and of other products in the *Star*s Family*) is to provide practical data which one seeks to always have at ones disposal. Over the years they have proved to not only be valuable auxiliaries for improving national and international relationships, but also to be efficient tools for helping laypersons and public bodies to contact organizations easily.

These directories have taken advantage of the experience gained with each successive edition, especially in the development of techniques for collecting, verifying and treating the data. To compile a directory of real value is a quite different venture to just reproducing and distributing, with comments of greater or lesser interest, data collected indiscriminately from all available sources (including the World-Wide Web which already contains too many obsolete pointers and too much outdated information). If professional file construction techniques are necessary, then they cannot spare the extensive background, unrewarding and very careful work which is indispensable for the compilation of a valuable directory.

StarGuides 2001 gathers together around 6,200 entries from over 100 countries. The information is given in an uncoded way for easy and direct use. For each entry, all practical data returned by the organizations themselves are listed: city, postal and electronic-mail addresses; telephone and telefax numbers; WWW sites; foundation years; numbers of members or staff; main activities; titles, frequencies, ISS-Numbers and circulations of periodicals produced; names and geographical coordinates of observing sites; names of planetariums; awards, prizes or distinctions granted; and so on. City reference coordinates are now provided for each entry (based on the location of the head office or of the main centre of activities). Organizations not answering several updating requests have been deleted, and numerous new ones have been introduced.

The entries are listed alphabetically in each country. An exhaustive index gives a breakdown not only by different designations and acronyms, but also by location and major terms in the names. A search for information in the directory would normally begin with consultation of this index.

Thematic subindices of academies, awards, bibliographical services, consultants, data centres, dealers and distributors, funding organizations, Internet service providers, ISS-Numbers, manufacturers, meteorological offices, norms and standards institutes, observatories, periodicals, planetariums, publishers, science museums, software producers, etc. are also provided as well as a list of telephone and telefax national codes.

The quality behind the master files of *StarGuides* has been recognized by their implementation as a separate database called *StarWays* by the European Space Information System (ESIS) group (see *e.g.* Heck *et al.* 1992) and as another independent database called *StarGates* (Albrecht & Heck 1994) at the European Southern Observatory (ESO). *StarGuides* files are also reachable at CDS as the database *StarWorlds* (Heck *et al.* 1994) via the URL http://vizier.u-strasbg.fr/starworlds.html.

An extensive dictionary of abbreviations, acronyms, contractions, and symbols used in astronomy, space sciences, and related fields is compiled in parallel and is available on paper in the sister publication *StarBriefs 2001* (ISBN 0-7293-6510-0). All these products are members of the *Star*s Family of Astronomy and Related Resources* (Heck 1997).

When compiling resources such as *StarGuides* and *StarBriefs*, one cannot but be impressed by the very broad spectrum of disciplines to which astronomy and related space sciences are linked, and by the very large variety of techniques applied in these fields.

The successive editions of the directories give fairly accurate global pictures of the active organizations in the fields covered. Their sequence testifies to the sometimes rapid evolution of scientific interests, of data collecting and handling techniques, as well as of communications in the broad sense. A few countries have also rearranged the structure of their national facilities in the course of the past years. With the rapid spreading of the WWW and the globalization of telecommunications, we now tend to include only the headquarters of commercial organizations since the data (often changing) of their subsidiaries and branches can be found on their main web sites.

Studies based on data extracted from the master files have been published. They provide geographical distributions and general characteristics of various categories of astronomy-related organizations (see Heck 1998a&b, 1999 and 2000a&b). Other investigations are currently under way. Figs. 1 to 3 (from Heck 2000c) illustrate the overall world distribution with blowups for Europe and North America of the geographical locations recorded in our files – in other words what could be called StarGuides/StarWorlds world (or *'Planet Astronomy'*) at the beginning of 2000.

Feedback from readers on possible modifications or additions to the data published here would be highly appreciated in order to ensure permanent accuracy of the master files (see updating form at the end of the volume). These are continually updated and new versions of *StarGuides* will be released regularly.

The information is provided in *StarGuides* 'bone fide'. The best is done to keep track of the modifications happening and to implement them as soon as they are confirmed or recognized by the international community. Any information received is included within a reasonably short time after reaching us.

Acknowledgements

Finally it is a very pleasant duty to express our gratitude to all persons and organizations who contributed over the past quarter of a century to the very substance of the master files used here, by returning the questionnaires, by providing the relevant documentation, by participating in the various procedures of maintenance, validation and verification of the information, or otherwise. The *Star*s Family* products have been conceived for them and for the vast community of users. We are looking forward to satisfying their needs in continually better ways.

The implementation, as databases, of the *Star*s Family* products by the European Space Agency, the European Southern Observatory and Strasbourg Astronomical Data Centre have been strong incentives to continue and always improve these time-consuming compilations for the benefit of the best possible communication within the astronomical community, as well as between it and the outside world.

June 2000.

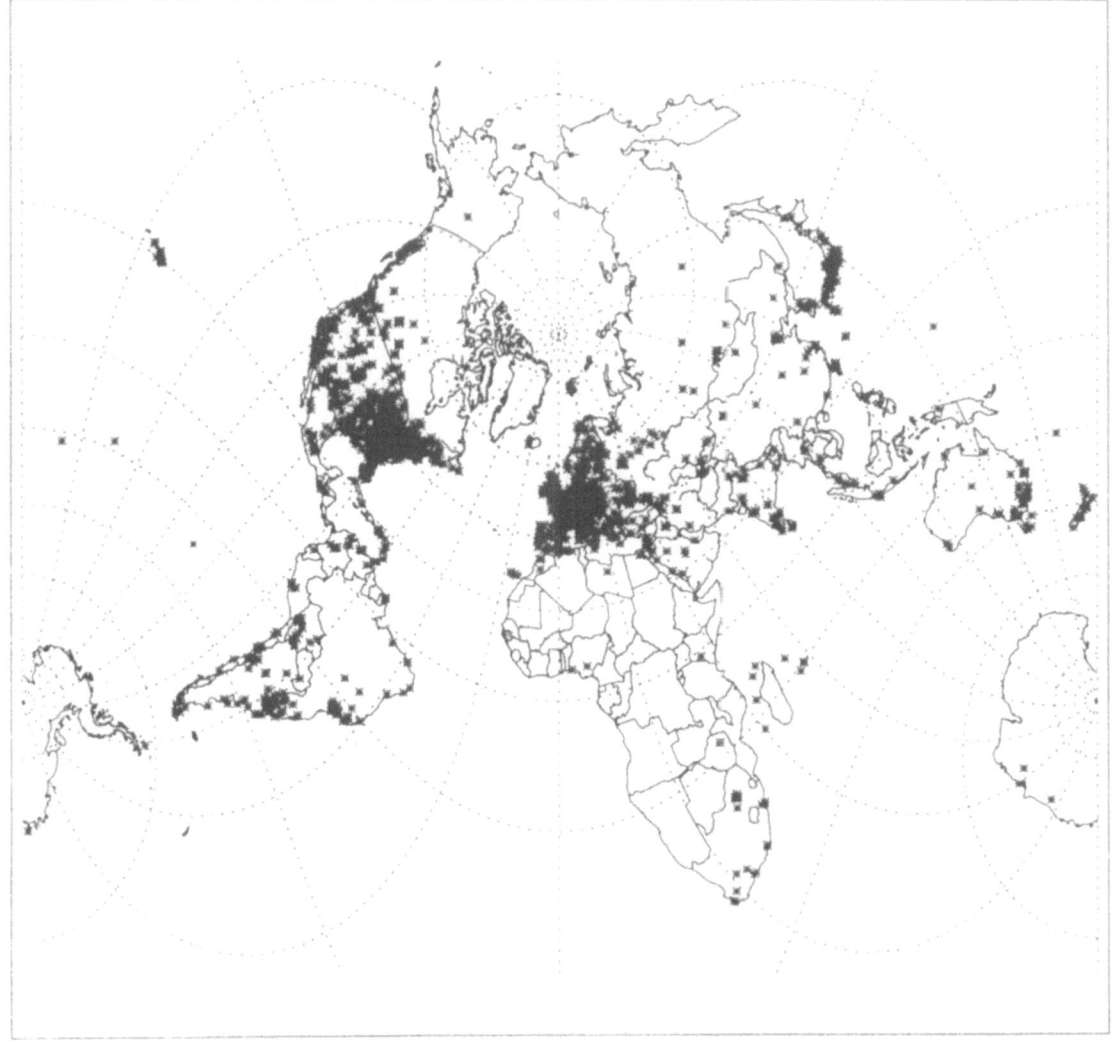

Fig. 1 – 'Planet Astronomy' – World distribution

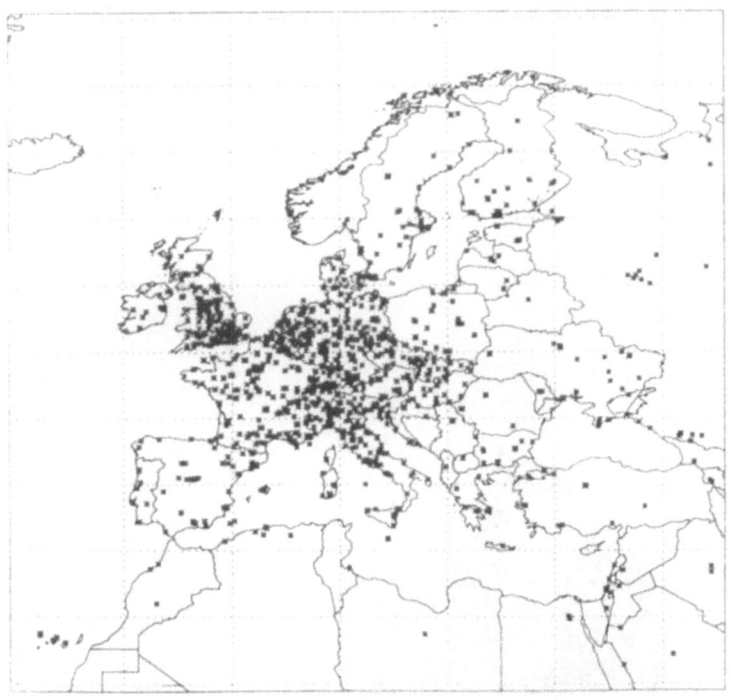

Fig. 2 – 'Planet Astronomy' – Western Europe

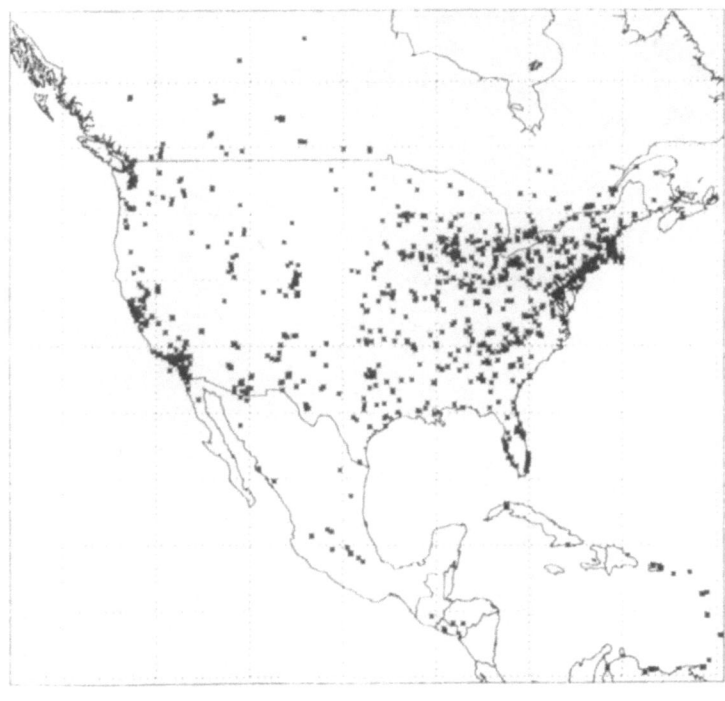

Fig. 3 – 'Planet Astronomy' – North America

REFERENCES

Albrecht, M. & Heck, A.: 1994, StarGates – An On-line Database of Astronomy, Space Sciences, and Related Organizations of the World, *Astron. Astrophys. Suppl.* **103**, 473-474.

Heck, A.: 1995, The Star*s Family: An Example of Comprehensive Yellow-Page Services, in *Information & On-Line Data in Astronomy*, Eds. D. Egret & M.A. Albrecht, Kluwer Acad. Publ., Dordrecht, 195-205

Heck, A.: 1997, Electronic Yellow-Page Services: The Star*s Family as an Example of Diversified Publishing, in *Electronic Publishing for Physics and Astronomy*, Ed. A. Heck, Kluwer Acad. Publ., Dordrecht, 211-220.

Heck, A.: 1998a, Geographical Distribution of Observational Activities in Astronomy, *Astron. Astrophys. Suppl.* **130**, 403-406

Heck, A.: 1998b, Astronomy-Related Organizations over the World, *Astron. Astrophys. Suppl.* **132**, 65-81

Heck, A.: 1999, The Ages of Astronomy-Related Organizations, *Astron. Astrophys. Suppl.* **135**, 467-475 and **136**, 615

Heck, A.: 2000a, Where Astronomers Are: A Stagnant Century, *Sky & Tel.* **99** (2000) 32-35

Heck, A.: 2000b, Characteristics of Astronomy-Related Organizations, *Astrophys. Sp. Sc.*, in press

Heck, A., Ciarlo, A. & Stokke, H.: 1992, StarWays – An On-line Database of Astronomy, Space Sciences, and Related Organizations of the World, *Astron. Astrophys. Suppl.* **96**, 565-566.

Heck, A., Egret, D. & Ochsenbein, F.: 1994, StarWorlds – StarBits (announcement of two databases), *Astron. Astrophys. Suppl.* **108**, 447-448.

Author's address: **André HECK**
Observatoire Astronomique
11, rue de l'Université
F-67000 Strasbourg
France
Telephone: (+33)(0) 388 150 743
Telefax: (+33)(0) 388 491 255
Electronic mail: starpages@astro.u-strasbg.fr
WWW: http://vizier.u-strasbg.fr/~heck

How to use this directory

A search for information in the directory would normally begin with consultation of the three-level index located at the end the volume (yellow pages). It may be tackled by the name or by any of the major terms appearing in the title of the organization. Acronyms may also be used, as well as geographical names (city, region, etc.).

Thematic subindices are available for academies, awards, bibliographic services, consultants, data centres, dealers and distributors, funding organizations, Internet service providers, ISS-Numbers, manufacturers, meteorological offices, norms and standards institutes, observatories, periodicals, planetariums, publishers, science museums, software producers, and so on.

In the directory itself, the information is given is an uncoded way for easy and direct use. A one-column layout has been selected to avoid cutting, across two lines, important data such as addresses, World-Wide Web URLs, coordinates, and so on. The ample blank space can also be used for annotations.

Besides the astronomy-related organizations (about 6,200), listed alphabetically in each country (over a hundred), we systematically included the national representative meteorological offices (routinely approached by observers world-wide) and the national institutions responsible for norms and standards (to be consulted regarding compatibility of instrumentation and similar technical issues, for instance).

If several addresses are listed for a given entry, the first one should be given preference for all contacts by post. Twelve entries have neither postal nor street addresses, but do have electronic contacts. More than a thousand cross-references (or pointers) are provided throughout the directory for accessing information via alternate denominations.

The individual telephone and telefax numbers are preceded, whenever possible, by the national inter-area first digits (0, 9, and so on). National telephone and telefax codes for countries and territories around the world are provided in a separate section. They should be preceded by the international access number of the country or territory from which the call is placed (*e.g.* 00, 011, etc.). Unless otherwise indicated, all electronic-mail addresses are given for Internet.

This directory includes about 6,000 Uniform Resource Locators (URLs), also called 'anchors' or 'links', that can be used for accessing the World-Wide Web (WWW) sites of organizations with an Internet presence. Users are however advised that such electronic addresses are still somewhat volatile and that the best is done to keep the master files as up-to-date as possible. The latest information in our possession can always be obtained from the electronic version of the directory, called *StarWorlds*, reachable at the web page http://vizier.u-strasbg.fr/starworlds.html.

The foundation years correspond to the current organizations and not to possible forerunners (unless clearly stated). The 'activities' listed have been homogenized and standardized as much as possible through keywords from the information delivered by the organizations themselves on the questionnaires returned and from the accompanying documentation.

As already mentioned, coding has been kept to a minimum. The annual frequencies of the periodicals and awards are indicated between parentheses. We insist that we have considered only the periodicals (not occasional books) currently published. At the request of librarians, we included sub-indices as per periodicals titles and ISS numbers.

The geographical coordinates of observing sites are self-explanatory. For each organization, we also entered a position based on the location of its head office or main center of activities. Only 12 entries (0.20%) were left without coordinates, in other words a negligible subset compare to the approximately 7,000 coordinate sets provided.

Finally let us recall that the information presented here is validated and authenticated (from signed and documented questionnaires). The data have been systematically compiled and presented, with a permament updating-process scheme. They give a faithful snapshot of the astronomy-related organizations world-wide at the end of the second millenium.

Feedback from readers on possible modifications or additions to the data published here would be highly appreciated in order to ensure permanent accuracy of the master files (see updating form at the end of the volume).

Albania

Albanian Physical Society
c/o M. Panariti (Secretary)
University of Tirana
Tirana
Telephone: (0)42-27669
Telefax: (0)42-27912
Membership: 60
City Reference Coordinates: 019°50'00"E 41°20'00"N

Directorate of Standardization
(Drejtoria e Standardizimit dhe Cilesise - DSC)
Rruga Mine Peza 143/3
Tirana
Telephone: (0)42-26255
Telefax: (0)42-26255
Electronic Mail: dsc@icc.al.eu.org
Staff: 25
City Reference Coordinates: 019°50'00"E 41°20'00"N

Hydrometeorological Institute
Rruga Durresi 219
Tirana
Telephone: (0)42-23518
Telefax: (0)42-23518
Founded: 1949
Staff: 120
Activities: meteorology * hydrology * climatology * oceanography
Periodicals: "Botime Hidrometeorologike"
Observatories: 110 sites for meteorology and 120 sites for hydrometry
City Reference Coordinates: 019°50'00"E 41°20'00"N

Algeria

Association d'Astronomie El-Battani (AAB)
c/o Centre Culturel Communal Ibn-Mahrez
6, rue de Lourmel
Boîte Postale 08 R.P.
Oran 31024
Telephone: (0)6-332168
Founded: 1985
Membership: 300
Activities: observing * lectures * workshop * library * education * computing * exhibitions * scientific expeditions
Periodicals: "Altaïr", "El-Fellek", "Science-Culture"
City Reference Coordinates: 000°43'00"W 35°43'00"N

Association Scientifique d'Astronomie (ASA) El-Bouzdjani
c/o Maison de la Culture
Medea 26000
Telephone: (0)3-585114
Telefax: (0)3-581208
Founded: 1988
Membership: 50
Activities: observing * meetings * lectures * popularization * exhibitions * meteorology * rocketry
Periodicals: (6) "El-Marsad"
Coordinates: 002°05'00"E 36°25'00"N H1,070m (Mont Tibarine)
City Reference Coordinates: 002°50'00"E 36°12'00"N

Association Sirius d'Astronomie
Cité du 20 Août 1955
Boîte Postale 18
25000 Constantine
Telephone: (0)4-900316
Electronic Mail: sirius_astro@hotmail.com
WWW: http://www.geocities.com/CapeCanaveral/Station/9720
Founded: 1995
Membership: 40
Activities: observing * lectures * exhibitions * radio broadcast * workshops * software * advising
Periodicals: (1-4) "Sirius Voice" (in Arabic)
Coordinates: 006°37'00"E 36°22'00"N H600m
City Reference Coordinates: 006°40'00"E 36°22'00"N

Centre de Recherches en Astronomie, Astrophysique et Géophysique (CRAAG)
Route de l'Observatoire
Boîte Postale 63
Bouzareah
16340 Alger
Telephone: (0)2-903160
(0)2-901424
(0)2-941572
(0)2-903267
Telefax: (0)2-903160
(0)2-901424
Electronic Mail: astro@ist.cerist.dz
Founded: 1980
Activities: seismology * geomagnetism * gravimetry * astronomy
Coordinates: 003°01'50"E 36°48'04"N H345m
City Reference Coordinates: 003°08'00"E 36°42'00"N

Institut Algérien de Normalisation et de Propriété Industrielle (INAPI)
5, rue Abou Hamou Moussa
Boîte Postale 403
Alger
Telephone: (0)2-635180
Telefax: (0)2-610971
City Reference Coordinates: 003°08'00"E 36°42'00"N

Observatoire d'Alger
● See "Centre de Recherches en Astronomie, Astrophysique et Géophysique (CRAAG)"

Office National de la Météorologie
Avenue Khemisti
Boîte Postale 153
Dar El Beïda

Alger
Telephone: (0)2-507393
Telefax: (0)2-508849
WWW: http://www.meteo.dz/
City Reference Coordinates: 003°08'00"E 36°42'00"N

Argentina

Asociación Argentina Amigos de la Astronomía (AAAA)
Avenida Patricias Argentinas 550
1405 Buenos Aires
Telephone: (0)11-48633366
Telefax: (0)11-48633366
Electronic Mail: postmaster@aaaa.org.ar
WWW: http://www.asaramas.com.ar/
Founded: 1929
Membership: 650
Activities: education * library * telescope making * visits * lectures * observing (comets, minor planets, variable stars, occultations, Sun, meteors) * astrometry
Periodicals: (4) "Revista Astronómica" (ISSN 0044-9253, circ.: 1,500)
Coordinates: 058°26'04"W 34°36'19"S H26m
City Reference Coordinates: 058°27'00"W 34°36'00"S

Asociación Argentina de Astronomía (AAA)
c/o Marta Rovira (President)
Instituto de Astronomía y Física del Espacio (IAFE)
Casilla de Correos 67 - Succ. 28
1428 Buenos Aires
or :
c/o Observatorio Astronómico de Córdoba
Laprida 854
5000 Córdoba
Telephone: (0)351-4331064
            (0)351-4331065
Telefax: (0)351-4331063
Electronic Mail: lapasset@mail.oac.uncor.edu.ar (Emilio Lapasset, Vice President)
               rovira@iafe.uba.ar
Founded: 1958
Membership: 260
Activities: organizing national scientific meetings * sponsoring international symposia and workshops in the country
Periodicals: (1) "Boletín de la AAA" (circ.: 300)
City Reference Coordinates: 058°27'00"W 34°36'00"S (Buenos Aires)
                                064°11'00"W 31°25'00"S (Córdoba)

Asociación Cordobesa Amigos de la Astronomía (ACAA)
Casilla de Correos 1105
5000 Córdoba
or :
c/o Observatorio Astronómico de Córdoba
Laprida 854
5000 Córdoba
Telephone: (0)51-246224
Telefax: (0)51-246224
Electronic Mail: acaa@unerin.edu.ar
WWW: http://www.cnba.uba.ar/acadr/astro/liada/liada.html
Founded: 1986
Membership: 20
Activities: popularization * radioastronomy * comets * Sun
Periodicals: "PC-Astro Magazine" (on diskettes)
City Reference Coordinates: 064°11'00"W 31°25'00"S

Centro Regional de Investigaciones Científicas y Tecnológicas (CRICYT)
Casilla de Correos 131
5500 Mendoza
or :
Avenida Ruiz Leal s/n
Parque General San Martin
Mendoza
Telephone: (0)61-288314
Telefax: (0)61-287370
Electronic Mail: ntcricyt@criba.edu.ar
WWW: http://www.cricyt.edu.ar/
Founded: 1973
Staff: 45
Activities: astronomy * mathematics * geology * space sciences * biology * social sciences
Periodicals: (1) "Memoria Anual"
Coordinates: 068°51'45"W 32°52'50"S
City Reference Coordinates: 068°52'00"W 32°48'00"S

Colegio Nacional de Buenos Aires, Observatorio
Bolívar 263
1066 Buenos Aires
WWW: http://www.cnba.uba.ar/acad/astro/
Founded: 1935
Staff: 10
Activities: education * observing (Sun, variable stars, occultations, binaries)
Coordinates: 058°22'25"W 34°36'39"S H45m
City Reference Coordinates: 058°27'00"W 34°36'00"S

Comisión Nacional de Actividades Espaciales (CONAE)
Avenida Dorrego 4010
1425 Buenos Aires
Telephone: (0)1-7762913
           (0)1-7762963
           (0)1-7749310
Telefax: (0)1-7745703
Electronic Mail: <userid>@conae.gov.ar
WWW: http://www.conae.gov.ar/
Founded: 1991
Staff: 100
Activities: peaceful uses of space
City Reference Coordinates: 058°27'00"W 34°36'00"S

Complejo Astronómico El Leoncito (CASLEO)
Avenida España 1512 Sur
Casilla de Correos 467
5400 San Juan
Telephone: (0)64-213653 (San Juan offices)
           (0)648-41088 (Observatory)
Telefax: (0)64-213693
Electronic Mail: <userid>@castec.edu.ar
WWW: http://www.fcaglp.unlp.edu.ar/casleo
        http://www.fcaglp.unlp.edu.ar/~scellone/casleo/
Founded: 1986
Staff: 50
Activities: observing * research * instrumentation
Periodicals: (4) "Noticias"
Coordinates: 069°18'00"W 31°47'57"S H2,550m
City Reference Coordinates: 068°30'00"W 31°30'00"S

Consejo Nacional de Investigaciones Científicas y Técnicas (CONICET)
Rivadavia 1917
1033 Buenos Aires
Telephone: (0)11-49537230
           (0)11-49537239
Telefax: (0)11-49518552
        (0)11-49544955
Electronic Mail: comunica@conicet.gov.ar
WWW: http://www.conicet.gov.ar/
Founded: 1958
Activities: promoting and organizing science and technology nationally
City Reference Coordinates: 058°27'00"W 34°36'00"S

Estación Astronómica Rio Grande (EARG)
Casilla de Correo 160
9420 Rio Grande
Electronic Mail: earg@earg.gov.ar
WWW: http://www.earg.gov.ar/
Founded: 1979
City Reference Coordinates: 067°40'00"W 53°50'00"S

Fundación José Maria Aragón
Avenida Felicia Moreau de Justo 1750
Piso 1 - Oficiba "C"
1107 Buenos Aires
Telephone: (0)11-43120055
Telefax: (0)11-43122299
Electronic Mail: info@aragon.org.ar
WWW: http://www.aragon.org.ar/
Founded: 1962
Staff: 15
Activities: information service * post-graduate grants * education * funding
Periodicals: (12) "Becas y Cursos" (circ.: 1,400); (1/2) "Fellowship Guide"
City Reference Coordinates: 058°27'00"W 34°36'00"S

Instituto Argentino de Racionalización de Materiales (IRAM)
Chile 1192
1098 Buenos Aires
Telephone: (0)1-3833751
Telefax: (0)1-3838463
Electronic Mail: iram2@sminter.com.ar
WWW: http://www.iram.org.ar/
Founded: 1935
Staff: 110
Activities: standardization and certification of products and quality systems
Periodicals: "Dinámica" (ISSN 3025-0733), "Guía de Licenciatorios"
City Reference Coordinates: 058°27'00"W 34°36'00"S

Instituto Argentino de Radioastronomía (IAR)
Casilla de Correos 5
1894 Villa Elisa
or :
Camino General Belgrano Km. 40
Parque Pereyra Iraola
Pdo. de Berazategui
Telephone: (0)21-254909
            (0)21-740230
Telefax: (0)21-254909
Electronic Mail: <userid>@irma.edu.ar
Founded: 1963
Staff: 40
Activities: radioastronomy * instrumentation
Coordinates: 058°08'12"W 34°52'06"S H11m
City Reference Coordinates: 058°25'00"W 32°10'00"S (Villa Elisa)
                                058°12'00"W 34°45'00"S (Berazategui)

Instituto Copernico
Casilla de Correos 51
1448 Buenos Aires
or :
Casilla de Correos 85
5600 San Rafael
Electronic Mail: <userid>@icoper.edu.ar
City Reference Coordinates: 058°27'00"W 34°36'00"S (Buenos Aires)
                                068°21'00"W 34°40'00"S (San Rafael)

Instituto de Astronomía y Física del Espacio (IAFE)
Casilla de Correos 67 - Succ. 28
1428 Buenos Aires
or :
Pabellón IAFE
Cuidad Universitaria
Intendente Guiraldes 2620
Buenos Aires
Telephone: (0)1-7832642
            (0)1-7816755
            (0)1-7868114
Electronic Mail: <userid>@iafe.uba.ar
WWW: http://www.iafe.uba.ar/
Founded: 1969
Staff: 50
Activities: globular clusters * SNR * symbiotic stars * interstellar dust * solar flares * solar prominences * stellar atmospheres * cool stars * solar and stellar winds * astrophysical plasmas * coronal heating * cosmology * relativity * quantum field theory * string theory * atomic collisons * photon-atom collisions * high-energy astrophysics
City Reference Coordinates: 058°27'00"W 34°36'00"S

Observatorio Astronómico de Córdoba (OAC)
● See "Universidad Nacional de Córdoba, Observatorio Astronómico de Córdoba (OAC)"

Observatorio Astronómico de La Plata
● See "Universidad Nacional de La Plata (UNLP), Observatorio Astronómico"

Observatorio Astronómico Felix Aguilar (OAFA)
● See "Universidad Nacional de San Juan, Observatorio Astronómico Felix Aguilar (OAFA)"

Observatorio Astronómico Municipal de Funes (OAMF)
Casilla de Correos 16
2132 Funes
or :
Estación Ferroviaria

Avenida Santa Fe 1689
2132 Funes
Telephone: (0)41-931234
Telefax: (0)41-931153
Electronic Mail: obsfunes@openware.com.ar
WWW: http://www.openware.com.ar/ObservatorioFunes/
Founded: 1991
Staff: 9
Activities: public education * variable stars * comets * meteors * sky patrol * portable planetarium
Periodicals: (1) "Equinoccio"; "Circulars"
Coordinates: 060°40'00"W 32°55'00"S H32m
City Reference Coordinates: 060°40'00"W 32°55'00"S

Observatorio Astronómico Municipal de Mercedes (OAMM)
Calle 35 N° 876
6600 Mercedes
Telephone: (0)2324-426775
Telefax: (0)2324-422442
Electronic Mail: oamm@satlink.com
                madel@iafe.edu.ar
Founded: 1974
Staff: 3
Activities: photoelectric photometry (eclipsing binaries, variable stars)
Periodicals: (1) "Manual Astronómico"
Coordinates: 059°25'59"W 34°37'55"S H34m
City Reference Coordinates: 059°30'00"W 34°42'00"S

Observatorio Astronómico, Planetario y Museo Experimental de Ciencias de Rosario
Parque Urquiza
Casilla de Correos 606
2000 Rosario
Telephone: (0)41-63084
Telefax: (0)41-257164
Electronic Mail: camussi@interactive.com.ar
WWW: http://www.rol.com.ar/camussi/museo.htm
Founded: 1970
Staff: 35
Activities: solar physics * planets * cosmology
Awards: "Luis C. Carballo"
City Reference Coordinates: 060°40'00"W 32°57'00"S

Observatorio del Colegio Cristo Rey
Laprida 1380
2000 Rosario
Telephone: (0)41-485421
            (0)41-213565
Founded: 1984
Activities: variable stars * double stars * planets * comets
Periodicals: (4) "Apex" (circ.: 20)
Coordinates: 060°37'55"W 32°57'20"S H50m
City Reference Coordinates: 060°40'00"W 33°00'00"S

Observatorio Nacional de Física Cosmica, San Miguel (ONFCSM)
Avenida Mitre 3100
1663 San Miguel
Telephone: (0)1-4557044
            (0)1-4651225
Telefax: (0)1-4651225
Founded: 1935
Staff: 15
Activities: sunspot monitoring
Periodicals: (12) "Sunspot Data"; (1) "General Catalog"; "Observaciones Solares"
Coordinates: 058°43'54"W 34°53'24"S H37m
City Reference Coordinates: 057°36'00"W 28°01'00"S

Observatorio Naval de Buenos Aires (ONBA)
Avenida España 2099
1107 Buenos Aires
Telephone: (0)1-3611162
Telefax: (0)1-3611162
Electronic Mail: <userid>@onba.mil.ar
                postmast@onba.mil.ar
Founded: 1881
Staff: 15
Activities: national time service * ephemerides for navigation and geodesy * research on time and reference frames

Periodicals: (1) "Almanaque Nautico y Aeronautico", "Suplemento al Almanaque"
Coordinates: 058°21'18"W 34°37'18"S H6m
City Reference Coordinates: 058°27'00"W 34°36'00"S

Observatorio San José (OSJ)
Azcuenaga 158
1029 Buenos Aires
Telephone: (0)951-4303
Electronic Mail: <userid>@sanjo.edu.ar
WWW: http://members.xoom.com/observatorio
Founded: 1913
Staff: 5
Activities: Sun * occultations * variable stars * comets * meteor streams * eclipses * photography * planets
Coordinates: 058°24'02"W 34°36'30"S H61m
City Reference Coordinates: 058°27'00"W 34°36'00"S

Planetario de la Ciudad de Buenos Aires Galileo Galilei
Avenidas Sarmiento y B. Roldán
1425 Buenos Aires
Telephone: (0)11-47716629
         (0)11-47729265
Telefax: (0)11-47716629
        (0)11-47729265
Electronic Mail: <userid>@bibpla.bib.cyt.edu.ar
WWW: http://www.earg.gov.ar/planetario/
Founded: 1967
Staff: 25
Activities: shows * lectures * exhibitions
City Reference Coordinates: 058°27'00"W 34°36'00"S

Servicio Meteorológico Nacional (SMN)
25 de Mayo 658
1002 Buenos Aires
Telephone: (0)1-3124481
         (0)1-3117176
Telefax: (0)1-3113968
Electronic Mail: dtec@udmeteor.mil.ar
WWW: http://www.meteofa.mil.ar/
Founded: 1872
Staff: 300
Activities: national meteorological office
Periodicals: (daily) "Boletín Meteorológico Diario"; (12) "Boletín Climatológico Mensual"; "Boletines Informativos"
Observatories: large network of stations
City Reference Coordinates: 058°27'00"W 34°36'00"S

Universidad Nacional de Córdoba, Observatorio Astronómico de Córdoba (OAC)
Laprida 854
5000 Córdoba
Telephone: (0)51-230491
         (0)51-236876
         (0)51-331064
         (0)51-331065
Telefax: (0)51-210613
        (0)51-331063
Electronic Mail: <userid>@uncbob.edu.ar
               <userid>%astro@ecord.gov.ar
               <userid>@oac.uncor.edu
WWW: http://www.oac.uncor.edu/
Founded: 1871
Staff: 70
Activities: open clusters * globular clusters * binary stars * intrinsic variable stars * celestial mechanics * galaxies * cosmology * astrometry
Periodicals: "Reprints"
Coordinates: 064°11'48"W 31°25'18"S H439m
           064°32'48"W 31°35'54"S H1,250m (Bosque Alegre Astrophysical Station)
City Reference Coordinates: 064°11'00"W 31°25'00"S

Universidad Nacional de La Plata (UNLP), Facultad de Ciencias Astronómicas y Geofísicas (FCAG)
Casilla de Correos 677
1900 La Plata
Telephone: (0)21-217308 (Switchboard)
         (0)21-38810 (Administration)
         (0)21-211761 (Academic Secretary)
         (0)21-216931 (Dean)
Electronic Mail: <userid>@fcaglp.fcaglp.unlp.edu.ar
WWW: http://www.unlp.edu.ar/fac_ciencias_astronomicas.html

Founded: 1883
Staff: 53
Activities: research and education in astronomy and geophysics
Periodicals: "Universidad Nacional de La Plata Series (Astronomy, Geophysics, Circulars and Special)"
Coordinates: 069°18'06"W 31°48'00"S (El Leoncito - see separate entry)
         067°45'06"W 53°47'12"S (Rio Grande - see also separate entry)
         057°41'24"W 35°00'01"S (Trelew)
         065°22'54"W 43°16'06"S (Observatorio Magnético Las Acacias)
City Reference Coordinates: 057°55'00"W 34°52'00"S

Universidad Nacional de La Plata (UNLP), Observatorio Astronómico

Paseo del Bosque s/n
1900 La Plata
Telephone: (0)21-217308
         (0)21-38810 (Switchboard)
         (0)21-211761 (Academic Secretary)
         (0)21-216931 (Dean)
Telefax: (0)21-211761
Electronic Mail: <userid>@fcaglp.edu.ar
WWW: http://www.fcaglp.unlp.edu.ar/
Founded: 1883
Staff: 160
Activities: research and education in astronomy and geophysics
Coordinates: 069°18'06"W 31°48'00"S (El Leoncito)
         067°45'06"W 53°47'12"S (Rio Grande)
         065°22'54"W 43°16'06"S (Trelew)
City Reference Coordinates: 057°55'00"W 34°52'00"S

Universidad Nacional de San Juan, Observatorio Astronómico Felix Aguilar (OAFA)

Avenida Benavidez 8175 Oeste
5400 San Juan
Telephone: (0)64-231467
Telefax: (0)64-238494
Staff: 50
Activities: astrometry * celestial mechanics * astrophysics
Periodicals: "Contributions"
Coordinates: 068°37'15"W 31°30'38"S H700m (Estación Carlos Cesco)
City Reference Coordinates: 068°30'00"W 31°30'00"S

Armenia

Armenian Academy of Sciences, Byurakan Astrophysical Observatory
378433 Byurakan
Telephone: (0)8852-283453
          (0)8852-284142
Telefax: (0)8852-523640
Electronic Mail: byur@mav.yerphi.am
              <userid>@bao.sci.am
Periodicals: "Soobshcheniya" (ISSN 0370-8691), "Reprints"
Coordinates: 044°17'30"E 40°20'06"N H1,500m
City Reference Coordinates: 044°18'00"E 40°20'00"N

Armenian Physical Society
c/o R. Avakian (President)
Alikhanian Brothers Street 2
375036 Yerevan
Telephone: (0)8852-341347
          (0)2-341347
Telefax: (0)8852-350030
        (0)2-350030
Electronic Mail: r.avakian@hermes.desy.de
Membership: 200
City Reference Coordinates: 044°30'00"E 40°11'00"N

Byurakan Astrophysical Observatory
● See "Armenian Academy of Sciences, Byurakan Astrophysical Observatory"

Department for Standardization, Metrology and Certification (SARM)
Komitas Avenue 49/2
375051 Yerevan
Telephone: (0)2-235600
Telefax: (0)2-285620
Electronic Mail: sarm@sarm.com
Founded: 1931
Staff: 450
City Reference Coordinates: 044°30'00"E 40°11'00"N

Garny Space Astronomy Institute
P.O. Box 370/15
375002 Yerevan
Telefax: (0)8852-647030
City Reference Coordinates: 044°30'00"E 40°11'00"N

Hydrometeorological Administration of Armenia
54 Leo Street
375002 Yerevan
Telephone: (0)2-532001
Telefax: (0)3-906863
Electronic Mail: meteo@mbox.amilink.net
Founded: 1930
Staff: 950
Activities: hydrology * meteorology * radiation * climate variability * weather services (public, aviation, warning)
City Reference Coordinates: 044°30'00"E 40°11'00"N

Yerevan Physics Institute, Department of Theoretical Physics
Alikhanian Brothers Street 2
375036 Yerevan
Telephone: (0)2-350150
          (0)2-341500
Telefax: (0)2-350030
Electronic Mail: <userid>@lx2.yerphi.am
WWW: http://www.yerphi.am/
Founded: 1943
Staff: 10
Activities: stellar dynamics * evolution of galaxies and clusters * cosmology
Periodicals: "Preprints"
City Reference Coordinates: 044°30'00"E 40°11'00"N

Yerevan State University (YSU), Department of Astrophysics
Al. Manukian Street 1
375025 Yerevan

Telefax: (0)8852-597939
Electronic Mail: <userid>@ysu.am
WWW: http://www.physic.ysu.am/Astrophys/
Founded: 1945
Staff: 3
City Reference Coordinates: 044°30'00"E 40°11'00"N

Australia

Adelaide Planetarium
University of South Australia
Levels Campus
Mawson Lakes Boulevard
Mawson Lakes
Adelaide, SA
Telephone: (0)8-8302-3138
              (0)8-8353-5762
WWW: http://ching.apana.org.au/~oliri/stars.html
         http://www.unisa.edu.au/erm/planetarium/pla-home.htm
Founded: 1972
City Reference Coordinates: 138°35'00"E 34°55'00"S

Advanced Telescope Supplies
P.O. Box 447
Engadine, NSW 2233
Telephone: (0)2-9541-1676
Telefax: (0)2-9541-4449
WWW: http://www.ozemail.com.au/~atsscope/
City Reference Coordinates: 151°01'00"E 34°04'00"S

Anglo-Australian Observatory (AAO), Coonabarabran
Siding Spring Mountain
Private Bag
Coonabarabran, NSW 2357
Telephone: (0)2-6842-6291
Telefax: (0)2-6884-2298
Electronic Mail: <userid>@aaocbn.aao.gov.au
WWW: http://www.aao.gov.au/
         http://www.aao.gov.au/ukst/ (UK Schmidt telescope)
Founded: 1974 (UK Schmidt: 1973)
Staff: 60
Activities: operating and maintaining the Anglo-Australian 3.9m Telescope (AAT) and the UK 1.2m Schmidt Telescope (UKST - see separate entry) * optical and near-IR astronomy
Periodicals: (4) "AAO Newsletter" (ISSN 0728-5833); (1) "AAO Annual Report" (ISSN 0728-6554)
Coordinates: 149°03'58"E 31°16'37"S H1,164m (AAT)
               149°04'12"E 31°16'24"S H1,130m (UKST)
City Reference Coordinates: 149°18'00"E 31°16'00"S

Anglo-Australian Observatory (AAO), Epping Laboratory
P.O. Box 296
Epping, NSW 2121
or :
167 Vimiera Road
Eastwood, NSW 2122
Telephone: (0)2-9372-4800
Telefax: (0)2-9372-4880
Electronic Mail: <userid>@aaoepp.aao.gov.au
WWW: http://www.aao.gov.au/
Founded: 1974
Staff: 50
Activities: research in all aspects of optical and near-IR astronomy * instrumentation * photography
Periodicals: (1) "Annual Report of Anglo-Australian Telescope Board" (ISSN 0728-6554); (4) "AAO Newsletter" (ISSN 0728-5833)
Coordinates: 149°03'58"E 31°16'37"S H1,164m (AAT)
City Reference Coordinates: 145°02'00"E 37°39'00"S (Epping)
                                151°05'00"E 33°48'00"S (Eastwood)

Astral Press
P.O. Box 107
Wembley, WA 6014
or :
46 Oceanic Drive
Floreat Park, WA
Telephone: (0)8-9387-4250
Telefax: (0)8-9361-4418
Electronic Mail: astral@psinet.net.au
Founded: 1985
Staff: 1
Periodicals: (2) "Journal of Astronomical History and Heritage" (ISSN 1440-2807)
Awards: "Astral Award"

● Publisher
City Reference Coordinates: 115°50'00"E 31°56'00"S (Perth)

AstroDomes
P.O. Box 52
Yandina, QLD 4561
Telephone: (0)7-5446-7449
Telefax: (0)7-5446-8544
Electronic Mail: astrodomes@astrodomes.com
WWW: http://www.astrodomes.com/
Activities: manufacturing astronomical domes
City Reference Coordinates: 152°52'00"E 26°36'00"S

Astronomical Society of Alice Springs Inc.
P.O. Box 739
Alice Springs, NT 0871
Telephone: (0)8-8952-8560 (Alan Viegas, President)
Electronic Mail: baney@topend.com.au
WWW: http://www.ozemail.com.au/~kkramer/astalice.html
Founded: 1973
Membership: 30
Activities: meetings * observing
Coordinates: 133°56'13"E 23°41'02"S H580m
City Reference Coordinates: 133°52'00"E 23°42'00"S

Astronomical Society of Australia (ASA)
c/o Department of Mathematics and Physics
University of Tasmania
G.P.O. Box 252-21
Hobart, TAS 7001
Telephone: (0)3-6226-2022
Telefax: (0)3-6223-3057
Electronic Mail: marc.duldig@utas.edu.au
WWW: http://www.atnf.csiro.au/asa_www/asa.html
        http://www.atnf.csiro.au/pasa/ (Electronic PASA)
Founded: 1965
Membership: 377
Activities: annual conference
Periodicals: (2) "Publications of the Astronomical Society of Australia (PASA)", "Newsletter", "Proceedings" (ISSN 0066-9997)
Awards: (1/2) "Page Medal"; (1) "Bok Prize", "Ellery Lectureship"
City Reference Coordinates: 149°19'00"E 42°31'00"S

Astronomical Society of Frankston (ASF) Inc.
P.O. Box 596
Frankston, VIC 3199
Electronic Mail: aggro@peninsula.starway.net.au (Peter Lowe)
WWW: http://peninsula.starway.net.au/~aggro
Founded: 1969
Membership: 100
Activities: minor-planet occultations * lunar grazes * variable stars * Jovian-satellite eclipses * Earth satellites * deep sky
Periodicals: (6) "Scorpius" (circ.: 150)
Coordinates: 145°02'25"E 38°16'28"S H60m (The Briars)
City Reference Coordinates: 145°07'00"E 38°08'00"S

Astronomical Society of Melbourne
P.O. Box 92
Bentleigh, VIC 3204
WWW: http://www.astromelb.i.net.au/
City Reference Coordinates: 144°58'00"E 37°49'00"S

Astronomical Society of New South Wales (ASNSW)
G.P.O. Box 1123
Sydney, NSW 2001
Electronic Mail: asnsw@ozemail.com.au
WWW: http://www.ozemail.com.au/~asnsw/
Founded: 1985
Membership: 250
Periodicals: (6) "Universe"
City Reference Coordinates: 151°13'00"E 33°52'00"S

Astronomical Society of South Australia (ASSA) Inc.
G.P.O. Box 199
Adelaide, SA 5001
Telephone: (0)8-8272-7352 (Secretary)

(0)8-8338-1231 (Information Officer)
Telefax: (0)8-8379-4145
Electronic Mail: assa@gist.net.au
                  assamail@camtech.net.au
WWW: http://www.assa.org.au/
Founded: 1892
Membership: 400
Activities: observing * public nights * education * photography * CCDs * variable stars
Periodicals: (12) "Bulletin of the Astronomical Society of South Australia"
City Reference Coordinates: 138°35'00"E 34°55'00"S

Astronomical Society of Tasmania, Inc. (AST)
c/o Martin George (Vice President)
Queen Victoria Museum and Art Gallery
Wellington Street
Launceston, TAS 7250
or :
c/o Norwood Avenue P.O.
Launceston, TAS 7250
Telephone: (0)3-6331-6777 (Vice President)
WWW: http://www.vision.net.au/~peter/AST/
Founded: 1934
Membership: 80
Activities: occultations * photography * public involvment * computing
Periodicals: (6) "Bulletin of the Astronomical Society of Tasmania Inc."; (1) "AST Ephemeris"
City Reference Coordinates: 147°08'00"E 41°26'00"S

Astronomical Society of the Hunter (ASH)
c/o Colin Maybury
P.O. Box 69
Kurri Kurri, NSW 2327
or :
P.O. Box 193
Wallsend, NSW 2287
Telephone: (0)2-4961-5448
                  (0)2-4937-4664
Founded: 1973
Membership: 20
Activities: observing (solar system, Sun, comets, deep sky, meteors) * instrumentation * photography * transient phenomena
Periodicals: (4) "Observations"
Observatories: 3 (Blackhill, Wallaroo State Forest, Kurri Technical College)
Awards: "Norther Instruments Trophy", "ASH Annual Achievement Award", "Jim Maybury Memorial Award"
City Reference Coordinates: 151°30'00"E 32°49'00"S (Kurri Kurri)
                  151°40'00"E 32°55'00"S (Wallsend)

Astronomical Society of the South West (ASSW) Inc.
P.O. Box 1100
Bunbury, WA 6231
Telephone: (0)8-9725-0036
                  (0)8-9721-1586
Founded: 1980
Membership: 46
Activities: observing * popularization * education
Periodicals: "The Celestial Onlooker"
Coordinates: 115°22'00"E 33°35'30"S
City Reference Coordinates: 115°38'00"E 33°20'00"S

Astronomical Society of Victoria (ASV) Inc.
Box 1059J - G.P.O.
Melbourne, VIC 3001
Telephone: (0)3-9877-3181
WWW: http://www.gsat.net.au/astrovic
Founded: 1922
Membership: 680
Activities: photography * aurorae * Sun * demonstrations * history of astronomy * instrumentation * activities for juniors * Moon * planets * meteors * variable stars * observing * planetarium
Periodicals: (4) "ASV Newsletter"; (1) "Astronomical Yearbook" (ISSN 0067-0006); "Journal of the Astronomical Society of Victoria" (ISSN 0044-9814)
Coordinates: 144°58'24"E 37°49'54"S H28m
City Reference Coordinates: 144°58'00"E 37°49'00"S

Astronomical Society of Western Australia (ASWA) Inc.
P.O. Box 421
Subiaco, WA 6008
Telephone: (0)8-9384-2264
Electronic Mail: aswa@cleo.murdoch.edu.au

WWW: http://cleo.murdoch.edu.au/gen/aswa/
Founded: 1950
Membership: 150
Activities: meetings * viewing nights * camps * public education
Periodicals: (6) "Sidereal Times" (circ.: 200)
City Reference Coordinates: 115°50'00"E 31°56'00"S (Perth)

Astrovisuals
6 Lind Street
Strathmore, VIC 3041
Telephone: (0)3-9379-5753
Telefax: (0)3-9379-5753
Electronic Mail: astrovis@ozemail.com.au
WWW: http://www.ozemail.com.au/~astrovis/
Founded: 1983
Staff: 1
Activities: producing and distributing astronomy posters, slide sets, videos, CD-ROMs and postcards
City Reference Coordinates: 142°32'00"E 17°48'00"S

Auspace Ltd.
50 Hoskins Street
P.O. Box 17
Mitchell, ACT 2911
Telephone: (0)6-6242-2611
Telefax: (0)6-6241-6664
Electronic Mail: admin@auspace.com.au
WWW: http://www.auspace.com.au/
Founded: 1984
Staff: 35
Activities: manufacturing and designing space surveillance and communications equipment
City Reference Coordinates: 147°58'00"E 26°29'00"S

Australasian Planetarium Society (APS)
P.O. Box 207
Dickson, ACT 2602
Telephone: (0)2-6249-7817
Telefax: (0)2-6249-7238
Electronic Mail: aps@ctuc.asn.au
WWW: http://www.cfmeu.asn.au/aps/
Founded: 1997
Periodicals: "The Australasian Planetarian"
City Reference Coordinates: 149°08'00"E 35°55'00"S

Australasian Society for General Relativity and Gravitation (ASGRG)
c/o Susan M. Scott (Treasurer)
Australian National University,
Department of Physics and Theoretical Physics
Canberra, ACT 0200
Telefax: (0)2-6249-0741
Electronic Mail: asgrg@physics.adelaide.edu.au
WWW: http://www.physics.adelaide.edu.au/ASGRG/
Founded: 1994
City Reference Coordinates: 149°08'00"E 35°55'00"S

Australian Academy of Science (AAS)
G.P.O. Box 783
Canberra, ACT 2601
or :
Ian Potter House
Gordon Street
Acton, ACT 2600
Telephone: (0)2-6247-3966
            (0)2-6247-5777
Telefax: (0)2-6257-4620
Electronic Mail: eb@science.org.au
WWW: http://www.science.org.au/aashome.htm
Founded: 1954
Staff: 20
Membership: 280
Periodicals: "Academy Newsletter"
City Reference Coordinates: 149°08'00"E 35°55'00"S

Australian Defence Force Academy (AFDA), School of Physics
Canberra, ACT 2600
Electronic Mail: <userid>@afda.edu.au

WWW: http://www.adfa.edu.au/physics/astro/astron.html
Founded: 1986
Staff: 13
Activities: infrared, optical, X-ray and gamma-ray astronomy * laboratory astrophysics
• Joint undertaking between the "University of New South Wales (UNSW)" and the "Australian Defence Force"
City Reference Coordinates: 149°08'00"E 35°55'00"S

Australian Institute of Physics (AIP)
1/21 Vale Street
North Melbourne, VIC 3051
Telephone: (0)3-9326-6669
Telefax: (0)3-9328-2670
Electronic Mail: vjrjm@cc.newcastle.edu.au
                physics@raci.org.au
WWW: http://www.physics.usyd.edu.au/aipaust/
        http://www.physics.usyd.edu.au/aipaust/ANZPhysicist.html (The Physicist)
Periodicals: (11) "The Physicist" (ISSN 1036-3831)
City Reference Coordinates: 144°58'00"E 37°49'00"S

Australian National University (ANU), Research School of Astronomy and Astrophysics (RSAA), Mount Stromlo and Siding Spring Observatories (MSSSO)
Private Bag
Weston Creek Post Office
Weston Creek, ACT 2611
Telephone: (0)2-6249-0230 (Mount Stromlo)
            (0)2-6842-6262 (Siding Spring)
Telefax: (0)2-6249-0233 (Mount Stromlo)
          (0)2-6842-6240 (Siding Spring)
Electronic Mail: <userid>@mso.anu.edu.au
WWW: http://msowww.anu.edu.au/home.html
        http://msowww.anu.edu.au/exploratory/ (Exploratory)
Founded: 1924
Staff: 104
Activities: galactic structure * stellar populations * stellar dynamics * interstellar physics * radio galaxies * galaxy clusters * QSOs * star formation * Magellanic Clouds * Hubble flow * clusters of galaxies * stellar pulsations * stellar evolution * abundances * cooling flows
Periodicals: (1) "Annual Report"
Coordinates: 149°03'42"E 31°16'24"S H1,149m (Siding Spring Observatory)
             149°00'30"E 35°19'12"S H767m (Mount Stromlo Observatory)
City Reference Coordinates: 149°08'00"E 35°55'00"S (Canberra)

Australian National University (ANU), Research School of Earth Sciences (RSES)
Canberra, ACT 0200
Telephone: (0)2-6249-3406
Telefax: (0)2-6249-0738
Electronic Mail: school.secretary.rses@anu.edu.au
WWW: http://wwwrses.anu.edu.au/
Founded: 1973
Staff: 50
Activities: geochronology and isotope geochemistry * petrochemistry and experimental petrology * geophysical fluid dynamics * environmenal geochemistry * geodynamics * ore genesis * petrophysics * seismology and geomagnetism
Periodicals: "Annual Report" (ISSN 0155-624x), "Research Papers" (ISSN 0084-7518), "Earth Sciences at ANU" (ISSN 1032-5999)
City Reference Coordinates: 149°08'00"E 35°55'00"S

Australian National University (ANU), Research School of Physical Sciences, Department of Theoretical Physics
G.P.O. Box 4
Canberra, ACT 2601
Telephone: (0)2-6249-2943
Telefax: (0)2-6249-1884
Electronic Mail: <userid>@rsphy2.anu.edu.au
WWW: http://rsphysse.anu.edu.au/theophys/
Founded: 1949
Staff: 17
Activities: theoretical physics * cosmology * strings * nuclear astrophysics
City Reference Coordinates: 149°08'00"E 35°55'00"S

Australian National University (ANU), Research School of Physical Sciences and Engineering, Plasma Research Laboratory (PRL)
Canberra, ACT 0200
Telephone: (0)2-6249-4680
Telefax: (0)2-6249-2575
Electronic Mail: prl@anu.edu.au
WWW: http://www.anu.edu.au/rsphyse/prl/PRL.html
Founded: 1980

Staff: 20
Activities: plasma physics * fusion * plasma processing * thin films * radio frequency heating * space plasma * helicon waves
City Reference Coordinates: 149°08'00"E 35°55'00"S

Australian National University (ANU), School of Mathematical Sciences, Astrophysical Theory Centre (ATC)
Canberra, ACT 0200
Electronic Mail: <userid>@maths.anu.edu.au
WWW: http://wwwmaths.anu.edu.au/atc/
City Reference Coordinates: 149°08'00"E 35°55'00"S

Australian Space Industry Chamber of Commerce (ASICC)
G.P.O. Box 7048
Sydney, NSW 2001
Telephone: (0)2-6228-1327
Telefax: (0)2-6233-2858
Founded: 1993
Staff: 1
Activities: biennial Australian space development conference * promotion of space development * space industry liaison
City Reference Coordinates: 151°13'00"E 33°52'00"S

Australian Space Research Institute (ASRI)
P.O. Box 20
Elizabeth, SA 5112
Telephone: (0)8-8259-5316
Electronic Mail: mark.blair@dsto.defence.gov.au (Mark Blair, Chairman)
WWW: http://www.asri.org.au/
Founded: 1993
Membership: 60
Activities: space technology education
Periodicals: (6) "ASRI News"
City Reference Coordinates: 138°40'00"E 34°43'00"S

Australia Telescope National Facility (ATNF)
● See "Commonwealth Scientific and Industrial Research Organisation (CSIRO), Australia Telescope National Facility (ATNF)"

Ballaarat Astronomical Society (BAS), Inc.
P.O. Box 284
Ballarat, VIC 3353
or :
Corner Magpie and Cobden Streets
Mount Pleasant
Ballarat, VIC
Electronic Mail: wfiddian@cbl.com.au
WWW: http://www.giant.net.au/astronomy/society.html
Founded: 1958 (Observatory: 1886)
Membership: 40
Activities: education * public viewing * maintenance of historical telescopes
Periodicals: (4) "Oddie-Baker Bulletin"
Coordinates: 143°51'30"E 37°34'59"S (Ballarat Observatory)
City Reference Coordinates: 143°52'00"E 37°34'00"S (Ballarat)

Bendigo District Astronomical Society (BDAS) Inc.
P.O. Box 123
Golden Square
Bendigo, VIC 3550
Telephone: (0)3-5448-4563
Electronic Mail: rbath@bendigo.net.au
WWW: http://www.bendigo.net.au/~rbath/
Founded: 1985
Membership: 15
Activities: monthly meetings * field nights * education
City Reference Coordinates: 144°17'00"E 36°46'00"S

Binocular and Telescope Shop
55 York Street
Sydney, NSW 2000
Telephone: (0)2-9262-1344
Telefax: (0)2-9262-1844
Electronic Mail: mike@bintel .com.au
WWW: http://www.bintel.com.au/
Founded: 1984
Staff: 2
Activities: retail sales of binoculars, telescopes, astronomy books and related material

Periodicals: (12) "The Night Sky" (circ.: 2,500)
City Reference Coordinates: 151°13'00"E 33°52'00"S

Brisbane Astronomical Society (BAS), Inc.
P.O. Box 204
Morningside, QLD 4170
Telephone: (0)7-3286-5807
Electronic Mail: nwilliam@b022.aone.net.au
WWW: http://www.ozemail.com.au/~nwilliam/bas/bas.html
Founded: 1986
Membership: 60
Activities: public star nights * school astronomy nights * meetings * observing * education * astrocamps
Periodicals: (6) "Brisbane Astronomical Society Newsletter"
City Reference Coordinates: 153°02'00"E 27°28'00"S (Brisbane)

Brisbane Planetarium (Sir Thomas_)
● See "Sir Thomas Brisbane Planetarium"

Bundaberg Astronomical Society Inc.
c/o Alloway Observatory
Goodwood Road
P.O. Box 4221
Bundaberg, QLD 4670
Telephone: (0)7-4159-7231
Electronic Mail: observe@interworx.com.au
WWW: http://www.angelfire.com/al/AstronDirectory
Founded: 1964
Membership: 20
Activities: observing (variable stars, planetary phenomena) * photography * public lectures and viewing * telescope making * occultation timing * computing
Coordinates: 152°22'38"E 24°56'35"S H28m (Alloway Observatory)
City Reference Coordinates: 152°21'00"E 24°52'00"S

Canberra Astronomical Society (CAS), Inc.
P.O. Box 1338
Woden, ACT 2606
Telephone: (0)2-6248-0552
Electronic Mail: dm@isd.canberra.edu.au
WWW: http://msowww.anu.edu.au/cas/
        http://www.mso.anu.edu.au/cas/
Founded: 1969
Membership: 140
Activities: observing (grazing occultations * deep sky) * photography * instrumentation * CCDs * education
Periodicals: (12) "The Southern Cross"
City Reference Coordinates: 149°08'00"E 35°55'00"S

Canberra Deep Space Communication Complex (CDSCC)
P.O. Box 4350
Kingston, ACT 2604
or :
Tidbindilla, ACT
Telephone: (0)2-6201-7800
Telefax: (0)2-6201-7808
        (0)2-6201-7975
WWW: http://tid.cdscc.nasa.gov/
Founded: 1965
Staff: 150
Activities: NASA Space Tracking Station * tracking of Earth orbiting and deep-space spacecraft * communication between the Jet Propulsion Laboratory (JPL) (see separate entry, USA-CA) and spacecraft
Coordinates: 148°59'00"E 35°24'00"S H660m
City Reference Coordinates: 149°08'00"E 35°55'00"S (Canberra)

Canberra Planetarium and Observatory (CPO)
P.O. Box 207
Dickson, ACT 2602
or :
Hawdon Place
Dickson, ACT
Telephone: (0)2-6248-5333
Telefax: (0)2-6249-7238
Electronic Mail: planetarium@cfmeu.asn.au
WWW: http://www.cfmeu.asn.au/planetarium/
Founded: 1991
Staff: 14
Activities: public education in astronomy

Periodicals: (12) "Canberra Observer"
City Reference Coordinates: 149°08'00"E 35°55'00"S

Commonwealth Bureau of Meteorology, Head Office
P.O. Box 1289 K
Melbourne, VIC 3001
or :
150 Lonsdale Street
Melbourne, VIC 3000
Telephone: (0)3-9669-4000
Telefax: (0)3-9669-4699
       (0)3-9669-4548
Electronic Mail: info@bom.gov.au
WWW: http://www.bom.gov.au/
Founded: 1908
Staff: 1600
Activities: national meteorological service * network of 60 observing sites
Periodicals: (4) "Australian Meteorological Magazine"; (1) "Annual Report"
City Reference Coordinates: 144°58'00"E 37°49'00"S

Commonwealth Bureau of Meteorology, Canberra Meteorological Office
G.P.O. Box 787
Canberra, ACT 2601
Telephone: (0)2-6247-0411
Electronic Mail: climate.act@bom.gov.au
WWW: http://www.bom.gov.au/
Founded: 1908 (Head Office)
City Reference Coordinates: 149°08'00"E 35°55'00"S

Commonwealth Bureau of Meteorology, NSW Regional Office
P.O. Box 413
Darlinghurst, NSW 2010
Telephone: (0)2-9296-1555
Electronic Mail: info@bom.gov.au
WWW: http://www.bom.gov.au/
Founded: 1908 (Head Office)
City Reference Coordinates: 151°13'00"E 33°52'00"S (Sydney)

Commonwealth Bureau of Meteorology, Northern Territory Regional Office
P.O. Box 735
Darwin, NT 0801
Telephone: (0)8-8982-2800
Electronic Mail: info@bom.gov.au
WWW: http://www.bom.gov.au/
Founded: 1908 (Head Office)
City Reference Coordinates: 130°44'00"E 12°23'00"S

Commonwealth Bureau of Meteorology, Queensland Regional Office
G.P.O. Box 413
Brisbane, QLD 4001
Telephone: (0)7-3864-8101
Electronic Mail: info@bom.gov.au
WWW: http://www.bom.gov.au/
Founded: 1908 (Head Office)
City Reference Coordinates: 153°02'00"E 27°28'00"S

Commonwealth Bureau of Meteorology, South Australia Regional Office
P.O. Box 421
Kent Town, SA 5071
Telephone: (0)8-8366-2600
Electronic Mail: info@bom.gov.au
WWW: http://www.bom.gov.au/
Founded: 1908 (Head Office)
City Reference Coordinates: 138°35'00"E 34°55'00"S (Adelaide)

Commonwealth Bureau of Meteorology, Tasmania and Antarctica Regional Office
G.P.O. Box 727G
Hobart, TAS 7001
Telephone: (0)3-6221-2000
Electronic Mail: info@bom.gov.au
WWW: http://www.bom.gov.au/
Founded: 1908 (Head Office)
City Reference Coordinates: 149°19'00"E 42°31'00"S

Commonwealth Bureau of Meteorology, Victoria Regional Office
P.O. Box 1636M
Melbourne, VIC 3001
Telephone: (0)3-9669-4900
Electronic Mail: info@bom.gov.au
WWW: http://www.bom.gov.au/
Founded: 1908 (Head Office)
City Reference Coordinates: 144°58'00"E 37°49'00"S

Commonwealth Bureau of Meteorology, Western Australia Regional Office
P.O. Box 1370
West Perth, WA 6872
or :
5/1100 Hay Street
West Perth, WA 6005
Telephone: (0)8-9263-2222
Telefax: (0)8-9263-2211
Electronic Mail: info@bom.gov.au
WWW: http://www.bom.gov.au/
Founded: 1908 (Head Office)
Staff: 75
Activities: meteorological forecasting * hydrological forecasting * climate data * climate studies
Coordinates: 115°50'43"E 31°57'03"S
City Reference Coordinates: 115°50'00"E 31°56'00"S

Commonwealth Scientific and Industrial Research Organisation (CSIRO), Atmospheric Research
Private Bag 1
Robertson Parade
Aspendale, VIC 3195
Telephone: (0)3-9586-7522
                     (0)3-9239-4400
Telefax: (0)3-9239-4444
Electronic Mail: <userid>@dar.csiro.au
WWW: http://www.dar.csiro.au/
               http://www.csiro.au/
Founded: 1949 (CSIRO)
Staff: 150
City Reference Coordinates: 145°07'00"E 38°02'00"S

Commonwealth Scientific and Industrial Research Organisation (CSIRO), Australia Telescope National Facility (ATNF), Headquarters
P.O. Box 76
Epping, NSW 2121
or :
c/o Radiophysics Laboratory
Vimiera and Pembroke Roads
Marsfield, NSW 2122
Telephone: (0)2-9372-4100
Telefax: (0)2-9372-4310
Electronic Mail: atnf@atnf.csiro.au
                              <userid>@atnf.csiro.au
WWW: http://www.atnf.csiro.au/overview/guide_epping.html
               http://www.atnf.csiro.au/
Founded: 1988
Staff: 75
Activities: operating the Australia Telescope as a national facility * galactic and extragalactic research
Coordinates: 151°05'42"E 33°46'42"S
City Reference Coordinates: 145°02'00"E 37°39'00"S (Epping)
                                                  151°07'00"E 33°47'00"S (Marsfield)

Commonwealth Scientific and Industrial Research Organisation (CSIRO), Australia Telescope National Facility (ATNF), Mopra Telescope
c/o Paul Wild Observatory
Locked Bag 194
Narrabri, NSW 2390
Telephone: (0)2-6849-1800
Telefax: (0)2-6849-18888
Electronic Mail: <userid>@atnf.csiro.au
WWW: http://www.atnf.csiro.au/overview/guide_mopra.html
               http://www.atnf.csiro.au/
Founded: 1988 (ATNF)
Activities: mm signle-dish observing and VLBI
Coordinates: 149°06'00"E 31°16'10"S
City Reference Coordinates: 149°47'00"E 30°19'00"S

Commonwealth Scientific and Industrial Research Organisation (CSIRO), Australia Telescope National Facility (ATNF), Narrabri
(Paul Wild Observatory)
Locked Bag 194
Narrabri, NSW 2390
Telephone: (0)2-6790-4000
Telefax: (0)2-6790-4090
Electronic Mail: narrabri@atnf.csiro.au
               <userid>@atnf.csiro.au
WWW: http://www.narrabri.atnf.csiro.au/www/public/site_info/site_info.html
        http://www.atnf.csiro.au/
Founded: 1988
Staff: 33
Activities: control centre of and site of Compact Array of Australia Telescope
City Reference Coordinates: 149°47'00"E 30°19'00"S

Commonwealth Scientific and Industrial Research Organisation (CSIRO), Australia Telescope National Facility (ATNF), Parkes Radio Observatory
P.O. Box 276
Parkes, NSW 2870
Telephone: (0)2-6861-1700
Telefax: (0)2-6861-1730
Electronic Mail: parkes@atnf.csiro.au
                 <userid>@atnf.csiro.au
WWW: http://www.parkes.atnf.csiro.au/
        http://www.atnf.csiro.au/
        http://wwwpks.atnf.csiro.au/visitors_centre/VCHomePage.html (Visitor Centre)
Founded: 1988 (ATNF)
Staff: 22
Activities: radio observing * VLBI
Coordinates: 148°15'44"E 33°00'00"S H392m
City Reference Coordinates: 148°13'00"E 33°10'00"S

Commonwealth Scientific and Industrial Research Organisation (CSIRO), Earth Observation Centre (EOC)
P.O. Box 3023
Canberra, ACT 2600
or :
Corner North and Daley Roads
ANU Campus
Acton, ACT 2601
Telephone: (0)2-6216-7200
Telefax: (0)2-6216-7222
Electronic Mail: <userid>@cossa.csiro.au
WWW: http://www.eoc.csiro.au/
Founded: 1984
Staff: 16
City Reference Coordinates: 149°08'00"E 35°55'00"S

Commonwealth Scientific and Industrial Research Organisation (CSIRO), Editorial Services
● See now "CSIRO Publishing"

Commonwealth Scientific and Industrial Research Organisation (CSIRO), Office of Space Science and Applications (COSSA)
P.O. Box 3023
Canberra, ACT 2600
or :
Corner North and Daley Roads
ANU Campus
Acton, ACT 2601
Telephone: (0)2-6216-7200
Telefax: (0)2-6216-7222
Electronic Mail: <userid>@thor.cbr.cossa.csiro.au
WWW: http://www.eoc.csiro.au/cossa/cossa.htm
Founded: 1984
Staff: 11
Activities: international liaison on space projects * information on space programs * remote sensing instruments and aircraft * managing CSIRO space projects
Periodicals: (6) "CSIRO Space Industry News" (ISSN 1037-5759)
City Reference Coordinates: 149°08'00"E 35°55'00"S

Commonwealth Scientific and Industrial Research Organisation (CSIRO), Paul Wild Observatory
● See "Commonwealth Scientific and Industrial Research Organisation (CSIRO), Australia Telescope National Facility, Narrabri"

Computer Transition Systems (CTS)
P.O. Box 4553
Melbourne, VIC 3001
or :
20 Dawson Avenue
Brighton, VIC 3186
Telephone: (0)3-9530-6633
Telefax: (0)3-9530-6644
Electronic Mail: info@cts.com.au
WWW: http://www.labyrinth.net.au/~ctrans/
Founded: 1984
Staff: 2
Activities: supplying Fortran compilers and related support for PCs and workstations
Periodicals: (3) "Computer Transition Systems Report"
City Reference Coordinates: 144°58'00"E 37°49'00"S (Melbourne)
                                                  145°00'00"E 37°55'00"S (Brighton)

CSIRO Publishing
150 Oxford Street
P.O. Box 1139
Collingwood, VIC 3066
Telephone: (0)3-9662-7500
Telefax: (0)3-9662-7555
WWW: http://www.publish.csiro.au/
                http://www.csiro.au/
Founded: 1949
Staff: 50
Activities: publishing books, journals and multimedia
Periodicals: among others: (12) "Australian Journal of Chemistry" (ISSN 0004-9425), "Australian Journal of Physics" (ISSN 0004-9506); (3) "Publications of the Astronomical Society of Australia" (ISSN 1323-3580)
City Reference Coordinates: 145°00'00"E 37°48'00"S

Federation of Australian Scientific and Technological Societies (FASTS)
c/o Toss Gascoigne (Executive Director)
P.O. Box 218
Deakin West, ACT 2601
Telephone: (0)2-6257-2891 (Work)
Telefax: (0)2-6257-2987
Electronic Mail: fasts@anu.edu.au
WWW: http://www.usyd.edu.au/su/fasts/
Founded: 1985
Membership: 50 (societies)
Activities: representing the interests of scientists and technologists throughout Australia
City Reference Coordinates: 149°08'00"E 35°55'00"S (Canberra)

GIO Australia
G.P.O. Box 1559
Sydney, NSW 2001
or :
2 Martin Place
Sydney, NSW 2000
Telephone: (0)2-9228-1327
Telefax: (0)2-9233-2958
WWW: http://www.gio.com.au/
Founded: 1926
Staff: 4000
Activities: satellite and general insurance
City Reference Coordinates: 151°13'00"E 33°52'00"S

Government Insurance Office (GIO)
● See "GIO Australia"

Grove Creek Observatory (GCO)
Unit 16
1-5 Stokes Street
Lane Cove, NSW 2066
Telephone: (0)2-9428-4334
                (0)2-9636-8611 (Observatory)
Telefax: (0)2-9558-2721
                (0)2-9418-6005
Electronic Mail: info@gco.org.au
WWW: http://www.gco.org.au/
Founded: 1978
Coordinates: 149°22'23"E 33°51'05"S H830m
City Reference Coordinates: 151°13'00"E 33°52'00"S (Sydney)

Hawkesbury Astronomical Association (HAA) Inc.
P.O. Box 670
Windsor, NSW 2756
Telephone: (0)2-4572-1568
WWW: http://www.rpi.net.au/~haa/
Founded: 1989
Membership: 30
City Reference Coordinates: 150°50'00"E 39°39'00"S

H.V. McKay Planetarium
• See now "Melbourne Planetarium at Scienceworks"

Institution of Engineers, Australia (IEAUST)
11 National Circuit
Barton, ACT 2600
Telephone: (0)2-6270-6552
Telefax: (0)2-6273-1488
Electronic Mail: contact_cust_service@ieaust.org.au
                    memberservices@ieaust.org.au
WWW: http://www.ieaust.org.au/
Founded: 1919
Membership: 60000
Staff: 100
Activities: professional society * National Committee on Space Engineering
Periodicals: (4) "Civil Engineering Transactions" (ISSN 0020-3319), "Mechanical Engineering Transactions", "Electrical Engineering Transactions"; (2) "Multi-Disciplinary Transactions"
City Reference Coordinates: 132°39'00"E 30°31'00"S

IPS Radio & Space Services
P.O. Box 1386
Haymarket, NSW 1240
or :
Level 6 - North Wing
477 Pitt Street
Sydney, NSW 2000
Telephone: (0)2-9213-8000
Telefax: (0)2-9213-8060
Electronic Mail: office@ips.gov.au
                    <userid>@ips.gov.au
WWW: http://www.ips.gov.au/
        http://www.ips.gov.au/culgoora/ (Culgoora Solar Observatory)
        http://www.ips.gov.au/learmonth/ (Learmonth Solar Observatory)
Founded: 1948
Staff: 40
Activities: providing space weather data, products and services
Periodicals: (12) "IPS Monthly Solar and Geophysical Summary"
City Reference Coordinates: 151°13'00"E 33°52'00"S (Sydney)

La Trobe University, Space Physics Group
Bundoora, VIC 3083
Telephone: (0)3-9479-2735
Telefax: (0)3-9479-1552
Electronic Mail: <userid>@latrobe.edu.au
WWW: http://www.latrobe.edu.au/www/physics/space/space.htm
Founded: 1967
Staff: 8
Activities: ionosphere * magnetosphere * relativity * atomic physics * liquid physics
Coordinates: 144°06'00"E 37°28'00"S (Beveridge)
City Reference Coordinates: 144°06'00"E 37°28'00"S

Launceston Planetarium
c/o Queen Victoria Museum and Art Gallery
Wellington Street
Launceston, TAS 7250
Telephone: (0)3-6323-3777
Telefax: (0)3-6323-3776
Electronic Mail: martin@qvmag.tased.edu.au (Martin George, Director)
WWW: http://www.vision.net.au/~peter/AST/launplan/launplan.htm
Founded: 1967
Staff: 2
Activities: public and school programs
City Reference Coordinates: 147°08'00"E 41°26'00"S

McKay Planetarium (H.V._)
• See now "Melbourne Planetarium at Scienceworks"

Melbourne Planetarium at Scienceworks
Museum of Victoria
P.O. Box 666E
Melbourne, VIC 3001
Telephone: (0)3-9392-4815
Telefax: (0)3-9399-2028
WWW: http://www.mov.vic.gov.au/planetarium/
Founded: 1965
Staff: 4
Activities: shows for general public and schools
City Reference Coordinates: 144°58'00"E 37°49'00"S
● Formerly known as "H.V. McKay Planetarium"

Molonglo Radio Observatory
● See "University of Sydney, Department of Astrophysics, Molonglo Radio Observatory"

Monash University, Department of Mathematics, Astrophysics Group
Melbourne, VIC 3168
Telephone: (0)3-9905-4431
Telefax: (0)3-9905-3867
Electronic Mail: <userid>@sci.monash.edu.au
                    astro_contact@hal.maths.monash.edu.au
WWW: http://www.maths.monash.edu.au/astro/
Founded: 1961
Staff: 10
Activities: solar physics * stellar evolution * star formation * numerical hydrodynamics * planetary physics
City Reference Coordinates: 144°58'00"E 37°49'00"S

Monash University, Department of Physics
Melbourne, VIC 3168
Telephone: (0)3-9905-3639
Telefax: (0)3-9905-3637
Electronic Mail: <userid>@sci.monash.edu.au
WWW: http://www.physics.monash.edu.au/
Founded: 1970
Staff: 2
Activities: photometry * astrometry * education
Coordinates: 145°07'52"E 37°54'08"S H50m
              145°29'39"E 37°58'37"S H311m (Mount Burnett)
City Reference Coordinates: 144°58'00"E 37°49'00"S

Mopra Observatory
● See "Commonwealth Scientific and Industrial Research Organisation (CSIRO), Australia Telescope National Facility (ATNF), Mopra Observatory"

Morel Astrographics
c/o Mati Morel
6 Blakewell Road
Thornton, NSW 2322
Telephone: (0)2-4966-2078
Telefax: (0)2-4966-2078
Electronic Mail: morel@ozemail.com.au
Founded: 1985
Staff: 1
Activities: collecting stellar data * preparing charts * compiling indices and directories of stellar data * variable stars
Periodicals: (6) "Variable Star Memorandum"; (1) "Stellar Data"
City Reference Coordinates: 151°39'00"E 32°47'00"S

Mount Pleasant Radio Astronomy Observatory
● See "University of Tasmania, Department of Physics"

Mount Stromlo and Siding Spring Observatories (MSSSO)
● See "Australian National University (ANU), Research School of Astronomy and Astrophysics (RSAA), Mount Stromlo and Siding Spring Observatories (MSSSO)"

Mount Tamborine Observatory (MTO)
● See "University of Queensland, Department of Physics"

Murdoch Astronomical Society (MAS)
c/o Physics and Energy Studies
Murdoch University
South Street
Murdoch, WA 6150
Telephone: (0)8-9360-2433
          (0)8-9360-2865

Telefax: (0)8-9310-1711
Electronic Mail: mas@cleo.murdoch.edu.au
WWW: http://fizzy.murdoch.edu.au/~walker/mas.html
         http://cleo.murdoch.edu.au/clubs/mas/
Founded: 1978
Membership: 36
Activities: meetings * dark-sky nights * public viewing nights
Periodicals: "Newsletter"
City Reference Coordinates: 115°50'00"E 31°56'00"S (Perth)

National Committee for Astronomy (NCA), Australia
G.P.O. Box 783
Canberra, ACT 2601
or :
Ian Potter House
Gordon Street
Acton, ACT 2600
Telephone: (0)2-6247-3966
Telefax: (0)2-6257-4620
Electronic Mail: ns@science.org.au
WWW: http://msowww.anu.edu.au/~jrm/NCA/
         http://www.science.org.au/
Membership: 11
City Reference Coordinates: 149°08'00"E 35°55'00"S

National Space Society of Australia Ltd. (NSSA)
G.P.O. Box 7048
Sydney, NSW 2001
Telephone: (0)2-9687-3301
WWW: http://www.nssa.com.au/
Founded: 1982
Membership: 300
Activities: promoting space development, space education and space industry liaison
Periodicals: (6) "Ad Astra" (circ.: 25,000)
City Reference Coordinates: 151°13'00"E 33°52'00"S

Nepean Astronomy Centre (NAC)
• See "University of Western Sydney (UWS), Nepean Astronomy Centre (NAC)"

Newcastle Astronomical Society (NAS)
c/o Department of Physics
University of Newcastle
Callaghan, NSW 2308
Telephone: (0)2-4942-6029
Electronic Mail: physpuls4@cc.newcastle.edu.au (Ian Dunlop)
                phmc@cc.newcastle.edu.au (Michael Cvetanovski)
WWW: http://www.newcastle.edu.au/department/ph/plasma/NAS/nas_home.html
City Reference Coordinates: 151°56'00"E 32°56'00"S (Newcastle)

Newcastle University, Physics Department
Callaghan, NSW 2308
Telephone: (0)2-4921-5440
           (0)2-4921-5451
Telefax: (0)2-4921-6907
Electronic Mail: phcslk@cc.newcastle.edu.au
WWW: http://www.newcastle.edu.au/department/ph/main.htm
Founded: 1965
Activities: electrophonic effects of large meteor fireballs
Coordinates: 151°41'49"E 32°53'43"S
City Reference Coordinates: 151°42'00"E 32°54'00"S

Northern Sydney Astronomical Society (NSAS)
P.O. Box 999
North Turramurra, NSW 2074
or :
P.O. Box 214
West Ryde, NSW 2114
Electronic Mail: johnc@fl.net.au (John Curdie, President)
WWW: http://www.users.fl.net.au/~johnc/NSAS/
City Reference Coordinates: 151°13'00"E 33°52'00"S (Sydney)

Parkes Radio Observatory
• See "Commonwealth Scientific and Industrial Research Organisation (CSIRO), Australia Telescope National Facility (ATNF), Parkes Radio Observatory"

Paul Wild Observatory
● See "Commonwealth Scientific and Industrial Research Organisation (CSIRO), Australia Telescope National Facility (ATNF), Narrabri"

Perth Observatory
Walnut Road
Bickley, WA 6076
Telephone: (0)8-9293-8255
            (0)8-9293-8109 (Information Line)
Telefax: (0)8-9293-8138
Electronic Mail: perthobs@calm.wa.gov.au
WWW: http://www.wa.gov.au/perthobs/
Founded: 1896
Staff: 11
Activities: photometry of solar system objects * education * SN search * minor planet and comet astrometry * stellar photometry
Periodicals: "Astronomical Handbook", "Communications of Perth Observatory" (ISSN 0079-1067)
Coordinates: 116°08'06"E 32°00'30"S H391m
City Reference Coordinates: 115°50'00"E 31°56'00"S (Perth)

Perth Omni Theatre and Planetarium
29 City West Centre
Railway Parade
West Perth, WA 6005
Telephone: (0)8-9481-6481
            (0)8-9481-5890
Telefax: (0)8-9481-6530
Electronic Mail: omni@wt.com.au
WWW: http://edsitewa.iinet.net.au/omni/
         http://edsitewa.iinet.net.au/omni/planet.html
City Reference Coordinates: 115°50'00"E 31°56'00"S

Phase Space Technology (PST)
P.O. Box 422
Mount Evelyn, VIC 3796
Telephone: (0)3-9735-2270
Telefax: (0)3-9735-2270
Electronic Mail: pst@phasespace.com.au
WWW: http://www.phasespace.com.au/
● Software producer
City Reference Coordinates: 145°23'00"E 37°47'00"S (Perth)

Port Macquarie Astronomical Association
P.O. Box 1453
Port Macquarie, NSW 2444
or :
Observatory
Rotary Park
Port Macquarie, NSW
Telephone: (0)2-6583-1933
Founded: 1962
Activities: lectures * observing
City Reference Coordinates: 152°55'00"E 31°26'00"S

Scitech Discovery Centre
P.O. Box 1155
West Perth, WA 6872
or :
City West Centre
Railway Parade
West Perth, WA 6005
Telephone: (0)8-9481-6295
Telefax: (0)8-9321-2869
Electronic Mail: <userid>@scitech.org.au
WWW: http://www.scitech.org.au/
Founded: 1987
Staff: 50
Activities: exhibitions * portable Starlab planetariums
City Reference Coordinates: 115°50'00"E 31°56'00"S

Siding Spring Observatory
● See "Australian National University (ANU), Research School of Astronomy and Astrophysics (RSAA), Mount Stromlo and Siding Spring Observatories (MSSSO)"

Sir Thomas Brisbane Planetarium (STBP)
Botanic Gardens
Mount Coot-tha
Brisbane, QLD
Telephone: (0)7-3403-2578
WWW: http://enterprise.powerup.com.au/~stbp/stbp_.html
         http://www.powerup.com.au/~stbp/stbp_.html
Founded: 1978
Activities: shows * observing
City Reference Coordinates: 153°02'00"E 27°28'00"S

Sky & Space Publishing
80 Ebley Street
Bondi Junction, NSW 2022
Telephone: (0)2-9369-3344
Telefax: (0)2-9369-3366
Electronic Mail: 100246.2000@compuserve.com
Founded: 1988
Periodicals: "Sky & Space" (6) (ISSN 1035-932x)
City Reference Coordinates: 151°13'00"E 33°52'00"S (Sydney)

South East Queensland Astronomical Society (SEQAS)
P.O. Box 516
Strathpine, QLD 4500
Electronic Mail: glong@eis.net.au (Graham Long, President)
WWW: http://www.ozemail.com.au/~mhorn/seqas.html
Founded: 1985
Membership: 57
Periodicals: (4) "Universal Times"
City Reference Coordinates: 152°59'00"E 27°19'00"S

Southern Astronomical Society (SAS) Inc.
P.O. Box 867
Beenleigh, QLD 4207
Telephone: (0)7-3209-1741 (Sue Dreves)
Electronic Mail: renato@odyssey.com.au (Renato Langersek)
WWW: http://www.sas.org.au/
Founded: 1986
City Reference Coordinates: 153°12'00"E 27°43'00"S

Space Association of Australia (SAA) Inc.
P.O. Box 351
Mulgrave North, VIC 3170
Telephone: (0)3-9729-5538
Electronic Mail: abdilla@rmit.edu.au (Andrew Rennie, President)
WWW: http://vicnet.net.au/~saa/
Founded: 1981
Activities: promoting astronautics * monthly public meetings * weekly radio programmes
City Reference Coordinates: 145°12'00"E 37°56'00"S

Space Centre for Satellite Navigation (SCSN)
c/o Queensland University of Technology
G.P.O. Box 2434
Brisbane, QLD 4001
or :
2 George Street
Brisbane, QLD 4000
Telephone: (0)7-3864-2392
Telefax: (0)7-3864-1361
WWW: http://www.scsn.bee.qut.edu.au/
Founded: 1993
Staff: 10
Activities: satellite navigation * vehicle positioning * sensor integration
City Reference Coordinates: 153°02'00"E 27°28'00"S

Standards Association of Australia (SAA)
P.O. Box 1055
Strathfield, NSW 2135
or :
1 The Crescent
Homebush, NSW 2140
or :
80-86 Arthur Street
P.O. Box 458

North Sydney, NSW 2059
Telephone: (0)2-9746-4700
Telefax: (0)2-9746-8450
Electronic Mail: intsect@standards.com.au
WWW: http://www.standards.com.au/
Founded: 1922
City Reference Coordinates: 151°56'00"E 33°52'00"S (Strathfield)
                                151°05'00"E 33°50'00"S (Homebush)
                                151°13'00"E 33°50'00"S (North Sydney)

Sutherland Astronomical Society Inc. (SASI)
P.O. Box 31
Sutherland, NSW 1499
Telephone: (0)2-9589-1014
Telefax: (0)2-9589-1014
Electronic Mail: sasi@ozemail.com.au
WWW: http://www.ozemail.com.au/~sasi/
Founded: 1962
Membership: 110
Activities: observing * education * star nights * field nights * junior group
Periodicals: (6) "Southern Observer"
Awards: "SASI Award"
Coordinates: 151°04'18"E 34°00'13"S H54m
City Reference Coordinates: 151°04'00"E 34°02'00"S

Swinburne University of Technology, Astrophysics and Supercomputing
Mail Number 31
P.O. Box 218
Hawthorn, VIC 3122
Telephone: (0)3-9214-8782
Telefax: (0)3-9819-0856
Electronic Mail: mbailes@swin.edu.au (Matthew Bailes)
WWW: http://www.swin.edu.au/astronomy/
City Reference Coordinates: 144°58'00"E 37°49'00"S (Melbourne)

Sydney Observatory
c/o Museum of Applied Arts and Sciences
P.O. Box K346
Haymarket, NSW 1238
or :
Observatory Hill
The Rocks
Sydney, NSW
Telephone: (0)2-9217-0485
Telefax: (0)2-9217-0489
WWW: http://www.phm.gov.au/observe/default.htm
Founded: 1858
Staff: 8
Activities: public education * public observing * astronomical museum
Periodicals: (1) "The Sydney Sky Guide" (ISSN 1039-3048)
Coordinates: 151°12'12"E 33°51'40"S H44m
City Reference Coordinates: 151°13'00"E 33°52'00"S (Sydney)

Sydney Outdoor Lighting Improvement Society (SOLIS) Inc.
P.O. Box 999
North Turramurra, NSW 2074
Electronic Mail: solislp@netscape.net
WWW: http://sites.netscape.net/solislp/homepage
City Reference Coordinates: 151°13'00"E 33°52'00"S (Sydney)

Sydney Space Frontier Society (SSFS)
c/o Wayne Short
G.P.O. Box 7048
Sydney, NSW 2001
Telephone: (0)2-9502-3063
Electronic Mail: wshor@doh.health.nsw.gov.au
                    ssfs@nssa.com.au
Founded: 1982
Membership: 100
Activities: promoting space development
Periodicals: (4) "Space Frontier News" (ISSN 1032-4739, circ.: 300)
City Reference Coordinates: 151°13'00"E 33°52'00"S

Taree Astronomical Society (TAS)
P.O.Box 111

Taree, NSW 2430
Telephone: (0)2-6550-2213
Electronic Mail: rosco@midcoast.com.au (Jim Ross)
WWW: http://www.midcoast.com.au/users/rosco/rosco.html
City Reference Coordinates: 152°26'00"E 31°54'00"S

Thomas Brisbane Planetarium (Sir.)
• See "Sir Thomas Brisbane Planetarium"

Tidbindilla Space Tracking Station
• See "Canberra Deep Space Communications Complex (CDSCC)"

United Kingdom Schmidt Telescope (UKST)
Private Bag
Coonabarabran, NSW 2357
Telephone: (0)2-6842-1622
Telefax: (0)2-6842-2288
Electronic Mail: schmidt@aaocbn3.aao.gov.au
                 <userid>@aaocbn3.aao.gov.au
WWW: http://www.aao.gov.au/schmidt.html
Founded: 1973
Activities: service facility providing deep-sky photography for systematic sky surveys and for special research projects
Coordinates: 149°04'12"E 31°16'24"S H1,130m
City Reference Coordinates: 149°18'00"E 31°16'00"S

University of Adelaide, Department of Physics, High-Energy Astrophysics Research Group
North Terrace
G.P.O. Box 498
Adelaide, SA 5001
Telephone: (0)8-8228-5996
Telefax: (0)8-8224-0464
Electronic Mail: <userid>@physics.adelaide.edu.au
WWW: http://www.physics.adelaide.edu.au/astrophysics/
       http://pilot.physics.adelaide.edu.au/astrophysics/home.html
       http://bragg.physics.adelaide.edu.au/astrophysics/home.html
Founded: 1966 (University: 1874)
Staff: 6
Activities: VHE, UHE and theoretical gamma-ray astronomy
Coordinates: 138°34'58"E 34°55'38"S H41m
City Reference Coordinates: 138°35'00"E 34°55'00"S

University of Melbourne, School of Physics, Astrophysics Group
Parkville, VIC 3010
Telephone: (0)3-9344-7670
Telefax: (0)3-9347-4783
Electronic Mail: <userid>@physics.unimelb.edu.au
WWW: http://www.ph.unimelb.edu.au/
       http://www.ph.unimelb.edu.au/astro/home.html
       http://astro.ph.unimelb.edu.au/
Founded: 1853
Staff: 25
Activities: education * research
City Reference Coordinates: 144°58'00"E 37°49'00"S (Melbourne)

University of New South Wales (UNSW), Centre for Remote Sensing
P.O. Box 1
Kensington, NSW 2033
Telephone: (0)2-9697-4964
Telefax: (0)2-9313-7493
Electronic Mail: <userid>@unsw.edu.au
WWW: http://www.geog.unsw.edu.au/~web/crsgis/info.html
Founded: 1981
Staff: 10
Activities: remote sensing * education * research * consulting
Periodicals: (1) "Annual Report"
City Reference Coordinates: 151°14'00"E 33°55'00"S

University of New South Wales (UNSW), Department of Astrophysics and Optics
Sydney, NSW 2052
Telephone: (0)2-9385-5649
Telefax: (0)2-9385-6060
WWW: http://www.phys.unsw.edu.au/astro.html
       http://newt.phys.unsw.edu.au/~mgb/jacara.html (Joint Australian Centre for Astrophysical Research in Antarctica)

Founded: 1962

Staff: 15
Activities: star formation * IR astronomy * PN * variable stars * telescope automation * extragalactic astronomy * Antarctic astronomy * mm astronomy
City Reference Coordinates: 151°13'00"E 33°52'00"S

University of New South Wales (UNSW), University College, Department of Physics
Canberra, ACT 2600
Telephone: (0)2-6268-8801
Telefax: (0)2-6268-8786
Electronic Mail: r-sood@adfa.edu.au (Ravi Sood, Head of School)
WWW: http://www.ph.adfa.edu.au/
Founded: 1986
Staff: 13
Activities: IR astronomy * laboratory astrophysics * high-energy astrophysics * nuclear magnetic resonance
Periodicals: (1) "Annual Report"
City Reference Coordinates: 149°08'00"E 35°55'00"S

University of Queensland, Department of Physics
Saint Lucia, QLD 4072
Telephone: (0)7-3365-3424
            (0)7-3365-2422 (MTO)
Telefax: (0)7-3371-5896
            (0)7-3365-1242 (MTO)
Electronic Mail: <userid>@physics.uq.edu.au
WWW: http://www.physics.uq.edu.au/
            http://www.physics.uq.edu.au/ap/ap.html (Astrophysics Group)
Founded: 1910 (Observatory: 1973)
Staff: 4
Activities: photoelectric photometry (chromospherically active stars, cataclysmic and eruptive variable stars, eclipsing binaries, Be stars) * transient planetary and cometary phenomena * catalogues * Nova search programme
Periodicals: "Nova Search" (ISSN 1031-508x)
Observatories: 1 (Mount Tamborine Observatory - MTO)
Coordinates: 153°12'48"E 27°58'21"S H560m
City Reference Coordinates: 153°02'00"E 27°28'00"S (Brisbane)

University of South Australia, School of Environmental and Recreation Management, Planetarium
Building P
Mawson Lakes Boulevard
Mawson Lakes, SA 5095
Telephone: (0)8-8302-3138
Telefax: (0)8-8302-5082
Electronic Mail: wayne.looker@unisa.edu.au (Wayne Looker, Manager)
WWW: http://ching.apana.org.au/~oliri/planet.html
            http://www.unisa.edu.au/erm/planetarium/pla-home.htm
Founded: 1972
Staff: 3
Activities: shows * education
Coordinates: 138°30'00"E 34°52'00"S
City Reference Coordinates: 138°35'00"E 34°55'00"S (Adelaide)

University of Sydney, Chatterton Astronomy Department
School of Physics A28
Sydney, NSW 2006
Telephone: (0)2-9351-2544
Telefax: (0)2-9351-7726
Electronic Mail: <userid>@astron.physics.usyd.oz.au
WWW: http://www.physics.usyd.edu.au/astron/astron.html
Founded: 1961
Staff: 5
Activities: high-resolution stellar interferometry * determination of fundamental stellar quantities * high-precision stellar photometry * astronomical seeing studies * Sydney University Stellar Interferometer (SUSI)
City Reference Coordinates: 151°13'00"E 33°52'00"S

University of Sydney, Department of Astrophysics
Sydney, NSW 2006
Telephone: (0)2-9351-2544
Telefax: (0)2-9351-7726
Electronic Mail: <userid>@physics.usyd.oz.au
WWW: http://www.physics.usyd.edu.au/astrop/astrop.html
            http://www.physics.usyd.edu.au/
Founded: 1960
Staff: 10
Activities: galactic and extragalactic continuum radio astronomy * optical identification of radio sources * QSO optical spectra * galactic SNRs
Coordinates: 149°25'25"E 35°22'19"S H732m (Molonglo Observatory Synthesis Telescope)
City Reference Coordinates: 151°13'00"E 33°52'00"S

University of Sydney, Department of Astrophysics, Molonglo Radio Observatory
R.M.B. 107
Hoskinstown, NSW 2621
Telephone: (0)6-2238-2262
Telefax: (0)6-2238-2290
WWW: http://www.physics.usyd.edu.au/astrop/astrop.html
       http://www.physics.usyd.edu.au/astrop/most/
Founded: 1964
Activities: operation of the 843 MHz continuum Molonglo Observatory Synthesis Telescope (MOST)
Coordinates: 149°25'25"E 35°22'19"S H732
City Reference Coordinates: 151°13'00"E 33°52'00"S (Sydney)

University of Sydney, Research Centre for Theoretical Astrophysics (RCfTA)
c/o School of Physics
Sydney, NSW 2006
Telephone: (0)2-9351-2542
Telefax: (0)2-9351-7726
Electronic Mail: rcfta@physics.usyd.edu.au
            <userid>@physics.usyd.edu.au
WWW: http://www.physics.usyd.edu.au/rcfta/
       http://www.physics.usyd.edu.au/rcfta/rcfta.html
       http://www.physics.usyd.edu.au/
Founded: 1991
Staff: 10
Activities: theoretical astrophysics
Periodicals: (1) "Annual Report"
City Reference Coordinates: 151°13'00"E 33°52'00"S

University of Sydney, School of Mathematics and Statistics
F07
Sydney, NSW 2006
Telephone: (0)2-9351-3039
Telefax: (0)2-9351-4534
WWW: http://www.maths.usyd.edu.au/
Founded: 1991
Staff: 70
Activities: mathematical research * solar and stellar astrophysics * geophysics
City Reference Coordinates: 151°13'00"E 33°52'00"S

University of Tasmania, Department of Physics
G.P.O. Box 252-21
Hobart, TAS 7001
Telephone: (0)3-6226-2401
Telefax: (0)3-6226-2410
Electronic Mail: <userid>@phys.utas.edu.au
            <userid>@physvax.phys.utas.edu.au
WWW: http://physics01.phys.utas.edu.au/physics/
       http://www-ra.phys.utas.edu.au/(Radio Astronomy Group)
       http://www.phys.utas.edu.au/physics/cosray/default.htm (Cosmic-Ray Astronomy)
       http://xray03.phys.utas.edu.au/ (Optical and X-ray Astronomy)
Founded: 1960 (Mount Pleasant Radio Astronomy Observatory)
Activities: radio, X-ray, cosmic-ray and optical astronomy * pulsars and collapsed objects * molecular lines * VLBI * geodesy
Periodicals: (1) "Research Report" (ISSN 1035-4751)
Observatories: 2 (Mount Pleasant, Ceduna)
Coordinates: 147°25'00"E 42°50'00"S H300m
           147°26'21"E 42°48'18"S H55m (Mount Pleasant)
City Reference Coordinates: 147°25'00"E 42°50'00"S

University of Western Australia (UWA), Department of Physics
Nedlands, WA 6907
Telephone: (0)8-9380-2738
Telefax: (0)8-9380-1014
Electronic Mail: head@physics.uwa.edu.au
WWW: http://www.pd.uwa.edu.au/Physics/Research/astro.html
Founded: 1987
Staff: 15
Activities: image processing * SN * flare stars * transient astronomical phenomena * gravitational collapse * pulsars * neutron stars * gravitational-wave detection
Observatories: 1 (Perth Observatory - see separate entry)
City Reference Coordinates: 115°49'00"E 31°59'00"S

University of Western Sydney (UWS), Nepean Astronomy Centre (NAC)
P.O. Box 10
Kingswood, NSW 2747
Telephone: (0)47-360-222
Telefax: (0)47-360-779

Electronic Mail: d.bailey@nepean.uws.edu.au (David Bailey)
WWW: http://www.nepean.uws.edu.au/astronomy/
City Reference Coordinates: 151°13'00"E 33°52'00"S

University of Wollongong (UoW), Department of Physics, Astronomy and Astrophysics Group
Northfields Avenue
Wollongong, NSW 2500
Telephone: (0)2-4221-3517
Telefax: (0)2-4221-5944
Electronic Mail: <userid>@uow.edu.au
WWW: http://www.uow.edu.au/eng/phys/astronomy.html
Founded: 1967
Staff: 4
Activities: image digitising and analysis * radio and IR observing of galactic star forming regions * cooling flows * hydrodynamics * planetary surfaces * education * asteroid mining * Illawara Science Centre Planetarium and Observatory

City Reference Coordinates: 150°54'00"E 34°25'00"S

Warsash Pty. Ltd.
P.O. Box 1685
Strawberry Hills, NSW 2012
or :
7/1 Marian Street
Redfern, NSW 2016
Telephone: (0)2-9319-0122
Telefax: (0)2-9318-2192
Electronic Mail: warsash@ozemail.com.au
WWW: http://www.ozemail.com.au/~warsash/
Founded: 1975
Staff: 5
Activities: distributing optical components, micropositioning systems, lasers, adaptive optics, off & on-axis parabolics, acousto-optic modulators, piezo-actuators, optical benches, servo control systems, and more
City Reference Coordinates: 151°13'00"E 33°52'00"S (Sydney)

Western Sydney Amateur Astronomy Group (WSAAG)
c/o Brett White
P.O. Box 400
Kingswood NSW 2747
Telephone: (0)2-4753-1041
WWW: http://physics.st.nepean.uws.edu.au/nac/wsaag.html
        http://www.physics.usyd.edu.au/~ptitze/wsaag/index2.html
        http://www.wsaag.base.org/
City Reference Coordinates: 151°13'00"E 33°52'00"S (Sydney)

Wild Observatory (Paul_)
• See "Commonwealth Scientific and Industrial Research Organisation (CSIRO), Australia Telescope National Facility, Narrabri"

Wollongong Amateur Astronomy Club (WAAC)
P.O. Box 398
Unanderra, NSW 2526
Telephone: (0)2-4272-4505
Electronic Mail: paul.b@bigpond.com (Paul Brown)
WWW: http://www.users.bigpond.com/paul.b/index.htm
Founded: 1999
Membership: 15
Activities: observing nights * education * visits
Periodicals: (6) "Wollongong Observer"
City Reference Coordinates: 150°52'00"E 34°25'00"S (Wollongong)

Wollongong Science Centre and Planetarium
c/o University of Wollongong
Northfields Avenue
Wollongong, NSW 2522
or :
Cowper Street
Fairy Meadow, NSW 2519
Telephone: (0)2-4221-5591
Electronic Mail: <userid>@davinci.sci.uow.edu.au
WWW: http://www.uow.edu.au/science_centre/
        http://www.uow.edu.au/public/science_centre/planetarium.html
Founded: 1989
Staff: 35
Activities: science and astronomy education for public and schools
City Reference Coordinates: 150°54'00"E 34°23'00"S (Fairy Meadow)
                            150°52'00"E 34°25'00"S (Wollongong)

Austria

Adalbert Jeszenkowitsch Gesellschaft (Dr.-)
- See "Dr. Adalbert Jeszenkowitsch Gesellschaft"

Anich-Planetarium Kufstein (Peter--)
- See "Peter-Anich-Planetarium Kufstein"

Arbeitsgruppe für Astronomie Haus der Natur
c/o Roland Primas
Hoefelgasse 6
A-5020 Salzburg
Telephone: (0)662-624119
Telefax: (0)662-847905
Founded: 1954
Membership: 140
Activities: lectures * meetings
Periodicals: "Astronomische Kurzinformationen und Eilnachrichten"
Awards: "Golden Saturn"
Coordinates: 013°02'24"E 47°52'14"N H597m (Volkssternwarte Voggenberg)
City Reference Coordinates: 013°03'00"E 47°48'00"N

Astronomische Beobachtungsstation Martin Stangl
Liebiggasse 26
A-8010 Graz
Electronic Mail: Martin.Stangl@sni.siemens.at
Founded: 1987
Activities: observing and photography of solar-system bodies
Coordinates: 015°27'21"E 47°04'48"N H390m (Graz)
                  015°05'07"E 47°08'05"N H773m (Kainach)
City Reference Coordinates: 015°27'00"E 47°05'00"N

Astronomischer Arbeitskreis Salzkammergut (AAS)
c/o Wolfgang Vogl
Am Pfarrerfeld 53
A-4840 Vöcklabruck
or :
Sachsenstraße 2
A-4863 Seewalchen-am-Attersee
Telephone: (0)7662-87182
              (0)7662-8297
Electronic Mail: aas.gahberg.aut@online.edvg.co.at
WWW: http://www.nf-team.co.at/aas/
         http://www.astronomie.at/
Founded: 1980
Membership: 400
Activities: comets * meteor streams * photography * planets
Periodicals: (10) "Astro-Info" (circ.: 550)
Coordinates: 013°36'33"E 47°54'47"N H860m (Gahberg Observatory)
City Reference Coordinates: 013°39'00"E 48°01'00"N (Vöcklabruck)
                                    013°35'00"E 47°57'00"N (Seewalchen)

Astronomischer Jugendclub Dingi-Vindemiatrix
Richard-Wagner-Platz 2/8
A-1160 Wien
Telephone: (0)1-4956340
Telefax: (0)1-4956340
Founded: 1969
Membership: 50
Activities: variable stars * photography
Periodicals: (4) "Die Sternenrundschau"
City Reference Coordinates: 016°20'00"E 48°13'00"N

Astronomisches Büro
- See "Österreichischer Astronomischer Verein"

Astronomische Vereinigung Kärntens (AVK)
Villacherstraße 239
A-9020 Klagenfurt
Telephone: (0)463-21700
Telefax: (0)463-21700
Electronic Mail: avk@uni-klu.ac.at

WWW: http://www.buk.ktn.gv.at/sterne/avk.htm
          http://www.buk.ktn.gv.at/sterne/planet/planet.htm (Planetarium Europapark)
Founded: 1961
Membership: 359
Activities: education
Periodicals: (11) "Sternenwelt"
Observatories: 1 (Kreubergl - see separate entry)
City Reference Coordinates: 014°18'00"E 46°38'00"N

Astroteam Mariazellerland
Hangweg 12
A-8630 Mariazell
Telephone: (0)3882-3540
Telefax: (0)3882-217817
WWW: http://ajax.kfunigraz.ac.at/~webers/
Founded: 1995
Membership: 57
Activities: popularization * education * observing * CCD-imaging
Observatories: 2 (Raiba Volkssternwarte Mariazellerland, Sternwarte Sankt Sebastian)
City Reference Coordinates: 015°19'00"E 47°47'00"N

Austrian Academy of Sciences
● See "Österreichische Akademie der Wissenschaften (ÖAW)"

Austrian Society for Aerospace Medicine (ASM)
Lustkandlgasse 52/3
A-1090 Wien
Telephone: (0)1-3155777
Telefax: (0)1-31557778
Electronic Mail: headoffice@asm.at
WWW: http://www.asm.at/
Founded: 1991
Activities: biomedical investigations into spaceflight medical issues
City Reference Coordinates: 016°20'00"E 48°13'00"N

AVL Gesellschaft für Verbrennungskraftmaschinen und Meßtechnik mbH
Kleitstraße 48
A-8020 Graz
WWW: http://www.avl.co.at/
Founded: 1946
Staff: 1000
Activities: engineering services * test systems and instruments * software producer
City Reference Coordinates: 015°27'00"E 47°05'00"N

Burgenländische Amateurastronomen (BAA)
Neusiedlerstraße 8
A-7000 Eisenstadt
Telephone: (0)2682-66207
Electronic Mail: baa@zlsm03.arcs.ac.at
WWW: http://www1.arcs.ac.at/baa
Founded: 1992
Membership: 50
Activities: observing * meteors * education
Periodicals: (4) "Alrukaba" (circ.: 100)
City Reference Coordinates: 016°32'00"E 47°51'00"N

Burgenländische Landessternwarte
● See "Dr. Adalbert Jeszenkowitsch Gesellschaft"

Dr. Adalbert Jeszenkowitsch Gesellschaft
c/o Norbert Hauser
Dr. Karl Rennerstraße 1
A-7000 Eisenstadt
Telephone: (0)2682-66819
Founded: 1979
Membership: 80
Activities: guided tours * library * lectures * exhibitions * photography
Coordinates: 016°31'21"E 47°50'39"N H194m (Burgenländische Landessternwarte)
City Reference Coordinates: 016°32'00"E 47°50'00"N

Eisner-Sternwarte
c/o Karl Silber
Cumberlandpark 16
A-4810 Gmunden
Telephone: (0)7612-67777

Founded: 1949
Membership: 3
Activities: lectures
Coordinates: 013°48'00"E 47°55'27"N H490m
City Reference Coordinates: 013°48'00"E 47°55'00"N

Figl-Observatorium für Astrophysik (Leopold-_)
● See "Universität Wien, Institut für Astronomie"

Fonds zur Förderung der Wissenschaftlichen Forschung (FWF)
(Austrian Science Fund)
Weyringergasse 35
A-1040 Wien
Telephone: (0)1-50567400
Telefax: (0)1-5056739
        (0)1-505674045
WWW: http://www.fwf.ac.at/index-en.html
Founded: 1967
Staff: 40
Activities: funding research * public relations
Periodicals: "FWF-Info", "FWF-Statistics", "Annual Report"
City Reference Coordinates: 016°20'00"E 48°13'00"N

Franz-Kroller-Sternwarte (FKS)
Hauptplatz 18
A-2514 Traiskirchen
WWW: http://www.astro.or.at/astro/orgs/fks/
        http://www.mycity.at/privat/98013062/fks/maksutov.htm
City Reference Coordinates: 016°18'00"E 48°01'00"N

Gnomonicae Societas Austriaca (GSA)
(Österreichischer Astronomischer Verein, Arbeitsgruppe Sonnenuhren)
Am Tigls 76A
A-6073 Sistrans
Telephone: (0)512-378868
Telefax: (0)512-378868
Electronic Mail: sundial@tirol.com
               k.schwarzinger@tirol.com (Karl Schwarzinger, Chairman)
WWW: http://www.tirol.com/sundial/
        http://www.sundials.co.uk/gsa.htm
Founded: 1990
Membership: 94
Activities: preservation of historical sundials * survey of Austrian sundials * meetings and conferences
Periodicals: (2) "Rundschreiben"
City Reference Coordinates: 011°27'00"E 47°14'00"N

Hotel und Sternwarte Heiligkreuz
Reimmichlstraße 18
A-6060 Hall in Tirol
Telephone: (0)5223-57114
Telefax: (0)5223-571145
Electronic Mail: heiligkreuz@aon.at
WWW: http://math1.uibk.ac.at/students/heiligkreuz/
Founded: 1994
Staff: 3
Activities: observing * photography
Coordinates: 011°29'55"E 47°17'10"N H572m
City Reference Coordinates: 014°28'00"E 47°28'00"N

International Atomic Energy Agency (IAEA)
Wagramerstraße 5
Postfach 100
A-1400 Wien
Telephone: (0)1-20600
Telefax: (0)1-20607
WWW: http://www.iaea.or.at/
Founded: 1957
Staff: 2300
Activities: accelerating and enlarging the contribution of atomic energy to peace, health and prosperity throughout the world
* verifying that States use fissionable material for peaceful purposes only
Periodicals: (24) "INIS Atom Index"; (12) "Nuclear Fusion"; (6) "IAEA Newsbriefs"; (4) "IAEA Bulletin", "Meetings in Atomic Energy"
City Reference Coordinates: 016°20'00"E 48°13'00"N

Jeszenkowitsch Gesellschaft (Dr. Adalbert_)
● See "Dr. Adalbert Jeszenkowitsch Gesellschaft"

Karl-Franzens-Universität Graz
● See "Universität Graz"

Kroller-Sternwarte (Franz-_)
● See "Franz-Kroller-Sternwarte"

Kuffner-Sternwarte (KSW)
Johann Staud-Straße 10
A-1160 Wien
Telephone: (0)1-9148130
Telefax: (0)1-914813031
Electronic Mail: admin@kuffner.ac.at
WWW: http://www.kuffner.ac.at/
Founded: 1884
Staff: 6
Activities: public observatory * museum * education * history of astronomy
Periodicals: (1/3) "Tätigkeitsbericht"
Coordinates: 016°17'47"E 48°12'47"N H302m
City Reference Coordinates: 016°20'00"E 48°13'00"N

Leopold-Figl-Observatorium für Astrophysik
● See "Universität Wien, Institut für Astronomie"

Leopold-Franzens-Universität Innsbruck
● See "Universität Innsbruck"

Linzer Astronomische Gemeinschaft (LAG)
Sternwarteweg 5
A-4020 Linz
Telephone: (0)732-674042
Telefax: (0)732-673105
Electronic Mail: herbert.raab@jk.uni-linz.ac.at (Herbert Raab, President)
WWW: http://www.planet.co.at/lag/
Founded: 1947
Membership: 200
Activities: education * amateur astronomy
Periodicals: (12) "WEGA"
Coordinates: 014°16'09"E 48°17'38"N H341m (Johannes-Kepler-Sternwarte)
City Reference Coordinates: 014°18'00"E 48°18'00"N

Observatorium Lustbühel
● See "Universität Graz, Institut für Astronomie, Observatorium Lustbühel"

Österreichische Akademie der Wissenschaften (ÖAW)
(Austrian Academy of Sciences)
Dr.-Ignaz-Seipel-Platz 2
A-1010 Wien
Telephone: (0)1-515810
Telefax: (0)1-5139541
Electronic Mail: <userid>@oeaw.ac.at
WWW: http://www.oeaw.ac.at/deutsch.html
        http://www.oeaw.ac.at/english.html
Founded: 1847
City Reference Coordinates: 016°20'00"E 48°13'00"N

Österreichische Akademie der Wissenschaften (ÖAW), Institut für Weltraumforschung (IWF)
Inffeldgasse 12
A-8010 Graz
Telephone: (0)316-4636960
Telefax: (0)316-463697
Electronic Mail: <userid>@oeaw.ac.at
WWW: http://iwf.oeaw.ac.at/
        http://www.iwf.oeaw.ac.at/
Founded: 1972
Staff: 60
Activities: ionosphere and magnetosphere of planets and comets * interplanetary space * satellites * rockets * balloons * time keeping and distribution
Coordinates: 015°29'40"E 47°04'05"N H494m (Lustbühel)
City Reference Coordinates: 015°27'00"E 47°05'00"N

Österreichische Akademie der Wissenschaften (ÖAW), Institut für Weltraumforschung (IWF), Abteilung für experimentelle Weltraumforschung
Inffeldgasse 12
A-8010 Graz
Telephone: (0)316-463696
Telefax: (0)316-463697
Electronic Mail: <userid>@iwf.tu-graz.ac.at
WWW: http://fiwfds01.tu-graz.ac.at/
Founded: 1970
Staff: 34
Activities: experimental investigation of plasma-physical processes in near-Earth and interplanetary medium * developing and constructing space flight hardware * data processing and analysis
City Reference Coordinates: 015°27'00"E 47°05'00"N

Österreichische Gesellschaft für Weltraumfragen GesmbH
(Austrian Space Agency - ASA)
Garnisongasse 7
A-1090 Wien
Telephone: (0)1-40381770
Telefax: (0)1-4058228
Founded: 1972
Staff: 6
Activities: coordinating space activities in Austria
Periodicals: (1/2) "Report of Activities"; (5) "ASA-Information Service"
City Reference Coordinates: 016°20'00"E 48°13'00"N

Österreichische Physikalische Gesellschaft (ÖPG)
c/o Atominstitut der Österreichischen Universitäten
Schüttelstraße 115
A-1020 Wien
Telephone: (0)1-588015578
              (0)1-588015575
Telefax: (0)1-5864203
Electronic Mail: oepg@kph.tuwien.ac.at
WWW: http://www.ati.ac.at/OePG/
Membership: 1350
Periodicals: (4) "ÖPG-Mitteilungsblatt"
Awards: (1) "Physik-Preis", "Roman-Ulrich-Sexl-Preis", "Viktor-Heß-Preis"
City Reference Coordinates: 016°20'00"E 48°13'00"N

Österreichischer Astronomischer Verein
(Astronomisches Büro)
Hasenwartgasse 32
A-1238 Wien
Telephone: (0)1-8893541
Telefax: (0)1-8893541
Electronic Mail: astbuero@astronomisches-buero-wien.or.at
WWW: http://members.ping.at/astbuero/
Founded: 1924 (Büro: 1908)
Membership: 1500
Staff: 5
Activities: astronomical phenomenology * lunar occultations * comets * minor planets * meteors * sundials (see separate entry) * education * Sterngarten
Periodicals: (12) "Der Sternenbote" (ISSN 0039-1271); (1) "Österreichischer Himmelskalender", "Seminarpapiere"
City Reference Coordinates: 016°20'00"E 48°13'00"N

Österreichischer Astronomischer Verein, Arbeitsgruppe Sonnenuhren
● See "Gnomonicae Societas Austriaca (GSA)"

Österreichisches Normungsinstitut (ÖN)
Heinestraße 38
Postfach 130
A-1021 Wien
Telephone: (0)1-213000
Telefax: (0)1-21300650
Electronic Mail: <userid>@on-norm.at
WWW: http://www.on-norn.at/
City Reference Coordinates: 016°20'00"E 48°13'00"N

Peter-Anich-Planetarium Kufstein
Festungsneuhof 1
A-6332 Kufstein
or :
Schützenstraße 16

A-6330 Kufstein
Telephone: (0)5372-65050
            (0)5372-62111
            (0)664-4319133
Telefax: (0)5372-61714
Electronic Mail: josef.s.koller@uibk.ac.at
WWW: http://ast5.uibk.ac.at/student/planetarium/plhome.html
        http://members.tripod.de/PlanetariumKufstein/
City Reference Coordinates: 012°10'00"E 47°35'00"N

Planetarium der Stadt Wien
Oswald-Thomas-Platz
A-1020 Wien
Telephone: (0)1-249432
Telefax: (0)1-8893541
Electronic Mail: planwien@planetarium-uraniasternwarte-wien.or.at
WWW: http://members.ping.at/sonne/
Founded: 1927
Staff: 4
Activities: popularization * shows
Periodicals: (1) "Seminars"
City Reference Coordinates: 016°20'00"E 48°13'00"N

Planetarium Europapark
• See "Astronomische Vereinigung Kärntens (AVK)"

Planetarium Kufstein
• See "Peter-Anich-Planetarium Kufstein"

Salzburger Sterngucker
Himmelreichstrasse 30
A-5071 Wals
Telephone: (0)662-852938
Electronic Mail: reifberger.jo@salzburg.co.at
WWW: http://www.pgliefering.asn-sbg.ac.at/herbert/sterngucker.htm
City Reference Coordinates: 013°03'00"E 47°48'00"N (Salzburg)

Sonnenobservatorium Kanzelhöhe
• See "Universität Graz, Sonnenobservatorium Kanzelhöhe"

Sport- und Kulturvereins Forschungszentrum Seibersdorf (SKVFZ), Sektion Amateurastronomie
c/o Erich Weber
Österreichisches Forschungszentrum Seibersdorf GesmbH
A-2444 Seibersdorf
Telephone: (0)2254-7803104
Telefax: (0)2254-72133
Electronic Mail: erich@zditf2.arcs.ac.at
WWW: http://www1.arcs.ac.at/skvfz/sekt-aa.html
Founded: 1994
City Reference Coordinates: 016°20'00"E 48°13'00"N (Wien)

Star Observer Verlag
Linzerstraße 63
A-3002 Purkersdorf bei Wien
Telephone: (0)1-223161580
Telefax: (0)1-223161581
Electronic Mail: info@starobserver.com
WWW: http://www.starobserver.com/
Founded: 1993
Staff: 5
Periodicals: (10) "Star Observer"
City Reference Coordinates: 016°20'00"E 48°13'00"N (Wien)

Steirischer Astronomen Verein (StAV)
Postfach 72
A-8051 Graz
WWW: http://www.kfunigraz.ac.at/astwww/stav/stav.html
        http://www.kfunigraz.ac.at/astwww/stav/stern.html (Observatory)
Founded: 1981
Membership: 120
Activities: occultations by Moon and minor planets * meteors * comets * photography * public observing * CCD imaging
Sections: occultations * meteors * comets * astrophotography
Periodicals: (2) "Nova" (circ.: 140); (12) "Monthly Newsletter"
Coordinates: 015°19'55"E 47°04'08"N H554m (Johannes-Kepler-Volkssternwarte, Steinberg)
            015°28'20"E 47°11'00"N H810m (Rinnegg - private obs.)

015°35'52"E 46°51'13"N H300m (Sankt Georgen - private obs.)
City Reference Coordinates: 015°27'00"E 47°05'00"N

Sternwarte Heiligkreuz
● See "Hotel und Sternwarte Heiligkreuz"

Sternwarte Königsleiten
c/o Josef Anzer
Königsleiten 29
A-5742 Wald im Pinzgau
Telephone: (0)6564-877042
Telefax: (0)6564-877043
Electronic Mail: sternwarte.kgl@aon.at
WWW: http://www.ipoc.net/sternwarte/
       http://www.tcs.co.at/tourismus/wald/plane.html (Planetarium)
City Reference Coordinates: 012°14'00"E 47°16'00"N

Sternwarte Kreuzbergl
Giordano Bruno Weg 1
A-9020 Klagenfurt
Telephone: (0)463-21700
Telefax: (0)463-21700
WWW: http://www1.arcs.ac.at/aucona/avk/volksstw.htm
Founded: 1965
Membership: 10
Activities: observing * education * guided celestial tours
Coordinates: 014°17'24"E 46°37'47"N H600m
City Reference Coordinates: 014°18'00"E 46°38'00"N

Swarovski Optik KG
A-6060 Absam
Telephone: (0)5223-5110
Telefax: (0)5223-41860
Electronic Mail: <userid>@swarovskioptik.com
       swarovski.optik@tyrol.at
WWW: http://www.swarovskioptik.com/
Founded: 1895
City Reference Coordinates: 011°30'00"E 47°18'00"N

Technische Universität Graz, Institut für Angewandte Geodäsie, Abteilung für Positionierung und Navigation
Steyrergasse 30
A-8010 Graz
Telephone: (0)316-8736830
Telefax: (0)316-8738888
Electronic Mail: <userid>@ftug.tu-graz.ac.at
Founded: 1930
Staff: 4
Activities: geodesy, including space geodesy
Periodicals: "Mitteilungen" (circ.: 400)
City Reference Coordinates: 015°27'00"E 47°05'00"N

United Nations Organization (UNO), Office for Outer Space Affairs (OOSA)
Room E-0953
Vienna International Centre
Postfach 500
A-1400 Wien
Telephone: (0)1-211314951
Telefax: (0)1-213455830
Electronic Mail: oosa@unov.un.or.at
WWW: http://www.un.or.at/OOSA/
Founded: 1962
Staff: 16
Activities: providing services to the Committee on the Peaceful Uses of Outer Space (COPUOS) and subsidiary bodies * administering Space Applications Programme * carrying out studies recommended by COPUOS
Periodicals: (12) "Monthly Survey of Selected Events in the Peaceful Exploration and Use of Outer Space"; (1) "Seminars of the United Nations Programme on Space Applications", "Highlights in Space: Progress in Space Science, Technology and Applications, International Cooperation and Space Law"
City Reference Coordinates: 016°20'00"E 48°13'00"N

Universität Graz, Institut für Astronomie
Universitätsplatz 5
A-8010 Graz
Telephone: (0)316-3805270
Telefax: (0)316-3809820

Electronic Mail: <userid>@bimgs1.kfunigraz.ac.at
             <userid>@balu.kfunigraz.ac.at
WWW: http://www.kfunigraz.ac.at/astwww/
Staff: 7
Activities: education * dynamics of solar photosphere * minor planets
Periodicals: "Mitteilungen der Universitäts-Sternwarte Graz"
Coordinates: 015°26'54"E 47°04'36"N H375m
● Full university name: "Karl-Franzens-Universität Graz"
City Reference Coordinates: 015°27'00"E 47°05'00"N

Universität Graz, Institut für Astronomie, Observatorium Lustbühel
Lustbühelstraße 46
A-8042 Graz
Telephone: (0)316-467367
Telefax: (0)316-467365
Staff: 11
Coordinates: 015°29'42"E 47°03'56"N H480m
● Full university name: "Karl-Franzens-Universität Graz"
City Reference Coordinates: 015°27'00"E 47°05'00"N

Universität Graz, Institut für Meteorologie und Geophysik (IMG)
Halbärthgasse 1
A-8010 Graz
Telephone: (0)316-380 x5255
Telefax: (0)316-384091
Electronic Mail: <userid>@kfunigraz.ac.at
             imgwww@kfunigraz.ac.at
WWW: http://www.kfunigraz.ac.at/imgwww/
Founded: 1924
Staff: 8
Activities: education * research * ionospheric physics * aeronomy (Earth and planets) * non-thermal planetary radio emissions * climatic changes
Coordinates: 015°29'42"E 47°03'56"N H480m (Lustbühel)
● Full university name: "Karl-Franzens-Universität Graz"
City Reference Coordinates: 015°27'00"E 47°05'00"N

Universität Graz, Sonnenobservatorium Kanzelhöhe
A-9521 Treffen
Telephone: (0)4248-27170
Telefax: (0)4248-271715
WWW: http://www.solobskh.ac.at/
Founded: 1943
Staff: 5
Activities: daily sunspot drawings * white-light integral pictures * solar flare patrol * solar instrumentation * solar activity * solar differential rotation
Periodicals: "Mitteilungen des Sonnenobservatoriums Kanzelhöhe"
Coordinates: 013°54'24"E 46°40'42"N H1,526m
● Full university name: "Karl-Franzens-Universität Graz"
City Reference Coordinates: 015°27'00"E 47°05'00"N

Universität Innsbruck, Institut für Astronomie
Technikerstraße 25
A-6020 Innsbruck
Telephone: (0)512-5076031
Telefax: (0)512-5072923
Electronic Mail: astro@uibk.ac.at
WWW: http://astro.uibk.ac.at/
Founded: 1892
Staff: 9
Activities: galactic structure * emission nebulae * reduction methods
Coordinates: 011°22'54"E 47°16'05"N
● Full university name: "Leopold-Franzens-Universität Innsbruck"
City Reference Coordinates: 011°25'00"E 47°17'00"N

Universität Wien, Institut für Astronomie
Türkenschanzstraße 17
A-1180 Wien
Telephone: (0)1-427751801
Telefax: (0)1-42779518
Electronic Mail: <userid>@astro.univie.ac.at
WWW: http://www.astro.univie.ac.at/
Founded: 1755
Staff: 16
Activities: stellar astrophysics * dynamical astronomy * extragalactic research * astrometry * history of astronomy
Coordinates: 016°20'12"E 48°13'54"N H241m
             015°55'24"E 48°05'00"N H890m (Leopold-Figl-Observatorium für Astrophysik)

City Reference Coordinates: 016°20'00"E 48°13'00"N

Universität Wien, Institut für Geochemie
Althanstraße 14
A-1090 Wien
Telephone: (0)1-313361714
                (0)1-313361725
Telefax: (0)1-31336781
Electronic Mail: geochemie@univie.ac.at
WWW: http://www.univie.ac.at/geochemistry/
Founded: 1985
Staff: 5
Activities: cosmochemistry (impact craters, meteorites, lunar rocks, cosmic dust)
City Reference Coordinates: 016°20'00"E 48°13'00"N

Universität Wien, Institut für Mathematik
Boltzmanngasse 5
A-1090 Wien
Telephone: (0)1-427750602
Telefax: (0)1-42779506
Electronic Mail: mathematik@univie.ac.at
WWW: http://www.mat.univie.ac.at/
                http://gauss.mat.univie.ac.at/
Staff: 2
Activities: numerical hydrodynamics * 3D convection modelling * stellar radial pulsations * stellar atmospheres
City Reference Coordinates: 016°20'00"E 48°13'00"N

Universität Wien, Institut für Meteorologie und Geophysik
Hohe Warte 38
A-1190 Wien
Telephone: (0)1-360263050
Telefax: (0)1-3685612
Electronic Mail: img-wien@univie.ac.at
WWW: http://www.univie.ac.at/IMG-Wien/mainimg.htm
Founded: 1851
Staff: 18
Activities: general meteorology * climatology * theoretical meteorology * gravimetry * magnetics * electromagnetics *seismology
Coordinates: 016°21'42"E 48°14'54"N
City Reference Coordinates: 016°20'00"E 48°13'00"N

Urania-Sternwarte, Wien
Oswald-Thomas-Platz
A-1020 Wien
or :
Uraniastraße 1
A-1010 Wien
Telephone: (0)1-249432
                (0)1-7126191
                (0)1-7295454
Electronic Mail: planwien@planetarium-uraniasternwarte-wien.or.at
WWW: http://members.ping.at/sonne/
                http://www.ping.at/members/sonne/
Founded: 1910
Staff: 8
Activities: popularization * phenomenology * astrometry * lunar occultations * minor planets * comets * Galilean satellites of Jupiter
Coordinates: 016°23'08"E 48°12'43"N H186m
City Reference Coordinates: 016°20'00"E 48°13'00"N

Vorarlberger Amateur Astronomen (VAA)
c/o Manfred Böhler
Hofsteigstraße 33
A-6890 Lustenau
Telephone: (0)5577-88647
WWW: http://www.vobs.at/astronomen/
Founded: 1989
Membership: 60
Activities: observing * lectures
City Reference Coordinates: 009°39'00"E 47°26'00"N

Wiener Arbeitsgemeinschaft für Astronomie (WAA)
Postfach 360
A-1070 Wien
or :

Dreyhausenstraße 11/53
A-1140 Wien
Telephone: (0)664-2561221
Electronic Mail: apikhard@ping.at (Alexander Pikhard, President)
             waa@eunet.at
WWW: http://members.eunet.at/waa/
Founded: 1998
Membership: 14
Activities: education * lectures * observing * star parties * camps * meetings
City Reference Coordinates: 016°20'00"E 48°13'00"N

Zentralanstalt für Meteorologie und Geodynamik (ZAMG)
Hohe Warte 38
A-1190 Wien
Telephone: (0)1-36026
Telefax: (0)1-3691233
Electronic Mail: <userid>@zamg.ac.at
WWW: http://www.zamg.ac.at/
Founded: 1851
Staff: 190
Activities: meteorology * climatology * environmental meteorology * geophysics
City Reference Coordinates: 016°20'00"E 48°13'00"N

Azerbaijan

Azerbaijan State Standardization and Metrology Centre
Mardanov Gardashlary 124
3700748 Baku
Telephone: (0)12-400396
Telefax: (0)12-403798
City Reference Coordinates: 049°51'00"E 40°23'00"N

Azerbaijan Academy of Sciences
Istiglaliyyat Street 10
370001 Baku
Telephone: (0)12-923529
(0)12-925883
Telefax: (0)12-925699
WWW: http://wwwinfo.cern.ch/∼mrashid/antaas.htm
Founded: 1916
City Reference Coordinates: 049°51'00"E 40°23'00"N

Azerbaijan Academy of Sciences, Shemakha Astrophysical Observatory
373243 Shemakha
Telephone: (0)8922-398248
Founded: 1960
Staff: 150
Activities: non-stationary processes in stars * celestial mechanics * Sun * planets * small solar-system bodies
Periodicals: "Circulars" (ISSN 0135-0420)
Coordinates: 048°35'04"E 40°46'20"N H1,580m
City Reference Coordinates: 048°37'00"E 40°38'00"N

Azerbaijanian National Aerospace Agency (ANASA)
159 Azadlyg Street
370106 Baku
Telephone: (0)12-629387
Telefax: (0)12-621738
Founded: 1975
Staff: 400
Activities: remote sensing * space image processing * developing hardware and software
City Reference Coordinates: 049°51'00"E 40°23'00"N

Shemakha Astrophysical Observatory
● See "Azerbaijan Academy of Sciences, Shemakha Astrophysical Observatory"

State Hydrometeorological Committee
Str. Rasul Rza 3
370000 Baku
Telephone: (0)12-892296
Telefax: (0)12-936937
City Reference Coordinates: 049°51'00"E 40°23'00"N

Bangladesh

Bangladesh Meteorological Department
Abhawa Bhaban
Agargaon
Sher-e-Bangla Nagar
Dhaka 7
Telephone: (0)2-311032
Telefax: (0)2-319251
City Reference Coordinates: 085°10'00"E 26°41'00"N

Bangladesh Standards and Testing Institution (BSTI)
116/A Tejgaon Industrial Area
Dhaka 1208
Telephone: (0)2-881462
Telefax: (0)2-885685
Electronic Mail: bsti@bangla.net
City Reference Coordinates: 085°10'00"E 26°41'00"N

Space Research and Remote Sensing Organization (SPARRSO)
Agargaon
Sher-e-Bangla Nagar
Dhaka 1207
Telephone: (0)2-327913
(0)2-319301
Telefax: (0)2-813080
Electronic Mail: sparrso@bangla.net
WWW: http://www.bangla.net/sparrso/
Founded: 1980
Staff: 142
Activities: R&D * applications of remote-sensing technology and GIS * resource management * environmental monitoring
Periodicals: (1) "Annual Report"; (4) "SPARRSO Newsletter"
City Reference Coordinates: 085°10'00"E 26°41'00"N

Barbados

Barbados Astronomical Society (BAS)
c/o Harry Bayley Observatory
P.O. Box 41-B
Brittons Hill
Saint-Michael
Telephone: 426-1317
          422-2394
Founded: 1956
Membership: 58
Activities: weekly meetings * education * open nights * limited research
Periodicals: (2) "Journal" (circ.: 70)
Coordinates: 059°35'10"W 13°05'07"N (Harry Bayley Observatory)
City Reference Coordinates: 059°37'00"W 13°06'00"N (Bridgetown)

Barbados National Standards Institution (BNSI)
"Flodden"
Culloden Road
Saint-Michael
Telephone: 426-3870
Telefax: 436-1495
Founded: 1973
Staff: 31
Activities: standards development and implementation * metrology * quality control * education * product development * product certification * quality management systems promotion
Periodicals: "BNSI Standards News", "National Standards and Codes of Practice"
City Reference Coordinates: 059°37'00"W 13°06'00"N (Bridgetown)

Caribbean Institute for Meteorology and Hydrology (CIMH)
P.O. Box 130
Bridgetown
or :
Husbands
Saint-James
Telephone: 425-1362
          425-1365
Telefax: 424-4733
WWW: http://inaccs.com.bb/carimet/top.htm
Founded: 1967
Staff: 40
Activities: training and research in meteorology and hydrology
City Reference Coordinates: 059°37'00"W 13°06'00"N (Bridgetown)

Meteorological Department
Grantley Adams International Airport
Christ Church
Telephone: 246-0910
Telefax: 246-1676
Founded: 1941
Activities: weather forecast for public and aviation * severe weather warnings
City Reference Coordinates: 059°37'00"W 13°06'00"N (Bridgetown)

Belarus

Belarusian Physical Society (BPS)
c/o Institute of Molecular and Atomic Physics
National Academy of Sciences of Belarus
70 F. Skaryna Avenue
220072 Minsk
Telephone: (0)172-685346
Telefax: (0)172-393064
Electronic Mail: imafbel@imaph.bas-net.by
WWW: http://imaph.bas-net.by/BPS/
Founded: 1990
Membership: 130
City Reference Coordinates: 027°34'00"E 53°54'00"N

Committee for Standardization, Metrology and Certification (BELST)
Starovilensky Trakt 93
220053 Minsk
Telephone: (0)0172-375213
Telefax: (0)0172-372588
Electronic Mail: belst@mcsm.belpak.minsk.by
City Reference Coordinates: 027°34'00"E 53°54'00"N

Belgium

Académie Royale des Sciences, des Lettres et des Beaux-Arts de Belgique
Palais des Académies
Rue Ducale 1
B-1000 Bruxelles
Telephone: (0)2-5502211
Telefax: (0)2-5502205
Founded: 1772
Membership: 300
Periodicals: "Bulletin de la Classe des Sciences" (ISSN 0001-4141), "Mémoires de la Classe des Sciences in 8°" (ISSN 0365-0936), "Mémoires de la Classe des Sciences in 4°" (ISSN 0365-0952)
Awards: (1) "Prix Léon et Henri Fredericq", "Prix Jean-Servais Stas", "Prix Adolphe Wetrems"; (1/2) "Prix Professeur Louis Baes", "Prix de Boelpaepe", "Prix Théophile Gluge", "Prix Charles Lemaire", "Prix Paul et Marie Stroobant", "Prix Frédéric Swarts", "Prix Baron van Ertborn"; (1/3) "Prix Albert Brachet", "Prix Henri Buttgenbach", "Prix Théophile De Donder", "Prix de la Fondation Jean-Marie Delwart", "Prix Jean De Meyer", "Prix Agathon De Potter", "Prix Léo Errera", "Prix Paul Fourmarier", "Prix Jean Lebrun", "Prix de l'Adjudant Hubert Lefebvre", "Prix Camille Liégeois", "Prix Joseph Schepkens", "Prix P.-J. et Édouard Van Beneden"; (1/4) "Prix François Deruyts", "Prix Jacques Deruyts", "Prix Dubois-Debauque", "Prix Charles Lagrange", "Prix Émile Laurent", "Prix Édouard Mailly", "Prix Louis Melsens", "Prix Pol et Christiane Swings", "Prix Georges Van der Linden"; (1/5) "Prix de la Belgica", "Prix Eugène Catalan", "Prix Edmond de Selys Longchamps", "Prix Lamarck"; (1/6) "Prix Auguste Sacré"; (1/8) "Prix Octave Dupont"
City Reference Coordinates: 004°20'00"E 50°50'00"N

Advanced Mechanical and Optical Systems (AMOS)
Parc de Recherches du Sart Tilman
Rue des Chasseurs Ardennais
B-4031 Angleur
Telephone: (0)4-3614040
Telefax: (0)4-3672007
WWW: http://www.reality.be/business/amos/
Founded: 1983
Staff: 26
Activities: mechanical and optical engineering * mirror polishing * telescope collimators * mechanical and optical ground support equipment * space simulator
City Reference Coordinates: 005°34'00"E 50°38'00"N

Alcatel, Études Techniques et Constructions Aérospatiales (ETCA)
Rue Chapelle Beaussart 101
B-6032 Mont-sur-Marchienne
Telephone: (0)71-442853
Telefax: (0)71-435778
WWW: http://www.charline.be/alcatelecom-etca/
Founded: 1963
Staff: 400
Activities: designing and developing i.a. satellite power conditioning systems, spectrum analyzers, digital processing circuits and camera control electronics
City Reference Coordinates: 004°26'00"E 50°25'00"N (Charleroi)

Astro CORAM
Rue des Haies 83
B-7712 Herseaux
or :
Rue de la Broche de Fer 101
B-7712 Herseaux
Telephone: (0)56-346242
Electronic Mail: 101544.2100@compuserve.com
WWW: http://ourworld.compuserve.com/homepages/OlivierDuquesne/coram.htm
        http://pegase.unice.fr/~skylink/pages_clubs/coram/coram.html
Founded: 1986
Membership: 19
Activities: observing * journeys * meetings * astronomy * space
Periodicals: (6) "Quasar/NOUT"
Coordinates: 003°12'45"E 50°43'35"N H30m
City Reference Coordinates: 003°14'00"E 50°43'00"N

AstroCosmique NGC 006
c/o Nathalie Heck
Rue Courtois 20/11
B-4000 Liège
Founded: 1985
Membership: 2
Activities: astronomical documentation * observing * field trips * promoting peaceful exploitation of space * lectures * colloquia

City Reference Coordinates: 005°34'00"E 50°38'00"N

Astronomie Centre Ardenne (ACA)
Chaussée de Bastogne 91
B-6840 Neufchâteau
Telephone: (0)61-279316
Electronic Mail: astroaca@france-mail.com
WWW: http://www.astrosurf.org/mercure/aca/
Founded: 1988
City Reference Coordinates: 005°26'00"E 49°51'00"N

Belgian Physical Society (BPS)
c/o J. Ingels (Secretary)
Belgisch Instituut voor Ruimte-Aeronomie
Ringlaan 3
B-1180 Brussel
Telephone: (0)2-3730378
Telefax: (0)2-3748423
WWW: http://www.ulg.ac.be/ipne/data/sbp.html
Membership: 450
Periodicals: "Physicalia"
City Reference Coordinates: 004°20'00"E 50°50'00"N

Belgian Space Industry Association (BELGOSPACE)
Boulevard Auguste Reyers 80
B-1030 Brussels
Telephone: (0)2-7067948
Telefax: (0)2-7067952
Electronic Mail: belgospace@fabrimetal.be
WWW: http://www.fabrimetal.be/
Founded: 1962
Membership: 12 (companies)
Activities: coordination between the members and with the public authorities * industrial forum for discussing common problems and major space options concerning Belgium
City Reference Coordinates: 004°20'00"E 50°50'00"N

Belgisch Instituut voor Ruimte-Aeronomie (BIRA)
Ringlaan 3
B-1180 Brussel
Telephone: (0)2-3730400
         (0)2-3730402
         (0)2-3730404
Telefax: (0)2-3748323
Electronic Mail: <userid>@bira-iasb.oma.be
WWW: http://www.oma.be/BIRA-IASB/
Founded: 1964
Staff: 90
Activities: public service * atmospheric physics and chemistry * solar-terrestrial physics * global change * space observations * data analysis
Periodicals: "Aeronomica Acta"
City Reference Coordinates: 004°20'00"E 50°50'00"N

BELGOSPACE
● See "Belgian Space Industry Association (BELGOSPACE)"

Canberra Semiconductor nv
Lammerdries 25
B-2250 Olen
Telephone: (0)14-221975
Telefax: (0)14-221991
Electronic Mail: csnv@canberra.com
Founded: 1981
Staff: 25
Activities: producing HPGe and Si detectors i.a. for space applications
Periodicals: "Information", "NPG-News"
City Reference Coordinates: 004°51'00"E 51°09'00"N

Centre National de Documentation Scientifique et Technique (CNDST)
c/o Bibliothèque Royale de Belgique
Boulevard de l'Empereur 4
B-1000 Bruxelles
Telephone: (0)2-5195640
Telefax: (0)2-5195679
Electronic Mail: cndst@kbr.be
WWW: http://www.belspo.be/stis/index_fr.htm

Founded: 1964
Staff: 11
Activities: on-line scientific, technical, economical, patent documentation and information * European Union National Awareness Partner * ESA-IRS National Aerospace Centre
City Reference Coordinates: 004°20'00"E 50°50'00"N

Centre Permanent d'Étude la Nature (CPEN), Observatoire
Route de Mons 52
B-6470 Sivry
Telephone: (0)60-455128
Telefax: (0)60-456142
WWW: http://www.vifcom.be/Local/Cpen.html
Founded: 1967
Membership: 12
Activities: meteorology * astronomy * education * planetarium
Coordinates: 004°13'45"E 50°10'50"N H220m
City Reference Coordinates: 004°11'00"E 50°10'00"N

Centre Spatial de Liège (CSL)
● See "Université de Liège (ULg), Centre Spatial de Liège (CSL)"

Cercle Astronomique de Bruxelles (CAB)
Allé D 5
B-6001 Marcinelle
Telephone: (0)71-434040
Telefax: (0)71-434040
Founded: 1963
Membership: 200
Electronic Mail: jean.schwaenen@worldonline.be (Jean Schwaenen, Secretary)
WWW: http://users.skynet.be/zmn/cab/cab.html
Activities: monthly lectures * observing (occultations, double stars, variable stars) * camps * instruments on loan
Periodicals: (10) "OBAFGKM Informations"; (4) "OBAFGKM" (ISSN 0775-0315)
Coordinates: 004°20'39"E 50°47'44"N H78m
City Reference Coordinates: 004°27'00"E 50°24'00"N

Cercle Astronomique Mosan (CAM)
Rue de Soulme 94
B-5620 Morville
Telephone: (0)82-688212
Electronic Mail: alain.gilson@skynet.be (Alain Gilson)
WWW: http://users.skynet.be/alagils/
Founded: 1977
Membership: 45
Activities: Sun * Moon * planets * nebulae
Periodicals: (6) "L'Univers du Namurois" (circ.: 45)
Coordinates: 004°55'39"E 50°14'56"N H215m (Herbuchenne, Dinant)
City Reference Coordinates: 004°45'00"E 50°14'00"N

Cercle d'Astronomie Olympus Mons
c/o Service de Physique Expérimentale et Biologique
Faculté de Médecine et de Pharmacie
Avenue du Champ de Mars 24
B-7000 Mons
Telephone: (0)65-373536
      (0)65-373468
Telefax: (0)65-373537
Electronic Mail: moiny@umh.ac.be (Francis Moiny)
WWW: http://olympus.umh.ac.be/
Periodicals: (6) "Galactée" (in collaboration with "Cercle d'Astronomie Tycho")
City Reference Coordinates: 003°56'00"E 50°27'00"N

Cercle d'Astronomie Tycho
c/o Pierre Dehombreux
Service de Mécanique Rationnelle, Dynamique et Vibrations
Boulevard Dolez 31
B-7000 Mons
Telephone: (0)65-374179
      (0)65-374184 (answering machine)
Telefax: (0)65-374183
Electronic Mail: tycho@fpms.ac.be
WWW: http://www.fpms.ac.be/~tycho/
Founded: 1997
Membership: 50
Periodicals: (6) "Galactée" (in collaboration with "Olympus Mons")
City Reference Coordinates: 003°56'00"E 50°27'00"N

Club d'Astronomie d'Amateurs "Orion"

Avenue Sart Paradis 68
B-5100 Wépion
or :
Chaussée de Waterloo 52
B-5002 Saint-Servais
Telephone: (0)81-731657
Electronic Mail: beatrice.gregoire@club.innet.be
Founded: 1996
Membership: 12
Periodicals: (4) "L'Orionide"; (12) "Bulletin d'Information"
Coordinates: 004°51'00"E 50°24'00"N
City Reference Coordinates: 004°52'00"E 50°25'00"N (Wépion)

Comité Belge pour l'Investigation Scientifique des Phénomènes Réputés Paranormaux (Comité PA-RA)

c/o Observatoire Royal de Belgique
Avenue Circulaire 3
B-1180 Bruxelles
Telephone: (0)2-3730241
WWW: http://www.comitepara.be/
Founded: 1949
Membership: 150
Activities: critical review of allegations by paranormal believers * research * experiments
Periodicals: "Nouvelles Brèves" (ISSN 0774-5834, circ.: 300), "Le Trait d'Union"
City Reference Coordinates: 004°20'00"E 50°50'00"N

École Royale Militaire (ERM), Département d'Astronomie, de Géodésie et de Topographie

Avenue de la Renaissance 30
B-1000 Bruxelles
Telephone: (0)2-7376120
(0)2-7376123
Electronic Mail: <userid>@elec.rma.ac.be
Founded: 1834
Staff: 3
Activities: (D)GPS * remote sensing
City Reference Coordinates: 004°20'00"E 50°50'00"N

European Association of Research and Technology Organisations (EARTO)

Rue du Luxembourg 3
B-1000 Bruxelles
Telephone: (0)2-5028698
Telefax: (0)2-5028693
Electronic Mail: info@earto.org
WWW: http://www.earto.org/
Activities: trade association of Europe's specialised research and technology organisations
City Reference Coordinates: 004°20'00"E 50°50'00"N

Europlanetarium vzw

Planetariumweg 19
B-3600 Genk
Telephone: (0)89-307990
Telefax: (0)89-307991
Electronic Mail: planetar@europlanetarium.com
WWW: http://www.europlanetarium.com/
Founded: 1977
Staff: 11
Membership: 350
Activities: popularization of astronomy, space sciences and meteorology
Periodicals: (12) "Info"; (1) "Jaarverslag"
● Earlier know as "Limburgse Volkssterrenwacht (LVS) vzw"
Coordinates: 005°32'10"E 50°57'30"N H90m
City Reference Coordinates: 005°30'00"E 50°58'00"N

Euro Space Center (ESC)

Rue Devant les Hêtres 1
B-6890 Transinne
Telephone: (0)61-656465
Telefax: (0)61-656461
Electronic Mail: escinfo@ping.be
WWW: http://www.ping.be/eurospace/
City Reference Coordinates: 005°12'00"E 50°00'00"N

Expert Software Systems (E2S) NV

Technologiepark 5

B-9052 Zwijnaarde
Telephone: (0)9-2210383
Telefax: (0)9-2203191
Electronic Mail: rma@e2s.be
Founded: 1984
Staff: 20
● Software producer
City Reference Coordinates: 003°43'00"E 51°00'00"N

Facultés Universitaires Notre-Dame de la Paix (FUNDP), Département de Mathématique
Rempart de la Vierge 8
B-5000 Namur
Telephone: (0)81-724903
Telefax: (0)81-724914
Electronic Mail: <userid>@fundp.ac.be
WWW: http://www.fundp.ac.be/sciences/math/dpt.html
Founded: 1971
Activities: celestial mechanics * dynamics of satellites and minor planets * perturbation methods
City Reference Coordinates: 004°52'00"E 50°28'00"N

Flemish Aerospace Group (FLAG)
Brouwersvliet 5 Box 4
B-2000 Antwerpen
Telephone: (0)3-2024448
Telefax: (0)3-2264334
Founded: 1980
Staff: 3
Activities: providing support for company-members active in aeronautics and aerospace
City Reference Coordinates: 004°25'00"E 51°13'00"N

Fonds de la Recherche Fondamentale Collective (FRFC)
Rue d'Egmont 5
B-1000 Bruxelles
Telephone: (0)2-5049211
Telefax: (0)2-5049292
Electronic Mail: admfnrs@bearn.bitnet
WWW: http://www.fnrs.be/Programmes_du_FRFC.htm
Founded: 1965
Activities: research council * funding basic research
City Reference Coordinates: 004°20'00"E 50°50'00"N

Fonds National de la Recherche Scientifique (FNRS)
Rue d'Egmont 5
B-1000 Bruxelles
Telephone: (0)2-5049211
Telefax: (0)2-5049292
Electronic Mail: mjsimoen@fnrs.be (M.J. Simoen, General Secretary)
WWW: http://www.fnrs.be/
Founded: 1928
Activities: research council * funding basic research
Periodicals: (1) "Annual Report"
Awards: (1/5) "Prix Dr. A. De Leeuw-Damry-Bourlart"; "Prix Scientifique Ernest-John Solvay", "Prix Scientifique Joseph Maisin"
City Reference Coordinates: 004°20'00"E 50°50'00"N

Fonds voor Kollektief en Fundamenteel Onderzoek (FKFO)
Egmontstraat 5
B-1000 Brussel
Telephone: (0)2-5129110
Telefax: (0)2-5125890
Electronic Mail: admfnrs@bearn.bitnet
Founded: 1965
Activities: research council * funding basic research
Periodicals: (1) "Annual Report"
City Reference Coordinates: 004°20'00"E 50°50'00"N

Groupe Astronomie de Spa (GAS)
Rue Amédée Hesse 3
B-4900 Spa
Telephone: (0)87-775699
Electronic Mail: jehun@astro.ulg.ac.be (Emmanuël Jehin, President)
WWW: http://www.ping.be/eurospace/astro.htm#GAS
Founded: 1989
Membership: 190
Activities: observing * star parties * lectures * exhibitions * photography

Periodicals: (6) "Formation Information Liaison (FIL)"
Coordinates: 005°52'00"E 50°29'00"N (Aérodrome de la Sauvenière)
City Reference Coordinates: 005°52'00"E 50°30'00"N

Groupe d'Astronomie Quasar
c/o Marc Lefort
Rue des Carrières 1
B-5353 Emptinne
Electronic Mail: quasar@trifide.com
WWW: http://www.trifide.com/quasar/index-2.html
City Reference Coordinates: 005°07'00"E 50°20'00"N

Institut Belge de Normalisation (IBN)
Avenue de la Brabançonne 29
B-1000 Bruxelles
Telephone: (0)2-7349205
          (0)2-7380111
Telefax: (0)2-7334264
Electronic Mail: <userid>@ibn.be
WWW: http://www.ibn.be/
City Reference Coordinates: 004°20'00"E 50°50'00"N

Institut d'Aéronomie Spatiale de Belgique (IASB)
Avenue Circulaire 3
B-1180 Bruxelles
Telephone: (0)2-3730400
          (0)2-3730402
          (0)2-3730404
Telefax: (0)2-3748323
Electronic Mail: <userid>@bira-iasb.oma.be
WWW: http://www.oma.be/BIRA-IASB/
Founded: 1964
Staff: 90
Activities: public service * atmospheric physics and chemistry * solar-terrestrial physics * global change * space observations * data analysis
Periodicals: "Aeronomica Acta"
City Reference Coordinates: 004°20'00"E 50°50'00"N

Institut Royal Météorologique de Belgique (IRMB)
Avenue Circulaire 3
B-1180 Bruxelles
Telephone: (0)2-3730611
Telefax: (0)2-3751259
Electronic Mail: <userid>@oma.be
WWW: http://www.oma.be/KMI-IRM/
          http://www.meteo.oma.be/IRM-KMI/
Founded: 1913
Staff: 190
Activities: climatology * hydrology * dynamic climatology * applied meteorology * aerology * aeronomy * internal/external geophysics * computing
Periodicals: (12) "Observations Climatologiques" (ISSN 0029-7682), "Observations Géophysiques" (ISSN 0020-2525), "Observations Ionosphériques - Rayonnement Cosmique" (ISSN 0020-2533); (4) "Observations de l'Ozone" (ISSN 0770-0164); (1) "Rayonnement Solaire" (ISSN 0524-7780), "Magnétisme Terrestre" (ISSN 0770-4569), "Rapport Annuel" (ISSN 0770-7371); "Publications Série A" (ISSN 0020-255x), "Publications Série B" (ISSN 0770-4615); "Hors Série" (ISSN 0772-4330); "Documentation Météorologique" (ISSN 0772-4349), "Miscellanea A" (ISSN 0770-0261), "Miscellanea B" (ISSN 0770-0318), "Miscellanea C" (ISSN 0772-4357), "Publications Scientifiques et Techniques"
Coordinates: 004°21'00"E 50°48'00"N H104m (Uccle)
          004°36'00"E 50°06'00"N H230m (Dourbes)
City Reference Coordinates: 004°21'00"E 50°48'00"N

International Astronomical Youth Camps (IAYC) Workshop for Astronomy (IWA)
c/o Sam Vereecke
Ter Weibroek 20
B-9880 Aalter
Telephone: (0)9-3742752
Electronic Mail: sv@cage.rug.ac.be
WWW: http://www.iayc.org/
Founded: 1977
Membership: 40
Activities: organizing international astronomical youth camps
Periodicals: (10) "Communicator"; (1) "IAYC Report"
City Reference Coordinates: 003°27'00"E 51°05'00"N

International Meteor Organization (IMO)
c/o Paul Roggemans (Secretary General)

Pijnboomstraat 25
B-2800 Mechelen
Telephone: (0)15-411225
WWW: http://www.imo.net/
        http://www.tu-chemnitz.de/~smo/imo/
Founded: 1988
Membership: 260
Activities: observing (visual, photographic, telescopic, radio and video) * databases * bibliography
Periodicals: (6) "WGN" (ISSN 1016-3115; circ.: 320); (1) "Annual Report", "IMO Proceedings"
Sections: (commissions) Radio * Telescopic * Visual * Camera Network
City Reference Coordinates: 004°27'00"E 51°02'00"N

International Union of Radio Science (URSI)

c/o INTEC
Universiteit Gent
Sint-Pieters-Nieuwstraat 41
B-9000 Gent
Telephone: (0)9-2643320
          (0)9-2643321
Telefax: (0)9-2644288
Electronic Mail: heleu@intec.rug.ac.be
Founded: 1919
Membership: 39 (countries)
Periodicals: (4) "URSI Information Bulletin"; (1/3) "Proceedings of URSI General Assemblies", "Review of Radio Science", "Modern Radio Science"
Sections: (commissions) A. Electromagnetic Metrology * B. Fields and Waves * C. Signals and Systems * D. Electronics and Photonics * E. Electromagnetic Noise and Interference * F. Wave Propagation and Remote Sensing * G. Ionospheric Radio and Propagation * H. Waves in Plasmas * J. Radio Astronomy * K. Electromagnetics in Biology and Medicine
Awards: (1/3) "Balth. van der Pol Gold Medal", "John Howard Dellinger Medal", "Issac Koga Gold Medal", "Appleton Prize" (together with the Royal Society)
City Reference Coordinates: 003°43'00"E 51°03'00"N

International Union of the History and Philosophy of Science (IUHPS)

c/o Centre d'Histoire des Sciences
Université de Liège
Avenue des Tilleuls 15
B-4000 Liège
Telephone: (0)4-3669479
Telefax: (0)4-3669447
Electronic Mail: chst@ulg.ac.be
               chstulg@vml.ulg.ac.be
WWW: http://www.icsu.org/Membership/SUM/iuhps.html
Founded: 1956 (1947 as "International Union of History of Science")
Membership: 80 (member committees)
City Reference Coordinates: 005°34'00"E 50°38'00"N

Jongerenvereniging voor Sterrenkunde (JvS)

H. d'Ydewallestraat 46
B-8730 Beernem
Telephone: (0)50-789705
Electronic Mail: info@vvs.be
WWW: http://www.vvs.be/
        http://www.uunet.be/vvs/
        http://www.innet.be/vvs/
Founded: 1971
Membership: 350
Activities: meetings * camps * observing
Periodicals: (6) "Astra"
City Reference Coordinates: 003°20'00"E 51°09'00"N

Katholieke Universiteit Leuven

● See "Universiteit Leuven"

Koninklijke Vlaamse Academie van België voor Wetenschappen en Kunsten

Paleis der Academiën
Hertogsstraat 1
B-1000 Brussel
Telephone: (0)2-5502323
Telefax: (0)2-5502325
Electronic Mail: kawlsk@skynet.be
Founded: 1938
Staff: 15
Periodicals: "Mededelingen", "Academiae Analecta"
Awards: "H.L. Vanderlinden Prize", "MacLeod Prize", "P. van Oye Prize", "H. Schouteden Prize", "J. Gillis Prize", "O. Callebaut Prize", "J. Coppens Prize", "C. de Clercq Prize", "F. Van der Mueren Prize", "R. Lenaerts Prize"
City Reference Coordinates: 004°20'00"E 50°50'00"N

Koninklijk Meteorologisch Instituut van België (KMIB)
Ringlaan 3
B-1180 Brussel
Telephone: (0)2-3730611
Telefax: (0)2-3751259
Electronic Mail: <userid>@oma.be
WWW: http://www.oma.be/KMI-IRM/
        http://www.meteo.oma.be/IRM-KMI/
Founded: 1913
Staff: 190
Activities: climatology * hydrology * dynamic climatology * applied meteorology * aerology * aeronomy * internal/external geophysics * computing
Periodicals: (12) "Klimatologische Waarnemingen" (ISSN 0029-7682), "Geofysche Waarnemingen" (ISSN 0020-2525), "Ionosferische Waarnemingen - Kosmische Straling" (ISSN 0020-2533); (4) "Ozon Waarnemingen" (ISSN 0770-0164); (1) "Zonnestraling" (ISSN 0524-7780), "Aardmagnetisme" (ISSN 0770-4569), "Jaarverslag" (ISSN 0772-3288); "Publikaties Reeks A" (ISSN 0020-255x), "Publikaties Reeks B" (ISSN 0770-4615); "Buiten Reeks" (ISSN 0772-4330); "Meteorologische Documentatie" (ISSN 0772-4349), "Miscellanea A" (ISSN 0770-0261), "Miscellanea B" (ISSN 0770-0318), "Miscellanea C" (ISSN 0772-4357), "Wetenschappelijke en Technische Publikaties"
Coordinates: 004°21'00"E 50°48'00"N H104m (Ukkel)
              004°36'00"E 50°06'00"N H230m (Dourbes)
City Reference Coordinates: 004°21'00"E 50°48'00"N

Koninklijk Sterrenkundig Genootschap van Antwerpen (KSGA)
Kapelsesteenweg 340
B-2930 Brasschaat
Telephone: (0)3-6642893
Founded: 1905
Membership: 110
Activities: lectures * public observing * library * excursions
Periodicals: (1) "Jaarverslag"; (12) "Astronomische Gazet"
Coordinates: 004°24'12"E 51°13'24"N H20m
City Reference Coordinates: 004°30'00"E 51°17'00"N

Koninklijke Sterrenwacht van België (KSB)
Ringlaan 3
B-1180 Brussel
Telephone: (0)2-3730211
Telefax: (0)2-3749822
Electronic Mail: rob@oma.be
WWW: http://www.oma.be/KSB-ORB/
Founded: 1826
Staff: 100
Activities: fundamental astronomy * time and frequency * satellite geodesy * geodynamics * seismology * gravimetry * physics of planetary interiors * astrometry * celestial mechanics * astrophysics * radioastronomy * solar physics * stellar statistics
Periodicals: (1) "Jaarboek" (ISSN 0524-7780); "Astronomisch Bulletin"; "Mededelingen"; "Annalen"
Coordinates: 004°21'29"E 50°47'51"N H105m
              005°15'19"E 50°11'31"N H293m (Humain Radioastronomy Station)
City Reference Coordinates: 004°20'00"E 50°50'00"N

Kot Astro
Place de l'Escholier 1/208
B-1348 Louvain-la-Neuve
Electronic Mail: olivier.duquesne@compuserve.com
WWW: http://www.student.info.ucl.ac.be:8080/HomePages/dumortie/kotastro/kotastro.htm
Telephone: (0)10-453008
Founded: 1989
Activities: observing * astronomy week * conferences * popularization
City Reference Coordinates: 004°37'00"E 50°40'00"N

Lichtenknecker Optics (LO) sa
Grote Breemstraat 21
B-3500 Hasselt
Telephone: (0)11-253052
Telefax: (0)11-250090
Founded: 1973
Activities: manufacturing astronomical optics and instrumentation (including optical components up to 600mm in diameter) * designing * prototype making * distributor for Belgium of most astronomical instruments
City Reference Coordinates: 005°20'00"E 50°56'00"N

Limburgse Volkssterrenwacht (LVS) vzw
• See now "Europlanetarium vzw"

Logis-Ciel
c/o Alphonse Pouplier
Chemin des Vignerons 47

B-5100 Wépion
Telephone: (0)81-460122
Telefax: (0)81-460567
WWW: http://users.skynet.be/alphonse
Founded: 1988
Staff: 1
● Software producer
City Reference Coordinates: 004°52'00"E 50°25'00"N

Nationaal Centrum voor Wetenschappelijke en Technische Dokumentatie (NCWTD)
Keizerslaan 4
B-1000 Brussel
Telephone: (0)2-5195640
Telefax: (0)2-5195645
(0)2-5195679
Electronic Mail: cnsdt@kbr.be
WWW: http://www.belspo.be/stis/index_nl.htm
Founded: 1964
Staff: 11
Activities: on-line scientific, technical, economical, patent documentation and information * European Union National Awareness Partner (EU-NAP) * ESA-IRS National Aerospace Centre (NAC)
City Reference Coordinates: 004°20'00"E 50°50'00"N

Nationaal Fonds voor Wetenschappelijk Onderzoek (NFWO)
Egmontstraat 5
B-1000 Brussel
Telephone: (0)2-5129110
Telefax: (0)2-5125890
Electronic Mail: <userid>@nfwo.be
jose.traest@nfwo.be (José Traest, General Secretary)
WWW: http://www.nfwo.be/
Founded: 1928
Activities: research council * funding basic research
Periodicals: (1) "Annual Report"
Awards: (1/5) "Prijs Dr. A. De Leeuw-Damry-Bourlart"
City Reference Coordinates: 004°20'00"E 50°50'00"N

North Atlantic Treaty Organization (NATO), Division of Scientific Affairs
(Organisation du Traité de l'Atlantique Nord, OTAN, Division des Affaires Scientifiques)
B-1110 Brussels
Telephone: (0)2-7284111
Telefax: (0)2-7284232
Electronic Mail: <userid>@hq.nato.int
natodoc@hq.nato.int
WWW: http://www.nato.int/
Founded: 1958
Staff: 20
Activities: funding international collaborative scientific research and training
Periodicals: (4) "NATO Science & Society Newsletter"; "NATO Scientific Publications" (ISSN 0770-3244), "NATO Yearbook"

City Reference Coordinates: 004°20'00"E 50°50'00"N

Observatoire Royal de Belgique (ORB)
Avenue Circulaire 3
B-1180 Bruxelles
Telephone: (0)2-3730211
Telefax: (0)2-3749822
Electronic Mail: rob@oma.be
WWW: http://www.oma.be/KSB-ORB/
Founded: 1826
Staff: 100
Activities: fundamental astronomy * time and frequency * satellite geodesy * geodynamics * seismology * gravimetry * physics of planetary interiors * astrometry * celestial mechanics * astrophysics * radioastronomy * solar physics * stellar statistics
Periodicals: (1) "Annuaire" (ISSN 0524-7780); "Bulletin Astronomique"; "Communications"
Coordinates: 004°21'29"E 50°47'51"N H105m
005°15'19"E 50°11'31"N H293m (Humain Radioastronomy Station)
City Reference Coordinates: 004°20'00"E 50°50'00"N

Organisation du Traité de l'Atlantique Nord (OTAN), Division des Affaires Scientifiques
● See "North Atlantic Treaty Organization (NATO), Division of Scientific Affairs"

Rijksuniversitair Centrum Antwerpen (RUCA)
● See "Universiteit Antwerpen"

Rijksuniversiteit Gent (RUG)
● See "Universiteit Gent"

Royal Observatory of Belgium, Planetarium
Boechoutlaan - Avenue Bouchout 10
B-1020 Brussels
Telephone: (0)2-4789526
          (0)2-4789513
Telefax: (0)2-4783026
WWW: http://www.oma.be/BIRA-IASB/SRBA/planetarium.html
Founded: 1973
Activities: education * lectures * colloquia * exhibitions
City Reference Coordinates: 004°20'00" E 50°50'00" N

Société Astronomique de Liège (SAL)
c/o Institut d'Astrophysique
Avenue de Cointe 5
B-4000 Liège
Telephone: (0)4-2533590
Telefax: (0)4-2547511
Electronic Mail: sal@astro.ulg.ac.be
WWW: http://www.astro.ulg.ac.be/~sal
        http://www.ulg.ac.be/musees/planetarium/ (Planetarium)
Founded: 1938
Membership: 850
Activities: education * library * observing * monthly lectures * instruments on loan * trips * planetarium
Periodicals: (10) "Le Ciel" (circ.: 1,100)
Observatories: 1 (Nandrin)
City Reference Coordinates: 005°34'00" E 50°38'00" N

Société Royale Belge d'Astronomie, de Météorologie et de Physique du Globe (SRBA)
Observatoire Royal de Belgique
Avenue Circulaire 3
B-1180 Bruxelles
Telephone: (0)2-3730253
Telefax: (0)2-3749822
Electronic Mail: srba@oma.be
WWW: http://www.oma.be/BIRA-IASB/SRBA/
Founded: 1884 ("Ciel et Terre": 1880)
Membership: 800
Activities: lectures * publishing * essentially Brussels-oriented
Periodicals: (6) "Ciel et Terre" (ISSN 0009-6709, circ.: 1,200)
City Reference Coordinates: 004°20'00" E 50°50'00" N

Société Royale des Sciences de Liège (SRSL)
c/o Institut de Mathématique
Université de Liège
Bâtiment B37
Grande Traverse 12
B-4000 Liège
Telephone: (0)4-3669371
Telefax: (0)4-3669547
Electronic Mail: n.naa@ulg.ac.be
WWW: http://www.ulg.ac.be/ipne/srsl/
Founded: 1835
Membership: 220
Activities: publishing monographs, proceedings, theses and original scientific papers * organizing lectures by foreign scientists * exchange of periodicals
Periodicals: (6) "Bulletin" (circ.: 500)
Awards: (1/5) "Prix Louis D'Or" (chemistry), "Prix Lucien Godeaux" (mathematics), "Prix Pol Swings" (physics), "Prix E. van Beneden" (biology)
City Reference Coordinates: 005°34'00" E 50°38'00" N

Société Scientifique de Bruxelles
Rue de Bruxelles 61
B-5000 Namur
Telephone: (0)81-724464
Telefax: (0)81-724502
Founded: 1875
Membership: 154
Periodicals: "Revue des Questions Scientifiques" (ISSN 0035-2160)
City Reference Coordinates: 004°52'00" E 50°28'00" N

Spacebel Instrumentation SA
Parc Industriel de Recherches du Sart Tilman

Rue des Chasseurs Ardennais
B-4031 Angleur
Telephone: (0)4-3666611
Telefax: (0)4-3660433
Electronic Mail: christine.mottard@spacebel.be (Public Relations)
Founded: 1988
Staff: 57
Activities: design and development of space-based optical equipment
City Reference Coordinates: 005°34'00"E 50°38'00"N

Sunspot Index Data Center (SIDC)
c/o Pierre Cugnon (Director)
Observatoire Royal de Belgique (ORB)
Avenue Circulaire 3
B-1180 Bruxelles
Telefax: (0)2-3730224
Electronic Mail: p.cugnon@oma.be (Pierre Cugnon
WWW: http://www.oma.be/KSB-ORB/SIDC/
Founded: 1981
City Reference Coordinates: 004°20'00"E 50°50'00"N

Union of International Associations (UIA)
Rue Washington 40
B-1050 Bruxelles
Telephone: (0)2-6401808
Telefax: (0)2-6460525
WWW: http://www.uia.org/
Electronic Mail: info@uia.be
Founded: 1907
Membership: 195
Staff: 20
Activities: collecting, analysing and publishing data on over 30,000 international governmental and non-governmental organizations, as well as on their meetings, publications and strategies, and on over 10,000 world problems with which they are preoccupied * maintaining on-line databases * carrying out surveys and studies
Periodicals: (6) "Transnational Associations"; (4) "International Congress Calendar"; (1) "Yearbook of International Organizations"; "Encyclopedia of World Problems and Human Potential"
City Reference Coordinates: 004°20'00"E 50°50'00"N

Universitaire Instelling Antwerpen (UIA)
● See "Universiteit Antwerpen"

Université Catholique de Louvain (UCL), Institut d'Astronomie et de Géophysique G. Lemaître
Chemin du Cyclotron 2
B-1348 Louvain-la-Neuve
Telephone: (0)10-473297
Telefax: (0)10-474722
Electronic Mail: <userid>@astr.ucl.ac.be
WWW: http://www.astr.ucl.ac.be/
Founded: 1966
Staff: 50
Activities: education * research in climatology and meteorology
Periodicals: "Contributions", "Bulletin Climatologique de la Station de Louvain-la-Neuve", "Scientific Reports", "Progress Reports"
Coordinates: 004°37'30"E 50°39'55"N
City Reference Coordinates: 004°38'00"E 50°40'00"N

Université de Liège (ULg), Centre Spatial de Liège (CSL)
Parc Scientifique du Sart Tilman
Avenue du Pré-Aily
B-4031 Liège-Angleur
Telephone: (0)4-3676668
Telefax: (0)4-3675613
Electronic Mail: cslulg@ulg.ac.be
WWW: http://www.ulg.ac.be/
Founded: 1988
Staff: 80
Activities: space qualification of instruments for optoelectronic observing * astronomical payloads * designing optical instrumentation
City Reference Coordinates: 005°34'00"E 50°38'00"N

Université de Liège (ULg), Institut d'Astrophysique et de Géophysique (IAGL)
Avenue de Cointe 5
B-4000 Liège
Telephone: (0)4-2547510
Telefax: (0)4-2547511

Electronic Mail: <userid>@astro.ulg.ac.be
WWW: http://www.astro.ulg.ac.be/
       http://vela.astro.ulg.ac.be/
       http://lpap.astro.ulg.ac.be/
Founded: 1880
Staff: 85
Activities: theoretical and observational astrophysics * spectroscopy * solar atmosphere * solar and atmospheric spectroscopy * planetary physics * geophysics * cosmology
Coordinates: 005°33'54"E 50°37'06"N H127m (Cointe)
Awards: "Prix Dehalu"
City Reference Coordinates: 005°34'00"E 50°38'00"N

Université de Mons-Hainaut, Service d'Astrophysique
Place du Parc 20
B-7000 Mons
Telephone: (0)65-373727/8
Telefax: (0)65-373054
Electronic Mail: <userid>@umh.ac.be
WWW: http://www.umh.ac.be/
Founded: 1968
Staff: 5
Activities: education in astronomy, astrophysics and atomic physics * research in astrophysics and atomic spectroscopy
Periodicals: "Mons Astrophysical Papers" (circ.: 300)
Coordinates: 016°30'00"W 28°18'00"N (Izaña, Teide, Islas Canarias)
City Reference Coordinates: 003°56'00"E 50°27'00"N

Universiteit Antwerpen, Rijksuniversitair Centrum Antwerpen (RUCA), Onderzoeksgroep Astrofysica
Groenenborgerlaan 171
B-2020 Antwerpen
Telephone: (0)3-2180355
Telefax: (0)3-2180204
Electronic Mail: astro@ruca.ua.ac.be
                <userid>@ruca.ua.ac.be
WWW: http://www.ruca.ua.ac.be/
Founded: 1985
Staff: 3
Activities: open clusters * associations * very young stellar systems * dynamical evolution and models * numerical N-body calculations * radial velocities * echelle spectroscopy
City Reference Coordinates: 004°25'00"E 51°13'00"N

Universiteit Antwerpen, Universitaire Instelling Antwerpen (UIA), Department of Physics
Universiteitsplein 1
B-2610 Antwerpen (Wilrijk)
Telephone: (0)3-8202457
Telefax: (0)3-8202245
WWW: http://nat-www.uia.ac.be/
Founded: 1972
Staff: 30
Activities: theoretical astrophysics * cosmology * relativity
City Reference Coordinates: 004°25'00"E 51°13'00"N

Universiteit Brussel, Astronomy Group
Pleinlaan 2
B-1050 Brussel
Telephone: (0)2-6293469
Telefax: (0)2-6293424
Electronic Mail: <userid>@vub.ac.be
WWW: http://www.vub.ac.be/STER/ster.html
Founded: 1990
Staff: 8
Activities: education * research * stellar evolution * binaries * stellar atmospheres * pulsating stars * photometry
Coordinates: 004°23'00"E 50°48'48"N H147m
City Reference Coordinates: 004°20'00"E 50°50'00"N
● Full university name: "Vrije Universiteit Brussel (VUB)"

Universiteit Gent, Sterrenkundig Observatorium
Gebouw S9
Krijgslaan 281
B-9000 Gent
Telephone: (0)9-2644798
Telefax: (0)9-2644989
Electronic Mail: <userid>@rug.ac.be
WWW: http://www.rug.ac.be/~hdejongh/astro/astro.html
Founded: 1907
Staff: 5

Activities: stellar dynamics * numerical astrophysics * radiation transfer * plasma astrophysics * popularization
Periodicals: "Mededelingen" (ISSN 0072-4432)
Coordinates: 003°04'00"E 51°02'00"N
• Full university name: "Rijksuniversiteit Gent (RUG)"
City Reference Coordinates: 003°04'00"E 51°02'00"N

Universiteit Leuven, Center for Plasma Astrophysics

Celestijnenlaan 200 B
B-3001 Heverlee
Telephone: (0)16-327007
Telefax: (0)16-327998
WWW: http://cpa.wis.kuleuven.ac.be/
Founded: 1992
Staff: 15
Activities: plasma astrophysics * solar physics * MHD * education
• Full university name: "Katholieke Universiteit Leuven"
City Reference Coordinates: 004°42'00"E 50°53'00"N

Universiteit Leuven, Instituut voor Sterrenkunde

Celestijnenlaan 200 B
B-3001 Heverlee
Telephone: (0)16-327033
Telefax: (0)16-327999
Electronic Mail: <userid>@ster.kuleuven.ac.be
WWW: http://www.ster.kuleuven.ac.be/
Staff: 14
Activities: astronomy and astrophysics
• Full university name: "Katholieke Universiteit Leuven"
City Reference Coordinates: 004°42'00"E 50°53'00"N

Université Libre de Bruxelles (ULB), Institut d'Astronomie et d'Astrophysique (IAA)

Case Postale 226
Boulevard du Triomphe 2
B-1050 Bruxelles
Telephone: (0)2-6502842
Telefax: (0)2-6504226
Electronic Mail: <userid>@astro.ulb.ac.be
WWW: http://astro.ulb.ac.be/iaa.htm
Founded: 1923
Staff: 11
Activities: theoretical astrophysics * theoretical and experimental nuclear astrophysics * stellar evolution * stellar structure * stellar abundances * CP stars * close binaries * nucleosynthesis * spectroscopy, photometry and radial velocities of red giant stars
City Reference Coordinates: 004°20'00"E 50°50'00"N

Université Libre de Bruxelles (ULB), Physique Nucléaire Théorique et Physique Mathématique (PNTPM)

Campus Plaine
Avenue F.D. Roosevelt
Case Postale 229
B-1050 Bruxelles
Telephone: (0)2-6505560
         (0)2-6505562
Telefax: (0)2-6505045
Electronic Mail: <userid>@ulb.ac.be
WWW: http://www.ulb.ac.be/recherche/cr/cr.html
Staff: 14
Activities: theoretical physics * nuclear reactions in general and of astrophysical interest * nuclear spectroscopy * nuclear and atomic physics * mathematical physics
City Reference Coordinates: 004°20'00"E 50°50'00"N

Urania, Volkssterrenwacht van Antwerpen

J. Mattheessensstraat 60
B-2540 Hove
Telephone: (0)3-4552493
Telefax: (0)3-4542297
Electronic Mail: info@urania.be
WWW: http://www.urania.be/
Founded: 1969
Membership: 350
Activities: education * lectures * visits * observing * meteorology * working groups * exhibitions * planetarium (Antwerp Zoo) * astronomy software * radio meteors
Periodicals: (15) "Urania in de Kijker" (circ.: 150); (10) "Oberonnieuws" (circ: 100); (4) "De Sterrenwachter" (circ.: 500)
Coordinates: 004°28'02"E 51°08'45"N H18m
City Reference Coordinates: 004°29'00"E 51°09'00"N

Vereniging voor Sterrenkunde, Meteorologie, Geofysica en Aanverwante Wetenschappen (VVS)
Brieversweg 147
B-8310 Brugge
Brugge
Telephone: (0)50-358872
Telefax: (0)50-355007
Electronic Mail: info@vvs.be
WWW: http://www.vvs.be/
Founded: 1944
Membership: 1900
Activities: amateur astronomy and meteorology
Periodicals: (12) "Heelal" (ISSN 0772-6422, circ.: 2,100), "Halo"; (4) "Distant Targets"
Sections: Astrophotography * Occultations * Computing Technology * Comets * Artificial Satellites * Meteors * Planets * Space * Variable Stars * Meteorology * Sun * Light Pollution * Solar Eclipses * Deep Sky
Awards: (1) "Galilei Prizes"
City Reference Coordinates: 003°14'00"E 51°13'00"N

Volkssterrenwacht Beisbroek vzw
Zeeweg 96
B-8200 Brugge
Telephone: (0)50-390566
Telefax: (0)50-393244
Electronic Mail: beisbroek@unicall.be
WWW: http://www.urania.be/beisbroek/
Founded: 1984
Staff: 1
Membership: 170
Activities: popularization * visits of schools * education * observing (meteors, variable stars) * investigation of selected astrophysical topics * planetarium
Periodicals: (1) "Jaarverslag"
Coordinates: 003°09'00"E 51°10'00"N H20m
City Reference Coordinates: 003°09'00"E 51°10'00"N

Volkssterrenwacht Mira vzw
Abdijstraat 20
B-1850 Grimbergen
Telephone: (0)2-2691280
Telefax: (0)2-2691075
Electronic Mail: info@mira.be
        http://www.mira.be/
Founded: 1967
Staff: 3
Membership: 300
Activities: popularization (astronomy, meteorology, space sciences) * lectures * exhibitions * library and documentation center * observing (Sun) * static planetarium
Periodicals: (4) "Mira Ceti"
Coordinates: 004°22'13"E 50°56'08"N
City Reference Coordinates: 004°22'00"E 50°56'00"N

Volkssterrenwacht Nysa
Pastorijstraat 1
B-9403 Neigem
Telephone: (0)54-328420
        (0)54-323003
        (0)95-333242
        (0)95-335236
Electronic Mail: nysa.volkssterrenwacht@skynet.be
WWW: http://users.skynet.be/volkssterrenwacht.nysa/
Founded: 1990
City Reference Coordinates: 004°04'00"E 50°48'00"N

Vrije Universiteit Brussel (VUB)
● See "Universiteit Brussel"

Wega vzw
Escoriallaan 18
B-3000 Leuven
Telephone: (0)16-250942
Electronic Mail: wega.leuven@usa.net
WWW: http://surf.to/wega
Founded: 1975
Membership: 4
Activities: observing * lectures * education * camps * instrumentation * star parties * meetings
City Reference Coordinates: 004°42'00"E 50°53'00"N

Bolivia

Academia Nacional de Ciencias de Bolivia (ANCB)
Avenida 16 de Julio 1732
Casilla 5829
La Paz
Telephone: (0)2-363990
Telefax: (0)2-379681
Electronic Mail: ancb@ancb.bo
Founded: 1960
Membership: 60
Activities: science policy studies * advising government * environmental sciences * astronomy
Periodicals: (4) "Revista"
City Reference Coordinates: 068°10'00"W 16°30'00"N

Asociación Boliviana de Astronomía (ABA)
Casilla 7707
La Paz
or :
Calle México 1771
La Paz
Telephone: (0)2-326922
               (0)2-352660
Telefax: (0)2-791696
Electronic Mail: lcabezas@entelsa.entelnet.bo (Luis Cabezas Tito, Vice President)
Founded: 1969
Membership: 160
Activities: observing * popularization
Sections: Meteors
Periodicals: (6) "Astromundo"; "Firmamento", "Alajpacha", "Meteoros"
Coordinates: 068°00'00"W 16°00'00"S (La Paz)
                066°00'00"W 20°00'00"S (Potosí)
                062°00'00"W 18°00'00"S (Santa Cruz)
                066°00'00"W 19°00'00"S (Sucre)
                066°00'00"W 21°00'00"S (Tarija)
City Reference Coordinates: 068°10'00"W 16°30'00"N

Astronomía Sigma Octante (ASO), Centro de Investigación y Estudio en Astronomía
Casilla 1491
Cochabamba
or :
Centro "Simón I. Patiño"
Avenida Potosí 1450
Cochabamba
Telephone: (0)42-80602
Telefax: (0)42-81099
Electronic Mail: german@bo.net (German Morales)
WWW: http://www.lostiempos.com/public/sigma/asondx.htm (Spanish version)
         http://www.lostiempos.com/public/sigma/asondxen.htm (English version)
Founded: 1977
Membership: 20
Activities: observing (Sun, planets, satellites, minor planets, variable stars, comets, meteors) * theoretical astronomy * computational astronomy * education * astrometry
Sections: positional astronomy * meteors * Sun * solar system * stellar observing * popularization
Periodicals: (12) "Boletín de los Amigos de ASO"; (2) "Sunspots Report"
Coordinates: 066°09'48"W 17°24'06"S H2,570m
City Reference Coordinates: 066°10'00"W 17°26'00"S

Instituto Boliviano de Normalización y Calidad (IBNORCA)
Avenida Camacho Esq. Bueno 1488
Casilla 5034
La Paz
Telephone: (0)2-317262
Telefax: (0)2-317262
City Reference Coordinates: 068°10'00"W 16°30'00"N

Liga Ibero-Americana de Astronomía (LIADA)
c/o Rodolfo Zalles (President)
Casilla 7090
La Paz
or :
c/o Mirko Raljevic (Secretary)
Casilla 7090

La Paz
or :
c/o Raul Cagigao (Treasurer)
Casilla 387
Sucre
WWW: http://www.cnba.uba.ar/acad/astro/liada/liada.html
Founded: 1982
Membership: 300
Activities: promoting observing (meteors, comets, Sun, planets, variable stars) * grouping amateur astronomers * developing respect and appreciation of Earth as a planet * publishing works by members
Periodicals: (4) "Universo" (ISSN 0012-9820), "La Red"
City Reference Coordinates: 068°10'00"W 16°30'00"N (La Paz)
071°08'00"W 08°36'00"N (Sucre)

Observatorio Astronómico Boliviano-Ruso
Santa Ana
Casilla 20
Tarija
Telephone: (0)1-860979
Telefax: (0)66-44238
Electronic Mail: obsrzb@uajms.bo
Founded: 1984
Staff: 4
Activities: astrometry * photometric catalogues
Coordinates: 064°36'00"W 21°35'00"S H2,000m
City Reference Coordinates: 064°36'00"W 21°35'00"S

Observatorio Astronómico de Patacamaya (OAP)
● See "Universidad Mayor de San Andrés, Observatorio Astronómico de Patacamaya (OAP)"

Observatorio San Calixto (OSC)
Indaburu 944
Casilla 12656
La Paz
Telephone: (0)2-356098
Telefax: (0)2-376805
Electronic Mail: sancalixto@mail.megalink.com
Founded: 1913
Staff: 6
Activities: astronomical demonstrations to visiting school groups * seismology * meteorology
Periodicals: (12) "Lecturas de la Estación de la Paz" (circ.: 60)
Observatories: network of seismic stations
City Reference Coordinates: 068°10'00"W 16°30'00"N

Planetario Dr. Max Schreier
Calle Frederico Zuazo 1976
Casilla 3164
La Paz
Telephone: (0)2-359522
Telefax: (0)2-316738
Electronic Mail: planetar@astro.bo
WWW: http://www.umsanet.edu.bo/org/astro/
Founded: 1976
Staff: 7
Activities: astronomy * archaeoastronomy * solar activity * geomagnetism
Periodicals: "Intijiwaña"; (1) "Calendario"
Observatories: 1 (see Estación Astronómica de Patacamaya)
City Reference Coordinates: 068°10'00"W 16°30'00"S

Servicio Nacional de Meteorología e Hidrología (SENAMHI)
Avenida Arce 2579
Casilla 10993
La Paz
Telephone: (0)2-326165
(0)2-355824
Telefax: (0)2-392413
Founded: 1944
Periodicals: "Boletines Meteorológicos", "Anuarios Meteorológicos"
Observatories: from 57°59'W to 69°20'W and from 11°5'S to 22°30'S
City Reference Coordinates: 068°10'00"W 16°30'00"S

Universidad Mayor de San Andrés (UMSA), Observatorio Astronómico de Patacamaya (OAP)
Calle Federico Zuazo 1976
Casilla 3164
La Paz

Telephone: (0)2-359522
Telefax: (0)2-316738
Electronic Mail: planetar@astro.bo
                    planetar@utama.bolnet.bo
WWW: http://www.umsanet.edu.bo/org/astro/
Founded: 1974
Activities: astrometry * photometry * special events * Sun * planetarium (see Planetario Dr. Max Schreier) * ethnoastronomy

Periodicals: (4) "Boletín de Actividad Solar"; "Calendario Astronómico"
Coordinates: 067°57'07"W 17°15'57"S H3,789m
City Reference Coordinates: 068°10'00"W 16°30'00"N

Brazil

Academia Brasileira de Ciencias (ABC)
Caixa Postal 229
20001-970 Rio de Janeiro - RJ
or :
Rua Anfilófio de Carvalho 29 - 3°
20030-060 Rio de Janeiro - RJ
Telephone: (0)21-2204794
Telefax: (0)21-2404695
Electronic Mail: <userid>@abc.org.br
WWW: http://www.abc.org.br/
Founded: 1916
Membership: 285
Activities: publishing * meetings * advising
Periodicals: (4) "Anais da Academia Brasileira de Ciências" (ISSN 0001-3765), "Revista Brasileira de Biologia" (ISSN 0034-7108)
Awards: (1/4) "Ruschi Award"
City Reference Coordinates: 043°15'00"W 22°54'00"S

Associação Astronômica da Poços de Caldas
Caixa Postal 340
37701-010 Poços de Caldas - MG
or :
Rua Assis Figueiredo 1.174 Conj. 34
37701-000 Poços de Caldas - MG
Telephone: (0)35-7222148
Founded: 1987
Membership: 20
Activities: education * celestial mechanics * astrophysics * Sun * Moon
Periodicals: (4) "As Plêiades" (circ.: 50)
Coordinates: 046°33'53"W 21°50'20"S H1,600m (Morro de São Domingo)
City Reference Coordinates: 046°33'00"W 21°48'00"S

Associação Brasileira de Normas Técnicas (ABNT)
Avenida 13 de Maio 13 - 27° andar
Caixa Postal 1680
20003-900 Rio de Janeiro - RJ
Telephone: (0)21-2103122
Telefax: (0)21-2408249
Electronic Mail: abnt@embratel.net.br
WWW: http://www.abnt.org.br/
Founded: 1940
City Reference Coordinates: 043°15'00"W 22°54'00"S

Centro de Divulgação da Astronomia (CDA)
c/o Centro de Divulgação Científica e Cultural
Rua 9 de Julho 1227
13569-590 São Carlos - SP
Telephone: (0)16-2739191
(0)16-2739772
Telefax: (0)16-2723910
Electronic Mail: cda@cdcc.sc.usp.br
WWW: http://www.cdcc.sc.usp.br/cda/observat.htm
Founded: 1986
Staff: 10
Activities: education * observing * exhibitions
Coordinates: 047°54'00"W 22°01'00"S H940m
City Reference Coordinates: 047°53'00"W 22°02'00"S

Centro de Estudios Astronômicos de Alagoas (CEAAL)
Caixa Postal 215
57.020-970 Maceió - AL
Electronic Mail: ceaal@fapeal.br
WWW: http://www.fapeal.br/ceaal/
Founded: 1978
Membership: 12
Periodicals: (12) "Apolo"
City Reference Coordinates: 035°43'00"W 09°40'00"S

Centro de Previsão de Tempo e Estudos Climáticos (CPTEC)
● See "Instituto Nacional de Pesquisas Espaciais (INPE), Centro de Previsão de Tempo e Estudos Climáticos (CPTEC)"

Centro de Radio-Astronomia e Aplicações Espaciais (CRAAE)
● See "Instituto Presbiteriano Mackenzie, Centro de Radio-Astronomia e Aplicações Espaciais (CRAAE)"

Clube de Astronomia do Rio de Janeiro (CARJ)
Caixa Postal 65.090
20072-970 Rio de Janeiro - RJ
Telephone: (0)21-3594790
Electronic Mail: carj@astronomia-carj.com.br
WWW: http://www.astronomia-carj.com.br/
Founded: 1976
Membership: 150
Activities: education * popularization
Periodicals: (12) "Guia do Amador de Astronomia", "Circular Astronômica"
City Reference Coordinates: 043°15'00"W 22°54'00"S

Clube Estudantil de Astronomia (CEA)
Caixa Postal 736
50001-000 Recife - PE
or :
Conjunto Residencial INOCOOP
Rua Azeredo Continho s/n
Várzea
50741-000 Recife - PE
WWW: http://www.cea.org.br/
Founded: 1972
Membership: 120
Activities: occultations by minor planets * comets * binaries * education
Periodicals: (4) "Boletim Astronômico" (circ.: 400)
Awards: "Messier Club Member Certificate"
Coordinates: 034°57'28"W 08°03'03"S H2m (Jorge Polman Observatory)
City Reference Coordinates: 034°53'00"W 08°06'00"S

Colégio Magno, Observatório
Rua Duque Costa 164
04671-160 São Paulo - SP
Telephone: (0)11-5486166
Electronic Mail: colmagno@eu.ansp.br
WWW: http://eu.ansp.br:80/~colmagno/
City Reference Coordinates: 046°37'00"W 23°32'00"S

Conselho Nacional Desenvolvimento Scientifico e Tecnologico
Edificio CNPq
SEPN 507 - Bloco "B"
Ed. Sede CNPq
70740-901 Brasília - DF
Telephone: (0)61-3489000
Telefax: (0)61-2741950
Electronic Mail: <userid>@cnpq.br
WWW: http://www.cnpq.br/
Founded: 1951
City Reference Coordinates: 044°26'00"W 16°12'00"S

Fundação de Amparo à Pesquisa do Estado de São Paulo (FAPESP)
Rua Pio XI 1500
Alto da Lapa
05468-901 São Paulo - SP
Telephone: (0)11-8384000
Telefax: (0)11-2614167
Electronic Mail: info@fapesp.br
WWW: http://www.fapesp.br/
Founded: 1960
City Reference Coordinates: 046°37'00"W 23°32'00"S

Fundação Planetário da Cidade do Rio de Janeiro
Avenida Padre Leonel Franca 240
Gávea
22451-000 Rio de Janeiro - RJ
Telephone: (0)21-2740046
          (0)21-2740096
Telefax: (0)21-2396927
Electronic Mail: planet@rdc.puc-rio.br
WWW: http://www.rio.rj.gov.br/planetario/
          http://io.rdc.puc-rio.br/planetario/
          http://io.rdc.puc-rio.br/planetario/ingles/
Founded: 1970

Staff: 40
Activities: shows * public observing * education * lectures
Coordinates: 043°10'00"W 22°53'00"S H10m
City Reference Coordinates: 043°10'00"W 22°53'00"S

Grupo de Estudos de Astronomia (GEA)
Planetário
Campus UFSC
88.000-000 Florianopolis - SC
Telephone: (0)48-3319241
Electronic Mail: geraldomattos@hotmail.com (Geraldo Mattos)
WWW: http://www.gea.org.br/
Founded: 1985
Activities: education * popularization * lectures
City Reference Coordinates: 048°31'00"W 27°35'00"S

Instituto Nacional de Meteorologia (INMET)
Eixo Monumental
Via S-1
Cruzeiro
70610-400 Brasília - DF
Telephone: (0)61-2247944
         (0)61-2251230
         (0)61-3443333
Telefax: (0)61-2266967
WWW: http://www.inmet.gov.br/
Founded: 1909
Activities: weather forecast * climatology
Periodicals: "Boletim de Radiação Solar", "Boletim Agrometeorológico"
City Reference Coordinates: 044°26'00"W 16°12'00"S

Instituto Nacional de Pesquisas Espaciais (INPE), Cachoeira Paulista
Rodovia Presidente Dutra - Km. 40
Caixa Postal 01
12630-000 Cachoeira Paulista - SP
Telephone: (0)125-609200
Telefax: (0)125-612088
WWW: http://www.dgi.inpe.br/
         http://www.inpe.br/
Founded: 1961 (INPE)
City Reference Coordinates: 045°01'00"W 22°40'00"S

Instituto Nacional de Pesquisas Espaciais (INPE), Centro de Previsão de Tempo e Estudos Climáticos (CPTEC)
Rodovia Presidente Dutra Km 40
Caixa Postal 01
12630-000 Cachoeira Paulista
Telephone: (0)12-5608400
Telefax: (0)12-5612835
Electronic Mail: clima@cptec.inpe.br
              <userid>@cptec.inpe.br
WWW: http://www.cptec.inpe.br/
Founded: 1986 (INPE: 1961)
Staff: 15
Activities: climatology
Periodicals: (12) "Climanálise" (ISSN 0103-0019)
City Reference Coordinates: 045°01'00"W 22°40'00"S

Instituto Nacional de Pesquisas Espaciais (INPE), São José dos Campos
Caixa Postal 515
12201-970 São José dos Campos - SP
or :
Avenida dos Astronautas 1758
Jardim da Granja
12227-010 São José dos Campos - SP
Telephone: (0)123-418977
         (0)12-3456000
Telefax: (0)12-3218743
Electronic Mail: <userid>@inpe.br
WWW: http://www.inpe.br/astro/home
         http://www.inpe.br/
Founded: 1961
Staff: 1400
Activities: astrophysics * space geophysics * meteorology * remote sensing * space sciences * spacecraft engineering
Periodicals: "Espacial" (ISSN 0103-0795), "INPE Space News", "Relatório de Atividades", "Climanálise"
City Reference Coordinates: 045°13'00"W 23°11'00"S

Instituto Presbiteriano Mackenzie, Centro de Radio-Astronomia e Aplicações Espaciais (CRAAE)
Rua Consolação 896
01302-000 São Paulo - SP
Telephone: (0)11-2368331
Telefax: (0)11-2142990
Electronic Mail: <userid>@craae.mackenzie.br
WWW: http://craae.mackenzie.br/
Founded: 1989 (Mackenzie: 1870)
Staff: 15
Activities: radioastronomy * VLBI * space geodesy * solar physics * STP
Coordinates: 046°33'48"W 23°11'03"S (ROI - see separate entry)
               038°25'35"W 03°52'40"S (ROEN - see separate entry)
City Reference Coordinates: 046°37'00"W 23°32'00"S

International Mathematical Union (IMU)
c/o Jacob Palis Jr.
IMPA
Estrada Dona Castorina 110
Jardim Botânico
22460-320 Rio de Janeiro - RJ
Telephone: (0)21-2949032
           (0)21-5111749
Telefax: (0)21-5124112
Electronic Mail: imu@impa.br
WWW: http://elib.zib.de/IMU/
Founded: 1951
Membership: 60 (countries)
Periodicals: "Bulletin" (ISSN 1015-8081), "World Directory of Mathematicians"
Awards: "Fields Medal", "Rolf Nevanlinna Prize"
City Reference Coordinates: 043°15'00"W 22°54'00"S

Laboratório Nacional de Astrofísica (LNA), Itajubá
Rua Estados Unidos 154
Bairro das Nações
Caixa Postal 21
37500-000 Itajubá - MG
Telephone: (0)35-6231500
           (0)35-6212121 (Observatório do Pico dos Dias)
Telefax: (0)35-6231544
           (0)35-6212137 (Observatório do Pico dos Dias)
Electronic Mail: <userid>@lna.br
             diretor@lna.br
             secret@lna.br
             hotel@elisa.lna.br (Observatório do Pico dos Dias)
WWW: http://www.lna.br/
Founded: 1980
Staff: 58
Activities: instrumentation * stellar and extragalactic astronomy * young and active stars * PN * image processing
Coordinates: 045°34'59"W 22°32'04"S H1,864m (Observatório do Pico dos Dias)
City Reference Coordinates: 045°27'00"W 22°26'00"S

Museu de Astronomia e Ciências Afins
Rua General Bruce 586
São Cristóvão
20921-000 Rio de Janeiro - RJ
Telephone: (0)21-5809432
           (0)21-5804531
Electronic Mail: mast@omega.lncc.br
WWW: http://pub2.lncc.br:80/mast/
       http://www.info.lncc.br/mast
Founded: 1985
Staff: 90
Periodicals: "Cadernos de Astronomia", "Perspicillum" (ISSN 0102-9495)
City Reference Coordinates: 043°15'00"W 22°54'00"S

Observatório Astronômico Antares (OAA)
Rua da Barra 925
Jardim Cruzeiro
44015-430 Feira de Santana - BA
Telephone: (0)75-6241921
Telefax: (0)75-6241921
Electronic Mail: vmartin@uefs.br (Vera Aparecida Fernandes Martin)
WWW: http://www.uefs.br/antares/default_htm.htm
Founded: 1971
Staff: 10
Activities: fundamental astronomy * solar physics * planetarium

City Reference Coordinates: 038°57'00"W 12°15'00"S

Observatório Astronômico de Piedade (OAP)
● See "Universidade Federal de Minas Gerais, Departamento de Física, Observatório Astronômico de Piedade (OAP)"

Observatório Astronômico de Uberlândia (OAU)
c/o Roberto F. Silvestre
Rua das Seriemas 475
Cidade Jardim
38412-158 Uberlândia - MG
Electronic Mail: silvestre@ufu.br
WWW: http://lfc.df.ufu.br/~silvestr/
Founded: 1996
Activities: private observatory
City Reference Coordinates: 048°17'00"W 18°57'00"S

Observatório Astronômico Herschel-Einstein (OAHE)
c/o Iolanda Siqueira Pamplona (General Coordinator)
Rua General Joaquim de Andrade 68
Aldeota
60150-030 Fortaleza - CE
Telephone: (0)85-2245496
          (0)85-2241741
Founded: 1965
Membership: 12
Activities: comets * meteors * Moon * Sun * eclipses * planets * meteorology
Periodicals: "Pleyades"
Coordinates: 038°30'36"W 03°43'53"S H16m
City Reference Coordinates: 038°30'00"W 03°43'00"S

Observatório Astronômico Monoceros (OAM)
Rua Luiz Marota
Bairro Santa Marta
36660-000 Além-Paraíba - MG
Telephone: (0)32-4624582
Electronic Mail: monocero@fusoes.com.br
Founded: 1975
Membership: 7
Activities: solar activity * education * popularization * meteorology
Periodicals: (12) "Circular"
Coordinates: 042°40'32"W 21°52'23"S H300m
City Reference Coordinates: 042°36'00"W 21°49'00"S

Observatório Copérnico
Praça da Bandeira 28
36680-000 São João Nepomuceno - MG
Telephone: (0)32-2612463
Telefax: (0)32-2612762
Founded: 1977
Staff: 7
Activities: education * lectures * exhibitions * investigations
Coordinates: 043°01'10"W 21°32'00"S
City Reference Coordinates: 043°04'00"W 21°33'00"S

Observatório Municipal de Americana (OMA)
Avenida Brasil 2525
13465-000 Americana - SP
Telephone: (0)19-4607026
Electronic Mail: oma@dglnet.com.br
WWW: http://www.dglnet.com.br/oma/
Founded: 1985
Staff: 2
Activities: solar research * public observing
Coordinates: 047°21'10"W 22°45'20"S H582m
City Reference Coordinates: 047°19'00"W 22°44'00"S

Observatório Municipal de Campinas (OMC)
Caixa Postal 27
Distrito de Sousa
13100-000 Campinas - SP
or :
Serra das Cabras
Avenida Anchieta 200 - 6°
13100-000 Campinas - SP
Telephone: (0)192-423252

Founded: 1977
Staff: 15
Activities: education * popularization * planets * comets * Sun * astrometry * photometry
Periodicals: "Boletim Astronômico"
Coordinates: 047°04'39"W 22°53'20"S
City Reference Coordinates: 047°06'00"W 22°54'00"S

Observatório Municipal de Piracicaba (OMP)
Rue Antonio Corrêa Barbosa 2.233 - 6° andar
13400-900 Piracicaba - SP
Telephone: (0)194-334233
Telefax: (0)194-342823
Founded: 1992
Staff: 3
Activities: education * popularization * observing (solar system, double stars)
City Reference Coordinates: 047°38'00"W 22°43'00"S

Observatório Nacional (ON)
Rua General José Cristino 77
20921-030 Rio de Janeiro - RJ
Telephone: (0)21-5806087 (Director)
          (0)21-5853215 (Switchboard)
Telefax: (0)21-5800332
          (0)21-5806041
Electronic Mail: <userid>@obsn.on.br
WWW: http://obsn.on.br/
Founded: 1827
Staff: 170
Activities: astronomy * geophysics * time & frequency service
Periodicals: (1) "Efemérides Astronômicas" (ISSN 0101-935x, circ.: 1,000); "Preprints", "Publicações do Observatório Nacional" (circ.: 500)
Coordinates: 043°13'22"W 22°53'42"S H33m
City Reference Coordinates: 043°15'00"W 22°54'00"S

Observatório Young
Instituto de Aeronautica e Espaço (IAE)
Praça Marechal Eduardo Gomes 50
12228-901 São José dos Campos
Electronic Mail: naee@iconet.com.br
Founded: 1960
Staff: 3
Activities: education
City Reference Coordinates: 045°13'00"W 23°11'00"S

Planetário da Cidade do Rio de Janeiro
● See "Fundação Planetário da Cidade do Rio de Janeiro"

Planetário Prof. José Baptista Pereira
● See "Universidade Federal do Rio Grande do Sul (UFRGS), Planetário Prof. José Baptista Pereira"

Rádio Observatório do Itapetinga (ROI)
Rodovia Municipal Engenharia do Mackenzie s/n
Bairro do Itapetinga
Caixa Postal 200
12940-000 Atibaia - SP
Telephone: (0)11-78711503
Telefax: (0)11-78711784
WWW: http://craae.ptr.usp.br/roi2.htm
Founded: 1971
Coordinates: 046°33'48"W 23°11'03"S
City Reference Coordinates: 046°33'00"W 23°07'00"S

Rádio Observátorio Espacial do Nordeste (ROEN)
Rua José Hipólito s/n
Bairro Tupuiù
Caixa Postal 21
61760-000 Fortaleza - CE
Telephone: (0)85-2609266
Telefax: (0)85-260959
WWW: http://craae.ptr.usp.br/roen.html
Coordinates: 038°25'35"W 03°52'40"S
City Reference Coordinates: 038°30'00"W 03°43'00"S

Sociedade Astronômica Brasileira (SAB)
Avenida Miguel Stefano 4200

04301-904 São Paulo - SP
Telephone: (0)11-5778599
Telefax: (0)11-2763848
Electronic Mail: sab@iagusp.usp.br
                 sab@orion.iagusp.usp.br
WWW: http://www.iagusp.usp.br/sab
Founded: 1974
Membership: 460
Activities: annual meetings
Periodicals: (3) "Boletim da Sociedade Astronômica Brasileira" (ISSN 0101-3440, circ.: 600)
City Reference Coordinates: 046°37'00"W 23°32'00"S

Sociedade Astronômica Maranhense de Amadores (SAMA)
Rua Arimatéia Cisme, 234 - Apeadouro
65030-000 São Luís - MA
Telephone: (0)98-2212008
           (0)98-2436022
WWW: http://www.elo.com.br/~cportela/sama01.html
       http://www.inf.puc-rio.br/~portela/sama01.html
Founded: 1976
Membership: 12
Activities: observing (Sun, sunspots) * telescope gatherings * videos shows * meetings
City Reference Coordinates: 044°16'00"W 02°31'00"S

Sociedade Brasileira dos Amigos da Astronomia (SBAA)
Avenida do Imperador 1330
Centro
60015-000 Fortaleza - CE
Telephone: (0)85-2525077
           (0)85-2525237
Telefax: (0)85-2525238
Electronic Mail: sbaa@sbaa.com.br
WWW: http://www.fortalnet.com.br/sbaa/
Founded: 1947
City Reference Coordinates: 038°30'00"W 03°43'00"S

Sociedade Brasileira para o Ensino da Astronomia (SBEA)
Rua Texas 1.177
Sala 04
04557-001 São Paulo - SP
Telefax: (0)11-5506-7838
Electronic Mail: sbea@mandic.com.br
WWW: http://pessoal.mandic.com.br/~sbea/
Periodicals: (12) "Astronomia Novae"
City Reference Coordinates: 046°37'00"W 23°32'00"S

Sociedade de Astronomia e Astrofísica de Diadema (SAAD)
Avenida Antônio Cunha Silva Bueno 1233
09970-160 Diadema - SP
Telephone: (0)11-4457974
Telefax: (0)11-7130901
Electronic Mail: joelfurlani@sti.com.br
                 siga@cepa.com.br
                 ozimar@uol.com.br
WWW: http://www.geocities.com/capcanaveral/2303
Founded: 1989
Membership: 60
Activities: observing (eclipses, occultations, variable stars, Moon, Sun, planets, meteors) * instrumentation * popularization
* education
Periodicals: (2) "Boletim da SAAD"
Coordinates: 046°36'38"W 23°41'10"S H820m
City Reference Coordinates: 046°37'00"W 23°42'00"S

Sociedade de Estudos Astronômicos de Ouro Preto (SEAOP)
Praça Tiradentes 20
35400-000 Ouro Preto - MG
Telephone: (0)31-5511533
Telefax: (0)31-5591528
Electronic Mail: seaop@em.ufop.br
WWW: http://www.seaop.em.ufop.br/
Founded: 1992
Membership: 6
Activities: education * public outreach
Coordinates: 043°30'16"W 20°22'59"S (Observatório Astronômico da Escuela de Minas)
City Reference Coordinates: 043°30'00"W 20°25'00"S

Universidade de São Paulo (USP), Instituto Astronômico e Geofísico (IAG), Departamento de Astronomia
Avenida Miguel Stéfano 4200
Água Funda
Caixa Postal 3386
01060-970 São Paulo - SP
Telephone: (0)11-5778599
Telefax: (0)11-5570270
Electronic Mail: <userid>@iagusp.usp.br
WWW: http://www.iagusp.usp.br/
Founded: 1932
Staff: 35
Activities: astrophysics (stars, ISM, galaxies, solar system) * cosmology * fundamental astronomy * dynamical astronomy
Periodicals: "Anuário Astronômico"
Coordinates: 046°58'00"W 23°06'00"S H850m (São Paulo)
                           046°37'22"W 23°39'07"S H800m (Observatório Abrahão de Moraes, Valinhos)
City Reference Coordinates: 046°58'00"W 23°06'00"S

Universidade Estadual de Campinas, Instituto de Física, Departamento Raios Cosmicos
Caixa Postal 6165
13083-970 Campinas - SP
Telephone: (0)192-398112
Telefax: (0)192-393127
Electronic Mail: <userid>@ifi.unicamp.br
WWW: http://www.ifi.unicamp.br/ifgw.html
Founded: 1967
Staff: 13
Activities: cosmic rays * high-energy physics * hadronic interactions
City Reference Coordinates: 047°05'00"W 22°54'00"S

Universidade Estadual de Campinas, Instituto de Física, Núcleo de Ciência, Aplicações e Tecnologia Espaciais (NUCATE)
Rua Roxo Moreira 1752
Barão Geraldo
Cidade Universitária "Zefirano Vaz"
Caixa Postal 6165
13083-592 Campinas - SP
Telephone: (0)19-2391301
                   (0)19-2397150
                   (0)19-3293125
Telefax: (0)19-2393680
                (0)19-2393127
                (0)19-3293125
Electronic Mail: <userid>@ifvax.unicamp.ansp.br
WWW: http://www.ifi.unicamp.br/ifgw.html
Activities: atmospheric sciences * space geophysics
City Reference Coordinates: 047°05'00"W 22°54'00"S

Universidade Estadual Paulista (UNESP), Instituto de Física Teorica
Rua Pamplona 145
01405-000 São Paulo - SP
Telephone: (0)11-2515155
Electronic Mail: <userid>@unesp.br
WWW: http://www.unesp.br/
Founded: 1950
Staff: 26
Activities: field theory * elementary particles * gravitation * cosmology * nuclear physics * mathematical physics
City Reference Coordinates: 046°37'00"W 23°32'00"S

Universidade Federal de Goiás (UFG), Planetário
Caixa Postal 131
74055-140 Goiânia - GO
Telephone: (0)62-2245787
Telefax: (0)62-2245787
WWW: http://www.ac-digital.com/abplanetarios/goiania/
              http://www.fis.ufg.br/
Founded: 1970
Staff: 5
Activities: education * shows
Coordinates: 049°15'29"W 16°41'21"S H802m
City Reference Coordinates: 049°16'00"W 16°40'00"S

Universidade Federal do Espírito Santo (UFES), Observatório Astronômico
Avenida Fernando Ferrari
Goiabeiras

Caixa Postal 19015
29060-970 Vitória - ES
Telephone: (0)27-3352484
             (0)27-3251711 x267
Electronic Mail: oaufes@cce.ufes.br
WWW: http://www.cce.ufes.br/~oaufes
Founded: 1986
City Reference Coordinates: 040°21'00"W 20°19'00"S

Universidade Federal de Minas Gerais (UFMG), Departamento de Física, Observatório Astronômico de Piedade (OAP)
Caixa Postal 702
30161-000 Belo Horizonte - MG
Telephone: (0)31-4412541
Telefax: (0)31-4481372
WWW: http://www.fisica.ufmg.br/
       http://www.fisica.ufmg.br/~astrof/
       http://www.fisica.ufmg.br/OAP/
Founded: 1972
Staff: 10
Activities: ISM * eclipsing binaries * photometry * stellar evolution
Coordinates: 043°30'43"W 19°49'20"S H1,746m
City Reference Coordinates: 043°56'00"W 19°55'00"S

Universidade Federal de Rio de Janeiro (UFRJ), Observatório do Valongo
Ladeira do Pedro Antônio 43 - Saŭde
20080-090 Rio de Janeiro - RJ
Telephone: (0)21-2630685
Telefax: (0)21-2630685
Electronic Mail: ov@ov.ufrj.br
WWW: http://www.ufrj.br/ov/
Founded: 1881
Staff: 11
Activities: education * fundamental astronomy * stellar astrophysics * ISM * extragalactic astronomy * laboratory astrophysics * PNs
Coordinates: 043°11'12"W 22°53'54"S H43m
City Reference Coordinates: 043°15'00"W 22°54'00"S

Universidade Federal de Santa Maria (UFSM), Núcleo de Estudos e Pesquisas Aeroespaciais (NE-PAE)
Campus UFSM
97119-900 Santa Maria - RS
Telephone: (0)55-2262013
Telefax: (0)55-2262013
Electronic Mail: <userid>@brufsm.bitnet
WWW: http://www.ufsm.br/
Founded: 1973 (reactivated: 1985)
Activities: radioastronomy * astrophysics * space geophysics * observing
City Reference Coordinates: 053°48'00"W 29°41'00"S

Universidade Federal do Rio Grande do Sul (UFRGS), Instituto de Física, Departamento de Astronomia
Avenida Bento Gonçalves 9500
Campus do Vale
Caixa Postal 15051
91501-970 Porto Alegre - RS
Telephone: (0)51-3167111
Telefax: (0)51-3191762
Electronic Mail: <userid>@if.ufrgs.br
WWW: http://www.if.ufrgs.br/ast
       http://sofia.if.ufrgs.br/ast/
       http://www.cesup.ufrgs.br/ufrgs.html
Founded: 1968
Staff: 10
Activities: education * research
Coordinates: 051°06'15"W 30°04'31"S H400m (Morro Santana)
City Reference Coordinates: 051°11'00"W 30°04'00"S

Universidade Federal do Rio Grande do Sul (UFRGS), Planetário Prof. José Baptista Pereira
Avenida Ipiranga 2000
90160-091 Porto Alegre - RS
Telephone: (0)51-3315434
Founded: 1972
Staff: 7
Activities: education * shows * school visits * popularization * sky of the month
Periodicals: (1) "Agenda Astronômica", "Cadernos de Astronomia"
City Reference Coordinates: 051°11'00"W 30°04'00"S

Bulgaria

Amateur Astronomical Club Canopus
Primorski Park 4
P.O. Box 108
BG-9000 Varna
Telephone: (0)52-222890
Electronic Mail: astro@ms3.tu-varna.acad.bg
Founded: 1963
Membership: 100
Activities: observing (meteors, variable stars, minor planets, comets, Sun)
City Reference Coordinates: 027°55'00"E 43°13'00"N

Astronomical Association - Sofia (AAS)
c/o Boriana Bontcheva (President)
Juri Gagarin Street
Bl.98 vh.B
BG-1113 Sofia
or :
49 Tzar Assen Street
BG-1463 Sofia
Telephone: (0)2-9811327 (Office)
           (0)2-9810898
Telefax: (0)2-9810510 (Office)
Electronic Mail: aas@bgcict.acad.bg
WWW: http://www.iccs.acad.bg/astro/www/
Founded: 1993
Membership: 40
Activities: popularization * observing * exhibitions
Periodicals: (6) "Andromeda" (ISSN 1310-3571)
City Reference Coordinates: 023°19'00"E 42°41'00"N

Astronomical Observatory "Slavey Zlatev"
P.O. Box 134
BG-6600 Kardjali
Telephone: (0)361-22595
           (0)361-26457
Telefax: (0)361-26457
Electronic Mail: obsk@yahoo.com
Founded: 1972
Staff: 9
Activities: meteors * Sun * planets * education
Coordinates: 023°25'23"E 41°38'28"N H330m
City Reference Coordinates: 026°19'00"E 42°40'00"N

Astronomical Observatory, Sliven
P.O. Box 7
BG-8800 Sliven
Telephone: (0)44-28353
Electronic Mail: <userid>@sliven.osf.acad.bg
           aosliven@sliven.osf.acad.bg
WWW: http://www.sliven.osf.acad.bg/
Founded: 1979
Staff: 10
Activities: variable stars * Sun * minor planets * meteors * globular clusters * photometry * instrumentation * history of astronomy * education
Periodicals: (1) "Calendars"
Coordinates: 026°15'14"E 42°40'50"N H259m (Sliven)
           026°22'00"E 42°43'00"N H1,150m (Haramiata)
• Alternate name: "Dr. Petur Beron Observatory"
City Reference Coordinates: 026°19'00"E 42°40'00"N

Astronomical Observatory and Planetarium N. Copernicus
Primorski Park 4
P.O. Box 120
BG-9000 Varna
Telephone: (0)52-222890
Electronic Mail: astro@ms3.tu-varna.acad.bg
WWW: http://aol.skyarchive.org/varna/astro
Founded: 1961 (Planetarium: 1968)
Staff: 8
Activities: popularization * lectures * comets * minor planets * meteors * Sun * education
Coordinates: 027°55'26"E 43°12'12"N H10m (Varna)

027°40'14"E 43°07'13"N H398m (Avren)
City Reference Coordinates: 027°55'00"E 43°13'00"N

Beron Observatory (Dr. Petur_)
● See "Astronomical Observatory, Sliven"

Bulgarian Academy of Sciences, Central Laboratory for Geodesy
Acad. G. Bonchev Street 1
BG-1113 Sofia
or :
Ul. 15 Noemvri 1
BG-1040 Sofia
Telephone: (0)2-720841
Telefax: (0)2-720841
Electronic Mail: <userid>@bgearn.bitnet
WWW: http://www.acad.bg/BulRTD/earth/geodesy2.html
Founded: 1948
Staff: 37
Activities: geodesy * satellite geodesy * optical astrometry * geodynamics * geodetic networks
Periodicals: (1) "Geodesy" (ISSN 0324-1114)
Coordinates: 023°24'37"E 42°29'55"N H1,200m (Plana)
City Reference Coordinates: 023°19'00"E 42°41'00"N

Bulgarian Academy of Sciences, Institute of Astronomy and National Astronomical Observatory, Smolyan
P.O. Box 136
BG-4700 Smolyan
Telephone: (0)3-021356
         (0)3-021357
         (0)3-0128902
Telefax: (0)3-021356
Electronic Mail: rozhen@tu-plovdiv.bg
WWW: http://www.astro.bas.bg/
Founded: 1980 (reestablsihed: 1995)
Staff: 40
Activities: comets * minor planets * stellar atmospheres * variable stars * stellar clusters * galaxies
Coordinates: 024°45'00"E 41°42'00"N H1,750m
● Alternate name: "Rozhen Observatory"
City Reference Coordinates: 024°45'00"E 41°42'00"N

Bulgarian Academy of Sciences, Institute of Astronomy and National Astronomical Observatory, Sofia
72 Tsarigradsko Chaussee Boulevard
BG-1784 Sofia
Telephone: (0)2-758927
         (0)2-7144614
Telefax: (0)2-758927
Electronic Mail: <userid>@astro.bas.bg
           kpanov@astro.bas.bg (Kiril Panteleev Panov, Director)
WWW: http://www.astro.bas.bg/
Founded: 1958
Staff: 90
Activities: astronomy * astrophysics * education
Periodicals: (1) "Astronomical Calendar" (ISSN 0861-1270)
Coordinates: 024°43'00"E 41°43'00"N H1,750m (Rozhen)
         022°40'30"E 43°37'35"N H630m (Belogradtchik)
City Reference Coordinates: 023°19'00"E 42°41'00"N

Bulgarian Academy of Sciences, Geophysical Institute
Acad. G. Bonchev Street
Block 3
BG-1113 Sofia
Telephone: (0)2-700128
Telefax: (0)2-700226
Electronic Mail: office@geophys.acad.bg
WWW: http://www.geophys.bas.bg/
Founded: 1961
Staff: 150
Activities: atmospheric and ionospheric physics * seismology * geomagnetism * gravimetry * network of seismic stations
Periodicals: (12) "Seismic Bulletin"; (4) "Bulgarian Geophysical Journal" (ISSN 0323-9918, circ.: 250); (1) "Ionospheric Bulletin", "Magnetic Bulletin"
City Reference Coordinates: 023°19'00"E 42°41'00"N

Bulgarian Academy of Sciences, National Institute of Meteorology and Hydrology (NIMH)
66 Tsarigradsko Chaussee Boulevard

BG-1784 Sofia
Telephone: (0)2-9733831
Telefax: (0)2-884494
Electronic Mail: office@meteo.bg
WWW: http://www.meteo.bg/
Founded: 1890
Observatories: 35 (plus 2000 observing stations)
City Reference Coordinates: 023°19'00"E 42°41'00"N

Bulgarian Academy of Sciences, Space Research Institute
6 Moskovska Street
BG-1000 Sofia
Telephone: (0)883503
(0)2-9793422
Telefax: (0)2-9813347
Founded: 1974
Staff: 310
Activities: space astronomy * space physics * space instrumentation * remote sensing * aerospace technologies
Periodicals: (12) "Aerospace Investigations in Bulgaria"
City Reference Coordinates: 023°19'00"E 42°41'00"N

Committee for Standardization and Metrology (BDS)
6th September Street 21
BG-1000 Sofia
Telephone: (0)2-8591
Telefax: (0)2-801402
Electronic Mail: csm@techno-link.com
City Reference Coordinates: 023°19'00"E 42°41'00"N

Dr. Petur Beron Observatory
● See "Astronomical Observatory, Sliven"

Petur Beron Observatory (Dr._)
● See "Astronomical Observatory, Sliven"

Rozhen Observatory
● See "Bulgarian Academy of Sciences, Department of Astronomy and National Astronomical Observatory, Smolyan"

Shoumen University, Department of Physics, Astronomical Group
BG-9700 Shoumen
Telephone: (0)54-63151 x289
Telefax: (0)54-63171
Electronic Mail: diana@uni-shoumen.bg (Diana Kjurkchieva)
Staff: 4
Activities: spotted stars * cataclysmic stars * close binaries * imaging
● Formerly "Higher Pedagogical Institute"
City Reference Coordinates: 026°55'00"E 43°16'00"N

Smolyan Planetarium
20 Bulgaria Boulevard
P.O. Box 132
BG-4700 Smolyan
Electronic Mail: arm@lgi-mtk.cit.bg
WWW: http://www.wfpa.acad.bg/smolyan/
Founded: 1975
Staff: 20
City Reference Coordinates: 024°41'00"E 41°35'00"N

Union of Physicists in Bulgaria
c/o I. Lalov (President)
Faculty of Physics
Sofia University
James Bourchier Boulevard 5
BG-1126 Sofia
Telephone: (0)2-627660
Telefax: (0)2-689085
Electronic Mail: rector@uni-sofia.bg
Founded: 1972
Membership: 100
City Reference Coordinates: 023°19'00"E 42°41'00"N

University of Sofia, Department of Astronomy
James Bourchier Boulevard 5
BG-1126 Sofia
Telephone: (0)2-688176

Telefax: (0)2-689085
WWW: http://www.phys.uni-sofia.bg/
Founded: 1891
Staff: 16
Activities: education * research
City Reference Coordinates: 023°19'00"E 42°41'00"N

Canada

ADR Spacelink & Commercialization Inc.
2426 Georgina Drive
Suite B
Ottawa, ON K2B 7M7
Telephone: 613-596-6032
Telefax: 613-596-2986
Electronic Mail: 102746.313@compuserve.com
WWW: http://adr.on.ca/
Activities: technology transfer
City Reference Coordinates: 075°42'00"W 45°25'00"N

Agence Spatiale Canadienne (ASC)
● See "Canadian Space Agency (CSA)"

Association Canadienne des Physiciens (ACP)
● See "Canadian Association of Physicists (CAP)"

Association des Astronomes Amateurs Abitibi-Témiscamingue (AAAAT)
1367 Chemin des Anciens
Taschereau, QC J0Z 3N0
Electronic Mail: aaaat@hotmail.com
WWW: http://astro.uqat.uquebec.ca/Astro/
City Reference Coordinates: 078°42'00"W 48°40'00"N

Astro-Club du CEGEP de Lévis-Lauzon
205 Monseigneur Bourget
Lévis, QC G6V 6Z9
Telephone: 418-833-5110
Electronic Mail: jfilion@zone.ca (Jacky Filion)
WWW: http://www2.zone.ca/~jfilion/
City Reference Coordinates: 071°11'00"W 46°49'00"N

Astro-Club du Collège de Lévis
9 Monseigneur Gosselin
Lévis, QC G6V 5K1
Telephone: 418-833-1249
Telefax: 418-833-1974
WWW: http://www.engsoc.carleton.ca/~wiswaud/acclf.htm
Founded: 1970
Membership: 30
Activities: observing * photography
Coordinates: 070°42'12"W 46°43'48"N H305m (Saint-Nérée-de-Bellechasse)
City Reference Coordinates: 071°11'00"W 46°49'00"N

Astronomical Society of Fort McMurray (ASFM)
c/o Josyane Cloutier
290 Ross Haven Drive
Fort McMurray, AB T9H 3P4
Telephone: 403-791-6372
Electronic Mail: josyane@telusplanet.net
WWW: http://www.realtime.ab.ca/~asfm/astro/astm.html
Founded: 1986
Membership: 12
Activities: observing * public nights
City Reference Coordinates: 111°07'00"W 56°40'00"N

Athena Community Astronomy Club
P.O. Box 1432
Summerside, PI C1N 4K2
Electronic Mail: egaudet@bud.peinet.pe.ca (Ed Gaudet)
WWW: http://www.rcch.com/athena
City Reference Coordinates: 063°47'00"W 46°24'00"N

Atlantic Space Sciences Foundation (TASSF) Inc. (The-)
● See "The Atlantic Space Sciences Foundation (TASSF) Inc."

Atomic Energy of Canada Ltd. (AECL), Physical Sciences
Chalk River, ON K0J 1J0
Telephone: 613-584-3311
Telefax: 613-584-4024

Electronic Mail: physics@crl.aecl.ca
WWW: http://www.sno.phy.queensu.ca/ (Sudbury Neutrino Observatory - SNO)
Founded: 1956
Staff: 300
Activities: condensed matter * theoretical, neutrino, nuclear and accelerator physics * analytical, physical and general chemistry * radiation applications * fusion
Periodicals: "Progress Reports" (ISSN 0067-0367)
Observatories: 1 (Sudbury Neutrino Observatory - SNO) (H-2300m)
City Reference Coordinates: 077°27'00"W 46°01'00"N

Bibliothèque Nationale du Canada (BNC)
Rue Wellington 395
Ottawa ON K1A 0N4
Telephone: 613-995-7969
Telefax: 613-991-9871
Electronic Mail: publications@nlc-bnc.ca
WWW: http://www.nlc-bnc.ca/
Founded: 1953
Staff: 500
Activities: national library
Periodicals: (10) "Nouvelles de la Bibliothèque Nationale"
● See also "National Library of Canada (NLC)"
City Reference Coordinates: 075°42'00"W 45°25'00"N

Bondar Planetarium (Roberta_)
● See "Seneca College, Roberta Bondar Planetarium"

Brandon University, Department of Physics and Astronomy
Brandon, MB R7A 6A9
Telephone: 204-727-9680
Telefax: 204-726-4573
Electronic Mail: <userid>@brandonu.ca
WWW: http://www.brandonu.ca/Departments/Science/Physics/Welcome.html
Founded: 1938
Staff: 4
Activities: stellar spectra * atmospheres * Be and Ap stars
City Reference Coordinates: 099°57'00"W 49°50'00"N

Brockville Astronomical Society
c/o Devin Fetter
14 Bramshot Avenue
Brockville, ON K6V 1Y5
Telephone: 613-342-2668
Electronic Mail: kfetter@geocities.com
WWW: http://www.geocities.com/CapeCanaveral/Lab/7855/
Membership: 4
City Reference Coordinates: 075°44'00"W 44°35'00"N

Bruce County Astronomical Society
c/o Charlie Szabototh
477 Johnston Avenue
Port Elgin, ON N0H 2C1
Telephone: 519-389-3150
Electronic Mail: csz@bmts.com
WWW: http://www.geocities.com/CapeCanaveral/Hall/6380/
City Reference Coordinates: 064°08'00"W 46°03'00"N

Bureau de Normalisation du Québec (BNQ)
70 Rue Dalhousie
Bureau 220
Québec, QC G1K 4B2
Telephone: 418-643-5114
Telefax: 418-646-3315
WWW: http://www.criq.qc.ca/bnq/
Founded: 1961
Staff: 28
Activities: normalizing * certifying * recording quality systems * informing * advising
City Reference Coordinates: 071°13'00"W 46°49'00"N

Burke-Gaffney Observatory (BGO) (Rev. M.W._)
● See "Saint Mary's University, Department of Astronomy and Physics, Rev. M.W. Burke-Gaffney Observatory (BGO)"

Calgary Centennial Planetarium
● See "Calgary Science Center"

Calgary Science Center
P.O. Box 2100
Station "M"
Location Code 73
Calgary, AB T2P 2M5
or :
701 - 11 Street SW
Calgary, AB
Telephone: 403-221-3700
Telefax: 403-237-0186
Electronic Mail: discover@calgaryscience.ca
WWW: http://www.calgaryscience.ca/
Founded: 1966
Staff: 30
Activities: education * shows * exhibitions * Centennial Planetarium
Periodicals: (6) "The Spark"
Coordinates: 114°05'21"W 51°02'51"N H1,084m (Public Observatory)
City Reference Coordinates: 114°05'00"W 51°03'00"N

Canadian Aeronautics and Space Institute (CASI)
(Institut Aéronautique et Spatial du Canada)
130 Slater
Suite 818
Ottawa, ON K1P 6E2
Telephone: 613-234-0191
Telefax: 613-234-9039
Electronic Mail: casi@asi.ca
WWW: http://www.casi.ca/
Founded: 1954
Staff: 4
Activities: technical symposia * publishing * branch programs
Periodicals: (4) "Canadian Aeronautics and Space Journal/Le Journal Aéronautique et Spatial du Canada" (ISSN 0008-2821); "Canadian Journal of Remote Sensing" (ISSN 0763-8992)
Awards: "McCurdy Award", "C.D. Howe Award", "Romeo Vachon Award", "F.W. (Casey) Baldwin Award", "Dr. Wilbur Franklin Award", "Remote Sensing Gold Medal"
City Reference Coordinates: 075°42'00"W 45°25'00"N

Canadian Association of Physicists (CAP)
(Association Canadienne des Physiciens - ACP)
c/o Gary D. Enright (Secretary/Treasurer)
Steacie Institute
100 Sussex Drive
Room 126
Ottawa, ON K1A 0R6
Telephone: 613-993-7393
Telefax: 613-954-5242
Electronic Mail: enright@ned1.sims.nrc.ca
                 cap@physics.uottawa.ca (General Inquiries)
WWW: http://www.cap.ca/
Founded: 1945
Staff: 3
Sections: (divisions) Aeronomy and Space Physics * Atomic and Molecular Physics * Condensed Matter Physics * Medical and Biological Physics * Nuclear Physics * Optics & Photonics * Particle Physics * Physics Education * Plasma Physics * Theoretical Physics * Industrial and Applied Physics * Surface Science
Periodicals: (1) "Annual Report" (ISSN 0068-8339); "Physics in Canada - La Physique au Canada" (ISSN 0031-9147)
Awards: (1) "CAP Medal of Achievement in Physics", "Herzberg Medal", "Lloyd G. Elliot Prize"; (12) "CAP Medal for Outstanding Achievement in Industrial and Applied Physics"
City Reference Coordinates: 075°42'00"W 45°25'00"N

Canadian Astronomical Society (CASCA)
(Société Canadienne d'Astronomie)
c/o Serge Demers (Secretary)
Département de Physique
Université de Montréal
Montréal, QC H3C 3J7
Telephone: 514-343-2364
Electronic Mail: casca@astro.umontreal.ca
WWW: http://www.astro.queensu.ca/~casca/
Founded: 1971
Membership: 360
Activities: support of astronomical research and education
Periodicals: "Cassiopeia"
Awards: "Carlyle S. Beals Award", "J.S. Plaskett Medal", "R.M. Petrie Prize", "Helen Sawyer Hogg Prize"
City Reference Coordinates: 073°34'00"W 45°31'00"N

Canadian Astronomy Data Centre (CADC)
• See "National Research Council of Canada (NRCC), Herzberg Institute of Astrophysics (HIA), Canadian Astronomy Data Centre (CADC)"

Canadian Coast Guard College Planetarium
P.O. Box 4500
Sydney, NS B1P 6L4
Telephone: 902-564-3660
Telefax: 902-564-3672
Electronic Mail: <userid>@cgc.ns.ca
WWW: http://www.cgc.ns.ca/
Founded: 1965
Staff: 105 (College)
Activities: education * astronomical navigation
City Reference Coordinates: 060°11'00"W 46°09'00"N

Canadian Council of Science Centres (CCSC)
(Conseil Canadien des Centres des Sciences - CCCS)
c/o Pacific Space Centre
1100 Chesnut Street
Vanier Park
Vancouver, BC V6J 3J9
Telephone: 604-738-STAR
        604-738-7827
Telefax: 604-736-5665 (Administration)
Electronic Mail: jdickens@pacific-space-centre.bc.ca (John Dickenson, President)
WWW: http://pacific-space-centre.bc.ca/
Founded: 1986
Membership: 26
Periodicals: "Newsletter"
City Reference Coordinates: 123°07'00"W 49°16'00"N

Canadian Institute for Advanced Research (CIAR)
180 Dundas Street West
Suite 1400
Toronto, ON M5G 1Z8
Telephone: 416-971-4251
Telefax: 416-971-6169
Electronic Mail: info@ciar.ca
WWW: http://www.ciar.ca/
Founded: 1982 (Programme in Cosmology and Gravity: 1986)
Staff: 14
City Reference Coordinates: 079°23'00"W 43°39'00"N

Canadian Institute for Scientific and Technical Information (CISTI)
• See "National Research Council of Canada (NRCC), Canadian Institute for Scientific and Technical Information (CISTI)"

Canadian Institute for Theoretical Astrophysics (CITA)
(Institut Canadien d'Astrophysique Théorique - ICAT)
c/o University of Toronto
McLennan Physical Laboratories
Room 1212A
60 Saint George Street
Toronto, ON M5S 3H8
Telephone: 416-978-6879
Telefax: 416-978-3921
Electronic Mail: office@cita.utoronto.ca
WWW: http://www.cita.utoronto.ca/
        http://www.physics.utoronto.ca/department/groups_partners/cita.html
Founded: 1984
Staff: 25
Activities: theoretical astrophysics * cosmology
City Reference Coordinates: 079°23'00"W 43°39'00"N

Canadian Meteorological Centre
4905 Dufferin Street
Downsview, ON M3H 5T4
Telephone: 416-739-4810
Telefax: 416-739-4999
Electronic Mail: <userid>@dow.on.doe.ca
WWW: http://cmits02.dow.on.doe.ca/
        http://www.ec.gc.ca/weather_e.html
        http://www.ec.gc.ca/weather_f.html
        http://www.tor.ec.gc.ca/cmc.html

http://www.tor.ec.gc.ca/forecasts/text/ (Regional Weather Forecasts)
City Reference Coordinates: 079°28'00"W 43°45'00"N

Canadian Space Agency (CSA), David Florida Laboratory (DFL)
(Agence Spatiale Canadienne - ASC, Laboratoire David Florida)
3701 Carling Avenue
Ottawa, ON K2H 8S2
Telephone: 613-998-2383
Telefax: 613-993-6103
Electronic Mail: <userid>@sp-agency.ca
WWW: http://www.dfl.doc.ca/
        http://www.space.gc.ca/ENG/Programs/DFL/dfl1.html
        http://maki.incen.doc.ca/
Founded: 1972
City Reference Coordinates: 075°42'00"W 45°25'00"N

Canadian Space Agency (CSA), Space Science Program
(Agence Spatiale Canadienne - ASC, Programme Scientifique Spatial)
100 Sussex Drive
P.O. Box 7275
Ottawa, ON K1L 8E3
Telephone: 613-990-0798
        613-990-0799
Telefax: 613-952-0970
        613-941-4294
Electronic Mail: <userid>@sp-agency.ca
WWW: http://www.space.gc.ca/
        http://www.space.gc.ca/ENG/Programs/Space_Science/sp-sci1.html
Founded: 1989
Staff: 30
Activities: selection, management, development, and operation of space experiments proposed by Canadian scientists
Periodicals: (4) "Materials Science in Microgravity/Science des Matériaux dans un Environnement de Microgravité"; (2) "Space Life Sciences/Sciences Spatiales de la Vie"; "Solar-Terrestrial Relations Newsletter/Bulletin des Relations Soleil-Terre", "Canadian Microgravity Program Update"
City Reference Coordinates: 075°42'00"W 45°25'00"N

Canadian Space Resource Centre, Atlantic
1593 Barrington Street
Halifax, NS B3J 1Z7
Telephone: 902-492-4422
            1-800-511-3500 (Canada and USA only)
Telefax: 902-492-3170
Electronic Mail: space@hercules.stmarys.ca
WWW: http://apwww.stmarys.ca/space
Founded: 1995
Staff: 2
Activities: space science educational outreach and resource centre
City Reference Coordinates: 063°36'00"W 44°39'00"N

Canadian Space Resource Centre, Ontario
c/o Marc Garneau Collegiate Institute
135 Overlea Boulevard
East York, ON M3C 1B3
Telephone: 416-396-2421
Telefax: 416-396-2423
Electronic Mail: csrs@interlog.com
WWW: http://www.interlog.com/~csrc
Staff: 2
Activities: space science educational outreach and resource centre
City Reference Coordinates: 076°43'00"W 39°58'00"N

Canadian Space Resource Centre, Pacific
● See "Pacific Space Centre"

Canadian Space Resource Centre, Prairies
● See "Western Space Education Network (WSEN) Inc."

Canadian Space Resource Centre, Québec
● See "Cosmodôme"

Cape Breton Astronomical Society (CBAS)
c/o John Reppa
131 Green Acres Drive
Sydney, NS B1S 1K5
Electronic Mail: ptech45@avtccb.nscc.ns.ca (John Reppa)

WWW: http://www.cbnet.ns.ca/cbnet/comucntr/astronomy/cbas.html
http://highlander.cbnet.ns.ca/~cbas/
Founded: 1986
City Reference Coordinates: 060°11'00"W 46°09'00"N

Carleton University, Centre for Research in Particle Physics (CRPP)
Ottawa, ON K1S 5B6
Telephone: 613-786-7552
Telefax: 613-786-7546
Electronic Mail: admin@crpp.carleton.ca
<userid>@physics.carleton.ca
WWW: http://www.crpp.carleton.ca/
http://www.physics.carleton.ca/
Founded: 1990
Staff: 15
Activities: research in particle physics and neutrino astrophysics
City Reference Coordinates: 075°42'00"W 45°25'00"N

Carleton University, Department of Physics, High-Energy Physics Group
Colonel By Drive
Herzberg Building
Ottawa, ON K1S 5B6
Telephone: 613-788-4377
Telefax: 613-788-4061
Electronic Mail: carnegie@physics.carleton.ca
WWW: http://www.hepnet.carleton.ca/
http://www.physics.carleton.ca/
<userid>@physics.carleton.ca
Staff: 40
Activities: particle physics * astrophysics * medical physics
City Reference Coordinates: 075°42'00"W 45°25'00"N

Centre Canadiens de Ressources Spatiales
• See the various "Canadian Space Resource Centres"

Centre for Climate and Global Change Research (C2GCR)
• See "McGill University, Centre for Climate and Global Change Research (C2GCR)"

Centre for Research in Earth and Space Technology (CRESTech)
4850 Keele Street
Floor 2
North York, ON M3J 3K1
Telephone: 416-665-3311
Telefax: 416-665-2032
Electronic Mail: <userid>@crestech.ca
<userid>@ists.ca
WWW: http://www.crestech.ca/
Founded: 1997
Activities: space science and technology * atmospheric studies * land resources * water resources * human performance in a aerospace environment * controlled environmental systems
• Results of the merger between the "Institute for Space and Terrestrial Science (ISTS)" and the "Waterloo Center for Groundwater Research (WCGR)"
City Reference Coordinates: 079°25'00"W 43°46'00"N

Climenhaga Observatory
• See "University of Victoria, Department of Physics and Astronomy"

Club d'Astronomie de Beloeil
c/o Gilles Ouellet
1020a Dupré
Beloeil, PQ J3G 4A8
Telephone: 514-461-0087
Electronic Mail: ouellet.gilles@ireq.ca
WWW: http://www.quebectel.com/beloeil/
City Reference Coordinates: 073°12'00"W 45°34'00"N

Club d'Astronomie de la Péninsule Acadienne
Case Postale 2000
Shippagan, NB E0B 2P0
Telephone: 506-336-3427
Telefax: 506-336-3477
Electronic Mail: jacques@admin.cus.ca (Jacques Robichaud, President)
WWW: http://www.cus.ca/Astronomie/
Founded: 1994
Membership: 25

Coordinates: 064°42'42"W 47°44'42"N
City Reference Coordinates: 064°42'00"W 47°45'00"N

Club d'Astronomie de l'Observatoire "L'Étoile d'Acadie"
98 Rue Marcoux
Gr. 37 - Boîte 20
Balmoral, NB E0B 1C0
Telephone: 506-826-9893
Electronic Mail: raudet@nbnet.nb.ca
Founded: 1995
Membership: 12
Activities: observing * meetings * photography * CCDs
Coordinates: 066°27'00"W 47°58'00"N H150m
City Reference Coordinates: 066°27'00"W 47°58'00"N

Club d'Astronomie de l'Université du Québec à Rimouski
c/o Jean-Paul Caron
300 Rue des Ursulines
Local E205
Rimouski, QC G5L 3A1
Telephone: 418-723-5197
             418-723-4402
Electronic Mail: astro_uqar@uqar.uquebec.ca
                 dlemay@quebectel.com
WWW: http://wwwb.uqar.uquebec.ca/astro/
       http://www.quebectel.com/astro/
City Reference Coordinates: 068°32'00"W 48°27'00"N

Club d'Astronomie Jupiter
c/o Thomas Collin
3601 Rue Pontoise
Trois-Rivières, QC G8Y 3K6
Telephone: 819-691-0726
Electronic Mail: thomas.collin@cgocable.ca
WWW: http://www.quebectel.com/jupiter/
City Reference Coordinates: 072°33'00"W 46°21'00"N

Club d'Astronomie Les Almucantars
425 Boulevard du Collège
Case Postale 1500
Rouyn-Noranda, QC J9X 5E5
Telephone: 819-762-0931 x1528
Electronic Mail: glesage@lino.com (Gilles Lesage)
Activities: observing * photography
City Reference Coordinates: 079°01'00"W 48°14'00"N

Club d'Astronomie Val d'Orion
c/o Jean-Guy Moreau
125 Self
Val d'Or, QC
Telephone: 819-738-7512
WWW: http://lino.com/astro/astro.html
City Reference Coordinates: 077°47'00"W 48°07'00"N

Club des Astronomes Amateurs de Laval (CAAL)
c/o Jean-Marc Richard (President)
Case Postale 214 - Succ. Saint-Martin
Laval, QC H7V 3P5
Telephone: 450-661-9390
Electronic Mail: astrolv@ca.org
WWW: http://www.cam.org/~astrolv/
Founded: 1986
Membership: 120
Activities: weekly meetings * star parties * education
Periodicals: "L'Orionide"
Coordinates: 073°39'00"W 45°36'00"N
City Reference Coordinates: 073°44'00"W 45°33'00"N

Club des Astronomes Amateurs de Longueuil (CAAL)
c/o Daniel Rompré
1279 Rue Brassard
Longueuil, QC J4M 1Z2
Telephone: 514-647-1072
Electronic Mail: drompre@vir.com
WWW: http://www.vir.com/~leia/caal.html

http://www.dsuper.net/~leia/caal/ecaal.html
City Reference Coordinates: 073°30'00"W 45°33'00"N

Club des Astronomes Amateurs de Sherbrooke (CAAS)
c/o Gisele Gilbert
Case Postale 352
Sherbrooke, QC J1H 5J1
Telephone: 819-821-1138
Electronic Mail: gisgil@videotron.ca
WWW: http://www.caas.sherbrooke.qc.ca/
Founded: 1980
Membership: 107
Activities: amateur astronomy * conferences * exhibitions * camps
Periodicals: (2) "Ciel de Nuit"
Coordinates: 072°01'53"W 45°12'10"N H345m (Les Sommets)
            071°53'53"W 45°27'17"N H240m (Beauvoir)
City Reference Coordinates: 071°54'00"W 45°24'00"N

Comox Valley Astronomy Club (CVAC)
c/o Keith Finnie
1467 Balmoral Avenue
Comox, BC V9N 7N9
Telephone: 604-339-3720
WWW: http://www.island.net/~dgraham/cvac.htm
Founded: 1989
Membership: 25
City Reference Coordinates: 124°55'00"W 49°40'00"N

Conceptron Associates
1195 Durant Drive
Coquitlam, BC V3B 6R3
Telephone: 1-800-871-4161 (USA & Canada only)
          604-945-5241
Telefax: 604-941-5562
Electronic Mail: info@conceptron.com
WWW: http://www.conceptron.com/
Founded: 1994
Activities: audio visual design consultants * planetarium technical systems consultants
City Reference Coordinates: 123°07'00"W 49°16'00"N (Vancouver)

Conseil Canadien des Centres des Sciences (CCCS)
• See "Canadian Council of Science Centres (CCSC)"

Conseil Canadien des Normes (CCN)
• See "Standards Council of Canada (SCC)"

Conseil National de Recherches du Canada (CNRC)
Montréal Road
Ottawa, ON K1A 0R6
Telephone: 613-993-9101 (general enquiry)
          613-993-2054 (NRC publications)
          613-745-1576 (time signal)
          613-745-9426 (horloge parlante)
          613-993-6549 (IAU contact)
Telefax: 613-952-6602
WWW: http://www.nrc.ca/
Founded: 1916
Periodicals: "Canadian Journal of Physics" (ISSN 0008-4204), "Canadian Journal of Earth Sciences"
• See also "National Research Council of Canada (NRCC)"
City Reference Coordinates: 075°42'00"W 45°25'00"N

Conseil National de Recherches du Canada (CNRC), Institut Canadien de l'Information Scientifique et Technique (ICIST)
• See "National Research Council of Canada (NRCC), Canadian Institute for Scientific and Technical Information (CISTI)"

Conseil National de Recherches du Canada (CNRC), Institut Herzberg d'Astrophysique, Observatoire Fédéral de Radioastrophysique (OFR)
• See "National Research Council of Canada (NRCC), Herzberg Institute of Astrophysics (HIA), Dominion Radio Astrophysical Observatory (DRAO)"

Conseil National de Recherches du Canada (CNRC), Institut Herzberg d'Astrophysique, Observatoire Fédéral d'Astrophysique (OFA)
• See "National Research Council of Canada (NRCC), Herzberg Institute of Astrophysics (HIA), Dominion Astrophysical Observatory (DAO)"

Coreco Inc.
6969 Trans-Canada Highway
Suite 113
Saint-Laurent, QC H4T 1V8
Telephone: 514-333-1301
Telefax: 514-333-1388
Electronic Mail: info@coreco.com
WWW: http://www.coreco.com/
Founded: 1979
Staff: 60
● Hardware and software manufacturer
City Reference Coordinates: 073°41'00"W 45°31'00"N

Cosmodôme
2150 Autoroute des Laurentides
Laval, QC H7T 2T8
Telephone: 514-978-3600
             514-978-3602 (Space Resource Centre)
             1-800-565-CAMP
Telefax: 514-978-3601
Electronic Mail: info@cosmodome.org
               document@cosmodome.org (Space Resource Centre)
WWW: http://www.cosmodome.org/
Founded: 1991 (Space Resource Centre: 1996)
Staff: 20
Activities: space science educational outreach and resource centre * hosting a Canadian Space Resource Centre (Québec)
City Reference Coordinates: 073°44'00"W 45°33'00"N

Cosmo-Logique Informatique Inc.
15 Rue Dollard
Vaudreuil-Dorion, QC J7V 8P5
Telephone: 450-455-1457
Telefax: 450-424-0677
               acoulombe@cosmo-logique.qc.ca (André Coulombe, Director)
Founded: 1997
Staff: 5
● Software producer
City Reference Coordinates: 074°02'00"W 45°24'00"N

CRESTech
● See "Centre for Research in Earth and Space Technology (CRESTech)"

Cyanogen Productions Inc.
25 Conover Street
Nepean, ON K2G 4C3
Telephone: 613-225-2732
Telefax: 613-225-9688
Electronic Mail: cyanogen@cyanogen.on.ca
               cyanogen@cyanogen.com
WWW: http://www.cyanogen.on.ca/
        http://www.cyanogen.com/
Founded: 1996
Activities: CCD camera control and image processing
City Reference Coordinates: 075°42'00"W 45°25'00"N (Ottawa)

David Dunlap Observatory (DDO)
● See "University of Toronto, Department of Astronomy"

David Florida Laboratory
● See "Canadian Space Agency (CSA), David Florida Laboratory"

Deep River Astronomy Club
c/o Harry Adams
P.O. Box 606
Deep River, ON K0J 1P0
Telephone: 613-584-3246
Electronic Mail: adams@intranet.on.ca
WWW: http://intranet.on.ca/~rbirchal/draco.html
        http://intranet.ca/~scarlisl/draco/draco.html
Founded: 1995
City Reference Coordinates: 077°30'00"W 46°06'00"N

Devon Astronomical Observatory
● See "University of Alberta, Department of Physics"

Dominion Astrophysical Observatory (DAO)
● See "National Research Council of Canada (NRCC), Herzberg Institute of Astrophysics (HIA), Dominion Astrophysical Observatory (DAO)"

Dominion Radio Astrophysical Observatory (DRAO)
● See "National Research Council of Canada (NRCC), Herzberg Institute of Astrophysics (HIA), Dominion Radio Astrophysical Observatory (DRAO)"

Doran Planetarium
● See "Laurentian University, Department of Physics and Astronomy"

Dunlap Observatory (DDO) (David_)
● See "University of Toronto, Department of Astronomy"

Edmonton Space and Science Centre (ESSC)
11211 - 142 Street
Edmonton, AB T5M 4A1
Telephone: 403-452-9100
Electronic Mail: essc@planet.eon.net
WWW: http://www.planet.eon.net/~essc/start.html
Founded: 1984
City Reference Coordinates: 113°28'00"W 53°33'00"N

Efston Science Inc.
3350 Dufferin Street
Toronto, ON M6A 3A4
Telephone: 416-787-4581
Telefax: 416-787-5140
Electronic Mail: efston@idirect.com
WWW: http://www.efstonscience.com/efsthome.html
City Reference Coordinates: 079°23'00"W 43°39'00"N

E.J. Tasso
3468 Drummond
Appartement 902
Montréal, QC H3G 1Y4
Telephone: 514-289-9577
Telefax: 514-842-9205
Founded: 1963
Staff: 2
Activities: distributing optical components
City Reference Coordinates: 073°34'00"W 45°31'00"N

Elginfield Observatory
● See "University of Western Ontario (UWO), Astronomy Department"

Eureka Software Inc.
110 Bagot Street
Kingston, ON K7L 3E5
Telephone: 613-546-4818
Electronic Mail: info@eureka.ca
WWW: http://www.eureka.ca/
Founded: 1994 (as "Logos Software")
● Software producer
City Reference Coordinates: 076°30'00"W 44°14'00"N
● Formerly "Logos Software"

Fédération des Astronomes Amateurs du Québec (FAAQ)
4545 Avenue Pierre-de-Coubertin
Casier Postal 1000
Station "M"
Montréal, QC H1V 3R2
Telephone: 514-252-3038
Telefax: 514-251-8038
Electronic Mail: faaq%cap@qcmedia.com
        ddurand@stsim.com (Danis Durand, President)
WWW: http://stsim.com/astro/faaq.html
      http://www.quebectel.com/faaq/
Founded: 1975
Membership: 650 (30 clubs)
Activities: gathering amateur astronomers * astronomical camps * lectures * popularization
Periodicals: (6) "Astronomie-Québec" (ISSN 1183-5362)
Awards: (1) "Prix Meritas", "Prix Les Pléiades"
City Reference Coordinates: 073°34'00"W 45°31'00"N

Fraser Valley Astronomers Society (FVAS)
2659 Valemont Crescent
Abbotsford, BC V2T 3V6
Telephone: 604-850-3931
Electronic Mail: victorp@uniserve.com (James Victor Pollock, President)
WWW: http://www.geocities.com/capecanaveral/hangar/6395/
Founded: 1987
Membership: 25
Activities: telescope making * stargazing
City Reference Coordinates: 122°17'00"W 49°03'00"N

Gemini Canadian Project Office (GCPO)
• See "National Research Council of Canada (NRCC), Herzberg Institute of Astrophysics (HIA), Gemini Canadian Project Office (GCPO)"

Glenlea Astronomical Observatory
• See "University of Manitoba, Department of Physics and Astronomy"

Gordon Southam Observatory
• See "Pacific Space Centre"

Great Island Science and Adventure Park
P.O. Box 430
Kensington, PE C0B 1M0
or :
Route 6
Stanley Bridge, PE C0A 1E0
Telephone: 902-836-3883
Electronic Mail: scifun@auracom.com
WWW: http://www.sciencefun.com/
Founded: 1986
Staff: 10
Activities: science park * planetarium
City Reference Coordinates: 063°39'00"W 46°26'00"N

Groupe Astronomie & CCD
c/o Luc Bellavance (President)
434 Des Passereaux
Rimouski, QC G5L 8K4
Telephone: 418-722-1529 (Office)
            418-723-5197 (Home)
Electronic Mail: astroccd@globetrotter.net
WWW: http://www.quebectel.com/astroccd/
City Reference Coordinates: 068°32'00"W 48°27'00"N

Hamilton Amateur Astronomers (HAA)
P.O. Box 65578
Dundas, ON L9H 6Y6
Telephone: 905-827-9105
Telefax: 905-627-3683
Electronic Mail: haa@www.science.mcmaster.ca
WWW: http://www.science.mcmaster.ca/HAA/
Founded: 1993
Membership: 110
Activities: monthly meetings * observing * education * instrumentation
Periodicals: (10) "Event Horizon"
City Reference Coordinates: 079°58'00"W 43°16'00"N

High Res Technologies (HRT)
40 Richview Road
Suite 409
Etobicoke, ON M9A 5C1
Telephone: 416-248-4473
Telefax: 416-248-4125
Electronic Mail: hrt@planeteer.com
WWW: http://toronto.planeteer.com/~hrt/
Activities: manufacturing and distributing video frame grabbers
City Reference Coordinates: 079°34'00"W 43°39'00"N

H.R. MacMillan Planetarium and Gordon Southam Observatory
• See now "Pacific Space Centre"

Institut Aéronautique et Spatial du Canada
• See "Canadian Aeronautics and Space Institute (CASI)"

Institut Canadien d'Astrophysique Théorique (ICAT)
• See "Canadian Institute for Theoretical Astrophysics (CITA)"

International Civil Aviation Organization (ICAO)
1000 Sherbrooke Street West
Montréal, QC H3A 2R2
Telephone: 514-285-8219
Telefax: 514-288-4772
Electronic Mail: icaohq@icao.org
WWW: http://www.icao.int/
Founded: 1947
Periodicals: "ICAO Journal", "Annual Report"
City Reference Coordinates: 073°34'00"W 45°31'00"N

International Federation of Institutes for Advanced Study (IFIAS)
39 Spadina Road
Toronto, ON M5R 2S9
Telephone: 416-926-7570
Telefax: 416-926-9481
WWW: http://www.ifias.ca/
Founded: 1971
Staff: 15
Activities: international non-governmental federation of research institutes * transdisciplinary, transnational projects on development and environment issues aimed at developing specific policy alternative for decision makers
City Reference Coordinates: 079°23'00"W 43°39'00"N

International Meteor Organization (IMO), North American Section
c/o Peter Brown
Department of Physics
University of Western Ontario
London, ON N6A 3K7
Telephone: 403-743-8625
Electronic Mail: peter@canton.physics.uwo.ca
WWW: http://www.imo.net/
            http://www.tu-chemnitz.de/~smo/imo/
Founded: 1988
Activities: see main entry in Belgium
Periodicals: see main entry in Belgium
City Reference Coordinates: 081°14'00"W 42°99'00"N

Khan Scope Centre
3243 Dufferin Street
Toronto, ON M6A 2T2
Telephone: 416-783-4140
Telefax: 416-783-7697
Electronic Mail: khan@globalserve.net
WWW: http://www.khanscope.com/
Founded: 1986
City Reference Coordinates: 079°23'00"W 43°39'00"N

Lakehead University, School of Mathematical Sciences
955 Oliver Road
Thunder Bay, ON P7B 5E1
Telephone: 807-343-8469
Telefax: 807-343-8821
Electronic Mail: math.enquiries@lakeheadu.ca
WWW: http://flash.lakeheadu.ca/~mathwww/mathwww.html
Founded: 1964
Activities: education * research
City Reference Coordinates: 080°03'00"W 44°48'00"N

Laurentian University, Department of Physics and Astronomy
Ramsey Lake Road
Sudbury, ON P3E 2C6
Telephone: 705-675-1151 x2220
            705-675-1151 x2227 (Doran Planetarium)
Telefax: 705-675-4868
Electronic Mail: <userid>@nickel.laurentian.ca
                plegault@nickel.laurentian.ca (Paul-Emile Legault, Director, Doran Planetarium)
WWW: http://www.laurentian.ca/www/physics/
            http://www.laurentian.ca/www/physics/planetarium/planetarium.html (Doran Planetarium)
Founded: 1960 (Doran Planetarium: 1968)
Staff: 17 (Doran Planetarium: 1)
Activities: research * education * shows for schools and public * Doran Planetarium
Coordinates: 081°00'00"W 46°30'00"N H900m

City Reference Coordinates: 081°00'00"W 46°30'00"N

Laurier University (WLU) (Wilfrid_)
● See "Wilfrid Laurier University (WLU)"

Les Observateurs de la Magnitude Absolue
c/o Gaetan Cormier
9070 Boivin
LaSalle, PQ H8R 2E4
Telephone: 514-623-7287
Electronic Mail: lafre@odyssee.net
WWW: http://www.microid.com/loma.htm.
City Reference Coordinates: 073°40'00"W 45°26'00"N

Lethbridge Astronomy Society (LAS)
P.O. Box 1104
Lethbridge, AB T1J 4A2
Telephone: 403-381-7827
Electronic Mail: lasa@telusplanet.net
WWW: http://www.telusplanet.net/public/lasa
Founded: 1988
Membership: 40
Activities: monthly meetings * public education
Periodicals: (12) "Newsletter"
Observatories: 1 (Oldman River Observatory)
City Reference Coordinates: 110°50'00"W 49°42'00"N

Lire La Nature Inc. (LLN)
1699 Chemin de Chambly
Longueuil, QC J4J 3X7
Telephone: 514-463-5072
Telefax: 514-463-3409
Electronic Mail: lirelanature@videotron.ca
WWW: http://www.stjeannet.ca/broquet/zindex.html
Founded: 1988
Staff: 2
Activities: distributing telescopes, binoculars, other optical instrumentation, books, other products
City Reference Coordinates: 073°30'00"W 45°33'00"N

London Regional Children's Museum
21 Wharncliffe Road South
London ON N6J 4G5
Telephone: 519-434-5726
Telefax: 519-434-1443
Founded: 1977
Activities: science gallery * space gallery * science experiment demonstrations * hands-on activities * planetarium
Periodicals: (12) "Thumbprint"
City Reference Coordinates: 081°14'00"W 42°59'00"N

Lumonics Inc.
105 Schneider Road
Kanata, ON K2K 1Y3
Telephone: 613-592-1460
Telefax: 613-592-7549
WWW: http://www.lumonics.com/
Founded: 1970
Staff: 740
Activities: manufacturing laser and laser-based systems
City Reference Coordinates: 081°14'00"W 42°59'00"N

MacMillan Planetarium (H.R._)
● See "Pacific Space Centre"

Maison de l'Astronomie P.L. Inc.
7974 Saint-Hubert
Montréal, QC H2R 2P3
Telephone: 514-279-0063
Telefax: 514-279-9628
Electronic Mail: rlotte@interlink.net
WWW: http://www.microid.com/maison.htm
Founded: 1987
Activities: distributing and renting telescopes and accessories * education
City Reference Coordinates: 073°34'00"W 45°31'00"N

Manitoba Astronomy Club
c/o Bert Valentin
Museum of Man and Nature
190 Rupert Avenue
Winnipeg, MB R3B 0N2
Telephone: 204-988-0625
Telefax: 204-942-3679
Electronic Mail: valentin@mbnet.mb.ca
WWW: http://www.manitobamuseum.mb.ca/cla.htm
Founded: 1982
Membership: 20
Activities: meetings * instrumentation * lectures * observing
Periodicals: (12) "The Nebula"
City Reference Coordinates: 097°09'00"W 49°53'00"N

Manitoba Planetarium
c/o Manitoba Museum of Man and Nature
190 Rupert Avenue
Winnipeg, MB R3B 0N2
Telephone: 204-956-2830
Telefax: 204-942-3679
Electronic Mail: info@manitobamuseum.mb.ca
WWW: http://www.manitobamuseum.mb.ca/planet.htm
Founded: 1968
Staff: 76
Activities: shows * lectures * exhibitions * education
City Reference Coordinates: 097°09'00"W 49°53'00"N

McGill University, Centre for Climate and Global Change Research (C2GCR)
805 Sherbrooke Street West
Montréal, QC H3A 2K6
Telephone: 514-398-3759
Telefax: 514-398-1381
Electronic Mail: mansi@felix.geog.mcgill.ca
WWW: http://www.mcgill.ca/ccgcr/
Founded: 1990
Staff: 15
City Reference Coordinates: 073°34'00"W 45°31'00"N

McGill University, Department of Atmospheric and Oceanic Sciences
805 Sherbrooke Street West
Montréal, QC H3A 2K6
Telephone: 514-398-3764
Telefax: 514-398-6115
Electronic Mail: <userid>@zephyr.meteo.mcgill.ca
WWW: http://zephyr.meteo.mcgill.ca/
Founded: 1992
Staff: 30
Activities: education * research * atmospheric and oceanic sciences
City Reference Coordinates: 073°34'00"W 45°31'00"N

McGill University, Department of Earth and Planetary Sciences
3450 University Street
Montréal, QC H3A 2A7
Telephone: 514-398-6767
Telefax: 514-398-4680
Electronic Mail: eps@stoner.eps.mcgill.ca
WWW: http://stoner.eps.mcgill.ca/
http://www.eps.mcgill.ca/
http://www.mcgill.ca/
Founded: 1992
City Reference Coordinates: 073°34'00"W 45°31'00"N

McGill University, Department of Mathematics and Statistics
805 Sherbrooke Street West
Montréal, QC H3A 2K6
Telephone: 514-398-3800
Telefax: 514-398-3899
Electronic Mail: chair@math.mcgill.ca
WWW: http://www.math.mcgill.ca/
Staff: 37
Activities: education * research
City Reference Coordinates: 073°34'00"W 45°31'00"N

McGill University, Department of Physics
Ernest Rutherford Physics Building
3600 University Street
Montréal, QC H3A 2T8
Telephone: 514-398-6485
Telefax: 514-398-8434
Electronic Mail: <userid>@physics.mcgill.ca
WWW: http://www.physics.mcgill.ca/
        http://www.mcgill.ca/
Founded: 1891
City Reference Coordinates: 073°34'00"W 45°31'00"N

McGill University, Institute and Centre of Air and Space Law (IASL)
3661 Peel Street
Montréal, QC H3A 1X1
Telephone: 514-398-5094
Telefax: 514-398-8197
Electronic Mail: <userid>@falaw.lan.mcgill.ca
WWW: http://www.mcgill.ca/iasl/
Founded: 1951
Staff: 7
Activities: education * research * air and space law
Periodicals: "Annals of Air and Space Law - Annales de Droit Aérien et Spatial"
City Reference Coordinates: 073°34'00"W 45°31'00"N

McLaughlin Planetarium
c/o Royal Ontario Museum
100 Queens Park
Toronto, ON M5S 2C6
Telephone: 416-586-5736
Telefax: 416-586-5887
Electronic Mail: tomc@rom.on.ca
WWW: http://www.rom.on.ca/
        http://ddo.astro.utoronto.ca/planetarium.html
Founded: 1968
Staff: 1
Activities: public and school programs * astronomy gallery
● Closed on 15 Dec. 1995 - reopening undefined
City Reference Coordinates: 079°23'00"W 43°39'00"N

McMaster University, Department of Physics and Astronomy
1280 Main Street West
Hamilton, ON L8S 4M1
Telephone: 905-525-9140
Telefax: 905-546-1252
Electronic Mail: physics@mcmaster.ca
                planetarium@physics.mcmaster.ca (William J. McCallion Planetarium)
WWW: http://www.physics.mcmaster.ca/
        http://www.physics.mcmaster.ca/Planetarium/Planetarium.html (Planetarium)
Founded: 1894 (Planetarium: 1949)
Staff: 20
Activities: undergraduate and graduate education * research * William J. McCallion Planetarium
City Reference Coordinates: 079°51'00"W 43°15'00"N

Memorial University of Newfoundland (MUN), Department of Physics
Saint John's, NF A1B 3X7
Telephone: 709-737-8735
            709-737-8736
            709-737-8737
            709-737-8738
Telefax: 709-737-8739
Electronic Mail: <userid>@kelvin.physics.mun.ca
WWW: http://www.physics.mun.ca/
Founded: 1960
Activities: education * research in molecular physics * emission spectroscopy of diatomic molecules * collision-induced absorption in hydrogen
City Reference Coordinates: 052°43'00"W 47°34'00"N

Merlan Scientific Ltd.
247 Armstrong Avenue
Georgetown, ON L7G 4X6
Telephone: 416-877-0171
Telefax: 416-877-0929
Electronic Mail: discount@merlan.ca
WWW: http://www.merlan.ca/
Founded: 1970

Staff: 25
Activities: science and technology sales to the educational market
City Reference Coordinates: 079°55'00"W 43°39'00"N

Micromedia Ltd.
20 Victoria Street
Toronto, ON M5C 2N8
Telephone: 1-800-387-2689 (Canada only)
        416-362-5211
WWW: http://www.micromedia.on.ca/
City Reference Coordinates: 079°23'00"W 43°39'00"N

Mount Allison University, Department of Physics, Engineering and Geoscience
Sackville, NB E0A 3C0
Telephone: 506-564-2580
Telefax: 506-364-2580
Electronic Mail: <userid>@mta.ca
WWW: http://aci.mta.ca/peg/
Founded: 1992 (University: 1838)
Staff: 15
Activities: meteors * meteorites * fireballs * aurorae
City Reference Coordinates: 064°22'00"W 45°54'00"N

Mount Kobau Star Party (MKSP)
c/o Caroline Wallace
P.O. Box 20119 TCM
Kelowna, BC V1Y 9H2
Telephone: 250-498-3244
Telefax: 250-498-3244
Electronic Mail: loveseth@vip.net
WWW: http://www.bcinternet.com/~mksp/
City Reference Coordinates: 119°29'00"W 49°50'00"N

Mount Royal College, Department of Mathematics, Physics and Engineering
4825 Richard Road SW
Calgary, AB T3E 6K6
Telephone: 403-240-6029
Telefax: 403-240-6505
Electronic Mail: <userid>@mtroyal.ab.ca
Founded: 1911
Activities: education * research
City Reference Coordinates: 114°05'00"W 51°03'00"N

M.W. Burke-Gaffney Observatory (BGO) (Rev._)
● See "Saint Mary's University, Department of Astronomy and Physics, Rev. M.W. Burke-Gaffney Observatory (BGO)"

National Library of Canada (NLC)
395 Wellington Street
Ottawa ON K1A 0N4
Telephone: 613-995-7969
Telefax: 613-991-9871
Electronic Mail: publications@nlc-bnc.ca
WWW: http://www.nlc-bnc.ca/
Founded: 1953
Staff: 500
Activities: national library
Periodicals: (10) "National Library News"
● See also "Bibliothèque Nationale du Canada (BNC)"
City Reference Coordinates: 075°42'00"W 45°25'00"N

National Museum of Science and Technology Corp. (NMSTC)
P.O. Box 9754
Ottawa Terminal
Ottawa, ON K1G 5A3
Telephone: 613-991-3044
Telefax: 613-990-3636
WWW: http://www.smnst.ca/
        http://www.nmstc.ca/
Founded: 1967
Staff: 200
Activities: science and technology education * public programs * related historical collections
Periodicals: "Sky News", "StarGazing", "Ciel Info"
City Reference Coordinates: 075°42'00"W 45°25'00"N

National Research Council of Canada (NRCC)
Montréal Road
Ottawa, ON K1A 0R6
Telephone: 613-993-9101 (general enquiry)
        613-993-2054 (NRC publications)
        613-745-1576 (time signal)
        613-745-9426 (horloge parlante)
        613-990-0928 (IAU contact)
        613-990-6091 (IAU contact)
Telefax: 613-952-6602
      613-952-9907
WWW: http://www.nrc.ca/
Founded: 1916
Periodicals: "Canadian Journal of Physics" (ISSN 0008-4204), "Canadian Journal of Earth Sciences"
● See also "Conseil National de Recherches du Canada (CNRC)"
City Reference Coordinates: 075°42'00"W 45°25'00"N

National Research Council of Canada (NRCC), Canadian Institute for Scientific and Technical Information (CISTI)
Ottawa, ON K1A 0S2
Telephone: 613-993-1600 (inquiries)
        613-993-1095 (administration)
        613-232-6727 (National Science Film Library)
Telefax: 613-952-9112
WWW: http://www.nrc.ca/icist/
Founded: 1974
Staff: 193
Activities: government agency providing scientific and technical information
Periodicals: "Annual Report", "CISTI News"
City Reference Coordinates: 075°42'00"W 45°25'00"N

National Research Council of Canada (NRCC), Herzberg Institute of Astrophysics (HIA)
c/o Donald C. Morton
5071 West Saanich Road
Victoria, BC V8X 4M6
Telephone: 250-363-0040 (Director General)
        250-363-0567 (Administrator)
        250-363-0007 (Director's office)
        250-363-0001 (Dominion Astrophysical Observatory - DAO)
        250-493-2277 (Dominion Radio Astrophysical Observatory - DRAO)
Telefax: 250-363-8483
Electronic Mail: don.morton@hia.nrc.ca
              <userid>@hia.nrc.ca
WWW: http://www.hia.nrc.ca/
Founded: 1975
Staff: 150
Activities: optical and radio astronomy
City Reference Coordinates: 123°22'00"W 48°25'00"N

National Research Council of Canada (NRCC), Herzberg Institute of Astrophysics (HIA), Canadian Astronomy Data Centre (CADC)
c/o Dominion Astrophysical Observatory
5071 West Saanich Road
Victoria, BC V8X 4M6
Telephone: 250-363-0023
        250-363-0024
        250-363-0025
        250-363-0052
Telefax: 250-363-0045
Electronic Mail: cadc@dao.nrc.ca
WWW: http://cadcwww.dao.nrc.ca/CADC-homepage.html (English Homepage)
      http://cadcwww.dao.nrc.ca/CADC-homepage_fr.html (French Homepage)
      http://www.dao.nrc.ca/
      http://www.hia.nrc.ca/
Founded: 1986
Staff: 7
Activities: astronomical data archiving * data analysis
City Reference Coordinates: 123°22'00"W 48°25'00"N

National Research Council of Canada (NRCC), Herzberg Institute of Astrophysics (HIA), Dominion Astrophysical Observatory (DAO)
5071 West Saanich Road
Victoria, BC V8X 4M6
Telephone: 250-363-0001 (receptionist/office)
Telefax: 250-363-0045

Electronic Mail: <userid>@hia.nrc.ca
WWW: http://www.hia.nrc.ca/facilities/dao/dao.html
Founded: 1916
Staff: 80
Activities: cosmology * galactic and extragalactic astronomy * interstellar clouds * stellar astronomy * star clusters * magnetic fields * minor planets * optical instrumentation * radio submillimeter instrumentation * public outreach * astronomical software * Canadian Astronomy Data Centre (CADC - see separate entry)
Coordinates: 123°25'01"W 48°31'12"N H238m
City Reference Coordinates: 123°22'00"W 48°25'00"N

National Research Council of Canada (NRCC), Herzberg Institute of Astrophysics (HIA), Dominion Radio Astrophysical Observatory (DRAO)
P.O. Box 248
Penticton, BC V2A 2S8
or :
White Lake Road
Regional District of Okanogan-Similkameen, BC
Telephone: 604-493-2277
Telefax: 604-493-7767
Electronic Mail: <userid>@drao.nrc.ca
WWW: http://www.drao.nrc.ca/
         http://www.hia.nrc.ca/DRAO/DRAO-homepage.html
         http://www.hia.nrc.ca/
Founded: 1959
Staff: 25
Activities: SNR * HII regions * ISM * PN * star formation * clusters of galaxies * aperture synthesis * radio instrumentation and software
Periodicals: "Preprint Series"
Coordinates: 119°37'12"W 49°19'12"N H545m
City Reference Coordinates: 119°35'00"W 49°30'00"N (Penticton)
                                119°43'00"W 48°06'00"N (Okanogan)
                                120°32'00"W 49°18'00"N (Similkameen)

National Research Council of Canada (NRCC), Herzberg Institute of Astrophysics (HIA), Gemini Canadian Project Office (GCPO)
c/o Dominion Astrophysical Observatory
5071 West Saanich Road
Victoria, BC V8X 4M6
Telephone: 250-363-0024
Telefax: 250-363-6970
Electronic Mail: <userid>@hia.nrc.ca
WWW: http://www.hia.nrc.ca/
Founded: 1991
Staff: 4
City Reference Coordinates: 123°22'00"W 48°25'00"N

Natural Resources Canada, Geodetic Survey Division
(Ressources Naturelles Canada - Division des Levés Géodésiques)
615 Booth Street
Ottawa, ON K1A 0E9
Telephone: 613-995-4410
           613-992-2061 (French)
Telefax: 613-995-3215
Electronic Mail: information@geod.nrcan.ca
WWW: http://www.geod.nrcan.gc.ca/
Founded: 1909
Staff: 94
Activities: providing and maintaining Canadian spatial reference system, standards and national networks of gravity and survey control points for Canada * ensuring availability of spatial referencing information, expertise and services
City Reference Coordinates: 075°42'00"W 45°25'00"N

New Brunswick Astronomy - Astronomie Nouveau Brunswick (NBANB) Inc.
c/o Adrien Bordage (President)
390 Kingsville Road
Saint John, NB E2M 4T2
Telephone: 506-635-3004
Electronic Mail: astro-l@admin.cus.ca
                 abordage@sjfn.nb.ca
                 rachett@nbnet.nb.ca
WWW: http://www.osco.nb.ca/nbanb/
Founded: 1995
Membership: 125
Activities: annual star party * observing * talks * instrumentation
Periodicals: (4) "Pleiades"
City Reference Coordinates: 066°03'00"W 45°16'00"N

Newfoundland Science Centre
The Murray Premises
5 Beck's Cove
P.O. Box 1312
Station C
Saint John's, NF A1C 5N5
Telephone: 709-754-0807
Telefax: 709-738-3276
WWW: http://www.nsc.nfld.com/
City Reference Coordinates: 052°43'00"W 47°34'00"N

North Bay Astronomy Club
c/o Merlin Clayton
50 Van Horne Crescent
North Bay, ON P1A 3L3
Telephone: 705-472-1182
Electronic Mail: brentt@efni.com
WWW: http://www.efni.com/@brentt/nbclub/nbclub.htm
City Reference Coordinates: 079°28'00"W 46°20'00"N

North York Astronomical Association (NYAA)
26 Chryessa Avenue
Toronto, ON M6N 4T5
Electronic Mail: walmac@spanit.com (Walter MacDonald)
WWW: http://www.interlog.com/~nyaa/
Activities: monthly meetings * photography observing (comets, meteors, planets, Sun, SN, variable stars, deep sky) * instrumentation
Periodicals: (4) "AstroTent"
City Reference Coordinates: 079°23'00"W 43°39'00"N

Nova Astronomics
P.O. Box 31013
Halifax, NS B3K 5T9
Telephone: 902-499-6196
Telefax: 902-826-7957
Electronic Mail: info@nova-astro.com
WWW: http://www.nova-astro.com/
Founded: 1991
Staff: 1
● Software producer
City Reference Coordinates: 063°36'00"W 44°39'00"N

Observateurs de la Magnitude Absolue (Les_)
● See "Les Observateurs de la Magnitude Absolue"

Observatoire Astronomique du Mont Mégantic (OAMM)
● See "Université de Montréal, Observatoire Astronomique du Mont Mégantic (OAMM)"

Observatoire "L'Étoile d'Acadie"
● See "Club d'Astronomie de l'Observatoire L'Étoile d'Acadie"

Ontario Science Centre (OSC)
770 Don Mills Road
North York, ON M3C 1T3
Telephone: 416-696-3127
        416-696-3147 (information in French)
Electronic Mail: <userid>@osc.on.ca
WWW: http://www.osc.on.ca/
Founded: 1969
City Reference Coordinates: 079°25'00"W 43°46'00"N

Ottawa Valley Astronomy and Observers Group (OAOG)
c/o Rock MAllin
P.O. Box 8285
Ottawa, ON K1K 0S1
Telephone: 613-741-1612
Electronic Mail: observers-group@home.com
WWW: http://www.members.home.net/observers-group/
City Reference Coordinates: 075°42'00"W 45°25'00"N

OZ Optics Ltd.
219 Westbrook Road
West Carleton Industrial Park
Carp, ON K0A 1L0

Telephone: 613-831-0981
Telefax: 613-836-5089
Electronic Mail: sales@ozoptics.com
WWW: http://www.ozoptics.com/
Founded: 1985
Activities: manufacturing fiber optic components
City Reference Coordinates: 076°02'00"W 45°21'00"N (Carp)

Pacific Space Centre

1100 Chesnut Street
Vanier Park
Vancouver, BC V6J 3J9
Telephone: 604-738-STAR
         604-738-7827
Telefax: 604-736-5665 (Administration)
         604-738-5086 (Planetarium)
Electronic Mail: <userid>@pacific-space-centre.bc.ca
         jdickens@axionet.com (John Dickenson, Managing Director)
         rsrs@pacific-space-centre.bc.ca
WWW: http://pacific-space-centre.bc.ca/
Founded: 1997 (Planetarium: 1968) (Observatory: 1978)
Staff: 46
Activities: planetarium shows * lectures * workshops * education * public observing * Starlab portable planetarium * space science educational outreach and resource centre * Canadian Space Resource Centre (Pacific)
Periodicals: (4) "The Starry Messenger"
Coordinates: 123°08'31"W 49°16'32"N H5m
● Formerly know as "H.R. MacMillan Planetarium and Gordon Southam Observatory"
City Reference Coordinates: 123°07'00"W 49°16'00"N

Pacific Space Centre Society (PSCS)

1100 Chesnut Street
Vancouver, BC V6J 3J9
Telephone: 604-738-7827
Telefax: 604-736-5665
Electronic Mail: ddodge@wimsey.ca
Founded: 1987 (Planetarium: 1968) (Observatory: 1972)
Membership: 3000
Staff: 37
Activities: administering "H.R. MacMillan Planetarium" and "Gordon Southam Observatory" * motion simulator * space flight simulations * "Canada in Space" exhibitions
Periodicals: (6) "The Starry Messenger"
● Formerly "British Columbia Space Sciences Society (BCSSS)"
City Reference Coordinates: 123°07'00"W 49°16'00"N

Physics and Astronomy Student Society (PASS)

c/o Department of Physics and Astronomy
University of Victoria
P.O. Box 3055 STN CSC
Victoria, BC V8W 3P6
Telephone: 250-721-7700
Telefax: 250-721-7715
Electronic Mail: uvpass@uvastro.phys.uvic.ca
WWW: http://astrowww.phys.uvic.ca/uvpass/pass.html
Founded: 1985
City Reference Coordinates: 123°22'00"W 48°25'00"N

Planetarium Association of Canada (PAC)

● See now "Canadian Council of Science Centres (CCSC)"

Planétarium de Montréal

1000 Rue Saint-Jacques Ouest
Montréal, QC H3C 1G7
Telephone: 514-872-4530
         514-872-3611
Telefax: 514-872-8102
Electronic Mail: chastenay@astro.umontreal.ca (Pierre Chastenay)
         planetarium@astro.umontreal.ca
WWW: http://www.planetarium.montreal.qc.ca/
Founded: 1966
Staff: 25
Activities: public and school shows * lectures * exhibits * workshops * star parties
● Alternate name: "Planétarium Dow"
City Reference Coordinates: 073°34'00"W 45°31'00"N

Prince George Astronomical Society (PGAS)
c/o Robert H. Nelson (President)
3330 - 22nd Avenue
Prince George, BC V2N 1P8
or :
Tedford Road
Prince George, BC
Telephone: 250-964-3600
Telefax: 250-561-5848
Electronic Mail: bnelson@pgonline.com
WWW: http://www.pgweb.com/astronomical
Founded: 1979
Membership: 30
Activities: education * star parties * photography * observing * photometry
Periodicals: (12) "PeGASus"
Coordinates: 122°50'52"W 53°45'29"N H600m (Prince George Astronomical Observatory)
City Reference Coordinates: 122°45'00"W 53°55'00"N

Queen's University, Department of Physics, Astronomy Research Group (QUARG)
Kingston, ON K7L 3N6
Telephone: 613-533-2707
               613-533-2711
Telefax: 613-533-6463
Electronic Mail: <userid>@qucdnast.queensu.cdn
                   <userid>@astro.queensu.ca
WWW: http://www.astro.queensu.ca/
Founded: 1960
Staff: 20
Activities: galactic structure * star formation * theoretical astrophysics
Observatories: 1 (Queen's Observatory)
City Reference Coordinates: 076°30'00"W 44°14'00"N

Relative Data Products
365 Sherwood Drive
Ottawa, ON K1Y 3X3
Telefax: 613-728-4240
Electronic Mail: dpatte@relativedata.com (David Patte)
WWW: http://www.relativedata.com/
Founded: 1996
● Software producer
City Reference Coordinates: 075°42'00"W 45°25'00"N

Resonance Ltd.
143 Ferndale Drive North
Barrie, ON L4M 4S4
Telephone: 705-733-3633
Telefax: 705-733-1388
Electronic Mail: res@bconnex.net
WWW: http://www.bconnex.net/~res/
Founded: 1980
Staff: 7
Activities: research and development and production of light sources, remote sensors and detector technology for commercial, industrial and space applications
City Reference Coordinates: 079°40'00"W 44°24'00"N

Ressources Naturelles Canada, Division des Levés Géodésiques
● See "Natural Resources Canada, Geodetic Survey Division"

Rev. M.W. Burke-Gaffney Observatory (BGO)
● See "Saint Mary's University, Department of Astronomy and Physics, Rev. M.W. Burke-Gaffney Observatory (BGO)"

Roberta Bondar Planetarium
● See "Seneca College, Roberta Bondar Planetarium"

Rothney Astrophysical Observatory (RAO)
● See "University of Calgary, Department of Physics and Astronomy, Rothney Astrophysical Observatory (RAO)"

Royal Astronomical Society of Canada (RASC)
136 Dupont Street
Toronto, ON M5R 1V2
Telephone: 416-924-7973
Telefax: 416-924-7973
Electronic Mail: rasc@vela.astro.utoronto.ca
WWW: http://www.rasc.ca/

Founded: 1890
Membership: 3000
Activities: observing * education * telescope making
Periodicals: (6) "Journal" (ISSN 0035-872x); (1) "Observer's Handbook" (ISSN 0080-4193), "Annual Report"
Awards: "Gold Medal", "Chant Medal", "Ken Chilton Prize", "Plaskett Medal", "Messier Certificate", "Service Award", "Simon Newcomb Award"
● See also "Société Royale d'Astronomie du Canada (SRAC)"
City Reference Coordinates: 079°23'00"W 43°39'00"N

Royal Astronomical Society of Canada (RASC), Calgary Centre
c/o Calgary Centennial Planetarium
P.O. Box 2100
Calgary, AB T2P 2M5
or :
c/o Susan Yeo
248 Second Street NW
Calgary, AB T2K 0X9
WWW: http://www.syz.com/rasc/
         http://www.rasc.ca/
Founded: 1890 (RASC)
Awards: See RASC awards
City Reference Coordinates: 114°05'00"W 51°03'00"N

Royal Astronomical Society of Canada (RASC), Edmonton Centre
c/o Edmonton Space and Science Centre
11211 - 142nd Street
Edmonton, AB T5M 4A1
Telephone: 403-452-9100
Electronic Mail: rasc@worldgate.com
WWW: http://planet10.v-wave.com/rasc
         http://www.rasc.ca/
Founded: 1932
Activities: lectures * observing
Periodicals: (10) "Stardust"
Coordinates: 112°55'00"W 53°32'00"N H800m (Waskahegan)
Awards: (1) "Angus Smith Telescope Making Award", "Astrophotography Award", "Service Award"; see also RASC awards

City Reference Coordinates: 113°28'00"W 53°33'00"N

Royal Astronomical Society of Canada (RASC), Halifax Centre
c/o Nova Scotia Museum
1747 Summer Street
Halifax, NS B3H 3A6
or :
c/o Patrick Kelly
Rural Route 2
Falmouth, NS B0P 1L0
Electronic Mail: dlane@husky1.stmarys.ca (David Lane)
WWW: http://halifax.rasc.ca/
         http://www.rasc.ca/
Founded: 1890
Membership: 165
Activities: public awareness of astronomy * observing
Periodicals: (6) "Nova Notes"
Coordinates: 063°42'00"W 44°57'00"N H50m (Beaverbank)
Awards: "Burke-Gaffney Award"; see also RASC awards
City Reference Coordinates: 063°36'00"W 44°39'00"N (Halifax)

Royal Astronomical Society of Canada (RASC), Hamilton Centre
P.O. Box 1223
Waterdown, ON L0R 2H0
Telephone: 905-689-0266
Electronic Mail: rasc@ad-here.com
WWW: http://www.homeroom.ca/rasc.html
         http://www.rasc.ca/
Founded: 1909
Membership: 50
Activities: observing (deep sky, planets, double stars, variable stars, occultations, meteors) * photography * photometry * public education * lectures
Periodicals: (11) "Orbit"
Coordinates: 079°55'12"W 43°23'24"N H271m (Leslie V. Powis Observatory)
Awards: see RASC awards
City Reference Coordinates: 079°51'00"W 43°15'00"N

Royal Astronomical Society of Canada (RASC), Kingston Centre
P.O. Box 1793

Kingston, ON K7L 5J6
or :
c/o Kim Hay
1462 Leland Road
R.R.#2 Perth Road
Kingston, ON K0H 2L0
Telephone: 613-353-7910
Electronic Mail: rascexec@cliff.path.queensu.ca
WWW: http://www1.kingston.net/~rasc
        http://www.rasc.ca/
Founded: 1961 (RASC: 1890)
Membership: 150
Activities: education * observing
Periodicals: "Regulus"
Awards: see RASC awards
City Reference Coordinates: 076°30'00"W 44°14'00"N

Royal Astronomical Society of Canada (RASC), Kitchener-Waterloo Centre
c/o Paul Bigelow
114 Westvale Drive
Waterloo, ON N2T 1J2
or :
c/o Jeff Brunton
123 Grand River Street North
Paris, ON N3L 2M4
Telephone: 519-888-7516
Electronic Mail: bigelow@waterloo.hp.com
WWW: http://kw.rasc.ca/
        http://www.rasc.ca/
Founded: 1970
Membership: 41
Activities: public star nights
Periodicals: (6) "Pulsar"
Coordinates: 080°29'49"W 43°17'01"N H320m (Dance Hill)
Awards: see RASC awards
City Reference Coordinates: 080°31'00"W 43°28'00"N (Waterloo)
                                080°23'00"W 43°12'00"N (Paris)

Royal Astronomical Society of Canada (RASC), London Centre
P.O. Box 842
Station "B"
London, ON N6A 4Z3
or :
c/o David MacMillan
15 Clenray Place
London, ON N6A 3Y9
Electronic Mail: svonesh@julian.uwo.ca (Susan Vonesh)
WWW: http://phobos.astro.uwo.ca/~rasc/
        http://www.rasc.ca/
Founded: 1922
Membership: 46
Activities: observing (planets, deep sky) * lectures * public "Star Nights"
Periodicals: (12) "Astronomy London"
Awards: "Messier Certificate", "Merit Award"; see also RASC awards
City Reference Coordinates: 081°14'00"W 42°59'00"N

Royal Astronomical Society of Canada (RASC), Montréal Centre
P.O. Box 1752
Station "B"
Montréal, QC H3B 3L3
or :
c/o Dominique MacKenzie
530 Cedar Street
Beaconsfield, QC J9W 3W1
Telephone: 514-845-2612
Electronic Mail: rasc.mtl%cap@qumedia.com
WWW: http://www.generation.net/~gang/mc/
        http://www.rasc.ca/
Founded: 1918
Membership: 120
Activities: lectures * observing
Periodicals: (12) "Skyward"
Awards: "Charles M. Good Award"; see also RASC awards
Coordinates: 073°34'53"W 45°30'42"N H94m
City Reference Coordinates: 073°34'00"W 45°31'00"N (Montréal)
                                073°51'00"W 45°26'00"N (Beaconsfield)

Royal Astronomical Society of Canada (RASC), Niagara Centre
P.O. Box 241
Niagara Falls, ON L2E 6T3
or :
c/o Ron Gasbarini
64 Roehampton Avenue
Saint Catherines, ON L2M 7P5
WWW: http://www.vaxxine.com/rascniag/
         http://www.rasc.ca/
Founded: 1890 (RASC)
Awards: see RASC awards
City Reference Coordinates: 079°04'00"W 43°06'00"N (Niagara Falls)
                                079°13'00"W 43°10'00"N (Saint Catherines)

Royal Astronomical Society of Canada (RASC), Okanogan Centre
c/o Ron Scherer
11450 Darlene Road
Winfield, BC V4V 1Y4
WWW: http://www.bcinternet.com/~rasc/
         http://www.rasc.ca/
Founded: 1890 (RASC)
Awards: see RASC awards
City Reference Coordinates: 119°29'00"W 49°50'00"N (Kelowna)
                                119°19'00"W 50°17'00"N (Vernon)

Royal Astronomical Society of Canada (RASC), Ottawa Centre
c/o Jane Lund
1974 Baseline Road
P.O. Box 33012
Nepean, ON K2C 0E0
Electronic Mail: george@sce.carleton.ca (Doug George)
WWW: http://ottawa.rasc.ca/
         http://www.rasc.ca/
Founded: 1890 (RASC)
Awards: see RASC awards
City Reference Coordinates: 075°42'00"W 45°25'00"N (Ottawa)

Royal Astronomical Society of Canada (RASC), Regina Centre
c/o Ross Parker (Secretary)
P.O. Box 20014
Cornwall Central Postal Outlet
Regina, SK S4P 4J7
Telephone: 306-565-0980 (Secretary)
WWW: http://www3.sk.sympatico.ca/kostaw/rasc/
         http://www.rasc.ca/
Founded: 1910
Membership: 35
Activities: meetings * star parties * lectures
Periodicals: (10) "The Stargazer"
Awards: see RASC awards
Coordinates: 104°36'00"W 50°26'00"N H577m (Regina Kalium Observatory)
                 104°06'00"W 50°23'00"N H670m (Davin Observing Compound)
City Reference Coordinates: 104°39'00"W 50°25'00"N

Royal Astronomical Society of Canada (RASC), Saint John's Centre
206 Frecker Drive
Saint John's, NF A1B 4G7
or :
c/o Dennis Ryan
74A Moores Drive
Mount Pearl, NF A1N 3R4
Telephone: 709-364-2924
               709-368-5677
Electronic Mail: randy@kean.ucs.mun.ca
WWW: http://www.infonet.st-johns.nf.ca/providers/rasc/rasc.html
         http://www.infonet.st-johns.nf.ca/rasc/
         http://www.rasc.ca/
Founded: 1965
Membership: 30
Activities: meetings * public relations * education * observing * Permanent Observing Site Search Committee
Periodicals: "All About Stars"
Awards: see RASC awards
City Reference Coordinates: 052°43'00"W 47°34'00"N (Saint John's)

Royal Astronomical Society of Canada (RASC), Sarnia Centre
c/o Alice Lester (President)

P.O. Box 394
587 Ontario Street
Apartment 204
Wyoming, ON N0N 1T0
or :
c/o Jim Selinger (Secretary)
160 George Street
Sarnia, ON N7T 7V4
Telephone: 519-845-0454 (President)
          519-337-6815 (Secretary)
WWW: http://www.rasc.ca/
Founded: 1983
Membership: 13
Activities: popularization * observing
Awards: See RASC Awards
City Reference Coordinates: 082°07'00"W 42°57'00"N (Wyoming)
                               082°23'00"W 42°58'00"N (Sarnia)

Royal Astronomical Society of Canada (RASC), Saskatoon Centre
P.O. Box 317
RPO University
Saskatoon, SK S7N 4J8
Telephone: 306-665-3392
Electronic Mail: huziak@sedsystems.ca
WWW: http://maya.usask.ca/~sarty/rasc/rasc.html
          http://www.rasc.ca/
Founded: 1968
Membership: 65
Activities: promoting amateur astronomy and related sciences to members, media and general public * meetings * public starnights
Periodicals: (12) "Saskatoon Skies" (circ.: 120)
Coordinates: 106°33'44"W 52°02'34"N H561m (Rystrom Observatory)
Awards: See RASC Awards
City Reference Coordinates: 106°38'00"W 52°07'00"N

Royal Astronomical Society of Canada (RASC), Thunder Bay Centre
c/o R.C. Bishop
135 Hogarth Street
Thunder Bay, ON P7A 7H1
Electronic Mail: bob.bishop@oln.com
WWW: http://www.rasc.ca/
Founded: 1890 (RASC)
Awards: See RASC Awards
City Reference Coordinates: 080°03'00"W 44°48'00"N

Royal Astronomical Society of Canada (RASC), Toronto Centre
c/o McLaughlin Planetarium
100 Queen's Park
Toronto, ON M5S 2C6
or :
Ontario Science Centre
770 Don Mills Road
North York, ON M3C 1T3
Telephone: 416-724-7827
          416-696-3127
Electronic Mail: bchou@watserv1.uwaterloo.ca
WWW: http://durham.durhamc.on.ca/ras/centre1.html
          http://www.rasc.ca/
Founded: 1868
Membership: 800
Activities: public education * meetings * observing
Periodicals: (6) "Scope"
Awards: (1) "Bert Winnearls Award", "Andrew Elvins Award", "Bertram J. Topham Award", "Jesse Ketchum Award"; see also RASC Awards
City Reference Coordinates: 079°23'00"W 43°39'00"N (Toronto)
                               079°25'00"W 43°46'00"N (North York)

Royal Astronomical Society of Canada (RASC), Vancouver Centre
c/o Pacific Space Centre
1100 Chesnut Street
Vancouver, BC V6J 3J9
or :
c/o Reid Hart
4976 Sherbrooke Street
Vancouver, BC V5W 3M2
WWW: http://pacific-space-centre.bc.ca/rasc.html

http://members.home.net/ronaldwp/rasc_vc/
http://www.rasc.ca/
Founded: 1890 (RASC)
Awards: see RASC awards
City Reference Coordinates: 123°07'00"W 49°16'00"N

Royal Astronomical Society of Canada (RASC), Victoria Centre
c/o F.W. Almond
354 Benhomer Drive
Victoria, BC V9C 2C6
Telephone: 250-478-8065
WWW: http://victoria.tc.ca/~rasc/
http://www.rasc.ca/
Founded: 1914
Membership: 145
Activities: meetings * observing * hobby show * field trips
Periodicals: (10) "Skynews"
Awards: see RASC awards
City Reference Coordinates: 123°22'00"W 48°25'00"N

Royal Astronomical Society of Canada (RASC), Windsor Centre
c/o Joady Ulrich
5450 Haig Avenue
Windsor, ON N8T 1K9
or :
c/o Steve Pellarin
3140 Parkwood Avenue
Windsor, ON N8W 2K4
Telephone: 519-966-0713
Electronic Mail: sheple@server.uwindsor.ca
WWW: http://www.wincom.net/rasc/
http://www.rasc.ca/
Founded: 1890 (RASC)
Awards: see RASC awards
City Reference Coordinates: 083°01'00"W 42°18'00"N

Royal Astronomical Society of Canada (RASC), Winnipeg Centre
x/o Chris Brown (Treasurer)
P.O. Box 2694
Winnipeg, MB R3C 4B3
or :
183 Canora Street
Winnipeg, MB R3G 1T1
Telephone: 204-775-6392
WWW: http://www.winnipeg.freenet.mb.ca/rasc/
http://www.rasc.ca/
Founded: 1890 (RASC)
Membership: 107
Activities: amateur astronomy * photometry * CCD photography
Periodicals: (6) "Winnicentrics"
Awards: see RASC awards
City Reference Coordinates: 097°09'00"W 49°53'00"N

Saint Mary's University (SMU), Department of Astronomy and Physics, Rev. M.W. Burke-Gaffney Observatory (BGO)
923 Robie Street
Halifax, NS B3H 3C3
Telephone: 902-420-5633
902-420-5828
Telefax: 920-420-5561
902-420-5141
Electronic Mail: <userid>@husky1.stmarys.ca
<userid>@mnbsun.stmarys.ca
WWW: http://mnbsun.stmarys.ca/smu_home.html
http://mnbsun.stmarys.ca/WWW/smu_home.html
http://apwww.stmarys.ca/
http://apwww.stmarys.ca/bgo/bgo.html
Founded: 1971
Staff: 4
Activities: education * research
Coordinates: 063°34'52"W 44°37'50"N H125m
City Reference Coordinates: 063°36'00"W 44°39'00"N

Science North Solar Observatory
100 Ramsey Lake Road
Sudbury, ON P3E 5S9

Telephone: 705-522-3701 x243
Telefax: 705-522-4954
Electronic Mail: yee@ramsey.cs.laurentian.ca
                    <userid>@sciencenorth.on.ca
WWW: http://sciencenorth.on.ca/
Founded: 1984
Staff: 9
Activities: interactive exhibits * solar observing * theatre programs
Coordinates: 080°59'46"W 46°28'12"N H260m
City Reference Coordinates: 081°00'00"W 46°30'00"N

Sciencetech Inc.
45 Meg Drive
London, ON N6E 2V2
Telephone: 519-668-0131
Telefax: 519-668-0132
Electronic Mail: scitech@execulink.com
WWW: http://www.execulink.com/~scitech/
Founded: 1985
Staff: 15
Activities: designing and manufacturing optical spectroscopy instruments
City Reference Coordinates: 081°14'00"W 42°59'00"N

Science World British Columbia
1455 Quebec Street
Vancouver, BC V6A 3Z7
Telephone: 604-268-6363 (24-hour information)
               604-443-7440 (Administration)
Telefax: 604-682-2923
Electronic Mail: question@scienceworld.bc.ca
WWW: http://www.scienceworld.bc.ca/
Founded: 1989
Staff: 100
Activities: hands-on science centre * exhibition galleries * Ominmax theater * 3D laser theater
Periodicals: (4) "Newsletter"; (1) "Annual Report"
City Reference Coordinates: 123°07'00"W 49°16'00"N

Scintrex Ltd.
222 Snidercroft Road
Concord, ON L4K 1B5
Telephone: 905-669-2280
Telefax: 905-669-6403
Electronic Mail: scintrex@scintrexltd.com
WWW: http://www.scintrexltd.com/
Founded: 1964
Staff: 140
Activities: manufacturing geophysical instrumentation
City Reference Coordinates: 079°29'00"W 43°48'00"N

Seneca College, Roberta Bondar Planetarium
1750 Finch Avenue East
North York, ON M2J 2X5
Telephone: 416-491-5050 x2227
Electronic Mail: stars@mars.senecac.on.ca
WWW: http://www.senecac.on.ca/bondar/planet.htm
Founded: 1972
City Reference Coordinates: 079°25'00"W 43°46'00"N

Sequiter Software Inc.
9644 - 54 Avenue
Suite 209
Edmonton, AB T6E 5V1
Telephone: 403-437-2410
Telefax: 403-436-2999
Electronic Mail: info@sequiter.com
WWW: http://www.sequiter.com/
Founded: 1985
Staff: 14
• Software producer
City Reference Coordinates: 113°28'00"W 53°33'00"N

Sienna Software Inc.
105 Pears Avenue
Toronto, ON M5R 1S9
Telephone: 1-800-252-5417 (Canada & USA only)

416-410-0259
Telefax: 416-410-0359
Electronic Mail: contact@siennasoft.com
WWW: http://www.siennasoft.com/
Founded: 1992
• Software producer
City Reference Coordinates: 079°23'00"W 43°39'00"N

SkyNews
c/o National Museum of Science and Technology Corp.
2421 Lancaster Road
P.O. Box 9724
Ottawa Terminal
Ottawa, ON K1G 5A3
Telephone: 613-991-2983
          1-800-267-3999
Telefax: 613-990-3636
Founded: 1995
• Periodical (6)
City Reference Coordinates: 075°42'00"W 45°25'00"N

Sky Optics
4031 Fairview Street
Unit 216B
Burlington, ON L7L 2A4
Telephone: 905-9631-9944
Electronic Mail: sbarnes@worldchat.com
WWW: http://www.worldchat.com/commercial/skyoptics/
City Reference Coordinates: 079°48'00"W 43°19'00"N

Snowbird Software Inc.
71A Leland Street
Hamilton, ON L8S 3A1
Telephone: 905-521-9667
Telefax: 905-521-9667
Electronic Mail: info@snowbirdsoftware.on.ca
WWW: http://www.snowbirdsoftware.on.ca/
Founded: 1984
Staff: 10
• Software producer
City Reference Coordinates: 079°51'00"W 43°15'00"N

Société Canadienne d'Astronomie (CASCA)
• See "Canadian Astronomical Society (CASCA)"

Société d'Astronomie de Montréal (SAM)
Casier Postal 206
Station Saint-Michel
Montréal, QC H2A 3L9
or :
7110 8e Avenue
Montréal, QC
Telephone: 514-728-4422
WWW: http://www.cam.org/~sam/
Founded: 1968
Membership: 150
Activities: lectures * library * telescope making * shop
Periodicals: (1) "Annuaire Astronomique" (ISSN 0825-9984, circ.: 1,300); (4) "Astro-Notes" (ISSN 0843-8978, circ.: 300)
Observatories: 1 (Saint-Valérien de Milton)
Awards: "Prix Georgette-Lemoyne", "Prix Étoile d'Argent"
City Reference Coordinates: 073°34'00"W 45°31'00"N

Société Royale d'Astronomie du Canada (SRAC)
• See also "Royal Astronomical Society of Canada (RASC)"

Société Royale d'Astronomie du Canada (SRAC), Centre de Québec
2000 Boulevard Montmorency
Québec, QC G1J 5E7
or :
c/o Clément Drolet
14 Rue du Moulin
Beaumont, QC G0R 1C0
Telephone: 418-660-2815 (answering machine)
WWW: http://www2.zone.ca:80/~marcelf/srac.htm
          http://www.rasc.ca/

Founded: 1942
Membership: 72
Activities: monthly lectures * observing * library * telescope making * photography * computing
Periodicals: (1) "Almanach Graphique" (ISSN 0384-7691); "La Chouette" (ISSN 1191-1220)
Awards: See RASC awards
Observatories: 1 (Saint-Nérée-de-Bellechasse, H330)
City Reference Coordinates: 071°13'00"W 46°49'00"N (Québec)

Société Royale d'Astronomie du Canada (SRAC), Centre Francophone de Montréal

Case Postale 206
Station Saint-Michel
Montréal, QC H3B 3L3
or :
c/o Alain Metras
868 Milot
Saint-Hubert, QC J3Y 8C2
WWW: http://www.cam.org/~sam/
          http://www.rasc.ca/
Founded: 1890 (RASC)
Awards: See RASC awards
City Reference Coordinates: 073°34'00"W 45°31'00"N (Montréal)
                              073°25'00"W 45°30'00"N (Saint-Hubert)

SoftQuad Inc.

20 Eglinton Avenue West
13th Floor
P.O. Box 2025
Toronto ON M4R 1K8
Telephone: 416-544-9000
             1-800-387-2777 (Canada and USA only)
Telefax: 416-544-0300
Electronic Mail: <userid>@sq.com
                     support@sq.com
                     sales@sq.com
                     mail@sq.com
WWW: http://www.sq.com/
● Software producer
City Reference Coordinates: 079°23'00"W 43°39'00"N

Solar Terrestrial Dispatch (STD)

P.O. Box 357
Stirling, AB T0K 2E0
Telephone: 403-756-3008
Electronic Mail: coler@solar.stanford.edu (Cary Oler)
                     oler@holly.cc.uleth.ca
WWW: http://solar.uleth.ca/solar
Founded: 1990
Activities: providing time-critical solar and geophysical information and data * daily summaries and reports of solar and geophysical activity * alerts and warnings of impending activity
City Reference Coordinates: 112°31'00"W 49°30'00"N

Southam Observatory (Gordon_)

● See "Pacific Space Centre"

Space Astrophysics Laboratory (SAL)

● See "York University, Department of Physics and Astronomy, Space Astrophysics Laboratory (SAL)"

Space West Science Activities (SWSA) Inc.

2115 McEown Avenue
P.O. Box 1811
Saskatoon, SK S7K 3S2
Telephone: 306-374-1395
Telefax: 306-374-6270
Electronic Mail: brooksws@the.link.ca (William S.C. Brooks, Executive Director)
Founded: 1997
Staff: 3
Activities: providing space camps for children and adults
City Reference Coordinates: 106°38'00"W 52°07'00"N

Standards Council of Canada (SCC)

(Conseil Canadien des Normes - CCN)
45 O'Connor Street
Suite 1200
Ottawa, ON K1P 6N7
Telephone: 613-238-3222

Telefax: 613-995-4564
Electronic Mail: info@scc.ca
WWW: http://www.scc.ca/
Founded: 1977
Staff: 79
Activities: coordinating voluntary standardization activities in Canada
Periodicals: (4) "Consensus" (English edition: ISSN 0380-1314, French edition: ISSN 0380-1322); (1) "Annual Report/Rapport Annuel"
Awards: (1) "Jean P. Carrière Award"
City Reference Coordinates: 075°42'00"W 45°25'00"N

Stargazer Steve
c/o Steve Dodson
1752 Rutherglen Crescent
Sudbury, ON P3A 2K3
Telephone: 705-566-1314
Electronic Mail: stargazr@isys.ca
WWW: http://ww2.isys.ca/stargazer/
Founded: 1994
Staff: 5
Activities: manufacturing and marketing innovative telescopes
City Reference Coordinates: 081°00'00"W 46°30'00"N

Stormy Weather Software Ltd.
Box 125
Picton, ON K0K 2T0
Telephone: 613-476-5779
Telefax: 613-476-7598
Electronic Mail: stormy@stormy.ca
                  astro@stormy.ca
WWW: http://www.stormy.ca/
Founded: 1988
Staff: 4
● Software producer
City Reference Coordinates: 077°08'00"W 44°00'00"N

Syzygy Research & Technology Ltd.
Box 83
Legal, AB T0G 1LO
Telephone: 780-961-2213
Electronic Mail: support@syz.com
                  sales@syz.com
WWW: http://www.syz.com/
Founded: 1995
Staff: 2
● Software producer
City Reference Coordinates: 113°40'00"W 53°58'00"N

Tasso (E.J._)
● See "E.J. Tasso"

The Atlantic Space Sciences Foundation (TASSF) Inc.
P.O. Box 31011
Halifax, NS B3K 5T9
Telephone: 902-864-7256
Telefax: 902-492-3170
Electronic Mail: tassf@hercules.stmarys.ca
WWW: http://halifax.rasc.ca/tassf/tassf.html
         http://halifax.rasc.ca/hp/ (Halifax Planetarium)
Founded: 1990
Staff: 8
Activities: astronomy education for teachers and students * portable and fixed planetariums
City Reference Coordinates: 063°36'00"W 44°39'00"N

Toronto Sidewalk Astronomers (TSA)
Electronic Mail: burns@astro.utoronto.ca (Chris Burns)
WWW: http://www.astro.utoronto.ca/~burns/TSA.html
Membership: 4
City Reference Coordinates: 079°23'00"W 43°39'00"N

Université de Moncton, Département de Physique
Moncton, NB E1A 3E9
Telephone: 506-858-4339
Telefax: 506-858-4541
Electronic Mail: <userid>@umoncton.ca

WWW: http://139.103.16.55/phys/dphys.htm
       http://www.umoncton.ca/leblanfn/observ.html
Activities: stellar astrophysics * optical instrumentation * micro-gravity * thin layers * astronomical * electroluminescence * observatory projet
City Reference Coordinates: 064°50'00"W 46°04'00"N

Université de Montréal, Département de Physique, Groupe d'Astronomie
2900 Ed. Montpetit
Casier Postal 6128
Station "A"
Montréal, QC H3C 3J7
Telephone: 514-343-6718
Telefax: 514-343-2071
Electronic Mail: <userid>@astro.umontreal.ca
WWW: http://www.astro.umontreal.ca/
       http://www.astro.umontreal.ca/index_eng.html (English Homepage)
Founded: 1968
Staff: 10
Activities: stellar and extragalactic astronomy
City Reference Coordinates: 073°34'00"W 45°31'00"N

Université de Montréal, Observatoire Astronomique du Mont Mégantic (OAMM)
2900 Ed. Montpetit
Casier Postal 6128
Station "A"
Montréal, QC H3C 3J7
Telephone: 514-343-6718
Telefax: 514-343-2071
Electronic Mail: megantic@astro.umontreal.ca
WWW: http://www.astro.umontreal.ca/omm/
Founded: 1978
Staff: 21
Activities: research * education * optical astronomy * instrumentation
Coordinates: 071°09'11"W 45°27'20"N H1,111m
City Reference Coordinates: 073°34'00"W 45°31'00"N

Université du Québec à Trois-Rivières (UQTR), Département de Physique
3351 Boulevard des Forges
Casier Postal 500
Trois-Rivières, QC G9A 5H7
Telephone: 819-376-5107
Telefax: 819-376-5164
Electronic Mail: <userid>@uqtr.uquebec.ca
WWW: http://www.uqtr.uquebec.ca/dphy/
Founded: 1969
Staff: 5
Activities: education * research
City Reference Coordinates: 072°33'00"W 46°21'00"N

Université Laval, Département de Physique, Groupe de Recherche en Astrophysique
Sainte-Foy, QC G1K 7P4
Telephone: 418-656-2652
Telefax: 418-656-2040
Electronic Mail: <userid>@phy.ulaval.ca
WWW: http://astrosun.phy.ulaval.ca/astro/
Founded: 1975
Staff: 5
Activities: evolution and dynamics of galaxies and ISM * stellar formation * stellar populations and synthesis * massive stars * starbursts * energetic phenomena (impacts on ISM) * instrumentation
Coordinates: 071°09'12"W 45°27'20"N H1,108m (Observatoire du Mont Mégantic)
City Reference Coordinates: 071°18'00"W 46°47'00"N

Université Laval, Département de Sciences Géomatiques
Sainte-Foy, QC G1K 7P4
Telephone: 418-656-2530
Telefax: 418-656-3177
       418-656-7411
Electronic Mail: scg@scg.ulaval.ca
WWW: http://www.scg.ulaval.ca/scg/fr/
Founded: 1994
Activities: positioning
Coordinates: 071°16'49"W 46°46'50"N H112m
● Formerly "Université Laval, Département de Sciences Géodésiques et de Télédétection"
City Reference Coordinates: 071°18'00"W 46°47'00"N

University of Alberta, Centre for Subatomic Research (CSR)
Edmonton, AB T6G 2N5
Telephone: 403-492-5011
Telefax: 403-492-3408
Electronic Mail: <userid>@physics.ualberta.ca
WWW: http://www.phys.ualberta.ca/csr/
City Reference Coordinates: 113°28'00"W 53°33'00"N

University of Alberta, Department of Physics
Edmonton, AB T6G 2J1
Telephone: 403-492-5286
            403-492-5410 (D. Hube)
            403-987-3933 (Devon Astronomical Observatory)
Telefax: 403-492-0714
Electronic Mail: <userid>@phys.ualberta.ca
                 dept@phys.ualberta.ca
WWW: http://www.phys.ualberta.ca/
Founded: 1968 (Devon Astronomical Observatory)
Staff: 2
Activities: spectroscopy and photometry of binary and variable stars * CCD imagery of SNRs
Coordinates: 113°45'31"W 53°23'27"N H708m (Devon Astronomical Observatory)
City Reference Coordinates: 113°28'00"W 53°33'00"N

University of Alberta, Institute of Geophysics, Meteorology and Space Physics
Edmonton, AB T6G 2J1
Electronic Mail: <userid>@phys.ualberta.ca
WWW: http://rubble.phys.ualberta.ca/inst/
City Reference Coordinates: 113°28'00"W 53°33'00"N

University of British Columbia (UBC), Department of Earth and Ocean Studies
6339 Stores Road
Vancouver, BC V6T 1Z4
Telephone: 604-822-2449
Telefax: 604-822-6088
Electronic Mail: <userid>@eos.ubc.ca
WWW: http://www.eos.ubc.ca/
Founded: 1996
Staff: 110
Activities: Earth interior * near-surface geological studies * environmental earth science * oceans * atmosphere
City Reference Coordinates: 123°07'00"W 49°16'00"N

University of British Columbia (UBC), Department of Physics and Astronomy
2219 Main Mall
Vancouver, BC V6T 1Z4
or :
6224 Agricultural Road
Vancouver, BC V6T 1Z1
Telephone: 604-822-2267
            604-822-3853
            604-822-6673
Telefax: 604-822-6047
        604-822-5324
Electronic Mail: <userid>@physics.ubc.ca
                 <userid>@geop.ubc.ca
WWW: http://www.physics.ubc.ca/Homepage.html
     http://www.astro.ubc.ca/
Founded: 1996
Staff: 55
Activities: variable sources * insterstellar molecules * star formation * instrumentation * radial velocities * globular clusters * close groups of galaxies * QSOs * observational cosmology * model atmospheres * ISM
Periodicals: (1) "Annual Report" (ISSN 0068-1725)
Coordinates: 123°15'24"W 49°15'30"N
City Reference Coordinates: 123°07'00"W 49°16'00"N

University of Calgary, Department of Physics and Astronomy, Rothney Astrophysical Observatory (RAO)
2500 University Drive NW
Calgary, AB T2N 1N4
Telephone: 403-220-5385 (Department Office)
            403-931-2366 (RAO)
            403-220-5410
Telefax: 403-289-3331
Electronic Mail: <userid>@ucalgary.ca
WWW: http://www.phas.ucalgary.ca/
     http://www.acs.ucalgary.ca/~milone/rao.html (RAO)

Founded: 1972
Staff: 12
Activities: variable-star photometry * analysis of eclipsing binary stars light and radial-velocity curves * solar IR spectroscopy * Be stars * IR and optical astronomy * optical imaging
Periodicals: "Publications of the Rothney Astrophysical Observatory"
Coordinates: 114°17'18"W 50°52'06"N H1,272m (Priddis)
City Reference Coordinates: 114°05'00"W 51°03'00"N

University of Guelph, Department of Physics
MacNaughton Building
Guelph, ON N1G 2W1
Telephone: 519-824-4120 x2261
Telefax: 519-836-9967
Electronic Mail: <userid>@physics.uoguelph.ca
WWW: http://www.physics.uoguelph.ca/physics.html
Founded: 1964
Activities: education
Coordinates: 080°13'42"W 43°31'52"N H364m
City Reference Coordinates: 080°15'00"W 43°33'00"N

University of Lethbridge, Department of Physics
4401 University Drive
Lethbridge, AB T1K 3M4
Telephone: 403-329-2280
Telefax: 403-329-2057
Electronic Mail: naylor@uleth.ca (David A. Naylor)
WWW: http://home.uleth.ca/fas/phy/
Founded: 1967
Staff: 700
Activities: education * high-resolution IR spectroscopy
Coordinates: 113°00'00"W 50°00'00"N H1,000m (Astrophysical Observatory)
City Reference Coordinates: 110°50'00"W 49°42'00"N

University of Manitoba, Department of Physics and Astronomy
Winnipeg, MB R3T 2N2
Telephone: 204-474-9817
         204-883-2567 (Glenlea Astronomical Observatory)
Telefax: 204-474-7622
Electronic Mail: physics@umanitoba.ca
WWW: http://www.physics.umanitoba.ca/
         http://www.umanitoba.ca/faculties/science/astronomy/
Founded: 1997 (Department of Physics: 1904)
Coordinates: 097°07'15"W 49°38'43"N (Glenlea Astronomical Observatory)
City Reference Coordinates: 097°09'00"W 49°53'00"N

University of New Brunswick, Department of Physics
IUC Building
Bailey Drive
P.O. Box 4400
Fredericton, NB E3B 5A3
Telephone: 506-453-4723
Telefax: 506-453-4581
Electronic Mail: physics@unb.ca
WWW: http://www.unb.ca/physics/dept/
City Reference Coordinates: 066°39'00"W 45°58'00"N

University of Regina, Department of Physics
Regina, SK S4S 0A2
Telephone: 306-585-4149
Telefax: 306-585-4890
Electronic Mail: <userid>@meena.cc.uregina.ca
WWW: http://www.uregina.ca/
Founded: 1974
Staff: 16
Activities: subatomic physics * theoretical physics * general relativity * low-temperature metal physics * molecular biophysics * meteoroid streams
Coordinates: 104°35'26"W 50°25'08"N H575m
City Reference Coordinates: 104°39'00"W 50°25'00"N

University of Saskatchewan, Department of Physics and Engineering Physics
116 Science Place
Saskatoon, SK S7N 5E2
Telephone: 306-966-6396
Telefax: 306-966-6400
Electronic Mail: <userid>@physics.usask.ca

WWW: http://physics.usask.ca/
          http://physics.usask.ca/observ.htm (Duncan Observatory)
Founded: 1911
Staff: 117
Activities: education * observing
Coordinates: 106°37'53"W 52°07'55"N H500m (Duncan Observatory)
City Reference Coordinates: 106°38'00"W 52°07'00"N

University of Saskatchewan, Institute of Space and Atmospheric Studies (ISAS)
116 Science Place
Saskatoon, SK S7N 5E2
Telephone: 306-966-6401
Telefax: 306-966-6428
WWW: http://www.usask.ca/physics/isas/
Founded: 1957
Staff: 35
Activities: research in atmospheric and space environments * satellite and ground-based systems * education
Coordinates: 103°70'00"W 58°20'00"N (Rabbit Lake)
          107°07'00"W 52°12'00"N (Park Site)
          106°27'00"W 52°15'00"N (Bakker's Farm)
          106°32'00"W 52°09'00"N (Kernen Farm)
City Reference Coordinates: 106°38'00"W 52°07'00"N

University of Toronto, Department of Astronomy
Room 1403
60 Saint George Street
Toronto, ON M5S 3H8
Telephone: 416-978-2016
          905-884-2112 (DDO)
Telefax: 416-971-2026
          905-971-2026 (DDO)
Electronic Mail: <userid>@vela.astro.utoronto.ca
WWW: http://www.astro.utoronto.ca/home.html
          http://ddo.astro.utoronto.ca/ddohome (David Dunlap Observatory)
          http://www.astro.utoronto.ca/~utso/ (Univ. Toronto Southern Obs.)
Founded: 1935
Staff: 30
Activities: education * research * stellar and extragalactic astrophysics * David Dunlap Observatory (DDO)
Coordinates: 079°30'00"W 43°54'00"N H244m (DDO, Richmond Hill, Ontario)
          070°42'00"W 29°00'00"S H2,280m (Las Campanas, Chile)
City Reference Coordinates: 079°30'00"W 43°54'00"N

University of Toronto, Department of Astronomy, Erindale Campus
Mississauga Road
Mississauga, ON L5L 1C6
Telephone: 905-828-5351
          905-828-3818
Telefax: 416-828-5425
Electronic Mail: <userid>@astro.erin.utoronto.ca
WWW: http://www.erin.utoronto.ca/~astro/
Founded: 1967
Staff: 3
Activities: education * studies of stellar atmospheres * variable stars * stellar spectroscopy * planetarium (Erindale Campus Planetarium)
City Reference Coordinates: 079°37'00"W 43°35'00"N

University of Toronto, Institute for Aerospace Studies (UTIAS)
4925 Dufferin Street
Downsview, ON M3H 5T6
Telephone: 416-667-7700
Telefax: 416-667-7799
Electronic Mail: info@utias.utoronto.ca
WWW: http://www.utias.utoronto.ca/
Founded: 1949
Staff: 70
Activities: aerospace engineering * applied physics * research * education
Periodicals: (1) "Annual Progress Report" (ISSN 0082-5239, circ.: 1,200); "UTIAS Reports" (ISSN 0082-5255, circ.: 350), "UTIAS Reviews" (ISSN 0082-5247, circ.: 350), "UTIAS Technical Notes" (ISSN 0082-5263, circ.: 350)
City Reference Coordinates: 079°28'00"W 43°45'00"N

University of Toronto Press Inc.
10 Saint Mary Street
Suite 700
Toronto ON M4Y 2W8
Telephone: 416-978-2239
Telefax: 416-978-4738

Electronic Mail: info@utpress.utoronto.ca
                   <userid>@utpress.utoronto.ca
WWW: http://www.utpress.utoronto.ca/
Founded: 1901
City Reference Coordinates: 079°30'00"W 43°54'00"N

University of Victoria, Department of Physics and Astronomy
P.O. Box 3055
Victoria, BC V8W 3P6
Telephone: 250-721-7700
           250-721-7750 (Climenhaga Observatory)
Telefax: 250-721-7715
Electronic Mail: <userid>@phys.uvic.ca
WWW: http://astrowww.phys.uvic.ca/
        http://info.phys.uvic.ca/
        http://info.phys.uvic.ca/uvphys_welcome.html
        http://almuhit.phys.uvic.ca/
Founded: 1963
Staff: 22
Coordinates: 123°18'30"W 48°27'48"N H74m (Climenhaga Observatory)
City Reference Coordinates: 123°22'00"W 48°25'00"N

University of Waterloo, Department of Physics, Astronomy Programme
Waterloo, ON N2L 3G1
Telephone: 519-885-1572 x2215
Telefax: 519-746-8115
Electronic Mail: observe@astro.uwaterloo.ca
                   <userid>@watsci.waterloo.edu
                   <userid>@watsci.waterloo.ca
WWW: http://astro.uwaterloo.ca/
Periodicals: "Contributions of the University of Waterloo"
City Reference Coordinates: 080°31'00"W 43°28'00"N

University of Western Ontario (UWO), Physics and Astronomy Department
London, ON N6A 3K7
Telephone: 519-661-3183
          519-225-2620 (Elginfield Observatory)
Telefax: 519-661-2033
Electronic Mail: <userid>@uwo.ca
WWW: http://www.physics.uwo.ca/
        http://www.astro.uwo.ca/ (Astronomy Group)
Founded: 1966
Staff: 12
Activities: education * research * high-resolution spectroscopy * polarimetry * photometry * theory of Be stars * Ap stars * cool stars * stellar interiors * cosmology * ISM
Coordinates: 081°19'00"W 43°12'00"N H323m (Elginfield Observatory)
               081°16'34"W 43°00'18"N (Hume Cronyn Observatory)
City Reference Coordinates: 081°14'00"W 42°59'00"N

Western Space Education Network (WSEN) Inc.
2115 McEown Avenue
P.O. Box 1811
Saskatoon, SK S7K 3S2
Telephone: 306-374-1375
Telefax: 306-374-1395
Electronic Mail: wsen@the.link.ca
WWW: http://www.sfn.saskatoon.sk.ca/science/wsen
Founded: 1993
Staff: 3
Activities: space science educational outreach and resource centre * Canadian Space Resource Centre (Prairies)
City Reference Coordinates: 106°38'00"W 52°07'00"N

Wilfrid Laurier University (WLU), Department of Physics and Computing
75 University Avenue West
Waterloo, ON N2L 3C5
Telephone: 519-884-1970
Telefax: 519-746-0677
Electronic Mail: <userid>@mach1.wlu.ca
                   jlit@mach1.wlu.ca (John Lit, Chair)
WWW: http://info.wlu.ca/~wwwphys/
Founded: 1963
Staff: 1100
Activities: academic courses in astronomy and optics
City Reference Coordinates: 080°31'00"W 43°28'00"N

York University Astronomy Physics Club
c/o Room 329
Petrie Science Building
4700 Keele Street
North York, ON M3J 1P3
Telephone: 416-736-2100 x77763
Electronic Mail: yuac@yorku.ca
WWW: http://www.yorku.ca/org/yuapc/
Founded: 1998 (1987 as "York University Astronomy Club - YUAC")
City Reference Coordinates: 079°25'00"W 43°46'00"N

York University, Department of Earth and Atmospheric Sciences
Room 101/102
Petrie Science Building
4700 Keele Street
North York, ON M3J 1P3
Telephone: 416-736-5245
Telefax: 416-736-5817
Electronic Mail: eas@science.yorku.ca
WWW: http://www.eas.yorku.ca/
Founded: 1983
Staff: 12
Activities: geophysics * atmospheric dynamics * Earth's core * ozone layer * boundary layer * laboratory fluid dynamics * mesoscale meteorology * aeronomy * remote sensing * space sciences * instrumentation
City Reference Coordinates: 079°25'00"W 43°46'00"N

York University, Department of Physics and Astronomy
4700 Keele Street
North York, ON M3J 1P3
Telephone: 416-736-5249
Telefax: 416-736-5516
Electronic Mail: <userid>@sol.yorku.ca
WWW: http://science.yorku.ca/units/phas/home.html
        http://www.physics.yorku.ca/home.htm
        http://aries.phys.yorku.ca/physics/home.html
Founded: 1965
Staff: 28
Activities: education * research
City Reference Coordinates: 079°25'00"W 43°46'00"N

York University, Department of Physics and Astronomy, Space Astrophysics Laboratory (SAL)
Petrie Science Building
Room 332
4850 Keele Street
North York, ON M3J 3K1
Telephone: 416-736-2100 x77721 (John Caldwell)
        416-665-3311
Telefax: 416-736-5516
        416-665-2032
Electronic Mail: enquiries@sal.phys.yorku.ca
                <userid>@sal.phys.yorku.ca
WWW: http://www.sal.phys.yorku.ca/Welcome.html
● Previously "Institute for Space and Terrestrial Science (ISTS), Space Astrophysics Laboratory (SAL)"
City Reference Coordinates: 079°25'00"W 43°46'00"N

Channel Islands

La Société Guernesiaise, Section Astronomique
Highcliffe
Avenue Beauvais
Ville au Roi
Saint Peter Port
Guernsey
Telephone: (0)1481-724101
Electronic Mail: vs76@dial.pipex.com (David Le Conte, Newsletter Editor)
WWW: http://ds.dial.pipex.com/nightsky/astro/
Founded: 1972
Membership: 45
Activities: amateur astronomy * meetings * lectures * observing
Periodicals: (12) "Sagittarius"; (1) "Transactions"
Coordinates: 002°38'09"W 49°26'57"N H46m
City Reference Coordinates: 002°32'00"W 49°27'00"N

Chile

Academia de Astronomía Armazones
● See "Universidad Católica del Norte (UCN), Academia de Astronomía Armazones"

Asociación Chilena de Astronomía y Astronáutica (ACHAYA)
Marcoleta 485
Departamento H
Santiago
Telephone: (0)2-6327556
WWW: http://macul.ciencias.uchile.cl/achaya/
        http://macul.ciencias.uchile.cl/achaya/pochoco.html (Cerro Pochoco Observatory)
City Reference Coordinates: 070°40'00"W 33°27'00"S

Carnegie Institution of Washington (CIW), Observatories, Las Campanas Observatory (LCO)
Casilla 601
La Serena
Telephone: (0)51-224680 (El Pino Office)
        (0)51-211254
        (0)51-211032 (Office)
        (0)51-211413 (LCO Site)
WWW: http://www.ociw.edu/lco.html
Founded: 1969
Activities: field station
Coordinates: 070°42'00"W 29°00'30"S H2,282m
City Reference Coordinates: 071°18'00"W 29°54'00"S

Centro de Estudios Científicos de Santiago (CECS)
Casilla 16443
Santiago 9
or :
Avenida Presidente Errazuriz 3132
Las Condes
Santiago
Telephone: (0)2-2338342
        (0)2-2060092
Telefax: (0)2-2338336
Electronic Mail: cecs@cecs.cl
WWW: http://www.cecs.cl/
Founded: 1984
Staff: 20
Activities: theoretical physics * biophysics
City Reference Coordinates: 070°40'00"W 33°27'00"S

Cerro Tololo Interamerican Observatory (CTIO)
● See "National Optical Astronomy Observatories (NOAO), Cerro Tololo Interamerican Observatory (CTIO)"

Comisión Nacional de Investigaciones Científicas y Técnologicas (CONICYT)
Casilla 297-V
Santiago 21
or :
Canada 308
Santiago
Telephone: (0)2-2744537
Telefax: (0)2-2096729
Electronic Mail: <userid>@daniel.conicyt.cl
WWW: http://www.conicyt.cl/
Founded: 1968
Staff: 120
Activities: scientific policy * financing * information * administration
City Reference Coordinates: 070°40'00"W 33°27'00"S

Dirección Meteorológica de Chile
Aeropuerto Arturo Merino Benitez
Casilla 717
Correo Central
Santiago
Telephone: (0)2-6019001
Telefax: (0)2-6019590
Electronic Mail: dimetchi@reuna.cl
                dimetchi@meteochile.cl
WWW: http://www.meteochile.cl/
Founded: 1880

Staff: 273
Activities: weather forecast * climatology * consulting * instrumentation * network of about 200 stations
Periodicals: (12) "Boletín Climatológico", "Boletín Agrometeorológico"; (1) "Anuarios Meteorológicos"
City Reference Coordinates: 070°40'00"W 33°27'00"S

Dirección Meteorológica de Chile, Centro Meteorológico Antártico (CMA) "Presidente Eduardo Frei"
Base Antártica Teniente Rodolfo Marsh
Antártica
Electronic Mail: dimetchi@reuna.cl
                dimetchi@meteochile.cl
WWW: http://www.meteochile.cl/
Founded: 1969
Staff: 6
Activities: meteorological observing * weather forecast
Periodicals: (12) "Boletín Climatológico"
City Reference Coordinates: 058°54'00"W 62°12'00"S

Dirección Meteorológica de Chile, Centro Meteorológico Regional Austral
Aeropuerto Carlos Ibañez del Campo
Punta Arenas
Telephone: (0)61-211003
              (0)61-231292
              (0)61-231371
Telefax: (0)61-214623
Electronic Mail: dimetchi@reuna.cl
                dimetchi@meteochile.cl
WWW: http://www.meteochile.cl/
Founded: 1880 (DMC)
Staff: 16
Activities: weather forecast * observing
City Reference Coordinates: 070°55'00"W 53°09'00"S

Dirección Meteorológica de Chile, Centro Meteorológico Regional del Pacifico
Aeropuerto Mataveri
Isla de Pascua
Telephone: (0)108-237
              (0)108-245
              (0)108-246
Electronic Mail: dimetchi@reuna.cl
                dimetchi@meteochile.cl
WWW: http://www.meteochile.cl/
Founded: 1880 (DMC)
Staff: 7
Activities: weather forecast * observing
City Reference Coordinates: 109°23'00"W 27°08'00"S

Dirección Meteorológica de Chile, Centro Meteorológico Regional Norte
Casilla 90
Antofagasta
or :
Aerodromo Cerro Moreno
Antofagasta
Telephone: (0)55-269077
Telefax: (0)55-225022
Electronic Mail: jduarte@reuna.cl (Julio Duarte)
                dimetchi@meteochile.cl
WWW: http://www.meteochile.cl/
Founded: 1880 (DMC)
Staff: 11
Activities: weather forecast * meteorological instrumental activities
Coordinates: 070°26'00"W 23°27'00"S H135m
City Reference Coordinates: 070°26'00"W 23°27'00"S

Dirección Meteorológica de Chile, Centro Meteorológico Regional Sur
Aeropuerto El Tepual
Puerto Montt
Telephone: (0)65-253205
Telefax: (0)65-253205
Electronic Mail: dimetchi@reuna.cl
                dimetchi@meteochile.cl
WWW: http://www.meteochile.cl/
Founded: 1880 (DMC)
Staff: 15
Activities: weather forecast * observing
City Reference Coordinates: 072°57'00"W 41°28'00"S

European Southern Observatory (ESO), Antofagasta Office
Casilla 540
Antofagasta
or :
Balmaceda 2536
Oficina 504
Antofagasta
Telephone: (0)55-260032
             (0)55-260048
Telefax: (0)55-260081
Electronic Mail: <userid>@ls.eso.org
WWW: http://www.ls.eso.org/
         http://www.eso.org/observing/vlt/ (Paranal Observatory)
         http://www.hq.eso.org/
Founded: 1962 (ESO)
Coordinates: 070°24'10"W 24°37'30"S H2,635m
• See main entry in Germany
City Reference Coordinates: 070°24'00"W 23°39'00"S

European Southern Observatory (ESO), Chile
Casilla 19001
Santiago 19
or :
Alonso de Córdova 3107
Vitacura
Santiago
Telephone: (0)2-2285006
Telefax: (0)2-2285132
Electronic Mail: <userid>@ls.eso.org
                     <userid>@sc.eso.org
WWW: http://www.ls.eso.org/
         http://www.hq.eso.org/
Founded: 1962 (ESO)
• See main entry in Germany
City Reference Coordinates: 070°40'00"W 33°27'00"S

European Southern Observatory (ESO), La Serena Office
Casilla 567
La Serena
or :
Avenida El Santo 1538
La Serena
Telephone: (0)51-215175
Telefax: (0)51-225387
Electronic Mail: <userid>@ls.eso.org
WWW: http://www.ls.eso.org/lasilla/lasilla-homepage.html (La Silla Observatory)
         http://www.ls.eso.org/lasilla/LaSilla.html
         http://www.hq.eso.org/
Founded: 1962 (ESO)
• See main entry in Germany
City Reference Coordinates: 071°16'00"W 29°54'00"S

European Southern Observatory (ESO), La Silla Observatory
Casilla 19001
Santiago 19
Telephone: (0)2-6988757 (through Santiago)
             (0)2-6993425 (through Santiago)
             (0)2-6954263 (through Santiago)
             (0)51-224527 (through La Serena)
             (0)51-224932 (through La Serena)
Telefax: (0)2-6954263 (through Santiago)
         (0)51-224932 (through La Serena)
Electronic Mail: lasilla@eso.org
                     <userid>@ls.eso.org
WWW: http://www.ls.eso.org/
         http://www.ls.eso.org/lasilla/lasilla-homepage.html
         http://www.ls.eso.org/lasilla/LaSilla.html
         http://www.hq.eso.org/
Founded: 1962
Activities: astronomical research in the southern hemisphere
Coordinates: 070°43'48"W 29°15'24"S H2,347m
• See main entry in Germany
City Reference Coordinates: 070°40'00"W 33°27'00"S (Santiago)

European Southern Observatory (ESO), Paranal Observatory
Telephone: (0)281291

Telefax: (0)285064
WWW: http://www.eso.org/observing/vlt/
Founded: 1962 (ESO)
Coordinates: 070°24'10"W 24°37'30"S H2,635m
● See main entry in Germany
City Reference Coordinates: 070°24'00"W 23°39'00"S (Antofagasta)
● Use Antofagasta office for mail
● See main entry in Germany

Instituto Isaac Newton (IIN)
Casilla 8-9
Santiago 9
or :
Fernández Concha 472
Las Condes
Santiago
Telephone: (0)2-2172013
Telefax: (0)2-2172352
Founded: 1978
Staff: 4
Activities: globular clusters * novae * archaeoastronomy * galaxies * compact binaries * CP stars * AGNs * blazars
City Reference Coordinates: 070°40'00"W 33°27'00"S

Instituto Nacional de Normalización (INN)
Matías Cousiño 64 - 6° piso
Casilla 995
Santiago 1
Telephone: (0)2-6968144
Telefax: (0)2-6960247
Electronic Mail: inn@reuna.cl
WWW: http://www.inn.cl/
Founded: 1973
City Reference Coordinates: 070°40'00"W 33°27'00"S

Las Campanas Observatory (LCO)
● See "Carnegie Institution of Washington (CIW), Observatories, Las Campanas Observatory (LCO)"

La Silla Observatory
● See "European Southern Observatory (ESO), La Silla Observatory"

National Optical Astronomy Observatories (NOAO), Cerro Tololo Interamerican Observatory (CTIO), La Serena
Casilla 603
La Serena
Telephone: (0)51-225415
Telefax: (0)51-205342
　　　　(0)51-214458
Electronic Mail: <userid>@noao.edu
WWW: http://ctios2.ctio.noao.edu/ctio.html
　　　　http://www.ctio.noao.edu/ctio.html
　　　　http://ctiot6.ctio.noao.edu/ (Southern Columbia Millimeter Telescope - SCMT)
Founded: 1963
Staff: 117
Activities: astronomical research * observing * instrumentation * data reduction
Periodicals: (4) "NOAO Newsletter", "NOAO Quarterly Report"; (1) "NOAO Annual Report"
Coordinates: 070°48'53"W 30°09'57"S H2,215m
City Reference Coordinates: 071°18'00"W 29°54'00"S

Observatorio Tololito
Casilla 558
La Serena
Electronic Mail: leiton@ctio.noao.edu
WWW: http://www.ctio.noao.edu/~leiton/tololito.d/tololito.html
Founded: 1977
City Reference Coordinates: 071°18'00"W 29°54'00"S

Pontificia Universidad Católica de Chile
● See "Universidad Católica de Chile"

Universidad Católica de Chile, Departamento de Astrofísica y Astronomía
Casilla 306
Santiago 22
or :
Vicuña Mackenna 4860
Santiago

Telephone: (0)2-6864940
Telefax: (0)2-5525692
      (0)2-6864948
Electronic Mail: daa@astro.puc.cl
             graf@astro.puc.cl
WWW: http://www.astro.puc.cl/
Founded: 1929
Staff: 12
Activities: galaxies * clusters of galaxies * observational cosmology * cataclysmic variable stars * Cepheid variable stars
Periodicals: "Preprints"
Coordinates: 070°37'48"W 33°25'06"S H840m (Observatorio M. Foster)
City Reference Coordinates: 070°40'00"W 33°27'00"S
● Full university name: "Pontificia Universidad Católica de Chile"

Universidad Católica del Norte (UCN), Academia de Astronomía Armazones
Avenida Angamos 0610
Antofagasta
Electronic Mail: <userid>@sanpedro.cecun.ucn.cl
WWW: http://www.ucn.cl/~astro/
Founded: 1994
City Reference Coordinates: 070°24'00"W 23°39'00"S

Universidad Católica del Norte (UCN), Instituto de Astronomía
Avenida Angamos 0610
Antofagasta
Telephone: (0)55-241300
Electronic Mail: astro@sanpedro.ucn.cl
WWW: http://www.ucn.cl/~astro/
City Reference Coordinates: 070°24'00"W 23°39'00"S

Universidad de Chile, Departamento de Astronomía
Casilla 36-D
Santiago
or :
Cerro Calan s/n
Las Condes
Santiago
Telephone: (0)2-6784731
      (0)2-2294101
Telefax: (0)2-2294002
Electronic Mail: oficina@das.uchile.cl
            <userid>@das.uchile.cl
WWW: http://www.das.uchile.cl/
Founded: 1852
Staff: 18
Activities: astrometry * radioastronomy * astrophysics
Coordinates: 070°32'40"W 33°23'50"S H860m
City Reference Coordinates: 070°40'00"W 33°27'00"S

Universidad de Concepción, Departamento de Física, Grupo de Astronomía
Casilla 4009
Concepción
Telephone: (0)41-204500
Telefax: (0)41-245622
Electronic Mail: <userid>@coma.cfm.udec.cl
WWW: http://phys.cfm.udec.cl/~rmennick/grupo.html
City Reference Coordinates: 073°03'00"W 36°50'00"S

China (PRC)

Academia Sinica
● See "Chinese Academy of Sciences"

Acta Astronomica Sinica
c/o Purple Mountain Observatory
2 Beijing Xilu
Nanjing 210008
Telephone: (0)25-3302147
Telefax: (0)25-3301459
● Journal (4) (ISSN 0001-5245)
City Reference Coordinates: 118°47'00"E 32°03'00"N

Acta Astrophysica Sinica
c/o Beijing Astronomical Observatory (BAO)
Zhongguancun
Beijing 100080
Telephone: (0)10-64877287
Telefax: (0)10-64888731
Electronic Mail: zjz@bao.ac.cn (Zhao Jing-zhi, Editor)
WWW: http://www.bao.ac.cn/bao/publ/aaps
Founded: 1981
Staff: 3
● Journal (4) (ISSN 0253-2379)
City Reference Coordinates: 116°25'00"E 39°55'00"N

Beijing Ancient Observatory
Jian Guo Men Street
Beijing
Telephone: (0)542202
WWW: http://china-window.com/beijing/tour/museum/guan/
Founded: 1442
Activities: exhibitions * ancient astronomy research
City Reference Coordinates: 116°25'00"E 39°55'00"N

Beijing Astronomical Observatory (BAO)
● See "Chinese Academy of Sciences, Beijing Astronomical Observatory (BAO)"

Beijing Astronomical Data Center (BADC)
● See "Chinese Academy of Sciences, Beijing Astronomical Data Center (BADC)"

Beijing Astrophysics Center (BAC)
Peking University
5501 Yifu No. 1 Building
Beijing 100871
WWW: http://vega.bac.pku.edu.cn/
● Jointly run by the "Chinese Academy of Sciences" and "Beijing University"
City Reference Coordinates: 116°25'00"E 39°55'00"N

Beijing Normal University, Department of Astronomy
Beijing 100875
Telephone: (0)1-2012255 x2618
Telefax: (0)1-2013929
WWW: http://www.bnu.edu.cn/
Founded: 1960
Staff: 35
Activities: education * research * Chinese Visiting Lecturer Programme * planetarium
Coordinates: 116°21'37"E 39°57'27"N H70m
City Reference Coordinates: 116°25'00"E 39°55'00"N

Beijing Planetarium
138 Xi Wai Street
Beijing 100044
Telephone: (0)10-8361691
WWW: http://www.bao.ac.cn/cas/pic/bp.gif
Telefax: (0)10-8353003
Founded: 1957
Staff: 130
Activities: shows * exhibitions * education * observing * research * instrumentation
Periodicals: "Amateur Astronomer"
City Reference Coordinates: 116°25'00"E 39°55'00"N

Beijing University, Astrophysics Division
Beijing 100871
Telephone: (0)1-2552471 x3929
Telefax: (0)1-2564095
WWW: http://www.pku.edu.cn/
City Reference Coordinates: 116°25'00"E 39°55'00"N

Beijing University, Department of Geophysics
Beijing 100871
Telephone: (0)1-2552471 x3929
Telefax: (0)1-2564095
WWW: http://www.pku.edu.cn/
City Reference Coordinates: 116°25'00"E 39°55'00"N

Changchun Artificial Satellite Observatory
• See "Chinese Academy of Sciences, Changchun Artificial Satellite Observatory"

China Meteorological Administration (CMA)
46 Baishiqiao Road
Beijing 100081
Telephone: (0)10-62174797
Telefax: (0)10-62174797
            (0)10-62173417
WWW: http://www.cma.gov.cn/
Founded: 1951
Staff: 65000 (whole country)
Activities: meteorological observing, telecommunications, data processing, analysis and prediction * research and education * training
Periodicals: "Meteorology", "Meteorological Knowledge", "Journal of Meteorology", "Journal of Tropical Meteorology", "Journal of Tropical Meteorology", "Journal of Applied Meteorology"
City Reference Coordinates: 116°25'00"E 39°55'00"N

China State Bureau of Technical Supervision (CSBTS)
P.O. Box 8010
Beijing 100088
or :
4 Zhi Chun Road
Haidian District
Beijing
Telephone: (0)1-2032424
            (0)10-62032424
Telefax: (0)1-2031010
            (0)10-62031010
Electronic Mail: intl@std.csbts.cn.net
City Reference Coordinates: 116°25'00"E 39°55'00"N

Chinese Academy of Sciences, Beijing Astronomical Data Center (BADC)
c/o Beijing Astronomical Observatory
Datun Road 20A
Chaoyang District
Beijing 100012
Telephone: (0)10-64888763
Telefax: (0)10-64888731
Electronic Mail: ghf@bdc.bao.ac.cn (Guo Hongfeng)
WWW: http://www.bao.ac.cn/bdc/
Founded: 1987
City Reference Coordinates: 116°25'00"E 39°55'00"N

Chinese Academy of Sciences, Beijing Astronomical Observatory (BAO)
Zhongguancun
Beijing 100080
Telephone: (0)1-281968
Telefax: (0)1-2561085
Electronic Mail: bmabao@ica.beijing.canet.cn
                lqb@bao01.bao.ac.cn (Li Qibin, Director)
WWW: http://www.bao.ac.cn/bao/
Founded: 1279
Staff: 400
Activities: solar physics * stellar physics * galaxies * AGNs * QSOs * cosmology * radio astronomy * stellar catalogues * atomic time
Periodicals: "Acta Astrophysica Sinica" (see separate entry), "Publications of the Beijing Astronomical Observatory", "Chinese Solar Geophysical Data", "Chinese Astronomy Abstracts"
Coordinates: 117°34'30"E 40°23'42"N H960m (Xinglong Station)
            116°45'54"E 40°33'24"N H160m (Miyun Station)
            116°19'42"E 40°06'06"N H40m (Shahe Station)

116°36'00" E 40°19'00" N H62m (Huairou Station)
City Reference Coordinates: 116°25'00" E 39°55'00" N

Chinese Academy of Sciences, Changchun Artificial Satellite Observatory

P.O. Box 1067
Changchun
Jilin 130117
Telephone: (0)431-4511337
Telefax: (0)431-4513550
Electronic Mail: zhxh@mail.jlu.edu.cn (Cui Douxing, Director)
Founded: 1957
Staff: 56
Activities: artificial satellite observing * satellite laser ranging * GPS * celestial mechanics
Coordinates: 125°26'36" E 43°47'26" N H268m
City Reference Coordinates: 126°33'00" E 43°51'00" N

Chinese Academy of Sciences, Graduate School, Department of Physics

19A Yuguan Lu
P.O. Box 3908
Beijing 100039
Telephone: (0)1-8217031 x713
Founded: 1978
Activities: high-energy astrophysics * relativistic astrophysics * AGNs * QSOs * galaxies * cosmology * IRAS galaxies
Periodicals: (4) "Journal of the Graduate School"
City Reference Coordinates: 116°25'00" E 39°55'00" N

Chinese Academy of Sciences, Institute of High-Energy Physics, Laboratory of Cosmic-Ray and High-Energy Astrophysics

P.O. Box 918-3
19 Yuquanlu
Beijing 100039
Telephone: (0)10-6821334 x2129
WWW: http://astrosv1.ihep.ac.cn/
Founded: 1977
Staff: 70
Activities: cosmic rays * high-energy astrophysics * manufacturing and launching balloons
City Reference Coordinates: 116°25'00" E 39°55'00" N

Chinese Academy of Sciences, Nanjing Astronomical Instruments Research Centre (NAIRC)

P.O. Box 846
Nanjing 210042
or :
188 Bancang Jie
Nanjing
Jiangsu 210042
Telephone: (0)25-5411476
Telefax: (0)25-5411872
WWW: http://www.bao.ac.cn/nairc/
Founded: 1958
Staff: 500
Activities: research and manufacturing of astronomical instruments
● Formerly "Nanjing Astronomical Instruments Factory (NAIF)"
City Reference Coordinates: 118°47'00" E 32°03'00" N

Chinese Academy of Sciences, National Astronomical Observatories

Beijing 100012
City Reference Coordinates: 116°25'00" E 39°55'00" N

Chinese Academy of Sciences, Purple Mountain Observatory

2 Beijing Xilu
Nanjing 210008
Telephone: (0)25-3303921
         (0)25-3308986
Telefax: (0)25-3300818
Founded: 1934
Staff: 250
Activities: astrophysics * celestial mechanics * astrometry * history of astronomy * instrumentation
Periodicals: (4) "Publications of the Purple Mountain Observatory" (ISSN 1000-3681), "Acta Astronomica Sinica" (see separate entry)
Coordinates: 118°49'18" E 32°04'00" N H267m (Purple Mountain)
         097°43'00" E 37°22'00" N H3,200m (Delingha - see separate entry)
City Reference Coordinates: 118°47'00" E 32°03'00" N

Chinese Academy of Sciences, Purple Mountain Observatory, Delingha Radio Astronomy Station

5 Qing Xin Lu

Delingha 817000
Telephone: (0)977-221935
Telefax: (0)25-3301459
Founded: 1986
Staff: 39
Activities: mm-wave astronomy
Coordinates: 097°43'00"E 37°22'00"N H3,200m
City Reference Coordinates: 097°43'00"E 37°22'00"N

Chinese Academy of Sciences, Shaanxi Astronomical Observatory (CSAO)
P.O. Box 18
Lintong
Xian
Shaanxi
Telephone: (0)29-3890207
Telefax: (0)29-3890344
Electronic Mail: kyc@ms.sxso.ac.cn
                <userid>@ms.sxso.ac.cn
Founded: 1966
Staff: 370
Activities: astrometry * astrophysics * STP * celestial mechanics * natural satellites * applied historical astronomy * time service
Periodicals: (12) "Time and Frequency Bulletin" (ISSN 1001-1811); (2) "Publications of the Shaanxi Astronomical Observatory" (ISSN 1001-1544)
Coordinates: 109°33'06"E 34°56'42"N H468m
City Reference Coordinates: 108°52'00"E 34°15'00"N

Chinese Academy of Sciences, Shanghai Astronomical Observatory (SHAO)
80 Nandan Road
Shanghai 200030
Telephone: (0)21-64386191
          (0)21-57651763 (Station)
Telefax: (0)21-64384618
Electronic Mail: <userid>@center.shao.ac.cn
WWW: http://center.shao.ac.cn/
       http://www.shao.ac.cn/
Founded: 1872
Staff: 300
Activities: astro-geodynamics * astrometry * stellar astronomy * astrophysics * satellite dynamics
Periodicals: "Annals of Shanghai Observatory"
Coordinates: 121°11'12"E 31°05'48"N H98m (Sheshan Station)
City Reference Coordinates: 121°28'00"E 31°14'00"N

Chinese Academy of Sciences, Urumqi Astronomical Observatory (UAO)
40 Suburb. 5
South Beijing Road
Urumqi
Xinjiang 830011
Telephone: (0)991-3838007
Telefax: (0)991-3838628
Electronic Mail: uao@public.wl.xj.cn
Founded: 1957
Staff: 92
Coordinates: 087°14'00"E 43°15'00"N H2000m
City Reference Coordinates: 087°38'00"E 43°43'00"N

Chinese Academy of Sciences, Yunnan Observatory
P.O. Box 110
Kunming 650011
Telephone: (0)871-3347087
          (0)871-3368497
Telefax: (0)871-3358437
        (0)871-3911845
Electronic Mail: ynao@public.km.yn.cn
Founded: 1972
Staff: 304
Activities: stellar theory * stellar spectra * photometry * extragalactic astrophysics * solar activity and corresponding forecast * solar radio observing * solar seismology * VLBI * astrometry * celestial mechanics * observational techniques * planetarium * popular science and educational center
Periodicals: (4) "Publications of Yunnan Observatory" (ISSN 1001-7526)
Coordinates: 102°47'18"E 25°01'32"N H2,002m
City Reference Coordinates: 102°41'00"E 25°04'00"N

Chinese Astronomical Society (CAS)
c/o Purple Mountain Observatory
2 Beijing Xilu

Nanjing 210008
Telephone: (0)25-3302147
Telefax: (0)25-3301459
Electronic Mail: lqb@bao01.bao.ac.cn (Li Qibin, President)
WWW: http://www.bao.ac.cn/cas/
Founded: 1922
Membership: 1800
Activities: promoting study and development of astronomy * disseminating knowledge of astronomy * sponsoring scientific meetings * improving astronomical education * directing amateur astronomical activities * developing international collaborations * acting as consultant for related decision-making bodies
Sections: (committees) Organising * Teaching of Astronomy * Astronomical Terminology * Popularization of Astronomy * Publications, Library and Information
Sections: (commissions) Stellar and Planetary Physics * Galaxies and Cosmology * Radio Astronomy * Solar Physics and Solar-Terrestrial Relationships * Catalogues and astronomical Constants * Astrogeodynamics * Time and Frequency * Celestial Mechanics * Instruments and Techniques * History of Astronomy * Satellite Dynamics * High-Energy Astrophysics * Astronomy from Space
Periodicals: (24) "Tianwen Aihaozhe" (Astronomical Amateur) (ISSN 0493-2285); (4) "Acta Astronomica Sinica" (see separate entry), "Acta Astrophysica Sinica" (see separate entry), "Progress in Astronomy" (see separate entry); (1-2) "CAS Bulletin" (ISSN 0001-5245); "Astronomical Circulars"
City Reference Coordinates: 118°47'00"E 32°03'00"N

Chinese Society of Astronautics (CSA)
P.O. Box 1408
Beijing 100013
Telephone: (0)10-68372905
(0)10-68373488
Telefax: (0)10-68372521
Electronic Mail: iaf@public.bta.net.cn
Founded: 1980
Membership: 20
City Reference Coordinates: 116°25'00"E 39°55'00"N

Delingha Radio Astronomy Station
● See "Chinese Academy of Sciences, Purple Mountain Observatory, Delingha Radio Astronomy Station"

Geocarto International Centre (GIC)
G.P.O. Box 4122
Hong Kong
or :
Wah Ming Centre
Room 217
421 Queen's Road West
Hong Kong
Telephone: (0)852-25464262
Telefax: (0)852-25593419
Electronic Mail: geocarto@hkstar.com
WWW: http://www.geocarto.com/
Founded: 1985
Activities: publishing and distributing maps, satellite images, journals and books on remote sensing, geoscience and cartography
Periodicals: (4) "Geocarto International" (ISSN 1010-6049)
City Reference Coordinates: 114°10'00"E 22°17'00"N

Hong Kong Astronomical Society (HKAS)
G.P.O. Box 2872
Hong Kong
or :
5/F, 315 Des Voeux Road West
Sai Ying Pun
Hong Kong
Telephone: (0)852-25474543
Telefax: (0)852-27152345
Electronic Mail: astronet@hk.super.net
<userid>@hkas.org
WWW: http://www.ee.cuhk.hk/klyip/aas.html
http://www.hk.super.net/~astronet/
http://www.hkas.org/
Founded: 1970
● Formerly "Hong Kong Amateur Astronomical Society (HKAAS)"
City Reference Coordinates: 114°10'00"E 22°17'00"N

Hong Kong Observatory (HKO)
134A Nathan Road
Kowloon
Hong Kong
Telephone: (0)852-29268200

Telefax: (0)852-27215034
Electronic Mail: mailbox@hko.gcn.gov.hk
WWW: http://www.weather.gov.hk/
Founded: 1883
Staff: 336
Activities: meteorological and geophysical services
Coordinates: 114°10'19"E 22°18'13"N
• Formerly "Royal Hong Kong Observatory (RHKO)"
City Reference Coordinates: 114°10'00"E 22°17'00"N

Hong Kong Space Museum, Planetarium
10 Salisbury Road
Tsim Sha Tsui
Kowloon
Hong Kong
Telephone: (0)852-27342722
Telefax: (0)852-23115804
Electronic Mail: hkspm@spider.net.hk
WWW: http://www.usd.gov.hk/hkspm/
Founded: 1980
Staff: 135 (Museum)
Activities: sky shows * omnimax shows * lectures * education * exhibitions
Periodicals: (4) "Newsletter"; "Astro Calendar"
City Reference Coordinates: 114°10'00"E 22°17'00"N

Hong Kong University (HKU), Astronomy Club
Pokfulam Road
Hong Kong
Telephone: (0)852-28583657
Electronic Mail: suastro@hkusua.hku.hk
WWW: http://www.hku.hk/suastro/home41/
City Reference Coordinates: 114°10'00"E 22°17'00"N

Nanjing Astronomical Instruments Research Centre (NAIRC)
• See "Chinese Academy of Sciences, Nanjing Astronomical Instruments Research Centre (NAIRC)"

Nanjing Normal University, Department of Physics
Nanjing 210097
City Reference Coordinates: 118°47'00"E 32°03'00"N

Nanjing University, Astronomy Department
22 Hankou Road
Nanjing 210093
Telephone: (0)25-3592882
          (0)25-3593510
Telefax: (0)25-3302728
Electronic Mail: <userid>@nju.edu.cn
WWW: http://www.nju.edu.cn/
Founded: 1952
Staff: 40
Activities: high-energy astrophysics * solar physics * astrometry * ERP * celestial mechanics * history of astronomy * radio astronomy * QSOs * AGNs * galactic astronomy
Periodicals: "Journal of Nanjing University (Natural Sciences Edition)" (ISSN 0465-7926)
Coordinates: 118°46'07"E 32°02'26"N H60m
          118°51'00"E 32°03'00"N H36m (solar tower)
City Reference Coordinates: 118°47'00"E 32°03'00"N

National Astronomical Observatories
• See "Chinese Academy of Sciences, National Astronomical Observatories"

Progress in Astronomy
c/o Editorial Office
80 Nandan Road
200030 Shanghai
Telephone: (0)21-64386191
Telefax: (0)21-64384618
Electronic Mail: twxjz@center.shao.ac.cn
Founded: 1983
Staff: 5
• Journal (4) (ISSN 1000-8349)
City Reference Coordinates: 121°28'00"E 31°14'00"N

Purple Mountain Observatory
• See "Chinese Academy of Sciences, Purple Mountain Observatory"

Shaanxi Astronomical Observatory
● See "Chinese Academy of Sciences, Shaanxi Astronomical Observatoryi (CSAO)"

Shanghai Astronomical Observatory (SHAO)
● See "Chinese Academy of Sciences, Shanghai Astronomical Observatory (SHAO)"

Shanghai Institute of Electron Physics
● See "Shanghai University, Shanghai Institute of Electron Physics"

Shanghai Jiao Tong University (SJTU), Institute for Space Astrophysics
Department of Applied Physics
Shanghai 200030
WWW: http://www.sjtu.edu.cn/
City Reference Coordinates: 121°28'00"E 31°14'00"N

Shanghai Teachers University, Department of Physics
100 Guilin Road
Shanghai 200234
Telephone: (0)21-64700700
Telefax: (0)21-64339150
Electronic Mail: shtunet@public.sta.net.cn
Founded: 1954
Staff: 25
Activities: black holes * cosmology * early universe * particle physics
City Reference Coordinates: 121°28'00"E 31°14'00"N

Shanghai University, Shanghai Institute of Electron Physics
Jia Ding
Shanghai 201800
Telephone: (0)21-59528602
Telefax: (0)21-59529932
Founded: 1966
Staff: 120
Activities: electronics * microwave engineering * remote sensing * ISM * VLBI * radioastronomical instrumentation
Periodicals: (4) "Journal of Shanghai University" (ISSN 0258-7041)
Coordinates: 121°15'00"E 31°23'00"N H23m
City Reference Coordinates: 121°15'00"E 31°23'00"N

Sky Observers' Association (SOA)
155 Fuk Wing Street
4th. Floor - Room 6
Shamshuipo
Hong Kong
Telephone: (0)852-3861658
Electronic Mail: skyweb@hk.super.net
WWW: http://www.vol.net/~slam/soainf.htm
        http://www.hk.super.net/~skyweb/
Founded: 1972
Membership: 410
Activities: observing * lectures * instrument making * photography * education * camps
Periodicals: "The Sky Observers"
Awards: "Astronomical Observation Award"
City Reference Coordinates: 114°10'00"E 22°17'00"N

Tianjin Normal University, Department of Physics, Astrophysics Group
Tianjin 300074
Telephone: (0)22-27372231
Electronic Mail: xyxia@aso.bao.ac.cn (Xiaoyang Xia)
Founded: 1984
Staff: 4
Activities: galaxies * cosmology
City Reference Coordinates: 117°12'00"E 39°08'00"N

University of Science and Technology of China (USTC), Center for Astrophysics
Hefei 230026
Telephone: (0)551-3601861
Telefax: (0)551-3631760
Electronic Mail: cfaoffice@ustc.edu.cn
WWW: http://www.cfa.ustc.edu.cn/
Founded: 1973
Staff: 16
Activities: research * education
City Reference Coordinates: 117°17'00"E 31°51'00"N

Urumqi Astronomical Observatory (UAO)
● See "Chinese Academy of Sciences, Urumqi Astronomical Observatory (UAO)"

World Data Center D for Astronomy
c/o Beijing Astronomical Observatory
Datun Road 20A
Chao Yang District
Beijing 100012
Telephone: (0)10-64888734
Telefax: (0)10-64888731
Electronic Mail: ghf@class1.bao.ac.cn (Guo Hongfong)
Founded: 1988
City Reference Coordinates: 116°25'00"E 39°55'00"N

Yunnan Observatory
● See "Chinese Academy of Sciences, Yunnan Observatory"

Colombia

Asociación de Astrónomos Autodidactos de Colombia (ASASAC)
Calle 116 40'64
Apartado 112
Bogotá 2 DE
or :
Carretera 32 # 71-A-31
Apartado Aéreo 59534
Bogotá 2 DE
Telephone: (0)2401-721
          (0)1-2151850
Telefax: (0)1-2181894
Electronic Mail: dc34930@unalcol.bitnet
Founded: 1965
Membership: 35
Activities: telescope making * observing (occultations, variable stars, comets) * ephemerides * education * planetarium
Sections: (commissions) Celestial Mechanics * Astrophotography * Instruments * Variable Stars * Comets and Meteors *
Occultations * Planets * Galactic Astronomy * Ephemerides
Periodicals: "Quasar"
Coordinates: 074°05'00"W 04°39'00"N
City Reference Coordinates: 074°05'00"W 04°39'00"N

Asociación Colombiana de Estudios Astronómicos (ACDA)
Apartado Aéreo 47731
Bogotá 2 DE
or :
Carretera 11 # 82-60
Apartamento 202
Bogotá 8 DE
Telephone: (0)1-2563030
          (0)2-225525
          (0)2-563030
Electronic Mail: acda@inter.net.co
WWW: http://www2.inter.net.co/~nr005000/acda.html
Founded: 1986
Membership: 44
Activities: weekly meetings * education * lectures * telescope making
Periodicals: (4) "Kajuyali"
Coordinates: 073°48'00"W 05°06'00"N H2,584 (Suesca)
          073°54'00"W 05°21'00"N H2,980 (Carmen de Carupa)
          073°34'00"W 05°38'00"N H2,143 (Villa de Leyva)
City Reference Coordinates: 073°48'00"W 05°06'00"N

Instituto Colombiano de Normas Técnicas (ICONTEC)
Carrera 37 N° 52-95
Apartado 14237
Santafé de Bogotá
Telephone: (0)1-3150377
Telefax: (0)1-2221435
Electronic Mail: sicontec@col1.telecom.com.co
WWW: http://www.icontec.org.co/
Founded: 1963
Periodicals: (1) "Memoria Anual"; "Normas y Calidad", "Boletín Informativo"
City Reference Coordinates: 074°05'00"W 04°39'00"N

Instituto de Hidrología, Meteorología y Estudios Ambientales (IDEAM)
Apartado Aéreo 20032
Bogotá 1 DE
or :
Carretera 5 N°15-80
Bogotá DE
Telephone: (0)1-2836937
          (0)1-2860658
Telefax: (0)1-2842402
          (0)1-2860658
Electronic Mail: <userid>@ideam.gov.co
WWW: http://www.ideam.gov.co/
City Reference Coordinates: 074°05'00"W 04°39'00"N

Observatorio Astronómico Nacional (OAN), Bogotá
Apartado Aéreo 2584
Bogotá 1 DE

or :
Carrera 8
Calle 8
Bogotá
Telephone: (0)1-3423786
                (0)1-3165222
Telefax: (0)1-3165383
Electronic Mail: observat@ciencias.ciencias.unal.edu.co
WWW: http://argos.observatorio.unal.edu.co/
Founded: 1803
Staff: 10
Activities: education * astrometry * celestial mechanics * open clusters * cosmology * stellar structures * history of astronomy
Periodicals: "Publicaciones del Observatorio Astronómico Nacional"; (1) "Annuario del Observatorio Astronómico Nacional"

Coordinates: 074°04'54"W 04°35'54"N H2,640m
City Reference Coordinates: 074°05'00"W 04°39'00"N

Croatia

Andrija Mohorovicic Geophysical Institute
● See "University of Zagreb, Andrija Mohorovicic Geophysical Institute"

Astronomical Astronautical Society Zadar (AASZ)
(Astronomsko Astronautičko Društvo Zadar)
Brne Karnarutica 2
HR-23000 Zadar
Telefax: (0)23-314052
Electronic Mail: aad@zadarnet.hr
WWW: http://pubwww.srce.hr/astro/zadaren.html
Founded: 1980
Membership: 28
City Reference Coordinates: 015°14'00"E 44°07'00"N

Astronomical Astronautical Society Zagreb
c/o Sanjin Kovacic
Opaticka 22
HR-10000 Zagreb
Telephone: (0)1-271418
Telefax: (0)1-271418
Electronic Mail: sanjin.kovacic@public.srce.hr
                astro@jagor.srce.hr
WWW: http://pubwww.srce.hr/astro/aadzen.html
Activities: popularization * observing (Sun, meteors)
Periodicals: "Journal of the Astronomical Astronautical Society Zagreb"
City Reference Coordinates: 015°58'00"E 45°48'00"N

Astronomical Society Ivan Štefek
Školska bb
HR-44320 Kutina
Telephone: (0)44-687952
Telefax: (0)44-687952
Electronic Mail: astro-kutina@sk.tel.hr
WWW: http://www.angelfire.com/sc2/ADIvanS/
City Reference Coordinates: 016°48'00"E 45°29'00"N

Astronomical Society Leo Brenner
Zabrebačka 2/II kat
HR-51550 Mali Lošinj
Telephone: (0)51-233203
Electronic Mail: dorian.bozicevic@aad.hr (Dorian Božičevic, Vice President)
                bozicevic.dorian@excite.com
Founded: 1994
Membership: 35
Activities: observing * popularization * education
Periodicals: "foton" (ISSN 1331-5765)
Coordinates: 014°31'24"E 44°27'48"N H100m (Veli Lošinj)
              014°28'42"E 44°29'30"N H200m (Mali Lošinj)
              014°27'48"E 44°30'12"N H90m (Mali Lošinj)
City Reference Coordinates: 014°25'00"E 44°35'00"N (Lošinj)

Astronomical Society Pitomaca
Trg Kralja Tomislava 9
HR-43405 Pitomača
Telephone: (0)33-782217
WWW: http://pubwww.srce.hr/astro/pitomacaen.html
Founded: 1981
Membership: 100
City Reference Coordinates: 017°14'00"E 45°57'00"N

Astronomical Society "Gea X"
Maruliceva 5
HR-55000 Slavonski Brod
Telephone: (0)35-441743
WWW: http://pubwww.srce.hr/astro/sbroden.html
Membership: 90
City Reference Coordinates: 018°00'00"E 45°09'00"N

Croatian Physical Society
Bijenička cesta 32

P.O. Box 162
HR-10000 Zagreb
Telephone: (0)41-432480
Telefax: (0)41-432525
Electronic Mail: hfd@phy.hr
                   <userid>@hfd.hr
WWW: http://www.hfd.hr/hfd/
Membership: 45
City Reference Coordinates: 015°58'00"E 45°48'00"N

Hvar Observatory
● See "University of Zagreb, Faculty of Geodesy, Hvar Observatory"

Mohorovicic Geophysical Institute (Andrija_)
● See "University of Zagreb, Andrija Mohorovicic Geophysical Institute"

State Hydrometeorological Institute
Grič 3
HR-41100 Zagreb
Telephone: (0)1-421222
Telefax: (0)1-278703
WWW: http://madhz.dhz.hr/
Founded: 1947
Staff: 450
Activities: official meteorological and hydrological service
City Reference Coordinates: 015°58'00"E 45°48'00"N

State Office for Standardization and Metrology (DZNM)
Ulica grada Vukovara 78
HR-10000 Zagreb
Telephone: (0)1-6133444
Telefax: (0)1-536668
Electronic Mail: <userid>@dznm.hr
WWW: http://www.dznm.hr/
City Reference Coordinates: 015°58'00"E 45°48'00"N

University of Rijeka, Physics Department
Omladinska 14
HR-51000 Rijeka
Telephone: (0)51-227474
Telefax: (0)51-515142
Electronic Mail: <userid>@mapef.pefri.hr
Founded: 1972
Staff: 8
Activities: research and education in physics and astronomy
City Reference Coordinates: 014°27'00"E 45°20'00"N

University of Zagreb, Andrija Mohorovicic Geophysical Institute
Horvatovac BB
HR-10000 Zagreb
Telephone: (0)1-420222
              (0)1-4605900
Telefax: (0)1-432462
            (0)1-4680331
Electronic Mail: seiszag@olimp.irb.hr
WWW: http://www.phy.hr/~moho/geophysi.htm
Founded: 1861
Staff: 27
Activities: physics of the atmosphere, the sea and Earth interior
Periodicals: (1) "Geofizika" (ISSN 0352-3659, circ.: 300)
Coordinates: 015°59'24"E 45°49'48"N H182m (Zagreb/Horvatovac)
                015°58'12"E 45°54'00"N H988m (Puntijarka)
                016°27'00"E 43°10'48"N H250m (Hvar - see separate entry)
                014°23'24"E 45°20'24"N H75m (Rijeka)
                014°32'24"E 45°18'00"N H2m (Bakar)
City Reference Coordinates: 015°58'00"E 45°48'00"N

University of Zagreb, Faculty of Geodesy, Hvar Observatory
Kaciceva 26
HR-10000 Zagreb
or :
P.O. Box 18
HR-21450 Hvar
Telephone: (0)1-4561222
              (0)21-525966

Telefax: (0)21-741427
          (0)1-4828081
Electronic Mail: <userid>@geodet.geof.hr
WWW: http://pubwww.srce.hr/geo/hrv/gf.htm
Founded: 1972
Activities: solar physics * photoelectric photometry of variable stars * observing station of the University of Zagreb, Faculty of Geodesy
Periodicals: (1) "Hvar Observatory Bulletin" (ISSN 0351-2657, circ.: 400)
Coordinates: 016°01'18"E 45°49'30"N H146m (Zagreb)
          016°27'00"E 43°10'00"N H245m (Hvar)
City Reference Coordinates: 015°58'00"E 45°48'00"N (Zagreb)
          016°27'00"E 43°10'00"N (Hvar)

University of Zagreb, Institute of Physics
Bijenička 46
P.O.Box 304
HR-10000 Zagreb
Telephone: (0)1-4605555
Telefax: (0)1-4680336
WWW: http://www.phys.hr/
City Reference Coordinates: 015°58'00"E 45°48'00"N

Visnjan Observatory
Istarska 5
HR-51463 Visnjan
Telephone: (0)52-449212
Telefax: (0)52-449212
Electronic Mail: korado@astro.hr (Korado Korlevic)
WWW: http://www.astro.hr/
Founded: 1976
Staff: 1
Membership: 32
Activities: meteors * minor planets * instrumentation * education
Periodicals: (12) "Nebeske Krijesnice" (ISSN 1330-4410); (1) "Acta Astronomica Visnianiensis" (ISSN 1330-2620)
Coordinates: 013°43'26"E 45°16'38"N H243m (Visnjan)
          013°46'00"E 45°11'00"N H297m (Rusnjak)
City Reference Coordinates: 013°43'00"E 45°16'00"N

Cuba

Academia de Ciencias de Cuba (ACC)
Capitolio Nacional
La Habana 2
Telephone: (0)6-8914
Electronic Mail: acc@ceniai.cu
WWW: http://www.ceniai.inf.cu/acc/ACADEMIA.HTM
Founded: 1962
City Reference Coordinates: 082°22'00"W 23°08'00"N

Comité Estatal de Normalización (CN)
Egido 610 entre Gloria y Apodaca
La Habana 10100
Telephone: (0)7-621503
Telefax: (0)7-338048
Electronic Mail: ncnorma@ceniai.inf.cu
Founded: 1976
Staff: 300
Activities: standardization * metrology * quality assurance
Periodicals: (2) "Normalización" (ISSN 0138-8118)
City Reference Coordinates: 082°22'00"W 23°08'00"N

Grupo de Aficionados a la Astronomía
c/o Universidad de Oriente
Apartado Postal 261
Santiago de Cuba 90100
Telephone: (0)226-23689
Telefax: (0)226-32263
Electronic Mail: astro@csd.uo.edu.cu
City Reference Coordinates: 075°49'00"W 20°00'00"N

Instituto de Geofísica y Astronomía (IGA)
Calle 212 No. 2906, entre 29 y 31
La Coronela
Lisa
La Habana 11600
Telephone: (0)218435
            (0)210644
            (0)214331
Telefax: (0)339497
Founded: 1964
Staff: 22
Activities: solar physics * International Sun Service daily observations * computing * astrometry
Periodicals: (1) "Datos Astronómicos para Cuba" (ISSN 0864-0645), "Ciencias de la Tierra y del Espacio" (ISSN 0138-6026)

Coordinates: 082°28'00"W 23°04'00"N H10m (Havana Radioastronomical Station)
            082°23'00"W 22°57'00"N H160m (Havana Optical Station)
City Reference Coordinates: 082°28'00"W 23°04'00"N

Instituto de Meteorología (INSMET)
La Habana
Electronic Mail: meteoro@ceniai.inf.cu
WWW: http://www.met.inf.cu/
Founded: 1965
City Reference Coordinates: 082°28'00"W 23°04'00"N

Sociedad Meteorológica de Cuba
Apartado Postal 4279
La Habana 11400
Telephone: 730-8996
Telefax: 733-3681
Electronic Mail: lulu@met.inf.cu
WWW: http://www.met.inf.cu/sometcub/default.htm
Membership: 500
Periodicals: "Boletín de SOMETCUBA" (ISSN 1025-921x)
City Reference Coordinates: 082°28'00"W 23°04'00"N

Czech Republic

Academy of Sciences of the Czech Republic
Národní třída 3
111 42 Praha 1
Telephone: (0)2-24240532
Telefax: (0)2-24240608
Electronic Mail: rihova@kav.cas.cz (Blanka Říhová, President of the Council for International Affairs)
WWW: http://www.cas.cz/
Founded: 1992
Staff: 6000
Activities: research * education
City Reference Coordinates: 014°26'00"E 50°05'00"N

Academy of Sciences of the Czech Republic, Astronomical Institute, Ondřejov
Fričova 1
251 65 Ondřejov
Telephone: (0)204-649201
Telefax: (0)204-620110
Electronic Mail: asu@asu.cas.cz
WWW: http://www.asu.cas.cz/
Founded: 1953
Staff: 127
Activities: stellar, solar and meteor astronomy * satellite dynamics * celestial mechanics
Periodicals: "Time and Latitude" (ISSN 1210-0463), "Solar Radio Data"
Coordinates: 014°47'00"E 49°54'40"N H528m (Ondřejov Observatory)
City Reference Coordinates: 014°47'00"E 49°55'00"N

Academy of Sciences of the Czech Republic, Astronomical Institute, Praha
Boční II 1401
141 31 Praha 4
Telephone: (0)2-67103111
Telefax: (0)2-769023
Electronic Mail: asu@asu.cas.cz
WWW: http://www.asu.cas.cz/
Founded: 1953
Staff: 133
City Reference Coordinates: 014°26'00"E 50°05'00"N

Academy of Sciences of the Czech Republic, Institute of Physics
Na Slovance 2
180 40 Praha 8
Telephone: (0)2-66052660
      (0)2-66051111
Telefax: (0)2-821227
      (0)2-8584569
Electronic Mail: <userid>@fzu.cz
      fzu@cas.cz
WWW: http://www.fzu.cz/
Founded: 1954
City Reference Coordinates: 014°26'00"E 50°05'00"N

Academy of Sciences of the Czech Republic, Institute of Plasma Physics
Za Slovankou 3
P.O. Box 17
182 00 Praha 8
Telephone: (0)2-66052037
      (0)2-66413030
      (0)2-66051111
Telefax: (0)2-8586389
Electronic Mail: ipp@ipp.cas.cz
      <userid>@ipp.cas.cz
WWW: http://tokamak.ipp.cas.cz/
Founded: 1959
Staff: 91
Activities: basic and applied research in high-temperature and low-temperature plasma physics (experimental and theoretical)
City Reference Coordinates: 014°26'00"E 50°05'00"N

Academy of Sciences of the Czech Republic, Optical Development Workshop
Skálova 89
511 01 Turnov
Telephone: (0)436-22718
      (0)436-22622

          (0)436-22587
Telefax: (0)436-22913
Founded: 1965
Staff: 24
Activities: research * development * manufacturing * optical components * glass * crystals * X-rays * thin films
City Reference Coordinates: 015°10'00"E 50°35'00"N

Association of Observatories and Planetariums
c/o Eva Marková (Chairman)
Úpice Observatory
P.O. Box 8
542 32 Úpice
Telephone: (0)439-881731
Telefax: (0)439-881289
Electronic Mail: obsupice@mbox.vol.cz
Founded: 1991
Membership: 19 (observatories and planetariums)
Activities: protecting common interests of observatories and planetariums * educating staff of public observatories * advising
Coordinates: 016°00'44"E 50°30'27"N H416m
City Reference Coordinates: 016°01'00"E 50°30'00"N

Astro Klub Kostkov
Návsí 645
739 92 Jablunkov
Telephone: (0)659-932623
Telefax: (0)659-25877
Electronic Mail: honza@elanor.sci.muni.cz
WWW: http://www.sci.muni.cz/~honza
          http://monoceros.physics.muni.cz/~honza/asko.htm
Founded: 1994
Membership: 41
Activities: observing * lectures * instrumentation * photography * CCDs * interplanetary matter * symbiotic stars * software
City Reference Coordinates: 018°47'00"E 49°35'00"N

Astronomers in Lelekovice
Lelekovice 393
664 31 Lelekovice
Telephone: (0)5-41232105
Electronic Mail: hornoch@astro.sci.muni.cz (Kamil Hornoch)
          xplsek01@dcse.fee.vutbr.cz
WWW: http://astro.sci.muni.cz/lelek/
Founded: 1996
Membership: 2
Activities: observing (comets, novae, SNs, meteors)
Coordinates: 016°39'18"E 49°21'15"N H365m
City Reference Coordinates: 016°39'00"E 49°21'00"N

Benátky nad Jizerou Popular Observatory
c/o Radovan Kováč
Platanová 647
294 71 Benátky nad Jizerou
Telephone: (0)326-762664
Founded: 1980
Staff: 3
Activities: popularization * observing
Coordinates: 014°50'09"E 50°17'24"N H210m
City Reference Coordinates: 014°51'00"E 50°17'00"N

Boskovice Observatory
Kulturní Zařízení Města Boskovice
Kpt Jaroše 15
680 01 Boskovice
Telephone: (0)501-453544
Telefax: (0)501-453253
Electronic Mail: kopecky@fch.vutbr.cz (Pavel Kopecký)
          nemecm@post.cz
Founded: 1960
Staff: 5
Activities: public observatory
Coordinates: 016°29'00"E 49°30'00"N H370m
City Reference Coordinates: 016°29'00"E 49°30'00"N

Brno N. Copernicus Observatory and Planetarium
Kraví Hora
616 00 Brno

Telephone: (0)5-41321287
Telefax: (0)5-41321287
       (0)5-744374
Electronic Mail: hvezdarn@sci.muni.cz
WWW: http://www.sci.muni.cz/obsbrno/
Founded: 1953
Staff: 21
Activities: education * popularization * lectures * Summer camps * telescope making * observing (eclipsing binaries, meteors, peculiar stars, comets, planets, eclipses)
Periodicals: (1) "Contributions of Brno N. Copernicus Observatory and Planetarium" (ISSN 0862-173x, circ.: 400); (4) "Perseus" (circ.: 150)
Coordinates: 016°35'18"E 49°12'15"N H304m (Brno)
              017°01'34"E 49°17'06"N H310m (Vyškov)
City Reference Coordinates: 016°37'00"E 49°12'00"N

České Budějovice Observatory and Planetarium

Zátkovo Nábřeží 4
370 01 České Budějovice
Telephone: (0)38-6352044
Telefax: (0)38-6352239
Electronic Mail: hap@ipex.cz
WWW: http://www.ipex.cz/HaP
       http://www.hvezcb.cz/index_e.html
Founded: 1937
Staff: 12
Activities: popularization * education * exhibitions * library
Coordinates: 014°28'20"E 48°58'23"N H394m
City Reference Coordinates: 014°29'00"E 48°58'00"N

Charles University, Astronomical Institute

V. Holešovičkách 2
180 00 Praha 8
Telephone: (0)2-21912572
Telefax: (0)2-21912577
       (0)2-6885095
       (0)2-299272
Electronic Mail: solc@mbox.cesnet.cz (Martin Solc, Director)
WWW: http://astro.mff.cuni.cz/
Founded: 1889
Staff: 6
Activities: spectrometry * photometry * cosmology * stellar astronomy * comets * minor planets * ISM * relativistic astrophysics
City Reference Coordinates: 014°26'00"E 50°05'00"N

Charles University, Centrum for Theoretical Study

Ovocny trh 3
116 36 Praha 1
Telephone: (0)2-263089
Telefax: (0)2-269466
Electronic Mail: <userid>@cuni.cz
WWW: http://www.cts.cuni.cz/
Founded: 1990
Staff: 10
Activities: research * seminars * workshops * mathematics * physics * astronomy * cognitive and computer sciences * political philosophy
Periodicals: "Research Reports"
City Reference Coordinates: 014°26'00"E 50°05'00"N

Czech Astronomical Society (CAS)

Královská Obora 233
170 21 Praha 7
Telephone: (0)2-370840 (answering machine)
Electronic Mail: borovic@asu.cas.cz (Jiří Borovička, President)
WWW: http://www.astro.cz/
Founded: 1917
Membership: 450
Activities: lectures * conferences * observing camps * press releases * observing (variable stars, meteors, comets, occultations, Sun)
Periodicals: "Kosmické Rozhledy"
City Reference Coordinates: 014°26'00"E 50°05'00"N

Czech Hydrometeorological Institute (CHMI)

Na Šabatce 17
Komořany
143 06 Praha 4

Telephone: (0)2-4095111
          (0)2-4016503
          (0)2-4019600 (Foreign Relations)
          (0)2-4019062 (Meteorology)
          (0)2-4016617 (Hydrology)
          (0)2-4016717 (Air Quality)
Telefax: (0)2-4010800
Electronic Mail: <userid>@chmi.cz
WWW: http://www.chmi.cz/
Founded: 1919
Staff: 905
Activities: meteorology * hydrology * climatology * air pollution
Periodicals: (6) "Meteorological Bulletin" (ISSN 0026-1173); (1-2) "Transactions" (ISSN 0232-0401)
City Reference Coordinates: 014°26'00"E 50°05'00"N

Czech Office for Standards, Metrology and Testing (COSMT)
Biskupský dvur 5
110 02 Praha 1
or :
Václavské Námestí 19
113 47 Praha 1
Telephone: (0)2-24224734
          (0)2-1802311
Telefax: (0)2-24224726
          (0)2-2324373
Electronic Mail: u30-csni@login.cz
City Reference Coordinates: 014°26'00"E 50°05'00"N

Czech Technical University, Astronomical Observatory
Thákurova 7
166 29 Praha 6
Telephone: (0)2-311279
Electronic Mail: <userid>@fsv.cvut.cz
Founded: 1926
Staff: 1
Activities: astrometry * celestial mechanics
Coordinates: 014°23'15"E 50°06'20"N
City Reference Coordinates: 014°26'00"E 50°05'00"N

Ďáblice Observatory
● See "Praha Observatory and Planetarium, Ďáblice Observatory"

European Astronomical Society (EAS)
c/o Jan Palouš (Secretary)
Astronomical Institute
Academy of Sciences of the Czech Republic
Boční II 1401
141 31 Praha 4
Telephone: (0)2-67103065
Telefax: (0)2-769023
Electronic Mail: palous@ig.cas.cz
WWW: http://www.iap.fr/eas/
Founded: 1991
Membership: 1500
Activities: coordinating and fostering astronomy in Europe * meetings
Periodicals: "EAS Newsletter"
City Reference Coordinates: 014°26'00"E 50°05'00"N

Group of Amateur Astronomers (GAMA)
c/o Rudolf Novak
Nicholas Copernicus Observatory
Kravi hora 2
616 00 Brno
Electronic Mail: gama@sci.muni.cz
WWW: http://www.sci.muni.cz/~gama/
          http://www.ian.cz/cba/
City Reference Coordinates: 016°37'00"E 49°12'00"N

Hradec Králové Observatory and Planetarium
Zamecek 456,
500 08 Hrádec Kralové
Telephone: (0)49-5264087
Telefax: (0)49-5267952
Electronic Mail: planethk@planethk.anet.cz
WWW: http://www.astrohk.cz/

http://www.astrohk.cz/indexen.html (English page)
City Reference Coordinates: 015°50'00"E 50°13'00"N

Jeseník School Observatory

Postovní 115
790 01 Jeseník
Telephone: (0)645-2477
Founded: 1965
Activities: lectures * observing * photography
Coordinates: 017°12'00"E 50°14'00"N H450m
City Reference Coordinates: 017°12'00"E 50°14'00"N

Jindřichuv Hradec Public Observatory

Gymnázium V. Nováka 333/II
377 01 Jindřichuv Hradec
Telephone: (0)331-22564
Telefax: (0)331-362248
Electronic Mail: strobl@port.troja.mff.cuni.cz
Founded: 1961
Membership: 15
Activities: popularization
Coordinates: 015°00'04"E 49°07'48"N H492m
City Reference Coordinates: 015°00'00"E 49°09'00"N

Karlovy Vary City Observatory

(Hvězdárna Úřadu Města Karlovy Vary)
c/o Miroslav Spurný
P.O. Box 175
360 01 Karlovy Vary
or :
Odbor Kulturya Školství
Moskevskaá 20
360 21 Karlovy Vary
Telephone: (0)17-25772
Telefax: (0)17-3222913 (Attn: Hvězdárna p. Spurný)
Electronic Mail: ok1msh@prgate.sci.muni.cz
WWW: http://www.karlovarsko.cz/hvezdar/
Founded: 1963
Staff: 1
Activities: popularization * observing (variable stars, comets, AGNs)
Coordinates: 012°54'29"E 50°12'52"N H615m (Hurky)
City Reference Coordinates: 012°52'00"E 50°11'00"N

Klet Observatory

Zátkovo Nábřeží 4
370 01 České Budějovice
Telephone: (0)337-711242
Electronic Mail: klet@klet.cz
              klet@asteroid.cz
WWW: http://www.klet.cz/
Founded: 1957
Staff: 4
Activities: minor planets * near-Earth objects * comets * astrometry
Coordinates: 014°17'17"E 48°51'46"N H1,070m
City Reference Coordinates: 014°28'00"E 48°59'00"N

Liberec Astronomy Club

P.O. Box 24
462 12 Liberec 25
Telephone: (0)48-29557
          (0)48-24553
Founded: 1963
Membership: 15
Activities: variable stars
Coordinates: 015°03'50"E 50°50'51"N H425m
            015°04'10"E 50°42'00"N H568m
City Reference Coordinates: 015°03'00"E 50°46'00"N

Masaryk University, Department of Theoretical Physics and Astrophysics

Kotlářská 2
611 37 Brno
Telephone: (0)5-41129251
Telefax: (0)5-41211214
Electronic Mail: <userid>@physics.muni.cz
WWW: http://astro.sci.muni.cz/

http://www.physics.muni.cz/
http://www.physics.muni.cz/mb/mb.htm
Founded: 1945
Staff: 12
Activities: education * relativity * quantum field theory * gravitation * spectroscopy * CCD photometry * close binary stars * carbon stars
Coordinates: 016°35'18"E 49°12'15"N H310m (Kraví Hora)
City Reference Coordinates: 016°37'00"E 49°12'00"N

Most Planetarium
P.O. Box 138
434 01 Most
or :
c/o Cultural Institution of Miners
Námestí VMS c. 4
434 25 Most
Telephone: (0)35-796381
          (0)35-42256
          (0)35-769384
WWW: http://www.mumost.cz/english/turisti/planetarium/eplanet.htm
Founded: 1984
Activities: shows for schools and public
City Reference Coordinates: 013°39'00"E 50°32'00"N

Ondřejov Observatory
● See "Czech Academy of Sciences, Astronomical Institute"

Palacky University, Department of Theoretical Physics
Trida Svobody 26
771 46 Olomouc
Telephone: (0)68-5222451 x374
Telefax: (0)68-5225737
Electronic Mail: <userid>@risc.upol.cz
Founded: 1954
Staff: 3
Activities: observing (Sun, meteors, occultations) * education * public parties
Coordinates: 017°13'06"E 49°34'14"N
City Reference Coordinates: 017°16'00"E 49°36'00"N

Planetarium Praha
● See "Praha Observatory and Planetarium, Planetarium Praha"

Plzeň Observatory and Planetarium
Dráhy 11
318 03 Plzeň
Telephone: (0)19-288400
Electronic Mail: hvezdarna@mmp.plzen-city.cz
Founded: 1954
Staff: 8
Activities: education * occultations
Periodicals: (12) "Observatory and Planetarium Reporter"; "Occultation Express"
Coordinates: 013°26'24"E 49°41'58"N H405m
City Reference Coordinates: 013°25'00"E 49°45'00"N

Praha Observatory and Planetarium, Planetarium Praha
Královská Obora 233
170 21 Praha 7
Telephone: (0)2-371746
          (0)2-377069
          (0)2-377746
Telefax: (0)2-375970
Electronic Mail: planetarium@planetarium.cz
WWW: http://www.planetarium.cz/
Founded: 1960
Staff: 30
Activities: school programmes * shows * education * exhibitions
City Reference Coordinates: 014°26'00"E 50°05'00"N

Praha Observatory and Planetarium
Petřín 205
118 46 Praha 1
Telephone: (0)2-57320540 (Observatory)
          (0)2-33376452 (Planetarium)
Telefax: (0)2-33376437
Electronic Mail: observat@ms.anet.cz

WWW: http://planet.infoserver.cz/
Founded: 1928
Membership: 45
Activities: popularization * education * publishing * ISM * maps * atlases * lunar occultations
Periodicals: (1) "Hvězdářská Ročenka" (ISSN 0373-8280)
Coordinates: 014°23'58"E 50°04'56"N H327m (Stefanik Observatory, Petrín)
               014°28'00"E 50°08'00"N H325m (Ďáblice - see separate entry)
               014°17'17"E 48°51'46"N H1,068m (Klet)
City Reference Coordinates: 014°26'00"E 50°05'00"N

Praha Observatory and Planetarium, Ďáblice Observatory
182 00 Praha 8
Telephone: (0)2-83910644
Electronic Mail: dabliceobs@planetarium.cz
WWW: http://www.planetarium.cz/
Founded: 1956
Staff: 5
Activities: occultations * variable stars
Coordinates: 014°28'00"E 50°08'00"N H325m
City Reference Coordinates: 014°26'00"E 50°05'00"N (Praha)

Přerov Astronomical Club
Hvězdárna
Macharova 21
750 02 Přerov
or :
c/o Miloš Slaný
Kosmákova 47
750 02 Přerov
or :
c/o Vladimir Kovarik
Zalátovská 22
750 00 Přerov
Telephone: (0)641-242361
Founded: 1961
Membership: 28
Activities: education * observing (occultations, meteors)
Periodicals: (1) "Annual Report"
Coordinates: 017°28'54"E 49°26'51"N H230m
City Reference Coordinates: 017°27'00"E 49°27'00"N

Prostějov Popular Observatory
Kolářovy Sady 3348
796 01 Prostějov
Telephone: (0)508-24130
Electronic Mail: hvezdarna.pv@mbox.vol.cz
WWW: http://www.oku-pv.cz/hvezdarna
Founded: 1962
Staff: 5
Activities: variable stars * solar activity * photography of planets
Periodicals: (12) "Zpravodaj"
Coordinates: 017°05'58"E 49°28'08"N H225m
City Reference Coordinates: 017°07'00"E 49°29'00"N

Research Institute of Geodesy, Topography and Cartography
250 66 Zdiby 98
Telephone: (0)2-6857351
Telefax: (0)2-6857056
Electronic Mail: vugtk@vugtk.cz
WWW: http://panurgos.fsv.cvut.cz/~odis/vugtk.shtml
Founded: 1954
Staff: 40
Activities: geodetic and cartographic research * astrometry
Periodicals: "Proceedings of Research Works"
Coordinates: 015°00'00"E 50°00'00"N
City Reference Coordinates: 015°00'00"E 50°00'00"N

Rokycany Astronomical Observatory
Voldusska 721/II
337 11 Rokycany
Telephone: (0)181-722622
Electronic Mail: halir@okn-ro.cz (Karel Halír, Director)
Founded: 1947
Staff: 5
Activities: Sun * occultations

Coordinates: 013°36'16"E 49°45'07"N H400m
City Reference Coordinates: 013°36'00"E 49°45'00"N

Rtyně v Podkrkonoší Observatory
Soukromá Hvězdárna 143
542 33 Rtyně v Podkrkonoší
Telephone: (0)439-787384
Electronic Mail: modr1@volny.cz
Founded: 1980
Staff: 2
Activities: private observatory * astronomical optics * mechanical manufacturing
City Reference Coordinates: 016°01'00"E 50°30'00"N (Úpice)

Uherský Brod Cultural House Observatory
c/o Rostislav Rajchl
Praksická 2222
688 11 Uherský Brod
Telephone: (0)633-634690
Founded: 1961
Membership: 15
Activities: solar activity * education * popularization * history of astronomy * archeoastronomy
Coordinates: 017°38'51"E 49°02'18"N H310m
City Reference Coordinates: 017°39'00"E 49°02'00"N

Union of Czech Mathematicians and Physicists, Physics Section
Na Slovance 2
180 40 Praha 8
Telephone: (0)2-66052910
(0)2-825459
Telefax: (0)2-8586389
(0)2-8584569
Electronic Mail: krejci@ipp.cas.cs
WWW: http://www-hep.fzu.cz/~www/
http://www.ujf.cas.cz:8080/fvs/fvs.html
Founded: 1968
Membership: 763
Activities: organizing activities of physicists in the Czech Republic * seminars * conferences * promoting knowledge of physics in the society
Periodicals: (6) "Bulletin of the Physics Section" (together with the "Czechoslovak Journal of Physics" - ISSN 0009-0700)
Awards: "Gold Medal", "Silver Medal"
City Reference Coordinates: 014°26'00"E 50°05'00"N

Úpice Observatory
P.O. Box 8
542 32 Úpice
Telephone: (0)439-932289
(0)439-881731
Telefax: (0)439-881289
Electronic Mail: obsupice@mbox.vol.cz
WWW: http://www.trutnov.vol.cz/obsupice/
Founded: 1959
Staff: 12
Activities: observing (solar activity, eclipses, comets) * popularization * optical counterparts of gamma-ray bursts * meteorology * seismology * air pollution
Periodicals: (4) "Kvart" (ISSN 1211-2453); "Proceedings", "Circulars"
Coordinates: 016°00'44"E 50°30'27"N H416m
City Reference Coordinates: 016°01'00"E 50°30'00"N

Valašské Meziříčí Astronomical Observatory
Vsetínská 78
757 01 Valašské Meziříčí
Telephone: (0)651-611928
Telefax: (0)651-611928
Electronic Mail: libor.lenza@vm.inext.cz
hvezdhvm@vm.inext.cz
WWW: http://www.inext.cz/hvezdarna/
Founded: 1955
Staff: 11
Activities: Sun * occultations * symbiotic stars * education * popularization
Periodicals: "Leaflets"
Coordinates: 017°58'32"E 49°27'51"N H338m
City Reference Coordinates: 017°58'00"E 49°28'00"N

Veselí nad Moravou Observatory
698 01 Veselí nad Moravou

Telephone: (0)631-2614
Electronic Mail: micek@vmb.cz
Founded: 1960
Activities: observing & photography (meteors, comets, occultations, sunspots, planets) * popularization of astronomy and astronautics
Coordinates: 017°22'17"E 48°57'17"N H163m
City Reference Coordinates: 017°22'00"E 48°58'00"N

Vlašim Astronomical Society (VAS)
B. Martinů 1341
258 01 Vlašim
Telephone: (0)303-42923
Telefax: (0)303-45169
Electronic Mail: jan.urban@csop.cz (Jan B. Urban, President)
WWW: http://www.vas.cz/
Founded: 1991
Membership: 53
Activities: solar activity * meteorology * gamma-ray burst sources * occultations * popularization * audio-visual programs
Periodicals: "Sunspot Report", "SEA Report", "Meteorological Report"
Coordinates: 014°53'47"E 49°41'37"N H423m
City Reference Coordinates: 014°54'00"E 49°42'00"N

VSB Technical University, Observatory and Planetarium
Listopadu 17
708 33 Ostrava-Poruba
Telephone: (0)69-6911007
Telefax: (0)69-6911009
Electronic Mail: thomas.graf@vsb.cz
                    planetarium@vsb.cz
WWW: http://www.vsb.cz/PLANET/welcome.htm
Founded: 1980
Staff: 6
Activities: popularization * solar activity * occultations * variable stars * education
Coordinates: 018°08'50"E 49°50'18"N H276m
• Full university name: "Vysoka Skola Banská Technical University"
City Reference Coordinates: 018°17'00"E 49°50'00"N

Vyškov Observatory
P.O. Box 43
682 00 Vyškov
Telephone: (0)507-21668
Electronic Mail: qhajek@fee.vutbr.cz (Petr Hájek)
Founded: 1970
Staff: 5
Activities: variable stars * planets * Sun
Coordinates: 017°01'34"E 49°17'06"N H254m
City Reference Coordinates: 017°01'00"E 49°17'00"N

Vysoka Skola Banská Technical University
• See "VSB Technical University"

Ždánice Public Observatory
Lovecká 678
696 32 Ždánice
Telephone: (0)629-633356
Founded: 1965
Staff: 2
Activities: variable stars
Coordinates: 017°02'27"E 49°03'58"N H226m
City Reference Coordinates: 017°02'00"E 49°04'00"N

Žebrák Popular Observatory
Lidova Hvezdárna
267 53 Žebrák
Founded: 1951
Staff: 3
Activities: observing (Moon, Sun, stars) * photography
Coordinates: 013°53'56"E 49°52'28"N H370m
City Reference Coordinates: 013°54'00"E 49°53'00"N

Zlín Astronomical Observatory
Hvezdárna
760 01 Zlín
Telephone: (0)67-36045
Electronic Mail: coufal@batahospital.iqnet.cz (Zdeněk Coufal, Chairman)

WWW: http://www.zas.cz/
Founded: 1993
Membership: 47
Activities: variable stars * occultations
Periodicals: (6) "Zpravodaj Zorné Pole"
Coordinates: 017°41'53"E 49°13'08"N H333m
City Reference Coordinates: 017°40'00"E 49°14'00"N

Denmark

Ålborg Amateur Radio Astronomy Observatory (AARAO)
Kalmanparken 20
DK-9200 Frejlev
Electronic Mail: i5sp@civil.auc.dk
               steffen.petersen@civil.auc.dk
WWW: http://www.protein.auc.dk/~i5sp/radioastronomy/AARAO.html
City Reference Coordinates: 009°56'00"E 57°03'00"N (Ålborg)

Amateur Astronomers of Eastern Jutland
(Østjyske Amatør Astronomer - ØAA)
c/o Elisabeth Siegel
Agertoften 2
DK-8340 Malling
Telephone: (0)86933458
Telefax: (0)86930015
Electronic Mail: wsiegel@daimi.au.dk
WWW: http://www.imf.au.dk/~kls/astro/oaa.html
Founded: 1977
Membership: 45
Activities: observing * instrumentation * meetings * computing
Periodicals: "Stjerneskuddet"
City Reference Coordinates: 010°12'00"E 56°02'00"N

Århus University, History of Science Department
Ny Munkegade
Building 521
DK-8000 Århus C
Telephone: (0)89423512
Telefax: (0)89423510
Electronic Mail: ievhlk@ifa.aau.dk
WWW: http://www.dfi.aau.dk/ievh/home.htm
Founded: 1965
Staff: 10
Activities: research in the history of science from Babylonian to modern times * museum for the history of science * archive with Hertzsprung and Strömgren documents
Periodicals: (4) "Centaurus" (ISSN 0008-8994)
City Reference Coordinates: 010°13'00"E 56°09'00"N

Århus University, Institute of Physics and Astronomy
Bygning 520
DK-8000 Århus C
Telephone: (0)86-128899
Telefax: (0)86-120740
Electronic Mail: <userid>@obs.aau.dk
WWW: http://www.dfi.aau.dk/
        http://www.aau.dk/uk/nat/fysik/
        http://www.obs.aau.dk/
Founded: 1911
Activities: minor planets * astro- and helioseismology * stellar abundances * CCD photometry and spectroscopy * formation and structure of galaxies * cosmology
Coordinates: 010°14'00"E 56°07'42"N H46m (Ole Rømer Observatory)
City Reference Coordinates: 010°13'00"E 56°09'00"N

Association Internationale de Géodésie (AIG)
• See "International Association of Geodesy (IAG)"

Astro Mekanik
Berenstorffsgade 32
DK-9000 Ålborg
Telephone: (0)98-134396
Telefax: (0)98-164226
Electronic Mail: .astro@pip.dknet.dk
WWW: http://pip.dknet.dk/~pip10846
Founded: 1965
Staff: 4
Activities: manufacturing and distributing astronomical instruments and accessories * representing Meade and GOTO
Coordinates: 009°56'32"E 57°02'16"N
City Reference Coordinates: 009°56'00"E 57°03'00"N

Astronomical Society of South Zealand
(Astronomisk Forening for Sydsjaelland - AFS)
c/o Ove Larsen (President)
Fasanvej 2
DK-4370 Store Merløse
Telephone: (0)57801812
Telefax: (0)57801812
Electronic Mail: otlastro@post.tele.dk
Founded: 1969
Membership: 50
Activities: education * observing * CCD imaging
Coordinates: 011°45'00"E 55°13'10"N H63m
City Reference Coordinates: 011°44'00"E 55°33'00"N

Astronomisk Forening for Sydsjaelland (AFS)
● See "Astronomical Society of South Zealand"

Astronomisk Forlag
c/o Tycho Brahe Planetarium
Gl. Kongevej 10
DK-1610 København V
Telephone: (0)1-33144666
Telefax: (0)1-33142888
Electronic Mail: tycho@tycho.dk
WWW: http://www.tycho.dk/
Founded: 1978
● Publisher
City Reference Coordinates: 012°35'00"E 55°40'00"N

Astronomisk Selskab i Danmark
● See "Danish Astronomical Association"

Bohr Institute (NBI) (Niels_)
● See "Copenhagen University, Niels Bohr Institute for Astronomy, Physics and Geophysics (NBIfAFG), Niels Bohr Institute (NBI)"

Brahe Planetarium (Tycho_)
● See "Tycho Brahe Planetarium"

Copenhagen Astronomical Society
(Københavns Astronomiske Forening - KAF)
c/o Claus Jensen (President)
Måløvvang 35 7B
DK-2760 Måløv
or :
c/o Gert Larsen (Secretary)
Auroravej 58
DK-2610 Rodovre
Telephone: (0)44-65-17-66
         (0)36-722011
Telefax: (0)36-72-20-11
WWW: http://www2.dk-online.dk/users/hpersson/kaf.htm
Founded: 1978
Membership: 175
Activities: general astronomy * all aspects of amateur astronomy
City Reference Coordinates: 012°35'00"E 55°40'00"N (Copenhagen)

Copenhagen Astronomical Observatory
● See "Copenhagen University, Niels Bohr Institute for Astronomy, Physics and Geophysics (NBIfAFG), Astronomical Observatory"

Copenhagen University, Niels Bohr Institute for Astronomy, Physics and Geophysics (NBIfAFG)
Blegdamsvej 17
DK-2100 København Ø
Telephone: (0)35325241
Telefax: (0)35431087
Electronic Mail: <userid>@nbi.dk
WWW: http://www.nbi.dk/NBIfAFG/
Founded: 1993
Staff: 145
City Reference Coordinates: 012°35'00"E 55°40'00"N

Copenhagen University, Niels Bohr Institute for Astronomy, Physics and Geophysics (NBIfAFG), Astronomical Observatory
Juliane Maries Vej 30
DK-2100 København Ø
Telephone: (0)35323999
Telefax: (0)35323989
Electronic Mail: <userid>@astro.ku.dk
WWW: http://www.astro.ku.dk/
Founded: 1993 (NBIfAFG)
Staff: 16
Activities: cosmology * extragalactic research * galactic research * stellar structure * astrometry
Coordinates: 012°34'45"E 55°41'12"N
City Reference Coordinates: 012°35'00"E 55°40'00"N

Copenhagen University, Niels Bohr Institute for Astronomy, Physics and Geophysics (NBIfAFG), Department of Geophysics
Juliane Maries Vej 30
DK-2100 København Ø
Telephone: (0)35320602
Telefax: (0)35365357
Electronic Mail: <userid>@gfy.ku.dk
WWW: http://www.gfy.ku.dk/
Founded: 1993 (NBIfAFG)
Staff: 18
Activities: geophysics * geodesy * solid-Earth physics * oceanography * glaciology * meteorology
Observatories: 1 (Greenland)
City Reference Coordinates: 012°35'00"E 55°40'00"N

Copenhagen University, Niels Bohr Institute for Astronomy, Physics and Geophysics (NBIfAFG), Ørsted Laboratory
Universitetsparken 5
DK-2100 København Ø
Telephone: (0)35-321818
Telefax: (0)35-320460
Electronic Mail: <userid>@fys.ku.dk
WWW: http://ntserv.fys.ku.dk/hco/
Founded: 1993 (NBIfAFG)
Staff: 20
Activities: research * education * solid-state physics * ion implantation * metastable phases * chaos * hydrodynamics * Mössbauer spectroscopy * laser spectroscopy * planetary and meteoritic sciences
City Reference Coordinates: 012°35'00"E 55°40'00"N

Copenhagen University, Niels Bohr Institute for Astronomy, Physics and Geophysics (NBIfAFG), Niels Bohr Institute (NBI)
Blegdamsvej 17
DK-2100 København Ø
Telephone: (0)31-421616
Telefax: (0)31-421016
Electronic Mail: <userid>@nbi.dk
WWW: http://www.nbi.dk/
Founded: 1993 (NBIfAFG)
Activities: extragalactic research * galactic dynamics
City Reference Coordinates: 012°35'00"E 55°40'00"N

Danish Astronautical Society
(Dansk Selskab for Rumfartsforksning - DSR)
Postboks 31
DK-1002 København K
or :
Branddamsvej 179
DK-2860 Soeborg
Telephone: (0)29677633
Telefax: (0)35362282
Electronic Mail: dsr@forening.dk (Thomas A.E. Andersen, President)
WWW: http://www.rumfart.dk/
Founded: 1949
Membership: 170
Activities: meetings * exhibitions
Periodicals: (4) "Dansk Rumfart" (ISSN 0905-2410)
City Reference Coordinates: 012°35'00"E 55°40'00"N (København)

Danish Astronomical Association
(Astronomisk Selskab i Danmark)
c/o Observatoriet
Juliane Maries Vej 30

DK-2100 København Ø
Telephone: (0)3532-5999
Electronic Mail: paldrich@inet.uni-c.dk
WWW: http://www.dsri.dk/AS/
         http://www.dsri.dk/AS/as-uk.html
Founded: 1916
Membership: 700
Activities: observing * computing * public lectures and presentations
Sections: planets * meteors * satellites * occultations * variable stars * telescopes and instrumentation * history of astronomy
Periodicals: (4) "Astronomisk Tidsskrift" (ISSN 0004-6345); "Knudepunktet"
City Reference Coordinates: 012°35'00"E 55°40'00"N

Danish Meteorological Institute (DMI)
Lyngbyvej 100
DK-2100 København
Telephone: (0)39-157500
Telefax: (0)39-271080
WWW: http://www.dmi.dk/
Founded: 1872
Staff: 400
Activities: monitoring and forecasting weather and climate, including the middle and upper atmosphere * research
City Reference Coordinates: 012°35'00"E 55°40'00"N

Danish Natural Science Research Council
Randergade 60
DK-2100 København
Telephone: (0)35446200
Telefax: (0)35446201
Electronic Mail: fr@forskraad.dk
WWW: http://www.forskraad.dk/
Founded: 1968
Membership: 6
City Reference Coordinates: 012°35'00"E 55°40'00"N

Danish Physical Society
(Dansk Fysisk Selskab - DFS)
c/o Jens Olaf Pepke Pedersen
Danish Center for Earth System Science (DCESS)
Juliane Maries Vej 30
DK-2100 København Ø
Telephone: (0)35320573
Telefax: (0)35320576
Electronic Mail: jopp@dcess.ku.dk
WWW: http://dfswww.fysik.dtu.dk/dfs/
Membership: 620
City Reference Coordinates: 012°35'00"E 55°40'00"N

Danish Space Research Advisory Board
Bredgade 43
DK-1260 København K
Telephone: (0)33929734
Telefax: (0)33150205
Electronic Mail: cbl@fsk.dk
WWW: http://www.fsk.dk/
Founded: 1996
Membership: 12
Activities: advisory board on space research
City Reference Coordinates: 012°35'00"E 55°40'00"N

Danish Space Research Institute (DSRI)
Juliane Maries Vej 30
DK-2100 København Ø
Telephone: (0)35325830
Telefax: (0)35362475
Electronic Mail: <userid>@dsri.dk
WWW: http://dsri.dk/
         http://www.dsri.dk/
Founded: 1966
Activities: developing balloon, rocket and satellite instrumentation for X- and gamma-ray astronomy * cosmic rays * electromagnetic and particle interactions in the magnetosphere
City Reference Coordinates: 012°35'00"E 55°40'00"N

Danish Standards Association (DS)
Baunegårdsvej 73
DK-2900 Hellerup

Telephone: (0)39-770101
Telefax: (0)39-770202
Electronic Mail: ds@ds.dk
WWW: http://www.ds.dk/
Founded: 1926
Staff: 150
Activities: standardization
Periodicals: "Dansk Standard"
City Reference Coordinates: 012°36'00"E 55°44'00"N

Dansk Fysisk Selskab (DFS)
● See "Danish Physical Society"

Dansk Selskab for Rumfartsforksning (DSR)
● See "Danish Astronautical Society"

Euromath Center (EmC)
Universitetsparken 5
DK-2100 København Ø
Telephone: (0)35320705
Telefax: (0)35320719
Electronic Mail: <userid>@euromath.dk
WWW: http://laurel.euromath.dk/
Founded: 1989
Staff: 9
Activities: development * documentation * integration * distribution * user support
Periodicals: (2) "Euromath Bulletin"
City Reference Coordinates: 012°35'00"E 55°40'00"N

Federation of Astronomical and Geophysical Data Analysis Services (FAGS)
c/o Niels Andersen (Secretary)
Kort & Matrikelstyrelsen
Rentemestervej 8
DK-2400 København NV
Telephone: (0)35875283
Telefax: (0)35875057
Electronic Mail: fags@kms.dk
WWW: http://www.kms.dk/fags
Founded: 1956
Activities: collecting and processing observations and data related to astronomy, geodesy, geophysics and related sciences
City Reference Coordinates: 008°24'00"E 49°03'00"N

International Association of Geodesy (IAG)
(Association Internationale de Géodésie - AIG)
c/o Department of Geophysics
Niels Bohr Institute for Astronomy, Physics and Geophysics
Juliane Maries Vej 30
DK-2100 København Ø
Telephone: (0)35320600
Telefax: (0)35365357
Electronic Mail: iag@gfy.ku.dk
WWW: http://www.gfy.ku.dk/~iag/
Founded: 1922
Membership: 78 (national committees)
Activities: promoting study of geodetic problems * encouraging geodetic research * coordinating international cooperation
Periodicals: (12) "Journal of Geodesy" (ISSN 0949-7714); (1/4) "Travaux de l'Association Internationale de Géodésie"
Awards: (1) "Best Paper Award"; (1/4) "Bomford Prize", "Levallois Medal"
City Reference Coordinates: 012°35'00"E 55°40'00"N

Københavns Astronomiske Forening (KAF)
● See "Copenhagen Astronomical Society"

Niels Bohr Institute (NBI)
● See "Copenhagen University, Niels Bohr Institute for Astronomy, Physics and Geophysics (NBIfAFG), Niels Bohr Institute (NBI)"

Niels Bohr Institute for Astronomy, Physics and Geophysics (NBIfAFG)
● See "Copenhagen University, Niels Bohr Institute for Astronomy, Physics and Geophysics (NBIfAFG)"

Nordic Institute for Theoretical Physics (NORDITA)
Blegdamsvej 17
DK-2100 København Ø
Telephone: (0)35325500
Telefax: (0)35389157
Electronic Mail: nordita@nordita.dk

WWW: http://www.nordita.dk/
Founded: 1957
Staff: 26
Activities: astrophysics * nuclear physics * condensed-matter physics * elementary-particle physics * complex systems
City Reference Coordinates: 012°35'00"E 55°40'00"N

Odense University, Physics Department
Campusvej 55
DK-5230 Odense M
Telephone: (0)66-158600
Telefax: (0)66-158760
Electronic Mail: <userid>@dou.dk
WWW: http://www.ou.dk/Nat/Fysik/
Founded: 1966
Staff: 13
City Reference Coordinates: 010°23'00"E 55°24'00"N

Ole Rømer Museet (ORM)
Kroppedals Allé 3
Vridsløsemagle
DK-2630 Tåstrup
Telephone: (0)43-529585
Telefax: (0)43-990163
Electronic Mail: ormus@page.dk
WWW: http://www.page.dk/ormus/
Founded: 1979
Staff: 8
Activities: exhibitions * history of astronomy
Periodicals: "Meddelelser fra Ole Rømer Venner"
Coordinates: 012°18'26"E 55°41'14"N
City Reference Coordinates: 012°19'00"E 55°39'00"N

Ole Rømer Observatory (ORO)
Observatorievej 3
DK-8000 Århus C
Telephone: (0)89423975
Telefax: (0)89423995
Electronic Mail: stenoplt@aau.dk
WWW: http://www.obs.aau.dk/oro/oro.html
Founded: 1910
Staff: 6
Activities: public shows * research
Coordinates: 010°14'00"E 56°07'42"N H46m
City Reference Coordinates: 010°13'00"E 56°09'00"N

Orion Planetarium
Sovej 36
Jels
DK-6630 Rødding
Telephone: (0)74552400
Telefax: (0)74552655
Electronic Mail: bacher@fys.ku.dk (Camilla Bacher, Director)
WWW: http://www.kulturnet.dk/museer/haandbog/mht485.html
Founded: 1993
Staff: 4
City Reference Coordinates: 009°06'00"E 55°23'00"N

Ørsted Laboratory
● See "Copenhagen University, Niels Bohr Institute for Astronomy, Physics and Geophysics (NBIfAFG), Ørsted Laboratory"

Østjyske Amatør Astronomer (ØAA)
● See "Amateur Astronomers of Eastern Jutland"

Risø National Laboratory
Frederiksborgvej 399
P.O. Box 49
DK-4000 Roskilde
Telephone: (0)4677-4677
Telefax: (0)4677-5688
Electronic Mail: risoe@risoe.dk
WWW: http://www.risoe.dk/
Founded: 1958
Staff: 900
City Reference Coordinates: 012°05'00"E 55°39'00"N

Rømer Museet (ORM) (Ole_)
● See "Ole Rømer Museet (ORM)"

Rømer Observatory (ORO) (Ole_)
● Ole Rømer Observatory (ORO)

Royal Danish Academy of Sciences and Letters
H.C. Andersens Boulevard 35
DK-1553 København V
Telephone: (0)33-435300
Telefax: (0)33-435301
Electronic Mail: kdvs@royalacademy.dk
WWW: http://www.royalacademy.dk/
Founded: 1742
Membership: 475
Activities: scientific meetings * public lectures * member of ICSU and adhering bodies * publishing * cooperation with international research groups * advising
Periodicals: "Matematisk-Fysiske Meddelelser", "Annual Report"
Awards: "Silver Medal"
City Reference Coordinates: 012°35'00"E 55°40'00"N

Steno Museum
(Steno Museet)
C.F. Møllers Allé
Building 100
University Park
DK-8000 Århus C
Telephone: (0)89423975
Telefax: (0)89423995
Electronic Mail: stenomus@aau.dk
                    stenoplt@aau.dk (Planetarium)
WWW: http://www.aau.dk/~stenomus/
                http://www.aau.dk/~stenomus/planetarium/planetarium.html
Founded: 1994
Staff: 30
Activities: exhibitions * herbal garden * planetarium * Ole Rømer Observatory (see separate entry) * lectures * guided tours * museum shop * class room
Coordinates: 010°14'00"E 56°07'42"N H46m (Ole Rømer Observatory)
City Reference Coordinates: 010°13'00"E 56°09'00"N

Theoretical Astrophysics Center (TAC)
Juliane Maries Vej 30
DK-2100 København Ø
Telephone: (0)35325900
Telefax: (0)35325910
Electronic Mail: <userid>@tac.dk
WWW: http://www.tac.dk/
Founded: 1994
Staff: 25
Activities: large-scale structure * formation and evolution of galaxies *AGNs * structure and oscillations of Sun and stars
City Reference Coordinates: 012°35'00"E 55°40'00"N

Tycho Brahe Planetarium
Gl. Kongevej 10
DK-1610 København V
Telephone: (0)33-121224 (Booking)
                (0)33-144888 (Administration)
Telefax: (0)33-142888
Electronic Mail: tycho@tycho.dk
WWW: http://www.tycho.dk/
Founded: 1989
Staff: 30
Activities: shows * OmniMax films
Periodicals: (4) "Aktuel Astronomi" (ISSN 0905-8958)
City Reference Coordinates: 012°35'00"E 55°40'00"N

World Association of Industrial and Technological Research Organisations (WAITRO)
c/o Secretariat
Danish Technological Institute
P.O. Box 141
DK-2630 Tåstrup
Telephone: (0)43507045
Telefax: (0)43507050
Electronic Mail: waitro@dti.dk
WWW: http://waitro.dti.dk/

Founded: 1970
Membership: 150 (from 76 countries)
Activities: international cooperation in industrial and technological R&D
Periodicals: (4) "WAITRO News"
Awards: (1/2) "WAITRO Honorary Award"
City Reference Coordinates: 012°29'00"E 55°39'00"N

Ecuador

Centro Cultural del Instituto Geográfico Militar (CCIGM)
Casilla 17-01-2435
Quito
or :
Seniergues y General Pazmiño
El Dorado
Quito
Telephone: (0)2-502091
           (0)2-522148 x3503-3550
Telefax: (0)2-569097
Electronic Mail: igm@igm-mil.ec
Founded: 1988
Staff: 16
Activities: cultural and scientific education
City Reference Coordinates: 078°30'00"W 00°13'00"S

Fundación Mundo Juvenil, Planetario
Avenida Los Shyris y Rusia s/n
Quito
Telephone: (0)2-244314
Founded: 1966
Staff: 12
City Reference Coordinates: 078°30'00"W 00°13'00"S

Instituto Ecuatoriano de Normalización (INEN)
A. Baquerizo Moreno 454 y 6 de Diciembre
Casilla 17-01-3999
Quito
Telephone: (0)2-524499
           (0)2-565626
           (0)2-544885
Telefax: (0)2-527561
        (0)2-567815
        (0)2-222223
Electronic Mail: inen1@inen.gov.ec
WWW: http://www4.ecua.net.ec/inen/
Founded: 1970
City Reference Coordinates: 078°30'00"W 00°13'00"S

Instituto Nacional de Meteorología e Hidrología (INAMHI)
Iñaquito 700 y Corea
Quito
Telephone: (0)2-433934
Telefax: (0)2-433934
Electronic Mail: inamhi@ecnet.ec
WWW: http://www.ecua.net.ec/inamhi
Founded: 1961
Staff: 4
Activities: weather and climate forecast * network of stations in the whole country
Periodicals: (12) "Meteorological Monthly Bulletin"; (1) "Meteorological and Hydrological Annual Report"
City Reference Coordinates: 078°30'00"W 00°13'00"S

Observatorio Astronómico de Quito (OAQ)
Parque Alameda
Apartado 17-01-165
Quito
Telephone: (0)2-570765
           (0)2-583541
Telefax: (0)2-567848
Electronic Mail: observaq@uio.satnet.net
WWW: http://www.satnet.net/observatorio/
Founded: 1873
Staff: 10
Activities: astronomy * meteorology * astrophysics * archaeoastronomy
Periodicals: "Boletín del Observatorio Astronómico de Quito"
Coordinates: 078°29'56"W 00°12'57"S H2,818m
City Reference Coordinates: 078°30'00"W 00°13'00"S

Egypt

Academy of Scientific Research and Technology (ASRT)
101 Kasr El-Eini Street
Cairo 11516
Telephone: (0)2-3557972
 (0)2-3542714
 (0)2-5940127
Telefax: (0)2-356280
 (0)2-3547807
WWW: http://www.asrt.sci.eg/
Founded: 1971
Membership: 2250
Activities: enhancing contribution of Egyptian scientific capabilities to development * ensuring high-level planning and coordinating among leading research workers, technologists and users
Periodicals: "Proceedings of the Mathematical and Physical Society of Egypt" and many other scientific journals in various fields
Awards: 32 annual prizes in mathematics, physics, geology, chemistry, biology, agriculture, engineering and medicine
City Reference Coordinates: 031°15'00"E 30°03'00"N

Astronomical Society of Egypt
c/o Cairo University
Faculty of Sciences
Astronomy Department
Geza
Telephone: (0)727022
Founded: 1975
Membership: 250
Activities: promoting astronomical research in Egypt * popularization
Periodicals: (6) "Journal of the Astronomical Society of Egypt"; "World of Astronomy and Space" (in arabic)
City Reference Coordinates: 031°15'00"E 30°03'00"N

Cairo University, Department of Astronomy and Meteorology
Gamaa Street
Geza
Telephone: (0)727022
Electronic Mail: <userid>@frcu.eun.eg
WWW: http://www.cairo.eun.eg/
Founded: 1937
Staff: 19
Activities: education * research * thesis supervision
City Reference Coordinates: 031°15'00"E 30°03'00"N

Egyptian Meteorological Authority (EMA)
P.O. Box 11784
Koubry El-Quobba
Cairo
Telephone: (0)2-830053
 (0)2-830069
 (0)2-2820790
 (0)2-2849858
Telefax: (0)2-849857
 (0)2-2849857
Electronic Mail: met@nwp.gov.eg
WWW: http://nwp.gov.eg/
Founded: 1971
City Reference Coordinates: 031°15'00"E 30°03'00"N

Egyptian Organization for Standardization and Quality Control (EOS)
2 Latin America Street
Garden City
Cairo
Telephone: (0)2-3549720
Telefax: (0)2-3557841
Electronic Mail: moi@idsc.gov.eg
Founded: 1957
Staff: 1000
Activities: standardization * quality control * testing * information * metrology
City Reference Coordinates: 031°15'00"E 30°03'00"N

Helwân Observatory
● See "National Institute of Astronomical and Geophysical Researches, Helwân Observatory"

National Institute of Astronomical and Geophysical Researches (NIAGR), Helwân Observatory
Helwân
Cairo 11421
or :
98.104 Street
Maadi
Telephone: (0)78-2683
            (0)78-0645
Telefax: (0)6221-405297
Electronic Mail: <userid>@frcu.eun.eg
Coordinates: 031°22'48"E 29°51'30"N H116m
City Reference Coordinates: 031°15'00"E 30°03'00"N

El Salvador

Asociación Salvadoreña de Astronomía
Apartado Postal 1440
San Salvador
or :
69 Avenida Norte #148
San Salvador
Telephone: (0)2986489
Telefax: (0)2242539
Electronic Mail: fegaram@es.com.sv
Founded: 1991
Membership: 60
Activities: observing * meetings * lectures * public discussions * education * popularization
Periodicals: (12) "Astro"
Coordinates: 089°04'59"W 13°29'41"N H175m
City Reference Coordinates: 089°12'00"W 13°42'00"N

Consejo Nacional de Ciencia y Tecnología (CONACYT)
Colonia Médica
Avenida Dr. E. Alvarez y Pje
Dr. Guillermo Rodriguez 51
(Edificio Espinoza)
San Salvador
Telephone: (0)2262800
Telefax: (0)2256255
Electronic Mail: iso@ns.conacyt.gob.sv
City Reference Coordinates: 089°12'00"W 13°42'00"N

Servicio de Meteorología e Hidrología Nacional (SMHN)
Apartado Postal 2265
San Salvador
Telephone: (0)2940566 x43
Telefax: (0)2944750
        (0)2940575
Founded: 1953
Staff: 120
Activities: operational forecasting * climatology * hydrology
City Reference Coordinates: 089°12'00"W 13°42'00"N

Estonia

Estonian Academy of Sciences
Kohtu 6
EE-10130 Tallinn
Telephone: (0)6442149
Telefax: (0)6451805
WWW: http://www.aca.ee/
Founded: 1938
City Reference Coordinates: 024°45'00"E 59°25'00"N

Estonian Physical Society
(Eesti Füüsika Selts - EFS)
Tähe 4
EE-51010 Tartu
Telephone: (0)7-383006
Telefax: (0)7-383033
Electronic Mail: piret@fi.tartu.ee (Piret Kuusk, Chairman)
WWW: http://www.physic.ut.ee/efs/
Founded: 1989
Membership: 200
Activities: organizing the "Estonian Days of Physics"
Periodicals: (1) "Annual Report of the Estonian Physical Society" (ISSN 1406-0574)
Awards: (1) "EFS Prize", "EFS Student Prize"
City Reference Coordinates: 026°44'00"E 58°20'00"N

National Standards Board of Estonia (EVS)
Aru 10
EE-10317 Tallinn
Telephone: (0)2-493572
Telefax: (0)2-492002
Electronic Mail: info@evs.ee
Founded: 1991
Staff: 24
Activities: standardization * metrology * accreditation
Periodicals: (12) "EVS Teataja"
City Reference Coordinates: 024°45'00"E 59°25'00"N

Tartu Observatory
EE-61602 Töravere
Telephone: (0)7-410265
Telefax: (0)7-410205
Electronic Mail: aai@aai.ee
　　　　　　　<userid>@aai.ee
WWW: http://www.aai.ee/
Founded: 1804
Staff: 65
Activities: stellar atmospheres * variable stars * galactic structure * large-scale structure * remote sensing * dynamic weather forecast
Periodicals: (1) "Calendar of Tartu Observatory" (ISSN 0202-2214), "Annual Report"
Coordinates: 026°43'18"E 58°22'47"N H67m
City Reference Coordinates: 026°44'00"E 58°20'00"N

Tartu Old Observatory
Tähetorn Toomel
EE-2400 Tartu
Electronic Mail: astro@obs.ee
　　　　　　　<userid>@obs.ee
WWW: http://www.obs.ee/
Coordinates: 026°43'18"E 58°22'47"N
City Reference Coordinates: 026°44'00"E 58°20'00"N

Tartu University, Institute of Physics
Riia 142
EE-51014 Tartu
Telephone: (0)27-383039
Telefax: (0)27-383033
Electronic Mail: dir@fi.tartu.ee
WWW: http://www.fi.tartu.ee/
Founded: 1973
Membership: 170
Activities: materials science * matter structure * laser physics * laser optical technologies * environmental science * biophysics
Periodicals: (1) "Annual Report"
City Reference Coordinates: 026°44'00"E 58°20'00"N

Finland

Academy of Finland
Hämeentie 68
P.O. Box 99
FIN-00501 Helsinki
or :
Vilhonvuorenkatu 6
FIN-00500 Helsinki
Telephone: (0)9-774881
Telefax: (0)9-77488299
WWW: http://www.aka.fi/
          http://www.aka.fi/eng/akatemia.htm
City Reference Coordinates: 024°58'00"E 60°10'00"N

Amateur Astronomers Society Seulaset
● See "Tähtitieteen Harrastajan Yhdistys Seulaset ry."

Astronomical Association Tampereen Ursa
● See "Tähtitieteellinen Yhdistis Tampereen Ursa ry."

Etelä-Karjalan Nova
(South Carelian Amateur Astronomical Association Nova)
P.O. Box 244
FIN-53101 Lappeenranta
Electronic Mail: nova@ursa.fi
WWW: http://www.funet.fi/pub/astro/html/yhd/novaeng.html
          http://www.ursa.fi/yhd/nova/novaeng.html
Founded: 1974
Membership: 45
Periodicals: (2) "Tähtiharrastustietoa" (ISSN 1238-0091)
City Reference Coordinates: 028°11'00"E 61°04'00"N

European Incoherent Scatter Facility (EISCAT), Sodankylä Station
c/o Geophysical Observatory
FIN-99600 Sodankylä
Telephone: (0)16-619880
Telefax: (0)16-610375
Electronic Mail: eiscat@sgo.fi
WWW: http://eiscat.sgo.fi/
Founded: 1975
Staff: 5
Activities: incoherent scatter radar receiving facility for ionospheric and magnetospheric research
Coordinates: 026°37'00"E 67°22'00"N H198m
City Reference Coordinates: 026°37'00"E 67°22'00"N

Finnish Academy of Science and Letters
Mariankatu 5
FIN-00170 Helsinki
Telephone: (0)9-636800
Telefax: (0)9-660117
WWW: http://www.acadsci.fi/
Founded: 1908
Membership: 450
Periodicals: "Annales Academiae Scientiarum Fennicae", "Finnish Academy Year Book" (ISSN 1238-9137)
Awards: (1) "Academy Award"
City Reference Coordinates: 024°58'00"E 60°10'00"N

Finnish Academy of Science and Letters, Sodankylä Geophysical Observatory (SGO)
FIN-99600 Sodankylä
Telephone: (0)16-619811
Telefax: (0)16-619875
WWW: http://www.sgo.fi/
Founded: 1913
Staff: 25
Activities: geomagnetism * ionosphere * magnetosphere * seismology * EISCAT site (see separate entry)
Periodicals: "Publications" (ISSN 0355-0826)
Coordinates: 026°37'55"E 67°22'05"N H180m
City Reference Coordinates: 026°37'00"E 67°22'00"N

Finnish Academy of Technology
Tekniikantie 12

FIN-02150 Espoo
Telephone: (0)9-4554565
Telefax: (0)9-4554626
Founded: 1957
Staff: 4
Activities: meetings * seminars * reports
Periodicals: "Acta Polytechnica Scandinavica"
Awards: "Craftsmans Award"
City Reference Coordinates: 024°40'00"E 60°13'00"N

Finnish Astronautical Society
● See "Suomen Avaruustutkimusseura"

Finnish Astronomical Society
(Suomen Tähtitieteilijäseura)
c/o University of Helsinki Observatory
P.O. Box 14
FIN-00014 Helsinki
or :
Tähtitorninmäki
FIN-00130 Helsinki 13
Telephone: (0)19122909
Telefax: (0)19122952
Electronic Mail: kimmo.lehtinen@helsinki.fi
WWW: http://cc.oulu.fi/tati/kauko/tts.html
Founded: 1969
Membership: 100
Activities: conferences * astronomy days * representing professional astronomers in Finland
Periodicals: "Proceedings of the Finnish Astronomy Days"
City Reference Coordinates: 024°58'00"E 60°10'00"N

Finnish Geodetic Institute (FGI)
Geodeetinrinne 2
FIN-02430 Masala
Telephone: (0)9-295550
Telefax: (0)9-29555200
Founded: 1918
Staff: 50
Activities: geodesy * gravimetry * photogrammetry * remote sensing * cartography * GIS
Periodicals: "Publications of the Finnish Geodetic Institute" (ISSN 0085-6932); "Reports of the Finnish Geodetic Institute" (ISSN 0355-1962)
City Reference Coordinates: 024°33'00"E 60°10'00"N H10m

Finnish Meteorological Institute (FMI)
Vuorikatu 24
P.O. Box 503
FIN-00101 Helsinki
Telephone: (0)9-19291
Telefax: (0)9-179581
Electronic Mail: <userid>@fmi.fi
WWW: http://www.fmi.fi/
Founded: 1838
Staff: 485
Activities: weather forecast * climatology * air pollution * geomagnetism * space phys cs
Periodicals: (1) "Annual Report"
Observatories: 3 (Nurmijärví, Jokioinen, Sodankylä)
City Reference Coordinates: 024°58'00"E 60°10'00"N

Finnish Meteorological Institute (FMI), Department of Geophysics
Vuorikatu 19
P.O. Box 503
FIN-00101 Helsinki
Telephone: (0)9-19291
Telefax: (0)9-1929539
      (0)9-19294603
Electronic Mail: <userid>@fmi.fi
WWW: http://www.geo.fmi.fi/
      http://www.fmi.fi/
      http://sumppu.fmi.fi/
Founded: 1838 (as Geomagnetic Observatory)
Staff: 45
Activities: geomagnetism * space physics
Periodicals: "Geophysical Publications" (ISSN 0782-6078)
City Reference Coordinates: 024°58'00"E 60°10'00"N

Finnish Meteorological Institute (FMI), Nurmijärví Geophysical Observatory
FIN-05100 Röykkä
Telephone: (0)9-8787030
Telefax: (0)9-87870350
Electronic Mail: <userid>@fmi.fi
WWW: http://www.fmi.fi/
Founded: 1953
Activities: geomagnetism * magnetometer calibration
Periodicals: (1) "Magnetic Results from Nurmijärví Geophysical Observatory" (ISSN 0782-6087)
Coordinates: 024°39'18"E 60°30'30"N H105m
City Reference Coordinates: 024°48'00"E 60°28'00"N (Nurmijärví)

Finnish Physical Society
c/o S. Nokkanen (Secretary)
P.O. Box 9
FIN-00014 Helsinki
Telephone: (0)9-1918375
Telefax: (0)9-1918378
Electronic Mail: nokkanen@phcu.helsinki.fi
                 finphys@pcu.helsinki.fi
WWW: http://www.physics.helsinki.fi/~sfs/
Founded: 1947
Membership: 775
City Reference Coordinates: 024°58'00"E 60°10'00"N

Finnish Standards Association (SFS)
P.O. Box 116
FIN-00241 Helsinki
or :
Maistraatinportli 2
FIN-00240 Helsinki
Telephone: (0)9-1499331
Telefax: (0)9-1464925
Electronic Mail: sfs@sfs.fi
WWW: http://www.sfs.fi/
Founded: 1924
Staff: 70
Periodicals: (6) "SFS-Hedatus" (ISSN 0356-1089, circ.: 1,900)
City Reference Coordinates: 024°58'00"E 60°10'00"N

Hämeenlinnan Tähtitieteen Harrastajlen Yhdistis Vega ry.
c/o Raimo Känkänen
Viialantie 5 D 16
FIN-13500 Hämeenlinna
or :
c/o Jari Vento
Viertokatu 18 A 1
FIN-13210 Hämeenlinna
Telephone: (0)3-6172696
Electronic Mail: vega@ursa.fi
WWW: http://www.helsinki.fi/~tmjlehto/engvega.htm
     http://www.htk.fi/public/vega/eindex.htm
Founded: 1981
Membership: 100
City Reference Coordinates: 024°27'00"E 61°61'00"N

Helsinki University of Technology (HUT), Institute of Photogrammetry and Remote Sensing
P.O. Box 1200
FIN-02015 HUT
or :
Otakaari 1
FIN-02150 Espoo
Telephone: (0)9-4513900
Telefax: (0)9-465077
WWW: http://foto.hut.fi/
     http://www.hut.fi/
Founded: 1957
Staff: 15
Activities: digital photogrammetry * digital image processing in remote sensing * real-time photogrammetry * projective transformations
Periodicals: "The Photogrammetric Journal of Finland" (ISSN 0557-1069), "Annual Report" (ISSN 0788-5474)
City Reference Coordinates: 024°40'00"E 60°13'00"N

Helsinki University of Technology (HUT), Laboratory of Space Technology
c/o Department of Electrical Engineering

Otakaari 5 A
FIN-02150 Espoo
Telephone: (0)9-4512378
Telefax: (0)9-460224
Electronic Mail: <userid>@delta.hut.fi
WWW: http://www.space.hut.fi/
City Reference Coordinates: 024°40'00"E 60°13'00"N

Helsinki University of Technology (HUT), Metsähovi Radio Observatory

Metsähovintie 114
FIN-02540 Kylmälä
Telephone: (0)9-2564831
Telefax: (0)9-2564531
Electronic Mail: <userid>@hut.fi
            vlbi@hut.fi (VLBI group)
            solar@hut.fi (solar group)
WWW: http://kurp-www.hut.fi/
Founded: 1974
Staff: 18
Activities: radio astronomy * microwaves * solar research * QSOs * molecular lines * VLBI
Periodicals: "Metsähovi Publications on Radio Science" (ISSN 1455-9587) "Metsähovi Radio Observatory Report" (ISSN 1455-9579)
Coordinates: 024°23'38"E 60°13'05"N H61m
City Reference Coordinates: 024°24'00"E 60°13'00"N

Jyväskylän Sirius ry

Sepänaukion Vapaa-Aikakeskus
Kyllikinkatu 1
FIN-40100 Jyväskylä
Telephone: (0)14-3731250
Electronic Mail: sirius@ursa.fi
WWW: http://www.ursa.fi/sirius
        http://www.ursa.fi/sirius/siriuseng.html
Founded: 1959
Membership: 180
Activities: meetings with lectures, slides, films * telescope making * optics * observing * photography * study trips
Periodicals: (4) "Valkoinen Kääpiö" (White Dwarf) (ISSN 0781-0466)
Awards: (1) "Amateur Astronomer of the Year"
Coordinates: 025°42'24"E 62°14'03"N H151m (Rihlaperä)
City Reference Coordinates: 025°44'00"E 62°14'00"N

Kuopion Tähtitieteellinen Seura Saturnus

• See "Saturnus Astronomical Association of Kuopio"

Lahden URSA ry

(URSA Astronomical Society)
c/o Yrjö Pullinen (Secretary)
Koulutie 8 A 25
FIN-15860 Hollola
or :
c/o Markku Pyykkönen (President)
Tenniläntie 301
FIN-16630 Hollola
Telephone: (0)3-7804534 (Yrjö Pullinen)
          (0)3-7885125 (Markku Pyykkönen)
          (0)3-7534002 (Observatory + answering machine)
Electronic Mail: yrjo.pullinen@pp.kolumbus.fi
Founded: 1948
Membership: 300
Activities: meetings * observing * public star shows
Coordinates: 025°35'31"E 60°59'03"N H193m (Pirttiharju)
City Reference Coordinates: 025°36'00"E 60°59'00"N

Metsähovi Radio Observatory

• See "Helsinki University of Technology (HUT), Metsähovi Radio Obsevatory"

Nordic Optical Telescope Scientific Association (NOTSA)

c/o The Director
Tuorla Observatory
Väisäläntie 20
FIN-21500 Piikkiö
Telephone: (0)2-2744274
Telefax: (0)2-2433767
Electronic Mail: piirola@astro.utu.fi (Vilppu Piirola, Director)
        <userid>@not.iac.es

WWW: http://www.not.iac.es/
http://www.astro.utu.fi/
Founded: 1984
Staff: 12
Activities: constructing and operating the Nordic Optical Telescope (NOT) and its auxiliary instrumentation in the Canary Islands
Coordinates: 017°53'00"E 28°45'21"N H2,382m
City Reference Coordinates: 022°31'00"E 60°26'00"N

Nurmijärví Geophysical Observatory
● See "Finnish Meteorological Institute, Nurmijärví Geophysical Observatory"

Olympos Astronomical Association
● See "Tähtitieteellinen Yhdistis Olympos ry."

Pollux
P.O. Box 69
FIN-02151 Espoo
Telephone: (0)9-4683124 (Chairman)
(0)40-5103523 (Secretary)
Telefax: (0)9-4683218
Electronic Mail: aapo.puhakka@hut.fi (Aapo Puhakka, Chairman)
jtmakela@c.hut.fi (Jussi Mäkelä, Secretary)
pollux@tky.hut.fi
pollux@otax.tky.hut.fi
WWW: http://www.tky.hut.fi/~pollux/
http://www.tky.hut.fi/~pollux/eng-index.html
Founded: 1994
Membership: 62
Activities: observing * visits
City Reference Coordinates: 024°40'00"E 60°13'00"N

Rauma Vocational College, Maritime Department, Planetarium
Suojantie 2
FIN-26100 Rauma
Telephone: (0)2-837721
Telefax: (0)2-83772222
Electronic Mail: skarlsson@rai.rauma.fi (Sune Karlsson)
Founded: 1880
Staff: 70
Activities: education
City Reference Coordinates: 021°30'00"E 61°08'00"N

Saturnus Astronomical Association of Kuopio
(Kuopion Tähtitieteellinen Seura Saturnus)
P.O. Box 293
FIN-70101 Kuopio
Founded: 1956
Membership: 67
Activities: observing (deep sky, planets, Sun) * meetings * education * photography
Coordinates: 027°39'03"E 62°53'11"N H159m (Huuhanmäki)
City Reference Coordinates: 027°40'00"E 62°54'00"N

Seinäjoen Ursa Astronomical Association
c/o Tapani Koskiniemi
Poutunkatu 6A
FIN-60320 Seinäjoki
Telephone: (0)6-4143149
Membership: 54
Activities: lectures * observing
City Reference Coordinates: 022°50'00"E 62°47'00"N

Sodankylä Geophysical Observatory (SGO)
● See "Finnish Academy of Science and Letters, Sodankylä Geophysical Observatory (SGO)"

South Carelian Amateur Astronomical Association Nova
● See "Etelä-Karjalan Nova"

Space Systems Finland (SSF) Ltd.
Keilaranta 8
FIN-02150 Espoo
Telephone: (0)9-61328600
Telefax: (0)9-61328699
WWW: http://www.ssf.fi/
Founded: 1990

Staff: 27
● Software producer and consultant
City Reference Coordinates: 024°40'00"E 60°13'00"N

Suomen Avaruustutkimusseura (ATS)
(Finnish Astronautical Society)
P.O. Box 507
FIN-00101 Helsinki
Telephone: (0)9-5874433
Electronic Mail: paul.stigell@vtt.fi (Paul Stigell)
WWW: http://www.vtt.fi/aut/pro/prg/staff/autpks/ats.htm
Founded: 1959
Membership: 100
Activities: space technology * rocketry * space research
Periodicals: (4) "Avaruusluotain" (ISSN 0356-021x, circ.: 150)
City Reference Coordinates: 024°58'00"E 60°10'00"N

Tähtitieteellinen Yhdistis Olympos ry.
(Olympos Astronomical Association)
P.O. Box 18
FIN-33501 Tampere
Electronic Mail: markku.nyfelt@nmp.nokia.com (Markku Nyfelt)
WWW: http://www.mikrolog.fi/olympos/
Founded: 1993
Membership: 28
Activities: deep-sky observing * telescope making * optics
Periodicals: (4) "Equatorial Dust Lane (EDL)"
Coordinates: 023°53'48"E 61°23'65"N H158m
City Reference Coordinates: 023°45'00"E 61°32'00"N

Tähtitieteellinen Yhdistis Tampereen Ursa ry.
(Astronomical Association Tampereen Ursa)
P.O. Box 18
FIN-33501 Tampere
Telephone: (0)3-2611005
Electronic Mail: risto.lehti@valmet.com (Risto Lehti)
         kakuure@sci.fi (Kari A. Kuure)
WWW: http://www.mikrolog.fi/olympos/ursa.htm
      http://www.sci.fi/~ursa/
Founded: 1950
Membership: 220
Activities: lectures * public star shows * observing
Periodicals: (4) "Radianti" (ISSN 0785-5672)
Coordinates: 023°47'38"E 61°30'44"N H162m
City Reference Coordinates: 023°45'00"E 61°32'00"N

Tähtitieteen Harrastajan Yhdistys Seulaset ry.
(Amateur Astronomers Society Seulaset)
c/o Pertti Pääkkönen
Department of Physics
University of Joensuu
P.O. Box 111
FIN-80101 Joensuu
Telephone: (0)13-2513238
Telefax: (0)13-2513290
Electronic Mail: pertti.paakkonen@joensuu.fi
WWW: http://cc.joensuu.fi/seulaset
Founded: 1973
Membership: 70
Activities: stargazing * public observing * education
Coordinates: 029°59'49"E 62°43'39"N H155m
City Reference Coordinates: 029°45'00"E 62°36'00"N

Tampereen Särkänniemi Oy, Planetarium
Särkänniemi
FIN-33230 Tampere
Telephone: (0)3-2488111
Telefax: (0)3-2121279
Electronic Mail: timo.rahunen@sarkanniemi.fi
WWW: http://www.sarkanniemi.fi/
      http://www.sarkanniemi.fi/oppimateriaali/tahtiakatemia/
Founded: 1969
Staff: 5
City Reference Coordinates: 023°45'00"E 61°32'00"N

Technical Research Centre of Finland
- See "VTT"

Technology Development Centre
- See "TEKES"

TEKES
(Technology Development Centre)
P.O.Box 69
FIN-00101 Helsinki
or :
Kyllikinportti 2
Helsinki
Telephone: (0)105-2151
Telefax: (0)9-6949196
Electronic Mail: tekes@tekes.fi
WWW: http://www.tekes.fi/
Founded: 1983
Staff: 200
Activities: coordinating and financing technological research in Finland
Periodicals: "Views on Finnish Technology"
City Reference Coordinates: 024°58'00" E 60°10'00" N

TEKES
- See "Technology Development Centre (TEKES)"

Teknofokus
P.O. Box 47
FIN-00711 Helsinki
or :
Pihlajistonkuja 4C
FIN-00710 Helsinki
Telephone: (0)9-370471
Telefax: (0)9-377388
Electronic Mail: hannu.maattanen@kolumbus.fi
Activities: manufacturing telescope mirrors and related accessories * supplying optical components and materials, abrasives and polishing compounds * designing instrumentation * prototype services
City Reference Coordinates: 024°58'00" E 60°10'00" N

Tuorla Observatory
- See "University of Turku, Tuorla Observatory"

Turun Ursa Astronomical Association
c/o Iso-Heikkilä Observatory
FIN-20200 Turku
Telephone: (0)2-302195
Electronic Mail: ursa@ursa.utu.fi
WWW: http://org.utu.fi/(ef,hr)/yhd/ursa/index.html
Founded: 1928
Membership: 200
Activities: public shows * photography * observing * popularization
Periodicals: (4) "Ceres" (ISSN 1235-1083)
Coordinates: 022°13'46" E 60°27'09" N H28m (Iso-Heikkilä Observatory)
                022°13'46" E 60°25'14" N H28m (Kevola Observatory)
City Reference Coordinates: 022°17'00" E 60°27'00" N

University of Helsinki, Observatory and Astrophysical Laboratory
P.O. Box 14
FIN-00014 University of Helsinki
or :
Tähtitorninmäki
FIN-00130 Helsinki 13
Telephone: (0)9-19122940
Telefax: (0)9-19122952
Electronic Mail: <userid>@cc.helsinki.fi
                       <userid>@helsinki.fi
WWW: http://gstar.helsinki.fi/
        http://www.astro.helsinki.fi/
Founded: 1834
Staff: 20
Periodicals: "Report" (ISSN 0355-9289)
Coordinates: 024°57'18" E 60°09'42" N H33m
City Reference Coordinates: 024°58'00" E 60°10'00" N

University of Oulu, Department of Physical Sciences, Astronomy Division
P.O. Box 3000
FIN-90401 Oulu
Telephone: (0)8-5531280
Telefax: (0)8-5531934
Electronic Mail: <userid>@oulu.fi
WWW: http://physics.oulu.fi/fysiikka/
Founded: 1964
Staff: 20
Activities: astrophysics * planetology * stellar dynamics
City Reference Coordinates: 025°54'00"E 65°05'00"N

University of Turku, Space Research Laboratory
Vesilinnantie 4
FIN-20004 Turku
Telephone: (0)2-3335751
Telefax: (0)2-3335993
Electronic Mail: <userid>@utu.fi
WWW: http://srl.utu.fi/
Founded: 1988
Staff: 14
Activities: solar and heliospheric particle physics
City Reference Coordinates: 022°17'00"E 60°27'00"N

University of Turku, Tuorla Observatory
Väisäläntie 20
FIN-21500 Piikkiö
Telephone: (0)2-2744244
Telefax: (0)2-2433767
Electronic Mail: <userid>@astro.utu.fi
WWW: http://www.astro.utu.fi/
Founded: 1927
Staff: 30
Activities: positional astronomy * celestial mechanics * optical photo-polarimetry * radio astronomy * dynamical astronomy * cosmology
Periodicals: "Tuorla Observatory Reports", "Informo" (ISSN 0789-6719)
Coordinates: 022°26'48"E 60°25'00"N H40m
City Reference Coordinates: 022°31'00"E 60°26'00"N

Ursa Astronomical Association
Raatimiehenkatu 3 A 2
FIN-00140 Helsinki
Telephone: (0)9-68404000
Telefax: (0)9-68404040
Electronic Mail: ursa@ursa.fi
WWW: http://www.ursa.fi/ursa/
       http://www.ursa.fi/ursa/planetaario/ (Planetarium)
Founded: 1921
Membership: 6000
Activities: publishing * meetings * lectures * courses * youth camps * topical sections * library * portable planetarium * public events
Periodicals: (6) "Tähdet ja Avaruus" (ISSN 0355-9467), "Ursa Minor" (ISSN 0780-7945); (1) "Tähdet 19xx" (ISSN 0355-9459)

Awards: "Stella Arcti"
Coordinates: 024°57'31"E 60°09'20"N H20m (Kaivopuisto Observatory)
City Reference Coordinates: 024°58'00"E 60°10'00"N

URSA Astronomical Society
● See "Lahden URSA ry"

Verne Theatre
Heureka - The Finnish Science Centre
Tiedepuisto 1
FIN-01300 Vantaa
Telephone: (0)9-85799
Telefax: (0)9-8734142
Electronic Mail: info@heureka.fi
WWW: http://www.heureka.fi/en/verne/
City Reference Coordinates: 024°59'00"E 60°30'00"N

VTT
(Technical Research Centre of Finland)
P.O. Box 105
FIN-02151 Espoo

or :
P.O. Box 1000
FIN-02044 VTT
or :
Vuorimiehentie 5
Espoo
FIN-02044 VTT
Telephone: (0)9-4561
          (0)9-4564330
Telefax: (0)9-4567011
          (0)9-4553349
Electronic Mail: <userid>@vtt.fi
WWW: http://www.vtt.fi/
Founded: 1942
Staff: 3000
Activities: technical research * space instrumentation * remote sensing * satellite communications * mechanics * electronics * computers * information handling * optics * Earth observing
City Reference Coordinates: 024°40'00"E 60°13'00"N

VTT, Automation ProTechno
(Technical Research Centre of Finland)
P.O. Box 1303
FIN-02044 VTT
or :
Metallimiehenkuja 10
FIN-02150 Espoo
Telephone: (0)9-4564363
          (0)9-4553349
Electronic Mail: <userid>@vtt.fi
WWW: http://www.vtt.fi/aut/pro
Founded: 1942
Staff: 70
Activities: spacecraft scientific experiments (instruments) * spacecraft mechanisms * mechanical design * structural engineering * thermal design * machining * welding * EMC testing * vibration testing * thermal testing * reliability analysis * vacuum tests
City Reference Coordinates: 024°40'00"E 60°13'00"N

France

Académie des Sciences
c/o Institut de France
23, quai de Conti
F-75006 Paris
Telephone: (0)143266621
Telefax: (0)143546399
WWW: http://www.acad-sciences.institut-de-france.fr/
Founded: 1666
Membership: 126
Activities: meetings * publishing
Periodicals: "Comptes-Rendus de l'Académie des Sciences - Série I" (ISSN 0764-4442), "Comptes-Rendus de l'Académie des Sciences - Série II" (ISSN 0764-4450), "Sciences de la Terre" (ISSN 1164-5873)
City Reference Coordinates: 002°20'00"E 48°52'00"N

Aerospatiale
37, boulevard de Montmorency
F-75781 Paris Cedex 16
Telephone: (0)142242424
Telefax: (0)145245414
WWW: http://www.aerospatiale.fr/
Founded: 1970
Activities: manufacturing aerospace equipment
Periodicals: (12) "Revue Aerospatiale"
● Now part of the "European Aeronautic, Defense and Space Co. (EADS)"
City Reference Coordinates: 002°20'00"E 48°52'00"N

Agence Spatiale Européenne (ASE)
● See "European Space Agency (ESA)"

Air et Cosmos
1bis, avenue de la République
F-75011 Paris
Telephone: (0)149293200
Telefax: (0)149293201
Founded: 1963
Staff: 30
Periodicals: (52) "Air & Cosmos"
City Reference Coordinates: 002°20'00"E 48°52'00"N

Albiréo - Astronomes Amateurs Tarnais
c/o Jean-Luc Floutard
24, rue des Acacias
F-81000 Albi
Telephone: (0)563607317
Founded: 1992
Membership: 32
Activities: observing * photography * mirror polishing
Coordinates: 001°43'10"E 43°52'30"N H196m (Saint-Caprais, Rabastens)
City Reference Coordinates: 002°09'00"E 43°56'00"N

Amis du Planétarium d'Aix-en-Provence (APAP)
c/o Maison des Associations
Place de l'Eglise
F-13540 Puyricard
Telephone: (0)442921546
WWW: http://www.astrsp-mrs.fr/www_root/private/malbu/sommaire.htm
Founded: 1989
Membership: 100
Activities: setting up a planetarium in Aix-en-Provence * popularization
Periodicals: "Pour un Planétarium à Aix" (ISSN 1152-8982)
City Reference Coordinates: 005°26'00"E 43°32'00"N

Andromède
c/o Observatoire de Marseille
2, place Le Verrier
F-13248 Marseille Cedex 4
Telephone: (0)495044100
Telefax: (0)491621190
Electronic Mail: duval@observatoire.cnrs-mrs.fr
WWW: http://www-obs.cnrs-mrs.fr/Andromede/
Founded: 1976

Membership: 40
Activities: school visits * observing * planetarium * lectures
City Reference Coordinates: 005°24'00"E 43°18'00"N

Angénieux
F-42570 Saint-Héand
Telephone: (0)477304210
Telefax: (0)477304875
Electronic Mail: angenieux@calva.net
WWW: http://www.angenieux.com/
Founded: 1935
Staff: 200
Activities: manufacturing optical components
City Reference Coordinates: 004°22'00"E 45°31'00"N

Annales Geophysicae
Editorial Office
Centre d'Étude Spatiale des Rayonnements
9, avenue du Colonel Roche
F-31029 Toulouse Cedex
Telephone: (0)561558370
Telefax: (0)561556535
Electronic Mail: anngeo@cesr.fr
Founded: 1983
Staff: 1
● Journal (12) (ISSN 0992-7689)
City Reference Coordinates: 001°26'00"E 43°36'00"N

Arcane
3, rue du Puits d'Argent
F-02240 Itancourt
Telephone: (0)323088842
Telefax: (0)323088875
Electronic Mail: arcane.astro@wanadoo.fr
WWW: http://perso.wanadoo.fr/arcane
Founded: 1986
Activities: manufacturing telescopes and pieces * studies * designs * modifications * adjustments * software * CCD
Coordinates: 003°21'51"E 49°48'33"N H115m
City Reference Coordinates: 003°28'00"E 49°48'00"N (Ribemont)

Arianespace, Head Office
Boulevard de l'Europe
Boîte Postale 177
F-91006 Évry Cedex
Telephone: (0)160876000
Telefax: (0)160876217
          (0)160876247
WWW: http://www.arianespace.com/
Founded: 1980
Staff: 255
Activities: satellite launching services
Periodicals: (1) "Annual Report"
City Reference Coordinates: 002°27'00"E 48°38'00"N

Association À Ciel Ouvert
Moulin du Roy
F-32500 Fleurance
Telephone: (0)562060976
Telefax: (0)562062499
Electronic Mail: etoiles.fleurance@mipnet.fr
WWW: http://www.gascogne.com/Ferme/ciel.htm
Founded: 1994
Membership: 7
Activities: public education * activities in schools * small planetarium
City Reference Coordinates: 000°40'00"E 43°51'00"N

Association Aéronautique et Astronautique de France (AAAF)
6, rue Galilée
F-75782 Paris Cedex 16
Telephone: (0)147230749
Telefax: (0)147230748
WWW: http://www.aaaf.asso.fr/
City Reference Coordinates: 002°20'00"E 48°52'00"N

Association Andromède Astronomie Aveyronnaise (4A)
Boîte Postale 25
F-12850 Onet-le-Château
Telephone: (0)5-65785861
Founded: 1983
Membership: 30
Activities: photography * initiation
Coordinates: 002°33'00"E 44°21'00"N H650m (Lioujas)
City Reference Coordinates: 002°43'00"E 44°21'00"N (Rodez)

Association Astronomie Tycho Brahe
59bis, avenue de Novel
F-74000 Annecy
Telephone: (0)450279221
            (0)606396628
Electronic Mail: tbrahe@cybercable.tm.fr
WWW: http://www.cybercable.tm.fr/~tbrahe/
Founded: 1992
City Reference Coordinates: 006°07'00"E 45°54'00"N

Association Astronomique d'Anjou (AAA)
15, rue Marc Sangnier
F-49000 Angers
Telephone: (0)241475294
Founded: 1979
Membership: 75
Activities: photography * radioastronomy * telescope making * popularization * solar observing
Periodicals: (6) "Pégase" (ISSN 0981-6410)
Observatories: 2
City Reference Coordinates: 000°33'00"W 47°28'00"N

Association Astronomique de Franche-Comté (AAFC)
Parc de l'Observatoire
34, avenue de l'Observatoire
F-25000 Besançon
Telephone: (0)381888788
WWW: http://www.iap.fr/saf/claafc.htm
Founded: 1978
Membership: 80
Activities: observing (minor planets, comets) * education * lectures * workshops * planetarium
Periodicals: (3) "Le Point Astro" (ISSN 1168-1195)
Coordinates: 006°00'00"E 47°00'00"N H309m
City Reference Coordinates: 006°02'00"E 47°15'00"N

Association Astronomique de la Côte d'Or (AACO)
c/o Hôtel de Ville
650, rue de Moirey
F-21850 Saint-Apollinaire
Telephone: (0)380572903 (Christian Nitschelm, President)
Electronic Mail: nitschel@u-bourgogne.fr
            nitschel@iap.fr
WWW: http://www.u-bourgogne.fr/c.nitschelm/aaco.html
Founded: 1996
Membership: 35
Activities: popularization * planetarium project * debunking astrology
Coordinates: 004°43'00"E 47°16'16"N H531m
City Reference Coordinates: 005°02'00"E 47°20'00"N (Dijon)

Association Astronomique de l'Ain (AAA)
c/o Maison des Sociétés
Boulevard Joliot-Curie
F-01000 Bourg-en-Bresse
Telephone: (0)474250447
Founded: 1966
Membership: 65
Activities: observing * photography * public events
Periodicals: (12) "Bulletin de Liaison"
Coordinates: 005°20'23"E 46°11'35"N
City Reference Coordinates: 005°13'00"E 46°12'00"N

Association Astronomique de la Vallée (AAV)
c/o Maison des Associations
Mairie d'Orsay
2, place du Général Leclerc

Boîte Postale 47
F-91401 Orsay Cedex
Electronic Mail: aav@mygale.org
WWW: http://www.mygale.org/∼aav
Founded: 1978
City Reference Coordinates: 002°17'00"E 48°46'00"N

Association Astronomique de l'Indre (AAI)
21, rue Paul-Verlaine
F-36000 Chateauroux
Telephone: (0)254495016
Founded: 1983
Membership: 30
Activities: observing * photography * spectroscopy * popularization
Observatories: 1 (Grosses Roches)
City Reference Coordinates: 001°42'00"E 46°49'00"N

Association Astronomique de Loir-et-Cher (AALC)
10, rue Alexandre-Dumas
F-41350 Vineuil
Telephone: (0)254421954
Founded: 1978
Membership: 200
Activities: observing * education * documentation
Periodicals: (4) "Astro 41"
City Reference Coordinates: 001°22'00"E 47°35'00"N

Association Astronomique du Soissonnais (AAS)
16, rue de la Congrégation
F-02200 Soissons
Telephone: (0)323597243
Founded: 1979
Membership: 40
Activities: lectures * observing * variable stars * mirror making * sundials
City Reference Coordinates: 003°20'00"E 49°22'00"N

Association Astronomique Picarde M80
3, Le Pré du Bois
F-80260 Rubempré
or :
Le Safran
Rue Georges Guynemer
F-80080 Amiens
Telephone: (0)322934902 (answering machine)
         (0)681494492 (President)
Electronic Mail: m80@mygale.org
             lesieur_e@hotmail.com (Emmanuel Lesieur)
WWW: http://www.astrosurf.com/m80
Founded: 1996
Membership: 5
Periodicals: "Bulletin Interne"
City Reference Coordinates: 002°20'00"E 50°00'00"N (Villers Bocage)
             002°18'00"E 49°54'00"N (Amiens)

Association Ciel d'Anjou (ACA)
58, rue Pierre-Blandin
F-49000 Angers
Telephone: (0)241481554
Electronic Mail: b.lailler@unimedia.fr (Bruno Lailler)
WWW: http://www.unimedia.fr/homepage/aca/
     http://welcome.to/CielAnjou
Founded: 1994
Membership: 65
Activities: popularization * education * exhibitions * slide shows
Periodicals: (6) "La Petite Ourse"
City Reference Coordinates: 000°33'00"W 47°28'00"N

Association Copernic (Gap Astronomie -_)
● See "Gap Astronomie - Association Copernic"

Association d'Astronomie Véga
Beffroi des Associations
50, Avenue de Saintonge
F-78450 Villepreux
or :

20, chemin de Rambouillet
F-78450 Villepreux
Telephone: (0)134623052
Founded: 1982
Membership: 16
Activities: observing * photography * lectures * slide shows * talks
Periodicals: "Lyre"
Coordinates: 001°56'45"E 48°50'12"N H60m
City Reference Coordinates: 002°01'00"E 48°50'00"N

Association de Recherche de Phénomènes Astronomiques (ARPA)

c/o Serge Westrich
Bourg
F-71240 Saint-Cyr
Telephone: (0)385442864
Founded: 1996
Membership: 8
Activities: SN search * observing (variable stars)
City Reference Coordinates: 002°04'00"E 48°48'00"N

Association des Amis du Planétarium, Nîmes

Avenue Peladan
Mont-Duplan
F-30000 Nîmes
Telephone: (0)466676094
Founded: 1982
Membership: 20
Activities: managing Nîmes planetarium
City Reference Coordinates: 004°21'00"E 43°50'00"N

Association des Amis du Planétarium, Strasbourg

c/o Observatoire Astronomique
11, rue de l'Université
F-67000 Strasbourg
or :
Rue de l'Observatoire
F-67000 Strasbourg
Telephone: (0)388212043
Telefax: (0)388212045
Electronic Mail: planetar@astro.u-strasbg.fr
WWW: http://astro.u-strasbg.fr/Obs/PLANETARIUM/planetarium.html
Founded: 1979
Membership: 68
Activities: promoting astronomy and planetarium activities * organizing colloquiums of European planetariums and Summer schools
Periodicals: "Bulletin de Liaison"
City Reference Coordinates: 007°45'00"E 48°35'00"N

Association des Astronomes Amateurs d'Auvergne (AAAA)

c/o Bernard Guillaud-Saumur (Secretary)
73, rue Viviani
F-63100 Clermont-Ferrand
or :
UFR Sciences - Complexe Scientifique des Cézeaux
F-63177 Aubière Cedex
Telephone: (0)473250395
Telefax: (0)473230345
Electronic Mail: bernard.guillaud-saumur@wanadoo.fr
WWW: http://aaaa.fr.eu.org/
            http://utopia.intelmatique.fr/~mpj/aaaa/
Founded: 1983
Membership: 105
Activities: observing * education * CCDs * deep-sky objects * visits * sundials
Periodicals: (12) "La Garandie"
Coordinates: 002°57'30"E 45°39'54"N (Aydat/La Garandie)
City Reference Coordinates: 003°05'00"E 45°47'00"N (Clermont-Ferrand)

Association des Planétariums de Langue Française (APLF)

c/o Planétarium de Strasbourg
Observatoire Astronomique
11, rue de l'Université
F-67000 Strasbourg
or :
Rue de l'Observatoire
F-67000 Strasbourg
Telephone: (0)388212042

Telefax: (0)388212045
WWW: http://astro.u-strasbg.fr/Obs/PLANETARIUM/APLF.html
Founded: 1989
Activities: promoting creation of planetariums in France and French-speaking countries
Periodicals: (1) "Comptes-Rendus de l'Association des Planétariums de Langue Française"
City Reference Coordinates: 007°45'00"E 48°35'00"N

Association des Utilisateurs de Détecteurs Électroniques (AUDE)
c/o Robert Delmas
19bis, rue des Lucioles
F-31700 Beauzelle
Electronic Mail: a.aude@wanadoo.fr
Founded: 1994
Membership: 390
Activities: CCDs * publishing * education * training * exchanges
Periodicals: (3) "CCD & Telescope"
City Reference Coordinates: 001°23'00"E 43°39'00"N (Blagnac)

Association Drômoise d'Astronomie (ADA) "Les Pléiades"
c/o Jean François Léoni
Quartier des Perrots
F-26760 Beaumont-lès-Valence
Telephone: (0)475595525
Founded: 1986
Membership: 51
Activities: education * photography * lectures * popularization * training * exhibitions * visits * instrument making * observing * radio talks * columns in regional newspaper * variable stars
Coordinates: 004°55'30"E 43°53'30"N H135m (Beaumont-lès-Valence)
            005°10'10"E 43°54'00"N H1,000m (Col de Bacchus)
City Reference Coordinates: 004°54'00"E 44°56'00"N (Valence)

Association Éducative des Amateurs d'Astronomie du Centre (AEAAC)
2, Cloître Saint-Pierre-le-Puellier
Boîte Postale 2341
F-45023 Orléans Cedex 1
Telephone: (0)238540571
Founded: 1963
Membership: 80
Activities: introduction to astronomy * education * observing * Sun
Periodicals: (4) "Le Point de Lagrange" (circ.: 200)
Coordinates: 001°57'40"E 47°54'40"N
City Reference Coordinates: 001°54'00"E 47°50'00"N

Association Française d'Astronomie (AFA)
c/o Observatoire du Parc Montsouris
17, rue Emile-Deutsch-de-la-Meurthe
F-75014 Paris
Telephone: (0)145898144
Telefax: (0)145650895
Electronic Mail: cieletespace@iap.fr
WWW: http://www.cieletespace.com/
        http://www.cieletespace.fr/
Founded: 1946
Membership: 2000
Activities: popularization * education * colloquia * publishing * observing * training * planetarium (Parc Montsouris)
Periodicals: (12) "Ciel et Espace" (ISSN 0373-9139, circ.: 85,000)
Coordinates: 003°63'11"E 43°41'01"N H176m (Aniane - see separate entry)
City Reference Coordinates: 002°20'00"E 48°52'00"N

Association Française de Normalisation (AFNOR)
Tour Europe
F-92049 Paris La Défense Cedex
Telephone: (0)142915555
Telefax: (0)142915656
Electronic Mail: Teletel: 3616 Code: AFNOR
                Teletel: 36290078 (NORIANE+)
                <userid>@email.afnor.fr
WWW: http://www.afnor.fr/
Founded: 1926
Staff: 9
Activities: French norms and standards agency * standardisation * certification * consulting * information * publishing
Periodicals: "Enjeux"
City Reference Coordinates: 002°20'00"E 48°52'00"N

Association Française des Observateurs d'Étoiles Variables (AFOEV)
c/o Observatoire Astronomique
11, rue de l'Université
F-67000 Strasbourg
Telephone: (0)388843711
Telefax: (0)388150760
Electronic Mail: afoev@astro.u-strasbg.fr
WWW: http://astro.u-strasbg.fr/afoev
Founded: 1921
Membership: 100
Activities: observing and studying variable stars * collaboration with professional astronomers
Periodicals: (4) "Bulletin de l'AFOEV" (ISSN 0153-9949); (6) "Gazette des Étoiles Variables"
City Reference Coordinates: 007°45'00"E 48°35'00"N

Association Française pour l'Avancement des Sciences (AFAS)
c/o Cité des Sciences et de l'Industrie
F-75930 Paris Cedex 19
Telephone: (0)140058201
Telefax: (0)140058202
Founded: 1872
Membership: 500
Activities: education * lectures * colloquia * conferences * scientific trips
Periodicals: (4) "Sciences" (ISSN 0151-0304, circ.: 1,000)
City Reference Coordinates: 002°20'00"E 48°52'00"N

Association Guadeloupéenne d'Astronomes Amateurs (AGAA)
Résidence Les Pléiades
Section Barbotteau/Vernou
F-97170 Petit-Bourg
(Guadeloupe)
Telephone: (0)590940552
Electronic Mail: agaa@netguacom.fr
Founded: 1990
Membership: 200
Activities: popularization * education
City Reference Coordinates: 061°36'00"W 16°12'00"N

Association I3A
(Informatique et Astronomie pour Astronomes Amateurs - I3A)
30, Chemin Guilhermy
F-31100 Toulouse
Telephone: (0)561862623
Electronic Mail: i3a@easynet.fr
WWW: http://www.astrosurf.com/
Activities: publishing CD-ROMs with astronomy material
City Reference Coordinates: 001°26'00"E 43°36'00"N

Association Jeunes Sciences, Section Astronomie
62, rue du 110e R.I.
Boîte Postale 1501
F-59381 Dunkerque Cedex
Telephone: (0)328668839
          (0)328216064
Electronic Mail: js@netinfo.fr
Activities: meetings * training * practical and theoretical optics * using and observing amateur satellites * video shows
City Reference Coordinates: 002°22'00"E 51°03'00"N

Association Marseillaise d'Astronomie (AMAS)
CAQ du Petit Bosquet
213, avenue de Montolivet
F-13012 Marseille
Telephone: (0)491661078
WWW: http://www.astrosurf.org/pluton/amas/amas.html
Founded: 1984
Membership: 54
Activities: photography * telescope making * public observing
Coordinates: 005°45'00"E 43°17'75"N H550m (Riboux)
City Reference Coordinates: 005°24'00"E 43°18'00"N

Association Méditerranéenne des Sciences de l'Environnement et de l'Espace (AMSEE)
c/o Patrice Poyet
182, route de Gairaut
Parc Château d'Azur
Villa la Chamaille

F-06100 Nice
Telephone: (0)493090651
Electronic Mail: poyet@cstb.fr
WWW: http://www.cstb.fr/ILC/amsee/
            http://cic.cstb.fr/amsee/projects/
Founded: 1990
Activities: observing (planets, double stars, deep sky) * CCDs
City Reference Coordinates: 007°15'00"E 43°42'00"N

Association Narbonnaise d'Astronomie Populaire (ANAP)
c/o Observatoire de Narbonne
31, rue de la Distillerie
F-11110 Vinassan
Telephone: (0)468453113
Founded: 1981
Membership: 30
Activities: popularization * research * CCD imaging
Coordinates: 002°58'00"E 43°09'00"N (Domaine de Montplaisir, Narbonne)
City Reference Coordinates: 002°58'00"E 43°09'00"N

Association Nationale pour l'Amélioration de la Vue (ASNAV)
F-92038 Paris La Défense Cedex
or :
39-41, rue Louis-Blanc
F-92400 Courbevoie
Telephone: (0)147176478
Telefax: (0)147176883
Founded: 1954
City Reference Coordinates: 002°20'00"E 48°52'00"N (Paris)
                                    002°15'00"E 48°54'00"N (Courbevoie)

Association Nationale Sciences Techniques Jeunesse (ANSTJ), Section Astronomique
c/o Secrétariat
16, place Jacques-Brel
F-91130 Ris-Orangis
or :
c/o Siège Social
Palais de la Découverte
Avenue Franklin Roosevelt
F-75008 Paris
Telephone: (0)169027610
Telefax: (0)169432143
Electronic Mail: cc@anstj.mime.univ-paris8.fr
WWW: http://anstj.mime.univ-paris8.fr/astronomie/pageastro.html
Founded: 1969
Membership: 1200 (including 160 clubs)
Activities: national coordination of astronomical clubs * exhibitions * education * training of teachers * holiday camps for young people
Periodicals: (5) "3, 2, 1 Infos"; "Microbe", "Newton: le journal qui tombe bien", "Les Filles d'Ariane"
City Reference Coordinates: 002°25'00"E 48°39'00"N (Ris-Orangis)
                                    002°20'00"E 48°52'00"N (Paris)

Association Normande d'Astronomie (ASNORA)
c/o Christophe Lagoude (Treasurer)
71, rue de la Falaise
F-14000 Caen
or :
52, rue de la Folie
F-14000 Caen
Telephone: (0)231478659
WWW: http://www.cpod.com/mohoweb/asnora
Founded: 1948
Membership: 50
Activities: meetings * observing * lectures * radioastronomy * outings
Periodicals: (4) "Capella" (ISSN 1266-7390)
Observatories: 1 (Laize-la-Ville)
City Reference Coordinates: 000°21'00"W 49°11'00"N

Association Novae
584, boulevard Jean-Ossola
F-06700 Saint-Laurent-du-Var
Telephone: (0)493075403
Electronic Mail: novae@castor.unice.fr
WWW: http://pegase.unice.fr/~novae/novae.html
            http://dico.unice.fr:80/~ramey/novae.html
Founded: 1988

Membership: 71
Activities: education * observing * photography * electronics * spectroscopy * mirror polishing * exchanges * image processing
* rocketry
Observatories: 1 (Observatoire de Nice)
Awards: (1/2) "Concours National de Poésie sur l'Astronomie"
City Reference Coordinates: 007°11'00"E 43°40'00"N

Association pour le Développement de l'Astronautique et de la Recherche Spatiale
● See "Nouveaux Mondes - New Worlds"

Association pour le Développement International de l'Observatoire de Nice (ADION)
c/o Observatoire de Nice
Boîte Postale 4229
F-06304 Nice Cedex 4
Telephone: (0)492003011
Electronic Mail: <userid>@obs-nice.fr
Founded: 1964
Activities: international relations * relations with local authorities * awarding prizes and medals
Periodicals: (1) "ADION Bulletin" (ISSN 0249-7522)
Awards: (1) "Médaille de l'ADION", "Prix pour Bénévoles"
City Reference Coordinates: 007°15'00"E 43°42'00"N

Association Sammielloise d'Astronomie (ASA)
c/o Sylvain Jannot
28, rue de Saint-Mihiel
F-55300 Dompcevrin
Telephone: (0)329901183
Founded: 1983
Membership: 25
Activities: observing * photography * planetarium
Coordinates: 005°29'22"E 48°54'52"N H304m (Les Paroches)
City Reference Coordinates: 005°30'00"E 48°56'00"N

Association Saranaise des Astronomes Amateurs (ASAA)
41, allée des Pyrénées
F-45770 Saran
Telephone: (0)238430507
Electronic Mail: asaa_h5@yahoo.com
WWW: http://www.astrosurf.org/asaa
Founded: 1988
Membership: 14
Activities: Sun * variable stars * photography * general observing * education * popularization
Periodicals: (10) "Astromag"
Coordinates: 001°52'00"E 48°00'00"N
City Reference Coordinates: 001°52'00"E 48°00'00"N

Association Sportive et Culturelle Toussaintaise (ASCT), Section Astronomie
3, rue du 19 Mars 1962
F-76400 Toussaint
Telephone: (0)235271940
Electronic Mail: philippe.ledoux@wanadoo.fr
Founded: 1995
Membership: 55
Activities: observing * photography * popularization * education
City Reference Coordinates: 000°23'00"E 49°45'00"N (Fécamp)

Association Stéphanoise d'Astronomie M 42
28, rue P. et D. Ponchardier
F-42100 Saint-Étienne
Founded: 1988
Membership: 50
Coordinates: 004°07'00"E 45°22'00"N H1,000m
City Reference Coordinates: 004°07'00"E 45°22'00"N

Association Sterenn, Groupement d'Études Astronomiques de Queven (GEAQ)
c/o Bernard Goumon
29, rue Chateaubriand
F-56530 Queven
Telephone: (0)297053040
Founded: 1983
Membership: 25
Activities: instrument making * photography * computing * introduction to astronomy * CCDs
Coordinates: 003°23'00"W 47°45'00"N
City Reference Coordinates: 003°23'00"W 47°45'00"N

Astro Club Aubagnais (ACA)
MJC l'Escale
Les Aires Saint-Michel
F-13400 Aubagne
or :
c/o Lionel Ruiz (President)
Quartier le Clos
F-13360 Roquevaire
Telephone: (0)442031532
Electronic Mail: kirstin@vulcain.u-3mrs.fr
WWW: http://pegase.unice.fr/~skylink/pages_clubs/aubagne/ACA.html
            http://www.astrosurf.org/pluton/aubagne/ACA.html
Founded: 1993
Membership: 25
Activities: observing * lectures * education
Observatories: 1 (La Sainte Baume)
City Reference Coordinates: 005°34'00"E 43°17'00"N (Aubagne)
                                              005°36'00"E 43°21'00"N (Roquevaire)

Astro Club de Beauce (ACB)
9 & 11, rue des Changes
F-28000 Chartres
or :
c/o Centre Socio-Educatif
131, avenue de la Résistance
F-28300 Mainvilliers
Telephone: (0)237215139
Founded: 1978
Membership: 15
Activities: observing & photography (deep sky) * introduction to observing * variable stars * comets
Coordinates: 001°34'17"E 48°25'13"N H150m (Brétigny)
                      001°35'06"E 48°34'18"N H130m (Changé)
City Reference Coordinates: 001°30'00"E 48°27'00"N (Chartres)
                                              001°28'00"E 48°28'00"N (Mainvilliers)

Astro-Club de Limagne-Sud
3, rue des Sources
F-63730 Les Marthes de Veyre
Telephone: (0)473392336
Electronic Mail: guy.berson@inserm.u-clermont1.fr (Guy Berson, President)
Founded: 1993
Membership: 15
Activities: observing * education * telescope making
City Reference Coordinates: 003°11'00"E 45°41'00"N

Astro Club Sirius
Maison des Associations
Place du 11 Novembre
Le Village
F-38090 Villefontaine
Telephone: (0)474963989
Founded: 1983
Membership: 25
Activities: observing * instrumentation * popularization
Coordinates: 005°07'55"E 45°37'41"N H345m (Relong)
City Reference Coordinates: 005°08'00"E 45°38'00"N

Astronomes Amateurs Tarnais
● See "Albiréo - Astronomes Amateurs Tarnais"

Astronomie en Chinonnaix
Le Vauroux
F-37500 Chinon
Telephone: (0)247588038
Electronic Mail: atco@wanadoo.fr
Founded: 1984
Membership: 25
Activities: popularization * lectures * slide shows * telescope making * meetings
Coordinates: 002°30'00"E 47°10'00"N H100m
City Reference Coordinates: 000°15'00"E 47°10'00"N

Astronomie en Touraine et Centre-Ouest (ATCO)
Le Vauroux
F-37500 Chinon
Telephone: (0)247932744

Electronic Mail: atco@wanadoo.fr
WWW: http://assoc.wanadoo.fr/.atco/
Founded: 1985
Membership: 120
Activities: linking together amateur astronomers and associations in the area * meetings
Periodicals: (4) "Astronomie en Touraine et Centre-Ouest" (ISSN 1243-8219, circ.: 200)
City Reference Coordinates: 000°15'00"E 47°10'00"N

Astronomie Magazine
18, boulevard Léon-Blum
F-02100 Saint-Quentin
Telephone: (0)323651919
Telefax: (0)323651909
Electronic Mail: astronomie.magazine@wanadoo.fr
WWW: http://pro.wanadoo.fr/astronomie.magazine/
● Periodical (11)
City Reference Coordinates: 003°17'00"E 47°51'00"N

Astronomie, Physique, Élaboration, Instrumentation et Observation (APHELIE)
Pavillon Colbert
35, rue Jean-Longuet
F-92290 Châtenay-Malabry
Telephone: (0)546612492 (Hervé Dole)
          (0)547021584 (Pascal Audureau)
Electronic Mail: dole@mesiom.obspm.fr
          p96dole@psiun.u-psud.fr
          audureau@ief-paris-sud.fr
WWW: http://wwwfirback.ias.fr/users/dole/aphelie
Founded: 1994
Membership: 20
Activities: meetings * lectures * instrumentation
Periodicals: "La Lettre d'Aphélie"
Observatories: 3 (Orsay, Thorigné-sur-Dué, Ablis)
City Reference Coordinates: 002°17'00"E 48°46'00"N

Astronomy and Astrophysics (A&A), Main Journal & Supplement Series
c/o C. Bertout (Editor-in-Chief)
Editorial Office
Observatoire de Paris
61, avenue de l'Observatoire
F-75014 Paris
Telephone: (0)143290541
Telefax: (0)143290557
Electronic Mail: aanda@obspm.fr
WWW: http://aanda.obspm.fr/
       http://link.springer.de/link/service/journals/00230/
       http://cdsweb.u-strasbg.fr/abstract/Abstractlist.html (Abstracts at CDS)
       http://www.ed-phys.fr/docinfos/OnlineAetA.html
Founded: 1969
Staff: 3
● Journal (24) (ISSN 0004-6361)
City Reference Coordinates: 002°20'00"E 48°52'00"N

Astronomy and Astrophysics Review
c/o L. Woltjer (Editor)
Observatoire de Haute Provence
F-04870 Saint-Michel-l'Observatoire
Telephone: (0)492706400
Telefax: (0)492766295
Founded: 1989
● Journal (4) (ISSN 0935-4956)
City Reference Coordinates: 005°43'00"E 43°56'00"N

Astronomy and Space Information Service (ASIS)
c/o Service RNGC-0006
5, rue des Messanges
F-67120 Duttlenheim
Telephone: (0)388150743
Telefax: (0)388491255
Founded: 1985
Staff: 1
Activities: directories * databases * dictionaries * WWW * compilations * documentation * astronomy and space sciences * lectures * trainings * colloquia * conferences
City Reference Coordinates: 007°39'00"E 48°33'00"N

Astroqueyras (AQ)
c/o Jean-Christophe Le Floch (Treasurer)
3, rue Jacques Daguerre
F-92500 Rueil-Malmaison
or :
Mairie de Saint-Véran
F-05350 Saint-Véran
Telephone: (0)142705442
          (0)147499255
Telefax: (0)142705442
          (0)147499255
Electronic Mail: jclf@magic.fr
WWW: http://www.bdl.fr/s2p/stveran
          http://perso.wanadoo.fr/astroqueyras
Founded: 1989
Membership: 103
Activities: observing
Periodicals: (4) "Lettre Astroqueyras"
Coordinates: 006°54'24"E 44°41'56"N H2930m
City Reference Coordinates: 002°11'00"E 48°53'00"N (Rueil-Malmaison)
                                006°52'00"E 44°42'00"N (Saint-Véran)

Astro-Terre
c/o Observatoire des Vallons
F-83630 Bauduen
Telephone: (0)494843919
WWW: http://www.ec-lille.fr/~astro/
          http://www.astrosurf.org/pluton/bauduen/home.html
Founded: 1989
Membership: 6
Activities: observing * photography * construction of telescopes and observing domes for amateur astronomers
Coordinates: 006°11'03"E 43°43'21"N
City Reference Coordinates: 006°10'00"E 43°44'00"N

Atelier d'Helios (L'_)
● See "L'Atelier d'Helios"

Bergoz Instrumentation
F-01170 Crozet
Telephone: (0)450410089
Telefax: (0)450410199
Electronic Mail: sales@bergoz.com
WWW: http://www.bergoz.com/
Founded: 1981
Staff: 10
Activities: manufacturing particle beam instrumentation
Awards: (1/2) "Faraday Cup"
Coordinates: 006°00'38"E 46°16'52"N H542m
City Reference Coordinates: 006°03'00"E 46°20'00"N (Gex)

Blénod Animation Loisirs, Section Astronomie
c/o Patrick Brandebourg
16, rue de la Fontaine
F-54700 Blénod-lès-Pont-à-Mousson
Telephone: (0)383823105
Founded: 1971
Membership: 15
Activities: photometry * popularization * activities in schools
City Reference Coordinates: 006°03'00"E 48°53'00"N

Bureau des Longitudes (BDL)
● See now "Observatoire de Paris, Institut de Mécanique Céleste et de Calcul des Éphémérides (IMCCE)"

Bureau Gravimétrique International (BGI)
c/o Observatoire Midi-Pyrénées
18, avenue Edouard-Belin
F-31055 Toulouse Cedex
Telephone: (0)561332889
Telefax: (0)561253098
Electronic Mail: balmino.uggi@cnes.fr (Georges Balmino, Director)
WWW: http://www.obs-mip.fr/uggi/bgi.html
Founded: 1951
Staff: 6
Activities: collecting, checking, archiving and distributing gravity measurements
Periodicals: (2) "Bulletin d'Information"

City Reference Coordinates: 001°26'00"E 43°36'00"N

Bureau International de l'Heure (BIH)
● See now "International Earth Rotation Service (IERS), Central Bureau"

Bureau International des Poids et Mesures (BIPM)
Pavillon de Breteuil
F-92312 Sèvres Cedex
Telephone: (0)145077070
Telefax: (0)145342021
Electronic Mail: info@bipm.fr
                    <userid>@bipm.fr
WWW: http://www.bipm.fr/
Founded: 1875
Staff: 60
Activities: unification of the basic units for physics, with the Time Section establishing the International Atomic Time and, jointly with the International Earth Rotation Service, the Coordinated Universal Time
Periodicals: (12) "Circular T"; (1) "Annual Report of the Time Section" (ISSN 1016-6114); "Comptes-Rendus des Séances de la Conférence Générale des Poids et Mesures", "Procès-Verbaux des Séances du Comité International des Poids et Mesures", "Sessions des Comités Consultatifs" (ISSN 0588-6228), "Metrologia"
City Reference Coordinates: 002°17'00"E 48°49'00"N

Bureau National de Métrologie (BNM), Laboratoire Primaire du Temps et des Fréquences (LPTF)
c/o Observatoire de Paris
61, avenue de l'Observatoire
F-75014 Paris
Telephone: (0)140512213
Telefax: (0)143255542
Electronic Mail: lptfop@obspm.fr
WWW: http://opdaf1.obspm.fr/www/lptf.html
Founded: 1975
Staff: 30
Activities: time and frequency metrology
Periodicals: (12) "Bulletin"
City Reference Coordinates: 002°20'00"E 48°52'00"N

Cassini Astronomie
18, avenue de l'Europe
(Piscine Municipale)
F-94190 Villeneuve-Saint-Georges
Telephone: (0)143896099
Founded: 1972
Membership: 30
Activities: instrumentation * photography * CCDs * observing * popularization
Periodicals: (12) "Albiréo"
● Previously called "Club Astronomique Cassini"
City Reference Coordinates: 002°27'00"E 48°44'00"N quest Maxime Nesnier (Secretary) (981014)

Centre Culturel de l'Astronomie (CCA)
Boîte Postale 1088
F-34007 Montpellier
Telephone: (0)467617401
Telefax: (0)467611008
Founded: 1984
Membership: 50
Activities: popularization * studies of astronomical heritage
Periodicals: (4) "Revue Montpelliéraine d'Astronomie Languedocienne" (ISSN 0246-1390)
City Reference Coordinates: 003°53'00"E 43°36'00"N

Centre d'Analyse des Images (CAI)
● See "Institut National des Sciences de l'Univers (INSU), Centre d'Analyse des Images (CAI)"

Centre d'Astronomie
Le Moulin à Vent
F-04870 Saint-Michel-l'Observatoire
Telephone: (0)492766969
Telefax: (0)492766767
Electronic Mail: centre.astro@wanadoo.fr
WWW: http://www.astrosurf.com/centre.astro/
Founded: 1998
Activities: observing * displays * education * festival * camps
City Reference Coordinates: 005°43'00"E 43°56'00"N

Centre de Culture Scientifique et Technique (CCST)
28, rue Albert 1er

F-17000 La Rochelle
Telephone: (0)546411825
Telefax: (0)546506365
Electronic Mail: jacques.vialle@wanadoo.fr (Jacques Vialle, Astronomy Advisor)
Founded: 1990
Staff: 3
Activities: public education * exhibitions * lectures * mobile planetarium
Coordinates: 001°11'00"W 46°10'12"N
City Reference Coordinates: 001°10'00"W 46°10'00"N

Centre de Culture Scientifique, Technique et Industrielle (CCSTI)
6, place des Colombes
F-35000 Rennes
Telephone: (0)299352823
Telefax: (0)299352821
Electronic Mail: lespace-des-sciences@wanadoo.fr
WWW: http://astro.u-strasbg.fr/Obs/PLANETARIUM/ccsti.html
Founded: 1995
Activities: travelling planetarium
City Reference Coordinates: 001°41'00"W 48°05'00"N

Centre de Données astronomiques de Strasbourg (CDS)
c/o Observatoire Astronomique
11, rue de l'Université
F-67000 Strasbourg
Telephone: (0)388150721 (Director)
         (0)388150720 (Secretary)
Telefax: (0)388150740
Electronic Mail: question@simbad.u-strasbg.fr
         <userid>@simbad.u-strasbg.fr
WWW: http://cdsweb.u-strasbg.fr/CDS.html (English Homepage)
      http://cdsweb.u-strasbg.fr/CDS-f.html (French Homepage)
Founded: 1972
Staff: 12
Activities: astronomical data center * bibliographical service * compilation, critical evaluation and distribution of astronomical data * on-line catalogue server * SIMBAD database * Aladin project * yellow-page services * research in classification, astronomical statistics, quality control
Periodicals: (2) "Bulletin d'Information du CDS" (ISSN 0242-6536); "Publications Spéciales du CDS" (ISSN 0764-9614)
● Operated under agreement between the "Institut National des Sciences de l'Univers (INSU)" and the "Université de Strasbourg I (Université Louis Pasteur - ULP)"
City Reference Coordinates: 007°45'00"E 48°35'00"N

Centre de Physique des Particules de Marseille (CPPM)
163, avenue de Luminy
Case 907
F-13288 Marseille Cedex 9
Telephone: (0)491827200
Telefax: (0)491827299
Electronic Mail: <userid>@cppm.in2p3.fr
WWW: http://cppm.in2p3.fr/
      http://antares.in2p3.fr/antares/ (Antares Project)
Founded: 1982
Staff: 120
Activities: particle physics * high-energy physics * astroparticle physics * astrophysics * neutrino astronomy * electroweak and strong interactions * conservation laws * standard model * supersymmetry * detectors * data acquisition
Coordinates: 005°57'00"E 42°50'00"N (Antares Neutrino Detector)
● Jointly run by the "Centre National de la Recherche Scientifique (CNRS)" , the "Institut National de Physique Nucléaire et de Physique des Particules (IN2P3)" and the "Université de la Méditerranée (Aix-Marseille II)"
City Reference Coordinates: 005°24'00"E 43°18'00"N

Centre de Recherche Astronomique de Lyon (CRAL)
9, avenue Charles André
F-69561 Saint-Genis-Laval Cedex
or :
46, allée d'Italie
F-69364 Lyon Cedex 07
Telephone: (0)478868383
         (0)472728000
Telefax: (0)478868386
         (0)472728080
Electronic Mail: <userid>@obs.univ-lyon1.fr
WWW: http://www-obs.univ-lyon1.fr/
      http://image.univ-lyon1.fr/
      http://www-obs.univ-lyon1.fr/base/leda-doc.html (LEDA)
Founded: 1995 (1878 as "Observatoire de Lyon")
Staff: 60

Activities: research * cosmology * nuclei of galaxies * stellar astrophysics * hydrodynamics * instrumentation * education * Lyon-Meudon Extragalactic Database (LEDA)
Periodicals: "Reprints"
Coordinates: 004°47'06"E 45°41'42"N H299m
● Merging of the "Observatoire de Lyon" and of the "Groupe d'Astrophysique" from the "École Normale Supérieure de Lyon (ENSL)"
City Reference Coordinates: 004°48'00"E 45°41'00"N (Saint-Genis-Laval)
004°51'00"E 45°45'00"N (Lyon)

Centre de Recherches en Physique de l'Environnement Terrestre et Planétaire (CRPE)
● See now "Centre d'Étude des Environnements Terrestre et Planétaires (CETP)"

Centre de Spectrométrie Nucléaire et de Spectrométrie de Masse (CSNSM)
● See "Université de Paris XI, Centre de Spectrométrie Nucléaire et de Spectrométrie de Masse (CSNSM)"

Centre d'Étude des Environnements Terrestre et Planétaires (CETP), Vélizy
10-12, avenue de l'Europe
F-78140 Vélizy
Telephone: (0)139254905
Telefax: (0)139254922
Electronic Mail: <userid>@cetp.ipsl.fr
WWW: http://balsa.cetp.ipsl.fr/
Founded: 1965
Staff: 170
Activities: atmosphere * ionosphere * magnetosphere * solar wind * planets * remote sensing * telecommunications * signal processing
● Formerly known as "Centre de Recherches en Physique de l'Environnement Terrestre et Planétaire (CRPE)"
City Reference Coordinates: 004°51'00"E 45°45'00"N

Centre d'Étude des Environnements Terrestre et Planétaires (CETP), Saint-Maur-des-Fossés
4, avenue de Neptune
F-94107 Saint-Maur-des-Fossés Cedex
Telephone: (0)145114200
(0)145114270
Telefax: (0)148894333
Electronic Mail: <userid>@cetp.ipsl.fr
WWW: http://balsa.cetp.ipsl.fr/
Founded: 1994
Activities: ionosphere * magnetosphere * plasma * meteorology * radar
● Formerly known as "Centre de Recherches en Physique de l'Environnement Terrestre et Planétaire (CRPE)"
City Reference Coordinates: 002°30'00"E 48°48'00"N

Centre d'Études et de Recherches de Toulouse (CERT)
● See "Office National d'Études et de Recherches Aérospatiales (ONERA), Centre d'Études et de Recherches de Toulouse (CERT)"

Centre d'Études et de Recherches en Géodynamique et Astrométrie (CERGA)
● See "Observatoire de la Côte d'Azur (OCA), Centre d'Études et de Recherches en Géodynamique et Astrométrie (CERGA)"

Centre d'Études Nucléaires (CEN)
● See "Commissariat à l'Énergie Atomique (CEA), Centre d'Études Nucléaires (CEN)"

Centre d'Étude Spatiale des Rayonnements (CESR)
9, avenue du Colonel Roche
Boîte Postale 4346
F-31028 Toulouse Cedex 4
Telephone: (0)561556666
Telefax: (0)561558692 (Director)
(0)561556701 (CESR)
Electronic Mail: <userid>@cesr.fr
WWW: http://www.cesr.fr/
Founded: 1964
Staff: 2
Activities: high-energy astrophysics (gamma-rays) * IR * space science (particles and fields) * magnetosphere * black holes * solar wind * plasmas * planetology
City Reference Coordinates: 001°26'00"E 43°36'00"N

Centre du Cadran Solaire (CCS)
Boîte Postale 1088
F-34007 Montpellier
Telephone: (0)467611095
Telefax: (0)467611008

Founded: 1992
Membership: 20
Activities: studies of the sundial heritage in the Languedoc-Roussillon region
City Reference Coordinates: 003°53'00"E 43°36'00"N

Centre Européen de Recherche et de Formation Avancée en Calcul Scientifique (CERFACS)
42, avenue Gaspard-Coriolis
F-31057 Toulouse Cedex 1
Telephone: (0)561193131
Telefax: (0)561193000
Electronic Mail: secretar@cerfacs.fr
              <userid>@cerfacs.fr
WWW: http://www.cerfacs.fr/
Founded: 1988
Staff: 65
Activities: climate * ocean modelling * data assimilation * computational fluid dynamics * electromagnetism * parallel algorithms * technology transfer
City Reference Coordinates: 001°26'00"E 43°36'00"N

Centre National de la Recherche Scientifique (CNRS)
3, rue Michel-Ange
F-75794 Paris Cedex 16
Telephone: (0)144964000
Telefax: (0)144965000
Electronic Mail: <userid>@cnrs-dir.fr
WWW: http://www.cnrs.fr/
      http://www.cnrs.org/
Founded: 1939
Membership: 26000
Activities: funding research, laboratories and institutions
City Reference Coordinates: 002°20'00"E 48°52'00"N

Centre National de la Recherche Scientifique (CNRS), Centre de Physique des Particules de Marseille (CPPM)
● See "Centre de Physique des Particules de Marseille (CPPM)"

Centre National de la Recherche Scientifique (CNRS), Centre de Physique Théorique (CPT)
Luminy
Case 907
Route Léon Lachamp
F-13288 Marseille Cedex 9
Telephone: (0)491269500
Telefax: (0)491269553
Electronic Mail: <userid>@cpt.uni-mrs.fr
WWW: http://www.cpt.univ-mrs.fr/
Founded: 1969
Staff: 100
Activities: algebraic methods * constructive field theory & statistical mechanics * elementary-particle physics * physics & geometry * functional analysis * stochastic physics * astronomy & cosmology * dynamical systems * solid-state physics * numerical methods in physics * non-commutative differential geometry * field theory & operator algebra * non-relativistic quantum mechanics * elementary-particle and quantum field theory * wavelet analysis
City Reference Coordinates: 005°24'00"E 43°18'00"N

Centre National de la Recherche Scientifique (CNRS), Délégation aux Systèmes d'Information (DSI)
1, place Aristide-Briand
F-92195 Meudon Cedex
Telephone: (0)145075051
Telefax: (0)145075191
Electronic Mail: <userid>@dsi.cnrs.fr
WWW: http://www.dsi.cnrs.fr/
      http://www.lmb.cnrs.fr/ (LMB Actu)
Founded: 1996 (1990 as "Service de l'Organisation et du Système d'Information - SOSI)
Staff: 90
Periodicals: (5) "Le Micro Bulletin"
City Reference Coordinates: 002°14'00"E 48°48'00"N

Centre National de la Recherche Scientifique (CNRS), Éditions
● See "CNRS Éditions"

Centre National de la Recherche Scientifique (CNRS), Images/media
● See now "CNRS Images/media FEMIS - CICT"

Centre National de la Recherche Scientifique (CNRS), Institut d'Astrophysique de Paris (IAP)
98bis, boulevard Arago

F-75014 Paris
Telephone: (0)144328000 (switchboard)
                   (0)144328016 (Director)
Telefax: (0)144328001
Electronic Mail: <userid>@iap.fr
WWW: http://www.iap.fr/
Founded: 1936
Staff: 130
Activities: research * cosmology * extragalactic astronomy * nucleosynthesis * stellar astronomy * interstellar astronomy * solar astronomy * planetary sciences * education * databases * numerical simulations * data processing
City Reference Coordinates: 002°20'00"E 48°52'00"N

Centre National de la Recherche Scientifique (CNRS), Institut d'Astrophysique Spatiale (IAS)
c/o Université de Paris XI
Bâtiment 121
F-91405 Orsay Cedex
Telephone: (0)169858500
Telefax: (0)169858675
WWW: http://www.ias.fr/
          http://www.ias.fr/cdp/Welcome.html (Centre de Données Planétaires - CDP)
Founded: 1968
Staff: 140
Activities: solar system * galaxies * solar and stellar physics
City Reference Coordinates: 002°11'00"E 48°48'00"N

Centre National de la Recherche Scientifique (CNRS), Institut de l'Information Scientifique et Technique (INIST)
2, allée du Parc de Brabois
F-54514 Vandoeuvre-lès-Nancy Cedex
Telephone: (0)383504600
              (0)383504664 (Customer Desk)
Telefax: (0)383504650
            (0)383504666 (Customer Desk)
Electronic Mail: infoclient@inist.fr
WWW: http://www.inist.fr/
Founded: 1988
Staff: 330
Activities: producing bibliographical databases PASCAL, FRANCIS and article@inist * providing primary documents
Periodicals: "Special'IST" (ISSN 1244-9148)
City Reference Coordinates: 006°12'00"E 48°41'00"N

Centre National de la Recherche Scientifique (CNRS), Institut des Sciences de la Terre, de l'Eau et de l'Espace de Montpellier (ISTEEM)
● See "Institut des Sciences de la Terre, de l'Eau et de l'Espace de Montpellier (ISTEEM)"

Centre National de la Recherche Scientifique (CNRS), Institut Gassendi pour la Recherche Astronomique en Provence (IGRAP)
● See "Institut Gassendi pour la Recherche Astronomique en Provence (IGRAP)"

Centre National de la Recherche Scientifique (CNRS), Laboratoire d'Astronomie Spatiale (LAS)
Boîte Postale 8
F-13376 Marseille Cedex 12
or :
Traverse du Siphon
Allée Peiresc
F-13012 Marseille
Telephone: (0)491055900
Telefax: (0)491661855
Electronic Mail: <userid>@astrsp-mrs.fr
WWW: http://www.astrsp-mrs.fr/
Founded: 1965
Staff: 120
Activities: astronomy * astrophysics * image processing
● See also "Institut Gassendi pour la Recherche Astronomique en Provence (IGRAP)"
City Reference Coordinates: 005°24'00"E 43°18'00"N

Centre National de la Recherche Scientifique (CNRS), Laboratoire de l'Accélérateur Linéaire (LAL)
● See "Laboratoire de l'Accélérateur Linéaire (LAL)"

Centre National de la Recherche Scientifique (CNRS), Observatoire de Haute Provence (OHP)
F-04870 Saint-Michel-l'Observatoire
Telephone: (0)492706400
Telefax: (0)492766295
Electronic Mail: <userid>@obs-hp.fr
WWW: http://www.obs-hp.fr/

Founded: 1936
Staff: 75
Activities: maintaining observing facilities for visiting astronomers * peculiar stars (cool, variable, X sources) * galaxies * QSOs * ISM * interferometry
Periodicals: (1) "Rapport Annuel d'Activités"; "La Lettre de l'OHP", "Pré-publications"
Coordinates: 005°42'47"E 43°55'53"N H665m
● See also "Institut Gassendi pour la Recherche Astronomique en Provence (IGRAP)"
City Reference Coordinates: 005°43'00"E 43°56'00"N

Centre National de la Recherche Scientifique (CNRS), Service d'Aéronomie
Boîte Postale 3
F-91371 Verrières-le-Buisson Cedex
Telephone: (0)164474245
Telefax: (0)169202999
Electronic Mail: <userid>@aero.jussieu.fr
WWW: http://www.aero.jussieu.fr/
Founded: 1958
Staff: 140
Activities: optical studies of Earth and planetary atmospheres (UV, visible, IR, lidars)
Observatories: 1 (Station Géophysique de l'Observatoire de Haute Provence - see separate entry)
City Reference Coordinates: 002°16'00"E 48°45'00"N

Centre National d'Enseignement à Distance (CNED)
Avenue du Téléport
Boîte Postale 200
F-86960 Futuroscope Cedex
Telephone: (0)549499494
Telefax: (0)549499696
Electronic Mail: Teletel/Minitel: 3615 Code CNED
accueil@cned.fr
WWW: http://www.cned.fr/
Founded: 1939
City Reference Coordinates: 000°20'00"E 46°35'00"N

Centre National d'Études Spatiales (CNES)
(French Space Agency)
2, place Maurice-Quentin
F-75039 Paris Cedex 01
Telephone: (0)144767500
Telefax: (0)144767676
Electronic Mail: <userid>@cnes.fr
WWW: http://www.cnes.fr/
Founded: 1962
Staff: 2500
Activities: developing French space activities (national projects, cooperative ventures, participation in ESA's programmes)
Periodicals: (6) "Calendrier des Manifestations Spatiales"; (4) "CNES Magazine" (ISSN 1283-9817, circ.: 9000); (2) "Qualité Espace" (ISSN 1556-6558, circ.: 3000); (1) "Rapport Annuel d'Activités"; (1/2) "Rapport au COSPAR" (circ.: 2,500)
City Reference Coordinates: 002°20'00"E 48°52'00"N

Centre National d'Études Spatiales (CNES), Centre Spatial de Toulouse (CST)
18, avenue Edouard-Belin
F-31055 Toulouse Cedex
Telephone: (0)561273131
Telefax: (0)561273179
WWW: http://www.cnes.fr/scripts/toulouse.html
Founded: 1968
Staff: 1650
Activities: managing space programmes * launch and maintenance of satellites * telecommunications * remote sensing * microgravity * optics * space medicine * computing * stratospheric balloons
Periodicals: "CST Information" (ISSN 1153-0073)
City Reference Coordinates: 001°26'00"E 43°36'00"N

Centre National d'Études Spatiales (CNES), Centre Spatial d'Évry
Rond Point de l'Espace
F-91023 Évry Cedex
Telephone: (0)160877111
Telefax: (0)160877397
WWW: http://www.cnes.fr/
Founded: 1962 (CNES)
City Reference Coordinates: 002°27'00"E 48°38'00"N

Centre National d'Études Spatiales (CNES), Centre Spatial Guyanais (CSG)
Boîte Postale 726
F-97387 Kourou Cedex

(Guyane Française)
Telephone: (0)594335111
               (0)594325111
Telefax: (0)594334766
WWW: http://www.cnes.fr/scripts/kourou.html
Founded: 1964
Staff: 1800
Activities: operational launch site
Periodicals: (4) "Latitude 5" (ISSN 0293-0072, circ.: 2,500)
City Reference Coordinates: 052°39'00"W 05°09'00"N

Centre National d'Études Spatiales (CNES), Délégation à la Communication
c/o Centre Spatial de Toulouse
18, avenue Edouard-Belin
F-31055 Toulouse Cedex
Telephone: (0)561273131
Telefax: (0)561281327
WWW: http://www.cnes.fr/
Founded: 1962 (CNES)
Activities: see main entry
Periodicals: see main entry
City Reference Coordinates: 001°26'00"E 43°36'00"N

Centre National d'Études Spatiales (CNES), Institut Gassendi pour la Recherche Astronomique en Provence (IGRAP)
• See "Institut Gassendi pour la Recherche Astronomique en Provence (IGRAP)"

Centre Régional de Promotion de la Culture Scientifique, Technique et Industrielle
Rue Vercord
F-59650 Villeneuve d'Ascq
Telephone: (0)320193600
Telefax: (0)320193601
Founded: 1989
Activities: research * creation of cultural products * exhibitions * lectures * tra_ning * information centre * planetarium project
• Alternate name: "ALIAS"
City Reference Coordinates: 003°10'00"E 50°37'00"N

Centre Spatial de Toulouse (CST)
• See "Centre National d'Études Spatiales (CNES), Centre Spatial de Toulouse (CST)"

Centre Spatial d'Évry
• See "Centre National d'Études Spatiales (CNES), Centre Spatial d'Évry"

Centre Spatial Guyanais (CSG)
• See "Centre National d'Études Spatiales (CNES), Centre Spatial Guyanais (CSG)"

Cépaduès Éditions
111, rue Nicolas-Vauquelin
F-31100 Toulouse
Telephone: (0)561405736
Telefax: (0)561417989
Electronic Mail: cepadues@editions-cepadues.fr
WWW: http://www.editions-cepadues.fr/
Founded: 1967
Staff: 8
• Publisher
City Reference Coordinates: 001°26'00"E 43°36'00"N

CERGA
• See "Observatoire de la Côte d'Azur (OCA), Centre d'Études et de Recherches en Géodynamique et Astrométrie (CERGA)"

Cité de l'Espace
Avenue Jean-Gonord
Boîte Postale 5855
F-31506 Toulouse Cedex 5
Telephone: (0)562716480
               (0)562714871 (Online infos)
Telefax: (0)561807470
WWW: http://www.cite-espace.com/
Founded: 1997
Staff: 60

Activities: hands-on exhibits * space exploration * planetarium
City Reference Coordinates: 001°26'00"E 43°36'00"N

Cité des Sciences et de l'Industrie de la Villette, Planétarium
F-75930 Paris Cedex 19
or :
30, avenue Corentin-Cariou
F-75019 Paris
Telephone: (0)140057022
Telefax: (0)140057118
Electronic Mail: planetar@world-net.sct.fr
WWW: http://www.club-internet.fr/cite-sciences/
Founded: 1986
Activities: shows * exhibitions
City Reference Coordinates: 002°20'00"E 48°52'00"N

Club Ajaccien des Amateurs d'Astronomie (C3A)
c/o Centre Scientifique de Vignola
Route des Sanguinaires
F-20000 Ajaccio
Telephone: (0)495218448
Founded: 1986
Membership: 74
Activities: observing * photography * popularization * education * Sun * planetarium
Coordinates: 008°39'03"E 41°54'44"N H50m (Observatoire de Vignola)
City Reference Coordinates: 008°44'00"E 41°55'00"N

Club Astro Alpha Centauri
c/o MJC de Carcassonne
91, rue Aimé-Ramond
F-11000 Carcassonne
Telephone: (0)468111700
Founded: 1981
Membership: 12
Activities: monthly lectures * visual observing * meteors * meetings
Coordinates: 002°08'40"E 43°04'10"N H255m (Pauligne, Limoux)
City Reference Coordinates: 002°21'00"E 43°13'00"N

Club Astro de Wittelsheim (CAW)
c/o Jean-Luc Garambois (President)
32, rue du Rempart
F-68190 Ensisheim
Telephone: (0)389810718
Founded: 1977
Membership: 66
Activities: observing * popularization * meetings
Periodicals: (6) "Procyon" (ISSN 0297-1038)
City Reference Coordinates: 007°20'00"E 47°51'00"N (Ensisheim)
                              007°20'00"E 47°49'00"N (Wittelsheim)

Club Astro Guynemer
c/o Comité d'Entreprise
Thomson CSF Optronique
Rue Guynemer
Boîte Postale 55
F-78283 Guyancourt Cedex
Founded: 1999
Membership: 56
Activities: instrument loan * education * circulating periodicals
City Reference Coordinates: 002°08'00"E 48°48'00"N (Versailles)

Club Astro Junior M 67
7, quai Finkwiller
F-67000 Strasbourg
Telephone: (0)388364166
Founded: 1986
Membership: 25
Activities: observing * meetings * photography * education
City Reference Coordinates: 007°45'00"E 48°35'00"N

Club Astronomie de Saint-Claude
16, route de Chaumont
F-39200 Saint-Claude
Telephone: (0)384453156
Founded: 1973

Membership: 15
Activities: observing * photography
Coordinates: 005°48'09"E 46°15'38"N H1,180m
City Reference Coordinates: 005°52'00"E 46°23'00"N

Club Astronomique de la Région Lilloise (CARL)
23, rue Gosselet
F-59000 Lille
Telephone: (0)320859919
Telefax: (0)320861556
Founded: 1976
Membership: 58
Activities: promoting astronomy and related fields * popularization * observing * exhibitions * workshops * mobile planetarium

Periodicals: (6) "Astronomie pour Tous"
Observatories: 1 (Ferme du Héran, Villeneuve d'Ascq)
City Reference Coordinates: 003°04'00"E 50°38'00"N

Club Astronomique Véga de la Lyre
15, avenue Juncarret
F-33870 Vayres
Telephone: (0)557748100
Founded: 1985
Membership: 80
Activities: lectures * observing * introduction to astronomy
Coordinates: 000°21'00"W 44°52'00"N H33m
City Reference Coordinates: 000°21'00"W 44°52'00"N

Club Copernic
Maison des Associations
642, rue des Batteries
F-83600 Fréjus
Telephone: (0)494828361
Telefax: (0)494828361
WWW: http://www.chez.com/pstj/Astro/copernic.htm
Founded: 1974
Membership: 60
Activities: amateur astronomy and astrophysics * talks * lectures * visits * observing * education * popularization
City Reference Coordinates: 006°44'00"W 43°26'00"N

Club d'Astronomie Alpha Centaure (CAAC)
c/o Elisabeth Margaron-Invernizzi
Résidence "Le Parc"
1, avenue Dr. Bonnet
F-26100 Romans-sur-Isère
Telephone: (0)475430290
Telefax: (0)475431905
Electronic Mail: alpha-centaure@wanadoo.fr
WWW: http://assoc.wanadoo.fr/alpha-centaure/
Founded: 1979
Membership: 50
Activities: theoretical study of astronomy * observing natural, terrestrial and celestial phenomena
Coordinates: 005°06'00"E 45°28'00"N H630m (Rochefort-Samson)
City Reference Coordinates: 005°06'00"E 45°28'00"N

Club d'Astronomie de Chamonix
c/o Maison pour Tous
94, promenade du Fori
F-74400 Chamonix
Telephone: (0)450534516
Founded: 1979
Membership: 21
Activities: public presentations * lectures * exhibitions * photography * observing at high-altitude sites (Brévent: 2525m, Aiguille du Midi: 3842m, Grands Montets: 3300m, Observatoire Vallot: 4350m, Mont Blanc: 4807m, Col de Balme: 2200m)
Periodicals: (12) "Albedo"; (1) "Plongée dans l'Univers"
Coordinates: 007°00'00"E 45°55'00"N H4,807m (Observatoire du Mont Blanc)
City Reference Coordinates: 006°52'00"E 45°55'00"N

Club d'Astronomie de Lyon-Ampère (CALA)
37, rue Paul-Cazeneuve
F-69008 Lyon
Telephone: (0)478012905
Founded: 1968
Membership: 150
Activities: education * observing * photography * instrument making * popularization * lectures * planetarium shows

Periodicals: (4) "Nouvelle Gazette du Club 69 (NGC 69)"
City Reference Coordinates: 004°51'00"E 45°45'00"N

Club d'Astronomie de Villemur

c/o Jean-Louis Prieur
Bondigoux
F-31340 Villemur
Electronic Mail: prieur@obs-mip.fr (Jean-Louis Prieur, President)
Founded: 1975
Membership: 12
Activities: observing * photography * instrumentation * popularization
Coordinates: 001°32'00"E 43°53'00"N
City Reference Coordinates: 001°32'00"E 43°53'00"N

Club d'Astronomie Janus

1, place Saint-Just
F-92230 Gennevilliers
WWW: http://www.multimania.com/astrojan/
Founded: 1979
Membership: 22
Activities: lectures in schools * observing * computing * photography * CCD camera
Coordinates: 002°17'15"E 48°55'31"N
City Reference Coordinates: 002°18'00"E 48°06'00"N

Club d'Astronomie "Randonnée Céleste"

c/o ACS de Boissy-Fresnoy
Mairie
F-60440 Boissy-Fresnoy
or :
18, rue Jean-Charron
F-60440 Boissy-Fresnoy
Telephone: (0)344888300
Telefax: (0)344888300
Electronic Mail: bdussart@bigfoot.com (Bernard Dussart)
WWW: http://www.astrosurf.org/rceleste/
Founded: 1993
Membership: 23
Activities: observing * lectures * exhibitions * popularization * sundials
City Reference Coordinates: 002°49'00"E 49°08'00"N (Nanteuil le Haudouin)

Club d'Astronomie Spica

Maison des Associations
4, avenue de Verdun
F-06800 Cagnes-sur-Mer
Telephone: (0)493244969
WWW: http://www.chez.com/pstj/Astro/spica.htm
City Reference Coordinates: 007°09'00"E 43°04'00"N

Club d'Astronomie Uranie

Maison des Jeunes et de la Culture
2A, avenue de la Libération
F-42400 Saint-Chamond
Telephone: (0)477317115
Telefax: (0)477220422
Electronic Mail: cbreisse@easynet.fr
WWW: http://perso.easynet.fr/~cbreisse/
City Reference Coordinates: 004°30'00"E 45°28'00"N

Club d'Information Scientifique (CIS) de la Poste et de France Telecom

57, rue de la Colonie
F-75013 Paris
Telephone: (0)143132151
Electronic Mail: cis-colonie@netcourrier.com
WWW: http://www.astrosurf.org/cis/
Founded: 1976
Membership: 230
Activities: lectures * observing * photography * CCDs * education
Periodicals: "Regard de l'Astronome"
City Reference Coordinates: 002°20'00"E 48°52'00"N

Club Éclipse

c/o Thierry Midavaine
102, rue Vaugirard
F-75006 Paris
or :

c/o André Bradel
26, boulevard Jean-Jaurès
F-92100 Boulogne-Billancourt
Telephone: (0)146040791 (A. Bradel)
Electronic Mail: club-eclipse@egroups.com
WWW: http://www.astroclub.net/club_eclipse
Founded: 1979
Membership: 14
Activities: QSOs * comets * CCDs * image intensifiers * photography * Messier cbjects * minor planets
Periodicals: (4) "La Lettre du Club Éclipse"
City Reference Coordinates: 002°20'00"E 48°52'00"N (Paris)
                               002°15'00"E 48°50'00"N (Boulogne-Billancourt)

Club Jean Perrin (CJP)
Palais de la Découverte
Avenue F.D. Roosevelt
F-75008 Paris
Telephone: (0)140748108
Electronic Mail: moreau@astra.astro.ulg.ac.be
Founded: 1970
Membership: 25
Activities: introduction to theoretical astronomy and astrophysics * observing * photography * workshops
City Reference Coordinates: 002°20'00"E 48°52'00"N

CNRS Éditions
(Siège Social)
15, rue Malebranche
F-75005 Paris
or :
(Librairie)
151 bis, rue Saint-Jacques
F-75005 Paris
Telephone: (0)153102700
          (0)153100505 (Librairie)
Telefax: (0)153102727
          (0)153100507 (Librairie)
Electronic Mail: editions@edition.cnrs.fr
WWW: http://www.cnrs.fr/Editions/accueil.html
• Publisher
City Reference Coordinates: 002°20'00"E 48°52'00"N

CNRS Images/media FEMIS - CICT
27, rue Paul-Bert
F-94204 Ivry-sur-Seine Cedex
Telephone: (0)149604120
Telefax: (0)149604156
Electronic Mail: images.media@dr1.cnrs.fr
WWW: http://www.cnrs.fr/
Founded: 1988
Staff: 12
Activities: multimedia * public relations * festivals * conferences * exhibitions
Periodicals: (1) "Image & Science"
• Mixed unit of the "Centre National de la Recherche Scientifique (CNRS)", the "Fondation Européenne des Métiers de l'Image et du Son (FEMIS)" and the "International Film, Television and Audiovisual Council" by UNESCO (CICT/IFTC)
City Reference Coordinates: 002°23'00"E 48°49'00"N

CODATA
• See "Committee on Data for Science and Technology (CODATA)"

Collège J. Valéri, Association du Planétarium
128, avenue Saint-Lambert
F-06100 Nice
Telephone: (0)492090924
Telefax: (0)492090924
Electronic Mail: planet.valeri@wanadoo.fr
Founded: 1967
Membership: 90
Activities: shows * workshops * observing * mirror polishing
Periodicals: "Le Gluon"
City Reference Coordinates: 007°15'00"E 43°42'00"N

Comité de Liaison Enseignants Astronomes (CLEA)
c/o Laboratoire d'Astronomie
Bâtiment 470
Université de Paris-Sud

F-91405 Orsay Cedex
Telephone: (0)169157766
Telefax: (0)169156381
Founded: 1979
Membership: 1000
Activities: training of teachers in astronomy * Starlab planetarium
Periodicals: (4) "Les Cahiers Clairaut" (ISSN 0758-234x, circ.: 1,800)
City Reference Coordinates: 002°11'00"E 48°48'00"N

Commissariat à l'Énergie Atomique (CEA), Centre d'Études de Bruyères-le-Chatel
Service PTN
Boîte Postale 12
F-91680 Bruyères-le-Chatel
Telephone: (0)169264140
Telefax: (0)169266094
Electronic Mail: <userid>@cea.fr
WWW: http://www.cea.fr/
Founded: 1945 (CEA)
Staff: 1800
Activities: ISM * star formation * stellar evolution * hydrodynamics
Periodicals: "Clefs CEA", "Les Défis du CEA", "Revue Chocs", "Les Échos du CEA"
City Reference Coordinates: 002°11'00"E 48°36'00"N

Commissariat à l'Énergie Atomique (CEA), Centre d'Études Nucléaires (CEN), Fontenay-aux-Roses
Boîte Postale 6
F-92265 Fontenay-aux-Roses Cedex
or :
60-68, avenue du Général Leclerc
F-92260 Fontenay-aux-Roses
Telephone: (0)146547080
Telefax: (0)146549027
Electronic Mail: <userid>@cea.fr
WWW: http://www.cea.fr/
Founded: 1945 (CEA)
Staff: 2000
Activities: nuclear safety * biomedical aspects * chemistry
City Reference Coordinates: 002°17'00"E 48°47'00"N

Commissariat à l'Énergie Atomique (CEA), Département d'Astrophysique, de Physique des Particules, de Physique Nucléaire et de l'Instrumentation Associée (DAPNIA)
Orme des Merisiers
Bâtiment 709
F-91191 Gif-sur-Yvette Cedex
Telephone: (0)169083912
          (0)169415218
Telefax: (0)169086577
Electronic Mail: <userid>@cea.fr
WWW: http://www-dapnia.cea.fr/
          http://www.cea.fr/
Founded: 1945 (CEA)
Staff: 120
Activities: space astronomy * high-energy astrophysics * cosmic rays * X-rays * gamma rays * IR and optical astronomy * theoretical astrophysics * nuclear synthesis
City Reference Coordinates: 002°08'00"E 48°42'00"N

Committee on Data for Science and Technology (CODATA)
51, boulevard de Montmorency
F-75016 Paris
Telephone: (0)145250496
Telefax: (0)142881466
Electronic Mail: codata@paris7.jussieu.fr
WWW: http://www.codata.org/codata
Founded: 1966
Membership: 43
Activities: ISCU interdisciplinary scientific committee * preparing key data sets for which consistent international use is desirable * coordinating multinational projects * establishing format standards to promote compatibility of databases * guidelines to presentation of data in the primary literature * supplying information on sources of reliable data * education and training * organizing conferences and workshops
Periodicals: "CODATA Bulletin" (ISSN 0366-757x), "CODATA Newsletter" (ISSN 0538-6918)
City Reference Coordinates: 002°20'00"E 48°52'00"N

Committee on Space Research (COSPAR)
51, boulevard de Montmorency
F-75016 Paris
Telephone: (0)145250679

Telefax: (0)140509827
Electronic Mail: cospar@paris7.jussieu.fr
WWW: http://cospar.itodys.jussieu.fr/
Founded: 1958
Staff: 4
Membership: 52 (unions and institutions) 4500 (individual associates)
Activities: furthering, on an international scale, the progress of all kinds of scientific investigations which are carried out with space vehicles, rockets, and balloons * biennial conference * colloquia
Sections: (scientific commissions) space studies of the Earth's surface, meteorology and climate * space studies of the Earth-Moon system, planets and small bodies of the solar system * space studies of the upper atmospheres of the Earth and planets including reference atmospheres * space plasmas in the solar system, including planetary magnetospheres * research in astrophysics from space * life sciences as related to space * material sciences in space * fundamental physics in space
Sections: (panels) satellite dynamics * potentially environmentally detrimental activities in space * space research in developing countries * scientific ballooning * standard radiation belts * space weather
Periodicals: (11) "Advances in Space Research" (ISSN 0273-1177); (3) "COSPAR Information Bulletin"; "COSPAR Colloquia Series"; "COSPAR Directory of Organizations and Associates"
Awards: (1/2) "COSPAR Award", "International Cooperation Medal", "William Nordberg Medal", "Massey Award" (jointly with the Royal Society, UK), "Vikram Sarabhai Award" (jointly with the Indian Space Research Organization, India), "Zel'-dovich Award" (jointly with the Russian Academy of Sciences), "COSPAR Distinguished Service Medal"
City Reference Coordinates: 002°20'00"E 48°52'00"N

Composants et Systèmes de Précision (CSP)

5ter, rue C. Ader
ZAC de Mercières
F-60471 Compiègne
Telephone: (0)344868121
Telefax: (0)344866991
Electronic Mail: csp@imaginet.fr
Founded: 1988
Staff: 10
Activities: designing and manufacturing astronomical telescope systems (high-precision mechanisms, control systems, metrology, user software)
City Reference Coordinates: 002°50'00"E 49°25'00"N

COSPAR

● See "Committee on Space Research (COSPAR)"

Dassault Aviation

9, rond-point des Champs-Elysées
F-75008 Paris
Telephone: (0)153769300
Telefax: (0)153769320
WWW: http://www.dassault-aviation.fr/
        http://www.dassault-aviation.com/
        http://dassault-industries.fr/
Founded: 1945
Staff: 12000
Activities: manufacturing business jets and military aircraft, electronic and electrical equipments, flight control systems, CAD-CAM software * maintenance and repair of aeronautical material * scientific research and studies * space systems
City Reference Coordinates: 002°20'00"E 48°52'00"N

Destination Univers

c/o Christian Legrand
260, route de la Bellevue
F-76160 Préaux
Telephone: (0)611225217
Electronic Mail: christian.legrand@drire.industrie.fr
Founded: 1987
Staff: 1
Activities: lectures * publishing * software production * astronomical accessories
City Reference Coordinates: 001°17'00"E 47°02'00"N

Deutsches Zentrum für Luft- und Raumfahrt (DLR) eV,Büro Paris

17, avenue de Saxe
F-75007 Paris
Telephone: (0)142199426
Telefax: (0)142199629
Electronic Mail: <userid>@dlr.de
WWW: http://www.dlr.de/
Founded: 1988 (DLR: 1969)
Activities: liaison office for the "Deutsches Zentrum für Luft- und Raumfahrt (DLR) eV" (see main entry in Germany)
City Reference Coordinates: 002°20'00"E 48°52'00"N

École Centrale de Paris (ECP), Club d'Astronomie

2, Avenue Sully Prudhomme

F-92290 Châtenay-Malabry
Electronic Mail: astro@ecp.fr
WWW: http://www.astro.ecp.fr/
Activities: lectures * education
City Reference Coordinates: 002°20'00"E 48°52'00"N

École et Observatoire de Physique du Globe de Strasbourg (EOPGS)
5, rue René Descartes
F-67084 Strasbourg Cedex
Telephone: (0)388416300
Telefax: (0)388616747
WWW: http://nelumbo.u-strasbg.fr/
Founded: 1919
Staff: 80
Activities: education * research * observing
Periodicals: "Bulletins Sismologiques", "Bulletins Magnétiques"
Observatories: 4 (Welschbruch, Griesheim-sur-Souffel, Strasbourg, Echery)
City Reference Coordinates: 007°45'00"E 48°35'00"N

École Nationale Supérieure des Sciences de l'Informatique et des Bibliothèques (ENSSIB)
17/21, boulevard du 11 Novembre 1918
F-69623 Villeurbanne Cedex
Telephone: (0)472444343
Telefax: (0)472442788
Electronic Mail: <userid>@enssib.fr
WWW: http://www.enssib.fr/
Founded: 1992 (1963 as "École Nationale Supérieure des Bibliothèques" in Paris then)
City Reference Coordinates: 004°53'00"E 45°46'00"N

École Normale Supérieure (ENS), Laboratoire de Radioastronomie Millimétrique
24, rue Lhomond
F-75231 Paris Cedex 05
Telephone: (0)143291225
Telefax: (0)143367204
WWW: http://www.phys.ens.fr/labo/lra/
Founded: 1978
Staff: 60
Activities: mm observing * IR * detectors * quantum astrochemistry * cosmology * galaxies
City Reference Coordinates: 002°20'00"E 48°52'00"N

École Normale Supérieure de Lyon (ENSL),Groupe d'Astrophysique
● See now "Centre de Recherche Astronomique de Lyon (CRAL)"

École Polytechnique, Centre de Physique Théorique
Plateau de Palaiseau
F-91128 Palaiseau Cedex
Telephone: (0)169334736
Telefax: (0)169333008
Electronic Mail: <userid>@orphee.polytechnique.fr
Founded: 1960
Staff: 50
Activities: astrophysical plasmas * cosmology * gravitation * planet formation * galaxies * compact stars * pulsars * radiation mechanisms
City Reference Coordinates: 002°15'00"E 48°43'00"N

École Polytechnique, Laboratoire de Météorologie Dynamique (LMD)
Route de Saclay
F-91128 Palaiseau Cedex
Telephone: (0)169334145
Telefax: (0)169333005
Electronic Mail: <userid>@lmd.polytechnique.fr
Founded: 1959
Staff: 150
Activities: dynamic meteorology * climatology * modelling * turbulence * atmospheric sciences * Lidar * wavelets * numerical simulation * satellites * balloons
City Reference Coordinates: 002°15'00"E 48°43'00"N

École Polytechnique, Laboratoire de Physique Nucléaire des Hautes Énergies (LPNHE)
● See "Laboratoire de Physique Nucléaire des Hautes Énergies (LPNHE)"

Éditions Burillier
● See "Librairie Uranie"

Éditions de Physique
● See "EDP Sciences"

Éditions du CNRS
● See "CNRS Éditions"

Éditions Hermes
8, quai du Marché-Neuf
F-75004 Paris
Telephone: (0)153101520
Telefax: (0)151101521
Electronic Mail: hermes@iway.fr
WWW: http://www.editions-hermes.fr/
● Publisher
City Reference Coordinates: 002°20'00"E 48°52'00"N

Étoiles et Toile - Canal 606
5, rue des Mésanges
F-67120 Dutlenheim
Telephone: (0)388150743
Telefax: (0)388491255
Founded: 1998
Membership: 12
Activities: electronic astronomy
City Reference Coordinates: 007°39'00"E 48°33'00"N

EDP Sciences
Z.I. de Courtaboeuf
Avenue du Hoggar
Boîte Postale 112
F-91944 Les Ulis Cedex A
Telephone: (0)169187575
Telefax: (0)169288491
Electronic Mail: infos@edpsciences.com (Information)
　　　　　　articles@edpsciences.com (Papers)
　　　　　　<userid>@edpsciences.com
WWW: http://www.edpsciences.com/
Founded: 1920
Staff: 24
Periodicals: "Astronomy and Astrophysics (A&A), Supplement Series" (see separate entry), "European Physical Journal B & D", "Europhysics Letters"
● Publisher
● Formerly "Éditions de Physique"
City Reference Coordinates: 002°11'00"E 48°48'00"N (Orsay)

Équinoxe-Astronomie
Rue des Remparts
F-15000 Aurillac
Telephone: (0)471474984
Electronic Mail: laurent.pacelli@meteo.fr (Laurent Pacelli, President)
Founded: 1985
Membership: 30
Activities: popularization * education * photography * education
City Reference Coordinates: 002°26'00"E 44°56'00"N

Eurastro
c/o Arnaud Zenden (President)
200, rue Jeann-d'Arc
F-54000 Nancy
Telephone: (0)383402941
Electronic Mail: salque@iecn.u-nancy.fr (Bruno Salque, Secretary)
Founded: 1991
WWW: http://pegase.unice.fr/~skylink/pages_clubs/eurastro/eurastro.html
City Reference Coordinates: 006°12'00"E 48°41'00"N

Eureka-Plus
3, avenue de l'Amiral Lemonnier
F-78160 Marly-le-Roi
Telephone: (0)139588792
Electronic Mail: eureka@ifrance.com
WWW: http://eureka.ifrance.com/
Founded: 1985
Membership: 50
Activities: experimental rocketry and ballooning
City Reference Coordinates: 002°05'00"E 48°52'00"N

Euroconsult
71, boulevard Richard-Lenoir
F-75011 Paris
Telephone: (0)143380600
Telefax: (0)143381240
Electronic Mail: econsult@pratique.fr
Founded: 1983
Staff: 15
Activities: consulting and market research analyses on economic and industrial aspects of space applications * publishing
Periodicals: (1) "World Space Markets Survey - Ten Year Outlook", "World Space Communications and Broadcasting Markets Survey - Ten Year Outlook", "Space Business in Europe - Prospects to 2005"
City Reference Coordinates: 002°20'00"E 48°52'00"N

European Centre for Space Law (ECSL)
● See "European Space Agency (ESA), European Centre for Space Law (ECSL)"

European Industrial Space Study Group
(EUROSPACE)
16, rue Hamelin
F-75116 Paris
Telephone: (0)147558300
Telefax: (0)147556330
Founded: 1961
Activities: promoting European space activity
Periodicals: (1) "European Space Directory"
City Reference Coordinates: 002°20'00"E 48°52'00"N

European Optical Society (EOS)
(Société Européenne d'Optique - SEO)
c/o Centre Universitaire d'Orsay
Bâtiment 503
Boîte Postale 147
F-91403 Orsay Cedex
Telephone: (0)169358720
Telefax: (0)169853565
Electronic Mail: francoise.chavel@iota.u-psud.fr
WWW: http://kon-hp.risoe.dk/eos/
Founded: 1991
Membership: 250
Activities: conventions
Periodicals: (6) "Journal of the European Optical Society: Part A - Pure and Applied Optics", "Journal of the European Optical Society: Part B: Quantum Optics"; "EOS Newsletter"
City Reference Coordinates: 002°11'00"E 48°48'00"N

European Science Foundation (ESF)
(Fondation Européenne de la Science)
1, quai Lezay-Marnésia
F-67080 Strasbourg Cedex
Telephone: (0)388767100
Telefax: (0)388370532
Electronic Mail: esf@esf.org
WWW: http://www.esf.org/
Founded: 1974
Membership: 59 (organizations in 21 European countries)
Staff: 40
Activities: association of research funding organizations * coordinating research in Europe * scientific programmes * scientific networks * standing, ad hoc, and associated committees * European Research Conferences
Periodicals: (1) "Annual Report - Rapport Annuel"; (2) "ESF Communications" (ISSN 0293-082x)
City Reference Coordinates: 007°45'00"E 48°35'00"N

European Space Agency (ESA), European Centre for Space Law (ECSL)
8-10, rue Mario-Nikis
F-75738 Paris Cedex 15
Telephone: (0)153697605
Telefax: (0)153697560
Electronic Mail: ecsl@hq.esa.fr
WWW: http://edms.esrin.esa.it/ecsl/
           http://edms.esrin.esa.it/esalex/login
           http://www.esa.int/
Founded: 1989
Membership: 200
Activities: Summer courses on space law and policy * Practitioners' Forum
Periodicals: "ECSL News"
City Reference Coordinates: 002°20'00"E 48°52'00"N

European Space Agency (ESA), Headquarters
8-10, rue Mario Nikis
F-75738 Paris Cedex 15
Telephone: (0)153697654
Telefax: (0)153697560
(0)153697236
Electronic Mail: <userid>@hq.esa.fr
WWW: http://www.esrin.esa.it/esa/descrip/estabs.htm
http://www.esrin.esa.it/
http://www.esa.int/
Founded: 1964
Staff: 2000
Activities: promoting cooperation among European states in space research and technology and their application for peaceful purposes
Periodicals: see "European Space Agency (ESA), ESTEC, ESA Publications Division (EPD)" (Netherlands)
City Reference Coordinates: 002°20'00"E 48°52'00"N

European Synchrotron Radiation Facility (ESRF)
Boîte Postale 220
F-38043 Grenoble Cedex
Telephone: (0)476882000
Telefax: (0)476882020
Electronic Mail: <userid>@esrf.fr
information@esrf.fr
WWW: http://www.esrf.fr/
Founded: 1988 (Opening: 1994)
Staff: 500
City Reference Coordinates: 005°43'00"E 45°10'00"N

Euroscience
c/o Françoise Praderie
Observatoire de Paris
61, avenue de l'Observatoire
F-75014 Paris
Telephone: (0)144964558
Telefax: (0)144964910
Electronic Mail: francoise.praderie@obspm.fr
WWW: http://www.iway.fr/sc/tribune/eurosc.htm
http://www.euroscience.org/
Founded: 1997
Activities: European association for the promotion of science and technology
City Reference Coordinates: 002°20'00"E 48°52'00"N

EUROSPACE
● See "European Industrial Space Study Group"

Fédération Laïque d'Éducation Populaire (FLEP), Section Astronomie
Foyer Laïque d'Éducation Populaire
Château des Izards
F-24660 Coulounieix-Chamiers
or :
c/o Cyrille Debray (President)
2, impasse Suzanne-Lacorre
F-24660 Coulounieix-Chamiers
Telephone: (0)553352405
(0)553467520
Electronic Mail: astronomie24@perigord.tm.fr
WWW: http://www.astrosurf.org/astroflep/
Founded: 1991
Membership: 60
Activities: solar observing * education * photography
Periodicals: (12) "Astro Passion"
Coordinates: 000°42'24"E 45°09'17"N (La Rampisole)
City Reference Coordinates: 000°43'00"E 45°11'00"N (Périgueux)

Ferme des Étoiles (La_)
● See "La Ferme des Étoiles"

Fondation Européenne de la Science
● See "European Science Foundation (ESF)"

Fondation Européenne des Métiers de l'Image et du Son (FEMIS), Images/media
● See "CNRS Images/media FEMIS"

Fondation Louis de Broglie

23, quai de Conti
F-75006 Paris
Telephone: (0)140460554
Telefax: (0)140510865
Electronic Mail: fond_broglie@compuserve.com
Founded: 1973
Staff: 12
Activities: theoretical physics
Periodicals: (4) "Annales de la Fondation Louis de Broglie" (ISSN 0182-4295)
City Reference Coordinates: 002°20'00"E 48°52'00"N

Galaxy Contact

7, rue Gustave-Cuvelier
Boîte Postale 26
F-62101 Calais Cedex
Telephone: (0)321352515
Telefax: (0)321351784
Electronic Mail: galaxycontact@netinfo.fr
WWW: http://www.spacephotos.com/
Founded: 1983
Staff: 8
Activities: documentation, photographs, video tapes, etc. on space
City Reference Coordinates: 001°50'00"E 50°57'00"N

Galerie Alain Carion

6, rue Jean-du-Bellay
F-75004 Paris
or :
92, rue Saint-Louis-en-l'Île
F-75004 Paris
Telephone: (0)143260116
Telefax: (0)143259233
Founded: 1972
Activities: sales of minerals, fossils, meteorites, tectites, ...
City Reference Coordinates: 002°20'00"E 48°52'00"N

Gap Astronomie - Association Copernic

c/o Léone Pellenq
Mairie de Gap
Rue du Colonel Roux
F-05000 Gap
or :
c/o École Copernic
Haute Corréo
F-05400 La Roche des Arnauds
Telephone: (0)492578317 (Observatory)
Telefax: (0)492578317 (Observatory)
Electronic Mail: asso.copernic@netcourrier.com
WWW: http://www.astrosurf.com/copernic
Founded: 1985
Membership: 100
Activities: lectures * observing * photography * CCDs
Coordinates: 006°00'00"E 44°33'46"N H1,200m (Haute Corréo)
City Reference Coordinates: 006°05'00"E 44°34'00"N (Gap)
005°57'00"E 44°34'00"N (La Roche des Arnauds)

Géospace Observatoire d'Aniane

929, rue d'Alco
F-34080 Montpellier
Telephone: (0)467034949
Telefax: (0)467752864
Electronic Mail: geospace@cnusc.fr
WWW: http://astro.u-strasbg.fr/Obs/PLANETARIUM/aniane.html
Founded: 1982
Membership: 500
Coordinates: 003°36'11"E 43°41'01"N H176m
City Reference Coordinates: 003°53'00"E 43°36'00"N

Groupe Astronomique Hague Querqueville (GAHQ)

Ferme de la Rocambole
61, rue Roger-Glinel
F-50460 Querqueville
Telephone: (0)233033709
Telefax: (0)333033709
Electronic Mail: gahqhag@aol.com

Founded: 1980
Membership: 100
Activities: introduction to astronomy * instrument making * photography * CCD * image processing * history of astronomy * spectroscopy
Periodicals: "Mira"
Coordinates: 001°45'00"W 49°38'00"N
City Reference Coordinates: 001°45'00"W 49°38'00"N

Groupe d'Astronomie M 31
10, rue Alphonse-Daudet
F-31200 Toulouse
Founded: 1984
Membership: 12
Activities: observing
City Reference Coordinates: 001°26'00"E 43°36'00"N

Groupe d'Entraînement et de Recherche pour les Méthodes d'Éducation Active (GERMEA)
Domaine du Pignada
1, allée de l'Empereur
F-64600 Anglet
Telephone: (0)559522254
Telefax: (0)559401422
Founded: 1983
Membership: 100
Activities: education * teacher training * activities in schools * popularization
City Reference Coordinates: 001°30'00"W 43°29'00"N

Groupe des Observateurs du Ciel Profond (GOCP)
c/o Yann Pothier
11, impasse Canart
F-75012 Paris
Telephone: (0)143414329
Electronic Mail: ypothier@abi.snv.jussieu.fr
Founded: 1996
Membership: 36
Activities: observing
Periodicals: (4) "Ciel Extrême"
Coordinates: 006°46'00"E 44°40'00"N H1650m (La Clapière, Ceillac-en-Queyras)
City Reference Coordinates: 002°20'00"E 48°52'00"N

Groupe Électronique Astronomie Informatique (GEAI)
c/o Claude Rivas
113, rue Paradis
F-13006 Marseille
Telephone: (0)491815585
            (0)684541584
Electronic Mail: claude.rivas@club.francetelecom.fr
WWW: http://www.multimania.com/geai13
Founded: 1985
Membership: 15
Activities: radioastronomy * spectrography * photography * observing * Sun * Moon * planets * double stars * computing * electronics
Coordinates: 005°37'47"E 43°17'57"N H150m (Gemenos)
City Reference Coordinates: 005°24'00"E 43°18'00"N

Groupe Européen d'Observation Stellaire (GEOS)
c/o Michel Dumont (Secretary)
3, promenade Vénézia
F-78000 Versailles
Telephone: (0)130219023
Founded: 1973
Membership: 96
Activities: observing variable stars * analysis of collected data
Periodicals: (30) "Notes Circulaires"; (5) "GEOS Circulars"
City Reference Coordinates: 002°08'00"E 48°48'00"N

Groupement Astronomique Populaire de la Région d'Antibes (GAPRA)
18, boulevard Chancel
F-06600 Antibes
Electronic Mail: brunetto@pacwan.fr (Laurent Brunetto)
WWW: http://perso.pacwan.fr/brunetto/GAPRA/
Founded: 1975
Membership: 50
City Reference Coordinates: 007°07'00"E 43°35'00"N

Groupement d'Astronomie Populaire Sottevillais (GAPS)

c/o Maison pour Tous
2, rue Thiremberg
F-76300 Sotteville-lès-Rouen
Telephone: (0)235723105
Telefax: (0)235722567
Founded: 1990
Membership: 16
Activities: observing (Sun, planets, comets, Moon, variable and double stars) * radioastronomy * photography * instrument making * education * meetings * lectures * planetarium
Periodicals: (12) "Le Ciel du Mois"; (1) "Le Ciel de vos Vacances"
Coordinates: 001°05'00"E 49°25'00"N H10m
City Reference Coordinates: 001°05'00"E 49°25'00"N

Groupement des Astronomes Amateurs de la Gâtine (GAAG)

18, avenue de la Maladrerie
F-79200 Parthenay
Telephone: (0)549642301 (President)
Founded: 1983
Membership: 29
Activities: popularization * weekly meetings * observing * instrumentation * exhibitions * mobile planetarium (Constellarium E4)
Observatories: 2 (Pompaire, Le Pin)
City Reference Coordinates: 000°15'00"W 46°39'00"N

Groupement des Industries Françaises Aéronautiques et Spatiales (GIFAS)

4, rue Galilée
F-75782 Paris Cedex 16
Telephone: (0)144431700
Telefax: (0)140709141
Electronic Mail: infogifas@gifas.asso.fr
WWW: http://www.gifas.asso.fr/
Founded: 1908
Membership: 200 (companies)
Activities: union of French aerospace manufacturers
Periodicals: (11) "GIFAS Letter" (ISSN 0399-4864); (1) "Rapport Annuel"
City Reference Coordinates: 002°20'00"E 48°52'00"N

Groupement des Industries Françaises de l'Optique (GIFO)

39-41, rue Louis-Blanc
F-92400 Courbevoie
Telephone: (0)147176400
Telefax: (0)147176398
Electronic Mail: gifo@gifo.org
WWW: http://www.gifo.org/
Founded: 1896
Staff: 6
Activities: representing French optical industry
Periodicals: (4) "Infos GIFO" (ISSN 1149-168x)
City Reference Coordinates: 002°15'00"E 48°54'00"N

Groupement Français pour l'Observation et l'Étude du Soleil (GFOES)

Le Vauroux
F-37500 Chinon
Telephone: (0)247932744
Electronic Mail: gredin@ccr.jussieu.fr (Patrick Gredin)
Founded: 1989
Membership: 50
Activities: Sun observing * data processing
Periodicals: (2) "Helios" (circ.: 70)
City Reference Coordinates: 000°30'00"E 47°10'00"N

Images, Reflets, Initiation Scientifique (IRIS)

20, rue Diffonty
F-13600 La Ciotat
Telephone: (0)442714019
Founded: 1991
Membership: 70
Activities: lectures * observing
City Reference Coordinates: 005°36'00"E 43°10'00"N

IMASTRO Rhône-Alpes

c/o Laurent Guillot
Les jardins de Brou
Bâtiment A3

152, boulevard de Brou
F-01000 Bourg-en-Bresse
Telephone: (0)474213983
Telefax: (0)474230732
Electronic Mail: guillot@star.ipl.fr
Founded: 1990
Membership: 25
Activities: CCDs * popularization * portable planetarium
City Reference Coordinates: 005°13'00"E 46°12'00"N

Informatique et Astronomie pour Astronomes Amateurs (I3A)
● See "Association I3A"

Institut d'Astrophysique de Paris (IAP)
● See "Centre National de la Recherche Scientifique (CNRS), Institut d'Astrophysique de Paris (IAP)"

Institut d'Astrophysique Spatiale (IAS)
● See "Centre National de la Recherche Scientifique (CNRS), Institut d'Astrophysique Spatiale (IAS)"

Institut de l'Information Scientifique et Technique (INIST)
● See "Centre National de la Recherche Scientifique (CNRS), Institut de l'Information Scientifique et Technique (INIST)"

Institut de Mécanique Céleste et de Calcul des Éphémérides (IMCCE)
● See now "Observatoire de Paris, Institut de Mécanique Céleste et de Calcul des Éphémérides (IMCCE)"

Institut de Physique du Globe de Paris (IPGP)
Tour 24 B89
4, place Jussieu
F-75282 Paris Cedex 05
Telephone: (0)144272404
Telefax: (0)144273373
WWW: http://www.ipgp.jussieu.fr/
Founded: 1990
Activities: space geodynamics * geophysics * geodesy * seismology * planetology * magnetism * remote sensing * oceanic altimetry
City Reference Coordinates: 002°20'00"E 48°52'00"N

Institut de Physique du Globe de Strasbourg (IPGS)
● See now "École et Observatoire de Physique du Globe de Strasbourg (EOPGS)"

Institut de Radioastronomie Millimétrique (IRAM), Grenoble
300, rue de la Piscine
Domaine Universitaire de Grenoble
F-38406 Saint-Martin-d'Hères Cedex
Telephone: (0)476824900
Telefax: (0)476515938
Electronic Mail: <userid>@iram.grenet.fr
WWW: http://iram.fr/
Founded: 1979
Staff: 96
Periodicals: (1) "Annual Report"; (6) "IRAM Newsletter"
Coordinates: 005°54'26"E 44°38'01"N H2,552m (Observatoire du Plateau de Bure)
003°23'58"W 37°04'06"N H2,870m (Pico Veleta, Spain)
● Joint facility of the "Centre National de la Recherche Scientifique (CNRS)", France, the "Max-Planck-Gesellschaft zur Förderung der Wissenschaften eV" (Max-Planck-Gesellschaft), Germany, and the "Instituto Geográfico Nacional (IGN)", Spain, to carry out mm-wavelength astronomy
City Reference Coordinates: 005°43'00"E 45°10'00"N (Grenoble)

Institut de Radioastronomie Millimétrique (IRAM), Observatoire du Plateau de Bure
L'Enclus
F-05250 Saint-Étienne-en-Devoluy
Telephone: (0)476538520
Telefax: (0)476538523
WWW: http://iram.fr/
Founded: 1979
Activities: mm-wavelength astronomy
Coordinates: 005°54'29"E 44°38'02"N H2,560m (interferometer)
● Mail should be sent to IRAM headquarters in Grenoble (see separate entry)
City Reference Coordinates: 005°43'00"E 45°10'00"N (Grenoble)

Institut des Sciences de la Terre, de l'Eau et de l'Espace de Montpellier (ISTEEM)
Place Eugène Bataillon
F-34095 Montpellier Cedex 5
Telephone: (0)467144595

Telefax: (0)467144785
WWW: http://www.dstu.univ-montp2.fr/
Founded: 1994
• Jointly run by the "Centre National de la Recherche Scientifique (CNRS)" and the "Université de Montpellier II Sciences et des Techniques du Languedoc"
City Reference Coordinates: 003°53'00"E 43°36'00"N

Institut des Sciences de la Terre, de l'Eau et de l'Espace de Montpellier (ISTEEM), Groupe de Recherche en Astronomie et Astrophysique du Languedoc (GRAAL)
Place Eugène Bataillon
F-34095 Montpellier Cedex 5
Telephone: (0)467143415
Telefax: (0)467144535
Electronic Mail: <userid>@graal.univ-montp2.fr
WWW: http://www.isteem.univ-montp2.fr/GRAAL/
Founded: 1989 (ISTEEM: 1994)
Staff: 14
Activities: observational cosmology * stellar physics * stellar atmospheres * late-type stars * analysis of large surveys
City Reference Coordinates: 003°53'00"E 43°36'00"N

Institut Français de Recherche pour l'Exploitation de la Mer (IFREMER)
Technopolis 40
155, rue Jean-Jacques Rousseau
F-92138 Issy-lès-Moulineaux Cedex
WWW: http://www.ifremer.fr/
Founded: 1984
Staff: 1700
City Reference Coordinates: 002°17'00"E 48°49'00"N

Institut Français du Pétrole, Club d'Astronomie
1 & 4, avenue du Bois Préau
F-92852 Rueil-Malmaison Cedex
Telephone: (0)147527040
Telefax: (0)147527095
Founded: 1973
Membership: 20
Activities: observing * photography
City Reference Coordinates: 002°11'00"E 48°53'00"N

Institut Gassendi pour la Recherche Astronomique en Provence (IGRAP)
c/o Observatoire de Marseille
2, place Le Verrier
F-13248 Marseille Cedex 4
Telephone: (0)491107483
Telefax: (0)491107484
Electronic Mail: igrap@obmara.cnrs-mrs.fr
                 obmara::igrap
WWW: http://www-obs.cnrs-mrs.fr/igrap/igrap.html
Founded: 1995
Activities: research federation gathering the "Laboratoire d'Astronomie Spatiale (LAS)", the "Observatoire de Haute Provence (OHP)" and the "Observatoire de Marseille" and jointly run by the "Centre National de la Recherche Scientifique (CNRS)", the "Centre National d'Études Spatiales (CNES)" and the "Université d'Aix-Marseille I"
City Reference Coordinates: 005°24'00"E 43°18'00"N

Institut Géographique National (IGN)
136bis, rue de Grenelle
F-75700 Paris
Telephone: (0)143988000
Telefax: (0)143988400
Electronic Mail: <userid>@ign.fr
WWW: http://www.ign.fr/
Founded: 1940
Staff: 2000
Activities: French national geographical organization * selling databases, maps, and related geographical products * network of about 200,000 points over the whole French territory
City Reference Coordinates: 002°20'00"E 48°52'00"N

Institut Géographique National (IGN), Espace
Parc Technologique du Canal
24, rue Hermès
F-31527 Ramonville-Saint-Agne
Telephone: (0)562191818
Telefax: (0)561750317
Electronic Mail: <userid>@ign.fr
WWW: http://www.ign.fr/

Founded: 1989 (IGN: 1940)
Staff: 30
Activities: digitalized images (SPOT, LANDSAT, ERS) * digitalized elevation modelling
City Reference Coordinates: 001°26'00"E 43°36'00"N (Toulouse)

Institut National de Physique Nucléaire et de Physique des Particules (IN2P3), Centre de Physique des Particules de Marseille (CPPM)
● See "Centre de Physique des Particules de Marseille (CPPM)"

Institut National de Physique Nucléaire et de Physique des Particules (IN2P3), Institut de Physique Nucléaire de Lyon
43, boulevard du 11 Novembre 1918
F-69622 Villeurbanne Cedex
Telephone: (0)472448457
Telefax: (0)472431540
Electronic Mail: <userid>@ipnl.in2p3.fr
WWW: http://lyoinfo.in2p3.fr/
Founded: 1971 (IN2P3)
City Reference Coordinates: 004°51'00"E 45°45'00"N

Institut National de Physique Nucléaire et de Physique des Particules (IN2P3), Institut des Sciences Nucléaires de Grenoble
53, avenue des Martyrs
F-38026 Grenoble Cedex
Telephone: (0)476284000
Telefax: (0)476284004
WWW: http://isnwww.in2p3.fr/
Founded: 1966
City Reference Coordinates: 005°43'00"E 45°10'00"N

Institut National de Physique Nucléaire et de Physique des Particules (IN2P3), Laboratoire d'Annecy-le-Vieux de Physique des Particules (LAPP), Laboratoire d'Annecy-le-Vieux de Physique Théorique (LAPT)
Chemin de Bellevue
Boîte Postale 110
F-74941 Annecy-le-Vieux
Telephone: (0)450091683
Electronic Mail: <userid>@lapp.in2p3.fr
WWW: http://lapp.in2p3.fr/
Founded: 1971 (IN2P3)
City Reference Coordinates: 006°08'00"E 45°55'00"N

Institut National de Physique Nucléaire et de Physique des Particules (IN2P3), Laboratoire de l'Accélérateur Linéaire (LAL)
● See "Laboratoire de l'Accélérateur Linéaire (LAL)"

Institut National de Physique Nucléaire et de Physique des Particules (IN2P3), Laboratoire de Physique Nucléaire des Hautes Énergies (LPNHE)
● See "Laboratoire de Physique Nucléaire des Hautes Énergies (LPNHE)"

Institut National de Recherche en Informatique et en Automatique (INRIA), Rocquencourt
Boîte Postale 105
Domaine de Voluceau
Rocquencourt
F-78153 Le Chesnay Cedex
Telephone: (0)139635511
Telefax: (0)139635330
Electronic Mail: <userid>@inria.fr
WWW: http://www.inria.fr/
Founded: 1967
Activities: computer science * control * applied mathematics
Periodicals: (6) "INédit", "Thésauria"
City Reference Coordinates: 002°07'00"E 48°50'00"N

Institut National de Recherche en Informatique et en Automatique (INRIA), Sophia Antipolis
2004, route des Lucioles
Boîte Postale 93
F-06902 Sophia-Antipolis Cedex
Telephone: (0)493657777
Telefax: (0)493657765
Electronic Mail: <userid>@inria.fr
WWW: http://www.inria.fr/
Founded: 1967
City Reference Coordinates: 007°00'00"E 43°33'00"N (Cannes)

Institut National des Sciences Appliquées (INSA), Club Astronomie
Bureau des Élèves
Maison des Étudiants
20, avenue Albert-Einstein
F-69621 Villeurbanne Cedex
Telephone: (0)472438229
Electronic Mail: cail@free.fr
WWW: http://www.insa-lyon.fr/Associations/ASTRO/
Founded: 1967
Membership: 25
Activities: photography * CCDs * observing (planets, deep sky) * lectures
City Reference Coordinates: 004°53'00"E 45°46'00"N

Institut National des Sciences de l'Univers (INSU)
3, rue Michel-Ange
Boîte Postale 287
F-75766 Paris Cedex 16
Telephone: (0)144964000 (switchboard)
           (0)144964338 (switchboard)
           (0)144964377
Telefax: (0)144965005
Electronic Mail: <userid>@cnrs-dir.fr
WWW: http://www.insu.cnrs-dir.fr/
Founded: 1985
City Reference Coordinates: 002°20'00"E 48°52'00"N

Institut National des Sciences de l'Univers (INSU), Centre d'Analyse des Images (CAI)
c/o Observatoire de Paris
Bâtiment Perrault
77, avenue Denfert-Rochereau
F-75014 Paris
Telephone: (0)140512098
Telefax: (0)140512090
Electronic Mail: jean.guibert@obspm.fr
WWW: http://dsmama.obspm.fr/
Founded: 1990
Staff: 8
Activities: digitalized processing of astronomical and remote-sensing images * data analysis and processing * automatic classification * astrometry * photometry * managing the "Machine Automatique à Mesurer pour l'Astronomie (MAMA)"
City Reference Coordinates: 002°20'00"E 48°52'00"N

Institut National des Sciences de l'Univers (INSU), Centre de Données astronomiques de Strasbourg (CDS)
● See "Centre de Données astronomiques de Strasbourg (CDS)"

International Academy of Astronautics (IAA)
Boîte Postale 1268-16
F-75766 Paris Cedex 16
or :
6, rue Galilée
F-75116 Paris
Telephone: (0)147238215
Telefax: (0)147238216
Electronic Mail: sgeneral@iaastronautics.org
WWW: http://www.iafastro.iplus.fr/academy/academy.htm
        http://www.iaanet.org/
Founded: 1960
Membership: 1160
Periodicals: (4) "Acta Astronautica"
Awards: "Theodore von Kármán Award"
City Reference Coordinates: 002°20'00"E 48°52'00"N

International Association of Universities (IAU)
c/o F. Eberhard
1, rue Miollis
F-75732 Paris Cedex 15
Telephone: (0)145682545
           (0)147832339
Telefax: (0)147347605
Electronic Mail: iau@citi2.fr
                iau@unesco.org
WWW: http://www.unesco.org/iau
Founded: 1950
Staff: 16

Activities: information * studies * research * meetings * IAU/UNESCO Information Centre * coordinating centre for WAD (World Academic Database)
Periodicals: (6) "IAU Newsletter" (ISSN 0020-6032); (4) "Higher Education Policy" (ISSN 0279-4631); "Issues in Higher Education"
City Reference Coordinates: 002°20'00"E 48°52'00"N

International Astronautical Federation (IAF)
3-5, rue Mario Nikis
F-75015 Paris
Telephone: (0)145674260
Telefax: (0)142732120
        (0)142737537
WWW: http://www.iafastro.iplus.fr/home.htm
Founded: 1950
Staff: 3
Membership: 143
Activities: fostering development of astronautics for peaceful purposes * encouraging widespread dissemination of technical and other information concerning astronautics * stimulating public interest in and support for the development of all aspects of astronautics through the various media of mass communication * encouraging participation in astronautical research or other relevant projects by international and national research institutions, universities, commercial firms and individual experts * creating and fostering as activities of the Federation academies, institutes and commissions dedicated to continuing research in, and the fostering on, all aspects of the natural and social sciences relating to astronautics and the peaceful use of outer space * convoking and organizing with support of its respective academies, institutes and commissions, international astronautical congresses, symposia, colloquia and other scientific meetings * cooperating and advising with appropriate international and national governmental and non-governmental organizations and institutions on all aspects of the natural, engineering and social sciences related to astronautics and the peaceful uses of outer space
Sections: (committees) Allan D. Emil Memorial Award * Finance * Publications * Liaison with International Organizations and Developing Nations * Astrodynamics * Communications * Earth Observations * Education * Student Activities * Materials and Structures * Power * Propulsion * Space Exploration * Space Processing and Microgravity Applications * Space Station * Space Systems * Space Transportation * Life Sciences * Space Physiology and Medicine * Space and Planetary Biology and Biophysics * Human Factors * Biotechnology and Life Support * Space and Natural Disaster Reduction
Awards: "Allan D. Emil Memorial Award", "Frank J. Malina Award", "Student Award", "L.G. Napolitano Award"
City Reference Coordinates: 002°20'00"E 48°52'00"N

International Astronomical Union (IAU)
(Union Astronomique Internationale - UAI)
Secrétariat
98bis, boulevard Arago
F-75014 Paris
Telephone: (0)143258358
Telefax: (0)143252616
Electronic Mail: iau@iap.fr (Office/Secretariat)
            iaupres@iau.org (President)
            iaugs@iau.org (General Secretary)
WWW: http://www.iau.org/
Founded: 1919
Membership: 8500
Staff: 2
Activities: promoting astronomy * providing a forum where astronomers can develop astronomy through international cooperation * General Assembly every three years * sponsoring many major meetings * assigning designations and names to celestial bodies and the surface features thereon
Sections: (divisions) I. Fundamental Astrometry * II. The Sun & Heliosphere * III. Planetary System Sciences * IV. Stars * V. Variable Stars * VI. Interstellar Matter * VII. Galactic System * VIII. Galaxies and the Universe * IX. Optical Techniques * X. Radio Astronomy * XI. Space and High Energy Astrophysics
Sections: (commissions) 4. Ephemeris * 5. Documentation and Astronomical Data * 6. Astronomical Telegrams * 7. Celestial Mechanics * 8. Positional Astronomy * 9. Instruments and Techniques * 10. Solar Activity * 12. Solar Radiation and Structure * 14. Atomic and Molecular Data * 15. Physical Study of Comets, Minor Planets and Meteorites * 16. Physical Study of Planets and Satellites * 19. Rotation of the Earth * 20. Positions and Motions of Minor Planets, Comets and Satellites * 21. Light of the Night Sky * 22. Meteors and Interplanetary Dust * 24. Photographic Astrometry * 25. Stellar Photometry and Polarimetry * 26. Double and Multiple Stars * 27. Variable Stars * 28. Galaxies * 29. Stellar Spectra * 30. Radial Velocities * 31. Time * 33. Structure and Dynamics of the Galactic System * 34. Interstellar Matter * 35. Stellar Constitution * 36. Theory of Stellar Atmospheres * 37. Star Clusters and Associations * 38. Exchange of Astronomers * 40. Radio Astronomy * 41. History of Astronomy * 42. Close Binary Stars * 44. Astronomy from Space * 45. Stellar Classification * 46. Teaching of Astronomy * 47. Cosmology * 48. High Energy Astrophysics * 49. The Interplanetary Plasma and the Heliosphere * 50. Protection of Existing and Potential Observatory Sites * 51. Bioastronomy: Search for Extraterrestrial Life
Periodicals: "IAU Transactions"; "Highlights of Astronomy" (ISSN 0080-1372); "IAU Symposia Proceedings" (ISSN 0074-1809); (2) "IAU Information Bulletin"
City Reference Coordinates: 002°20'00"E 48°52'00"N

International Council for Scientific and Technical Information (ICSTI)
51, boulevard de Montmorency
F-75016 Paris
Telephone: (0)145256592
Telefax: (0)142151262
Electronic Mail: icsti@dial.oleane.com
WWW: http://www.icsti.nrc.ca/icsti

Founded: 1952 (1984 with current name)
Staff: 2
Membership: 53
Activities: studies, meetings, symposia aiming at improving the collection, storage, organization and dissemination of scientific and technical information
Periodicals: (4) "ICSTI Forum" (ISSN 1018-9580)
City Reference Coordinates: 002°20'00"E 48°52'00"N

International Council of Scientific Unions (ICSU)

51, boulevard de Montmorency
F-75016 Paris
Telephone: (0)145250329
Telefax: (0)142889431
Electronic Mail: secretariat@icsu.org
WWW: http://www.icsu.org/
Founded: 1931 (1919 as "International Research Council - IRC")
Membership: 95 multidisciplinary bodies and 25 scientific unions
Activities: non-governmental organization set up to promote international scientific activity in the different branches of science and their applications
Periodicals: (4) "Science International" (ISSN 1011-6257)
City Reference Coordinates: 002°20'00"E 48°52'00"N

International Earth Rotation Service (IERS)

c/o Observatoire de Paris
61, avenue de l'Observatoire
F-75014 Paris
Telephone: (0)140512226
Telefax: (0)140512291
Electronic Mail: iers@obspm.fr
WWW: http://hpiers.obspm.fr/
Founded: 1988
Staff: 12
Activities: Earth rotation * terrestrial and celestial reference systems
Periodicals: (1) "Annual Report" (ISSN 0068-4236); "Technical Notes", "Circulars"
City Reference Coordinates: 002°20'00"E 48°52'00"N

International Institute of Space Law (IISL)

3-5, rue Mario Nikis
F-75015 Paris
Telephone: (0)145674260
Telefax: (0)142732120
Electronic Mail: jtmasson@cyberway.com.sg
WWW: http://www.iafastro.iplus.fr/iisl/iisl_fra.htm
Founded: 1960
Activities: annual colloquium on the Law of Outer Space
Periodicals: "IISL Proceedings"
City Reference Coordinates: 002°20'00"E 48°52'00"N

International Space University (ISU)

Parc d'Innovation
Boulevard Gonthier d'Andernach
F-67400 Illkirch-Graffenstaden
Telephone: (0)388655430
Telefax: (0)388655447
Electronic Mail: info@isu.isunet.edu
WWW: http://www.isunet.edu/
Founded: 1987
Staff: 28
Activities: education in space-linked disciplines
Periodicals: (6) "The Universe"
City Reference Coordinates: 007°43'00"E 48°32'00"N

International Union of Geodesy and Geophysics (IUGG)

(Union Géodésique et Géophysique Internationale - UGGI)
c/o Georges Balmino
Bureau Gravimétrique International
Observatoire Midi-Pyrénées
18, avenue Edouard-Belin
F-31055 Toulouse Cedex
Telephone: (0)561332889
Telefax: (0)561253098
Electronic Mail: balmino.uggi@cnes.fr
WWW: http://www.obs-mip.fr/uggi/
Founded: 1919
Activities: scientific study of the Earth and planets and applications to the needs of society * mineral resources * effects of natural hazards * environmental preservation

Sections: International Association of Geodesy (IAG) * International Association of Seismology and Physics of the Earth's Interior (IASPEI) * International Association of Volcanology and Chemistry of the Earth's Interior (IAVCEI) * International Association of Geomagnetism and Aeronomy (IAGA) * International Association of Meteorology and Atmospheric Physics (IAMAP) * International Association of Hydrological Sciences (IAHS) * International Association for the Physical Sciences of the Ocean (IAPSO)
City Reference Coordinates: 001°26'00"E 43°36'00"N

International Union of the History and Philosophy of Science (IUHPS)
c/o Robert Halleux (Secretary General)
Institut d'Histoire des Sciences - IRFEST
Université Louis Pasteur
7, rue de l'Université
F-67000 Strasbourg
Telephone: (0)388528029
Telefax: (0)388528030
Electronic Mail: chst@ulg.ac.be
WWW: http://www.icsu.org/Membership/SUM/iuhps.html
Founded: 1956 (1947 as "International Union of History of Science")
Membership: 1800 (member committees: 80)
City Reference Coordinates: 007°45'00"E 48°35'00"N

Intespace
18, avenue Edouard-Belin
Boîte Postale 4356
F-31029 Toulouse Cedex 4
Telephone: (0)561281111
Telefax: (0)561281112
Electronic Mail: marketing@intespace.fr
WWW: http://www.intespace.fr/
Activities: consultancy in environmental engineering
City Reference Coordinates: 001°26'00"E 43°36'00"N

Laboratoire d'Astronomie Spatiale (LAS)
● See "Centre National de la Recherche Scientifique (CNRS), Laboratoire d'Astronomie Spatiale (LAS)"

Laboratoire de Glaciologie et de Géophysique de l'Environnement (LGGE)
● See "Université de Grenoble I, Laboratoire de Glaciologie et de Géophysique de l'Environnement (LGGE)"

Laboratoire de l'Accélérateur Linéaire (LAL)
Bâtiment 200
Centre Universitaire d'Orsay
F-91405 Orsay Cedex
Telephone: (0)164468300
Telefax: (0)169071526
WWW: http://www.lal.in2p3.fr/
Founded: 1955
Staff: 400
Activities: particle physics * particle astrophysics
● Jointly run by the "Centre National de la Recherche Scientifique (CNRS)" and the "Institut National de Physique Nucléaire et de Physique des Particules (IN2P3)"
City Reference Coordinates: 002°11'00"E 48°48'00"N

Laboratoire de Physique Nucléaire des Hautes Énergies (LPNHE)
Plateau de Palaiseau
F-91128 Palaiseau Cedex
Telephone: (0)169334136
Telefax: (0)169333002
Electronic Mail: <userid>@poly.in2p3.fr
                     <userid>@in2p3.fr
WWW: http://polywww.in2p3.fr/
Founded: 1936
Staff: 54 (Astrophysics: 12)
Activities: particle physics * astrophysics
Coordinates: 001°58'00"E 42°30'00"N H1,650m (Themis, French Pyrenees)
● Jointly run by the "École Polytechnique" and the "Institut National de Physique Nucléaire et de Physique des Particules (IN2P3)"
City Reference Coordinates: 002°15'00"E 48°43'00"N

Laboratoire de Physique Théorique et des Hautes Énergies (LPTHE)
● See "Université de Paris VI, Laboratoire de Physique Théorique et des Hautes Énergies (LPTHE)"
● See also "Université de Paris XI, Laboratoire de Physique Théorique et des Hautes Énergies (LPTHE)"

Laboratoire Primaire du Temps et des Fréquences (LPTF)
● See "Bureau National de Métrologie (BNM), Laboratoire Primaire du Temps et des Fréquences (LPTF)"

Laboratoire René Bernas
● See "Université de Paris XI, Laboratoire René Bernas"

Lancexport
84, boulevard de la République
F-92420 Vaucresson
Telephone: (0)147951701
Telefax: (0)147412520
Electronic Mail: lmoussy@lancexport.com
WWW: http://www.lancexport.com/
Founded: 1987
Staff: 1
Activities: consultancy for space programmes
City Reference Coordinates: 002°09'00"E 48°50'00"N

La Ferme des Étoiles
Moulin du Roy
F-32500 Fleurance
Telephone: (0)562060976
Telefax: (0)562062499
Electronic Mail: etoiles.fleurance@mipnet.fr
WWW: http://www.gascogne.com/Ferme/ferme.htm
Staff: 3
City Reference Coordinates: 000°40'00"E 43°51'00"N

La Recherche
c/o Société d'Éditions Scientifiques
57, rue de Seine
F-75280 Paris Cedex 06
Telephone: (0)153737979
Telefax: (0)146347508
Electronic Mail: courrier@larecherche.fr
WWW: http://www.larecherche.fr/
Founded: 1970
● Journal (12) (ISSN 0029-5671)
City Reference Coordinates: 002°20'00"E 48°52'00"N

L'Atelier d'Helios
La Provosté
F-72130 Saint-Georges-le-Gaultier
Telephone: (0)243973192
Founded: 1986
Staff: 1
Activities: designer * sculptor * sundials * lectures
Coordinates: 000°07'40"W 48°18'23"N H160m
City Reference Coordinates: 000°02'00"E 48°16'00"N (Fresnay/Sarthe)

Lavoisier Technique et Documentation
14, rue de Provigny
F-94236 Cachan Cedex
Telephone: (0)147406700
Telefax: (0)147406702
Electronic Mail: info@lavoisier.fr
      magasin@lavoisier.fr
WWW: http://www.lavoisier.fr/
Founded: 1947
Staff: 130
Activities: publisher * bookseller * subscription agent + database distributor
City Reference Coordinates: 002°20'00"E 48°48'00"N

Librairie Uranie
(Éditions Burillier)
Place Lucien-Laroche
F-56000 Vannes
Telephone: (0)297470997
Telefax: (0)297426067
Founded: 1996
Staff: 1
Activities: publishing books on astronomy and history of sciences * bookshop
City Reference Coordinates: 002°46'00"W 47°39'00"N

Loisir Art Culture, Section Astronomie
c/o Eurocopter
F-13700 Marignane
Telephone: (0)442857316

Founded: 1972
Membership: 12
Activities: observing * photography
Coordinates: 005°13'00"E 43°25'00"N H600m
City Reference Coordinates: 005°13'00"E 43°25'00"N

M53 Mayenne Astronomie
19, rue du Bois-de-l'Huisserie
F-53000 Laval
Telephone: (0)243535958
Electronic Mail: m53mayenneastronomie@wanadoo.fr
WWW: http://perso.wanadoo.fr/m53mayenneastro/
Founded: 1994
City Reference Coordinates: 000°46'00"W 48°04'00"N

Maison de l'Astronomie
33-35, rue de Rivoli
F-75004 Paris
Telephone: (0)142779955
Telefax: (0)148874087
Electronic Mail: info@maison-astronomie.com
WWW: http://www.maison-astronomie.com/
Founded: 1928
Staff: 17
Activities: selling astronomical instruments and accessories, books, maps, atlases, pictures and posters * lectures * workshops * travels * Club Privilège
Periodicals: (12) "Astro News" (ISSN 0990-2862, circ.: 4,000)
City Reference Coordinates: 002°20'00"E 48°52'00"N

Maison des Jeunes et de la Culture (MJC), Section Astronomie, Imphy
c/o Philippe Vignier
82ter, rue des Commes
F-58160 Imphy
Telephone: (0)386383681
Electronic Mail: astronomie.imphy@wanadoo.fr
Founded: 1983
Membership: 15
Activities: variable stars * planets * photography * observing * CCDs
Coordinates: 003°18'00"E 46°55'00"N H220m
City Reference Coordinates: 003°18'00"E 46°55'00"N

Maison des Jeunes et de la Culture (MJC), Section Astronomie, Rambouillet
32, rue Gambetta
F-78120 Rambouillet
Telephone: (0)130888904
Founded: 1979
Membership: 15
Activities: instrument making * observing * photography
Coordinates: 001°45'49"E 48°40'42"N H145m (Obs. Aquarius, Poigny-la-Forêt)
City Reference Coordinates: 001°50'00"E 48°39'00"N

Maison pour Tous, Section Astronomie, Vitrolles
(Uranomania)
c/o Thierry Sabuco
Maison pour Tous
6, rue Pierre et Marie Curie
Boîte Postale 151
F-13127 Vitrolles
Telephone: (0)442898077
Telefax: (0)442750410
Founded: 1983
Membership: 25
Activities: photography * CCDs
City Reference Coordinates: 005°15'00"E 43°28'00"N

Masson Éditeur
120, boulevard Saint-Germain
F-75280 Paris Cedex 06
Telephone: (0)146342160
(0)140466000
Telefax: (0)143296340
(0)140466001
Electronic Mail: infos@masson.fr
WWW: http://www.masson.fr/
Founded: 1804

● Publisher
City Reference Coordinates: 002°20'00"E 48°52'00"N

Matra Marconi Space, Headquarters
37, avenue Louis-Bréguet
Boîte Postale 1
F-78146 Vélizy-Villacoublay
Telephone: (0)134883000
Telefax: (0)134884343
WWW: http://www.matra-marconi-space.com/
        http://taps.com/matra_marconi
Founded: 1962
Activities: manufacturing space equipment
City Reference Coordinates: 004°51'00"E 45°45'00"N

Matra Marconi Space
Z.I. du Palays
31, rue des Cosmonautes
F-31077 Toulouse Cedex
Telephone: (0)561396139
Telefax: (0)561545710
WWW: http://www.matra-marconi-space.com/
        http://taps.com/matra_marconi
Founded: 1962
Staff: 2000
Activities: manufacturing space equipment
City Reference Coordinates: 001°26'00"E 43°36'00"N

Météo France
1, quai Branly
F-75340 Paris Cedex 07
Telephone: (0)145567171
        (0)836680808 (on-line weather forecast)
Telefax: (0)145567005
Electronic Mail: <userid>@meteo.fr
WWW: http://www.meteo.fr/
Founded: 1945
Staff: 3500
Activities: data acquisition and processing * weather forecast * climatology * documentation * research * training * network of climatic stations all over the country
Periodicals: (52) "Météo-Hebdo"; (12) "Bulletin Climatique"; (4) "Met Mar" (ISSN 0222-5123, circ.: 2,000); (1) "Rapport d'Activité"; "meteo.fr", "Atmosphère et Climat"
City Reference Coordinates: 002°20'00"E 48°52'00"N

Météo France, Centre de Météorologie Spatiale (CMS)
Avenue de Lorraine
Boîte Postale 147
F-22302 Lannion Cedex
Telephone: (0)296056700
Telefax: (0)296056737
WWW: http://www.meteo.fr/
Founded: 1963
Staff: 80
Activities: meteorological satellite data collection, processing and distribution * operational meteorology * archive centre * research
Periodicals: (12) "SATMER"
City Reference Coordinates: 003°28'00"W 48°44'00"N

Mission Astronomie Jeunesse (MAJ)
142, rue Demay
F-45650 Saint-Jean-le-Blanc
Telephone: (0)238510757
Founded: 1985
Membership: 15
Activities: observing * education * popularization * instrumentation
Periodicals: (6) "L'Obs'session"
Coordinates: 001°56'04"E 47°52'04"N
City Reference Coordinates: 001°53'00"E 47°53'00"N

Musée de l'Air et de l'Espace
Aéroport du Bourget
Boîte Postale 173
F-93352 Le Bourget Cedex
Telephone: (0)149927199
        (0)149927171

Telefax: (0)149927095
WWW: http://www.mae.org/
Founded: 1919
Staff: 100
Activities: exhibitions * visits * planetarium shows (closed until End 1998)
Periodicals: (4) "Pégase"
City Reference Coordinates: 002°26'00"E 48°56'00"N

Musée de l'Instrumentation Optique
Le Capitole
Place de la Mairie
F-68600 Biesheim
Telephone: (0)389720159
Telefax: (0)389721449
Activities: museum of optical instrumentation * exhibitions * astronomy club
City Reference Coordinates: 007°33'00"E 48°01'00"N (Neuf-Brisach)

Musée des Arts et Métiers (MAM)
292, rue Saint-Martin
F-75003 Paris
Telephone: (0)153018220
                (0)153018200
                (0)140272331 (answering machine)
Electronic Mail: <userid>@cnam.fr
WWW: http://www.cnam.fr/museum/
Founded: 1794
Staff: 65
● Temporarily closed - Reopening scheduled for End 1999
City Reference Coordinates: 002°20'00"E 48°52'00"N

Muséum National d'Histoire Naturelle (MNHN)
61, rue Buffon
F-75005 Paris
Telephone: (0)140793534
Telefax: (0)140793524
Electronic Mail: meteor@mnhn.fr
WWW: http://www.mnhn.fr/
Founded: 1974 (MNHN)
Staff: 5
Activities: meteorites * solar-system origin and evolution * minor planets * cosmic-ray irradiation
City Reference Coordinates: 002°20'00"E 48°52'00"N

Musicosmic
c/o Ch. Bruneau/UGC006
3, rue Wimphelinge
F-67000 Strasbourg
Telephone: (0)388150720
Founded: 1990
Membership: 8
Activities: astronomy-related music * musical historiography * contemporeanous creations
City Reference Coordinates: 007°45'00"E 48°35'00"N

National Space Development Agency of Japan (NASDA), Paris Office
3, avenue Hoche
F-75008 Paris
Telephone: (0)146224983
Telefax: (0)146224932
WWW: http://www.nasda.go.jp/welcome_e.html
Founded: 1986
Staff: 1
Activities: NASDA representative office in Paris (see main entry in Japan)
Periodicals: "NASDA Report"
City Reference Coordinates: 002°20'00"E 48°52'00"N

Nouveaux Mondes - New Worlds
(Association pour le Développement de l'Astronautique et de la Recherche Spatiale)
17, rue Rosette
F-36200 Argenton-sur-Creuse
Telephone: (0)254244774
Telefax: (0)254245174
Electronic Mail: nouvmondes@aol.com
WWW: http://members.aol.com/NadIarshan/
Founded: 1994
Membership: 12
Activities: documentation * research on astronautics and space sciences

City Reference Coordinates: 001°31'00" E 46°35'00" N

Novespace SA
15, rue des Halles
F-75001 Paris
Telephone: (0)142334141
Telefax: (0)140260860
Electronic Mail: espace@novespace.fr
WWW: http://www.novespace.fr/
Founded: 1986
Staff: 12
Activities: technology transfer
City Reference Coordinates: 002°20'00" E 48°52'00" N

Observatoire Antarès
193, chemin des Eaux
F-83500 La-Seyne-sur-Mer
Telephone: (0)494877447
Telefax: (0)494877447
Electronic Mail: antaresobservatoireanta@minitel.net
            sgmax.ob@caramail.com (Grégory Schiavone)
Founded: 1964
City Reference Coordinates: 005°53'00" E 43°06'00" N

Observatoire Astronomique d'Aniane
● See "Géospace Observatoire d'Aniane"

Observatoire Astronomique des Pises (OAP)
F-30750 Dourbies
Telephone: (0)467735558
Electronic Mail: jmlopez@pratique.fr (Jean-Marie Lopez, President)
WWW: http://www.bdl.fr/pises/pises.html
        http://www.bdl.fr/s2p/pises/
Founded: 1988
Membership: 50
Activities: deep-sky observing * photography * SN and comet search
Coordinates: 003°27'00" E 44°02'22" N H1,300m
City Reference Coordinates: 003°28'00" E 44°05'00" N

Observatoire Astronomique de Strasbourg
● See "Université de Strasbourg I, Observatoire Astronomique"

Observatoire de Besançon
Boîte Postale 1615
F-25010 Besançon Cedex
or :
41bis, avenue de l'Observatoire
F-25000 Besançon
Telephone: (0)381666900
Telefax: (0)381666944
Electronic Mail: <userid>@obs-besancon.fr
WWW: http://www.obs-besancon.fr/
Founded: 1882
Staff: 35
Activities: galactic structure and dynamics * time metrology * planetology
Coordinates: 005°59'18" E 47°14'58" N H315m
City Reference Coordinates: 006°02'00" E 47°15'00" N

Observatoire de Bordeaux
● See "Université de Bordeaux 1, Observatoire"

Observatoire de Grenoble
● See "Université de Grenoble I, Observatoire de Grenoble, Groupe d'Astrophysique"

Observatoire de Haute Provence (OHP)
● See "Centre National de la Recherche Scientifique (CNRS), Observatoire de Haute Provence (OHP)"

Observatoire de la Côte d'Azur (OCA)
Boîte Postale 229
F-06304 Nice Cedex 4
Telephone: (0)492003011
Telefax: (0)492003033
Electronic Mail: <userid>@obs-nice.fr
WWW: http://www.obs-nice.fr/

Founded: 1988 (1887 as Observatoire de Nice)
Staff: 220
Activities: astrometry * geodynamics * dynamics * fluid mechanics * stellar, galactic, and extragalactic astronomy * solar physics
Coordinates: 007°18'06"E 43°43'24"N H372m
City Reference Coordinates: 007°15'00"E 43°42'00"N

Observatoire de la Côte d'Azur (OCA), Centre d'Études et de Recherches en Géodynamique et Astrométrie (CERGA)
Avenue Copernic
F-06130 Grasse
or :
c/o Observatoire de Nice
Boîte Postale 229
F-06004 Nice Cedex 4
Telephone: (0)493405353
Telefax: (0)493405333
Electronic Mail: <userid>@obs-azur.fr
WWW: http://www.obs-nice.fr/
Founded: 1974 (1988 as part of OCA)
Staff: 60
Activities: astrometry * celestial mechanics * reference frames * Earth-Moon system * space geodesy * geodynamics * laser techniques * history of sciences * time standards
Coordinates: 006°55'30"E 43°44'54"N H1,270m (Calern - see separate entry)
City Reference Coordinates: 006°55'00"E 43°40'00"N (Grasse)
                              007°15'00"E 43°42'00"N (Nice)

Observatoire de la Côte d'Azur (OCA), Station de Calern
2130, route de l'Observatoire
Caussols
F-06460 Saint-Vallier-de-Thiey
Electronic Mail: <userid>@obs-azur.fr
WWW: http://www.obs-nice.fr/
Founded: 1974 (1988 as part of OCA)
Staff: 40
Coordinates: 006°55'30"E 43°44'54"N H1,270m
City Reference Coordinates: 006°15'00"E 43°42'00"N

Observatoire de l'Association Culturelle de Dax (OACD)
7, rue des Chênes
F-40100 Dax
Telephone: (0)558561447
WWW: http://www.iap.fr/saf/cldax.htm
         http://ourworld.compuserve.com/homepages/DUPOUY_Philippe/
Founded: 1968
Membership: 50
Activities: introduction to astronomy * observing * photography * radioastronomy * meteorological imaging * image processing * planetarium
Coordinates: 001°01'46"W 43°41'35"N H35m
City Reference Coordinates: 001°03'00"W 43°43'00"N

Observatoire de Lyon
• See now "Centre de Recherche Astronomique de Lyon (CRAL)"

Observatoire de Marseille
2, place Le Verrier
F-13248 Marseille Cedex 04
Telephone: (0)491959088
Telefax: (0)491621190
WWW: http://alpha0.cnrs-mrs.fr/
Staff: 35
Activities: stellar and interstellar studies of our Galaxy and of nearby galaxies * kinematics and dynamics of galaxies * optical instrumentation
• See also "Institut Gassendi pour la Recherche Astronomique en Provence (IGRAP)"
City Reference Coordinates: 005°24'00"E 43°18'00"N

Observatoire de Meudon
• See "Observatoire de Paris"

Observatoire de Nice
• See now "Observatoire de la Côte d'Azur (OCA)"

Observatoire de Paris
61, avenue de l'Observatoire
F-75014 Paris

or :
5, place Jules-Janssen
F-92195 Meudon Cedex
Telephone: (0)140512221
Telefax: (0)143541804
Electronic Mail: <userid>@obspm.fr
WWW: http://www.obspm.fr/
Founded: 1667
Staff: 700
Activities: astrophysical and physical research * cosmology * Sun * planets * stars * galaxies * ISM * radioastronomy * space-time reference frames * space research * instrumentation * education
Periodicals: "Cartes Synoptiques de la Chromosphère Solaire", "Bulletin de l'IERS"
Coordinates: 002°20'14"E 48°50'10"N H67m
            002°13'53"E 48°48'18"N H162m (Meudon - see separate entry)
            002°11'51"E 47°22'24"N H183m (Nançay - see separate entry)
City Reference Coordinates: 002°20'00"E 48°52'00"N (Paris)
            002°14'00"E 48°48'00"N (Meudon)

Observatoire de Paris, Institut de Mécanique Céleste et de Calcul des Éphémérides (IMCCE)

77, avenue Denfert Rochereau
F-75014 Paris
Telephone: (0)140512128
            (0)140512270
Telefax: (0)146332834
Electronic Mail: <userid>@bdl.fr
            bdl@bdl.fr
WWW: http://www.bdl.fr/
Founded: 1795
Staff: 43
Activities: celestial mechanics * solar system * ephemeris * fundamental astronomy * astrometry
Periodicals: (1) "Éphémérides Astronomiques - Annuaire du Bureau des Longitudes", "La Connaissance des Temps" (ISSN 0181-3048), "Phénomènes et Configurations des Satellites Galiléens de Jupiter" (ISSN 0769-1033), "Configurations des Huit Premiers Satellites de Saturne" (ISSN 0769-1025), "Éphémérides des Satellites de Jupiter, Saturne et Uranus", "Éphémérides des Satellites Faibles de Jupiter et de Saturne" (ISSN 0769-1041), "Éphémérides Nautiques"; "Notes Scientifiques et Techniques du BDL"
● Formerly "Bureau des Longitudes (BDL)"
City Reference Coordinates: 002°20'00"E 48°52'00"N

Observatoire de Paris, Station de Radioastronomie de Nançay

F-18330 Neuvy-sur-Barangeon
Telephone: (0)148518241
            (0)145077609 (Director)
            (0)145077754 (Secretary)
Telefax: (0)148518318
Electronic Mail: vandriel@obspm.fr (Willem van Driel, Director)
WWW: http://www.obs-nancay.fr/
        http://www.obspm.fr/
Founded: 1953
Staff: 50
Activities: radioastronomical research
Coordinates: 002°11'48"E 47°22'48"N H150m
City Reference Coordinates: 002°15'00"E 47°19'00"N

Observatoire de Physique du Globe de Clermont-Ferrand (OPGC), Département des Sciences de la Terre

5, rue Kessler
F-63038 Clermont-Ferrand Cedex
Telephone: (0)473406363
            (0)473346703 (Library)
Telefax: (0)473346744
Electronic Mail: <userid>@opgc.univ-bpclermont.fr
WWW: http://wwwobs.univ-bpclermont.fr/
Activities: research * education * volcanology * experimental magnetology * petrology * magnetism * geodynamics * isotope geochemistry
City Reference Coordinates: 003°05'00"E 45°47'00"N

Observatoire de Physique du Globe de Clermont-Ferrand (OPGC), Département des Sciences de l'Atmosphère

12, avenue Landais
F-63000 Clermont-Ferrand
Telephone: (0)473407380
Telefax: (0)473271657
Electronic Mail: <userid>@opgc.univ-bpclermont.fr
WWW: http://wwwobs.univ-bpclermont.fr/
City Reference Coordinates: 003°05'00"E 45°47'00"N

Observatoire de Puimichel

c/o Danny Cardoen
Le Haut de la Chapelle
F-04700 Puimichel
Telephone: (0)492787922
(0)492787969
WWW: http://www-eurinsa.insa-lyon.fr/Associations/astronomie/fr/puimichel.html
Coordinates: 006°01'00"E 43°58'30"N H750m
City Reference Coordinates: 006°01'00"E 43°59'00"N

Observatoire de Rouen (OR)

3, impasse Adrien-Auzout
F-76000 Rouen
Telephone: (0)235880196
Founded: 1884
Membership: 60
Activities: observing * popularization (public, schools) * photography * workshops for teachers * public sessions * video recording * planetarium shows
Periodicals: (12) "Bulletin de l'Observatoire"
Coordinates: 001°14'33"E 49°26'29"N H46m
City Reference Coordinates: 001°05'00"E 49°26'00"N

Observatoire des Alpes Maritimes (OAM)

● See now "Observatoire de la Côte d'Azur (OCA)"

Observatoire de Strasbourg

● See "Université de Strasbourg I, Observatoire Astronomique"

Observatoire de Toulouse

● See "Observatoire Midi-Pyrénées (OMP)"

Observatoire du Pic du Midi

● See "Observatoire Midi-Pyrénées (OMP), Observatoire du Pic du Midi"

Observatoire Midi-Pyrénées (OMP)

14, avenue Edouard-Belin
F-31400 Toulouse
Telephone: (0)561332929
Telefax: (0)561332888
(0)561536722
Electronic Mail: <userid>@obs-mip.fr
WWW: http://www.obs-mip.fr/omp/
http://webast.ast.obs-mip.fr/ (Laboratoire d'Astrophysique)
Founded: 1878
Staff: 350
Activities: extragalactic astronomy * theoretical astrophysics * solar physics * space oceanography and geophysics * gravimetry * space geodesy * planetology * aerology * geochemistry
City Reference Coordinates: 001°26'00"E 43°36'00"N

Observatoire Midi-Pyrénées (OMP), Centre de Recherches Atmosphériques (CRA)

Campistrous
F-65300 Lannemezan
Telephone: (0)562406100
Telefax: (0)562406101
Electronic Mail: <userid>@aero.obs-mip.fr
WWW: http://www.omp.obs-mip.fr/
Founded: 1960
Staff: 20
Activities: atmospheric chemistry and physics
City Reference Coordinates: 000°23'00"E 43°08'00"N

Observatoire Midi-Pyrénées (OMP), Observatoire du Pic du Midi

9, rue du Pont de la Moulette
F-65200 Bagnères-de-Bigorre
Telephone: (0)562951969
(0)562919106 (La Mongie)
Telefax: (0)62951070
Founded: 1878 (OMP)
Coordinates: 000°08'42"E 42°56'12"N H2,861m (Pic du Midi)
City Reference Coordinates: 000°09'00"E 43°04'00"N

Observatoire Populaire de Laval (OPL)

33bis allée du Vieux-Saint-Louis
Boîte Postale 1424
F-53014 Laval

Telephone: (0)243670506
Telefax: (0)243670173
Electronic Mail: sciences-fal53@laligue.org
WWW: http://www.fal53.laligue.org/opl/
Founded: 1986
Membership: 55
Activities: observing * photography * education * mobile planetarium
Periodicals: (12) "La Revue de l'Observatoire Populaire de Laval"
Coordinates: 000°47'41"W 48°03'20"N H80m
City Reference Coordinates: 000°46'00"W 48°04'00"N

Office National d'Études et de Recherches Aérospatiales (ONERA)
Boîte Postale 72
F-92322 Châtillon Cedex
or :
29, avenue de la Division Leclerc
F-92320 Châtillon
Telephone: (0)141734040
Telefax: (0)146734141
Electronic Mail: <userid>@onera.fr
WWW: http://www.onera.fr/
Founded: 1946
Staff: 2300
Activities: fundamental and applied research * technical assistance * energy * aerodynamics * structural resistance * physics
fluid mechanics * tests * materials * computing
Periodicals: (6) "La Recherche Aérospatiale" (ISSN 0034-1223, circ.: 600)
City Reference Coordinates: 002°17'00"E 48°48'00"N

Office National d'Études et de Recherches Aérospatiales (ONERA), Centre d'Études et de Recherches de Toulouse (CERT)
2, avenue Edouard-Belin
Boîte Postale 4025
F-31055 Toulouse Cedex
Electronic Mail: <userid>@cert.fr
WWW: http://www.cert.fr/
City Reference Coordinates: 001°26'00"E 43°36'00"N

Optilas, Composants
Z.I. de la Petite Montagne Sud
4, allée du Cantal
C.E. 1834
F-91018 Évry Cedex
Telephone: (0)160795900
Telefax: (0)160869633
Electronic Mail: sale_france@optilas.com
WWW: http://www.optilas.com/
Founded: 1973
Staff: 13
Activities: distributing high-technology equipment (optoelectronics, optical components, optical fibers & telecomms)
Periodicals: (2) "Opto Composants"
City Reference Coordinates: 002°27'00"E 48°38'00"N

Optilas, Systèmes
Z.I. de la Petite Montagne Sud
4, allée du Cantal
C.E. 1834
F-91018 Évry Cedex
Telephone: (0)160795900
Telefax: (0)160869633
Electronic Mail: sale_france@optilas.com
WWW: http://www.optilas.com/
Founded: 1973
Staff: 28
Activities: distributing high-technology equipment (lasers and instrumentation, spectroscopy, detection, high voltage, imaging, industrial measurement, ...)
Periodicals: (4) "Laser Actualités" (circ.: 15,000); (3) "Images & Mesures"
City Reference Coordinates: 002°27'00"E 48°38'00"N

Optique Unterlinden
Galerie du Rempart
F-68000 Colmar
Telephone: (0)389241605
(0)389233454
Telefax: (0)389416083
Founded: 1976

Staff: 9
Activities: optics * distributing telescopes, microscopes and accessories
City Reference Coordinates: 007°22'00"E 48°05'00"N

Palais de la Découverte, Planétarium
Avenue Franklin D. Roosevelt
F-75008 Paris
Telephone: (0)140748028
Telefax: (0)140748181
Electronic Mail: planetarium@palais-decouverte.fr
WWW: http://www.palais-decouverte.fr/
Founded: 1937
Membership: 220
Activities: promoting astronomy, Earth sciences and other sciences through exhibitions, shows, lectures etc.
Periodicals: (10) "Revue du Palais de la Découverte" (ISSN 0180-3344)
City Reference Coordinates: 002°20'00"E 48°52'00"N

Parc aux Étoiles
2, rue de la Chapelle
F-78510 Triel-sur-Seine
Telephone: (0)139747510
Telefax: (0)139749274
Electronic Mail: parc-aux-etoiles@hd.fr
Founded: 1991
Staff: 20
Activities: astronomy and space museum * popularization * observing
Periodicals: (4) "Objectif Univers"
Coordinates: 002°04'00"E 49°00'00"N H120m
City Reference Coordinates: 002°04'00"E 49°00'00"N

Parsec
18, avenue Maréchal Foch
F-06000 Nice
Telephone: (0)493412304
            (0)493412304 (Astrorama)
Telefax: (0)493856285
Electronic Mail: parsec@obs-nice.fr
WWW: http://www.obs-nice.fr/parsec
Founded: 1987
Membership: 30
Activities: education * popularization * portable planetariums * observing
Periodicals: "Astrorama" (ISSN 1256-7406)
Observatories: 1 (Astrorama, La Trinité)
City Reference Coordinates: 007°15'00"E 43°42'00"N

Planétarium d'Aix-en-Provence
● See "Amis du Planétarium d'Aix-en-Provence (APAP)"

Planétarium de Bourbon-Lancy
Place Sénateur Turlier
F-71140 Bourbon-Lancy
Telephone: (0)385890978
Electronic Mail: verdenetmichel@minitel.net (Michel Verdenet)
WWW: http://astro.u-strasbg.fr/Obs/PLANETARIUM/bourbon/bourbon.htm
Founded: 1993
City Reference Coordinates: 003°46'00"E 46°37'00"N

Planétarium de Cholet
c/o Maison des Sciences
1, rue Lamarque
F-49300 Cholet
Telephone: (0)241624036
Telefax: (0)241719461
Electronic Mail: sla@mygale.org (Société des Sciences, Lettres et Arts de Cholet)
WWW: http://www.mygale.org/04/sla
Founded: 1973
Membership: 400
Staff: 20
Activities: shows * observing * instrument making * lectures
City Reference Coordinates: 000°53'00"W 47°04'00"N

Planétarium de Montpellier
Jardin des Plantes
Boîte Postale 1088
F-34007 Montpellier

Telephone: (0)467617401
Telefax: (0)467611008
Founded: 1989
Staff: 50
Activities: shows * education
Periodicals: "Le Planétarium" (ISSN 1243-4833)
City Reference Coordinates: 003°53'00"E 43°36'00"N

Planétarium de Nantes
8, rue des Acadiens
F-44100 Nantes
Telephone: (0)240739923
Telefax: (0)240419239
Electronic Mail: Teletel: 3614 Code: TELEM
WWW: http://www.pageszoom.com/pt/sites/htm_ang/reg_nant/nantes/planetar/acc.htm
Founded: 1981
Staff: 5
Activities: shows related to astronomy and to space sciences and techniques
City Reference Coordinates: 001°33'00"W 47°13'00"N

Planétarium de Nîmes
• See "Association des Amis du Planétarium, Nîmes"

Planétarium de Parthenay
20, rue de la Citadelle
F-79200 Parthenay
Telephone: (0)549642301
Founded: 1996
Staff: 1
Activities: shows * workshops * observing
City Reference Coordinates: 000°15'00"W 46°39'00"N

Planétarium de Poitiers
• See "Planétarium - Lasérium"

Planétarium de Reims
1, place Museux
F-51100 Reims
Telephone: (0)326855150
Founded: 1980
Activities: education * public shows
City Reference Coordinates: 004°02'00"E 49°15'00"N

Planétarium de Saint-Étienne
Espace Fauriel
28, rue Pierre-et-Dominique Ponchardier
F-42100 Saint-Étienne
Telephone: (0)477255492
Telefax: (0)477333570
Electronic Mail: 100751.737@compuserve.com
WWW: http://ourworld.compuserve.com/homepages/planetarium/
Founded: 1990
Staff: 6
City Reference Coordinates: 004°24'00"E 45°26'00"N

Planétarium de Strasbourg
• See "Association des Amis du Planétarium, Strasbourg"

Planétarium de Vaulx-en-Velin
Place de la Nation
Boîte Postale 166
F-69512 Vaulx-en-Velin Cedex
Telephone: (0)478795010
         (0)478795013
City Reference Coordinates: 004°56'00"E 45°47'00"N

Planétarium du Trégor
c/o Société d'Economie Mixte
Cosmopolis
Parc Scientifique du Trégor
F-22560 Pleumeur-Bodou
Telephone: (0)296918378
         (0)296918378
Telefax: (0)296239891

(0)296158031
Electronic Mail: saem0694@eruobretagne.fr
atlas4@nu7eris96.isdnet.net
WWW: http://astro.u-strasbg.fr/Obs/PLANETARIUM/tregor.html
Founded: 1988
Staff: 6
Activities: popularization * shows
City Reference Coordinates: 003°32'00"W 48°46'00"N

Planétarium - Lasérium
c/o Espace Pierre Mendès France
1, place de la Cathédrale
F-86000 Poitiers
Telephone: (0)549415625
(0)549503300
Telefax: (0)549413856
Founded: 1992
Activities: shows * exhibitions * laser shows
City Reference Coordinates: 000°20'00"E 46°35'00"N

Planétarium Observatoire de Montredon-Labessonnié
Mairie
F-81360 Montredon-Labessonnié
Telephone: (0)563756312
Telefax: (0)563751811
Electronic Mail: planetarn@wanadoo.fr
WWW: http://perso.wanadoo.fr/planetarn/
Founded: 1993
Membership: 11
Activities: popularization * shows * observing * education
Coordinates: 002°17'50"E 43°44'25"N
City Reference Coordinates: 002°18'00"E 44°45'00"N

Pour la Science
8, rue Férou
F-75006 Paris
Telephone: (0)146342142
Telefax: (0)143251829
Founded: 1977
● Journal (12) (ISSN 0163-4092) (circ.: 70,000) (French edition of "Scientific American')
City Reference Coordinates: 002°20'00"E 48°52'00"N

Prospace
34, rue des Bourdonnais
F-75001 Paris
Telephone: (0)144889930
Telefax: (0)144889939
Electronic Mail: prospace@prospace-fr.com
WWW: http://www.prospace-fr.com/
Founded: 1974
Membership: 52 (societies)
Staff: 6
Activities: promoting activities, products, means, and services of French space industry for export * prospecting space markets
Periodicals: "News from Prospace"
City Reference Coordinates: 002°20'00"E 48°52'00"N

Provence Sciences Techniques Jeunesse (PSTJ)
1257, route de Grasse
F-06580 Pégomas
Telephone: (0)493609177
Telefax: (0)493609277
Electronic Mail: pstj@chez.com
WWW: http://www.chez.com/pstj
http://wwwrc.obs-azur.fr/pstj/pstj.htm
Founded: 1992
Membership: 200
Activities: education * astronomy * rocketry * robotics * computer science * environment
Periodicals: (6) "Scoop" (Circ.: 50)
City Reference Coordinates: 007°00'00"E 43°33'00"N (Cannes)

Recherche (La_)
● See "La Recherche"

REOSC, Recherche Étude Optique
(Groupe SFIM)

Avenue de la Tour Maury
F-91280 Saint Pierre du Perray
Telephone: (0)169897200
Telefax: (0)169897220
Electronic Mail: reosc@sfim.fr
WWW: http://www.reosc.com/
Founded: 1937
Staff: 95
Activities: studying and manufacturing optical instrumentation for space, astronomy and laser systems * figuring, lightweighting and polishing large mirrors
City Reference Coordinates: 002°18'00"E 48°42'00"N (Longjumeau)

R.S. Automation Industrie (RSAI)
Z.I. de la Vaure
Rue des Mineurs
Boîte Postale 40
F-42290 Sorbiers
Telephone: (0)477533048
Telefax: (0)477533861
Electronic Mail: rs.automation@wanadoo.fr
WWW: http://perso.wanadoo.fr/rsautomation/
Founded: 1979
Staff: 42
Activities: automation for industry and public works * manufacturing planetariums and equipment for public museums
City Reference Coordinates: 004°30'00"E 45°29'00"N (Izieux)

Salons Internationaux de l'Aéronautique et de l'Espace
4, rue Galilée
F-75116 Paris
Telephone: (0)153233333
Telefax: (0)147200086
Electronic Mail: info@salon-du-bourget.fr
WWW: http://www.salon-du-bourget.fr/
Founded: 1909
Staff: 15
Activities: organizing the biennial Paris air show
City Reference Coordinates: 002°20'00"E 48°52'00"N

SAT
● See "Société Anonyme de Télécommunications (SAT)"

Scot
8-10 rue Hermès
Parc Technologique du Canal
F-31526 Ramonville-Saint-Agne Cedex
Telephone: (0)561394600
Telefax: (0)561394610
Electronic Mail: contact@scot.cnes.fr
WWW: http://www.scot-sa.com/
Activities: design and consultancy in Earth resources
City Reference Coordinates: 001°26'00"E 43°36'00"N (Toulouse)

Service Hydrographique et Océanographique de la Marine (SHOM)
Boîte Postale 5
F-00307 Armées
or :
3, avenue Octave-Gréard
F-75007 Paris
Telephone: (0)144384116
Telefax: (0)140659998
Electronic Mail: <userid>@shom.fr
Founded: 1720
Staff: 862
Activities: hydrography * oceanography * marine mapping
Periodicals: "Épidécides Lunaires" (ISSN 0240-8376), "Épiménides Nautiques" (ISSN 0240-8368), "Annales Hydrographiques" (ISSN 0373-3629), "Rapport Annuel" (ISSN 0989-5876)
City Reference Coordinates: 002°20'00"E 48°52'00"N (Paris)

Société Anonyme de Télécommunications (SAT)
58B, rue du Dessous-des-Berges
Boîte Postale 326
F-75625 Paris Cedex 13
Telephone: (0)145823111
Telefax: (0)145823113
WWW: http://www.sagem-sat.co.uk/

Founded: 1932
Staff: 5000
Activities: manufacturing telecommunication and optronic IR equipment
Periodicals: (1) "Rapport Annuel"
City Reference Coordinates: 002°20'00"E 48°52'00"N

Société Anonyme de Télécommunications (SAT), Club Astro du Comité d'Établissement

51, rue du Chevaleret
F-75013 Paris
Telephone: (0)145823499
Telefax: (0)145823228
Electronic Mail: satdod@pobox.oleane.com (Attn.: Midavaine)
Founded: 1975
Membership: 41
Activities: instrument loan * CCD * computing * library
City Reference Coordinates: 002°20'00"E 48°52'00"N

Société Astronomique de Bordeaux (SAB)

Hotel des Sociétés Savantes
1, place Bardineau
F-33000 Bordeaux
Telephone: (0)556516932
Electronic Mail: sab33@wanadoo.fr
 guy-libante@wanadoo.fr (Guy Libante, Editor)
WWW: http://www.astrosurf.org/sab33
Founded: 1909
Membership: 70
Activities: general astronomy * meteorology * observing * lectures * slide shows * films
Periodicals: "Astronomie Passion" (ISSN 1283-3339)
City Reference Coordinates: 000°34'00"W 44°30'00"N

Société Astronomique de Bourgogne (SAB)

4, rue Chancelier de l'Hospital
F-21000 Dijon
Telephone: (0)380364413
Electronic Mail: sab@astrosurf.com
WWW: http://www.astrosurf.com/sab
Founded: 1975
Membership: 120
Activities: popularization * observing * education * lectures * Sun * Moon * deep sky * CCDs * light pollution
Periodicals: (4) "L'Observation du Ciel"
Coordinates: 004°57'44"E 47°18'11"N H429m (Hautes-Plates)
City Reference Coordinates: 005°02'00"E 47°20'00"N

Société Astronomique de France (SAF)

3, rue Beethoven
F-75016 Paris
Telephone: (0)142241374
Electronic Mail: saf@calva.net
WWW: http://www.iap.fr/saf/
Founded: 1887
Membership: 2500
Activities: amateur astronomy
Periodicals: (12) "L'Astronomie" (ISSN 0004-6302, circ.: 6,000); (4) "Observations et Travaux" (ISSN 0769-0878)
Coordinates: 002°20'41"E 48°50'55"N (Sorbonne, Paris)
Awards: (1) "Prix Janssen"
City Reference Coordinates: 002°20'00"E 48°52'00"N

Société Astronomique de France (SAF), Groupe d'Alsace (SAFGA)

c/o Observatoire Astronomique de Strasbourg
11, rue de l'Université
F-67000 Strasbourg
WWW: http://www.iap.fr/saf/
Founded: 1931
Membership: 120
Activities: lectures * education * colloquiums * observing * CCDs
Periodicals: (12) "Alsace Astronomie"
Coordinates: 007°46'04"E 48°35'02"N H142m
City Reference Coordinates: 007°45'00"E 48°35'00"N

Société Astronomique de Haute-Marne (SAHM)

Lotissement Le Chiny
F-52100 Valcourt
Telephone: (0)325060476
 (0)325062218 (Observatory)

Founded: 1985
Membership: 40
Activities: observing * lectures * exhibitions * education * meetings
Coordinates: 004°54'00"E 48°38'00"N H139m
City Reference Coordinates: 004°54'00"E 48°38'00"N

Société Astronomique de la Montagne de Lure
Route de Lure
F-04230 Saint-Étienne-des-Orgues
Telephone: (0)492731798
Founded: 1992
Activities: observing * popularization
City Reference Coordinates: 005°47'00"E 44°02'00"N

Société Astronomique de Lyon (SAL)
c/o Observatoire de Lyon
Avenue Charles André
F-69230 Saint-Genis-Laval
Telephone: (0)478595839
Founded: 1931
Membership: 150
Activities: observing * workshops * lectures * visits of observatories * mirror polishing
Periodicals: (4) "Société Astronomique de Lyon"
City Reference Coordinates: 004°51'00"E 45°45'00"N

Société Astronomique de Montpellier (SAM) "Pierre Vauriot"
66, boulevard de l'Observatoire
F-34000 Montpellier
Telephone: (0)467661214
Electronic Mail: nicolas.montviloff@wanadoo.fr (Nicolas Montviloff, President)
Founded: 1979
Membership: 30
Activities: popularization * observing * photography * CCDs * software * education
Observatories: 1 (Observatoire Astronomique des Pises - see separate entry)
City Reference Coordinates: 003°53'00"E 43°36'00"N

Société Astronomique du Havre
5, passage Lenormand
F-76600 Le Havre
Telephone: (0)235515611
Electronic Mail: philippe.baudouin@wanadoo.fr
WWW: http://perso.wanadoo.fr/philippe.baudouin/
Founded: 1971
Membership: 30
Activities: telescope making * introduction to observing * video observing * double stars * variable stars * photography * computing
City Reference Coordinates: 000°08'00"E 49°30'00"N

Société Astronomique Populaire du Centre (SAPC)
Observatoire d'Arçay
40, Grande Rue
F-18340 Arçay
or :
31, avenue Marcel-Haegelen
F-18000 Bourges
Telephone: (0)248251001
Founded: 1986
Membership: 80
Activities: education * observing * planets * comets * double stars
City Reference Coordinates: 002°24'00"E 47°05'00"N (Bourges)

Société d'Astronomie de Cannes (SACA)
Boîte Postale 125
F-06405 Cannes Cedex
Telephone: (0)493649884
            (0)493990345
Founded: 1991
Membership: 50
Activities: observing * lectures
City Reference Coordinates: 007°01'00"E 43°33'00"N

Société d'Astronomie de Nantes (SAN)
35, boulevard Louis-Millet
F-44300 Nantes
Telephone: (0)240689120

Telefax: (0)240938123
Electronic Mail: san@oceanet.fr
WWW: http://www.oceanet.fr/Associations/san/
Founded: 1971
Membership: 100
Activities: meetings * library * observing * computing * lectures * popularization
City Reference Coordinates: 001°33'00"W 47°13'00"N

Société d'Astronomie de Saône et Loire (SASL)
25, rue Pierre Bridet
F-71100 Chalon-sur-Saône
Telephone: (0)385461407
Electronic Mail: camille.fevrat@wanadoo.fr
Founded: 1984
Membership: 50
Activities: observing * instrumentation * lectures
Coordinates: 004°40'32"E 46°43'05"N H420m (Buxy)
City Reference Coordinates: 004°51'00"E 46°47'00"N

Société d'Astronomie Populaire (SAP)
c/o Observatoire de Jolimont
1, avenue Camille Flammarion
F-31500 Toulouse
Telephone: (0)561584201
Electronic Mail: sap@easynet.fr
WWW: http://www.astrosurf.com/
Founded: 1910
Membership: 1000
Activities: observing * popularization
Periodicals: (6) "Pulsar" (ISSN 0154-4101); (1) "Éphémérides"
Observatories: 1 (Observatoire de Jolimont)
City Reference Coordinates: 001°26'00"E 43°36'00"N

Société d'Astronomie Populaire Poitevine (SAPP)
c/o Centre Socio-Culturel
Clos Gauthier
7, rue Vallée Monnaie
F-86000 Poitiers
Telephone: (0)549013423
Founded: 1976
Membership: 80
Activities: observing * photography * lectures * video
Coordinates: 000°24'30"E 46°33'30"N H129m
City Reference Coordinates: 000°20'00"E 46°35'00"N

Société Européenne de Propulsion (SEP)
Boîte Postale 303
F-92156 Suresnes Cedex
or :
24, rue Salomon-de-Rotschild
F-92150 Suresnes
Telephone: (0)147286500
　　　　　　(0)147280884
Telefax: (0)147280580
　　　　　(0)147280536
　　　　　(0)147280813
Electronic Mail: info@sep.fr
WWW: http://www.sep.fr/
Founded: 1969
Staff: 1000
Activities: manufacturing rocket and satellite boosters and engines
City Reference Coordinates: 002°14'00"E 48°52'00"N

Société Européenne d'Optique (SEO)
● See "European Optical Society (EOS)"

Société Française de Physique (SFP)
33, rue Croulebarbe
F-75013 Paris
Telephone: (0)147073298
Telefax: (0)143317426
WWW: http://www.in2p3.fr/SFP/
Founded: 1873
Membership: 3200
Periodicals: (5) "Bulletin" (ISSN 0037-9360); (1) "Annuaire" (ISSN 0081-1076)

Awards: "Gentner-Kastler" (jointly with the Deutsche Physikalische Gesellschaft), "Holweck" (jointly with the Institute of Physics), "Rammal Rammal", "Jean Ricard", "Félix Robin", "Jean Perrin", "Louis Ancel", "Aimé Cotton", "Paul Langevin", "Joliot-Curie", "Esclangon", "Foucault", "Daniel Guinier"
City Reference Coordinates: 002°20'00"E 48°52'00"N

Société Française des Spécialistes d'Astronomie (SFSA)
● See now "Société Française d'Astronomie et d'Astrophysique (SF2A)"

Société Française d'Astronomie et d'Astrophysique (SF2A)
c/o Institut d'Astrophysique de Paris
98bis, boulevard Arago
F-75014 Paris
Electronic Mail: sf2a@iap.fr
WWW: http://www.iap.fr/sf2a/
Founded: 1978
Membership: 520
Activities: organizing colloquia, courses, meetings * yearly "École de Goutelas"
Periodicals: (3) "Journal des Astronomes Français (JAF)" (ISSN 0021-8979); (1/2) "Annuaire"; "Comptes-Rendus de l'École de Goutelas"
Awards: (1) "Prix Compaq"
● Formerly known as "Société Française des Spécialistes d'Astronomie (SFSA)"
City Reference Coordinates: 002°20'00"E 48°52'00"N

Société Française d'Optique (SFO)
Bâtiment 503
Centre Universitaire d'Orsay
Boîte Postale 147
F-91403 Orsay Cedex
Telephone: (0)169358816
Telefax: (0)169853565
Electronic Mail: joelle.bourges@iota.u-psud.fr
Founded: 1983
Membership: 1360
Periodicals: (4) "Optique et Photonique"; (1) "Annuaire" (ISSN 1154-4317)
City Reference Coordinates: 002°11'00"E 48°48'00"N

Société Lorraine d'Astronomie (SLA)
c/o Lycée Saint Joseph
413, avenue de Boufflers
F-54520 Laxou
Telephone: (0)383933568
WWW: http://perso.wanadoo.fr/isabelle.berquand/
City Reference Coordinates: 006°08'00"E 48°41'00"N

Société Météorologique de France (SMF)
1, quai Branly
F-75340 Paris Cedex 07
Telephone: (0)145567364
Telefax: (0)145567363
WWW: http://www.meteo.fr/publications/smf.html
Periodicals: (4) "La Météorologie"
City Reference Coordinates: 002°20'00"E 48°52'00"N

SPOT Image
Boîte Postale 4359
F-31030 Toulouse Cedex
or :
5, rue des Satellites
F-31030 Toulouse Cedex
Telephone: (0)562194040
Telefax: (0)562194011
WWW: http://www.spotimage.com/
Founded: 1982
Staff: 190
Activities: distributing worldwide geographical data collected by SPOT satellites
Periodicals: (4) "SPOT Flash" (ISSN 1161-3289 & 1161-3297); (2) "SPOT Magazine" (ISSN 0764-048x)
City Reference Coordinates: 001°26'00"E 43°36'00"N

Station Astronomique Jansky (SAJ)
Prairie de Mérignan
F-45240 La Ferté Saint-Aubin
Telephone: (0)254988797
Telefax: (0)254988797
Electronic Mail: bertrand.flouret@obs-nancay.fr (Bertrand Flouret, President)
observateur@obs-nancay.fr

WWW: http://www.obs-nancay.fr/saj/
Founded: 1987
Membership: 15
Activities: observing
Coordinates: 001°55'00"E 47°49'00"N H110m
City Reference Coordinates: 001°55'00"E 47°49'00"N

Station de Radioastronomie de Nançay
● See "Observatoire de Paris, Station de Radioastronomie de Nançay"

Telas
100, boulevard du Midi
F-06150 Cannes
Telephone: (0)493906643
Telefax: (0)493906689
Founded: 1987
Activities: manufacturing telescopes, mirrors and decay lines
City Reference Coordinates: 007°01'00"E 43°33'00"N

UNESCO
● See "United Nations Educational, Scientific and Cultural Organization"

Union Astronomique Internationale (UAI)
● See "International Astronomical Union (IAU)"

Union Internationale des Associations et Organismes Techniques (UATI)
c/o Maison de l'UNESCO
1, rue Miollis
F-75732 Paris Cedex 15
Telephone: (0)145669410
Telefax: (0)143062927
Founded: 1951
Staff: 10
Activities: promoting and coordinating scientific and technical activities of concern to members
Periodicals: (2) "Bulletin"; (1) "Annuaire"
City Reference Coordinates: 002°20'00"E 48°52'00"N

Union Rhône-Alpes des Clubs d'Astronomie (URACA)
c/o Club d'Astronomie de Lyon-Ampère
37, rue Paul-Cazeneuve
F-69008 Lyon
Telephone: (0)478012905
Founded: 1985
Membership: 25 (clubs)
Activities: promoting astronomy in the area
City Reference Coordinates: 004°51'00"E 45°45'00"N

United Nations Educational, Scientific and Cultural Organization (UNESCO)
7, place de Fontenoy
F-75700 Paris
Telephone: (0)145681000
            (0)145681681 (Office of Public Information)
            (0)145681682 (Office of Public Information)
WWW: http://www.unesco.org/
Founded: 1946
Membership: 170 (states)
Staff: 2680
Activities: promoting collaboration among the nations through education, science and culture in order to further universal respect for justice, for the rule of law and for the human rights and fundamental freedoms, without distinction of race, sex, language or religion
Periodicals: (12) "UNESCO Courier"; "New Trends in Physics Teaching" (ISSN 0077-8907), "UNESCO Sources", "Impact of Science on Society"
City Reference Coordinates: 002°20'00"E 48°52'00"N

Université d'Aix-Marseille I, Institut Gassendi pour la Recherche Astronomique en Provence (IGR-AP)
● See "Institut Gassendi pour la Recherche Astronomique en Provence (IGRAP)"

Université de Bordeaux 1, Observatoire
2, rue de l'Observatoire
Boîte Postale 89
F-33270 Floirac
Telephone: (0)557776100
Telefax: (0)557776110
Electronic Mail: <userid>@observ.u-bordeaux.fr

WWW: http://www.observ.u-bordeaux.fr/
Founded: 1879
Staff: 20
Activities: astrometry (automatic meridian circle, stars, planetary satellites, minor planets) * stellar kinematics and dynamics * radioastronomy (molecular clouds, star formation, VLBI) * solar convection * interplanetary matter * planetary atmospheres * Earth
Coordinates: 000°31'42"W 44°50'06"N H73m
City Reference Coordinates: 000°32'00"W 44°49'00"N

Université de Grenoble I, Laboratoire de Glaciologie et de Géophysique de l'Environnement (LGGE)

54, rue Molière
Boîte Postale 96
F-38402 Saint-Martin-d'Hères Cedex
Telephone: (0)476824200
Telefax: (0)476824201
Electronic Mail: <userid>@glaciog.grenet.fr
WWW: http://www.ujf-grenoble.fr/COM/labujf/lggefran.htm
Staff: 60
Activities: glaciology * climatology * atmospheric chemistry * planetology * experimental astrophysics * IR spectroscopy
● Alternate university name: "Université Joseph Fourier"
City Reference Coordinates: 005°43'00"E 45°10'00"N

Université de Grenoble I, Observatoire de Grenoble, Laboratoire d'Astrophysique (LAOG)

Boîte Postale 53X
F-38041 Grenoble Cedex 9
or :
414, rue de la Piscine
F-38042 Saint-Martin d'Hères
Telephone: (0)476514788
            (0)476514981
Telefax: (0)476448821
Electronic Mail: <userid>@obs.ujf-grenoble.fr
WWW: http://laog.obs.ujf-grenoble.fr/
Founded: 1985 (1979 as "Groupe d'Astrophysique")
Staff: 49
Activities: mm-wavelength astronomy * molecular clouds * star formation * astrochemistry * molecular excitation * ISM * QSOs * circumstellar envelopes * IR astronomy * high-angular resolution * very-low-mass stars * planetary disks * high-energy astrophysics
Coordinates: 005°54'26"E 44°38'01"N H2,552m (Observatoire du Plateau de Bure)
● Alternate university name: "Université Joseph Fourier"
City Reference Coordinates: 005°43'00"E 45°10'00"N

Université de Lille, Laboratoire d'Astronomie

1, impasse de l'Observatoire
F-59000 Lille
Telephone: (0)320524424
Telefax: (0)320580328
Electronic Mail: <userid>@gat.univ-lille1.fr
Founded: 1930
Staff: 4
Activities: celestial mechanics * astrometry * Saturn's satellites
Coordinates: 003°04'14"E 50°36'57"N H32m
● Full university name: "Université des Sciences et Technologies de Lille"
City Reference Coordinates: 003°04'00"E 50°38'00"N

Université de Montpellier II Sciences et des Techniques du Languedoc, Institut des Sciences de la Terre, de l'Eau et de l'Espace de Montpellier (ISTEEM)

● See "Institut des Sciences de la Terre, de l'Eau et de l'Espace de Montpellier (ISTEEM)"

Université de Montpellier II Sciences et des Techniques du Languedoc, Laboratoire de Physique Mathématique (LPM)

Place Eugène Bataillon
F-34095 Montpellier Cedex 5
Telephone: (0)467143567
Telefax: (0)467544850
Electronic Mail: <userid>@lpm.univ-montp2.fr
Founded: 1967
Staff: 25
Activities: mathematical physics * modelling * inverse problems * field and particle theory * inverse problems
City Reference Coordinates: 003°53'00"E 43°36'00"N

Université de Nice-Sophia-Antipolis, Département d'Astrophysique

Parc Valrose
F-06108 Nice Cedex 2
Telephone: (0)493529806

Telefax: (0)493529004
Electronic Mail: <userid>@ayalga.unice.fr
　　　　　　　　 <userid>@obs-nice.fr
WWW: http://boulega.unice.fr/da.html
　　　　 http://pegase.unice.fr/
Founded: 1968
Staff: 32
Activities: Sun * solar and stellar seismology * atmospheric optics * imagery * speckle interferometry
City Reference Coordinates: 007°15'00"E 43°42'00"N

Université de Paris-Sud
● See "Université de Paris XI"

Université de Paris VI, Laboratoire de Physique Théorique et des Hautes Énergies (LPTHE)
Tour 16 - 1er Étage - Boîte 126
4, place Jussieu
F-75252 Paris Cedex 05
Telephone: (0)144274121
Telefax: (0)144277088
Electronic Mail: login@lpthe.jussieu.fr
WWW: http://qcd.th.u-psud.fr/
● Alternate university name: "Université Pierre et Marie Curie"
City Reference Coordinates: 002°20'00"E 48°52'00"N

Université de Paris VI, Physique des Milieux Condensés
Tour 13 E4
Boîte Postale 77
4, place Jussieu
F-75252 Paris Cedex 05
Telephone: (0)144275897
　　　　　　 (0)144274461
Telefax: (0)144274469
Electronic Mail: <userid>@pmc.jussieu.fr
Founded: 1982
Staff: 20
Activities: high-pressure physics
● Alternate university name: "Université Pierre et Marie Curie"
City Reference Coordinates: 002°20'00"E 48°52'00"N

Université de Paris XI, Centre de Spectrométrie Nucléaire et de Spectrométrie de Masse (CSNSM)
Bâtiments 104-108
F-91405 Orsay Cedex
Telephone: (0)169415284
Telefax: (0)169415268
Electronic Mail: <userid>@csnsm.in2p3.fr
WWW: http://www-csnsm.in2p3.fr/
Founded: 1962
Staff: 100
Activities: research
Periodicals: (1) "Annual Report"
● Alternate university name: "Université de Paris-Sud"
City Reference Coordinates: 002°11'00"E 48°48'00"N

Université de Paris XI, Laboratoire d'Astronomie
Bâtiment 470
F-91405 Orsay Cedex
Telephone: (0)169417766
Telefax: (0)169416380
Electronic Mail: michele.presse@df.cso.u-psud.fr
WWW: http://www.u-psud.fr/
● Alternate university name: "Université de Paris-Sud"
City Reference Coordinates: 002°11'00"E 48°48'00"N

Université de Paris XI, Laboratoire de Physique des Gaz et des Plasmas (LPGP)
Bâtiment 212
F-91045 Orsay Cedex
Telephone: (0)169417251
Telefax: (0)169417844
Electronic Mail: deut@psisun.u-psud.fr
WWW: http://www.u-psud.fr/
Founded: 1959
Staff: 117
Activities: basic research * hot plasmas * strongly coupled plasmas * weakly ionized discharges * molecular plasmas * GAPHYOR database
● Alternate university name: "Université de Paris-Sud"

City Reference Coordinates: 002°11'00"E 48°48'00"N

Université de Paris XI, Laboratoire de Physique Théorique et des Hautes Énergies (LPTHE)
Bâtiment 211
F-91045 Orsay Cedex
Telephone: (0)169416353
Telefax: (0)169419551
Electronic Mail: adm@qcd.th.u-psud.fr
WWW: http://qcd.th.u-psud.fr/
Activities: elementary particles * theoretical physics * mathematical physics * plasma physics * statistical physics * physics of liquids
• Alternate university name: "Université de Paris-Sud"
City Reference Coordinates: 002°11'00"E 48°48'00"N

Université de Picardie Jules Verne (UPJV), Laboratoire de Physique Théorique et d'Astrophysique
33, rue Saint-Leu
F-80039 Amiens Cedex
Telephone: (0)322827633
Electronic Mail: <userid>@u-picardie.fr
WWW: http://www.u-picardie.fr/
Founded: 1982
Staff: 2
Activities: education * PN * HII regions
City Reference Coordinates: 002°18'00"E 49°54'00"N

Université des Sciences et Technologies de Lille
• See Université de Lille

Université de Strasbourg I, Centre de Données astronomiques de Strasbourg (CDS)
• See "Centre de Données astronomiques de Strasbourg (CDS)"

Université de Strasbourg I, Observatoire Astronomique
11, rue de l'Université
F-67000 Strasbourg
Telephone: (0)388150711 (Director)
            (0)388150710 (Secretary)
Telefax: (0)388150760
Electronic Mail: <userid>@astro.u-strasbg.fr
WWW: http://astro.u-strasbg.fr/Obs-e.html (English Homepage)
        http://astro.u-strasbg.fr/Obs.html (French Homepage)
Founded: 1881
Staff: 50
Activities: education * research * statistical astronomy * statistical methodology * galactic evolution * stellar populations * high-energy astrophysics * cosmology * databases * hosting the "Centre de Données astronomiques de Strasbourg (CDS)" (see separate entry) * planetarium (see separate entry)
Coordinates: 007°46'04"E 48°35'02"N H142m
• Alternate university name: "Université Louis Pasteur (ULP)"
City Reference Coordinates: 007°45'00"E 48°35'00"N

Université Denis Diderot
• See "Université de Paris VII"

Université Joseph Fourier
• See "Université de Grenoble I"

Université Louis Pasteur (ULP)
• See "Université de Strasbourg I"

Université Pierre et Marie Curie
• See "Université de Paris VI"

Uranoscope de France
94, rue de la Glacière
F-75013 Paris
Telephone: (0)164420002
Telefax: (0)164078604
Electronic Mail: uranos@club-internet.fr
WWW: http://perso.club-internet.fr/uranos/
Founded: 1995
Staff: 15
Activities: developing international relationships between amateur astronomers
City Reference Coordinates: 002°20'00"E 48°52'00"N

Uranoscope de l'Île de France
7, avenue Carnot
F-77220 Gretz-Armainvilliers
Telephone: (0)164420002
Telefax: (0)164078604
Electronic Mail: uranos@club-internet.fr
WWW: http://perso.club-internet.fr/uranos/
Founded: 1983
Membership: 125
Activities: observing * photography * introduction to astronomy * planetarium
Periodicals: (4) "Cosmos Express"
Observatories: 1 (Observatoire Astronomique de la Brie)
City Reference Coordinates: 002°44'00"E 48°44'00"N

World Intellectual Property Law Agency (WIPLA)
c/o Société CEDAT
26, rue George Sand
F-75016 Paris
Electronic Mail: cedat@wipla.com
WWW: http://www.wipla.com/
Founded: 1997
City Reference Coordinates: 002°20'00"E 48°52'00"N

Georgia

Abastumani Astrophysical Observatory (AbAO)
• See "Georgian Academy of Sciences, Abastumani Astrophysical Observatory (AbAO)"

Astronomical Society of Georgia (ASG)
Kazbegi Avenue 2a
380060 Tbilisi
Electronic Mail: gsal@dtapha.kheta.ge
WWW: http://www.gcci.org.ge/asgeo.htm
Founded: 1996
Membership: 60
Periodicals: "Letters in Astronomy"
City Reference Coordinates: 042°50'00"E 41°46'00"N

Georgian Academy of Sciences, Abastumani Astrophysical Observatory (AbAO)
Mount Kanobili
383762 Abastumani
Telephone: (0)32-955367
Electronic Mail: tenat@dtapha.kheta.ge
Founded: 1932
Staff: 185
Activities: Sun * planetary systems * variable stars * flare stars * spectral classification * Galaxy * galaxies * plasma astrophysics * upper atmosphere * interplanetary dust * history of astronomy
Periodicals: "Abastumanskaya Astrofizicheskaya Observatorija Byulleten" (ISSN 0375-6644)
Coordinates: 042°49'16"E 41°45'15"N H1,650m
City Reference Coordinates: 042°51'00"E 41°44'00"N

Georgian Academy of Sciences, Abastumani Astrophysical Observatory (AbAO), Town Department
Kazbegi Avenue 2a
380060 Tbilisi
Telephone: (0)32-375226
              (0)32-375228
Electronic Mail: tenat@dtapha.kheta.ge
Founded: 1967
Activities: see main entry
Periodicals: see main entry
City Reference Coordinates: 042°50'00"E 41°46'00"N

State Department for Standardization, Metrology and Certification of Georgia
67 Chargali Street
380092 Tbilisi
Telephone: (0)32-612530
Telefax: (0)32-612530
Electronic Mail: gestand@caucasus.net
City Reference Coordinates: 042°50'00"E 41°46'00"N

Germany

Abbe-Stiftung (Ernst-)
● See "Zeiss-Planetarium der Ernst-Abbe-Stiftung"

ABBS Astro-Mail
c/o Peter Bluhm
Ginsterweg 7
D-21368 Dahlenburg
Telephone: (0)5851-1514
Telefax: (0)5851-7230
Electronic Mail: pbluhm@abbs.heide.de
Founded: 1987
Activities: astronomical bulletin board (call 5851-7896, 1200-28,800 bps)
City Reference Coordinates: 010°44'00"E 53°11'00"N

Akademie der Wissenschaften und der Literatur Mainz
Geschwister-Scholl-Straße 2
D-55131 Mainz
Telephone: (0)6131-5770
Telefax: (0)6131-577111
Founded: 1949
Membership: 213
Activities: interdisciplinary research * basic research * publishing * symposiums
City Reference Coordinates: 008°16'00"E 50°01'00"N

Albert-Einstein-Institut (AEI)
● See "Max-Planck-Institut für Gravitationsphysik"

Albert-Ludwigs-Universität Freiburg-im-Breisgau
● See "Universität Freiburg-im-Breisgau"

Allgäuer Volkssternwarte Ottobeuren (AVSO) eV
Schwabenstraße 13
D-87724 Ottobeuren
Telephone: (0)8332-93186
Telefax: (0)8332-95145
Electronic Mail: info@avso.de
WWW: http://www.avso.de/
Founded: 1966
Membership: 100
Activities: education * photography * computing * observing
Periodicals: (4) "Astro-Amateur"
Coordinates: 010°17'18"E 47°55'48"N H746m
City Reference Coordinates: 010°19'00"E 47°56'00"N

Altec
Ostallee 41
D-54290 Trier
Telephone: (0)651-9940820
Telefax: (0)651-9940821
Electronic Mail: altectrier@aol.com
                   support@altec.de
WWW: http://www.altec.de/home.htm
               http://www.altec.de/astro.htm
Founded: 1994 (in telescope distribution)
City Reference Coordinates: 006°28'00"E 49°45'00"N

Amateurastronomische Vereinigung Göttingen (AVG) eV
Im Kolke 27
D-37083 Göttingen
Telephone: (0)551-7906881 (Rüdiger Rohrig)
Electronic Mail: rrohrig@t-online.de
WWW: http://www.gwdg.de/~unolte/AVG/AVG.html
City Reference Coordinates: 009°55'00"E 51°32'00"N

Amateur- und Präzisionsoptik-Mechanik Markus Ludes
● See now "APM-Telescopes Markus Ludes"

Ambrosius Barth (Johann-)
● See "Johann Ambrosius Barth"

Andromeda
Markenkamp 16
D-45721 Haltern
Telephone: (0)2364-8003
Telefax: (0)2364-169211
Electronic Mail: jelitte.andromeda@cityweb.de
Founded: 1988
Staff: 4
Activities: mail order dealer (NASA publications, English and German astronomical books, planetary maps, slides, videos, NASA decals, patches, ...)
City Reference Coordinates: 007°10'00"E 51°46'00"N

APM-Telescopes Markus Ludes
Kapellenstraße 1
D-66507 Reifenberg
Telephone: (0)6375-6345
Telefax: (0)6375-6397
Electronic Mail: apm-telescopes@saar-pfalz.net
WWW: http://www.apm-ludes-telescopes.de/
Founded: 1990
Staff: 3
Activities: designing and manufacturing telescopes and computer-controlled mounts * importing/exporting and distributing components and accessories
City Reference Coordinates: 007°22'00"E 49°15'00"N (Zweibrücken)
• Formerly "Amateur- und Präzisionsoptik-Mechanik Markus Ludes"

Arbeitsgemeinschaft Walter-Hohmann-Sternwarte Essen eV
Wallneyer Straße 159
D-45133 Essen
Telephone: (0)201-493941 (answering machine)
Telefax: (0)201-493941
Electronic Mail: info@walter-hohmann-sternwarte.de
WWW: http://www.walter-hohmann-sternwarte.de/
Founded: 1969
Membership: 85
Activities: public lectures * photography * optical and radio observing * CCDs * minor planets
Periodicals: (4) "Nova"
Coordinates: 006°58'46"E 51°23'41"N
City Reference Coordinates: 007°00'00"E 51°45'00"N

Arbeitskreis Astronomie Freiburg eV (AKAF)
Basler Landstraße 74b
D-79111 Freiburg-im-Breisgau
Telephone: (0)761-46700
Founded: 1981
Membership: 48
Activities: Sun * Moon * planets * galaxies * popularization
Periodicals: (4) "Astronomisches Informationsblatt der Volkssternwarte Freiburg" (circ.: 1,000)
Coordinates: 007°50'00"E 48°00'00"N (Volkssternwarte Freiburg)
City Reference Coordinates: 007°50'00"E 48°00'00"N

Arbeitskreis Meteore eV (AKM)
Postfach 60 01 18
D-14401 Potsdam
or :
Mehlbeerenweg 5
D-14469 Potsdam
Telephone: (0)331-520707
Electronic Mail: jrendtel@aip.de
WWW: http://www.tu-chemnitz.de/~smo/meteore/akm.html
Founded: 1978
Membership: 70
Activities: observing and analyzing data on meteors, haloes, noctilucent clouds, aurorae * photographic meteor patrol
Periodicals: (12) "Meteoros" (ISSN 1435-0424)
City Reference Coordinates: 013°04'00"E 52°24'00"N

Arbeitskreis Sternfreunde Lübeck (ASL) eV
Postfach 22 09
D-23510 Lübeck
Telephone: (0)451-898547 (Uwe Freitag, President)
Founded: 1977 (as former "Arbeitskreis Sternwarte Lübeck")
Membership: 50
Activities: education * observing (Sun, planets, occultations, comets, variable stars, aurorae) * photography
Periodicals: (3) "Polaris" (ISSN 0930-4916)
Observatories: 1 (Fliegenfelde)
City Reference Coordinates: 010°40'00"E 53°52'00"N

Arbeitskreis Volkssternwarte Recklinghausen (AVR)
c/o Franz Stark
Tannenstraße 17
D-45661 Recklinghausen
Telephone: (0)2361-7782
Electronic Mail: sternwarte.re@t-online.de
WWW: http://members.aol.com/WStrickli/avrhome.htm
http://home.t-online.de/home/Sternwarte.RE/set.htm
Founded: 1950
City Reference Coordinates: 007°11'00"E 51°37'00"N

Archenhold-Sternwarte
Alt-Treptow 1
D-12435 Berlin
Telephone: (0)30-5348080
Telefax: (0)30-5348083
WWW: http://www.cs.tu-berlin.de/~davidi/arch0000.html
http://www.tu-chemnitz.de/~smo/asw/
http://www.snafu.de/~astw/
Founded: 1896
Staff: 14
Activities: hosting the Arbeitskreis Geschichte der Astronomie
Periodicals: (1-2) "Vorträge und Schriften der Archenhold-Sternwarte"; (3-5) "Mitteilungen der Archenhold-Sternwarte",
"Veröffentlichungen der Archenhold-Sternwarte"; (12) "Monatsprogramm und Mitteilungen für Sternfreunde"
Coordinates: 013°28'42"E 52°29'12"N H32m
City Reference Coordinates: 013°24'00"E 52°31'00"N

Astroclub Radebeul eV (ACR)
Auf den Ebenbergen 10a
D-01445 Radebeul
Telephone: (0)351-8381907
Telefax: (0)351-8381907
Electronic Mail: rattei@chemie.rmhs1.tu-dresden.d400.de
WWW: http://ctch06.chm.tu-dresden.de/acr/
Founded: 1959
Membership: 40
Activities: Sun * meteors * planets * photography
Periodicals: (6) "Der Sternfreund" (co-publisher)
Coordinates: 013°37'20"E 51°06'59"N H185m
City Reference Coordinates: 013°43'00"E 51°13'00"N

Astro-Club Schwerin Friedrichsthal-Herrensteinfeld
Pingelshäger Straße 146
D-19057 Schwerin
Telephone: (0)385-4844137
Telefax: (0)385-4844138
Electronic Mail: contact@astro-online.de
WWW: http://www.astro-online.de/
City Reference Coordinates: 011°25'00"E 53°38'00"N

Astrocom GmbH
Lochhamer Schlag 5
D-82116 Gräfelfing
Telephone: (0)89-89889600
Telefax: (0)89-89889601
WWW: http://www.astrocom.de/
Founded: 1984
Activities: manufacturing and distributing telescopes and accessories
City Reference Coordinates: 011°35'00"E 48°08'00"N (München)

Astro-Mail (ABBS_)
● See "ABBS Astro-Mail"

Astronomie Arbeitsgruppe der Universität Oldenburg
Carl von Ossietzky Universität Oldenburg
Fachschule Physik
FB 8 Physik
Postfach
D-26111 Oldenburg
or :
Ammerländer Heerstraße 114-118
D-26129 Oldenburg (for parcel delivery)
Telephone: 441-7980
Telefax: 441-7983000
Electronic Mail: michael.uhlemann@informatik.uni-oldenburg.de

WWW: http://www.infodrom.north.de/~muh/Astronomie/Astro-AG/
Founded: 1992
City Reference Coordinates: 010°52'00"E 54°17'00"N

Astronomie Software Service
c/o Daniel Roth
Strundener Straße 79
D-51069 Köln
Telephone: (0)221-687112
Telefax: (0)221-687112
Electronic Mail: roth@ph-cip.uni-koeln.de
Founded: 1986
Staff: 1
● Software distributor
City Reference Coordinates: 006°57'00"E 50°56'00"N

Astronomiestation Demmin
An den Tannen
Postfach 11 46
D-19101 Demmin
Telephone: (0)3998-222410
Electronic Mail: c.fischer.demmin@t-online.de
WWW: http://home.t-online.de/home/c.fischer.demmin/Astrostation_Demmin.html
Founded: 1978
Staff: 1
Activities: education * observing
Coordinates: 013°03'18"E 53°54'18"N H34m
City Reference Coordinates: 013°03'00"E 53°55'00"N

Astronomische Arbeitsgemeinschaft der Liebigschule
c/o Werner Ziegs
Kollwitzstraße 3
D-60488 Frankfurt/Main
Telefax: (0)69-21239480
Electronic Mail: astro@jugendnetz-ffm.de
WWW: http://www.jugendnetz-ffm.de/astro
Founded: 1974
Membership: 25
Activities: theoretical and practical astronomy, astrophysics and physics for school children * observing
City Reference Coordinates: 008°40'00"E 50°07'00"N

Astronomische Arbeitsgemeinschaft der Volkssternwarte Singen eV
c/o Elmar Nestlen
Thurgauer Straße 9
D-78224 Singen
or :
Rielasingerstraße 37
D-78224 Singen
Telephone: (0)7731-63722
Founded: 1983
Membership: 41
Activities: meetings * education * slide and video shows * public observing * computing * photography
Coordinates: 008°50'40"E 47°45'15"N H422m
City Reference Coordinates: 008°50'00"E 47°45'00"N

Astronomische Arbeitsgemeinschaft im Kulturring Heuchelheim eV
Bachstraße 61
D-35452 Heuchelheim
Electronic Mail: frank.h.leiter@physik.uni-giessen.de
WWW: http://wwwstud.uni-giessen.de/~s608/AAG.html
Founded: 1975
City Reference Coordinates: 008°24'00"E 50°35'00"N (Giessen)

Astronomische Arbeitsgemeinschaft Mainz eV
Postfach 11 64
D-55001 Mainz
Telephone: (0)6764-3660
           (0)6131-236940 (Observatory)
Telefax: (0)6764-3660
WWW: http://iphcip1.physik.uni-mainz.de/~astro/pop/AAG.html
        http://iphcip1.physik.uni-mainz.de/~astro/pop/VSW.html (Observatory)
Founded: 1970
Membership: 110
Activities: observing (deep sky, occultations, comets, eclipses, haloes, ...) * CCDs
Periodicals: (6) "Mitteilungen astronomischer Vereinigungen Rhein-Main-Nahe" (circ.: 500)

City Reference Coordinates: 008°16'00"E 50°01'00"N

Astronomische Arbeitsgemeinschaft Pfaffenwinkel (AAP)
c/o Magnus Zwick
Döbendorferstraße 3
D-86956 Schongau
WWW: http://ourworld.compuserve.com/homepages/DBecker_AAP/
Coordinates: 010°47'10"E 47°50'11"N
City Reference Coordinates: 010°54'00"E 47°49'00"N

Astronomische Arbeitsgemeinschaft Waldhügel
c/o Georg Neumann
Birkhahnweg 8
D-48429 Rheine
Telephone: (0)5971-7406
Founded: 1988
Membership: 10
Activities: observing * photography * computing * lectures
Coordinates: 007°25'12"E 52°15'39"N (Waldhügel)
          007°28'00"E 52°18'23"N H38m (Altenrheine)
City Reference Coordinates: 007°26'00"E 52°17'00"N

Astronomische Arbeitsgemeinschaft Wanne-Eickel/Herne eV
c/o Dieter Rösener
Tupenweg 48
D-44651 Herne
or :
Am Böckenbusch 2a
D-44652 Herne
Telephone: (0)2325-77202
Telefax: (0)2325-51864
WWW: http://homepage.ruhr-uni-bochum.de/Bernd.A.Brinkmann/vsw.html
Founded: 1981
Membership: 40
Activities: observing * lectures * seminars
Coordinates: 007°10'34"E 51°31'40"N H55m
City Reference Coordinates: 007°12'00"E 51°32'00"N

Astronomische Arbeitsgruppe Laufen (AAL) eV
Goethestraße 8
D-83410 Laufen
or :
c/o Klaus Eder
Kohlhaasstraße 4
D-83410 Laufen
WWW: http://hiris.anorg.chemie.tu-muenchen.de/AAL/
Founded: 1986
City Reference Coordinates: 012°55'00"E 47°56'00"N

Astronomische Gesellschaft (AG)
c/o Reinhard E. Schielicke (Secretary)
Universitäts-Sternwarte Jena
Schillergäßchen 2
D-07745 Jena
Telephone: (0)3641-947526
Telefax: (0)3641-947502
Electronic Mail: schie@astro.uni-jena.de (Reinhard E. Schielicke, Secretary)
WWW: http://www.astro.uni-jena.de/Astron_Ges/ag0home.html
Founded: 1863
Membership: 800
Activities: scientific meetings * publication of scientific papers * support of young astronomers
Periodicals: (1) "Mitteilungen der Astronomischen Gesellschaft" (ISSN 0374-1958, circ.: 1,200); "Astronomische Gesellschaft Abstract Series" (ISSN 0934-4438), "Reviews in Modern Astronomy" (ISSN 0941-1445)
Awards: "Karl-Schwarzschild-Medaille", "Ludwig-Biermann-Förderpreis", "Bruno-H.-Bürgel-Preis", "Hans-Ludwig-Neumann-Preis"
City Reference Coordinates: 011°35'00"E 50°56'00"N

Astronomische Gesellschaft (AG), Arbeitskreis Astronomiegeschichte
c/o Wolfgang R. Dick (Secretary)
Otterkiez 14
D-14478 Potsdam
Telephone: (0)331-316618
Telefax: (0)331-316602
Electronic Mail: wdi@potsdam.ifag.de
WWW: http://www.astro.uni-bonn.de/~pbrosche/aa/aa.html

http://www.astro.uni-bonn.de/~pbrosche/astoria.html
Founded: 1992
Membership: 150
Activities: history of astronomy
Periodicals: (2) "Mitteilungen zur Astronomiegeschichte" (ISSN 0944-1999, circ.: 300); "Elektronische Mitteilungen zur Astronomiegeschichte / Electronic Newsletter for the History of Astronomy" (circ.: 250)
City Reference Coordinates: 013°04'00"E 52°24'00"N

Astronomische Gesellschaft Buchloe eV

Gansbichlstraße 10
D-86807 Buchloe
Telephone: (0)8241-7924
Electronic Mail: b_koch@t-online.de (Bernd Koch)
WWW: http://www.schwabmuenchen.de/~schenk
Founded: 1997 (Observatory: 1981)
Membership: 60
Activities: popularization * observing (comets, minor planets, meteors, atmospheric optics, variable stars, deep sky) * photography * fighting light pollution
Periodicals: "Faszinierendes Universum", "Buchloer Astronomisches Zirkular"
Coordinates: 010°43'58"E 48°01'02"N H636m
City Reference Coordinates: 010°45'00"E 48°04'00"N

Astronomische Gesellschaft Urania eV, Wiesbaden

c/o Sternwarte Wiesbaden
Bierstadter Straße 47
D-65189 Wiesbaden
or :
c/o Horst-Rainer Schneider
Feldbergstraße 22
D-65527 Niedernhausen
Telephone: (0)6127-7410 (H.R. Schneider)
            (0)611-317438 (Observatory - Mondays, after 20:00)
Electronic Mail: horst.schneider@arcor.net
WWW: http://iphcip1.physik.uni-mainz.de/~astro/urania/
Founded: 1925
Membership: 100
Activities: observing (Sun, comets, minor planets, occultations) * photography * photometry * CCD * astrometry
Periodicals: (6) "Mitteilungen astronomischer Vereinigungen Rhein-Main-Nahe" (circ.: 500)
Coordinates: 008°15'47"E 50°04'58"N H198m
City Reference Coordinates: 008°14'00"E 50°05'00"N (Wiesbaden)
                              008°20'00"E 50°09'00"N (Niedernhausen)

Astronomische Instrumente Stefan Thiele (AIT)

Walkmühlstraße 4
D-65195 Wiesbaden
Telephone: (0)611-407226
Telefax: (0)611-407226
Electronic Mail: office@ait-trading.com
WWW: http://www.ait-trading.com
Founded: 1980
Staff: 2
Activities: distributing telescopes, accessories, software and filters for H-alpha solar observing and photography * GPS handheld navigators
City Reference Coordinates: 008°14'00"E 50°05'00"N

Astronomische Interessengemeinschaft Brackwede

c/o H. Warnek
Brackweder Gymnasium
Beckumer Straße 10
D-33647 Bielefeld
Telephone: (0)521-444225
Founded: 1972
Membership: 15
Activities: education
Coordinates: 008°30'00"E 51°59'30"N
City Reference Coordinates: 008°32'00"E 52°02'00"N

Astronomische Nachrichten (AN)

Astrophysikalisches Institut Potsdam
An der Sternwarte 16
D-14482 Potsdam
Telephone: (0)331-762202
Telefax: (0)331-762200
Electronic Mail: kfritze@aip.de
WWW: http://www.wiley-vch.de/berlin/journals/an/
Founded: 1821

Staff: 3
● Journal (6) (ISSN 0004-6337)
City Reference Coordinates: 013°04'00"E 52°24'00"N

Astronomischer Arbeitskreis der Heimvolksschule Schloß Dhaun

c/o Heimvolksschule Schloß Dhaun
D-55606 Hochstetten-Dhaun
Telephone: (0)6752-5374
Electronic Mail: wichb000@mzdmza.zdv.uni-mainz.de (Burkhard Wiche)
WWW: http://iphcip1.physik.uni-mainz.de/~astro/pop/Dhaun.html
Membership: 20
Periodicals: "Mitteilungen astronomischer Vereinigungen Rhein-Main-Nahe" (circ.: 550)
Coordinates: 007°29'44"E 49°48'53"N H365m
City Reference Coordinates: 007°31'00"E 49°48'00"N

Astronomischer Arbeitskreis Ingolstadt (AAI) eV

c/o Dieter Leistritz
Lilienthalstraße 137
D-85077 Manching
Telephone: (0)8459-6587
Electronic Mail: aai@bingo.baynet.de
WWW: http://www.bingo.baynet.de/~aai/
          http://www.bingo-ev.de/~aa156/
          http://www.bingo.baynet.de/region/vereine/aai/
Coordinates: 011°25'26"E 48°44'47"N
City Reference Coordinates: 011°31'00"E 48°43'00"N

Astronomischer Arbeitskreis Kassel eV (AAK)

c/o K.P. Haupt
Wilhelmshöher Allee 300a
D-34131 Kassel
Telephone: (0)561-311116
Telefax: (0)561-311116
Electronic Mail: matsim@physik.uni-kassel.de (Matthias Simon)
WWW: http://www.kassel.de/ask/aak/
Founded: 1972
Membership: 185
Activities: observing (planets, Sun) * photography * popularization
Periodicals: (4) "Sternzeit" (co-publisher) (ISSN 0721-8168, circ.: 2,000); (3) "Korona" (circ.: 350)
Observatories: 1 (Sternwarte Calden)
City Reference Coordinates: 009°29'00"E 51°19'00"N

Astronomischer Arbeitskreis Merkur/Venus Göttingen (AAMVG)

c/o Detlev Niechoy
Bertheaustraße 26
D-37075 Göttingen
Telephone: (0)551-33830
Telefax: (0)551-33871
Electronic Mail: dniechoy@t-online.de
WWW: http://www.geocities.com/~dniechoy/
Founded: 1994
Membership: 4
Activities: inner planets * Sun * photometry * telescope drawings
Coordinates: 009°56'36"E 51°31'00"N
City Reference Coordinates: 009°55'00"E 51°32'00"N

Astronomischer Arbeitskreis Pforzheim (AAP) 1982 eV

Parkstraße 25
D-75223 Niefern-Öschelbronn
Telephone: (0)7233-81515
Telefax: (0)7233-81515
WWW: http://www.s-direktnet.de/homepages/aap/
Founded: 1982
Membership: 43
Activities: education * running a public observatory
Periodicals: (4) "Astro-News"
Observatories: 1 (Bieselsberg)
City Reference Coordinates: 008°42'00"E 48°54'00"N

Astronomischer Arbeitskreis Wetzlar eV

Lindenstraße 11
D-35606 Solms/Lahn
Telephone: (0)6442-1039
          (0)6442-927640 (astronomical information)
Telefax: (0)6085-970384

Electronic Mail: martin.pfeil@metronet.de
WWW: http://home.t-online.de/home/Sternwarte.Burgsolms/ (Sternwarte Burgsolms)
City Reference Coordinates: 008°29'00"E 50°33'00"N (Wetzlar)

Astronomischer Freundeskreis Ostsachsen (AFO)
c/o Volks- und Schulsternwarte Bruno H. Bürgel
Zöllnerweg 12
D-02689 Sohland
Telephone: (0)35936-37270
WWW: http://ctch06.chm.tu-dresden.de/afo/afo_home.htm
Founded: 1992
Membership: 25
Periodicals: (6) "Der Sternfreund"
City Reference Coordinates: 014°25'00"E 51°02'00"N

Astronomischer Verein der Grafschaft Bentheim (AVGB) eV
Am Westhang 19
D-48455 Bad Bentheim
Telephone: (0)5924-5157
Electronic Mail: avgb@eure.de
WWW: http://privat.eure.de/avgb/
         http://privat.eure.de/avgb/index2.htm
Founded: 1990
City Reference Coordinates: 007°10'00"E 52°19'00"N

Astronomischer Verein der Volkssternwarte Papenburg eV
Wilhelm-Leuschnerstraße 48
D-26871 Papenburg
Telephone: (0)4961-1694
Founded: 1984
Membership: 50
Activities: observing * photography * instrument making * activities in schools * seminars * consulting * journeys
Periodicals: (4) "Sternzeit" (co-publisher) (ISSN 0721-8168, circ.: 2,000); "Astro-Nachrichten"
Coordinates: 007°23'30"E 53°03'20"N
City Reference Coordinates: 007°25'00"E 53°05'00"N

Astronomischer Verein Dortmund (AVD) eV
Postfach 10 01 20
D-44001 Dortmund
WWW: http://www.astronomie.org/avd/
Telephone: (0)231-104076
Founded: 1913
Membership: 60
Activities: observing (variable stars, Sun, planets)
Periodicals: (4) "Sternzeit" (co-publisher) (ISSN 0721-8168, circ.: 2,000), "Raum und Zeit"
Observatories: 1 (Volkssternwarte Dortmund)
City Reference Coordinates: 007°28'00"E 51°31'00"N

Astronomischer Verein Hoyerswerda eV
c/o Peter Schubert
Jan-Arnost-Smoler-Straße 3
D-02977 Hoyerswerda
Telephone: (0)3571-417020
Electronic Mail: arlev@t-online.de
WWW: http://www.germany.net/teilnehmer/100/142601/astro.htm
City Reference Coordinates: 014°14'00"E 51°26'00"N

Astronomischer Verein Remscheid (AVRS) eV
Postfach 10 01 03
Palmstraße 29
D-42801 Remscheid
Telephone: (0)2191-75607
Electronic Mail: dick.huetzluff@t-online.de (Dick Hützluff)
WWW: http://www.sds.de/remscheid/vereine/avrs/
         http://central.sds.de/remscheid/vereine/avrs/
Founded: 1982
Membership: 75
Activities: observing * photography * popularization
Periodicals: (4) "Infoheft"
Coordinates: 007°12'00"E 51°12'00"N
City Reference Coordinates: 007°10'00"E 51°11'00"N

Astronomische Schulstation Adolph Diesterweg
G.-Petri-Straße 3
D-38855 Wernigerode

Telephone: (0)3943-632270
Telefax: (0)3943-632421
Founded: 1967
Staff: 10
Activities: observing * education * actualities
Coordinates: 010°47'00"E 51°50'00"N H240m
City Reference Coordinates: 010°47'00"E 51°50'00"N

Astronomisches Rechen-Institut (ARI)

Mönchhofstraße 12-14
D-69120 Heidelberg
Telephone: (0)6221-4050
Telefax: (0)6221-405297
Electronic Mail: <userid>@urz.uni-heidelberg.de
WWW: http://www.ari.uni-heidelberg.de/
         http://www.ari.uni-heidelberg.de/index.eng.htm
Founded: 1700
Staff: 70
Activities: astrometry * stellar dynamics * ephemerides * bibliography
Periodicals: (2) "Astronomy and Astrophysics Abstracts (AAA)" (ISSN 0067-0022); (1) "Apparent Places of Fundamental Stars", "Astronomische Grundlagen für den Kalender" (ISSN 0067-0014); "Veröffentlichungen des Astronomisches Rechen-Institut", "Mitteilungen des Astronomisches Rechen-Institut - Serie A", "Mitteilungen des Astronomisches Rechen-Institut - Serie B"
City Reference Coordinates: 008°43'00"E 49°25'00"N

Astronomische Station Heinrich S. Schwabe

c/o Walter-Gropius-Gymnasium
Peterholzstraße 58
D-06849 Dessau
Telephone: (0)340-8581858
WWW: http://lbs.st.schule.de/lbs/pl-dessau.htm
Founded: 1967
Staff: 1
Activities: observing * planetarium shows * education
City Reference Coordinates: 012°14'00"E 51°50'00"N

Astronomische Station Tycho Brahe

Nelkenweg 6
D-18057 Rostock
Telephone: (0)381-4934068
Founded: 1965
Staff: 3
Activities: education
City Reference Coordinates: 012°07'00"E 54°05'00"N

Astronomische Sternwarte Nessa

Dorfstraße 11
D-06682 Nessa
Telephone: (0)34443-20960
Telefax: (0)34443-20961
Electronic Mail: nessa@t-online.de
WWW: http://www.nessa.via.t-online.de/
Founded: 1969
Staff: 1
Activities: CCD observing (variable stars, faint planet satellites)
Coordinates: 012°01'15"E 51°09'05"N H169m
City Reference Coordinates: 012°01'00"E 51°09'00"N

Astronomisches Zentrum Bruno-H.-Bürgel

Im Neuen Garten 6
D-14469 Potsdam
Telephone: (0)331-2702721
           (0)331-2702724
Telefax: (0)331-292447
Electronic Mail: planetarium.potsdam@t-online.de
Founded: 1971
Staff: 2
Activities: popularization * teacher training * sundials * photography * computing * planetarium
Coordinates: 013°04'06"E 52°25'12"N
City Reference Coordinates: 013°04'00"E 52°24'00"N

Astronomisches Zentrum Burg

c/o Elke Sommer
Kirchofstraße 3
D-39288 Burg

Telephone: (0)3921-3014
WWW: http://lbs.st.schule.de/lbs/pl-burg.htm
City Reference Coordinates: 011°51'00"E 52°17'00"N

Astronomisches Zentrum Halberstadt (AZH)

Wilhelm-Trautewein-Straße 19
D-38820 Halberstadt
Telephone: (0)3941-600039
WWW: http://lbs.st.schule.de/lbs/pl-halb.htm
Founded: 1990
Activities: education * observing * planetarium shows
Coordinates: 011°06'00"E 51°09'00"N
City Reference Coordinates: 011°06'00"E 51°09'00"N

Astronomisches Zentrum Magdeburg (AZM)

Pablo-Picasso-Straße 21
D-39128 Magdeburg
Telephone: (0)391-2523975
Telefax: (0)391-2523974
Electronic Mail: astromd@gmx.de
WWW: http://www.now-online.de/astro
Founded: 1977
Activities: observing (Sun) * planetarium
Periodicals: (4) "Sternzeit" (co-publisher) (ISSN 0721-8168, circ.: 2,000)
Coordinates: 011°37'30"E 52°10'12"N H60m (Johannes Kepler Observatory)
City Reference Coordinates: 011°38'00"E 52°07'00"N

Astronomisches Zentrum Schkeuditz

Postfach 11 29
D-04431 Schkeuditz
or :
Bergbreite 1
D-04435 Schkeuditz
Telephone: (0)34204-62616
Telefax: (0)34204-62616
Electronic Mail: stern@rzaix530.rz.uni-leipzig.de
WWW: http://www.uni-leipzig.de/~stern/
Founded: 1978
Staff: 1
Activities: education * planetarium shows * public observatory
Coordinates: 012°13'30"E 51°23'45"N H119m
City Reference Coordinates: 012°13'00"E 51°24'00"N

Astronomische Vereinigung Albstadt (AVA) eV

Hartmannstraße 140
D-72458 Albstadt-Ebingen
Telephone: (0)7431-72881
Telefax: (0)7431-72881
Electronic Mail: 0743172881-0001@t-online.de
WWW: http://home.t-online.de/home/0743172881-0001@t-online.de/
Founded: 1973
Membership: 15
Activities: photography * Moon * planets * Sun * comets * observing * planetarium shows (see separate entry) * CCD * instrumentation
Coordinates: 008°59'59"E 48°12'37"N H737m
City Reference Coordinates: 009°02'00"E 48°13'00"N

Astronomische Vereinigung Augsburg eV (AVA)

c/o Volkssternwarte
Pestalozzistraße
D-86420 Diedorf
Telephone: (0)8238-7344
WWW: http://www.serve.com/ava/
        http://www.serve.com/ava/sternwarte_e.html
        http://www.intercon-spacetec.com/ava/
Founded: 1965
Membership: 180
Activities: public observing * education * photography * planetarium shows * star parties
Periodicals: (3) "Uranus" (circ.: 250)
Coordinates: 010°47'11"E 48°21'15"N H504m
City Reference Coordinates: 010°53'00"E 48°23'00"N (Augsburg)

Astronomische Vereinigung Karlsruhe (AVK) eV

Krokusweg 49
D-76199 Karlsruhe

WWW: http://www.uni-karlsruhe.de/~lh34/avkhomep.html
      http://www.karlsruhe.de/Umwelt/Volkssternwarte/ (Sternwarte Karlsruhe)
Founded: 1974
Coordinates: 008°24'55"E 48°58'39"N H125m
City Reference Coordinates: 008°24'00"E 49°03'00"N

Astronomische Vereinigung Nürtingen (AVN) eV
c/o Hans-Dieter Haas
Birkenweg 7
D-72622 Nürtingen
Telephone: (0)7022-33678
Telefax: (0)7022-31408
Electronic Mail: hdhaas@nuertingen.netsurf.net
WWW: http://www.avnev.nepustil.net
City Reference Coordinates: 009°20'00"E 48°37'00"N

Astronomische Vereinigung Tübingen (AVT) eV
Waldhäuser Straße 64
D-72076 Tübingen
Telephone: (0)7071-297-8607
Electronic Mail: avt@magellan.tat.physik.uni-tuebingen.de
WWW: http://magellan.tat.physik.uni-tuebingen.de/~avt/
Founded: 1972
City Reference Coordinates: 009°02'00"E 48°31'00"N

Astronomische Vereinigung Weikersheim eV
c/o Albert Hammer
Marienstraße 38
D-97980 Bad Mergentheim
Telephone: (0)7931-2120
Electronic Mail: astro@hamox.com
WWW: http://hamox.com/astro/
      http://www.hamox.com/astro/
Founded: 1978
Membership: 72
Activities: observing * photography * lectures
Periodicals: (4) "Sternzeit" (co-publisher) (ISSN 0721-8168, circ.: 2,000)
City Reference Coordinates: 009°46'00"E 49°30'00"N

Astronomische Vereinigung West-München (AVWM)
c/o Bruno Wagner
Egelseestraße 21
D-86949 Windach
or :
Grasslfinger Straße
D-82194 Gröbenzell
Telephone: (0)8193-366
Electronic Mail: info@avwm.org
WWW: http://www.avwm.org/
Founded: 1978
Membership: 55
Activities: observing * meteors
Periodicals: (3) "Ganymed"
Coordinates: 011°02'18"E 48°03'33"N H595m
City Reference Coordinates: 011°03'00"E 48°04'00"N (Windach)
                              011°22'00"E 48°11'00"N (Gröbenzell)
                              011°35'00"E 48°08'00"N (München)

Astronomy and Astrophysics (A&A), Letters
c/o P. Schneider (Letter Editor)
Max-Planck-Institut für Astrophysik
Postfach 1523
D-85740 Garching
or :
Karl-Schwarzschild-Straße 1
D-85748 Garching
Telephone: (0)89-32993296
Telefax: (0)89-32993596
Electronic Mail: aal@mpa-garching.mpg.de
WWW: http://link.springer.de/link/service/journals/00230/
Founded: 1969
● Journal (24) (ISSN 0004-6361)
City Reference Coordinates: 011°40'00"E 48°14'00"N

Astronomy Study Unit (ASU)
c/o Eckehard Schmidt
Postfach 46 16
D-90025 Nürnberg
or :
Brunhildstraße 1a
D-90461 Nürnberg
Telephone: (0)911-4720978
Telefax: (0)911-5865549
Founded: 1972
Membership: 61
Activities: gathering stamps with astronomical themes
Periodicals: (4) "Astrofax"
City Reference Coordinates: 011°04'00"E 49°27'00"N

Astrophysikalisches Institut Potsdam (AIP)
An der Sternwarte 16
D-14482 Potsdam
Telephone: (0)331-74990
Telefax: (0)331-7499200
Electronic Mail: <userid>@aip.de
WWW: http://www.aip.de:8080/
Founded: 1700
Staff: 78
Activities: extragalactic astrophysics * cosmology * cosmic magnetic fields * X-ray astrophysics
Coordinates: 013°04'00"E 52°22'54"N H74m (Babelsberg)
City Reference Coordinates: 013°04'00"E 52°24'00"N

Astrophysikalisches Institut Potsdam (AIP), Observatorium für Solare Radioastronomie (OSRA)
D-14552 Tremsdorf
Telephone: (0)33205-2261
Telefax: (0)33205-2393
Electronic Mail: <userid>@aip.de
WWW: http://aipsoe.aip.de/~det/online_spectra.html
Founded: 1700 (AIP)
City Reference Coordinates: 013°04'00"E 52°24'00"N (Potsdam)

Astrophysikalisches Institut Potsdam (AIP), Sonnenobservatorium Einsteinturm (SOE)
Telegrafenberg
D-14473 Potsdam
Telephone: (0)331-2880
Telefax: (0)331-2882310
Electronic Mail: <userid>@aip.de
WWW: http://aipsoe.aip.de/
Founded: 1920
Staff: 10
Activities: solar physics * solar magnetic fields * sunspots * solar activity
Coordinates: 013°04'00"E 52°22'54"N H100m
City Reference Coordinates: 013°04'00"E 52°24'00"N

astro-shop
Hindenburgstraße Ö1
D-22303 Hamburg
Telephone: (0)40-5114348
Telefax: (0)40-5114594
Electronic Mail: e.vesting@astro-shop.com (Eric-Sven Vesting)
WWW: http://www.astro-shop.com/
Founded: 1978
City Reference Coordinates: 009°59'00"E 53°33'00"N

Astro-Versand
Birkenstraße 14
D-72145 Hirrlingen
Telephone: (0)7478-261613
Telefax: (0)7478-261614
Electronic Mail: versand1@aol.com
WWW: http://www.astro-versand.com/
Founded: 1979
Staff: 3
Activities: astronomical mail order company
City Reference Coordinates: 009°02'00"E 48°31'00"N (Tübingen)

Baader Planetarium GmbH
Zur Sternwarte

D-82291 Mammendorf
Telephone: (0)8145-8802
Telefax: (0)8145-8805
Founded: 1966
Staff: 12
Activities: manufacturing and distributing observatory domes, planetariums, spectroscopes, coronagraphs, H-alpha filters, telescopes, optical pieces and accessories
City Reference Coordinates: 011°09'00"E 48°12'00"N

Barth (Johann Ambrosius)
● See "Johann Ambrosius Barth"

Bayerische Akademie der Wissenschaften
Marstallplatz 8
D-80539 München
Telephone: (0)89-230310
Telefax: (0)89-23031240
          (0)89-23031100
Founded: 1759
Membership: 90
Periodicals: (1) "Jahrbuch (ISSN 0084-6090); "Nova Kepleriana - Abhandlungen - Neue Folge" (ISSN 0078-2246), "Sitzungsberichte der Philosophische-historische Klasse" (ISSN 0342-5991), "Abhandlungen der Philosophische-historische Klasse" (ISSN 0005-710x), "Sitzungsberichte der Mathematische-naturwissenschaftliche Klasse" (ISSN 0340-7586), "Abhandlungen der Mathematische-naturwissenschaftliche Klasse" (ISSN 0005-6995)
City Reference Coordinates: 011°35'00"E 48°08'00"N

Bayerische Julius-Maximilians-Universität Würzburg
● See "Universität Würzburg"

Bayerische Volkssternwarte München eV
Rosenheimer Straße 145a
D-81671 München
Telephone: (0)89-406239
Telefax: (0)89-494987
WWW: http://www.lrz-muenchen.de/t/t7121bl/astro/vsw.html (German)
          http://www.lrz-muenchen.de/t/t7121bl/astro/vsw_e.html (English)
          http://www.lrz-muenchen.de/~t7121bl/astro/vsw_e.html
Founded: 1947
Membership: 440
Staff: 3
Activities: popularization * planetarium shows
Periodicals: (6) "Blick ins All"
Coordinates: 011°36'31"E 48°07'21"N H572m (München)
               011°10'00"E 48°00'00"N H600m (Herrsching)
City Reference Coordinates: 011°35'00"E 48°08'00"N

Bayerische Volkssternwarte Neumarkt/Opf. eV
c/o Hans-Werner Neumann
Moorstraße 5
D-92318 Neumarkt
Telephone: (0)9181-44592
WWW: http://www.neumarkt.net/sternw/
Founded: 1969
Membership: 250
Activities: weekly observing evenings * lectures * open-door days * excursions meteors * solar protuberances
Sections: Astrophotography * Space * Meteor Observatory * Solar Protuberances
Periodicals: (1) "Jahresbericht"
Coordinates: 011°27'40"E 49°16'50"N
City Reference Coordinates: 011°28'00"E 49°17'00"N

Berliner Sternfreunde
c/o Andreas Reinhard
Ettersburger Weg 4
D-13086 Berlin
Telephone: (0)30-9246778
Telefax: (0)30-9246778
WWW: http://www.astronomie.org/best/
Membership: 35
City Reference Coordinates: 013°24'00"E 52°31'00"N

Bild der Wissenschaft
Postfach 10 60 12
D-70049 Stuttgart
or :
Neckarstraße 121

D-70190 Stuttgart
Telephone: (0)711-26310
Telefax: (0)711-2631292
        (0)711-2631102
WWW: http://www.wissenschaft.de/
Founded: 1964
Activities: popularization
● Journal (12) (ISSN 0006-2375)
City Reference Coordinates: 009°11'00"E 48°46'00"N

Bismarckschule Hannover, Planetarium
An der Bismarckschule 5
D-30173 Hannover
Telephone: (0)511-16843456
WWW: http://www.h.shuttle.de/h/bimsch/
City Reference Coordinates: 009°44'00"E 52°24'00"N

Bremerhavener Sternfreunde eV
c/o Günter Neumann
Buschkämpen 29
D-27576 Bremerhaven
Founded: 1982
Membership: 20
Activities: observing * public lectures
City Reference Coordinates: 008°34'00"E 53°33'00"N

Bruno-H.-Bürgel-Sternwarte, Berlin
Heerstraße 531
D-13593 Berlin
Telephone: (0)30-3636242
Telefax: (0)30-36801180
Founded: 1982
Membership: 120
Activities: education * lectures * exhibitions * shows * radioastronomy
Periodicals: "Space"
Coordinates: 013°09'11"E 52°31'09"N (Hahneberg)
City Reference Coordinates: 013°24'00"E 52°31'00"N

Bruno-H.-Bürgel-Sternwarte, Hartha
Töpelstraße 43
D-04746 Hartha
Telephone: (0)34328-3158
WWW: http://home.t-online.de/home/03432839310-0001/STWHOME.HTM
Founded: 1956
Activities: variable stars
Periodicals: "Mitteilungen der Bruno-H. Bürgel-Sternwarte Hartha", "Harthaer Beobachtungszirkulare", "Sonderdrucke"
Coordinates: 012°57'37"E 51°06'11"N H324m
City Reference Coordinates: 013°00'00"E 51°06'00"N

Bundesamt für Seeschiffahrt und Hydrographie (BSH), Hamburg
Postfach 30 12 20
D-20305 Hamburg
or :
Bernhard-Nocht-Straße 78
D-20359 Hamburg
Telephone: (0)40-31900
Telefax: (0)40-31905000
Periodicals: (1) "Nautisches Jahrbuch" (ISSN 0077-6211)
City Reference Coordinates: 009°59'00"E 53°33'00"N

Bundesamt für Seeschiffahrt und Hydrographie (BSH), Rostock
Dierkower Damm 45
D-18146 Rostock
Telephone: (0)381-45635
Telefax: (0)381-4563948
City Reference Coordinates: 012°07'00"E 54°05'00"N

Bundesdeutsche Arbeitsgemeinschaft für Veränderliche Sterne eV (BAV)
Munsterdamm 90
D-12169 Berlin
Telephone: (0)30-7900930
Electronic Mail: u7x11az@lrz-muenchen.de (Thorsten Lange)
WWW: http://www.lrz-muenchen.de/~u7x11az/www/bav.html
Founded: 1950
Membership: 200

Activities: observing (variable stars)
Periodicals: (4) "BAV-Rundbrief" (ISSN 0405-5497, circ.: 320); (2) "BAV-Mitteilungen" (circ.: 320); (1) "BAV-Circular" (circ.: 280)
City Reference Coordinates: 013°24'00"E 52°31'00"N

Carl Zeiss Jena GmbH, Astronomische Geräte/Weltraumtechnik
Tatzend Promenade 1a
Postfach 125
D-07740 Jena
Telephone: (0)3641-640
          (0)3641-642542
Telefax: (0)3641-642023
Electronic Mail: deczjkhr@ibmmail.com (Peter Koehler, Product Manager/Astronomical Instruments)
          astro@zeiss.de
          <userid>@zeiss.de
WWW: http://www.zeiss.de/
Founded: 1889
Staff: 13000
Activities: designing and manufacturing astronomical instrumentation * CCD cameras * professional telescopes * space technology projects * optical, mechanical and electronic components * gratings * planetariums
Periodicals: "Zeiss Information with Jena Review" (ISSN 0941-7559 & 0941-7567)
City Reference Coordinates: 011°35'00"E 50°56'00"N

Carl Zeiss Jena GmbH, Oberkochen
D-73446 Oberkochen
Telephone: (0)7364-200
          (0)7364-203418
Telefax: (0)7364-6808
       (0)7364-204343
Electronic Mail: <userid>@zeiss.de
WWW: http://www.zeiss.de/
Founded: 1889
Staff: 13000
Activities: see "Carl Zeiss Jena GmbH, Astronomische Geräte/Weltraumtechnik"
Periodicals: see "Carl Zeiss Jena GmbH, Astronomische Geräte/Weltraumtechnik"
City Reference Coordinates: 010°06'00"E 48°47'00"N

Carl-Zeiss-Planetarium Stuttgart
Mittlerer Schloßgarten
D-70173 Stuttgart
Telephone: (0)711-1629215
          (0)711-162920
Telefax: (0)711-2163912
WWW: http://www.s.shuttle.de/hk1000/plan.html
Founded: 1977
Staff: 13
Activities: education * shows
Periodicals: (8-12) "Planetarium Stuttgart, Program"
Coordinates: 009°11'51"E 48°47'01"N H344m
          009°35'49"E 48°52'30"N H542m (Welzheim)
City Reference Coordinates: 009°11'00"E 48°46'00"N

Christian-Albrechts-Universität zu Kiel
● See "Universität Kiel"

Copernicus Planetarium (Nicolaus_)
● See "Nicolaus Copernicus Planetarium"

Daimler Benz Aerospace
Postfach 80 11 09
D-81663 München
Telephone: (0)89-6070
          (0)89-60734361
Telefax: (0)89-60726481
       (0)89-60734373
WWW: http://www.dasa.de/
Activities: manufacturing i.a. space modules and instrumentation
● Formerly "Deutsche Aerospace AG (DASA)"
● Now part of the "European Aeronautic, Defense and Space Co. (EADS)"
City Reference Coordinates: 011°35'00"E 48°08'00"N

Deutsch (Verlag Harri_)
● See "Verlag Harri Deutsch"

Deutsche Aerospace AG (DASA)
● See now "Daimler-Benz Aerospace"

Deutsche Agentur für Raumfahrtangelegenheiten (DARA) GmbH
● Now part of the "Deutsches Zentrum für Luft- und Raumfahrt (DLR) eV"

Deutsche Forschungsanstalt für Luft- und Raumfahrt (DLR) eV
● See now "Deutsches Zentrum für Luft- und Raumfahrt (DLR) eV"

Deutsche Forschungsgemeinschaft (DFG)
Kennedyallee 40
D-53175 Bonn
Telephone: (0)228-8851
Telefax: (0)228-8852777
        (0)228-8852180 (PR)
WWW: http://www.dfg.de/
Founded: 1920 (refounded: 1949)
Membership: 87 (organizations)
Activities: promoting research in all fields * counseling parliaments * fostering international academic relationships
Periodicals: (4) "forschung - Mitteilungen der DFG" (ISSN 0172-1518); (3) "german research - Reports of the DFG" (ISSN 0172-1526); (1) "Jahresbericht"; "Statistik",
City Reference Coordinates: 007°05'00"E 50°44'00"N

Deutsche Gesellschaft für Chronometrie (DGC) eV
Ziehrerweg 8
D-71254 Ditzingen
Telephone: (0)7156-951640
Telefax: (0)7156-951640
WWW: http://www.ph-cip.uni-koeln.de/~roth/dgc.html
        http://192.41.20.246/dgc/
Founded: 1949
City Reference Coordinates: 009°03'00"E 48°49'00"N

Deutsche Gesellschaft für Luft- und Raumfahrt (DGLR) eV
Godesberger Allee 70
D-53175 Bonn
Telephone: (0)228-376726
        (0)228-376727
Telefax: (0)228-374755
WWW: http://www.kp.dlr.de/DGLR/
Founded: 1967
Membership: 3300
Activities: meetings * fostering aeronautical and space research
Periodicals: (6) "Luft- und Raumfahrt" (ISSN 0173-6264), "Mitteilungen aus der DGLR"
Sections: (commissions) FG 1. Systemtechnik und Systemplanung * FA 1.1 Basismaterialien * FA 1.2 Projektführung * FA 1.3 Unterstützungsfunktionen * FA 2A. Luftfahrtzeuge * FA 2A.1 Transport- und Hochgeschwindigkeitsflugzeuge * FA 2A.2 Drehflüger * FA 2A.3 Segelflug und Sportflugzeuge * FA 2A.4 Flugbetrieb für Luftfahrtzeuge * FA 2A.5 Flugsysteme leichter als Luft * FG 2B. Flugkörper und Raumfahrt * FA 2B.1 Flugkörper, RPV's und Drohnen * FA 2B.2 Bemannte Raumfahrtsysteme * FA 2B.3 Satelliten und Sonden * FA 2B.4 Transportsysteme (einsch. Wiedereintrittssysteme) * FA 2B.5 Bodeneinrichtungen für Raumfahrtzeuge * FG 3. Fluid- und Thermodynamik * FA 3.1 Hydro-, Aero- und Gasdynamik * FA 3.2 Strömungsakustik/Fluglärm * FA 3.3 Versuchswesen der Fluid- und Thermodynamik * FG 4. Flugmechanik und Flugführung * FA 4.1 Flugleistungen und Bahnen * FA 4.2 Flugeigenschaften * FA 4.3 Flugregelung * FA 4.4 Ortung und Navigation * F14.5 Anthropotechnik * FA 4.6 Flugversuchstechnik * FA 4.7 Begriffsbestimmungen * FG 5. Antriebe * FA 5.1 Luftatmende Antriebe * FA 5.2 Raketenantriebe * FA 5.3 Elektrische und unkonventionelle Flugantriebe * FG 6. Energieversorgung, Schutzsysteme, Automation * FA 6.1 Energieversorgung * FA 6.2 Lebenserhaltungssysteme, Wärmeschutzsysteme * FA 6.3 Robotik, Automation, RVD, EVA * FA 6.4 Technologie der rechnergestützten Simulation * FG 7. Festigkeit, Bauweisen, Werkstoffe * FA 7.1 Festigkeit und Bauweisen * FA 7.2 Aeroelastik und Strukturdynamik * FA 7.3 Werkstoffe * FG 8. Elektronik-Nachrichtentechnik * FA 8.1 Telemetrie * FA 8.2 Avionik * FA 8.3 EMV/EMP * FA 8.4 Nachrichtenübertragung einsch. Signal- und Informationsverarbeitung * FA 8.5 Sensoren, Informationserfassung, Datenauswertung * FG 9. Flugmedizin und Biologie * FG 10. Luftraum- und Weltraumkunde * FA 10.1 Extraterrestrische Systeme * FA 10.2 Geowissenschaftliche Systeme * FA 10.3 Werkstoffforschung unter Schwerelosigkeit * FA 10.4 Medizin-/Biologieforschung unter Schwerelosigkeit * FG 11. Luftrecht und Weltraumrecht * FG 12. Geschichte der Luft- und Raumfahrt * FG 13. Dokumentation
Awards: (1) "Ludwig-Prandtl-Ring"
City Reference Coordinates: 007°05'00"E 50°44'00"N

Deutsche Meteorologische Gesellschaft (DMG) eV
c/o Sekretariat DMG
Institut für Meteorologie
Freie Universität Berlin
Carl-Heinrich-Becker-Weg 6-10
D-12165 Berlin
Telephone: (0)30-83871197
Telefax: (0)30-7919002
Electronic Mail: dmg@bibo.met.fu-berlin.de

WWW: http://www.met.fu-berlin.de/dmg
Founded: 1883
Activities: furthering meteorology in Germany
Periodicals: "Zeitschrift für Meteorologie"
Coordinates: 013°18'00"E 52°28'00"N H37m
City Reference Coordinates: 013°24'00"E 52°31'00"N

Deutsche Physikalische Gesellschaft (DPG) eV
c/o Volker Häselbarth (Chief Executive)
Hauptstraße 5
D-53604 Bad Honnef
Telephone: (0)2224-92320
Telefax: (0)2224-923250
Electronic Mail: arias@snhonnef1.pbh.uni-bonn.de
WWW: http://www.dpg-physik.de/
Founded: 1845
Membership: 30000
Periodicals: (12) "Physikalische Blätter" (ISSN 0031-9279); "Verhandlungen der Deutsche Physikalische Gesellschaft"
Awards: "Max-Planck-Medaille", "Stern-Gerlach-Preis", "Max-Born-Preis" (together with the "Institute of Physics"), "Gentner-Kastler-Preis" (together with the "Société Française de Physique"), "Robert-Wichard-Pohl-Preis", "Gustav-Hertz-Preis", "Walter-Schottky-Preis"
City Reference Coordinates: 007°13'00"E 50°39'00"N

Deutscher Wetterdienst (DWD)
Frankfurter Straße 135
D-63067 Offenbach/Main
Telephone: (0)69-80620
Telefax: (0)69-80622488
Electronic Mail: <userid>@dwd.d400.de
WWW: http://www.dwd.de/
Founded: 1952
Staff: 3200
Periodicals: (daily) "Europäischer Wetterbericht" (ISSN 0341-2970), "Wetterkarte" (ISSN 0936-5818); (52) "Agrarmeteorologischer Wochenhinweis" (ISSN 0172-0570); (12) "Monatlicher Witterungsbericht" (ISSN 0435-7965), "Großwetterlagen Europas" (ISSN 0017-4645); (4) "Promet - meteorologische Fortbildung" (ISSN 0340-4552); (1) "Agrarmeteorologische Bibliographie" (ISSN 0515-6831), "Deutsches Meteorologisches Jahrbuch" (ISSN 0724-7125), "Jahresbericht" (ISSN 0433-8251)
• Meteorological office
City Reference Coordinates: 008°47'00"E 50°08'00"N

Deutscher Wetterdienst (DWD), Berlin
Lindenberger Weg 24
D-13125 Berlin
Telephone: (0)30-9400940
Telefax: (0)30-9497324
Electronic Mail: <userid>@dwd.d400.de
WWW: http://www.dwd.de/
Founded: 1952 (DWD)
Activities: hydrometeorology
City Reference Coordinates: 013°24'00"E 52°31'00"N

Deutscher Wetterdienst (DWD), Klima- und Umweltberatung Weimar
Heinrich-Jäde-Straße 12
D-99425 Weimar
Telephone: (0)3643-54000
Telefax: (0)3643-540040
Electronic Mail: <userid>@dwd.d400.de
WWW: http://www.dwd.de/
Founded: 1945
Staff: 33
Activities: weather observing * climatology
Coordinates: 011°18'36"E 50°58'36"N H264m (Weimar)
City Reference Coordinates: 011°20'00"E 50°59'00"N

Deutscher Wetterdienst (DWD), Meteorologisches Observatorium Potsdam (MOP)
Postfach 60 05 52
D-14405 Potsdam
or :
Telegrafenberg
D-14473 Potsdam
Telephone: (0)331-316500
Telefax: (0)331-316591
Electronic Mail: <userid>@mop.dwd.d400.de
                <userid>@dwd.d400.de
WWW: http://www.dwd.de/
Founded: 1893
Staff: 38

Activities: radiation fluxes in the atmosphere and at the ground * passive remote sensing of the atmosphere from the ground * modelling of radiation processes and validation
Coordinates: 013°04'00"E 52°23'00"N H81m
City Reference Coordinates: 013°04'00"E 52°24'00"N

Deutsches Elektronen-Synchrotron (DESY), Hamburg
Notkestraße 85
D-22603 Hamburg
Telephone: (0)40-89980
Telefax: (0)40-89983282
Electronic Mail: desyinfo@desy.de
                    <userid>@desy.de
WWW: http://www.desy.de/
Founded: 1959
• Use postal code D-22607 for parcels
City Reference Coordinates: 009°59'00"E 53°33'00"N

Deutsches Elektronen-Synchrotron (DESY), Zeuthen
Platanenallee 6
D-15738 Zeuthen
Telephone: (0)33762-770
Telefax: (0)33762-77330
Electronic Mail: desyinfo@ifh.de
                    <userid>@ifh.de
WWW: http://www.ifh.de/
Founded: 1992 (DESY Hamburg)
Staff: 120
Activities: particle physics * detector development * neutrino astrophysics * parallel computing
• Formerly "Institut für Hochenergiephysik"
City Reference Coordinates: 013°37'00"E 52°20'00"N

Deutsches Fernerkundungsdatenzentrum (DFD)
• See "Deutsches Zentrum für Luft- und Raumfahrt (DLR) eV, Deutsches Fernerkundungsdatenzentrum (DFD)"

Deutsches Geodätisches Forschungsinstitut (DGFI)
Marstallplatz 8
D-80539 München
Telephone: (0)89-23031106
                (0)89-23031107
Telefax: (0)89-23031240
Electronic Mail: mailer@dgfi.badw-muenchen.de
Founded: 1952
Staff: 26
Activities: geodesy * theoretical geodesy * reference systems * space geodesy
Periodicals: "Veröffentlichungen der Deutschen Geodätischen Kommission"
City Reference Coordinates: 011°35'00"E 48°08'00"N

Deutsches Hydrographisches Institut (DHI)
• See now "Bundesamt für Seeschiffahrt und Hydrographie (BSH)"

Deutsches Institut für Normung (DIN)
D-10772 Berlin
or :
Burggrafenstraße 6
D-10787 Berlin
Telephone: (0)30-26010
Telefax: (0)30-26011231
Electronic Mail: <userid>@din.de
WWW: http://www.din.de/
Founded: 1917
Staff: 800
Membership: 6200
Activities: standardization
Periodicals: (12) "DIN Mitteilungen", "Elektronorm" (ISSN 0722-2912)
City Reference Coordinates: 013°24'00"E 52°31'00"N

Deutsches Klimarechenzentrum (DKRZ) GmbH
Bundesstraße 55
D-20146 Hamburg
Telephone: (0)40-411730 (Switchboard)
                (0)40-41173334 (Secretariat)
                (0)40-41173275 (User Information)
Telefax: (0)40-41173270
Electronic Mail: beratung@dkrz.de
                    <userid>@dkrz.de

WWW: http://www.dkrz.de/
Founded: 1987
City Reference Coordinates: 009°59'00"E 53°33'00"N

Deutsches Museum München, Abteilung Astronomie
Museumsinsel 1
D-80538 München
Telephone: (0)89-21791
Telefax: (0)89-2179324
WWW: http://www.lrz-muenchen.de/a/kdq01ai/www/
       http://www.deutsches-museum.de/astro.htm
Founded: 1903
Staff: 6
Activities: exhibitions * planetarium * public observatory * history of scientific publishing * popularization
Periodicals: "Kultur und Technik" (Museum publication)
Coordinates: 011°34'50"E 48°07'51"N H545m
City Reference Coordinates: 011°35'00"E 48°08'00"N

Deutsches Zentrum für Luft- und Raumfahrt (DLR) eV
Linder Höhe
D-51147 Köln
Telephone: (0)2203-6010
Telefax: (0)2203-67310
Electronic Mail: <userid>@dlr.de
WWW: http://www.dlr.de/
       http://www.dlr.de/institute_d.html (DLR Institutes)
Founded: 1969
Staff: 4500
Activities: national research establishment performing scientific and technical research for the development and utilization of future aircraft and spacecraft * research in energy technology * constructing and operating large-scale test and simulation equipment, as well as space operation facilities * operating research centres with local branches and liaison offices in Paris and washington (see separate entries) * acting as German space agency
Periodicals: (8) "Aerospace, Science and Technology (AST)" (ISSN 1270-9638, circ.: 1,000); (4) "DLR-Nachrichten" (ISSN 0937-0420, circ.: 8,000)
Awards: "Wissenschaftliche Preise der DLR"
City Reference Coordinates: 006°57'00"E 50°56'00"N

Deutsches Zentrum für Luft- und Raumfahrt (DLR) eV, Deutsches Fernerkundungsdatenzentrum (DFD)
(German Remote Sensing Data Center)
Kalkhorstweg
D-17235 Neustrelitz
Telephone: (0)2-99917481
Telefax: (0)2-99917485
Electronic Mail: <userid>@dlr.de
WWW: http://www.dfd.dlr.de/
       http://www.dlr.de/
Founded: 1992
Staff: 44
Activities: reception and processing of satellite signals * archiving and data distribution * microwave spectroscopy * regional oecology * atmospheric physics
City Reference Coordinates: 013°04'00"E 53°21'00"N

Deutsches Zentrum für Luft- und Raumfahrt (DLR) eV, Institut für Aeroelastik
Bunsenstraße 10
D-37073 Göttingen
Telephone: (0)551-7091
Telefax: (0)551-7092101
Electronic Mail: <userid>@dlr.de
WWW: http://www.ae.go.dlr.de:8088/Welcome.html
       http://www.ae.go.dlr.de:8088/AE/AEinfo/AEinfo_engl.html
       http://www.dlr.de/
Founded: 1969
Staff: 475
Activities: aeroelasticity * fluid mechanics * see also main entry
Periodicals: see main entry
City Reference Coordinates: 009°55'00"E 51°32'00"N

Deutsches Zentrum für Luft- und Raumfahrt (DLR) eV, Institut für Aeroelastik
Rüttelprüfstand MAVIS 1
Gebäude 9.8
Stetternicher Forst
D-52428 Jülich
Telephone: (0)2461-7024
Telefax: (0)2461-58601
Electronic Mail: <userid>@dlr.de

WWW: http://www.dlr.de/
Founded: 1969 (DLR)
Activities: see main entry
Periodicals: see main entry
City Reference Coordinates: 006°21'00"E 50°55'00"N

Deutsches Zentrum für Luft- und Raumfahrt (DLR) eV, Institut für Antriebtechnik
Linder Höhe
D-51147 Köln
Telephone: (0)2203-6012249
(0)2203-6012250
Telefax: (0)2203-64395
Electronic Mail: <userid>@dlr.de
WWW: http://www.kp.dlr.de/EN-AT/
http://www.kp.dlr.de/en/at/index_e.html
http://www.dlr.de/
Founded: 1969
Staff: 1470
Activities: propulsion technology * see also main entry
Periodicals: see main entry
City Reference Coordinates: 006°57'00"E 50°56'00"N

Deutsches Zentrum für Luft- und Raumfahrt (DLR) eV, Institut für Luft- und Raumfahrtmedizin
D-51140 Köln
or :
Linder Höhe
D-51147 Köln
Telephone: (0)2203-6013115
Telefax: (0)2203-695211
Electronic Mail: <userid>@dlr.de
WWW: http://www.me.kp.dlr.de/
http://www.dlr.de/
Founded: 1934
Staff: 100
Activities: aerospace medicine and psychology * space biology * biophysics
Periodicals: see main entry
City Reference Coordinates: 006°57'00"E 50°56'00"N

Deutsches Zentrum für Luft- und Raumfahrt (DLR) eV, Institut für Optoelectronik
NE-OE-PE
Oberpfaffenhofen
D-82234 Weßling
Telephone: (0)8153-28730
(0)8153-28731
Telefax: (0)8153-2476
Electronic Mail: <userid>@dlr.de
WWW: http://www.op.dlr.de/ne-oe/
http://www.dlr.de/
Founded: 1969 (DLR)
Activities: spectral imaging * photometry and spectroscopy of solid planetary bodies * processing and evaluation of imagery of planetary missions * instrument development for ground-based and planetary missions
Periodicals: see main entry
City Reference Coordinates: 011°17'00"E 48°07'00"N (Gilching)

Deutsches Zentrum für Luft- und Raumfahrt (DLR) eV, Institut für Physik der Atmosphäre (IPA)
Postfach 11 16
D-82230 Weßling
Telephone: (0)8153-282520
Telefax: (0)8153-281841
Electronic Mail: <userid>@dlr.de
WWW: http://www.op.dlr.de/NE-PA/
http://www.dlr.de/
Founded: 1962
Staff: 90
Activities: climate research * cloud physics * trace gases measurements
Periodicals: see main entry
Coordinates: 011°18'00"E 48°05'00"N H580m
City Reference Coordinates: 011°18'00"E 48°05'00"N

Deutsches Zentrum für Luft- und Raumfahrt (DLR) eV, Institut für Planetenerkundung
Rudower Chaussee 5
D-12489 Berlin
Telephone: (0)30-69545300
Telefax: (0)30-69545303
Electronic Mail: <userid>@terra.pe.ba.dlr.de
WWW: http://www.ba.dlr.de/NE-PE/

http://www.ba.dlr.de/
http://www.dlr.de/
Founded: 1992
Staff: 98
Activities: planetary surfaces and atmospheres * physics of planets, comets and minor bodies * comparative planetology * modelling of planetary processes * image processing * designing and developing sensor electronics for cameras and spectrometers
City Reference Coordinates: 013°24'00"E 52°31'00"N

Deutsches Zentrum für Luft- und Raumfahrt (DLR) eV, Institut für Raum-Simulation

Linder Höhe
D-51147 Köln
Telephone: (0)2203-6012331
Telefax: (0)2203-61768
Electronic Mail: <userid>@dlr.de
WWW: http://www.kp.dlr.de/WB-RS/willkommen.html
http://www.kp.dlr.de/WB-RS/welcome.html
http://www.dlr.de/
Founded: 1964
Staff: 55
Activities: material research under microgravity * metallic undercooling * nucleation * phase separation * space environment * detection of microgravity * comet simulation * microgravity user support
Periodicals: see main entry
City Reference Coordinates: 006°57'00"E 50°56'00"N

Deutsches Zentrum für Luft- und Raumfahrt (DLR) eV, Institut für Strömungsmechanik

Bunsenstraße 10
D-37073 Göttingen
Telephone: (0)551-7091
Telefax: (0)551-7092101
Electronic Mail: <userid>@dlr.de
WWW: http://www.sm.go.dlr.de/
http://www.sm.go.dlr.de/sm-sk_info/dlr_research_center_sm-sk_d.html
http://www.dlr.de/
Founded: 1969 (DLR)
Activities: see main entry
Periodicals: see main entry
City Reference Coordinates: 009°55'00"E 51°32'00"N

Deutsches Zentrum für Luft- und Raumfahrt (DLR) eV, Institut für Weltraumsensorik

Rudower Chaussee 5
D-12489 Berlin
Telephone: (0)30-67043481
Telefax: (0)30-67045768
Electronic Mail: <userid>@dlr.de
WWW: http://www.ba.dlr.de/NE-WS/
http://www.dlr.de/
Founded: 1992
Staff: 102
Activities: sensor development * data processing * Earth observing * artificial intelligence * expert systems * environmental conditioning
City Reference Coordinates: 013°24'00"E 52°31'00"N

Deutsches Zentrum für Luft- und Raumfahrt (DLR) eV, Standort Braunschweig

Am Flughafen
Postfach 32 67
D-38022 Braunschweig
Telephone: (0)531-2950
Telefax: (0)531-2952105
Electronic Mail: <userid>@dlr.de
WWW: http://www.dlr.de/
http://www.dlr.de/dlr_research_center_bs_d.html
Founded: 1969
Staff: 710
Activities: see main entry
Periodicals: see main entry
City Reference Coordinates: 010°31'00"E 52°16'00"N

Deutsches Zentrum für Luft- und Raumfahrt (DLR) eV, Standort Göttingen

Bunsenstraße 10
D-37073 Göttingen
Telephone: (0)551-7091
Telefax: (0)551-7092101
Electronic Mail: <userid>@dlr.de
WWW: http://www.dlr.de/
http://www.dlr.de/go_d.html
Founded: 1969

Staff: 380
Activities: aeroelasticity * fluid mechanics * see also main entry
Periodicals: see main entry
City Reference Coordinates: 009°55'00"E 51°32'00"N

Deutsches Zentrum für Luft- und Raumfahrt (DLR) eV, Standort Köln
Linder Höhe
D-51140 Köln
Telephone: (0)2203-6010
Telefax: (0)2203-67310
Electronic Mail: <userid>@dlr.de
WWW: http://www.dlr.de/
         http://www.dlr.de/dlr_research_center_kp_d.html
Founded: 1969
Staff: 1470
Activities: centre for manned space missions * microgravity sciences * see also main entry
Periodicals: see main entry
City Reference Coordinates: 006°57'00"E 50°56'00"N

Deutsches Zentrum für Luft- und Raumfahrt (DLR) eV, Standort Lampoldshausen
D-74239 Hardthausen
Telephone: (0)6298-280
Telefax: (0)6298-28408
Electronic Mail: <userid>@dlr.de
WWW: http://www.dlr.de/
Founded: 1969 (DLR)
Staff: 200
Activities: centre for space propulsion and technology * see also main entry
Periodicals: see main entry
City Reference Coordinates: 009°14'00"E 49°08'00"N (Heilbronn)

Deutsches Zentrum für Luft- und Raumfahrt (DLR) eV, Standort Oberpfaffenhofen
Postfach 11 16
D-82230 Weßling
Telephone: (0)8153-280
Telefax: (0)8153-281243
Electronic Mail: <userid>@dlr.de
WWW: http://www.dlr.de/
         http://www.op.dlr.de/dlr_research_center_op_d.html
         http://www.op.dlr.de/ne-hf/Welcome.html
Founded: 1969
Staff: 1070
Activities: space missions * telecommunications * radio frequency technology * atmospheric physics * central data processing * optoelectronics * robotics for space applications * dynamic systems research * research flight facilities * satellite data acquisition, processing and distribution * see also main entry
Periodicals: see main entry
City Reference Coordinates: 011°18'00"E 48°05'00"N

Deutsches Zentrum für Luft- und Raumfahrt (DLR) eV, Standort Stuttgart
Pfaffenwaldring 38/40
D-70569 Stuttgart
Telephone: (0)711-68620
Telefax: (0)711-6862349
Electronic Mail: <userid>@dlr.de
WWW: http://www.dlr.de/
         http://www.dlr.de/dlr_research_center_st_d.html
Founded: 1969
Staff: 420
Activities: see main entry * lasers * energy techniques
Periodicals: see main entry
City Reference Coordinates: 009°11'00"E 48°46'00"N

Die Sterne
● Now merged with "Sterne und Weltraum"

Dornier GmbH
Postfach 14 20
D-88039 Friedrichshafen
or :
An der Bundesstraße 31
D-88090 Immenstaad
Telephone: (0)7545-80
Telefax: (0)7545-84411
WWW: http://www.dasa.com/dasa/index.htm?/dasa/g/dornier.htm
Founded: 1914

Staff: 4300
Activities: space technology manufacturer
City Reference Coordinates: 009°28'00"E 47°39'00"N (Friedrichshafen)
009°22'00"E 47°40'00"N (Immenstaad)

Dr. Remeis-Sternwarte
● See "Universität Erlangen-Nürnberg, Astronomisches Institut, Dr. Remeis-Sternwarte"

Dr. Vehrenberg KG
Postfach 14 05 51
D-40075 Düsseldorf
or :
Schillerstraße 17
D-40237 Düsseldorf
Telephone: (0)211-672080
(0)211-672089
Telefax: (0)211-667726
Electronic Mail: service@vehrenberg.de
WWW: http://www.vehrenberg.de/
Founded: 1912
Staff: 9
Activities: publishing * distributing Celestron, Vixen and Tele Vue products in Germany, Austria and Luxemburg
City Reference Coordinates: 006°47'00"E 51°12'00"N

Eberhard-Karls-Universität Tübingen
● See "Universität Tübingen"

Erhard Friedrich Verlag GmbH & Co. KG
Postfach 10 01 50
D-30926 Seelze
Telephone: (0)511-400040
Telefax: (0)511-4000419
Periodicals: (6) "Astronomie + Raumfahrt im Unterricht"
City Reference Coordinates: 009°25'00"E 52°24'00"N
● Publisher

Ernst-Abbe-Stiftung
● See "Zeiss-Planetarium der Ernst-Abbe-Stiftung"

Ernst-Mach-Institut (EMI)
● See "Fraunhofer-Institut für Kurzzeitdynamik, Ernst-Mach-Institut (EMI)"

ESOC
● See "European Space Agency (ESA), European Space Operations Centre (ESOC)"

EUMETSAT
● See "European Organisation for the Exploitation of Meteorological Satellites"

European Association for Astronomy Education (EAAE)
c/o European Southern Observatory
Karl-Schwarzschild-Straße 2
D-85748 Garching
Electronic Mail: dps@eugenides_found.edu.gr (Dionysios P. Simopoulos, Chairman)
reichen@obs.unize.ch (Michael Reichen, Editor)
anders@astro.su.se (Anders Västerberg)
WWW: http://www.algonet.se/~sirius/eaae.htm
http://www.rz.uni-frankfurt.de/EAAE/
Founded: 1995
City Reference Coordinates: 011°40'00"E 48°14'00"N

European Asteroidal Occultation Network (EAON)
c/o Francis Delahaye
Font Darlan
Paillet
F-33550 Langoiran
Electronic Mail: francis.delahaye@wanadoo.fr
WWW: http://www.xcom.it/cana/EAON/welcome.htm
Founded: 1988
Membership: 200
Activities: observing (occultations by minor planets)
Periodicals: (3) "EAON Informations and Asteroidal Results"
City Reference Coordinates: 000°21'00"W 44°42'00"N

European Astronauts Centre (EAC)
● See "European Space Agency (ESA), European Astronauts Centre (EAC)"

European Council of Skeptical Organizations (ECSO)
c/o Amardeo Sarma
Kirchgasse 4
D-64380 Rossdorf
Telephone: (0)6154-695028
Telefax: (0)6154-695029
Electronic Mail: info@ecso.org
WWW: http://www.ecso.org/
Founded: 1995
Membership: 10
Activities: promoting science and critical thinking * organizing European conferences * exchange and dissemination of ideas and research results
City Reference Coordinates: 008°45'00"E 49°51'00"N

European Geophysical Society (EGS)
Max-Planck-Straße 13
D-37191 Katlenburg-Lindau
Telephone: (0)5556-1440
Telefax: (0)5556-4709
Electronic Mail: egs@copernicus.org
WWW: http://www.copernicus.org/EGS/EGS.html
Founded: 1971
Membership: 6000
Activities: pursuit of excellence in the geosciences and space sciences * organizing annual general assemblies, topical conferences and courses * publishing journals and books * supporting young scientists and colleagues from Eastern Europe
Periodicals: (25) "Physics and Chemistry of the Earth"; (12) "Annales Geophysicae" (see separate entry in France), "Geophysical Journal International", "Planetary and Space Science" (9) "Journal of Atmospheric Chemistry" (ISSN 0167-7764); (8) "Climate Dynamics" (ISSN 0930-7575), "Journal of Geodynamics"; (6) "Surveys in Geophysics" (ISSN 0169-3298), "Tectonics"; (4) "Newsletter"; "Hydrology and Earth System Sciences", "Nonlinear Processes in Geophysics" (ISSN 1023-5809)
City Reference Coordinates: 010°06'00"E 51°41'00"N

European Group of Astronomy Librarians (EGAL)
c/o Astrobibliothek
Max-Planck-Institut für Astrophysik
Karl-Schwarzschild-Straße 1
D-85748 Garching
Telephone: (0)89-32993306
Telefax: (0)89-32993235
Electronic Mail: emc@mpa-garching.mpg.de
Founded: 1989
Membership: 80
Activities: informal cooperation and communication between astronomy librarians in all European countries
Periodicals: (3) "EGAL Bulletin"; "EGAL Directory"
City Reference Coordinates: 011°40'00"E 48°14'00"N

European Organisation for the Exploitation of Meteorological Satellites
(EUMETSAT)
Am Kavalleriesand 31
D-64295 Darmstadt
Telephone: (0)6151-8077
Telefax: (0)6151-807555
Electronic Mail: <userid>@eumetsat.de
WWW: http://www.eumetsat.de/
Founded: 1986
Staff: 140
Membership: 17 (States)
Activities: European intergovernmental organisation * exploitation of meteorological satellites
Periodicals: (1) "Annual Report" (ISSN 1013-3410); (2) "Image"; "Image Bulletin", "Special Publications", "Proceedings of Conferences and Workshops",
City Reference Coordinates: 008°40'00"E 49°53'00"N

European Patent Office (EPO)
(Office Européen des Brevets - OEB)
Erhardtstraße 27
D-80331 München
Telephone: (0)89-23990
Telefax: (0)89-23994560
WWW: http://www.european-patent-office.org/
Founded: 1977
Staff: 3800
City Reference Coordinates: 011°35'00"E 48°08'00"N

European Planetarium Network (EuroPlaNet)
c/o Klaus Wörle
Hofweg 32
D-93053 Regensburg
Telephone: (0)941-944254
Telefax: (0)941-9449233
Electronic Mail: kw@fan.net
WWW: http://www-nw.uni-regensburg.de/∿.wok13927.augen.klinik.uni-regensburg.de/epn.htm
        http://www.artofsky.com/epn/
Founded: 1995
Staff: 3
Activities: planetarium resources * education
City Reference Coordinates: 012°06'00"E 49°01'00"N

European Southern Observatory (ESO), Headquarters
Karl-Schwarzschild-Straße 2
D-85748 Garching
Telephone: (0)89-320060 (Switchboard)
        (0)89-32006226 (Director General)
        (0)89-32006223 (Visiting Astronomers Office)
        (0)89-32006221 (Administration)
        (0)89-32006276 (Education and Public Relations)
        (0)89-32006252 (VLT Division)
        (0)89-32006356 (Instrumentation Division)
        (0)89-32006509 (Data Management Division)
Telefax: (0)89-3202362
Electronic Mail: <userid>@eso.org
WWW: http://www.eso.org/
Founded: 1962
Activities: intergovernmental organization (Member States: Belgium, Denmark, France, Germany, Italy, Netherlands, Sweden & Switzerland) * theoretical and observational astrophysics * instrumentation * observatories at La Silla and Paranal, Chile (see separate entries)
Periodicals: (4) "The Messenger" (ISSN 0722-6691); (1) "Annual Report" (ISSN 0531-4496); "ESO-MIDAS Courier" (ISSN 1018-3051), "ESO Conference and Workshop Proceedings"
City Reference Coordinates: 011°40'00"E 48°14'00"N

European Space Agency (ESA), European Astronauts Centre (EAC)
Linder Höhe
D-51147 Köln
Telephone: (0)2203-60010
Telefax: (0)2203-600166
WWW: http://www.estec.esa.nl/spaceflight/astronaut/
        http://www.esa.int/
Founded: 1990
Staff: 25
Activities: astronaut selection, recruitment, training and coordination
City Reference Coordinates: 006°57'00"E 50°56'00"N

European Space Agency (ESA), European Space Operations Centre (ESOC)
Robert-Bosch-Straße 5
D-64293 Darmstadt
Telephone: (0)6151-900
        (0)6151-902300
Telefax: (0)6151-90495
Electronic Mail: <userid>@esoc.esa.de
WWW: http://www.esoc.esa.de/
        http://www.esa.int/
Founded: 1967
Staff: 320
Activities: ground control of space operations * tracking stations in a worldwide network
City Reference Coordinates: 008°39'00"E 49°52'00"N

European Space Operations Centre (ESOC)
● See "European Space Agency (ESA), European Space Operations Centre (ESOC)"

Eurospace Technische Entwicklungen GmbH
Postfach 22 07
D-09551 Flöha
Telephone: (0)3726-783300
Telefax: (0)3726-712378
Electronic Mail: ri@eurospace.fg.eunet.de
        eurospace@t-online.de
WWW: http://home.t-online.de/home/eurospace/
City Reference Coordinates: 013°04'00"E 50°51'00"N

Fachhochschule Mannheim, Astronomische Arbeitsgemeinschaft
Windeckstraße 110
D-68163 Mannheim
Telephone: (0)621-2926413
Telefax: (0)621-2926360
Founded: 1976
Membership: 25
City Reference Coordinates: 008°29'00"E 48°29'00"N

Fachinformationszentrum (FIZ) Karlsruhe
(Gesellschaft für Wissenschaftlich-Technische Information mbH)
Postfach 24 65
D-76012 Karlsruhe
Telephone: (0)7247-8080
Telefax: (0)7247-808131
Electronic Mail: hlpdeskk@fiz-karlsruhe.de
WWW: http://www.fiz-karlsruhe.de/
Founded: 1977
Staff: 330
Activities: providing information and documentation in fields of science and technology including astronomy and astrophysics
Periodicals: (24) "Mathematics Abstracts"; (6) "STNews", "International Reviews on Mathematical Education", "Bibliographie Informatik für Schule, Hochschule und Weiterbildung"; (4) "Reports in the Fields of Science and Technology", "High-Energy Physics Index"; (1) "Jahresberichte des Bundesministers für Forschung und Entwicklung in der Biologie, Ökologie, Energie"; "Computer Theoretikum und Praktikum für Physiker", "Energy Data", "FIZ-KA Referenzserie", "FIZ-KA Berichte", "Physics Data", "STN International: Databases in Science and Technology" and others
• For parcels use: D-76344 Eggenstein-Leopoldshafen
City Reference Coordinates: 008°24'00"E 49°03'00"N

FAST COMTEC GmbH
Grünwalder Weg 28
D-82041 Oberhaching
Telephone: (0)89-66518050
Telefax: (0)89-66518040
Electronic Mail: <userid>@fastcomtec.com
WWW: http://www.fastcomtec.com/
Founded: 1984
Staff: 12
Activities: manufacturing scanning photon counting systems and transient recorders
City Reference Coordinates: 011°37'00"E 48°02'00"N

Fehrenbach-Planetarium (Richard-_)
• See "Richard-Fehrenbach-Planetarium"

Fireball Data Center (FIDAC)
• See "International Meteor Organization (IMO), Fireball Data Center (FIDAC)"

Förderkreis Planetarium Göttingen (FPG) eV
c/o Karsten Bischoff
Universitäts-Sternwarte Göttingen
Geismarlandstraße 11
D-37083 Göttingen
Telephone: (0)551-395068
Telefax: (0)551-395043
Electronic Mail: fpg@uni-sw.gwdg.de
WWW: http://www.uni-sw.gwdg.de/FPG
Founded: 1994
Membership: 118
Activities: construction of a planetarium
City Reference Coordinates: 009°55'00"E 51°32'00"N

Förderverband Astronomische Bildung in Thüringen eV (FABITH)
Postfach 10 05 19
D-07705 Jena
Founded: 1990
Membership: 35
Activities: supporting astronomical education for public and schools in Thuringia
Periodicals: (4) "Mitteilungsblatt"
City Reference Coordinates: 011°35'00"E 50°56'00"N

Forum der Technik, Planetarium
Postfach 26 02 61
D-80059 München
Telephone: (0)89-21125253
Telefax: (0)89-21125255

WWW: http://www.fdt.de/
City Reference Coordinates: 011°35'00"E 48°08'00"N

Forum Weltraumforschung Aachen
● See "Technische Hochschule Aachen, Forum Weltraumforschung"

Fraunhofer Gesellschaft (FhG)
Leonrodstraße 54
D-80636 München
Telephone: (0)89-1205577
(0)89-1205544
Telefax: (0)89-1205317
Electronic Mail: info@zv.fhg.de
<userid>@zv.fhg.de
WWW: http://www.fhg.de/
http://www.fhg.de/english.html
Founded: 1949
Activities: leading organization of applied research in Germany (47 institutes)
City Reference Coordinates: 011°35'00"E 48°08'00"N

Fraunhofer-Institut für Atmosphärische Umweltforschung
Postfach 13 43
D-82453 Garmisch-Partenkirchen
or :
Kreuzeckbahnstraße 19
D-82467 Garmisch-Partenkirchen
Telephone: (0)8821-1830
Telefax: (0)8821-73573
Electronic Mail: info@ifu.fhg.de
<userid>@ifu.fhg.de
WWW: http://www.fhg.de/depts/ifu-e.html
http://www.fhg.de/german/profile/ifu.html
http://www.fhg.de/
Founded: 1974
Staff: 80
Activities: biogeochemical cycles of trace constituents in the atmosphere * climatic changes * pollution
Observatories: 4 (Garmisch-Partenkirchen - H740; Wank-Gipfel - H1,780; Zugspitze - H2,964; Cape Point - South Africa)
City Reference Coordinates: 011°05'00"E 47°29'00"N

Fraunhofer-Institut für Kurzzeitdynamik, Ernst-Mach-Institut (EMI)
Eckerstraße 4
D-79104 Freiburg-im-Breisgau
Telephone: (0)761-27140
Telefax: (0)761-2714316
Electronic Mail: info@emi.fhg.de
<userid>@emi.fhg.de
WWW: http://www.emi.fhg.de/
Founded: 1949 (FhG)
Staff: 190
Activities: propulsion * impact physics * safety technology * numerical simulation * fluid dynamics * detonics * high-speed measurement techniques * system studies and analysis
City Reference Coordinates: 007°51'00"E 47°59'00"N

Fraunhofer-Institut für Physikalische Meßtechnik (IPM)
Heidenhofstraße 8
D-79110 Freiburg-im-Breisgau
Telephone: (0)761-88570
Telefax: (0)761-8857224
Electronic Mail: info@ipm.fhg.de
<userid>@ipm.fhg.de
WWW: http://www.ipm.fhg.de/
http://www.fhg.de/
Founded: 1973
Membership: 100
Activities: optical measurement systems * optical spectroscopy * measurement systems for space science * optical intersatellite communication * contract research and development
● Formerly "Institut für Physikalische Weltraumforschung"
City Reference Coordinates: 007°51'00"E 47°59'00"N

Freie Universität Berlin
● See "Universität Berlin"

Freundeskreis der Himmelskunde Bad Salzschlirf eV
c/o Michael Passarge
Dr.-Martiny-Straße 1

D-36364 Bad Salzschlirf
Telephone: (0)6648-3272
Electronic Mail: 066483236-0001@t-online.de
WWW: http://observatoriumbs.rhoen.de/
Founded: 1991 (Observatory: 1983) (Solar Observatory: 1995)
Membership: 40
Activities: observing (Sun, planets, comets, deep sky) * photography * lectures
Periodicals: "Himmelskundliches aus Bad Salzschlirf"
Coordinates: 009°29'57"E 50°37'37"N H280m
009°29'57"E 50°38'06"N H367m (Solar Observatory)
City Reference Coordinates: 009°30'00"E 50°38'00"N

Friedrich-Alexander-Universität Erlangen-Nürnberg
• See "Universität Erlangen-Nürnberg"

Friedrich Ebert Visual And Radiotelescope Observatory (FEVARO)
c/o Friedrich Ebert Gymnasium
Alter Postweg 30-38
D-21075 Hamburg
Telephone: (0)40-771702048
Telefax: (0)40-7659275
Electronic Mail: feg@feg.hh.schule.de
WWW: http://www.hh.shuttle.de/hh/feg/FEVARO.htm
Founded: 1996
Membership: 18
Activities: comets * nebulae * planets * Sun * Moon
City Reference Coordinates: 009°59'00"E 53°33'00"N

Friedrich-Schiller-Universität Jena
• See "Universität Jena"

Friedrichs-Gymnasium der Stadt Herford, Sternwarte
Werrestraße 9
D-32049 Herford
Telephone: (0)5221-72924
(0)5221-189358
Telefax: (0)5221-189763
Founded: 1963
Staff: 3
Activities: popularization
Coordinates: 008°40'27"E 52°07'18"N H78m
City Reference Coordinates: 008°40'00"E 52°07'00"N

Friedrich Verlag GmbH & Co. KG (Erhard_)
• See "Erhard Friedrich Verlag GmbH & Co. KG"

Friedrich Vieweg & Sohn Verlagsgesellschaft mbH
Postfach 15 46
D-65005 Wiesbaden
or :
Abraham-Lincoln-Straße 46
D-65189 Wiesbaden
Telephone: (0)611-7878357
Telefax: (0)611-7878420
Electronic Mail: wolfgang.schwarz@bertelsmann.de (Wolgang Schwarz, Editor for Physics & Applied Mathematics)
WWW: http://www.vieweg.de/
Founded: 1786
Staff: 50
• Publisher
City Reference Coordinates: 008°14'00"E 50°05'00"N

Georg-August-Universität Göttingen
• See "Universität Göttingen"

Gesellschaft für astronomische Bildung (GAB) eV
Peißnitzinsel 4a
D-06108 Halle
Telephone: (0)345-2028776
Electronic Mail: rfp@physik.uni-halle.de
WWW: http://www.planetarium.halle-aktuell.de/Html/GABeV/start_gabev.htm
Founded: 1990
Membership: 15
Activities: popularization * observing
Observatories: 1 (Weißenschirmbach)
City Reference Coordinates: 011°58'00"E 51°28'00"N

Gesellschaft für astronomische Bildung in Mecklenburg-Vorpommern

c/o Claus Fischer
Asternweg 13
D-17109 Demmin
Telephone: (0)3998-222410
Telefax: (0)3998-222410
Electronic Mail: c.fischer.demmin@t-online.de
WWW: http://www.physik.uni-greifswald.de/~sterne/Gab/
Founded: 1990
Membership: 30
Activities: popularization * education * teacher training * "Tag der Astronomie"
Periodicals: "Astroblick"
City Reference Coordinates: 013°02'00"E 53°54'00"N

Gesellschaft für volkstümliche Astronomie eV (GvA)

c/o Planetarium Hamburg
Hindenburgstraße Ö1
D-22303 Hamburg
Telephone: (0)40-516560
WWW: http://www.astrophysik.uni-kiel.de/gva/gva.html
       http://home.t-online.de/home/040295292-0002/steki.htm
       http://home.t-online.de/home/gva.hamburg/
Founded: 1964
Membership: 600
Activities: deep-sky observing * photography * CCDs * popularization
Periodicals: (4) "Sternkieker"
Observatories: 3 (Repsold-Sternwarte Hamburg, Max-Koch-Sternwarte Cuxhaven)
City Reference Coordinates: 009°59'00"E 53°33'00"N

Gesellschaft für Weltallkunde (GfW) eV

Postfach 20 51
D-40677 Erkrath
or :
Am Stadtweiher 5
D-40699 Erkrath
Telephone: (0)2104-47564
Telefax: (0)2104-47564
Founded: 1956
Membership: 1500
Activities: education * observing
Periodicals: "Occiale", "Weltallkunde", "Geophys"
City Reference Coordinates: 006°55'00"E 51°13'00"N

Gesellschaft zur wissenschaftlichen Untersuchung von Parawissenschaften eV (GWUP)

Postfach 12 22
D-64374 Rossdorf
or :
Kirchgasse 4
D-64380 Rossdorf
Telephone: (0)6154-695021
Telefax: (0)6154-695022
Electronic Mail: 100042.322@compuserve.com
           info@gwup.org
WWW: http://pc1502.geographie.uni-regensburg.de:80/html/gwup.htm
       http://www.gwup.org/
Founded: 1987
Membership: 330
Activities: conferences * investigations * press releases * publishing
Periodicals: (4) "Skeptiker" (ISSN 0936-9244)
City Reference Coordinates: 008°45'00"E 49°51'00"N

Grasweg Sternwarte

c/o Rudolf A. Hillebrecht
Heinrichstraße 4
D-37581 Bad Gandersheim
Telephone: (0)5382-3020
Telefax: (0)5382-3020
Electronic Mail: 053822297-0001@t-online.de
Founded: 1980
Activities: observing (Sun, Moon, planets, comets, stars) * photography * CCDs
City Reference Coordinates: 010°01'00"E 51°22'00"N

Gymnasium Philippinum Sternwarte

Leopold-Lucas-Straße 18
D-35037 Marburg

Telephone: (0)6421-201359
Telefax: (0)6421-201543
Founded: 1989
Membership: 30
Activities: popularization * observing (Sun, satellites, variable stars, planets) * photography * planetarium
Coordinates: 008°08'00"E 50°08'00"N H146m
City Reference Coordinates: 008°08'00"E 50°08'00"N

Hamburger Sternwarte
● See "Universität Hamburg, Hamburger Sternwarte"

Hans-Nüchter Sternwarte
Domänenweg 2
D-36037 Fulda
Telephone: (0)661-65037
         (0)661-969400
Founded: 1977
Membership: 30
Activities: education * photography * observing * planetarium
Periodicals: (2) "Ganymed" (circ.: 100)
Coordinates: 009°41'57"E 50°33'21"N H300m
City Reference Coordinates: 009°41'00"E 50°33'00"N

Harri Deutsch (Verlag_)
● See "Verlag Harri Deutsch"

Harzplanetarium
Walter-Rathenau-Straße 11
D-38855 Wernigerode
Telephone: (0)927-32277
         (0)43-602096
WWW: http://www.harzart.de/wernigerode/sehenswe.htm
Founded: 1972
Staff: 2
Activities: education * lectures
Coordinates: 010°45'00"E 51°49'00"N H240m
City Reference Coordinates: 010°45'00"E 51°49'00"N

Hermann-von-Helmholtz-Gemeinschaft Deutscher Forschungszentren (HGF)
Ahrstraße 45
D-53175 Bonn
Telephone: (0)228-308180
Telefax: (0)228-3081830
WWW: http://www.helmholtz.de/
Founded: 1970
Membership: 16 (research centres)
Periodicals: "Forschungsthemen", "HGF-Mitteilungen", "Handbuch der Helmholtz-Zentren"
● Formerly known as "Arbeitsgemeinschaft der Großforschungseinrichtungen"
City Reference Coordinates: 007°05'00"E 50°44'00"N

Hirzel Verlag GmbH & Co. (S._)
● See "S. Hirzel Verlag GmbH & Co."

Hüthig Buch Verlag GmbH
Postfach 10 28 69
D-69018 Heidelberg
or :
Im Weiher 10
D-69121 Heidelberg
Telephone: (0)6221-489250
          (0)6221-489261
          (0)6221-489334
Telefax: (0)6221-489450
         (0)6221-489205
Electronic Mail: info@huethig.de
WWW: http://www.huethig.de/
Founded: 1925
● Publisher
● See also "Johann Ambrosius Barth"
City Reference Coordinates: 008°43'00"E 49°25'00"N

Institut für Physikalische Weltraumforschung, Freiburg-im-Breisgau
● See now "Fraunhofer-Institut für Physikalische Meßtechnik"

Intercon Spacetec GmbH
Gablinger Weg 9
D-86154 Augsburg
Telephone: (0)821-414081
Telefax: (0)821-414085
Electronic Mail: info@intercon-spacetec.com
WWW: http://www.intercon-spacetec.com/
Founded: 1981
Staff: 7
Activities: distributing telescopes, accessories, and astronomy software * manufacturing ICS Newtonian telescopes from 8" to 25"
City Reference Coordinates: 010°53'00"E 48°23'00"N

Interessengemeinschaft Astronomie Crimmitschau (IGAC) eV
c/o Sternwarte Johannes Kepler
Lindenstraße 8
D-08451 Crimmitschau
Telephone: (0)3762-3730
Electronic Mail: can@stw-cri.wda.sn.schule.de
WWW: http://www.fh-zwickau.de/~fa/
Founded: 1960
Membership: 25
Activities: Sun (Wolf number) * lunar occultations * comets * deep sky * computing
Coordinates: 012°23'00"E 50°49'00"N H273m
City Reference Coordinates: 012°23'00"E 50°49'00"N

Interessengemeinschaft Astrofotografie Bochum (IAB)
Grillostraße 70
D-44799 Bochum 1
Telephone: (0)234-382771
WWW: http://homepage.ruhr-uni-bochum.de/Volker.Mette/
Founded: 1984
Activities: photography (deep sky, comets, zodiacal light, counterglow phenomenon) * image processing and analysis * darkroom techniques * hypersensitizing * excursions * education * amateur research projects
City Reference Coordinates: 007°13'00"E 51°28'00"N

Interessen Gemeinschaft Sternfreunde an der Ruhr (IG StaR)
c/o Ewald Goitowski
Kirschbaumsweg 8
D-45149 Essen
Telephone: (0)201-718903
Electronic Mail: egoi@myweb.de
WWW: http://hp.citystar.de/goitowski/ig_star/
Founded: 1994
Membership: 12
City Reference Coordinates: 007°01'00"E 51°28'00"N

International Meteor Organization (IMO), Fireball Data Center (FIDAC)
c/o André Knöfel
Saarbrücker Straße 8
D-40476 Düsseldorf
Telephone: (0)211-450719
Telefax: (0)211-450736
Electronic Mail: fidac@imo.net
WWW: http://www.imo.net/
Founded: 1988
Membership: 6
Activities: collecting data on fireballs, meteorite falls and accompanying phenomena * archiving * dissemination of data
Periodicals: "FIDAC News" (ISSN 1021-3228; circ.: 100), "Fireball Reports"
City Reference Coordinates: 006°47'00"E 51°12'00"N

International Union of Pure and Applied Physics (IUPAP)
c/o Wolfgang Heinicke (Secretary-General)
Deutsche Physikalische Gesellschaft
Hauptstraße 5
D-53604 Bad Honnef
Telephone: (0)2224-92320
Telefax: (0)2224-923250
Founded: 1922
Activities: stimulating and promoting international cooperation in physics * sponsoring international conferences
Sections: (commissions) 1. Finance * 2. Symbols, Units, Nomenclature, Atomic Masses and Fundamental Constants * 3. Statistical Physics * 4. Cosmic Rays * 5. Low-Temperature Physics * 6. Biological Physics * 7. Acoustics * 8. Semiconductors * 9. Magnetism * 10. Structure and Dynamics of Condensed Matter * 11. Particles and Fields * 12. Nuclear Physics * 13. Physics for Development * 14. Physics Education * 15. Atomic and Molecular Physics and Spectroscopy * 16. Plasma Physics * 17. Quantum Electronics * 18. Mathematical Physics * 19. Astrophysics

Periodicals: "News Bulletin", "General Report"
City Reference Coordinates: 007°13'00"E 50°39'00"N

Jenoptik AG
Carl-Zeiss-Straße 1
D-07743 Jena
Telephone: (0)3641-650
Telefax: (0)3641-424514
Founded: 1991 (1846 as "Carl Zeiss")
Staff: 7000
Activities: development, production and distribution of laser systems, optics and equipment * optical medical instruments * engineering metrology equipment * engineering of semiconductors * equipment for the telecommunication industry * facility management * clean-room automation * electromechanical systems
Awards: "Ernst-Abbe-Preis", "Otto-Schott-Preis"
City Reference Coordinates: 011°35'00"E 50°56'00"N

Johann Ambrosius Barth
Salomonstraße 18b
D-04103 Leipzig
Telephone: (0)70131
Telefax: (0)291979
Founded: 1780
● Publisher
● See also "Hüthig Buch Verlag GmbH"
City Reference Coordinates: 012°20'00"E 51°19'00"N

Johann-Wolfgang-Goethe-Universität Frankfurt/Main
● See "Universität Frankfurt/Main"

Joint Organization for Solar Observations (JOSO)
c/o P.N. Brandt
Kiepenheuer-Institut für Sonnenphysik
Schöneckstraße 6
D-79104 Freiburg-im-Breisgau
Telephone: (0)761-3198250
Telefax: (0)761-3198111
Electronic Mail: pnb@kis.uni-freiburg.de
　　　　　　　　pbrandt@solar.stanford.edu
WWW: http://joso.oat.ts.astro.it/
Founded: 1969
Membership: 26 (countries)
Activities: annual meetings * exchange of information on solar research in Europe and abroad
Periodicals: (1) "Annual Report" (circ.: 300)
City Reference Coordinates: 007°51'00"E 47°59'00"N

Karl-Schwarzschild-Observatorium (KSO)
● See "Thüringer Landessternwarte (TLS) - Tautenburg, Karl-Schwarzschild-Observatorium (KSO)"

Kayser-Threde GmbH
Wolfratshauser Straße 48
D-81379 München
Telephone: (0)89-724950
Telefax: (0)89-72495291
Electronic Mail: info@kayser-threde.de
　　　　　　　　<userid>@kayser-threde.de
WWW: http://www.kayser-threde.de/
Founded: 1967
Staff: 170
Activities: manufacturing microgravity research instruments, small satellites, Earth observation and optical instrumentation, small reentry capsules, sounding rockets payloads, navigation, space telescopes and space science instrumentation, on-board data handling systems
City Reference Coordinates: 011°35'00"E 48°08'00"N

Kiepenheuer-Institut für Sonnenphysik (KIS)
Schöneckstraße 6
D-79104 Freiburg-im-Breisgau
Telephone: (0)761-31980
　　　　　　(0)7602-226 (Schauinsland Observatory)
Telefax: (0)761-3198111
　　　　　(0)761-382280 (Schauinsland Observatory)
Electronic Mail: secr@kis.uni-freiburg.de
　　　　　　　　<userid>@kis.uni-freiburg.de
WWW: http://www.kis.uni-freiburg.de/
　　　　http://www.kis.uni-freiburg.de/kiswwwe.html (English page)
Founded: 1943

Staff: 15
Activities: solar research * observing * theoretical studies * instrumentation
Coordinates: 007°54'22"E 47°54'52"N H1,240m (Schauinsland)
  016°30'30"W 28°17'50"N H2,395m (Izaña, Teide, Islas Canarias)
City Reference Coordinates: 007°51'00"E 47°59'00"N

Kometen, Planetoiden, Meteore (KPM)
c/o Jost Jahn
Neustädter Straße 11
D-29389 Bodenteich
Telephone: (0)5824-3197
Telefax: (0)581-14824
Electronic Mail: j.jahn@abbs.heide.de
WWW: http://www.uni-essen.de/initiative/vds/jjkpminf.html
Founded: 1985
Staff: 4
● Journal (3) (ISSN 0930-102x, circ.: 120)
City Reference Coordinates: 010°41'00"E 52°50'00"N

Landessternwarte Heidelberg
Königstuhl
D-69117 Heidelberg
Telephone: (0)6221-5090
Telefax: (0)6221-509202
Electronic Mail: <userid>@lsw.uni-heidelberg.de
WWW: http://www.lsw.uni-heidelberg.de/
Founded: 1898
Staff: 45
Activities: observational and theoretical astrophysics
Coordinates: 008°43'18"E 49°23'54"N H564m
City Reference Coordinates: 008°43'00"E 49°25'00"N

LOBO electronic GmbH
Robert-Bosch-Straße 100
D-73437 Aalen
Telephone: (0)7361-968710
Telefax: (0)7361-968730
Electronic Mail: mail@lobo.de
WWW: http://www.lobo.de/
Founded: 1982
Staff: 25
Activities: manufacturing transputer-controlled laser animation systems for i.a. planetariums
City Reference Coordinates: 010°05'00"E 48°50'00"N

Lohrmann-Observatorium
● See "Technische Universität Dresden, Institut für Planetare Geodäsie und Lohrmann-Observatorium"

Max-Planck-Gesellschaft (MPG)
Residenzstraße 1a
D-80333 München
Telephone: (0)89-21081
Telefax: (0)89-229850
WWW: http://www.mpg.de/
  http://www.ipp-garching.mpg.de/mpi.html (WWW and Gopher servers of Max-Planck-Institutes)
  http://www.gwdg.de/Allgemeines/mpg.html
Founded: 1948
Periodicals: "MPG Spiegel" (ISSN 0341-7727), "Berichte und Mitteilungen" (ISSN 0341-7778)
City Reference Coordinates: 011°35'00"E 48°08'00"N

Max-Planck-Gesellschaft (MPG), Arbeitsgruppe Gravitationstheorie
Max-Wien-Platz 1
D-07743 Jena
or :
Fröbelstieg 1
D-07743 Jena
Telephone: (0)3641-636643
Telefax: (0)3641-636728
Electronic Mail: agg@gravi.physik.uni-jena.de
  agg@tpi.uni-jena.de
WWW: http://einstein.physik.uni-jena.de/mpg_gra2.html
Founded: 1991
Staff: 6
Activities: general relativity * gravitational field of rotating bodies * numerical relativity * relativistic celestial mechanics * accretion processes
Periodicals: "Preprints"

City Reference Coordinates: 011°35'00"E 50°56'00"N

Max-Planck-Gesellschaft (MPG), Arbeitsgruppe Staub in Sternentstehungsgebieten
Schillergäßchen 3
D-07745 Jena
Telephone: (0)3641-449874
Telefax: (0)3641-449875
Electronic Mail: <userid>@astro.uni-jena.de
                 mail@astro.uni-jena.de
                 mail@fred.stro.uni-jena.de
WWW: http://www.astro.uni-jena.de/
Founded: 1992
Staff: 12
Activities: star formation * interstellar dust * laboratory astrophysics
City Reference Coordinates: 011°35'00"E 50°56'00"N

Max-Planck-Institut für Aeronomie (MPAe), Katlenburg-Lindau
Postfach 20
D-37189 Katlenburg-Lindau
or :
Max-Planck-Straße 2
D-37191 Katlenburg-Lindau
Telephone: (0)5556-9790
Telefax: (0)5556-979240
Electronic Mail: <userid>@linmpi.mpae.gwdg.de
WWW: http://www.mpae.gwdg.de/
       http://www.mpae.gwdg.de/mpae_projects/SOUSY/sousy_home.html (SOUSY project)
       http://www.mpae.gwdg.de/mpae_projects/STARE/STARE.html (STARE project)
Founded: 1936
Staff: 300
Activities: atmospheric physics * magnetospheric physics * planets * Sun * interplanetary medium
City Reference Coordinates: 010°06'00"E 51°41'00"N

Max-Planck-Institut für Astronomie (MPIA), Heidelberg
Königstuhl 17
D-69117 Heidelberg
Telephone: (0)6221-5280
Telefax: (0)6221-528246
Electronic Mail: <userid>@mpia-hd.mpg.de
WWW: http://www.mpia-hd.mpg.de/
       http://www.mpia-hd.mpg.de/MPIA/
Founded: 1969
Staff: 150
Activities: star forming regions * circumstellar matter * ISM * AGNs * evolution of galaxies * clusters of galaxies * IR observing from space
Periodicals: (1) "Annual Report"
Coordinates: 008°43'07"E 49°23'49"N H560m
City Reference Coordinates: 008°43'00"E 49°25'00"N

Max-Planck-Institut für Astrophysik (MPA)
Postfach 1523
D-85740 Garching
or :
Karl-Schwarzschild-Straße 1
D-85748 Garching
Telephone: (0)89-329900
Telefax: (0)89-32993235
Electronic Mail: <userid>@mpa-garching.mpg.de
WWW: http://www.mpa-garching.mpg.de/
Founded: 1958
Staff: 68
Activities: astrophysical research
City Reference Coordinates: 011°40'00"E 48°14'00"N

Max-Planck-Institut für extraterrestrische Physik (MPE)
Postfach 16 03
D-85740 Garching
or :
Giessenbachstraße
D-85748 Garching
Telephone: (0)89-329900
Telefax: (0)89-32993569
Electronic Mail: mpe@mpe.mpg.de
                 <userid>@mpe.mpg.de
WWW: http://www.mpe.mpg.de/
Founded: 1962

Staff: 355
Activities: in situ space data collection * astrophysical data collection * data analysis * theory
Periodicals: (1) "Jahresbericht" (ISSN 0947-8787)
Coordinates: 011°40'23"E 48°15'46"N H476m
City Reference Coordinates: 011°40'00"E 48°14'00"N

Max-Planck-Institut für Gravitationsphysik

(Albert-Einstein-Institut - AEI)
Schlaatzweg 1
D-14473 Potsdam
Telephone: (0)331-275370
Telefax: (0)331-2753798
Electronic Mail: office@aei-potsdam.mpg.de
            <userid>@aei-potsdam.mpg.de
WWW: http://www.aei-potsdam.mpg.de/
        http://www.livingreviews.org/ (Living Reviews in Relativity)
Founded: 1995
Staff: 40
Activities: gravitation * relativity * astrophysics * quantum gravity * numerical relativity * black holes * mathematical relativity * gravitational waves * superconducting
Periodicals: "Living Reviews in Relativity" (ISSN 1433-8351)
City Reference Coordinates: 013°04'00"E 52°24'00"N

Max-Planck-Institut für Kernphysik (MPI-K)

Postfach 10 39 80
D-69029 Heidelberg
Telephone: (0)6221-5161
Telefax: (0)6221-516540
WWW: http://www.mpi-hd.mpg.de/
Founded: 1958
Staff: 270
Activities: particle physics * astrophysics * atmospheric physics * nuclear physics * atomic physics
Periodicals: (1) "Jahresbericht"
City Reference Coordinates: 008°43'00"E 49°25'00"N

Max-Planck-Institut für Kernphysik (MPI-K), Bereich Astrophysik

Saupfercheckweg 1
D-69117 Heidelberg
Telephone: (0)6221-516295
Telefax: (0)6221-516549
        (0)6221-516324
Electronic Mail: <userid>@mpi-hd.mpg.de
WWW: http://www.mpi-hd.mpg.de/
Founded: 1958
Staff: 22
Activities: gamma-ray astronomy * neutrino astronomy * ISM * IR astronomy * star formation * cosmic-ray physics
Periodicals: (1) "Jahresbericht"
City Reference Coordinates: 008°43'00"E 49°25'00"N

Max-Planck-Institut für Physik (MPP)

(Werner-Heisenberg-Institut - WHI)
Föhringer Ring 6
D-80805 München
Telephone: (0)89-323081
Telefax: (0)89-3226704
Electronic Mail: <userid>@mppmu.mpg.de
WWW: http://iws132a.mppmu.mpg.de/library/welcome.html
Founded: 1917 (as "Kaiser Wilhelm-Institut für Physik")
City Reference Coordinates: 011°35'00"E 48°08'00"N

Max-Planck-Institut für Radioastronomie (MPIfR)

Auf dem Hügel 69
D-53121 Bonn
Telephone: (0)228-5250
        (0)2257-3010 (Effelsberg)
Telefax: (0)228-525229
        (0)2257-30169 (Effelsberg)
Electronic Mail: <userid>@mpifr-bonn.mpg.de
WWW: http://www.mpifr-bonn.mpg.de/
Founded: 1967
Staff: 180
Activities: radioastronomy * spectroscopy * continuum radiation * VLBI * ISM structure and physics * extended galaxies * extragalactic sources * optical interferometry * developing new observing techniques
Coordinates: 006°53'06"E 50°31'36"N H369m (Bad Münstereifel/Effelsberg)
City Reference Coordinates: 007°05'00"E 50°44'00"N

Menke Planetarium und Sternwarte
Institut für Physik
Kanzleitstraße 91-93
D-24943 Flensburg
or :
Fördestraße 35
D-24960 Glücksburg
Telephone: (0)461-805269
            (0)461-805273
Telefax: (0)461-805300
Electronic Mail: uwe.roose@fh-flensburg.de (Uwe Roose)
WWW: http://www.fh-flensburg.de/ph/planetarium.html
Founded: 1969
Staff: 5
Activities: public shows * observing (variable stars)
Coordinates: 009°31'41"E 54°50'23"N
City Reference Coordinates: 009°26'00"E 54°47'00"N (Flensburg)
                            009°33'00"E 54°50'00"N (Glücksburg)

Meteorologisches Observatorium Hohenpeißenberg
Albin-Schwaiger-Weg 10
D-82383 Hohenpeißenberg
Telephone: (0)8805-92000
Telefax: (0)8805-920046
Founded: 1781
Activities: radar meteorology * atmospheric trace constituents * ozone * atmospheric chemistry
Periodicals: (2) "Ozone Bulletin"
Coordinates: 011°00'38"E 47°48'08"N H975m
City Reference Coordinates: 011°01'00"E 47°48'00"N

Moerser Astronomische Organisation eV (MAO)
Postfach 10 18 11
D-47408 Moers
Telephone: (0)2841-55995 (Helmut Gröll)
            (0)841-23187
Telefax: (0)841-23187
Electronic Mail: uwe.reimann@mpie-duesseldorf.mpg.d400.de (Uwe Reimann, Secretary)
                 uwe.reimann@t-online.de
                 juergen.hueneborn@uni.muenster.de
                 mao@physik.de
WWW: http://user.cs.tu-berlin.de/~davidi/mao.html
     http://members.aol.com/maowebpage/
     http://www.physik.de/MAO/
Founded: 1969
Membership: 50
Activities: popularization * Einstein-planetarium
Periodicals: (4) "Astro-Kurier" (circ.: 90), "Sternzeit" (co-publisher) (ISSN 0721-8168, circ.: 2,000)
Coordinates: 006°34'00"E 51°27'00"N
City Reference Coordinates: 006°34'00"E 51°27'00"N

Motivgruppe Astronomie & Philatelie
c/o Eckehard Schmidt
Postfach 46 16
D-90025 Nürnberg
or :
Brunhildstraße 1a
D-90461 Nürnberg
Telephone: (0)911-4720978
Telefax: (0)911-5865549
Founded: 1977
Membership: 61
Activities: gathering stamps with astronomical themes * popularization
Periodicals: (4) "Astronomie & Philatelie" (ISSN 0948-4418)
City Reference Coordinates: 011°04'00"E 49°27'00"N

MST Aerospace GmbH
Eupener Straße 150
D-50933 Köln
Telephone: (0)221-9498920
Telefax: (0)221-4912443
Electronic Mail: office@mst-aerospace.de
WWW: http://www.mst-aerospace.de/
Activities: technology transfer
City Reference Coordinates: 006°57'00"E 50°56'00"N

Museum am Schölerberg, Natur und Umwelt, Planetarium
Am Schölerberg 8
D-49082 Osnabrück
Telephone: (0)541-560030
                      ahaenel@rz.uni-osnabrueck.de (Andreas Hänel)
WWW: http://www.physik.uni-osnabrueck.de/astro
Founded: 1986
Staff: 1
Activities: planetarium shows * public star parties
Periodicals: "Osnabrücker Naturwissenschaftliche Mitteilungen" (ISSN 0340-4781)
Coordinates: 008°04'18"E 52°15'00"N
City Reference Coordinates: 008°03'00"E 52°17'00"N

Museum für Astronomie und Technikgeschichte, Planetarium
Postfach 41 04 20
D-34066 Kassel
or :
Orangerie
An der Karlsaue 20c
D-34121 Kassel
Telephone: (0)561-71543
Telefax: (0)561-7846222
WWW: http://www.kassel.de/ask/projekt/t-museum/
Founded: 1992
Staff: 5
City Reference Coordinates: 009°29'00"E 51°19'00"N

Naturwissenschaftliche Rundschau
Birkenwaldstraße 44
D-70191 Stuttgart
Telephone: (0)711-2582309
                (0)711-2582310
Telefax: (0)711-2582390
Electronic Mail: wvg.nr@t-online.de
Founded: 1948
Staff: 3
• Journal (12) (ISSN 0028-1050)
City Reference Coordinates: 009°11'00"E 48°46'00"N

Naturwissenschaftlicher Verein für Bielefeld und Umgegend eV, Arbeitsgemeinschaft Astronomie
c/o Holger Sturm
Kreuzstraße 48
D-33602 Bielefeld
Telephone: (0)521-172434 (Thursdays)
                (0)521-883982 (home)
Electronic Mail: spieweck@physik.uni-bielefeld.de (Michael Spieweck)
WWW: http://www.uni-bielefeld.de/biologie/Oekologie/NWV/apu/apu.html
Founded: 1991
Membership: 39
Activities: popularization * monthly lectures * public meetings * working groups
Periodicals: "Astronomie in Bielefeld"
City Reference Coordinates: 008°31'00"E 52°01'00"N

Nicolaus Copernicus Planetarium
Am Plärrer 41
D-90317 Nürnberg
Telephone: (0)911-265467
                (0)911-9296554
Electronic Mail: planetarium@osn.de
                      planet@nuernberg.de
WWW: http://www.nuremberg.de/ver/bz/level1/level2/plan.htm
                http://www.nuremberg.de/ver/him/planet.htm
Founded: 1961
Staff: 4
Activities: lectures * public shows * concerts * exhibitions
Coordinates: 011°06'42"E 49°27'56"N H338m (Regiomontanus Observatory)
City Reference Coordinates: 011°04'00"E 49°27'00"N

Nüchter Sternwarte (Hans-_)
• See "Hans-Nüchter Sternwarte"

Nürnberger Astronomische Arbeitsgemeinschaft (NAA) eV
c/o Regiomontanus Sternwarte
Regiomontanusweg 1

D-90491 Nürnberg
Telephone: (0)911-9593538
Telefax: (0)911-523791
Electronic Mail: info@naa.net
WWW: http://www.naa.net/
Founded: 1962
Membership: 230
Activities: public education * debunking pseudosciences * photoelectric observing of variable stars (see "Sternwarte Nürnberg")
Periodicals: (4) "Regiomontanusbote" (ISSN 0938-0205, circ.: 500)
Coordinates: 011°06'42"E 49°27'56"N H338m
City Reference Coordinates: 011°04'00"E 49°27'00"N

Oberbayerische Volkssternwarte Berg (OVSB) eV
Lindenallee
D-82335 Berg-Aufkirchen
Telephone: (0)8151-51112 (Christian Jutz)
Electronic Mail: t7121bl@sunmail.lrz-muenchen.de (Evi Hummel)
WWW: http://www.lrz-muenchen.de/~t7121bl/astro/vswberg.html
Founded: 1992
Membership: 30
Activities: public observing
City Reference Coordinates: 011°21'00"E 47°58'00"N

Observatorium für Solare Radioastronomie (OSRA)
● See "Astrophysikalisches Institut Potsdam (AIP), Observatorium für Solare Radioastronomie (OSRA)"

Observatorium Hoher List
● See "Universität Bonn, Astronomische Institute, Observatorium Hoher List"

Observatorium Ravenstein
c/o Gisbert Krause
Goethe-Straße 16
D-74747 Ravenstein
Telephone: (0)6297-95033
Founded: 1979
Coordinates: 009°33'52"E 49°25'52"N H360m
City Reference Coordinates: 009°34'00"E 49°26'00"N

Observatorium Wendelstein
● See "Universität München, Institut für Astronomie und Astrophysik, Observatorium Wendelstein"

Office Européen des Brevets (OEB)
● See "European Patent Office (EPO)"

Olbers-Planetarium
c/o Hochschule Bremen
Fachbereich Nautik
Werderstraße 73
D-28199 Bremen
or :
Feldstraße 26
D-28203 Bremen
Telephone: (0)421-5905678
           (0)421-706882
Telefax: (0)421-706882
Electronic Mail: dieter.vornholz@t-online.de
                 prichter@physik.uni-bremen.de
WWW: http://www.rz.hs-bremen.de/planetarium/
     http://www-theo.physik.uni-bremen.de/og/og.ger.html
     http://www-theo.physik.uni-bremen.de/og/og.html
Founded: 1952
Staff: 2
Activities: shows for students and public
City Reference Coordinates: 008°49'00"E 53°04'00"N

Optische und Electronische Systeme (OES) GmbH
Dr. Neumeyer Straße 240
D-91349 Egloffstein
Telephone: (0)9197-698980
Telefax: (0)9197-698982
Electronic Mail: fo0107@forchheim.baynet.de (Frank Fleischmann)
Founded: 1988
Staff: 4
Activities: manufacturing and distributing CCD cameras, image intensifiers, photon-counting devices, speckle-interferometric devices, telescope control systems and astronomical image processing software

Coordinates: 011°16'00" E 49°42'00" N H380m
City Reference Coordinates: 011°16'00" E 49°42'00" N

Physikalischer Verein Frankfurt, Volkssternwarte
Robert-Mayer-Straße 2-4
D-60054 Frankfurt/Main
Telephone: (0)69-704630
Telefax: (0)69-97981342
Electronic Mail: piehler@phys-verein.uni-frankfurt.de (Georg Piehler)
WWW: http://isis.phys-verein.uni-frankfurt.de/
        http://earth.astro.uni-frankfurt.de/vstw/
Founded: 1824 (Verein)
Membership: 700
Activities: popularization * observing (PN, deep sky, Sun, planets, Moon, double stars)
Periodicals: (4) "Mitteilungen astronomischer Vereinigungen Rhein-Main-Nahe" (circ.: 550); (1) "Jahresbericht"
Awards: (1) "Philipp-Siedler-Preis", "Christian-Ernst-Neeff-Preis", "Samuel-Thomas-von Soemmering-Preis", "Eugen-Hart-mann-Didaktik-Preis"
Coordinates: 008°39'12" E 50°07'05" N H115m
              008°26'43" E 50°13'22" N H825m (Feldberg)
City Reference Coordinates: 008°40'00" E 50°07'00" N

Physikalisch-Technische Bundesanstalt (PTB)
Postfach 33 45
D-38023 Braunschweig
or :
Bundesallee 100
D-38116 Braunschweig
Telephone: (0)531-5920
Telefax: (0)531-5929292
Electronic Mail: <userid>@ptb.de
WWW: http://www.ptb.de/
Founded: 1887
Staff: 1600
Activities: legal metrology * realization and dissemination of SI-units * determination of physical constants
Periodicals: (1) "Jahresbericht" (ISSN 0340-4366)
Awards: (1/3) "Helmholtz-Preis"
City Reference Coordinates: 010°31'00" E 52°16'00" N

Physik Instrumente (PI) GmbH & Co.
Polytec-Platz 5-7
D-76337 Waldbronn
Telephone: (0)7243-604100
Telefax: (0)7243-604145
WWW: http://www.physikinstrumente.com/
Founded: 1969
Staff: 40
Activities: manufacturing piezoelectric translators, tilting mirrors, positioning systems, step- and DC-motor controllers
Periodicals: "Movement & Positioning"
City Reference Coordinates: 008°25'00" E 48°56'00" N (Ettlingen)

Planetarium Aschersleben
Im Tierpark
Auf der Alten Burg 40
D-06449 Aschersleben
Telephone: (0)3473-2592
Telefax: (0)3473-3841
WWW: http://lbs.st.schule.de/lbs/pl-asch.htm
City Reference Coordinates: 011°28'00" E 51°46'00" N

Planetarium der Fachhochschule Kiel
Alte Chaussee 32
D-24107 Kiel
Telephone: (0)431-5198211
Electronic Mail: postfach@planetarium.fh-kiel.de
WWW: http://www.fh-kiel.de/planet/fh_planetarium.html
        http://www.planetarium.fh-kiel.de/planetarium/
        http://www.planetarium.fh-kiel.de/planet/fh_planetarium.html
Founded: 1969
Staff: 10
Activities: shows for schools and public
City Reference Coordinates: 010°08'00" E 54°20'00" N

Planetarium der Fachhochschule Stralsund
Hainholzstraße 59
D-18435 Stralsund

Telephone: (0)3831-456528
(0)3831-456529
Electronic Mail: rudi.wendorf@fh-stralsund.de
WWW: http://www-httpd.fh-stralsund.de/Allgemein/planet.html
City Reference Coordinates: 013°05'00"E 54°19'00"N

Planetarium der Stadt Wolfsburg GmbH
Uhlandweg 2
D-38440 Wolfsburg
Telephone: (0)5361-21939
Telefax: (0)5361-21272
WWW: http://enjoy-wolfsburg.com/DE/freizeit/planetarium.html
http://www.wolfsburg.de/wobline/kultur/planet/
Founded: 1983
Activities: planetarium shows * concerts * narrations * education * lectures * seminars
City Reference Coordinates: 010°47'00"E 52°25'00"N

Planetarium Hamburg
Hindenburgstraße Ö1
D-22303 Hamburg
Telephone: (0)40-5149850
Telefax: (0)40-51498510
WWW: http://www.hamburg-cityguide.de/planeta.html
http://www.hamburg.de/Behoerden/Kulturbehoerde/Planetarium/
Founded: 1930
City Reference Coordinates: 009°59'00"E 53°33'00"N

Planetarium Hoyerswerda
3. Mittelschule "Mittelschule am Planetarium" (WK VI)
Collins-Straße 29
D-02977 Hoyerswerda
Telephone: (0)3571-417020
WWW: http://www.germany.net/teilnehmer/101/57799/planetar.htm
Founded: 1969
Coordinates: 014°15'00"E 51°26'30"N
City Reference Coordinates: 014°14'00"E 51°26'00"N

Planetarium im Vonderau Museum
Jesuitenplatz 2
D-36037 Fulda
Telephone: (0)661-928350
WWW: http://www.fulda.com/fis/museen/planet.htm
http://www.fulda-online.de/db/rubdef/artikel_detail.phtml?id=2241
Founded: 1990
Staff: 5
Activities: popularization
City Reference Coordinates: 009°41'00"E 50°33'00"N

Planetarium Lübz
c/o Norbert Karsten
Stadtverwaltung
Am Markt 22
D-19386 Lübz
Telephone: (0)38731-23581
WWW: http://www.physik.uni-greifswald.de/~sterne/Planetarium_luebz_de/
Founded: 1980
Coordinates: 012°02'56"E 53°27'38"N H61m
City Reference Coordinates: 012°03'00"E 53°28'00"N

Planetarium Mannheim Gem. mbH
Wilhelm-Varnholt-Allee 1
D-68165 Mannheim
Telephone: (0)621-419420
(0)621-415692
Telefax: (0)621-412411
WWW: http://www.mannheim.de/planetarium/
Founded: 1984
Staff: 9
Activities: shows for public and schools * cultural events
City Reference Coordinates: 008°29'00"E 48°29'00"N

Planetarium Merseburg
c/o Helmut Conrad
Schulstraße 1a
D-06242 Braunsbedra

Telephone: (0)34633-20909
WWW: http://lbs.st.schule.de/lbs/pl-mersb.htm
Founded: 1966
Staff: 2
Activities: education
City Reference Coordinates: 011°53'00"E 51°18'00"N (Braunsbedra)
                                        012°00'00"E 51°22'00"N (Merseburg)

Planetarium Senftenberg
An der Ingenieurschule
D-01968 Senftenberg
Telephone: (0)3573-2112
Founded: 1966
Membership: 18
Activities: popularization * training * lectures * shows
Coordinates: 013°59'00"E 51°22'00"N H104m
City Reference Coordinates: 013°59'00"E 51°22'00"N

Rat Deutscher Sternwarten (RDS)
(Council of German Observatories)
c/o Gregor Morfill (Chairman)
Max-Planck-Institut für extraterrestrische Physik
Postfach 16 03
D-85740 Garching
Telephone: (0)89-23993567
Telefax: (0)89-23993399
Electronic Mail: gem@mpc-garching.mpg.de
Founded: 1959
Membership: 35 (institutes, observatories and research groups)
Activities: representing nationally and internationally research at astronomical observatories in Germany * coordinating activities between these institutions
Periodicals: (1) "Annual Report"
City Reference Coordinates: 011°40'00"E 48°14'00"N

Raumflugplanetarium Cottbus
Lindenplatz 21
D-03042 Cottbus
Telephone: (0)355-713109
Telefax: (0)355-7295822
WWW: http://www-user.tu-cottbus.de/~embergb/planetarium/
Founded: 1996
City Reference Coordinates: 014°21'00"E 51°43'00"N

Raumflugplanetarium Halle (RFP)
Peißnitzinsel 4a
D-06108 Halle
Telephone: (0)345-8060317
Telefax: (0)345-8060317
Electronic Mail: rfp@physik.uni-halle.de
WWW: http://www.planetarium.halle-aktuell.de/
          http://www.halle.de/DEUTSCH/4/2/01625/01625.HTM
Founded: 1978
Staff: 2
Activities: education * shows * popularization
Coordinates: 011°56'59"E 51°29'47"N
City Reference Coordinates: 011°58'00"E 51°28'00"N

Regionale Astronomische Arbeitsgemeinschaft Euskirchen (RAAGE)
c/o Heinz-Jürgen Schäfer
Danziger Straße 29
D-53879 Euskirchen
WWW: http://home.t-online.de/home/willi-graf-realschule/raage.htm
Founded: 1982
Membership: 15
Activities: lectures * photography * computing
City Reference Coordinates: 006°47'00"E 50°39'00"N

Remeis-Sternwarte (Dr._)
● See "Universität Erlangen-Nürnberg, Astronomisches Institut, Dr. Remeis-Sternwarte"

Rheinische Friedrich-Wilhelms-Universität Bonn
● See "Universität Bonn"

Rheinische-Westfälische Technische Hochschule Aachen
● See "Technische Hochschule Aachen"

Richard-Fehrenbach-Planetarium
Friedrichstraße 51
D-79098 Freiburg-im-Breisgau
Telephone: (0)761-276099
WWW: http://home.t-online.de/home/yachtschule-spittler/planetar.htm
Founded: 1975
Activities: shows for schools and public
City Reference Coordinates: 007°51'00"E 47°59'00"N

Robert-Mayer-Volks-und-Schulsternwarte Heilbronn eV
Bismarckstraße 10
D-74072 Heilbronn
Telephone: (0)7131-81299
Telefax: (0)7131-677777
Electronic Mail: info@sternwarte.org
WWW: http://www.sternwarte.org/
Founded: 1914 (Association: 1987)
Membership: 171
Activities: education * public observing * lectures * youth activities * working groups * trips
Coordinates: 009°13'40"E 49°08'25"N H188m
City Reference Coordinates: 009°14'00"E 49°08'00"N

Römer-Sternwarte Rheinhausen (RRS) eV (Rudolf-_)
● See "Rudolf-Römer-Sternwarte Rheinhausen (RRS) eV"

Rudolf-Römer-Sternwarte Rheinhausen (RRS) eV
Postfach 14 18 07
D-47208 Duisburg
or :
Scwarzenberger Straße 147
D-47226 Duisburg-Rheinhausen
Telephone: (0)2065-75012
Founded: 1971
Membership: 80
Activities: observing * photography * education
Periodicals: (4) "Duisburger Komet" (ISSN 0179-0730)
City Reference Coordinates: 006°46'00"E 51°25'00"N

Ruhr-Universität Bochum
● See "Universität Bochum"

Ruprecht-Karls-Universität Heidelberg
● See "Universität Heidelberg"

Rüsselsheimer Sternfreunde 1975 eV
c/o Horst Tremel
Am borngraben 40
D-65428 Rüsselsheim
Telephone: (0)6142-59789
Electronic Mail: 100665.1407@compuserve.com (Jürgen Bommarius)
WWW: http://iphcip1.physik.uni-mainz.de/~astro/pop/rues/
Founded: 1975
Periodicals: "Mitteilungen astronomischer Vereinigungen Rhein-Main-Nahe" (circ.: 550)
City Reference Coordinates: 008°25'00"E 50°00'00"N

Schulplanetarium Chemnitz
Nikolaus-Kopernikus-Mittelschule
Albert-Köhler-Straße 48
D-09122 Chemnitz
Telephone: (0)371-229111
Telefax: (0)371-229111
WWW: http://members.aol.com/ugbsoft/point/11_1.htm
        http://www.tu-chemnitz.de/~wth/11_1.htm
Founded: 1981
Staff: 2
Activities: education * shows * lectures
City Reference Coordinates: 012°55'00"E 50°50'00"N

Schulsternwarte und Planetarium Rodewisch
Rützengrüner Straße 41A
D-08228 Rodewisch
Telephone: (0)3744-32313
Telefax: (0)3744-32815
WWW: http://home.t-online.de/home/n.hornfischer/sternw.htm
        http://www.freiepresse.de/sternwarte.rodewisch/

Founded: 1950
Activities: education * observing * planetarium shows
Coordinates: 012°24'56" E 50°31'42" N H498m
City Reference Coordinates: 012°25'00" E 50°32'00" N

Schul- und Volkssternwarte Aalen

c/o Schubart-Gymnasium Aalen
Rombacher Straße 30
D-73430 Aalen
Telephone: (0)7361-95610
Telefax: (0)7361-956120
Founded: 1969
Staff: 2
Activities: popularization * education
Coordinates: 010°04'51" E 48°50'08" N H440m
City Reference Coordinates: 010°07'00" E 48°50'00" N

Schul- und Volkssternwarte K.E. Ziolkowski

Postfach 505
D-98504 Suhl
or :
Auf dem Hoheloh 1
D-98527 Suhl
Telephone: (0)3681-723556
Telefax: (0)3681-723556
WWW: http://granit.maschinenbau.tu-ilmenau.de/mb/wwwirp/projekte/schulen/testfile.htm
Founded: 1969
Activities: education * popularization * observing * astronautics * planetarium
Coordinates: 010°42'08" E 50°36'14" N H527m
City Reference Coordinates: 010°43'00" E 50°37'00" N

Schul- und Volkssternwarte Leinfelden-Echterdingen eV

Immanuel-Kant-Gymnasium
Anemonenstraße 15
D-70771 Leinfelden-Echterdingen
Telephone: (0)711-1600500
Telefax: (0)711-1600503
Electronic Mail: sternwarte-le@informatik.uni-stuttgart.de
WWW: http://www.informatik.uni-stuttgart.de/sternwarte/
Founded: 1987
Membership: 50
City Reference Coordinates: 009°08'00" E 48°41'00" N

Schwäbische Sternwarte (SSW) eV

(Geschäftsstelle)
Seestraße 59A
D-70174 Stuttgart
or :
Zur Uhlandshöhe 41
D-70188 Stuttgart
Telephone: (0)711-2260893 (Geschäftsstelle)
          (0)711-281871 (Observatory)
Telefax: (0)711-2260895 (Geschäftsstelle)
        (0)711-2624546 (Observatory)
WWW: http://ix.urz.uni-heidelberg.de/~mziegler/SSW.html
        http://home.t-online.de/home/farago@t-online.de/aktuell.htm
        http://www.rzuser.uni-heidelberg.de/~mziegler/SSW.html
        http://schwaebische-sternwarte.de/
Founded: 1922 (Association: 1920)
Membership: 30
Activities: occultations * variable stars * photography * CCDs
Periodicals: "Sternwarte Stuttgart"
Coordinates: 009°11'51" E 48°47'01" N H354m
City Reference Coordinates: 009°11'00" E 48°46'00" N

Schwarzschild-Observatorium (KSO) (Karl-_)

● See "Thüringer Landessternwarte (TLS) - Tautenburg, Karl-Schwarzschild-Observatorium (KSO)"

Scientific and Technical Information Network (STN) International

c/o Fachinformationszentrum Karlsruhe
Postfach 24 65
D-76012 Karlsruhe
Telephone: (0)7247-808555
Telefax: (0)7247-808131
Electronic Mail: hlpdeskk@fiz-karlsruhe.de

WWW: http://www.fiz-karlsruhe.de/
Founded: 1977
Staff: 330
Activities: operating the STN Karlsruhe centre * databanks including information in the fields of astronomy and astrophysics
Periodicals: (6) "STNews"
● For parcels use: D-76344 Eggenstein-Leopoldshafen
City Reference Coordinates: 008°24'00"E 49°03'00"N (Karlsruhe)
008°23'00"E 49°04'00"N (Eggenstein)

S. Hirzel Verlag GmbH & Co.

Postfach 10 10 61
D-70009 Stuttgart
or :
Birkenwaldstraße 44
D-70191 Stuttgart
Telephone: (0)711-25820
Telefax: (0)711-2582290
WWW: http://www.geist.spacenet.de/hirzel-steiner/verlag-D.html
Founded: 1853
● Publisher
City Reference Coordinates: 009°11'00"E 48°46'00"N

Sonnenobservatorium Einsteinturm (SOE)

● See "Astrophysikalisches Institut Potsdam (AIP), Sonnenobservatorium Einsteinturm (SOE)"

Spacetec GmbH (Intercon_)

● See "Intercon Spacetec GmbH"

Space Telescope - European Coordinating Facility (ST-ECF)

c/o E.S.O.
Karl-Schwarzschild-Straße 2
D-85748 Garching
Telephone: (0)89-32006291
Telefax: (0)89-32006480
Electronic Mail: <userid>@eso.org
WWW: http://www.stecf.org/
Founded: 1984
Staff: 18
Activities: European focal point of ST-related activities * coordinating the development of data analysis software in Europe and with the "Space Telescope Science Institute (STScI)" (USA-MD) * maintaining a copy of the ST archive * supporting European astronomers in the preparation of ST proposals * organizing HST-related meetings
Periodicals: "ST-ECF Newsletter"
● Joint facility of the "European Space Agency (ESA)" and the "European Southern Observatory (ESO)"
City Reference Coordinates: 011°40'00"E 48°14'00"N

Sparkassen-Planetarium Augsburg

Im Thäle 3
D-86152 Augsburg
Telephone: (0)821-3246740 (Information)
(0)821-3246762 (Management)
(0)821-314936
Telefax: (0)821-314946
Electronic Mail: s-planetarium@a-city.de
WWW: http://www.augsburg.baynet.de/s-planetarium/
http://www.a-city.de/s-planetarium/
City Reference Coordinates: 010°53'00"E 48°23'00"N

Spektrum Akademischer Verlag

Vangerowstraße 20
D-69115 Heidelberg
Telephone: (0)6221-91260
(0)6221-504743
Telefax: (0)6221-912638
(0)6221-504751
Electronic Mail: online@spektrum.com
WWW: http://www.spektrum.de/
Founded: 1990
Staff: 35
Periodicals: "Spektrum der Wissenschaft"
● Publisher
City Reference Coordinates: 008°43'00"E 49°25'00"N

Springer-Verlag

Postfach 10 52 80
D-69042 Heidelberg

or :
Tiergartenstraße 17
D-69121 Heidelberg
Telephone: (0)6221-487360
Telefax: (0)6221-487150
Electronic Mail: springer@vax.ntp.springer.de
WWW: http://www.springer.de/
         http://science.springer.de/
Founded: 1842
Staff: 1230
● Publisher
City Reference Coordinates: 008°43'00"E 49°25'00"N

SPSS Science Software GmbH
Postfach 41 07
D-40688 Erkrath
Telephone: (0)2104-9540
         (0)800-903755 (Numéro Vert - France)
Telefax: (0)2104-95410
Electronic Mail: euroscience@spss.com
WWW: http://www.spss.com/software/science
Founded: 1968
Staff: 15
● Software distributor
City Reference Coordinates: 006°55'00"E 51°13'00"N

Starkenburg-Sternwarte eV Heppenheim
Kleine Bach 3
D-64646 Heppenheim
Telephone: (0)6252-4247
Electronic Mail: e.schwab@gsi.de (Erwin Schwab)
WWW: http://www.gsi.de/~kaos/html/sternw/sternw-home.html
         http://www.fse.fh-darmstadt.de/ssw-hp/
         http://www.regio-info.de/sternwarte-heppenheim/
Founded: 1970
Membership: 180
Activities: minor planets * radioastronomy * Sun * photography
Periodicals: (4) "Sirius"
Coordinates: 008°39'11"E 49°38'53"N H256m
City Reference Coordinates: 008°39'00"E 49°38'00"N

Stefan Thiele (Astronomische Instrumente_)
● See "Astronomische Instrumente Stefan Thiele (AIT)"

Sterne (Die_)
● Now merged with "Sterne und Weltraum"

Sterne und Weltraum (SuW)
c/o H.J. Staude (Editor)
Redaktion SuW
MPI für Astronomie
Königstuhl 17
D-69117 Heidelberg
or :
Verlag Sterne und Weltraum
Hüthig GmbH
Im Weiher 10
D-69121 Heidelberg
Telephone: (0)6221-528229 (Editor)
         (0)6221-4890 (Publisher)
Telefax: (0)6221-528246 (Editor)
         (0)6221-489279 (Publisher)
Electronic Mail: quetz@mpia-hd.mpg.de
WWW: http://www.mpia-hd.mpg.de/suw/suw
         http://www.mpia-hd.mpg.de/MPIA/Projects/PUBREL/SuW/suw-welcome.html
         http://www.sterne-und-weltraum.de
Founded: 1962
Staff: 6
● Journal (11) (ISSN 0039-1263, circ.: 40,000)
City Reference Coordinates: 008°43'00"E 49°25'00"N

Sternfreunde Breisgau (SFB) eV
c/o Karl-Ludwig Bath
Geranienstraße 2
D-79312 Emmendingen

Telephone: (0)7641-3492
Telefax: (0)7641-3492
WWW: http://www.kis.uni-freiburg.de/~ps/SFB/
Founded: 1973
Membership: 60
Activities: occultations * general amateur astronomy * popularization
Periodicals: (3) "Mitteilungen"
Coordinates: 007°54'20"E 47°54'53"N H1,240m (Schauinslandsternwarte)
City Reference Coordinates: 007°51'00"E 48°07'00"N

Sternfreunde Donzdorf eV

Gmünder Straße 12
D-73072 Donzdorf
or :
Beim Schulzentrum
D-73072 Donzdorf
Telephone: (0)7162-24713
Founded: 1985
Membership: 140
Activities: observing * photography * education * lectures
Periodicals: (4) "Galaxis"
Coordinates: 009°49'12"E 48°41'10"N H456m (Messelberg-Sternwarte)
City Reference Coordinates: 009°49'00"E 48°41'00"N

Sternfreunde Durmersheim und Umgebung eV

c/o Jürgen Lindner
Würmersheimer-Straße 25
D-76448 Durmersheim
Electronic Mail: Juergen.Linder@t-online.de
WWW: http://home.t-online.de/home/Juergen.Linder/sternfr.htm
Founded: 1988
Membership: 40
City Reference Coordinates: 008°17'00"E 48°56'00"N

Sternfreunde Franken (SFF) eV

Dr. Neumeyer Straße 240
D-91349 Egloffstein
Telephone: (0)9197-698980
Telefax: (0)9197-698982
Electronic Mail: fo0107@fonline.de (Frank Fleischmann, President)
WWW: http://sff.coolworld.de/
Founded: 1995
Membership: 28
Activities: popularization * observing * photography * CCDs * instrumentation
Coordinates: 011°15'30"E 49°44'00"N H450m
City Reference Coordinates: 011°16'00"E 49°42'00"N

Sternfreunde im FEZ (SiFEZ)

c/o Steffen Janke
Eichgestell
D-12459 Berlin
Telephone: (0)30-53071445
Telefax: (0)30-53071445
Electronic Mail: astro@sjanke.in-berlin.de
WWW: http://www.sifez.de/
Founded: 1979
City Reference Coordinates: 013°24'00"E 52°31'00"N

Sternwarte Bautzen

● See "Sternwarte Johannes Franz Bautzen"

Sternwarte der Volkshochschule Aachen

Peterstraße 21-25
D-52062 Aachen
Telephone: (0)241-47920
Telefax: (0)241-406023
Electronic Mail: vhsac1@aol.com
WWW: http://www.rat.de/apd/STERNWT.HTM
        http://members.tripod.com/~apd2/sternwt.htm
Founded: 1935
Activities: education
Coordinates: 006°04'15"E 50°45'35"N
City Reference Coordinates: 006°05'00"E 50°47'00"N

Sternwarte des Bruder-Klaus-Heim

Sankt-Michael-Straße 15
D-86450 Altenmünster-Violau
Telephone: (0)8295-1097
Telefax: (0)8295-499
Electronic Mail: kcmayer@dillingen.baynet.de (Christoph Mayer, Manager)
WWW: http://home.t-online.de/home/082951097/stern.htm
Founded: 1962
Staff: 4
Activities: public observatory * popularization * photography * meteors * planetarium
Periodicals: (1) "Boundless Universe" (calendar)
Coordinates: 010°34'29"E 48°27'14"N H480m
City Reference Coordinates: 010°40'00"E 48°27'00"N (Welden)

Sternwarte des Max-Born-Gymnaiums

Johann-Sebastian-Bach-Straße 8
D-82110 Germering
Telephone: (0)89-843111
Telefax: (0)89-845790
Electronic Mail: hammer@max-born-gymn.by.schule.de
WWW: http://www.m.shuttle.de/ffb/max-born-gymn/
       http://members.xoom.com/astrombg/
Founded: 1967
Membership: 35
Activities: education * popularization * observing
Coordinates: 015°23'09"E 48°06'18"N H535m
City Reference Coordinates: 015°21'00"E 48°07'00"N

Sternwarte des Rottmayr-Gymnasiums

c/o Gerardo Inhester
Barbarossastraße 16
D-83410 Laufen
Telephone: (0)8682-9091
Telefax: (0)8682-95989
WWW: http://hiris.anorg.chemie.tu-muenchen.de/AAL/RGL/sternw.htm
Founded: 1970
City Reference Coordinates: 012°55'00"E 47°56'00"N

Sternwarte Greifswald

Domstraße 10a
D-17489 Greifswald
Telephone: (0)3834-63372
Telefax: (0)3834-63269
Electronic Mail: kersten@physik.uni-greifswald.de
         sternwarte@physik.uni-greifswald.de
WWW: http://www.physik.uni-greifswald.de/~sterne/Sternwarte/
       http://www.physik.uni-greifswald.de/~sterne/Observatory/
       http://www.greifswald-online.de/vv/sternwarte/
Founded: 1924
Staff: 1
Activities: popularization * education
Periodicals: "Mitteilungen der Greifswalder Sternwarte", "MV-Astroblick"
Coordinates: 013°22'34"E 54°05'38"N H35m
City Reference Coordinates: 013°24'00"E 54°06'00"N

Sternwarte Höfingen

Oberes Ende der Uhlandstraße
D-71229 Leonberg Höfingen
Electronic Mail: gdietze@uni-hohenheim.de (Gerald Dietze)
WWW: http://www.uni-hohenheim.de/~gdietze/astro.html
Founded: 1970
Membership: 15
City Reference Coordinates: 009°01'00"E 48°48'00"N

Sternwarte Johannes Kepler

Lindenstraße 8
D-08451 Crimmitschau
Telephone: (0)3762-3730
Founded: 1929
Electronic Mail: can@stw-cri.wda.sn.schule.de
WWW: http://www.tu-chemnitz.de/~sontag/ods/can1-95.html
       http://www.fh-zwickau.de/~fa/
Activities: observing (Sun, planets, Moon, comets) * education * computing
Coordinates: 012°23'00"E 50°49'00"N H273m
City Reference Coordinates: 012°23'00"E 50°49'00"N

Sternwarte Johannes Franz Bautzen

Postfach 11 09
D-02607 Bautzen
or :
Czornebohstraße 82
D-02625 Bautzen
Telephone: (0)3591-47126
Telefax: (0)3591-44071
WWW: http://www.chm.tu-dresden.de/bio/private/astro/bzn_home.htm
http://ctch06.chm.tu-dresden.de/afo/bzn_home.htm
Founded: 1872
Staff: 2
Activities: training for teachers * education * popularization * lectures * planetarium * meteorology
Periodicals: (4) "Der Sternfreund"
Coordinates: 014°27'30" E 51°09'48" N H207m
City Reference Coordinates: 014°29'00" E 51°11'00" N

Sternwarte Kronshagen

Hofbrook 64
D-24119 Kronshagen
Telephone: (0)431-581632
Founded: 1980
Staff: 1
Activities: education * planetarium
Coordinates: 010°04'44" E 54°20'33" N
City Reference Coordinates: 010°05'00" E 54°20'00" N

Sternwarte Lauenstein

c/o Kay Hardelt
Schloßstraße 8
D-01778 Geising
Telephone: (0)35054-25217
Telefax: (0)35054-25206
Electronic Mail: hardelt@t-online.de
WWW: http://pollux.hrz.tu-freiberg.de/~jkress/
City Reference Coordinates: 009°33'00" E 52°04'00" N (Lauenstein)

Sternwarte Neanderhöhe Hochdahl eV

Postfach 22 45
D-40679 Erkrath
Telephone: (0)2104-46888
Telefax: (0)2104-46887
Electronic Mail: office@snh.rp-online.de
WWW: http://snh.rp-online.de/
Founded: 1967
Staff: 3
Activities: public observing * planetarium shows * education
Periodicals: (4) "Sternzeit" (co-publisher) (ISSN 0721-8168, circ.: 2,000)
Coordinates: 006°58'59" E 51°12'37" N H135m
City Reference Coordinates: 006°55'00" E 51°13'00" N

Sternwarte-Planetarium Bochum

Castroper Straße 67
D-44777 Bochum
Telephone: (0)234-516060
(0)234-5160611
Telefax: (0)234-5160651
Electronic Mail: planetarium@bochum.de
WWW: http://bochum-info.ruhr.de/sci/planetar.htm
http://www.bochum.de/
Founded: 1947
Staff: 12
Activities: education * public observing * planetarium shows
Periodicals: (4) "Programm"
Coordinates: 007°13'24" E 51°27'54" N H132m
City Reference Coordinates: 007°13'00" E 51°28'00" N

Sternwarte Solingen

● See "Walter-Horn-Gesellschaft eV (WHG)"

Sternwarte Sonneberg

Sternwartestraße 32
D-96515 Sonneberg
Telephone: (0)3675-81210
Telefax: (0)3675-81219

Electronic Mail: cld@stw.tu-ilmenau.de (Constanze la Dous, Director)
WWW: http://www.stw.tu-ilmenau.de/∼web/public/indexd.html
Founded: 1925
Staff: 10
Activities: sky patrol * variable stars
Coordinates: 011°11'33"E 50°22'41"N H640m
City Reference Coordinates: 011°10'00"E 50°22'00"N

Sternwarte Torgelow

c/o Frau Spelly
Kopernikus Gymnasium Torgelow
D-17358 Torgelow
Telephone: (0)3976-202331
WWW: http://www.physik.uni-greifswald.de/∼sterne/Sternwarte_torgelow/
Coordinates: 014°00'00"E 53°38'00"N H0m
City Reference Coordinates: 014°00'00"E 53°38'00"N

Sternwarte und Planetarium Albstadt-Ebingen

Hartmannstraße 140
D-72458 Albstadt-Ebingen
Telephone: (0)7431-72881
Telefax: (0)7431-72881
Electronic Mail: 0743172881-0001@t-online.de
WWW: http://home.t-online.de/home/0743172881-0001@t-online.de/homepage.htm
Founded: 1973
Coordinates: 008°59'59"E 48°12'37"N H737m
City Reference Coordinates: 009°02'00"E 48°13'00"N

Sternwarte und Planetarium Köln-Nippes

Blücherstraße 15-17
D-50733 Köln
Telephone: (0)221-7761448
              (0)221-7761361
Electronic Mail: a0196@rrz.uni-koeln.de
WWW: http://www.uni-koeln.de/∼a0196/
Founded: 1960
Activities: education
Coordinates: 006°57'26"E 50°58'04"N H44m
City Reference Coordinates: 006°57'00"E 50°56'00"N

Sternwarte und Planetarium Schneeberg

Heinrich-Heine-Straße 13a
D-08289 Schneeberg
Telephone: (0)3772-22439
Telefax: (0)3772-22440
Electronic Mail: 0377222439@t-online.de
WWW: http://www.freiepresse.de/ORTE/Schneeberg/planetarium.htm
City Reference Coordinates: 012°38'00"E 50°36'00"N

Stiftung Volkssternwarte Trebur

c/o Michael Adrian Observatorium
Fichtenstraße 7
D-65468 Trebur
Telephone: (0)6147-50000
Telefax: (0)6147-50001
Electronic Mail: 100665.1407@compuserve.com (Jürgen Bommarius)
              gerd.koellner@t-online.de (Gerd Köllner)
              jomo@monet.fh-friedberg.de (Johannes M. Ohlert)
WWW: http://www.fh-friedberg.de/users/jomo/t1t.htm
Founded: 1997
Staff: 6
Activities: education * research
Coordinates: 008°24'42"E 49°55'35"N H100m
City Reference Coordinates: 008°25'00"E 50°00'00"N (Rüsselsheim)

STN International

● See "Scientific and Technical Information Network (STN) International"

Technische Hochschule Aachen, Forum Weltraumforschung

Templergraben 55
D-52062 Aachen
Telephone: (0)241-803513
Telefax: (0)241-404472
         (0)241-8888260
WWW: http://www.rwth-aachen.de/Einrichtungen/Fwf/fwf.html

Founded: 1987
Staff: 75
Activities: planning and coordinating common interests in space R&D * education * advising * industry-research interface
● Full institution name: "Rheinische-Westfälische Technische Hochschule Aachen"
City Reference Coordinates: 006°05'00"E 50°47'00"N

Technische Hochschule Aachen, Institut für Physikalische Chemie, Centrum für Physikalische Chemie unter Mikrogravitation
Templergraben 59
D-52056 Aachen
Telephone: (0)241-804743
Telefax: (0)241-36840
Electronic Mail: <userid>@rwth-aachen.de
WWW: http://www.rwth-aachen.de/pci/Ww/richter/cpc.html
Founded: 1988
Staff: 20
Activities: physical chemistry under microgravity
● Full institution name: "Rheinische-Westfälische Technische Hochschule Aachen"
City Reference Coordinates: 006°05'00"E 50°47'00"N

Technische Universität Berlin, Institut für Astronomie und Astrophysik
PN 8-1
Hardenbergstraße 36
D-10623 Berlin
Telephone: (0)30-31423783
Telefax: (0)30-31424885
Electronic Mail: <userid>@physik.tu-berlin.de
WWW: http://export.physik.tu-berlin.de/
          http://export.physik.tu-berlin.de/german.html
Founded: 1968
Staff: 21
Activities: theoretical astrophysics * ISM * astrochemistry * dust formation * physics of dusty objects * optical and X-ray astronomy * cataclysmic variable stars * active galaxies
City Reference Coordinates: 013°24'00"E 52°31'00"N

Technische Universität Braunschweig, Institut für Flugmechanik und Raumfahrttechnik
Hans-Sommer-Straße 5
D-38106 Braunschweig
Telephone: (0)531-3917880
Telefax: (0)531-3915193
Electronic Mail: d.rex@tu-bs.de
WWW: http://www.tu-bs.de/institute/fmrt/
Founded: 1996
Staff: 11
Activities: aerospace education * aerospace research * orbital mechanics * reentry * space debris * orbital debris * satellite constellations * flight mechanics * flying quality characteristics * flight performance * flight stability * flight dynamics
City Reference Coordinates: 010°31'00"E 52°16'00"N

Technische Universität Darmstadt (TUD), Institut für Physikalische Geodäsie (IPG)
Petersenstraße 13
D-64287 Darmstadt
Telephone: (0)6151-163109
Telefax: (0)6151-164512
Electronic Mail: <userid>@ipgs.ipg.verm.tu-darmstadt.de
Founded: 1970
Staff: 12
Activities: education and research in geodesy (physical, astronomical, satellite)
Coordinates: 008°40'38"E 49°51'48"N
City Reference Coordinates: 008°39'00"E 49°52'00"N

Technische Universität Dresden, Institut für Planetare Geodäsie und Lohrmann-Observatorium
Mommsenstraße 13
D-01062 Dresden
Telephone: (0)351-4634097
Telefax: (0)351-4637019
Electronic Mail: lohrmobs@rcs.urz.tu-dresden.de
WWW: http://www.tu-dresden.de/fghgipg/homepage.html
Founded: 1956
Activities: Earth rotation * photographic astrometry * star occultations * geodetical astronomy
Periodicals: "Mitteilungen des Lohrmann-Observatoriums" (ISSN 0323-8180)
Coordinates: 013°52'18"E 51°03'00"N H324m (Gönnsdorf)
              013°43'46"E 51°01'51"N H168m (University Refractor)
City Reference Coordinates: 013°44'00"E 51°03'00"N

Technische Universität München, Institut für Astronomische und Physikalische Geodäsie (IAPG)
Arcisstraße 21
D-80333 München
Telephone: (0)89-21053190
Telefax: (0)89-21053178
Founded: 1961
Staff: 11
Activities: theoretical geodesy * satellite geodesy * gravimetry
Periodicals: "Mitteilungen", "Jahresbericht" (ISSN 0938-846x)
Coordinates: 011°34'00"E 48°09'00"N H530m
City Reference Coordinates: 011°35'00"E 48°08'00"N

Teleskoptechnik Halfmann
Gessertshausener Straße 8
D-86356 Neusäß-Vogelsang
Telephone: (0)821-483070
Telefax: (0)821-485999
Founded: 1990
Staff: 25
Activities: manufacturing professional telescopes and instrumentation
City Reference Coordinates: 006°27'00"E 50°35'00"N

Thiele (Astronomische Instrumente Stefan_)
• See "Astronomische Instrumente Stefan Thiele"

Thüringer Landessternwarte (TLS) - Tautenburg, Karl-Schwarzschild-Observatorium (KSO)
Sternwarte 5
D-07778 Tautenburg
Telephone: (0)36427-8630
Telefax: (0)36427-86329
Electronic Mail: <userid>@tls-tautenburg.de
WWW: http://www.tls-tautenburg.de/
Founded: 1960
Staff: 24
Activities: star formation * YSOs * extrasolar planets * galaxy evolution * AGNs * gamma-ray bursts * magnetic stars
Coordinates: 011°42'48"E 50°58'54"N H331m
City Reference Coordinates: 011°43'00"E 50°59'00"N

Turtle Star Observatory (TSO)
c/o Axel Martin
Friedhofstraße 15
D-45478 Mülheim/Ruhr
Telephone: (0)208-55151
Electronic Mail: axelm@bph.ruhr-uni-bochum.de
WWW: http://www.bph.ruhr-uni-bochum.de/~axelm/tso/tso.htm
Founded: 1995
Staff: 4
Activities: astrometry * minor planets * astrophotography * CCD
Periodicals: (4) "Spica"
Coordinates: 006°50'39"E 51°25'43"N
City Reference Coordinates: 006°54'00"E 51°24'00"N

Universität Berlin, Institut für Meteorologie
Carl-Heinrich-Becker-Weg 6-10
D-12165 Berlin
Telephone: (0)30-83871172
           (0)30-83871171
           (0)30-83871169
Telefax: (0)838-71128
    (0)838-7919002
Electronic Mail: <userid>@bibo.met.fu-berlin.de
                <userid>@strat.met.fu-berlin.de
WWW: http://www.met.fu-berlin.de/
       http://www.met.fu-berlin.de/english
Founded: 1949
Membership: 105
Activities: atmospheric environmental research * radioactivity * radiation * synoptic weather maps and climate maps * observation systems * urban and regional meteorology * stratospheric research * meteorological satellite research * climatology * publishing reports on weather and climate * long-range weather forecasting
Periodicals: (daily) "Berliner Wetterkarte"; "Beilagen"
Coordinates: 013°18'00"E 52°28'00"N H51m (Berlin-Dahlem)
                013°18'45"E 52°27'35"N H51m (Berlin-Steglitz)
• Full university name: "Freie Universität Berlin"
City Reference Coordinates: 013°24'00"E 52°31'00"N

Universität Bochum, Astronomisches Institut

Universitätsstraße 150
Gebäude NA 7
D-44780 Bochum
Telephone: (0)234-7003454
    (0)234-7005802
Telefax: (0)234-7094169
Electronic Mail: <userid>@astro.ruhr-uni-bochum.dbp.de
WWW: http://www.astro.ruhr-uni-bochum.de/
Founded: 1966
Staff: 21
Activities: stellar, galactic and extragalactic astronomy
Coordinates: 007°15'51"E 51°26'41"N H179m
    070°44'16"W 29°15'17"S H2,340m (La Silla)
● Full university name: "Ruhr-Universität Bochum"
City Reference Coordinates: 007°13'00"E 51°28'00"N

Universität Bochum, Institut für Experimentalphysik V

Universitätsstraße 150
D-44780 Bochum
Telephone: (0)234-7005785
Telefax: (0)234-7094175
Staff: 20
Activities: spectroscopy of hot plasmas
WWW: http://www.ep5.ruhr-uni-bochum.de/
Founded: 1972
● Full university name: "Ruhr-Universität Bochum"
City Reference Coordinates: 007°13'00"E 51°28'00"N

Universität Bochum, Institut für Hochfrequenztechnik (IHFT)

Universitätsstraße 150
D-44780 Bochum
Telephone: (0)234-7002842
Telefax: (0)234-7094167
Electronic Mail: he@hf.ruhr.uni-bochum.de
WWW: http://www.hf.ruhr-uni-bochum.de/
Founded: 1965
Staff: 45
Activities: antennas and wave propagation * biomedical engineering * remote sensing * industrial testing and automation * microwave measurement * ultrasound medical imaging * magnetic resonance imaging
● Full university name: "Ruhr-Universität Bochum"
City Reference Coordinates: 007°13'00"E 51°28'00"N

Universität Bochum, Institut für Theoretische Physik, Lehrstuhl IV

Universitätsstraße 150
D-44780 Bochum
Telephone: (0)234-7004728
Telefax: (0)234-7094177
Electronic Mail: <userid>@tp4.ruhr-uni-bochum.de
WWW: http://www.tp4.ruhr-uni-bochum.de/
● Full university name: "Ruhr-Universität Bochum"
City Reference Coordinates: 007°13'00"E 51°28'00"N

Universität Bonn, Astronomische Institute, Institut für Astrophysik und Extraterrestrische Forschung

Auf dem Hügel 71
D-53121 Bonn
Telephone: (0)228-733676
    (0)228-733671
Telefax: (0)228-733672
Electronic Mail: <userid>@astro.uni-bonn.de
WWW: http://www.astro.uni-bonn.de/~webiaef
    http://www.astro.uni-bonn.de/
Founded: 1964
Staff: 11
Activities: ionosphere * exosphere * thermosphere * solar wind * interplanetary matter * ISM * SN physics * pulsars * QSOs * black holes * jets * cosmology * rocket and satellite experiments
● Full university name: "Rheinische Friedrich-Wilhelms-Universität Bonn"
City Reference Coordinates: 007°05'00"E 50°44'00"N

Universität Bonn, Astronomische Institute, Observatorium Hoher List

D-54550 Daun
Telephone: (0)6592-2150
Telefax: (0)6592-985140
Electronic Mail: <userid>@astro.uni-bonn.de
WWW: http://www.astro.uni-bonn.de/~webstw

http://www.astro.uni-bonn.de/
Founded: 1952
Staff: 15
Activities: optical observing * minor planets * stars and stellar systems
Coordinates: 006°51'02"E 50°09'49"N H533m
● Full university name: "Rheinische Friedrich-Wilhelms-Universität Bonn"
City Reference Coordinates: 006°50'00"E 50°11'00"N

Universität Bonn, Astronomische Institute, Radioastronomisches Institut (RAIUB)
Auf dem Hügel 71
D-53121 Bonn
Telephone: (0)228-733658
Telefax: (0)228-733672
Electronic Mail: <userid>@astro.uni-bonn.de
WWW: http://www.astro.uni-bonn.de/~webrai
http://www.astro.uni-bonn.de/
Founded: 1962
Staff: 12
Activities: solar, planetary, galactic and extragalactic radioastronomy * STP * technology, image and data processing in radioastronomy
Coordinates: 006°43'24"E 50°34'12"N H435m (Radioobservatorium Stockert)
● Full university name: "Rheinische Friedrich-Wilhelms-Universität Bonn"
City Reference Coordinates: 007°05'00"E 50°44'00"N

Universität Bonn, Astronomische Institute, Sternwarte
Auf dem Hügel 71
D-53121 Bonn
Telephone: (0)228-733655
Telefax: (0)228-733672
Electronic Mail: <userid>@astro.uni-bonn.de
WWW: http://www.astro.uni-bonn.de/~webstw
http://www.astro.uni-bonn.de/
Founded: 1846
Staff: 15
Activities: astrophysics * photometry * spectroscopy * stars * Magellanic clouds * ISM * planetoids * astrometry
Periodicals: "Veröffentlichungen der Astronomischen Institute der Universität Bonn"
Coordinates: 007°04'05"E 50°43'52"N H91m
● Full university name: "Rheinische Friedrich-Wilhelms-Universität Bonn"
City Reference Coordinates: 007°05'00"E 50°44'00"N

Universität Bonn, Geodätisches Institut
Nußallee 17
D-53115 Bonn
Telephone: (0)228-732621
(0)228-732622
Telefax: (0)228-732988
Electronic Mail: <userid>@sn-geod-1.geod.uni-bonn.de
WWW: http://giub.geod.uni-bonn.de/vlbi.html (VLBI Group)
http://giub.geod.uni-bonn.de/
City Reference Coordinates: 007°05'00"E 50°44'00"N

Universität Erlangen-Nürnberg, Astronomisches Institut, Dr. Remeis Sternwarte
Sternwartestraße 7
D-96049 Bamberg
Telephone: (0)951-952220
Telefax: (0)951-9522222
Electronic Mail: <userid>@sternwarte.uni-erlangen.de
WWW: http://www.sternwarte.uni-erlangen.de/
Founded: 1889
Staff: 8
Activities: variable stars * stellar atmospheres
Coordinates: 010°53'24"E 49°53'06"N H288m
● Full university name: "Friedrich-Alexander-Universität Erlangen-Nürnberg"
City Reference Coordinates: 010°53'00"E 49°53'00"N

Universität Frankfurt/Main, Institut für Theoretische Physik/Astrophysik
Robert-Mayer-Straße 10
D-60054 Frankfurt/Main
Telephone: (0)69-79822357
Telefax: (0)69-79828350
Electronic Mail: <userid>@astro.uni-frankfurt.de
WWW: http://www.astro.uni-frankfurt.de/
http://earth.astro.uni-frankfurt.de/
Founded: 1908
Staff: 12
Activities: ISM

● Full university name: "Johann-Wolfgang-Goethe-Universität Frankfurt/Main"
City Reference Coordinates: 008°40'00"E 50°07'00"N

Universität Göttingen, Institut für Geophysik
Herzberger Landstraße 180
D-37075 Göttingen
Telephone: (0)551-397451
Telefax: (0)551-397459
Electronic Mail: <userid>@willi.uni-geophys.gwdg.de
WWW: http://www.geo.physik.uni-goettingen.de/
● Full university name: "Georg-August-Universität Göttingen"
City Reference Coordinates: 009°55'00"E 51°32'00"N

Universität Hamburg, Hamburger Sternwarte
Gojenbergsweg 112
D-21029 Hamburg
Telephone: (0)40-72524112
Telefax: (0)40-72524198
Electronic Mail: <userid>@hs.uni-hamburg.de
WWW: http://www.physnet.uni-hamburg.de/fb12/stw.html
           http://www.hs.uni-hamburg.de/
Founded: 1833
Staff: 18
Activities: theoretical astrophysics * stellar structure * stellar evolution * stellar atmospheres * astrometry * ISM * QSO survey * gravitational lenses
Periodicals: "Abhandlungen aus der Hamburger Sternwarte" (ISSN 0374-1583)
Coordinates: 010°14'30"E 53°28'54"N H45m
City Reference Coordinates: 009°59'00"E 53°33'00"N

Universität Hannover, Institut für Erdmessung (IfE), Astronomische Station
Schneiderberg 50
D-30167 Hannover
Telephone: (0)511-7622475
Telefax: (0)511-7624006
Electronic Mail: <userid>@mbox.ife.uni-hannover.de
Founded: 1963
Staff: 16
Activities: determining astronomical latitude, longitude and vertical deflections by portable zenith cameras * satellite geodesy * GPS * geoid and gravity field determination
Coordinates: 009°42'45"E 52°23'15"N H123m
City Reference Coordinates: 009°44'00"E 52°24'00"N

Universität Hannover, Institut für Meteorologie und Klimatologie
Herrenhäuser Straße 2
D-30419 Hannover
Telephone: (0)511-7622677
Telefax: (0)511-7624418
Electronic Mail: <userid>@muk.uni-hannover.dbp.de
WWW: http://www.muk.uni-hannover.de/
Founded: 1949
Staff: 25
Activities: antarctic research * VHF-MST radar measurements * atmospheric turbulence * eolian energy * numerical simulation
Periodicals: "Berichte des Instituts für Meteorologie und Klimatologie der Universität Hannover"
City Reference Coordinates: 009°44'00"E 52°24'00"N

Universität Heidelberg, Institut für Theoretische Astrophysik (ITA)
Tiergartenstraße 15
D-69121 Heidelberg
Telephone: (0)6221-544837
Telefax: (0)6221-544221
Electronic Mail: ita@ita.uni-heidelberg.de
                <userid>@ita.uni-heidelberg.de
WWW: http://www.ita.uni-heidelberg.de/
     http://www.urz.uni-heidelberg.de/institute/d/53/
Founded: 1976
Activities: theoretical astrophysics * stellar spectra * chromospheres and coronae * star formation * accretion disks * evolution of galaxies * astrochemistry * galactic centre
● Full university name: "Ruprecht-Karls-Universität Heidelberg"
City Reference Coordinates: 008°43'00"E 49°25'00"N

Universität Jena, Astrophysikalisches Institut und Universitäts-Sternwarte
Schillergäßchen 2
D-07745 Jena
Telephone: (0)3641-947501
Telefax: (0)3641-947502

Electronic Mail: obs@astro.uni-jena.de
WWW: http://www.astro.uni-jena.de/Sternwarte/observ.html
Founded: 1813
Staff: 10
Activities: ISM * star formation * early stellar phases * photometry * laboratory astrophysics
Coordinates: 011°29'00"E 50°55'48"N H356m (Großschwabhausen)
● Full university name: "Friedrich-Schiller-Universität Jena"
City Reference Coordinates: 011°35'00"E 50°56'00"N

Universität Kiel, Institut für Theoretische Physik und Astrophysik
Olshausenstraße 40
D-24098 Kiel
Telephone: (0)431-8804110
Telefax: (0)431-8804100
Electronic Mail: <userid>@astrophysik.uni-kiel.de
WWW: http://www.astrophysik.uni-kiel.de/
        http://www.astrophysik.uni-kiel.de/e-home.html
Founded: 1872
Staff: 35
Activities: stellar atmospheres * stellar spectroscopy * stellar evolution * Sun and solar-type stars * white dwarfs * galactic evolution * ISM evolution * chemo-dynamical evolution of galaxies * hydrodynamics * stellar dynamics
● Use postal code D-24118 for parcels.
● Formerly "Institut für Astronomie und Astrophysik"
● Full university name: "Christian-Albrechts-Universität zu Kiel"
City Reference Coordinates: 010°08'00"E 54°20'00"N

Universität Köln, Erstes Physikalisches Institut
Zülpicher Straße 77
D-50937 Köln
Telephone: (0)221-4703562
Telefax: (0)221-4705162
Electronic Mail: <userid>@ph1.uni-koeln.de
WWW: http://www.uni-koeln.de/math-nat-fak/
        http://www.ph1.uni-koeln.de/kosma.html (KOSMA - see separate entry in Switzerland)
        http://www.ph1.uni-koeln.de/kosma_observatory.html (KOSMA - see separate entry in Switzerland)
Founded: 1980
Staff: 50
Activities: molecular spectroscopy * microwave-IR, sub-mm interstellar molecular spectroscopy
Coordinates: 007°47'04"E 45°59'04"N H3,130m (KOSMA, Gornergrat)
City Reference Coordinates: 006°57'00"E 50°56'00"N

Universität Köln, Institut für Geophysik und Meteorologie
Albertus-Magnus-Platz/Kerpener Straße 13
D-50923 Köln
Telephone: (0)221-4703682
Telefax: (0)221-4705161
Electronic Mail: <userid>@meteo.uni-koeln.de
WWW: http://www.uni-koeln.de/math-nat-fak/geomet/
Founded: 1961
Staff: 20
Activities: atmospheric circulation * mesoscale modelling * wind energy * transport of pollutants
Periodicals: "Mitteilungen aus dem Institut für Geophysik und Meteorologie Köln" (ISSN 0069-5882)
City Reference Coordinates: 006°57'00"E 50°56'00"N

Universität München, Institut für Astronomie und Astrophysik, Observatorium Wendelstein
Wendelsteingipfel
D-83735 Bayrischzell
Telephone: (0)8023-406
Telefax: (0)8023-9141
Electronic Mail: <userid>@usm.uni-muenchen.de
WWW: http://bigbang.usm.uni-muenchen.de:8002/USM/WDST/
Founded: 1940
Staff: 4
Activities: photometry * spectrophotometry * direct imaging * spectroscopy
Coordinates: 012°00'49"E 47°42'16"N H1,841m
City Reference Coordinates: 012°01'00"E 47°42'00"N

Universität München, Institut für Astronomie und Astrophysik, Universitäts-Sternwarte München
Scheinerstraße 1
D-81679 München
Telephone: (0)89-9220940
Telefax: (0)89-92209427
Electronic Mail: <userid>@usm.uni-muenchen.de
        adis@usm.uni-muenchen.de
WWW: http://www.usm.uni-muenchen.de:8001/
Founded: 1816

Staff: 48
Activities: stellar spectroscopy * model atmospheres * stellar winds * stellar evolution * UV spectroscopy * cataclysmic variable stars * high-speed photometry * hydrodynamics * atomic physics * instrumentation * galaxies * large-scale structure * plasma astrophysics
Coordinates: 011°36'30"E 48°08'48"N H529m (München)
              012°00'50"E 47°42'16"N H1,841m (Wendelstein)
City Reference Coordinates: 011°35'00"E 48°08'00"N

Universität Münster, Institut für Planetologie (IfP)
Wilhelm-Klemm-Straße 10
D-48149 Münster
Telephone: (0)251-8333496
Telefax: (0)251-8339083
          (0)251-8336301
Electronic Mail: <userid>@uni-muenster.de
WWW: http://ifp.uni-muenster.de/
Founded: 1986
Staff: 50
Activities: analytical planetology * planetary physics
● Full university name: "Westfälische Wilhelms-Universität Münster"
City Reference Coordinates: 007°37'00"E 51°58'00"N

Universität Potsdam, Lehrstuhl Astrophysik
Postfach 60 15 53
D-14415 Potsdam
Telephone: (0)331-9771054
Telefax: (0)331-9771107
Electronic Mail: office@astro.physik.uni-potsdam.de
WWW: http://www.astro.physik.uni-potsdam.de/
Founded: 1995
Staff: 6
City Reference Coordinates: 013°04'00"E 52°24'00"N

Universitäts-Sternwarte Bonn (USB)
● See "Universität Bonn, Astronomische Institute, Sternwarte"

Universitäts-Sternwarte Göttingen (USG)
Geismarlandstraße 11
D-37083 Göttingen
Telephone: (0)551-395042
          (0)551-395053
Telefax: (0)551-395043
Electronic Mail: <userid>@uni-sw.gwdg.de
WWW: http://www.uni-sw.gwdg.de/
        http://www.uni-sw.gwdg.de/Welcome_de.html (German version)
Founded: 1749
Staff: 40
Activities: solar physics * extragalactic research * theoretical astrophysics * stellar atmospheres * stellar evolution * high-energy astrophysics
Coordinates: 009°56'33"E 51°31'48"N H159m
              009°58'34"E 51°31'32"N H347m (Hainberg)
              016°30'00"W 28°18'00"N H2,409m (Izaña, Islas Canarias)
● Full university name: "Georg-August-Universität Göttingen"
City Reference Coordinates: 009°55'00"E 51°32'00"N

Universitäts-Sternwarte Jena (USJ)
● See "Universität Jena, Astrophysikalisches Institut und Universitäts-Sternwarte"

Universitäts-Sternwarte München (USM)
● See "Universität München, Institut für Astronomie und Astrophysik, Universitäts-Sternwarte München"

Universität Tübingen, Institut für Astronomie und Astrophysik, Abteilung Astronomie
Waldhäuser Straße 64
D-72076 Tübingen
Telephone: (0)7071-2972486
Telefax: (0)7071-2973458
Electronic Mail: <userid>@astro.uni-tuebingen.de
                <userid>@ait.physik.uni-tuebingen.de
WWW: http://astro.uni-tuebingen.de/
Founded: 1949
Staff: 24
Activities: optical, UV, EUV, X-ray, and gamma-ray astronomy * instrumentation * binaries * PN * AGNs * ISM
Coordinates: 009°03'30"E 48°32'18"N H470m
● Full university name: "Eberhard-Karls-Universität Tübingen"
City Reference Coordinates: 009°02'00"E 48°31'00"N

Universität Tübingen, Lehr- und Forschungsbereich Theoretische Astrophysik
Auf der Morgenstelle 10c
D-72076 Tübingen
Telephone: (0)7071-292487
Telefax: (0)7071-295889
Electronic Mail: <userid>@tat.physik.uni-tuebingen.de
WWW: http://www.tat.physik.uni-tuebingen.de/
Founded: 1968
Staff: 60
Activities: theoretical astrophysics * plasma physics * atomic physics in strong fields * cosmic X-sources * computational physics * solar physics * scientific visualization * biomechanics * neural networks
● Full university name: "Eberhard-Karls-Universität Tübingen"
City Reference Coordinates: 009°02'00"E 48°31'00"N

Universität Würzburg, Astronomisches Institut
Am Hubland
D-97074 Würzburg
Telephone: (0)931-8885031
Telefax: (0)931-8884603
Electronic Mail: <userid>@astro.uni-wuerzburg.de
WWW: http://www.astro.uni-wuerzburg.de/
Founded: 1967
Staff: 16
Activities: solar physics * galactic and extragalactic astronomy * theoretical astrophysics
Observatories: 1 (Izaña, Teide, Islas Canarias)
● Full university name: "Bayerische Julius-Maximilians-Universität Würzburg"
City Reference Coordinates: 009°56'00"E 49°48'00"N

Urania Verlag
Germaniastraße 18-20
D-12099 Berlin
Telephone: (0)30-75082300
Telefax: (0)30-75082400
Founded: 1924
● Publisher
City Reference Coordinates: 013°24'00"E 52°31'00"N

VCH Verlagsgesellschaft mbH
Postfach 10 11 61
Pappelallee 3
D-69469 Weinheim
Telephone: (0)6201-6060
Telefax: (0)6201-606328
WWW: http://www.wiley-vch.de/
Founded: 1921
Staff: 300
● Publisher
City Reference Coordinates: 008°39'00"E 49°33'00"N

Vehrenberg KG (Dr._)
● See "Dr. Vehrenberg KG"

Verein der Amateurastronomen des Saarlandes (VAS) eV
Postfach 18 06
D-66409 Homburg
or :
Händelstraße 1
D-66538 Neunkirchen
Telephone: (0)821-22803 (Michael Risch/private)
Electronic Mail: risch@activeminds.de
WWW: http://fsinfo.cs.uni-sb.de/~monz/vas.htm
      http://www.astronomie.com/vas/
      http://www.astronomie.com/vas/sternwar.htm (Sternwarte Peterberg)
Founded: 1977
Membership: 155
City Reference Coordinates: 008°06'00"E 50°32'00"N
                    007°20'00"E 49°20'00"N (Neunkirchen)

Verein für Himmelskunde Dresden eV
c/o Hans-Jörg Mettig
Böhmische Straße 11
D-01099 Dresden
Telephone: (0)351-801115
WWW: http://ctch06.chm.tu-dresden.de/afo/vfh_home.htm
      http://home.t-online.de/home/Badicke/dvfh.htm

Founded: 1991
Membership: 20
Activities: meetings * excursions * observing
Periodicals: (6) "Der Sternfreund" (co-publisher)
City Reference Coordinates: 013°44'00"E 51°03'00"N

Verein für Volkstümliche Astronomie Essen (VVA) eV
Weberplatz 1
D-45127 Essen
Telephone: (0)201-510401
　　　　　　(0)201-554018
Founded: 1981
Membership: 25
Activities: observing * photography * annual "Astrobörse"
Periodicals: (1) "Aus Astronomie und Raumfahrt" (circ.: 2,000); (4) "Sternzeit" (co-publisher) (ISSN 0721-8168, circ.: 2,000)
City Reference Coordinates: 007°01'00"E 51°28'00"N

Verein Historische Sternwarten Gotha eV
c/o O. Schwarz
Uthmannstraße 8
D-99867 Gotha
Electronic Mail: sternwgth@aol.com
WWW: http://members.aol.com/sternwGTH/
City Reference Coordinates: 010°43'00"E 50°57'00"N

Vereinigte Amateur-Astronomen Eschwege 1975 eV (VAAE)
c/o Tibor Kaulitzki
Hauptstraße 36
D-36205 Sontra-Ulfen
Telephone: (0)6221-509256
Electronic Mail: mdietric@mail.lsw.uni-heidelberg.de
WWW: http://www.eschwege.de/astro/vaae.htm
Founded: 1975
Membership: 15
Activities: meetings * observing (planets, Sun, deep sky) * popularization * trips * lectures
Periodicals: (4) "Sirius" (ISSN 0941-2352, circ.: 200), "Sternzeit" (co-publisher) (ISSN 0721-8168, circ.: 2,000)
Coordinates: 010°02'00"E 51°15'00"N (Meinhard-Hitzelrode)
　　　　　　010°00'06"E 51°10'00"N (Witzenhausen)
City Reference Coordinates: 010°03'00"E 51°11'00"N (Eschwege)
　　　　　　　　　　009°56'00"E 51°05'00"N (Sontra)

Vereinigung der Nordenhamer Sternfreunde eV
Lutherplatz 2
D-26954 Nordenham
Telephone: (0)4731-23251
WWW: http://www.nordenham.de/sterne.htm
City Reference Coordinates: 008°29'00"E 53°30'00"N

Vereinigung der Sternfreunde eV (VdS)
c/o Bayerische Volkssternwarte München
Rosenheimer Straße 145a
D-81671 München
Telephone: (0)89-406239
WWW: http://www.uni-essen.de/initiative/vds/
　　　　　http://www.vds-astro.de/
　　　　　http://neptun.uni-sw.gwdg.de/sonne.html (Solar Section)
Founded: 1953
Membership: 2500
Activities: meetings * observing * popularization
Sections: (Fachgruppen) Amateurteleskope * astrophotographie * Atmosphärische Phänomene * CCD-Technik * Geschichte * Jugendarbeit * Kleine Planeten * Kometen * Dark Sky - Bewegung gegen Lichtverschmutzung * Meteore * Planeten * Pseudowissenschaften, Esoterik und Astrologie * UFOs * Radioastronomie * Rechnende Astronomie * Sonne * Spektroskopie * Sternbedeckungen * Veränderliche * Visuelle Deep-Sky Beobachtung * Volkssternwarten und Planetarien
Periodicals: (12) "VdS-Nachrichten" (in "Sterne und Weltraum" - see separate entry) "Provisional Sunspot Numbers"; (4) "Sonne" (ISSN 0721-0094, circ.: 500); (1) "Sonne Datenblatt" (circ.: 200); "New Sunspot Indices Bulletin" (ISSN 0934-8220), "Sonne Tageskarten", "Sonne Zirkular", "Astro Fax Zirkular"
City Reference Coordinates: 011°35'00"E 48°08'00"N

Vereinigung Gandersheimer Sternfreunde (VGS)
c/o Uwe Schmidtmann
Eschenweg 8
D-37547 Kreiensen
Telephone: (0)5563-910060
Telefax: (0)5563-910060
Electronic Mail: uschmidtmann@solar.stanford.edu

Founded: 1974
Activities: observing * CCD imaging
Periodicals: (4) "Sternzeit" (co-publisher) (ISSN 0721-8168, circ.: 2,000)
Coordinates: 010°01'48"E 51°52'54"N H150m (Schulsternwarte)
City Reference Coordinates: 009°58'00"E 51°52'00"N

Vereinigung Krefelder Sternfreunde eV
Postfach 29 64
D-47729 Krefeld
or :
Waldhofstraße 132
D-47800 Krefeld
Telephone: (0)2151-503116
WWW: http://www.krefeld-city.de/vks/sternfreunde.html
Founded: 1966
Membership: 90
Activities: popularization
Periodicals: (4) "Sternenbote"
Coordinates: 006°34'00"E 51°19'00"N H80m
City Reference Coordinates: 006°34'00"E 51°19'00"N

Verkehr Raumfarht und Systemtechnik (VRS) GmbH
Walter-Köhn-Straße 1b
D-04356 Leipzig
Telephone: (0)341-526230
Telefax: (0)341-5262356
Electronic Mail: info@vrs.de
WWW: http://www.vrs.de/
Founded: 1993
Staff: 16
Activities: studies, analysis and development of subsystems for space research in the field of Earth observation, microgravity and small satellites
City Reference Coordinates: 012°20'00"E 51°19'00"N

Verlag Harri Deutsch
Gräfstraße 47
D-60486 Frankfurt/Main
Telephone: (0)69-775021
Telefax: (0)69-7073739
Electronic Mail: buchhandlung@harri-deutsch.de
                verlag@harri-deutsch.de
WWW: http://www.harri-deutsch.de/
Founded: 1961
Staff: 20
Activities: publishing * bookshop * software
City Reference Coordinates: 008°40'00"E 50°07'00"N

Viersener Astronomischer Arbeitskreis (VAA)
c/o Stefan Adler
Weiherstraße 10A
D-41748 Viersen
Telephone: (0)2162-24119
Founded: 1986
Membership: 21
Activities: popularization * lectures
City Reference Coordinates: 006°23'00"E 51°15'00"N

Vieweg & Sohn Verlagsgesellschaft mbH (Friedrich_)
● See "Friedrich Vieweg & Sohn Verlagsgesellschaft mbH"

Volkshochschule der Stadt Fürstenfeldbruck eV, Arbeitsgemeinschaft Astronomie
c/o Michael A. Rappenglück
Heinrich-Feller-Straße 15
D-82275 Emmering
or :
Theodor-Heuss-Straße 2
D-82256 Fürstenfeldbruck
Telephone: (0)8141-4827
Telefax: (0)8141-4827
Electronic Mail: 100639.637@compuserve.com (Michael A. Rappenglück)
Founded: 1981
Membership: 30
Activities: popularization * photography * planets * history of astronomy * computing * observing (Sun * deep sky)
Coordinates: 011°14'10"E 48°09'52"N H535m
City Reference Coordinates: 011°15'00"E 48°10'00"N (Fürstenfeldbruck)

Volkshochschule der Stadt Soest, Astronomische Arbeitsgemeinschaft
c/o Ernst Fleischer
Steinkuhlenweg 6
D-59494 Soest
Telephone: (0)2921-77490
Founded: 1966
Membership: 35
Activities: education * observing
Coordinates: 008°05'37"E 51°34'23"N H96m (Volkssternwarte Soest)
City Reference Coordinates: 008°06'00"E 51°34'00"N

Volkssternwarte Adolph Diesterweg Radebeul
Auf den Ebenbergen 10a
D-01445 Radebeul
Telephone: (0)351-8305905
Telefax: (0)351-8381906
Electronic Mail: rattei@chemie.rmhs1.tu-dresden.d400.de
WWW: http://ctch06.chm.tu-dresden.de/stw-rdbl/
Founded: 1959
Staff: 1
Activities: observing * popularization * planetarium
Coordinates: 013°37'20"E 51°06'59"N H185m
City Reference Coordinates: 013°43'00"E 51°13'00"N

Volkssternwarte Bonn (VSB), Astronomische Vereinigung eV
Poppelsdorfer Allee 47
D-53115 Bonn
Telephone: (0)228-222270
WWW: http://www.comedia.de/VSB/vsb.html
Founded: 1972
Membership: 130
Activities: education * public readings * observing (Sun, planets, deep sky)
Periodicals: (4) "Telescopium" (ISSN 0723-1121, circ.: 450), "Sternzeit" (co-publisher) (ISSN 0721-8168, circ.: 2,000)
Coordinates: 007°05'56"E 50°43'44"N H59m
City Reference Coordinates: 007°05'00"E 50°44'00"N

Volkssternwarte Darmstadt (VSD) eV
Herdweg 45
D-64285 Darmstadt
Telephone: (0)6151-61108
Electronic Mail: stwda@igd.fhg.de
WWW: http://www.igd.fhg.de/~stwda/
Founded: 1969
Activities: observing * popularization
Periodicals: (12) "Mitteilungen der Volkssternwarte Darmstadt"
Coordinates: 008°39'47"E 49°50'38"N
City Reference Coordinates: 008°39'00"E 49°52'00"N

Volkssternwarte Ennepetal eV
Am Hinnenberg 80
D-58247 Ennepetal
Telephone: (0)2333-62646
Electronic Mail: rolke@avu.de
City Reference Coordinates: 007°23'00"E 51°18'00"N

Volkssternwarte Erich Bär
Stolpener Straße 48
D-01454 Radeberg
WWW: http://ctch06.chm.tu-dresden.de/afo/rbg_home.htm
http://www.radeberg.de/sternwarte/stwhome.htm
Founded: 1964
Membership: 20
Activities: lectures * observing (occultations, Sun) * photography
Periodicals: (6) "Der Sternfreund" (co-publisher)
Coordinates: 013°55'59"E 51°06'51"N H218m
City Reference Coordinates: 013°56'00"E 51°08'00"N

Volkssternwarte Erich Scholz
Hochwaldstraße 21c
D-02763 Zittau
WWW: http://ctch06.chm.tu-dresden.de/afo/zit_home.htm
Founded: 1968
City Reference Coordinates: 014°47'00"E 50°54'00"N

Volkssternwarte Geschwister Herschel Hannover eV
Am Lindener Berge 27
D-30449 Hannover
Telephone: (0)511-456290
Telefax: (0)511-456290
Founded: 1968
Membership: 100
Activities: popularization * lectures * observing
Coordinates: 009°42'25"E 52°21'50"N
City Reference Coordinates: 009°44'00"E 52°24'00"N

Volkssternwarte Hagen eV
Postfach 146
D-58001 Hagen
or :
Eugen-Richter-Turm
D-58135 Hagen
Telephone: (0)2331-590790
Telefax: (0)2331-590791
Electronic Mail: 100276.1140@compuserve.com
          http://www.sternwarte.bnet.de/
Founded: 1955
Membership: 45
City Reference Coordinates: 007°28'00"E 51°22'00"N

Volkssternwarte Hof
Egerländerweg 25
D-95032 Hof
Telephone: (0)9281-95278
Telefax: (0)8281-95278
Electronic Mail: ba108@fim.uni-erlangen.de
            a0492@mz.uni-bayreuth.de
            hof-08@hof.baynet.de
WWW: http://www.hof.baynet.de/~ho1019/22e_ster.htm
        http://www.hof.baynet.de/~ho1019/vsw-www/22e_for.htm
Founded: 1971
Staff: 15
Activities: education * observing * digital image processing * radioastronomy * photography (H-alpha, deep sky) * weather satellite station * automatic telescope positioning * software * meteors * CCD * tours * meetings * lectures * Zeiss-Starlab planetarium
Periodicals: (4) "Sternzeit" (co-publisher) (ISSN 0721-8168, circ.: 2,000); (2) "Program"; (1) "Annual Report"
Coordinates: 011°54'57"E 50°18'07"N H500m
            011°52'36"E 50°14'56"N H623m (remote station)
City Reference Coordinates: 011°56'00"E 50°19'00"N

Volkssternwarte im Volksbildungswerk Hofheim-Marxheim
c/o Hermann Minor
Lessingstraße 56
D-65719 Hofheim/Taunus
Telephone: (0)6192-3599
Electronic Mail: lingnau@tm.informatik.uni-frankfurt.de (Anselm Lingnau)
Founded: 1977
Membership: 60
Periodicals: "Mitteilungen astronomischer Vereinigungen Rhein-Main-Nahe" (circ.: 550)
City Reference Coordinates: 008°26'00"E 50°07'00"N

Volkssternwarte Jonsdorf
An der Sternwarte 3
D-02796 Jonsdorf
WWW: http://ctch06.chm.tu-dresden.de/afo/jon_home.htm
Founded: 1962
City Reference Coordinates: 014°43'00"E 50°51'00"N

Volkssternwarte Köln
c/o Schiller-Gymnasium
Nikolausstraße 55
D-50937 Köln
Telephone: (0)221-415467
WWW: http://www.koeln.netsurf.de/~Volkssternwarte.Koeln/
Founded: 1922
City Reference Coordinates: 006°57'00"E 50°56'00"N

Volkssternwarte Laupheim eV und Planetarium
Leibnizstraße 35

D-88471 Laupheim
Telephone: (0)7392-18055
Telefax: (0)7392-17464
WWW: http://home.t-online.de/home/vstw.laupheim/
http://home.t-online.de/home/vstw.laupheim/planet.htm
http://www.planetarium-laupheim.de/
Founded: 1975
Membership: 100
Activities: popularization * amateur astronomical fair * exhibitions * planetarium
Coordinates: 009°52'56"E 48°13'37"N H518m
009°52'56"E 48°43'35"N H506m
City Reference Coordinates: 009°55'00"E 48°13'00"N

Volkssternwarte Manfred von Ardenne
c/o Lars Stephan
Klenzestraße 3
D-17424 Seebad Heringsdorf
WWW: http://www.physik.uni-greifswald.de/~sterne/Volkssternwarte_heringsdorf/
http://ctch06.chm.tu-dresden.de/MvAObserv/
Founded: 1960
Staff: 1
Activities: observing (occultations, planets) * photography
Coordinates: 014°10'27"E 53°57'22"N H5m
City Reference Coordinates: 014°10'00"E 53°58'00"N

Volkssternwarte Marburg eV
c/o Ursula von Geyr
Deutschhausstraße 27a
D-35037 Marburg
or :
Dresdener Straße 18
D-35274 Kirchhain
Telephone: (0)6422-7599
WWW: http://www.dsk.de/rds/24482.htm
Founded: 1975
Membership: 45
Activities: education * lectures * observing * popularization * photography
Coordinates: 008°55'27"E 50°49'51"N H225m
City Reference Coordinates: 008°36'00"E 50°49'00"N (Marburg)
008°55'00"E 50°49'00"N (Kirchhain)

Volkssternwarte Norderstedt eV
c/o Harald Prahl
Breslauer Straße 7
D-22850 Norderstedt
Telephone: (0)40-5285383
Electronic Mail: fornasiero@physnet.uni-hamburg.de (Lirio Fornasiero, Chairman)
Founded: 1973
Membership: 26
Activities: photography * observing (occultations, comets) * lectures
Periodicals: (4) "Sternzeit" (co-publisher) (ISSN 0721-8168, circ.: 2,000)
City Reference Coordinates: 008°59'00"E 53°43'00"N

Volkssternwarte Paderborn eV
Postfach 11 42
D-33041 Paderborn
or :
Schloß Neuhaus
Im Schloßpark
Marstallstraße 13
D-33104 Paderborn
Telephone: (0)5254-85157
Telefax: (0)5293-8132
Electronic Mail: astroobspb@aol.com
intersolpb@aol.com
WWW: http://members.aol.com/astroobspb/INDEX.HTM
Founded: 1971
Membership: 120
Activities: INTER-SOL programme * radioastronomy * lectures * popularization (press, radio, TV) * debunking astrology * photography
Periodicals: (3) "INTER-SOL Reports"
Coordinates: 008°42'39"E 51°44'50"N H119m
City Reference Coordinates: 008°44'00"E 51°43'00"N

Volkssternwarte Passau
c/o Rainer Klemm

Anton-Poetzl-Straße 6
D-94034 Passau
Telephone: (0)851-45453
Electronic Mail: kalz@passau.baynet.de
WWW: http://www.passau.baynet.de/~pa000127/vsw/
City Reference Coordinates: 013°28'00"E 48°35'00"N

Volkssternwarte Prenzlau
c/o Robert Ehrlich
Kirchstraße 6
D-17291 Roepersdorf
Telephone: (0)3984-802039
Electronic Mail: sternwarte-prenzlau@t-online.de
WWW: http://home.t-online.de/home/sternwarte-prenzlau/
Founded: 1962
Staff: 2
Activities: occultations
Coordinates: 013°51'51"E 53°18'38"N H57m
City Reference Coordinates: 013°52'00"E 53°19'00"N (Prenzlau)

Volkssternwarte Rothwesten
c/o Friedel Spitzer
Brüder-Grimm-Straße 24
D-34233 Fuldatal
Telephone: (0)5607-459
Electronic Mail: hmai@namu01.gwdg.de
                    hmai@euromail.de
                    pge12@rz.uni-kiel.dbp.de
WWW: http://www.ts.go.dlr.de/~hmai/vsw/
          http://www.sm.go.dlr.de/~hmai/vsw/
Founded: 1963
Activities: observing * photography
Coordinates: 009°30'53"E 51°23'26"N H319m
City Reference Coordinates: 009°31'00"E 51°23'00"N

Volkssternwarte Tirschenreuth
c/o Peter Postler
Blankenbühlstraße 28
D-95643 Tirschenreuth
or :
Großenseeser Straße
D-95643 Tirschenreuth
Telephone: (0)9631-6212 (after 18:00)
WWW: http://www.sternwarte-tir.de/
Founded: 1964
Activities: general observing * slide shows * education
Coordinates: 012°21'00"E 49°54'00"N
City Reference Coordinates: 012°21'00"E 49°54'00"N

Volkssternwarte und Planetarium Drebach
Straße der Jugend
D-09430 Drebach
Telephone: (0)37341-7435
Electronic Mail: stw@stw-drebach.zp.sn.schule.de
WWW: http://www.tu-chemnitz.de/~mwei/stw-drebach/stw-drebach.html
          http://www.tu-chemnitz.de/home/ods/stw_drebach/
Founded: 1969
Activities: photography * occultations * popularization
City Reference Coordinates: 013°01'00"E 50°40'00"N

Volkssternwarte und Planetarium Reutlingen
Karlstraße 40
D-72764 Reutlingen
Telephone: (0)7121-3360
WWW: http://home.t-online.de/home/B.Augustin/stwrtpln.htm
Founded: 1956
Membership: 18
Activities: education * photography * observing (Sun, lunar occultations)
Coordinates: 009°12'58"E 48°29'55"N H395m
City Reference Coordinates: 009°11'00"E 48°29'00"N

Volkssternwarte Urania Jena eV
Schillergäßchen 2a
D-07745 Jena
Telephone: (0)3641-423414 (Wilfried Weise - after 17:00)

WWW: http://www.uni-jena.de/~ows/Urania/
Founded: 1909
Membership: 50
Activities: observing (Sun, planets, deep sky, comets, variable stars) * photography * photometry * lectures * popularization
Observatories: 2 (Urania-Volkssternwarte & Forst-Sternwarte)
City Reference Coordinates: 011°35'00"E 50°56'00"N

Volkssternwarte Wetterau eV

c/o Walter Groening
Gartenfeldstrasse 16
D-61231 Bad Nauheim
WWW: http://ourworld.compuserve.com/homepages/wetterau/
Coordinates: 008°43'41"E 50°21'59"N H265m
City Reference Coordinates: 008°44'00"E 50°21'00"N

Volkssternwarte Würzburg eV

c/o Josef Laufer
Peter-Wagner-Straße 4
D-97230 Estenfeld
or :
Cronthalstraße 25
D-97074 Würzburg
Telephone: (0)9305-993216
          (0)931-73020
Electronic Mail: vstw@gmx.de
WWW: http://www.wuerzburg.de/vstw/
Founded: 1984
Membership: 85
Activities: observing * lectures
Coordinates: 009°57'34"E 49°46'33"N H271m
City Reference Coordinates: 009°56'00"E 49°48'00"N (Würzburg)

Volks- und Schulsternwarte Bartholomäus Scultetus

An der Sternwarte 1
D-02827 Görlitz
Telephone: (0)3581-78222
WWW: http://www.chm.tu-dresden.de/bio/private/astro/grl_home.htm
          http://ctch06.chm.tu-dresden.de/afo/grl_home.htm
Founded: 1947
Activities: education * popularization * Sun * variable stars * occultations * local astronomy history * planetarium
Periodicals: (4) "Der Sternfreund"
Coordinates: 014°57'05"E 51°08'08"N H235m
City Reference Coordinates: 014°59'00"E 51°09'00"N

Volks- und Schulsternwarte Bruno H. Bürgel eV

Zöllnerweg 12
D-02689 Sohland
Telephone: (0)35936-37270
Electronic Mail: starklabk@aol.com (Matthias Stark)
WWW: http://ctch06.chm.tu-dresden.de/afo/shl_home.htm
          http://members.aol.com/stwsohland/
Founded: 1963
Membership: 38
Activities: photography * lectures * education * variable stars
Coordinates: 014°25'00"E 51°12'00"N H335m
City Reference Coordinates: 014°25'00"E 51°12'00"N

Volks- und Schulsternwarte Geretsried eV

Adalbert-Stifter-Straße 14
D-82538 Geretsried
Telephone: (0)8171-932532
Telefax: (0)8171-90412
Electronic Mail: sternwarte@wor.de
WWW: http://www.ilo.baynet.de/sternwarte/
          http://www.wor.de/sternwarte/
Founded: 1976
Membership: 55
Activities: education * observing * photography * computing * weather satellite receiving system
Coordinates: 011°28'34"E 47°52'18"N H600m
City Reference Coordinates: 011°28'00"E 47°51'00"N

Volks- und Schulsternwarte Juri Gagarin

Mansberg 18
D-04838 Eilenburg
Telephone: (0)3423-603153

Telefax: (0)3423-603153
Founded: 1964
Staff: 2
Activities: popularization * education * observing (sunspots, lunar occultations) * planetarium
Coordinates: 012°37'39"E 51°27'05"N H120m
City Reference Coordinates: 012°37'00"E 51°27'00"N

VRS GmbH
● See "Verkehr Raumfarht und Systemtechnik (VRS) GmbH"

Walter-Horn-Gesellschaft eV (WHG)
c/o Sternwarte Solingen
Postfach 19 05 50
D-42705 Solingen
or :
Sternstraße 5
D-42719 Solingen
Telephone: (0)212-335555
Telefax: (0)212-331091
Electronic Mail: sternwarte@solingen.de
WWW: http://www.solingen.de/sternwarte
Founded: 1921
Membership: 75
Activities: education * multimedia theater * observing
Periodicals: (4) "Sternzeit" (co-publisher) (ISSN 0721-8168, circ.: 2,000)
Coordinates: 007°01'19"E 51°10'32"N H145m
City Reference Coordinates: 007°05'00"E 51°10'00"N

Weltastronomie Studienreisen (WAS)
c/o Eckehard Schmidt
Postfach 46 16
D-90025 Nürnberg
or :
Maxfeldstraße 50
D-90409 Nürnberg
Telephone: (0)911-586550
Telefax: (0)911-5865549
Electronic Mail: eckehard@orion.franken.de (Eckehard Schmidt)
Founded: 1990
Staff: 2
Activities: organizing travels to astronomy and space sites * study trips
Periodicals: "Astrotrotter"
City Reference Coordinates: 011°04'00"E 49°27'00"N

Werner-Heisenberg-Institut (WHI)
● See "Max-Planck-Institut für Physik (MPP)"

Westfälisches Museum für Naturkunde, Planetarium
Sentruper Straße 285
D-48161 Münster
Telephone: (0)251-59105
Telefax: (0)251-5916098
WWW: http://www.lwl.org/naturkundemuseum/
City Reference Coordinates: 007°37'00"E 51°58'00"N

Westfälische Volkssternwarte und Planetarium
Stadtgarten 6
D-45657 Recklinghausen
Telephone: (0)2361-23134
Telefax: (0)2361-23134
Electronic Mail: sternwarte.re@t-online.de
WWW: http://home.t-online.de/home/Sternwarte.RE/set.htm
        http://home.t-online.de/home/Sternwarte.RE/foerderv.htm
Founded: 1953
Staff: 4
Activities: public lectures * planetarium shows * public observing
Periodicals: (2) "Program of Activities" (circ.: 15,000)
Coordinates: 007°10'54"E 51°37'30"N H118m
City Reference Coordinates: 007°11'00"E 51°37'00"N

Westfälische Wilhelms-Universität Münster
● See "Universität Münster"

Wilhelm-Foerster-Sternwarte (WFS) eV
Munsterdamm 90

D-12169 Berlin
Telephone: (0)30-7900930
Telefax: (0)30-79009312
Electronic Mail: wfs@bics.be.schule.de
WWW: http://www.be.schule.de/schulen/wfs/homepage.html
        http://wwwwbs.cs.tu-berlin.de/~jensd/WFS/
Founded: 1953
Staff: 13
Activities: education * observing
Periodicals: "Veranstaltungsprogramm"
City Reference Coordinates: 013°24'00"E 52°31'00"N

Wittenberger Planetarium
Falkstraße 83
D-06886 Lutherstadt Wittenberg
Telephone: (0)3491-403455
Telefax: (0)3491-403465
WWW: http://lbs.st.schule.de/lbs/pl-witt.htm
Founded: 1987
Staff: 1
Activities: lectures
City Reference Coordinates: 012°39'00"E 51°53'00"N

Workgroup for Spectrography (WfS)
c/o Walter Diehl
Braunfelser Straße 81
D-35578 Wetzlar
Telephone: (0)6441-28377
Founded: 1990
Activities: spectrography of all kind of objects, with emphasis on the Sun
Coordinates: 008°30'00"E 50°36'00"N
City Reference Coordinates: 008°30'00"E 50°36'00"N

Zeiss Jena GmbH (Carl_)
● See "Carl Zeiss Jena GmbH

Zeiss-Grossplanetarium Berlin (ZGP)
Prenzlauer Allee 80
D-10405 Berlin
Telephone: (0)30-4218450
Telefax: (0)30-4251252
WWW: http://www.cs.tu-berlin.de/~davidi/zgp_0000.html
        http://www.zgp.be.schule.de/
        http://www.snafu.de/~astw/zgp/
Founded: 1987
Staff: 12
Activities: shows * exhibitions * life concerts
Periodicals: (12) "Program"
City Reference Coordinates: 013°24'00"E 52°31'00"N

Zeiss-Planetarium der Ernst-Abbe-Stiftung
Am Planetarium 5
D-07743 Jena
Telephone: (0)3641-885488
Telefax: (0)3641-885420
WWW: http://www.jena.de/tourism/planet.htm
Founded: 1926
Staff: 8
Activities: public shows
Periodicals: "Programmheft"
City Reference Coordinates: 011°35'00"E 50°56'00"N

Zeiss-Planetarium und Schulsternwarte Herzberg
Lugstraße 3
D-04916 Herzberg/Elster
Telephone: (0)3535-70057
Telefax: (0)3535-70057
WWW: http://www.herzberg-elster.de/A_Z/p/Pob1.htm
City Reference Coordinates: 013°14'00"E 51°42'00"N

Gibraltar

Gibraltar Astronomical Society (GAS)
c/o J. M. Reyes
13 Silver Birch Lodge
Montagu Gardens
Telephone: 75763
Electronic Mail: gasjjgon@gibnet.gi
WWW: http://www.gibnet.gi/~gasjjgon/
City Reference Coordinates: 005°21'00"E 36°09'00"N

Greece

Academy of Athens, Research Center for Astronomy and Applied Mathematics
14 Anagnostopoulou Street
GR-106 73 Athens
Telephone: (0)1-3613589
Telefax: (0)1-3631606
Founded: 1959
Staff: 8
Activities: solar activity * variable stars * astrophysics * stars * planets
Periodicals: "Contributions", "Annual Report"
City Reference Coordinates: 023°43'00"E 37°58'00"N

Aristotle University of Thessaloniki
● See "University of Thessaloniki"

Dionysos Satellite Observatory
● See "National Technical University of Athens, Dionysos Satellite Observatory"

Dourouti Observatory
● See "University of Ionnina, Department of Physics, Section of Astro-Geophysics"

Eugenides Foundation, Planetarium
87 Leof. Syggrou
GR-175 64 Athens
Telephone: (0)9411181
Telefax: (0)9417372
Electronic Mail: library@eugenides_found.edu.gr
WWW: http://www.eugenides_found.edu.gr/
        http://www.eugenfound.edu.gr/
City Reference Coordinates: 023°43'00"E 37°58'00"N

Greek Amateur Astronomers Society
46 Pentelis Street
Maroussi
GR-151 26 Athens
Telephone: (0)1-8053046
Electronic Mail: mzoulias@prometheus.hol.gr (Manolis Zoulias)
City Reference Coordinates: 023°43'00"E 37°58'00"N

Hellenic Aerospace Industry Ltd.
P.O. Box 23
GR-320 09 Schimatari
Telephone: (0)1-8836711
          (0)262-52185
          (0)262-52145
Telefax: (0)1-8838714
          (0)262-52170
Electronic Mail: marketing@haicorp.com
WWW: http://www.haicorp.com/
Founded: 1975
City Reference Coordinates: 023°43'00"E 37°58'00"N (Athens)

Hellenic Astronomical Society (HELAS)
c/o Section of Astrophysics, Astronomy and Mechanics
Department of Physics
National University of Athens
Panepistimiopolis
GR-157 84 Zografos
Telephone: (0)31-998173
Telefax: (0)31-995384
Electronic Mail: elaset@astro.auth.gr
WWW: http://www.hri.org/elaset/
        http://www.astro.auth.gr/elaset/
Founded: 1993
Membership: 140
Activities: promoting astronomy * distributing astronomical information
Periodicals: (4) "Hellenic Astronomical Society Newsletter"
City Reference Coordinates: 023°43'00"E 37°58'00"N

Hellenic National Meteorological Service (HNMS)
14 E. Venizelou Street

GR-167 77 Hellinikon
Telephone: (0)1-9629415
Telefax: (0)1-9628952
Electronic Mail: director@hnms.gr
Founded: 1931
Staff: 700
Activities: meteorological support of national defence, national economy, safety of lives and properties
City Reference Coordinates: 023°43'00"E 37°58'00"N

Hellenic Organization for Standardization (ELOT)
313 Acharnon Street
GR-111 45 Athens
Telephone: (0)1-2280001
Telefax: (0)1-2020776
Electronic Mail: elotinfo@elot.gr
WWW: http://www.elot.gr/
Founded: 1976
Staff: 71
Activities: standardization * certification * quality control * testing * information on standards and technical regulations
Periodicals: (1) "Catalogue of Hellenic Standards"; "Information Bulletin" (in Greek)
City Reference Coordinates: 023°43'00"E 37°58'00"N

Hellenic Physical Society
c/o C. Helmis (President)
6 Grivaion Street
GR-106 80 Athens
Telephone: (0)1-3635701
Telefax: (0)1-3610690
Electronic Mail: chelmis@atlas.uoa.gr
Membership: 65
City Reference Coordinates: 023°43'00"E 37°58'00"N

National Observatory of Athens (NOA), Astronomical Institute
P.O. Box 20048
GR-118 10 Athens
or :
Lofos Nymfon
Thission
Telephone: (0)1-3461191
         (0)1-8040619
Telefax: (0)1-3463803
      (0)1-8040453
      (0)1-3421019
Electronic Mail: <userid>@leon.ariadne-t.gr
              ourania@astro.noa.gr
WWW: http://www.astro.noa.gr/
Founded: 1853
Staff: 25
Activities: extragalactic astronomy * variable stars * solar system * solar physics
Periodicals: "Memoirs of the National Observatory of Athens, Series I, Astronomy"
Coordinates: 022°37'18"E 37°58'24"N H1,005m (Kryonerion Station)
City Reference Coordinates: 023°43'00"E 37°58'00"N (Athinai)

National Observatory of Athens (NOA), Institute of Meteorology and Physics of the Atmospheric Environment (IMPAE)
P.O. Box 200 48
GR-118 10 Athens
Telephone: (0)1-3456257
Telefax: (0)1-3464566
      (0)1-3421019
Electronic Mail: <userid>@inachos.hydro.ntua.civil.gr
WWW: http://www.softlab.ntua.gr/~retal/noa.html
Founded: 1860
Staff: 12
Activities: meteorological, solar radiation and daylight data collection and processing * research on atmospheric boundary layer, air pollution, modelling, wind field modelling, solar radiation modelling, atmospheric electricity, renewable forms of energy, indoor air pollution, and energy efficiency
Periodicals: (1) "Climatological Bulletin"
Coordinates: 023°43'30"E 37°58'12"N H107m (Athens)
             023°51'48"E 38°02'54"N H509m (Pentele)
City Reference Coordinates: 023°43'00"E 37°58'00"N

National Technical University of Athens (NTUA), Dionysos Satellite Observatory
9 Iroon Polytechneiou Street
GR-157 80 Zografos
Telephone: (0)1-7773613

(0)1-7721875
Telefax: (0)1-7708550
       (0)1-7721866
Electronic Mail: <userid>@ntua.gr
WWW: http://www.ntua.gr/
Founded: 1968
Staff: 10
Activities: satellite geodesy * satellite laser ranging * geodetic astronomy
Coordinates: 023°56'00"E 38°04'40"N H480m
City Reference Coordinates: 023°43'00"E 37°58'00"N (Athinai)

National University of Athens, Department of Physics, Section of Astrophysics, Astronomy and Mechanics
Panepistimiopolis
GR-157 84 Zografos
Telephone: (0)1-7243211
         (0)1-7243414
         (0)1-7235122
Telefax: (0)1-7238413
Electronic Mail: <userid>@atlas.uao.ariadne-t.gr
WWW: http://www.uoa.gr/departs/physics/sectc_en.htm
Founded: 1896
Staff: 24
Activities: astronomy & astrophysics * nonlinear mechanics * cosmology * relativity * stellar structure and evolution * solar physics * stellar dynamics * observing
Sections: Laboratory of Astronomy * Laboratory of Astrophysics
Observatories: 1 (Athens University Observatory)
City Reference Coordinates: 023°43'00"E 37°58'00"N

University of Crete, Department of Physics, Astrophysics Group
P.O. Box 1527
GR-711 11 Iraklion
or :
P.O. Box 2208
GR-710 03 Heraklion
Telephone: (0)81-235014
Telefax: (0)81-239735
Electronic Mail: <userid>@physics.uch.gr
WWW: http://www.physics.uch.gr/
Founded: 1986
Activities: compact X-ray sources * gamma-ray sources * ISM * star formation * jets * solar physics * Earth magnetosphere * ionosphere
City Reference Coordinates: 025°09'00"E 35°20'00"N

University of Ionnina, Department of Physics, Section of Astro-Geophysics
GR-451 10 Ionnina
Telephone: (0)651-98471
         (0)651-98480
Telefax: (0)651-45697
      (0)651-98682
Electronic Mail: calissar@cc.uoi.gr (C.E. Alissandrak)
WWW: http://www.uoi.gr/
      http://dioni.sci.uoi.gr/
      http://dioni.sci.uoi.gr/seci/tomeas.html
Founded: 1968
Activities: solar physics * chromospherically active dwarf stars * flare stars * galaxies * cosmology * education
Coordinates: 020°51'00"E 39°40'00"N H700m (Dourouti Observatory)
City Reference Coordinates: 020°51'00"E 39°40'00"N

University of Patras, Department of Physics, Astronomical Laboratory
GR-261 10 Rion
Telephone: (0)61-997571
         (0)61-997636
Telefax: (0)61-997571
WWW: http://www.apel.ee.upatras.gr/
Founded: 1970
Staff: 6
Activities: education * stellar dynamics * SETI * galactic nebulae * Seyfert galaxies
City Reference Coordinates: 021°44'00"E 38°15'00"N

University of Thessaloniki, Department of Physics, Section of Astrophysics, Astronomy and Mechanics
GR-540 06 Thessaloniki
Telephone: (0)31-998407
      (0)31-995384
Electronic Mail: <userid>@astro.auth.gr

WWW: http://www.astro.auth.gr/
Founded: 1943
Staff: 9
Activities: education * observing * theoretical astronomy
Periodicals: "Annual Report"
Coordinates: 022°57'30"E 40°37'00"N H28m
● Full university name: "Aristotle University of Thessaloniki"
City Reference Coordinates: 022°58'00"E 40°38'00"N

Guam

University of Guam (UoG), Planetarium
UoG Station
Mangilao, GU 96923
Telephone: 735-2783
        734-9273
Telefax: 734-1299
      734-4582
Electronic Mail: pameastl@uog9.uog.edu (Pam Eastlick, Director)
WWW: http://www.guam.net/planet/
Founded: 1992
City Reference Coordinates: 144°47'00"E 13°28'00"N

Guatemala

Asociación Astronómica de Guatemala (AAG)
Apartado Postal 307 "I"
01907 Guatemala
or :
14 Avenida 39-39
Zona 8
01008 Guatemala
Telephone: (0)472-1277
Electronic Mail: ecastro@guate.net (Edgar Castro, President)
WWW: http://www.icaiti.org.gt/astro/aag.htm
Founded: 1983
Membership: 100
Activities: observing * lectures * popularization
Periodicals: (24) "El Cyberastronomo Centroamericano"; (12) "Cometa"
City Reference Coordinates: 090°31'00"W 14°38'00"N

Comisión Guatemalteca de Normas (COGUANOR)
8a Avenida 10-43
Zona 1
Guatemala, CA
Telephone: (0)2-533547
          (0)2-383330
          (0)2-306086
Telefax: (0)2-533547
City Reference Coordinates: 090°31'00"W 14°38'00"N

Instituto Nacional de Sismología, Vulcanología, Meteorología e Hidrología (INSIVUMEH)
7a Avenida 14-57
Guatemala 13
Telephone: (0)2-314967
          (0)2-324722
          (0)2-314986
Telefax: (0)2-315005
City Reference Coordinates: 090°31'00"W 14°38'00"N

Honduras

Servicio Meteorológico Nacional (SMN)
Apartado 30 145
Tegucigalpa
Telephone: (0)338075
             (0)331112
             (0)331113
             (0)331114
Telefax: (0)333683
City Reference Coordinates: 088°13'00"W 14°06'00"N

Universidad Nacional Autónoma de Honduras (UNAH), Sección de Astrofísica
Apartado Postal 3023
Tegucigalpa MDC
Telephone: (0)322110 x230
Telefax: (0)327196
             (0)314686
Electronic Mail: huracan!<userid>@uunet.uu.net
                     <userid>@ns.hondunet.net
Founded: 1990
Staff: 2
Activities: education * theoretical research * observatory under construction
Coordinates: 087°09'00"W 14°05'00"N H1,063m
City Reference Coordinates: 087°09'00"W 14°05'00"N

Hungary

Albireo Amateur Astronomy Society
Nemzetör út. 8
H-8900 Zalaegerszeg
Telephone: (0)92-313490
Electronic Mail: albireo@alpha.dfmk.hu
WWW: http://alpha.dfmk.hu/~albireo
Founded: 1971
Membership: 100
Activities: solar system * deep sky * double stars * eclipsing binaries * amateur meteorology
Periodicals: (4) "Albireo"
Coordinates: 016°50'18"E 46°28'12"N H120m
City Reference Coordinates: 016°51'00"E 46°53'00"N

Astronomical Foundation of Nógrád County
Móricz Zs. út. 9
H-3100 Salgótarján
Telephone: (0)32-310464
           (0)32-310250
           (0)30-9108868
Founded: 1992
Staff: 7
Activities: assisting local and national amateur astronomy activities * public observatory
Periodicals: (6) "A Csillagvizsgáló"
Coordinates: 019°48'00"E 48°40'09"N H255m
              019°48'35"E 48°41'28"N H356m (Urania Observatory)
              019°53'20"E 48°41'46"N H555m (ALFA Station)
City Reference Coordinates: 019°48'00"E 48°07'00"N

Astrotech Instruments & Computers KKT
P.O. Box 116
H-6501 Baja
or :
Szegedi út. III/70
H-6500 Baja
Telephone: (0)20-9370042
Telefax: (0)79-427001
Electronic Mail: hege@electra.bajaobs.hu
               virtual@emitel.hu
WWW: http://www.bajaobs.hu/baja/astrotch.htm
Founded: 1993
Staff: 3
Activities: import-export of astronomical telescopes, accessories, CCD cameras, filters, educational material, computers and
software
City Reference Coordinates: 018°57'00"E 46°11'00"N

Attila University (József_)
• See "József Attila University"

Baja Astronomical Observatory
Szegedi út. III/70
P.O. Box 766
H-6500 Baja
Telephone: (0)79-424027
Telefax: (0)79-427001
Electronic Mail: hege@electra.bajaobs.hu
                baja@electra.bajaobs.hu
WWW: http://www.bajaobs.hu/
Founded: 1955
Staff: 6
Activities: photoelectric photometry * variable stars * binaries * cataclysmic variables * CCD imaging * lightcurve analysis
* period variations
Coordinates: 018°57'35"E 46°10'52"N H110m
City Reference Coordinates: 018°58'00"E 46°11'00"N

Budapest Planetarium
Ker. Népliget
P.O. Box 46
H-1476 Budapest
Telephone: (0)1-2650725
Telefax: (0)1-2633104
WWW: http://www.planetarium.hu/

Founded: 1977
Staff: 11
Activities: lectures * shows
City Reference Coordinates: 019°05'00"E 47°30'00"N

Debrecen Heliophysical Observatory
● See "Hungarian Academy of Sciences, Debrecen Heliophysical Observatory"

Eötvös Loránd Physical Society
P.O. Box 433
H-1371 Budapest
Telephone: (0)1-2018682
Telefax: (0)1-2018682
Electronic Mail: elft@rmk530.rmki.kfki.hu
　　　　　　　　　mail.elft@mtesz.hu
WWW: http://www.kfki.hu/~elfthp/
Membership: 700
City Reference Coordinates: 019°05'00"E 47°30'00"N

Eötvös Loránd University, Department of Astronomy
Ludovika tér 2
H-1083 Budapest
Telephone: (0)1-3141019
Telefax: (0)1-3141019
Electronic Mail: <userid>@innin.elte.hu
WWW: http://innin.elte.hu/
Founded: 1753
Staff: 10
Activities: stellar astronomy * celestial mechanics * solar MHD * dynamo theory * star formation * planetology
Periodicals: (3) "Publications of the Astronomy Department of the Eötvös Loránd University" (ISSN 0238-2423)
City Reference Coordinates: 019°05'00"E 47°30'00"N

Eötvös Loránd University, Department of Meteorology
Ludovika tér 2
H-1083 Budapest
Telephone: (0)1-1138617
　　　　　　(0)1-1334160
Telefax: (0)1-1343953
WWW: http://nimbus.elte.hu/
Founded: 1945
Staff: 9
Activities: education * research * meteorology * geography
Periodicals: "Annales" (ISSN 0237-2738), "Notes" (ISSN 0865-7920)
City Reference Coordinates: 019°05'00"E 47°30'00"N

Eötvös Loránd University, Gothard Astrophysical Observatory (GAO)
Szent Imre Herceg út. 112
H-9707 Szombathely
Telephone: (0)94-313871
Telefax: (0)94-328324
Electronic Mail: obs@gothard.hu
WWW: http://www.gothard.hu/
　　　　http://innin.elte.hu/
Founded: 1881
Staff: 4
Activities: high-dispersion spectroscopy of emission-line stars * education * history of astronomy
Coordinates: 016°36'30"E 47°15'40"N H223m
City Reference Coordinates: 016°38'00"E 47°14'00"N

Gothard Astrophysical Observatory (GAO)
● See "Eötvös Loránd University, Gothard Astrophysical Observatory (GAO)"

Gyula Observing Station
● See "Hungarian Academy of Sciences, Debrecen Heliophysical Observatory"

Haynald Observatory
Hunyadi János út. 23-25
H-6300 Kalocsa
Founded: 1878
Activities: education * public demonstrations
Coordinates: 018°58'30"E 46°31'42"N H117m
City Reference Coordinates: 019°00'00"E 46°31'00"N

Hungarian Academy of Sciences
Roosevelt tér 9

P.O. Box 6
H-1361 Budapest
Telephone: (0)1-3319353
        (0)1-3327176
Telefax: (0)1-3328943
Electronic Mail: <userid>@ella.hu
WWW: http://www.mta.hu/
Founded: 1825
Membership: 300
City Reference Coordinates: 019°05'00"E 47°30'00"N

Hungarian Academy of Sciences, Debrecen Heliophysical Observatory
P.O. Box 30
H-4010 Debrecen
Telephone: (0)52-311015
        (0)66-361553 (Gyula)
WWW: http://fenyi.sci.klte.hu/
Founded: 1958 (Gyula: 1972)
Staff: 13 (Gyula: 3)
Activities: sunspot positions and proper motions * solar flares * solar magnetic fields * photography (solar photosphere)
Periodicals: "Publications of the Debrecen Observatory" (ISSN 0209-7567)
Coordinates: 021°37'24"E 47°33'36"N H132m
        021°16'12"E 46°39'12"N H135m (Gyula Observing Station)
City Reference Coordinates: 021°38'00"E 47°32'00"N

Hungarian Academy of Sciences, Geodetical and Geophysical Research Institute
P.O. Box 5
H-9401 Sopron
or :
Csatkai Endre út. 6-8
H-9400 Sopron
Telephone: (0)99-314292
        (0)99-314293
Telefax: (0)99-313267
WWW: http://www.ggki.hu/ggki-ggri.html
Founded: 1972
City Reference Coordinates: 016°36'00"E 47°41'00"N

Hungarian Academy of Sciences, Institute for Particle and Nuclear Physics
P.O. Box 49
H-1525 Budapest
or :
Konkoly-Thege út. 29-33
H-1121 Budapest
Telephone: (0)1-1551682
Telefax: (0)1-1696567
Electronic Mail: <userid>@rmk520.rmki.kfki.hu
WWW: http://www.kfki.hu/
Founded: 1991
Staff: 253
Activities: particle and nuclear physics * space physics * plasma physics * materials science * biophysics
City Reference Coordinates: 019°05'00"E 47°30'00"N

Hungarian Academy of Sciences, Konkoly Observatory
P.O. Box 67
H-1525 Budapest
Telephone: (0)1-755866
        (0)1-754122
Telefax: (0)1-156940
Electronic Mail: <userid>@ella.hu
WWW: http://www.konkoly.hu/
Founded: 1871
Periodicals: "Communications" (ISSN 0238-2091), "Information Bulletin on Variable Stars" (see separate entry)
Coordinates: 018°57'54"E 47°30'00"N H474m
        019°54'00"E 47°55'00"N H946m (Piszkéstetö Mountain Station)
City Reference Coordinates: 019°05'00"E 47°30'00"N

Hungarian Amateur Astronomical Society
P.O. Box 36
H-1387 Budapest 62
or :
Ujhegyi út. 4
H-8624 Kötcse
Electronic Mail: ngus@sc.bme.hu
Founded: 1987
Membership: 40

Activities: observing * popularization
Periodicals: (12) "Amatörcsillagászati Courier"
Coordinates: 017°52'45"E 46°45'10"N H290m (Kötcse)
City Reference Coordinates: 019°05'00"E 47°30'00"N (Budapest)

Hungarian Astronautical Society

Fö út. 68
H-1027 Budapest
or :
P.O. Box 433
H-1371 Budapest
Telephone: (0)1-2018443
Telefax: (0)1-3561215
Founded: 1956
Membership: 450
Activities: conferences * lectures * tours * contest * popularization * international relationships
Periodicals: "Ürkaleidoszkóp", "Asztronautikai Tájékoztató"
Awards: (1) "Fonó Albert Award", "Nagy Ernö Award"
City Reference Coordinates: 019°05'00"E 47°30'00"N

Hungarian Astronomical Association

P.O. Box 219
H-1461 Budapest
or :
Bartók Béla út. 11-13
H-1114 Budapest
Telephone: (0)1-1862313
Electronic Mail: mzs@mcse.hu
           tepi@mcse.hu
           sky@mcse.hu
WWW: http://www.mcse.hu/
Founded: 1946 (refounded: 1989)
Membership: 2200
Activities: organizing amateur activities * fostering collaborations between amateur and professional astronomers
Sections: History of Astronomy * Moon * Astronomical Computing * Meteors * Variable Stars
Periodicals: (12) "Meteor" (ISSN 0133-249x): (1) "Yearbook" (ISSN 0866-2851)
Coordinates: 017°47'00"E 47°12'27"N H502m (Ráktanya)
City Reference Coordinates: 019°05'00"E 47°30'00"N

Hungarian Meteorological Service

P.O. Box 38
H-1525 Budapest
or :
Kitaibel Pál út. 1
H-1024 Budapest
Telephone: (0)1-2122699
Telefax: (0)1-2125153
Electronic Mail: intrel@met.hu
           mets@.met.hu
WWW: http://www.met.hu/
      http://www.met.hu/index-e.html
Founded: 1870
Staff: 307
Activities: observing * telecommunications * weather forecast * aeronautical meteorology * climatology * agroclimatology * satellite meteorology * environmental protection
Periodicals: (4) "Idöjárás" (ISSN 0324-6329)
City Reference Coordinates: 019°05'00"E 47°30'00"N

Hungarian Space Office (HSO)

Szervita tér 8
H-1052 Budapest
Telephone: (0)1-3178717
Telefax: (0)1-2666728
Electronic Mail: elod.both@omfb.x400gw.itb.hu (Elöd Both, Director)
Periodicals: "Space Activities in Hungary" (ISSN 1217-7725)
City Reference Coordinates: 019°05'00"E 47°30'00"N

Hungarian Standards Institution

(Magyar Szabványügyi Testület - MSZT)
Üllöi út. 25
P.O. Box 24
H-1450 Budapest
Telephone: (0)1-2183011
Telefax: (0)1-2185125
WWW: http://www.mszt.hu/

Founded: 1921
Staff: 120
Activities: standards of development * training * certification * information centre
Periodicals: (12) "MSZT Bulletin"
City Reference Coordinates: 019°05'00"E 47°30'00"N

Information Bulletin on Variable Stars (IBVS)
c/o Konkoly Observatory
P.O. Box 67
H-1525 Budapest
Telephone: (0)1-3754122
Telefax: (0)1-2754668
Electronic Mail: ibvs@ogyalla.konkoly.hu
WWW: http://www.konkoly.hu/IBVS/IBVS.html
            http://www.konkoly.hu:80/IAUC27/
Founded: 1961
Staff: 5
● Journal (ISSN 0374-0676) of IAU Commissions 27 and 42
City Reference Coordinates: 019°05'00"E 47°30'00"N

Institute for Geodesy, Cartography and Remote Sensing, Satellite Geodetic Observatory (SGO)
P.O. Box 546
H-1373 Budapest
or :
Bosnyak tér 5
H-1149 Budapest
Telephone: (0)27-310980
Telefax: (0)27-310982
Electronic Mail: <userid>@sgo.fomi.hu
WWW: http://www.sgo.fomi.hu/
Founded: 1972
Staff: 15
Activities: satellite geodesy
Coordinates: 019°16'55"E 47°47'26"N H245m (Penc)
City Reference Coordinates: 019°05'00"E 47°30'00"N

József Attila University, Department of Physics, Astronomical Observatory
Dóm tér 9
H-6720 Szeged
Telephone: (0)62-311154
Telefax: (0)62-311154
Electronic Mail: <userid>@physx.u-szeged.hu
                    k.szatmary@physx.u-szeged.hu (Károly Szatmáry, Observatory Head)
WWW: http://www.jate.u-szeged.hu/jate/central/obs/
           http://www.jate.u-szeged.hu/obs/
Founded: 1992
Activities: variable stars * pulsating variable stars * binary stars * photoelectric photometry * automation of telescopes * education * astronomical software
Coordinates: 020°09'31"E 46°14'12"N H85m
City Reference Coordinates: 020°09'00"E 46°15'00"N

Kecskemét Planetarium
Lánchíd utca 18/A
H-6000 Kecskemét
Telephone: (0)76-478994
Electronic Mail: kecsplan@externet.hu
WWW: http://planetarium.silicondreams.hu/
Founded: 1983
Staff: 7
Activities: shows * education
City Reference Coordinates: 019°43'00"E 46°46'00"N

Kiskunhalas Popular Observatory
Kossuth út. 43
H-6400 Kiskunhalas
Telephone: (0)77-423355
Founded: 1972
Staff: 1
Membership: 31
Activities: education * solar-activity impact
Coordinates: 019°00'00"E 46°06'00"N H136m
City Reference Coordinates: 019°00'00"E 46°06'00"N

Konkoly Observatory
● See "Hungarian Academy of Sciences, Konkoly Observatory"

Lajos Terkán Public Observatory
Fürdö sor 3
H-8000 Székesfehérvár
Telephone: (0)22-314456
            (0)22-313028
            (0)22-311001
Electronic Mail: <userid>@mars.iif.hu
Founded: 1967
Membership: 50
Activities: popularization * education
Periodicals: "Telapo"
Coordinates: 018°24'10"E 47°09'19"N H129m
City Reference Coordinates: 018°22'00"E 47°11'00"N

Loránd University (Eötvös_)
• See "Eötvös Loránd University"

Scientific Educational Society of Nógrádmegye
Mérleg út. 2
P.O. Box 156
H-3100 Salgótarján
Telephone: (0)32-314282
            (0)32-311734
Founded: 1984
Membership: 2
Activities: observing (Sun, Moon, comets, deep sky) * photography
Coordinates: 019°44'04"E 48°07'58"N H356m
City Reference Coordinates: 019°48'00"E 48°07'00"N

Terkán Public Observatory (Lajos_)
• See "Lajos Terkán Public Observatory"

Urania Observatory
Sánc út. 3/B
H-1016 Budapest
Telephone: (0)1-1869171
            (0)1-1869233
Telefax: (0)1-2671391
Electronic Mail: kondor@ludens.elte.hu
Founded: 1947
Membership: 21
Activities: popularization * lectures * telescope making * polishing mirrors * optical accessories
City Reference Coordinates: 019°05'00"E 47°30'00"N

Iceland

Amateur Astronomical Society of Seltjarnarnes
c/o Gudni G. Sigurdsson
Valhusaskola
170 Seltjarnarnes
Telephone: 561-2424
Electronic Mail: aquila@ismennt.is
WWW: http://rvik.ismennt.is/~aquila/
Membership: 70
City Reference Coordinates: 022°01'00"W 64°09'00"N

Icelandic Astronomical Society
Dunhaga 3
IS-107 Reykjavík
Telephone: (0)5254800
Telefax: (0)5528911
Electronic Mail: gulli@raunvis.hi.is
Founded: 1988
Membership: 20
Activities: promoting astronomy and astrophysics in Iceland
City Reference Coordinates: 021°51'00"W 64°09'00"N

Icelandic Council for Standardization (STRÍ)
c/o Technological Institute of Iceland
Keldnaholt
IS-112 Reykjavík
Telephone: (0)5877000
Telefax: (0)5877409
Electronic Mail: stri@stri.is
WWW: http://www.stri.is/
Founded: 1987
Staff: 10
Activities: Icelandic standards * representation of Iceland in international and regional standardization bodies
Periodicals: "Stadlatídindi"
City Reference Coordinates: 021°51'00"W 64°09'00"N

Icelandic Meteorological Office
Bústadavegi 9
IS-150 Reykjavík
Telephone: (0)600600
Telefax: (0)28121
Electronic Mail: <userid>@vedur.is
WWW: http://www.vedur.is/
　　　　http://www.vedur.is/index_eng.html
Founded: 1920
Staff: 64
Activities: meteorological service (about 130 weather stations in Iceland)
Periodicals: (12) "Vedráttan" (ISSN 0258-3836)
Coordinates: 021°51'00"W 64°08'00"N H52m
City Reference Coordinates: 021°51'00"W 64°08'00"N

Icelandic Physical Society
c/o Sveinbjorn Bjornsson (President)
Science Institute
University of Iceland
Dunhaga 3
IS-107 Reykjavík
Telephone: (0)694942
Telefax: (0)28911
Electronic Mail: svb@os.is
WWW: http://www.os.is/ei/
Founded: 1977
Membership: 80
City Reference Coordinates: 021°51'00"W 64°09'00"N

Icelandic Research Council
(Rannsóknarrád Íslands)
Laugavegi 13
IS-101 Reykjavík
Telephone: (0)5621320
Telefax: (0)5529814
Electronic Mail: rannis@rannis.is

WWW: http://www.rannis.is/english/
Founded: 1994
Membership: 11
City Reference Coordinates: 021°51'00"W 64°09'00"N

University of Iceland, Science Institute
Dunhaga 3
IS-107 Reykjavík
Telephone: (0)5254800
Telefax: (0)5528911
Electronic Mail: <userid>@raunvis.hi.is
WWW: http://www.raunis.hi.is/
Founded: 1966
Activities: almanac * STP * high-energy astrophysics * cosmology * cosmic rays
Coordinates: 021°57'24"W 64°08'24"N
City Reference Coordinates: 021°51'00"W 64°09'00"N

India

Aligarh Muslim University (AMU), Physics Department, Astrophysics Group
Aligarh 202 002
Telephone: (0)571-401001
Electronic Mail: agphys@amu.ernet.in
Founded: 1969
Staff: 5
Activities: theoretical astrophysics * ISM * history of astronomy in India and Islamic countries
City Reference Coordinates: 078°05'00"E 27°54'00"N

Amateur Astronomers' Association (Bombay)
c/o Physics Department
Saint Xavier's College
Mahapalika Marg
Bombay 400 001
Telephone: (0)22-4306519 (c/o Aadil Desai)
Electronic Mail: aaa_bombay@hotmail.com
Founded: 1977
Membership: 400
Activities: monthly lectures * field trips * conferences * instrumentation * photography * film and slide shows * library * visits * observing * popularization * research * data collection and distribution
Periodicals: (6) "M 51" (circ.: 500)
Awards: (1) "R. Kundaji Trophy"
Coordinates: 072°46'00"E 19°00'00"N H35m
City Reference Coordinates: 072°50'00"E 19°11'00"N

Amateur Astronomers Association of Delhi (AAAD)
c/o Nehru Planetarium
Teen Murti House
New Delhi 110 011
Electronic Mail: aaadel@nebulacorp.com
WWW: http://www.nebulacorp.com/org/aaadel/
        http://www.nebulacorp.com/aaa/
City Reference Coordinates: 077°12'00"E 28°37'00"N

Astronomical Society of India (ASI)
c/o Dipankar Bhattacharya (Secretary)
Raman Research Institute
Sadashivanagar
Bangalore 560 080
Telephone: (0)80-3340122
Telefax: (0)80-3340492
Electronic Mail: dipankar@rri.ernet.in
WWW: http://www.rri.res.in/asi/
Founded: 1972
Membership: 500
Activities: holding scientific meetings * publishing * encouraging amateur astronomers
Periodicals: (4) "Bulletin of the Astronomical Society of India" (ISSN 0304-9523, circ.: 500)
Awards: "Vainu Bappu Memorial Award", "Young Astronomers Award"
City Reference Coordinates: 077°36'00"E 12°58'00"N (Bangalore)

Astronomy and Space Science Association (ASSA)
c/o Astrophysics Group
Physics Department
Gauhati University
Guwahati 781 014
Telephone: (0)361-570412
Telefax: (0)361-570133
Electronic Mail: vcgu@gulib.iitg.ernet.in
Founded: 1993
Membership: 320
Activities: observing * popularization * research
Periodicals: (12) "The ASSA Newsletter"
Coordinates: 091°45'00"E 26°10'00"N H75m
City Reference Coordinates: 091°45'00"E 26°10'00"N

Bappu Observatory (VBO) (Vainu_)
• See "Indian Institute of Astrophysics (IIA)"

Birla Planetarium (B.M._)
• See "B.M. Birla Planetarium, Chennai"
• See also "B.M. Birla Planetarium, Hyderabad"

● See also "M.P. Birla Planetarium, Calcutta"

B.M. Birla Planetarium, Chennai
c/o Tamil Nadu Science and Technology Centre
Gandhi Mandapam Road
Engineering College P.O.
Chennai 600 025
Telephone: (0)44-416751
         (0)44-4915250
Telefax: (0)44-4918787
Founded: 1988
Staff: 8
Activities: programmes for public and students * education * seminars * workshops * observing * outreach activities
City Reference Coordinates: 080°16'00"E 13°04'00"N

B.M. Birla Planetarium, Hyderabad
c/o Birla Archaeological, Cultural and Research Institute
Adarsh Nagar
Hyderabad 500 463
Telephone: (0)842-235081
         (0)842-241067
Telefax: (0)842-222483
WWW: http://www.andhratoday.com:80/tourism/planet.htm
Founded: 1985
Staff: 20
Activities: shows * camps * lectures * education * seminars * films
Awards: (1) "B.M. Birla Science Prize"
City Reference Coordinates: 078°29'00"E 17°23'00"N

Breakthrough Science Society (BSS)
9 Creek Row
Calcutta 700 014
Telephone: (0)33-2460563
Telefax: (0)33-2465114
Electronic Mail: breakthrough@ieee.org
WWW: http://tnp.saha.ernet.in/~basak/bss.html
Founded: 1995
Membership: 500
Activities: amateur astronomy * popularization * pollution * promoting scientific approach
Periodicals: (4) "Breakthrough: A Journal on Science and Society", "Prakriti: Vigyan O Samaj Vishayak Patrika"
City Reference Coordinates: 088°20'00"E 22°34'00"N

Bureau of Indian Standards (BIS)
Manak Bhavan
9 Bahadur Shah Zafar Marg
New Delhi 110 002
Telephone: (0)11-3311375
Telefax: (0)11-3314062
Electronic Mail: bisind@de12.vsn1.net.in
Founded: 1947
Staff: 2380
Activities: formulating standards * certifications
Periodicals: (12) "Standard India" (ISSN 0970-2628), "Standards Monthly Additions" (ISSN 0970-3985); (1) "Annual Report"; "Manakdoot", "Standards Worldover", "Current Published Information on Standardization", "EEC Norm Scan"
City Reference Coordinates: 077°15'00"E 28°36'00"N

Committee on Science and Technology in Developing Countries (COSTED)
24 Gandhi Mandap Road
Chennai 600 025
Membership: 20
Activities: promoting science and technology in developing countries
Periodicals: (3) "COSTED Newsletter"
City Reference Coordinates: 080°16'00"E 13°04'00"N

Current Science Association
C.V. Raman Avenue
Sadashivanagar
P.O. Box 8001
Bangalore 560 080
Telephone: (0)80-3342310
Telefax: (0)80-3346094
Electronic Mail: currsci@ias.ernet.in
Founded: 1932
Staff: 7
Periodicals: (24) "Current Science" (ISSN 0011-3891, circ.: 3,200) (in collaboration with the Indian Academy of Sciences)

City Reference Coordinates: 077°36'00"E 12°58'00"N

C.Z. Instruments India Pvt. Ltd.
• See now "Gordhandas Desai Pvt. Ltd."

Federation of Asian Scientific Academies and Societies (FASAS)
c/o Indian Academy of Sciences
C.V. Raman Avenue
P.O. Box 8005
Sadashivanagar
Bangalore 560 080
Telephone: (0)80-3342546
(0)80-3344592
Telefax: (0)80-3346094
Electronic Mail: office@ias.ernet.in
Founded: 1934
City Reference Coordinates: 077°36'00"E 12°58'00"N

Gauhati University, Physics Department, Astrophysics Group
Guwahati 781 014
Telephone: (0)361-570531
Telefax: (0)361-570133
Electronic Mail: <userid>@iucaa.ernet.in
Founded: 1962
Activities: theoretical nuclear astrophysics * neutron stars * gravitational collapse * black holes * early universe * variable stars
City Reference Coordinates: 091°45'00"E 26°10'00"N

Goodwill Cryogenics Enterprises
213 Nirman Vyapar Kendra
Sector 17
Vashi
New Bombay 400 703
Telephone: (0)22-7890642
Telefax: (0)22-7822769
Electronic Mail: vinochop@bomb2.vsnl.net.in
Founded: 1979
Activities: cryogenics vacuum and refrigeration * temperature and magnetic-field measurement instrumentation
City Reference Coordinates: 072°50'00"E 19°11'00"N

Gordhandas Desai Pvt. Ltd.
(C.Z. Instruments India Pvt. Ltd.)
P.O. Box 11108
35-A New Marine Lines
Bombay 400 020
Telephone: (0)22-2004208
Telefax: (0)22-2000581
Electronic Mail: gdpl@lwbbs.net
WWW: http://www.zeiss.de/
Founded: 1924
Staff: 51
Activities: supplying, erecting and maintaining planetariums, telescopes and related equipment * Carl Zeiss representative
City Reference Coordinates: 072°50'00"E 19°11'00"N

GOTO Optical (India) Pvt. Ltd.
53 Syed Amir Ali Avenue
3 CD Shivam Chambers
Calcutta 700 019
Telephone: (0)33-2478656
(0)33-2478313
(0)33-2473120
Telefax: (0)33-2477844
Electronic Mail: <userid>@goto.co.jp
WWW: http://www.goto.co.jp/
Founded: 1979 (reorganised: 1989)
Staff: 50
Activities: representative of GOTO Optical Mfg. Co., Japan (see separate entry) * producing planetarium programmes
City Reference Coordinates: 088°20'00"E 22°34'00"N

India Meteorological Department (IMD), New Delhi
Mausam Bhavan
Lodi Road
New Delhi 110 003
Telephone: (0)11-618242
(0)11-616602

Telefax: (0)11-699216
WWW: http://www.nic.in/snt/c9imd.htm
Founded: 1875
Staff: 8400
Activities: weather forecasting * aviation weather service * hydrology * seismology
Periodicals: "MAUSAM"
City Reference Coordinates: 077°15'00"E 28°36'00"N

India Meteorological Department (IMD), Weather Forecasting
Observatory
Shivajinagar
Pune 411 005
Telephone: (0)212-325211
          (0)212-323403 x510
Telefax: (0)212-323201
WWW: http://www.nic.in/snt/c9imd.htm
Founded: 1875
Staff: 250
Activities: weather forecasting * training * R&D
Periodicals: "All India Weather Summary", "India Daily Weather Report", "India Weekly Weather Report", "Atlas of Tracks
of Storms and Depressions in the Bay of Bengal and Arabian Sea"
City Reference Coordinates: 073°54'00"E 18°31'00"N

Indian Academy of Sciences (IASc)
C.V. Raman Avenue
P.O. Box 8005
Sadashivanagar
Bangalore 560 080
Telephone: (0)80-3342546
          (0)80-3344592
Telefax: (0)80-3346094
Electronic Mail: office@ias.ernet.in
Founded: 1934
Staff: 20
Activities: publishing scientific journals * organising scientific meetings
Periodicals: (24) "Current Science" (ISSN 0011-3891, circ.: 4,500) (in collaboration with the Current Science Association);
(12) "Pramana - Journal of Physics" (ISSN 0304-4289, circ.: 1,000); (6) "Proceedings (Chemical Sciences)" (ISSN 0253-4134,
circ.: 900); (5) "Bulletin of Materials Science" (ISSN 0250-6327, circ.: 1,800); (4) "Journal of Astrophysics and Astronomy"
(ISSN 0250-6335, circ.: 900), "Sadhana (Engineering Science)" (ISSN 0256-2499, circ.: 700); (3) "Proceedings (Earth and
Planetary Sciences)" (ISSN 0253-4126, circ.: 900), "Proceedings (Mathematical Sciences)" (ISSN 0253-4142, circ.: 1,000));
"Resonance" (ISSN 0971-8044, circ.: 5,000); other journals in the field of life sciences
City Reference Coordinates: 077°36'00"E 12°58'00"N

Indian Institute of Astrophysics (IIA)
Koramangala
Bangalore 560 034
Telephone: (0)80-5530672
          (0)80-5530673
          (0)80-5530674
          (0)80-5530675
          (0)80-5530676
Telefax: (0)80-5534043
Electronic Mail: <userid>@iiap.ernet.in
WWW: http://www.iiap.ernet.in/~website/home_page.html
       http://www.iiap.ernet.in/~website/vbo/vbo.html (Vainu Bappu Observatory)
Founded: 1786 (Kodaikanal Observatory)
Staff: 337
Activities: solar chromospheres * convection and magnetic fields * plasma processes in the corona * solar cycles and related
topics * solar system, comets, minor planets * stellar atmospheres * abundances * Ap stars * binary stars * SN * photometry
of globular and galactic clusters * photometry of galaxies * galactic dynamics * pulsars * neutron stars * cosmology * ISM *
radioastronomy (decametric wavelengths) * flux distributions
Periodicals: (4) "IIA Newsletter"; "Kodaikanal Observatory Bulletin"
Coordinates: 077°36'00"E 12°58'00"N H921m
            078°49'54"E 12°24'32"N H725m (Vainu Bappu Observatory - VBO, Kavalur)
            077°28'00"E 10°03'08"N H2,343m (Kodaikanal Observatory)
            077°26'07"E 13°36'12"N H687m (Radiotelescope Station, Gauribidanur)
City Reference Coordinates: 077°36'00"E 12°58'00"N

Indian Institute of Science (IISc), Joint Astronomy Programme
Karnataka
Bangalore 560 012
Telephone: (0)812-344411
Telefax: (0)812-341683
WWW: http://www.iisc.ernet.in/
Founded: 1909 (IISc)

Activities: graduate programme in astronomy in collaboration with the other astronomical institutions in India * theoretical astrophysics * galactic dynamics * formation and evolution of galaxies * solar physics
Periodicals: "Journal"
City Reference Coordinates: 077°36'00"E 12°58'00"N

Indian Skeptics
c/o B. Premanand (Convener)
11/7 Chettipalayam Road
Podanur 641 023
Telephone: (0)422-872423
Founded: 1976
Activities: parascience debunking * education
Periodicals: (12) "Indian Skeptics"
City Reference Coordinates: 077°00'00"E 10°57'00"N

Indian Space Research Organization (ISRO)
Antariksh Bhavan
New BEL Road
Bangalore 560 094
Telephone: (0)80-3415474
         (0)80-3425275
Telefax: (0)80-3412253
Electronic Mail: <userid>@isro.ernet.in
WWW: http://www.isro.org/
Founded: 1972
Staff: 16700
Activities: providing operational space services in telecommunications, TV broadcast, radio networking, meteorology and remote sensing for natural resources survey and management * developing satellites, launch vehicles and associated ground systems
Periodicals: (4) "Space India"; (1) "Annual Report"
Awards: "Vikram Sarabhai Award"
City Reference Coordinates: 077°36'00"E 12°58'00"N

Indian Space Research Organization (ISRO), Satellite Centre (ISAC)
Airport Road
Vimanapura Post
Bangalore 560 017
Telephone: (0)80-5266251
Telefax: (0)80-5265407
Electronic Mail: <userid>@isac.ernet.in
WWW: http://www.isro.org/cen_isac.htm
Founded: 1972
Staff: 800
Activities: spacecraft design and development * space astronomy
Periodicals: (2) "Journal of Spacecraft Technology" (ISSN 0971-1600)
City Reference Coordinates: 077°36'00"E 12°58'00"N

Inter-University Centre for Astronomy and Astrophysics (IUCAA)
Post Bag 4
Ganeshkhind
Pune 411 007
or :
Meghnad Saha Road
Pune University Campus
Ganeshkhind
Pune 411 007
Telephone: (0)20-565414
Telefax: (0)20-565760
Electronic Mail: root@iucaa.ernet.in
         <userid>@iucaa.ernet.in
WWW: http://www.iucaa.ernet.in/
Founded: 1988
Staff: 100
Activities: fundamental research * Sun * planets * galactic and extragalactic astronomy * high-energy astrophysics * theoretical cosmology * gravitation and relativity * quantum gravity * quantum cosmology * education * observing * popularization
Periodicals: (4) "Khagol: The IUCAA Bulletin"
Coordinates: 073°58'00"E 18°34'00"N
City Reference Coordinates: 073°54'00"E 18°31'00"N

Mount Abu Infrared Observatory
● See "Physical Research Laboratory (PRL), Mount Abu Infrared Observatory"

M.P. Birla Planetarium, Calcutta
96 J. L. Nehru Road
Calcutta 700 071

Telephone: (0)33-2481515
(0)33-2486610
WWW: http://host.westbengal.com/misc/brlaplnt/
Founded: 1962
Periodicals: "The Journal of Birla Planetarium"
City Reference Coordinates: 088°20'00"E 22°34'00"N

National Centre for Radio Astrophysics (NCRA), Bangalore
P.O. Box 1234
I.I.Sc. Campus
Bangalore 560 012
Telephone: (0)812-363138
(0)812-364062
(0)812-362816
Electronic Mail: root@gmrt.ernet.in
WWW: http://www.ncra.tifr.res.in/
Activities: radio astronomy
Coordinates: 074°03'00"E 19°05'30"N H650m (Junnar Taluk)
City Reference Coordinates: 077°36'00"E 12°58'00"N

National Centre for Radio Astrophysics (NCRA), Junnar Taluk
Khodad Village
Narayangaon
Junnar Taluk 410 504
Electronic Mail: root@gmrt.ernet.in
WWW: http://www.ncra.tifr.res.in/
Activities: radio astronomy
Coordinates: 074°03'00"E 19°05'30"N H650m (Junnar Taluk)
City Reference Coordinates: 073°58'00"E 19°15'00"N (Junnar)

National Centre for Radio Astrophysics (NCRA), Ootacamund
P.O. Box 8
Ootacamund 643 001
Telephone: (0)423-4049 (Offices)
(0)423-2032 (Offices)
(0)423-2065 (Residences)
(0)423-2588 (Residences)
(0)423-2621 (Residences)
Electronic Mail: root@gmrt.ernet.in
WWW: http://www.ncra.tifr.res.in/
Founded: 1970
Staff: 15
Activities: observing at 327 MHZ * aperture synthesis * radio source structure * observational cosmology * pulsars * IPS * ISM * VLBI * SNR * galaxy clusters * galactic plane survey * protocluster search * image processing
Coordinates: 076°40'02"E 11°22'56"N H2,150m (Ooty Radio Telescope)
City Reference Coordinates: 076°43'00"E 11°25'00"N

National Centre for Radio Astrophysics (NCRA), Pune
c/o Pune University Campus
Post Bag 3
Ganeshkhind
Pune 411 007
Telephone: (0)212-337107
Telefax: (0)212-335149
Electronic Mail: root@gmrt.ernet.in
WWW: http://www.ncra.tifr.res.in/
Founded: 1990
Staff: 250
Activities: radio astronomy * cosmology * pulsars * interplanetary scintillations * aperture synthesis * meterwavelength observations (50-1500 MHz)
Coordinates: 074°03'00"E 19°06'00"N H650m (Giant Meterwave Radio Telescope)
City Reference Coordinates: 073°54'00"E 18°31'00"N

National Physical Laboratory (NPL), Radio and Atmospheric Sciences Division (RASD)
Dr. K.S. Krishnan Road
New Delhi 110 012
Telephone: (0)11-5787657
Telefax: (0)11-5752678
(0)11-5764189
Electronic Mail: mahajan@csnpl.ren.nic.in (K.K. Mahajan, Director)
Founded: 1947
Staff: 80
Activities: planetary atmospheres * aeronomy * tropospheric and ionospheric communications * middle atmosphere * solar-terrestrial physics * upper atmosphere * ionosphere * global change
City Reference Coordinates: 077°15'00"E 28°36'00"N

National Remote Sensing Agency (NRSA)
Balanagar
Hyderabad 500 037
Telephone: (0)40-279572
           (0)40-279573
           (0)40-279574
           (0)40-279575
           (0)40-279576
           (0)40-278360
Telefax: (0)40-278648
Electronic Mail: info@nrsa.gov.in
WWW: http://www.nrsa.gov.in/
Founded: 1975
Staff: 1025
Activities: satellite and aerial data acquisition * data processing and dissemination * remote sensing * R&D
Periodicals: (4) "Interface"
City Reference Coordinates: 078°29'00"E 17°23'00"N

Nehru Planetarium
Teen Murti House
New Delhi 110 011
City Reference Coordinates: 077°12'00"E 28°37'00"N

Optical Society of India
c/o Department of Applied Physics
University of Calcutta
92 Acharya Prafulla Chandra Road
Calcutta 700 009
Founded: 1965
Staff: 4
Membership: 350
Activities: promotion of optics in India * publishing * symposia
Periodicals: "Journal of Optics"
City Reference Coordinates: 088°20'00"E 22°34'00"N

Osmania University, Centre of Advanced Study in Astronomy (CASA)
Hyderabad 500 007
Telephone: (0)7018951 x247
Founded: 1908
Staff: 13
Activities: education * galactic dynamics * photometry & spectroscopy of variable and binary stars * radio observing (Sun)
Periodicals: "Contributions from the Nizamiah and Japal-Rangapur Observatories"
Coordinates: 078°27'12"E 17°25'59"N H554m (Nizamiah Observatory)
               078°43'42"E 17°05'54"N H695m (Japal-Rangapur Observatory)
City Reference Coordinates: 078°29'00"E 17°23'00"N

Physical Research Laboratory (PRL), Mount Abu Infrared Observatory, Headquarters
Navrangpura
Ahmedabad 380 009
Telephone: (0)79-462129
          (0)79-6425037
Telefax: (0)79-6560502
Electronic Mail: <userid>@prl.ernet.in
WWW: http://www.prl.ernet.in/astronomy/
       http://www.prl.ernet.in/astronomy/irtel.shtml
Founded: 1947
Staff: 85
Activities: AGNs * starburst galaxies * PNs * stellar physics * novae * star formation * HII regions * lunar occultations * comets * solar eclipses
Coordinates: 072°46'47"E 24°39'09"N H1,680m
City Reference Coordinates: 072°35'00"E 23°02'00"N

Physical Research Laboratory (PRL), Udaipur Solar Observatory (USO)
P.O. Box 198
Bari Road
Dewali
Udaipur 313 001
Telephone: (0)294-560626
Telefax: (0)294-526325
Electronic Mail: root@uso.ernet.in
WWW: http://www.prl.ernet.in/~shibu/uso.html
       http://www.prl.ernet.in/
Founded: 1975
Staff: 18
Activities: solar astronomy
Coordinates: 073°42'45"E 24°35'08"N H301m

City Reference Coordinates: 073°47'00"E 24°36'00"N

Punjabi University, Department of Astronomy and Space Sciences
Patiala 147 002
Telephone: (0)175-822161 x96
Founded: 1978
Staff: 16
Activities: stellar populations * space physics * Be stars * HII regions * ISM reddening
Coordinates: 076°45'00"E 30°45'00"N H253m
City Reference Coordinates: 076°45'00"E 30°45'00"N

Raman Research Institute (RRI)
C.V. Raman Avenue
Sadashivanagar
Bangalore 560 080
Telephone: (0)812-340122
         (0)80-3340122
Telefax: (0)812-340492
      (0)80-3340492
Electronic Mail: <userid>@rri.ernet.in
WWW: http://www.rri.res.in/
Founded: 1948
Activities: theoretical astrophysics * decametric- and mm-wavelength radioastronomy
Coordinates: 077°35'00"E 13°00'45"N (mm, Bangalore)
         077°26'07"E 13°36'12"N H500m (dkm, Gauribidanur)
City Reference Coordinates: 077°36'00"E 12°58'00"N

Tata Institute of Fundamental Research (TIFR)
Homi Bhabha Road
Colaba
Bombay 400 005
Telephone: (0)22-2152971
         (0)22-2152311
Telefax: (0)22-2152110
Electronic Mail: <userid>@tifr.res.in
WWW: http://www.tifr.res.in/
Founded: 1945
Staff: 700
Activities: fundamental research in mathematics * theoretical and experimental physics and biology, including astronomy, nuclear and atomic physics, high-energy physics, cosmic-ray physics, condensed-matter physics, chemical physics, solid-state electronics and computer science
Periodicals: (1) "Annual Report"
Coordinates: 076°40'02"E 11°22'50"N H2,150m (RAC, Ooty - see separate entry)
         074°03'00"E 19°06'00"N H655m (GMRT, Pune - see separate entry)
City Reference Coordinates: 072°50'00"E 19°11'00"N

Udaipur Solar Observatory (USO)
● See "Physical Research Laboratory (PRL), Udaipur Solar Observatory (USO)"

University of Delhi, Department of Physics and Astrophysics
Delhi 110 007
Telephone: (0)11-7257793
Founded: 1922
Staff: 50
Activities: education * research
City Reference Coordinates: 077°12'00"E 28°37'00"N

Uttar Pradesh State Observatory (UPSO)
Manora Peak
Naini Tal 263 169
Telephone: (0)5942-35136
         (0)5942-35583
         (0)5942-35053
Electronic Mail: <userid>@upso.ernet.in
Founded: 1954
Staff: 41
Activities: photometry of variable stars * spectrophotometry * open clusters * eclipsing binaries * occultations * comets * Sun * stellar atmospheres * Be stars * site survey
Coordinates: 079°27'24"E 29°21'36"N H1,950m
City Reference Coordinates: 079°26'00"E 29°22'00"N

Vainu Bappu Observatory (VBO)
● See "Indian Institute of Astrophysics (IIA)"

Vikram Sarabhai Space Centre (VSSC)
Trivandrum 695 022
Telephone: (0)471-562444
(0)471-562555
Telefax: (0)471-79795
Electronic Mail: root%vssct@sirnetm.ernet.in
Founded: 1963
Staff: 5600
Activities: launch vehicle technology development
City Reference Coordinates: 076°57'00"E 08°28'00"N

Indonesia

A. Heck

Bandung Institute of Technology, Bosscha Observatory
Lembang 40391
Telephone: (0)22-2786001
                   (0)22-2786027
Telefax: (0)22-2786001
                (0)22-2787289
Electronic Mail: bambang@as.itb.ac.id (Bambang Hidayat, Director)
WWW: http://www.itb.ac.id/
Founded: 1923
Staff: 11
Activities: binary stars * galactic structure * photometry * variable stars
Periodicals: "Contributions", "Publications", "Annals"
Coordinates: 107°37'00"E 06°49'30"S H1,400m
City Reference Coordinates: 107°37'00"E 06°50'00"S

Bandung Institute of Technology, Department of Astronomy
Jalan Ganesha 10
Bandung 40132
Telephone: (0)22-2501645 x780
                   (0)22-2501645 x781
Telefax: (0)22-2505442
Electronic Mail: <userid>@as.itb.ac.id
WWW: http://www.itb.ac.id/
Founded: 1956
Staff: 18
Activities: galactic structure * cosmology * stellar physics * solar systems
Observatories: 1 (Bosscha Observatory - see separate entry)
City Reference Coordinates: 107°36'00"E 06°54'00"S

Bandung Institute of Technology, Department of Geophysics and Meteorology
Jalan Ganesha 10
Bandung 40132
Telephone: (0)22-438078 x778
                   (0)22-438078 x779
Telefax: (0)22-438338
WWW: http://www.itb.ac.id/
Founded: 1961
Staff: 25
Activities: education * research
City Reference Coordinates: 107°36'00"E 06°54'00"S

Bosscha Observatory
● See "Bandung Institute of Technology, Bosscha Observatory"

Indonesian Institute of Sciences (LIPI)
Gedung Widya Graha
Jalan Jend. Gatot Subroto 10
Selatan 12710
Jakarta
Telephone: (0)22-5225711
Telefax: (0)22-5207226
WWW: http://www.lipi.go.id/
Founded: 1967
Staff: 4750
Activities: research * development
City Reference Coordinates: 106°48'00"E 06°10'00"S

Meteorological and Geophysical Agency
(Badan Meteorologi dan Geofisika - BMG)
Jalan Arief Rakhman Hakim 3
Tromolpos 3540
Jakarta
Telephone: (0)21-3906482
Telefax: (0)22-85163
Electronic Mail: bmg@cbn.net.id
WWW: http://www.cbn.net.id/commerce/bmg/
Founded: 1980
City Reference Coordinates: 106°48'00"E 06°10'00"S

National Institute of Aeronautics and Space (LAPAN)
Jalan Pemuda Persil 1

Jakarta 13220
Telephone: (0)21-4892802
          (0)21-4894941
Telefax: (0)21-4894815
WWW: http://www.lapan.go.id/
Founded: 1963
Staff: 1350
Activities: remote sensing * space technology * atmosphere * ionosphere * Sun
Periodicals: (4) "Majalah Lapan" (ISSN 0126-0480), "Warta Lapan" (ISSN 0216-9754)
Coordinates: 107°10'00"E 07°40'00"S (Pameungpeuk)
          107°47'00"E 06°54'00"S H760m (Tanjung Sari)
          113°30'00"E 06°55'00"S (Watukosek)
          135°47'00"E 01°15'00"S (Biak)
City Reference Coordinates: 106°48'00"E 06°10'00"S

Standardization Council of Indonesia
(Dewan Standardisasi Nasional - DSN)
Jalan Jenderal Gatot Subroto 10
P.O. Box 3123
Jakarta 12710
Telephone: (0)21-5221686
Telefax: (0)21-5206574
Electronic Mail: pustan@rad.net.id
Founded: 1984
Staff: 55
Activities: services in standardization
Periodicals: "Warta Standardisasi"
City Reference Coordinates: 106°48'00"E 06°10'00"S

Iran

Biruni Observatory
● See "Shiraz University, Physics Department, Biruni Observatory"

Institute of Standards and Industrial Research of Iran (ISIRI)
(Headquarters)
P.O. Box 31585-163
Karaj
or :
(Central Office)
14 Shahamati Alley
Veli-e-Asr Avenue
Tehran 15946
Telephone: (0)261-227045 (Headquarters)
            (0)261-226031 (Public and International Relations)
            (0)21-899308 (Central Office)
Telefax: (0)261-225015 (Headquarters)
          (0)21-8802276 (Central Office)
Founded: 1960
Staff: 1125
City Reference Coordinates: 050°59'00"E 35°48'00"N (Karaj)
                              051°26'00"E 35°40'00"N (Tehran)

Islamic Republic of Iran Meteorological Organization (IRIMO)
P.O. Box 13185-461
Mehrabad Airport
Tehran
Telephone: (0)21-6004026/8
Telefax: (0)21-6000417
          (0)21-6469044
Founded: 1955
Staff: 2200
Activities: meteorology * agrometeorology * climatology * statistics
Periodicals: (52) "Bulletin of Meteorological Data"; (4) "Journal of NIVAR"
Observatories: 154 synoptic stations
City Reference Coordinates: 051°26'00"E 35°40'00"N

Shiraz University, Physics Department, Biruni Observatory
Shiraz 71454
Telephone: (0)71-24609
Telefax: (0)71-20027
WWW: http://www.physics.susc.ac.ir/Biruni.html
Founded: 1975
Activities: photoelectric photometry of variable stars * theoretical study of stellar systems * cosmology
Coordinates: 052°30'00"E 29°35'00"N H1,600m
City Reference Coordinates: 052°34'00"E 29°38'00"N

Tabriz University, Center for Astronomical Research
Tabriz 51664
Telephone: (0)41-342564
Founded: 1973
Activities: education * observing * solar physics * photometry of double stars * planetarium * Khadjeh Nassir Addin Toussi Observatory
Coordinates: 046°19'57"E 37°52'10"N (Gazan Mountains)
City Reference Coordinates: 046°18'00"E 38°05'00"N

Zarvan Co. Ltd.
P.O. Box 15875-1487
Tehran
or :
4 Seventh Alley
Balbochestan Street
Nasr Avenue
14469 Tehran
Telephone: (0)21-8271363
Telefax: (0)21-8272507
Electronic Mail: nojum@apadana.com
Founded: 1991
Staff: 18
Activities: publishing * popularization * manufacturing domes * telescope dealer
Periodicals: "Nojum Magazine" (ISSN 1019-584x)
City Reference Coordinates: 051°26'00"E 35°40'00"N

Iraq

Central Organization for Standardization and Quality Control (COSQC)
c/o Ministry of Planning
P.O. Box 13032
Aljadiria
Baghdad
Telephone: (0)1-7765180
Telefax: (0)1-7765781
City Reference Coordinates: 044°25'00"E 33°21'00"N

Iraqi Meteorological Organization
Almansoor
P.O. Box 6078
Baghdad
Telephone: (0)1-5560070
City Reference Coordinates: 044°25'00"E 33°21'00"N

Scientific Research Council, Space and Astronomy Research Center (SARC)
P.O. Box 2441
Jadiriyah
Baghdad
Telephone: (0)1-7756127
Founded: 1980
Activities: galaxies * ISM * variable stars * solar system
Periodicals: "Journal of Space and Astronomy Research"
Coordinates: 044°24'00"E 33°48'00"N H100m
City Reference Coordinates: 044°24'00"E 33°48'00"N

Ireland

Astronomy Ireland
P.O. Box 2888
Dublin 1
Telephone: (0)1-4598883
Telefax: (0)1-4599933
Electronic Mail: ai@iol.ie
WWW: http://www.astronomy.ie/
Founded: 1990
Membership: 4
Activities: meetings
Periodicals: (6) "Astronomy & Space Magazine" (ISSN 0781-8062, circ.: 5,000)
City Reference Coordinates: 006°15'00"W 53°20'00"N

Birr Castle
Demesne
Birr, Co. Offaly
Telephone: (0)509-20056
         (0)509-20336
Telefax: (0)509-21583
Electronic Mail: info@birrcastle.com
WWW: http://www.birrcastle.com/
Founded: 1984 (Birr Scientific & Heritage Foundation)
Staff: 1
Activities: Ireland's Historic Science Centre * museum * exhibitions * displays * lectures
Periodicals: "Focus"
City Reference Coordinates: 007°54'00"W 53°05'00"N

Cork Astronomy Club
c/o Charles Coughlan (Secretary)
12 Forest Ridge Crescent
Wilton, Cork
Telephone: (0)21-543669
Electronic Mail: chas@indigo.ie
WWW: http://indigo.ie/~chas/astronomy.html
City Reference Coordinates: 008°28'00"W 51°54'00"N

Cork Institute of Technology, Department of Applied Physics and Instrumentation
Rossa Avenue
Bishopstown, Co. Cork
Telephone: (0)21-326100
Telefax: (0)21-545343
WWW: http://www.rtc-cork.ie/faculty/science/Physics/PHYMAI.html
City Reference Coordinates: 008°28'00"W 51°54'00"N (Cork)

Dublin Institute for Advanced Studies (DIAS), School of Cosmic Physics, Astronomy Section
c/o Dunsink Observatory
Castleknock
Dublin 15
Telephone: (0)1-387911
         (0)1-387959
Telefax: (0)1-387090
Electronic Mail: iae@starlink.tl.ac.uk
           rlvad::iae.starlink
           astro@dunsink.diea.ie
WWW: http://www.dunsink.dias.ie/
      http://www.dias.ie/astro/
Founded: 1947
Activities: stellar variability (cepheids, symbiotic stars) * solar system dynamics (comets, minor planets) * interpretation of QSO spectra
Periodicals: "Contributions of Dunsink Observatory"; "Dunsink Observatory Reprints"
Coordinates: 006°20'18"W 53°23'13"N H86m
City Reference Coordinates: 006°15'00"W 53°20'00"N

Dublin Institute for Advanced Studies (DIAS), School of Cosmic Physics, Cosmic-Ray Section
5 Merrion Square
Dublin 2
Telephone: (0)1-6621333
Telefax: (0)1-6621477
Electronic Mail: <userid>@cp.dias.ie
WWW: http://www.cp.dias.ie/
      http://www.dias.ie/astro/

Founded: 1947
Staff: 11
Activities: cosmic-ray abundances * cosmic-ray acceleration * IR * star formation
Periodicals: "Reprints"
Coordinates: 006°25'00"W 53°24'00"N
City Reference Coordinates: 006°25'00"W 53°24'00"N

Dublin Institute for Advanced Studies (DIAS), School of Theoretical Physics
10 Burlington Road
Dublin 4
Telephone: (0)1-6140100
Telefax: (0)1-6680561
Electronic Mail: physics@stp.dias.ie
WWW: http://www.stp.dias.ie/
       http://www.cp.dias.ie/theory.html
Founded: 1940
Staff: 6
Activities: theoretical physics * applied statistics
Periodicals: "Communications"
City Reference Coordinates: 006°15'00"W 53°20'00"N

Dunsink Observatory
• See "Dublin Institute for Advanced Studies (DIAS), School of Cosmic Physics, Astronomy Section"

EG&G Optoelectronics
Bay TS3
Shannon Free Zone, Co. Clare
Telephone: (0)61-472577
Telefax: (0)61-472390
Electronic Mail: eod@egginc.com
WWW: http://www.egginc.com/optogrp
Founded: 1995
Staff: 7
Activities: European sales and support centre for high-energy, high-voltage components
Periodicals: (1) "Annual Report"
City Reference Coordinates: 007°54'00"W 53°05'00"N

Electronic Space Systems Corp. (ESSCO) - Collins
• See now "L-3 Communications ESSCO Collins Ltd."

Farran Technology Ltd. (FTL)
Ballincollig
Cork
Telephone: (0)21-872814
Telefax: (0)21-873892
Electronic Mail: farran@iol.ie
WWW: http://ireland.iol.ie/farran/
Founded: 1984
Staff: 19
Activities: manufacturing mm waveguide components, mm-wavelength sources, Schottky barrier mixer and varactor diodes for mm/sub-mm wavelengths, mm/sub-mm radiometer systems, and quasi-optical 300-3000 GHz components
City Reference Coordinates: 008°28'00"W 51°54'00"N

FORBAIRT
(Irish Science and Technology Agency)
Glasnevin
Dublin 9
Telephone: (0)1-8370101
          (0)1-8082000
Telefax: (0)1-8370172
         (0)1-8082020
Electronic Mail: <userid>@forbairt.ie
WWW: http://www.forbairt.ie/
Founded: 1994
Staff: 480
Activities: state agency responsible for the development, application, and promotion of science and technology in Ireland * coordinating Ireland's industrial and scientific involvment in the European Space Agency (ESA) space programmes
Periodicals: "Technology Ireland" (ISSN 0040-1676), "S+T News" (ISSN 0079-2735)
Awards: (1) research grants
City Reference Coordinates: 006°15'00"W 53°20'00"N

Irish Astronomical Society (IAS)
P.O. Box 2547
Dublin 14
Telephone: (0)1-2981268 (John O'Neill, Secretary)

(0)1-2980181 (James O'Connor, Secretary)
Electronic Mail: jgoneill@indigo.ie
WWW: http://indigo.ie/~stepryan/ias.htm
Founded: 1937
Membership: 120
Activities: lectures * observing * exhibitions * astronomy weekends * stars parties * beginners' classes * telescope making classes * visits * sidewalk astronomy * relativity classes
Periodicals: (6) "Orbit" (circ.: 580); (1) "Sky-High"
Coordinates: 006°13'00"W 53°14'00"N H300m
City Reference Coordinates: 006°15'00"W 53°20'00"N

Irish Meteorological Service
● See "Met Éireann"

Irish National Committee for Astronomy and Space Research
c/o Royal Irish Academy
Academy House
19 Dawson Street
Dublin 2
Telephone: (0)1-6762570
Telefax: (0)1-6762346
Electronic Mail: admin@ria.ie
WWW: http://www.ria.ie/
Founded: 1969
Membership: 17
Activities: international collaboration * promotion of astronomy and space science within Ireland
City Reference Coordinates: 006°15'00"W 53°20'00"N

Irish National Committee for Physics
c/o Royal Irish Academy
Academy House
19 Dawson Street
Dublin 2
Telephone: (0)1-6762570
Telefax: (0)1-6762346
Electronic Mail: admin@ria.ie
WWW: http://www.ria.ie/
Founded: 1963
Membership: 17
Activities: international collaboration * promotion of physics within Ireland
City Reference Coordinates: 006°15'00"W 53°20'00"N

Irish Science and Technology Agency
● See "FORBAIRT"

L-3 Communications ESSCO Collins Ltd.
Kilkishen, Co. Clare
Telephone: (0)61-367244
Telefax: (0)61-311044
Electronic Mail: moconn@indigo.ie
WWW: http://www.esscoradomes.com/
Founded: 1976
Staff: 52
Activities: designing, manufacturing and installing radiotelescopes, antennae and radomes
City Reference Coordinates: 007°54'00"W 53°05'00"N

Met Éireann
(Irish Meteorological Service)
Glasnevin Hill
Dublin 9
Telephone: (0)1-8064234
Telefax: (0)1-8064247
Electronic Mail: meteireann@met.ie
Founded: 1936
Membership: 100
Activities: weather forecast * climatic data * geophysical data
Periodicals: "Met Éireann Technical Note Series"
Observatories: anout 14 stations throughout the country
Coordinates: 006°18'00"W 53°23'00"N
City Reference Coordinates: 006°15'00"W 53°20'00"N

National Standards Authority of Ireland (NSAI)
Glasnevin
Dublin 9
Telephone: (0)1-8073800

Telefax: (0)1-8073838
Electronic Mail: nsai@nsai.ie
WWW: http://www.nsai.ie/
Founded: 1996
Staff: 130
Periodicals: "Standards Bulletin"
City Reference Coordinates: 006°15'00"W 53°20'00"N

National University of Ireland, Galway, Department of Applied Physics and Electronics
Galway
Telephone: (0)9-124411
Telefax: (0)9-125700
Electronic Mail: <userid>@epona.physics.ucg.ie
WWW: http://www.ucg.ie/
Founded: 1981 (University, as "University College Galway - UCG": 1845)
City Reference Coordinates: 009°03'00"W 53°16'00"N

National University of Ireland, Galway, Department of Physics, Astrophysics and Applied Imaging Research Group
Galway
Telephone: (0)9-124411
Telefax: (0)9-125700
Electronic Mail: <userid>@epona.physics.ucg.ie
WWW: http://www.physics.ucg.ie/airg/
Founded: 1845 (University, as "University College Galway - UCG")
City Reference Coordinates: 009°03'00"W 53°16'00"N

National University of Ireland, Maynooth, Experimental Physics Department
Maynooth, Co. Kildare
Telephone: (0)1-7083641
Telefax: (0)1-6289277
Electronic Mail: physicsec@may.ie
WWW: http://www.may.ie/academic/physics/
Founded: 1795
Staff: 12
Activities: space science * mm and radio astronomy * ground-based gamma-ray astronomy
● Formerly "Saint Patrick's College"
City Reference Coordinates: 006°35'00"W 53°23'00"N

Royal Irish Academy
Academy House
19 Dawson Street
Dublin 2
Telephone: (0)1-6762570
Telefax: (0)1-6762346
Electronic Mail: admin@ria.ie
WWW: http://www.ria.ie/
Founded: 1785
Membership: 250
Activities: learned society of Ireland including IAU contact through the "Irish National Committee for Astronomy and Space Research" and IUPAP contact through the "Irish National Committee for Physics" (see separate entries)
Periodicals: "Proceedings. Sect. A (Mathematical, Astronomical and Physical Sciences)" and others
City Reference Coordinates: 006°15'00"W 53°20'00"N

Schull Planetarium
de 'Leyns'
Colla Road
Schull, Co. Cork
Telephone: (0)28-28552
City Reference Coordinates: 009°33'00"W 51°32'00"N

Shannonside Astronomy Club (SAC)
26 Ballycannon Heights
Meelick, Co. Clare
Telephone: (0)61-453322
Telefax: (0)61-453322
WWW: http://members.xoom.com/shnastroclub/home.htm
Founded: 1986
Membership: 50
Activities: public educational programmes * annual Whirlpool Star Party * observing
Periodicals: (6) "Constellation"
Coordinates: 008°11'00"W 52°42'00"N H315m (Wood-Cock Hill)
City Reference Coordinates: 008°11'00"W 52°42'00"N

Trinity Astronomy and Space Society (TASS)
Box 8
The Atrium
Trinity College
Dublin 2
Telephone: (0)1-7021827
Telefax: (0)1-6778996
Electronic Mail: tass@maths.tcd.ie
WWW: http://www.csc.tcd.ie/~tass/Welcome.html
Founded: 1995
Membership: 130
Activities: talks * lectures * space/astronomy-related films * classes for beginners * field trips * planetarium shows * observing
City Reference Coordinates: 006°15'00"W 53°20'00"N

Trinity College Dublin (TCD), Department of Physics
College Green
Dublin 2
Telephone: (0)1-6081675
Telefax: (0)1-6711759
Electronic Mail: physics@tcd.ie
WWW: http://www2.tcd.ie/Physics/
Founded: 1592
Staff: 31
Activities: optoelectronics * condensed-matter physics * magnetism * computational physics (including astrophysics)
City Reference Coordinates: 006°15'00"W 53°20'00"N

Tullamore Astronomical Society (TAS)
145 Arden Vale
Tullamore, Co. Offaly
Telephone: (0)506-41983
Electronic Mail: seanmck@iol.ie
WWW: http://www.iol.ie/~seanmck/tas.htm
Founded: 1986
Membership: 45
Activities: lectures * education * observing * telescope and observatory building * star parties * Irish Astrofest * visits * photography * radiotelescope
Periodicals: (4) "Réalta"
Coordinates: 007°31'00"E 53°17'00"N H100m
City Reference Coordinates: 007°31'00"E 53°17'00"N

University College Dublin (UCD), Experimental Physics Department
Stillorgan Road
Belfield
Dublin 4
Telephone: (0)1-693244
Telefax: (0)1-837275
          (0)1-694409
Electronic Mail: <userid>@irlearn.bitnet
WWW: http://www.ucd.ie/~physics/main.html
          http://www.ucd.ie/~physics/research/hea.html (High-Energy Astrophysics)
Staff: 4
Activities: gamma-ray and optical astronomy
City Reference Coordinates: 006°15'00"W 53°20'00"N

Israel

Asher Space Research Institute
• See "Israel Institute of Technology, Asher Space Research Institute"

Florence and George Wise Observatory
• See "Tel Aviv University, Florence and George Wise Observatory"

Hebrew University, Racah Institute of Physics
Jerusalem 91904
Telephone: (0)2-6584605
Telefax: (0)2-5611519
Electronic Mail: <userid>@vms.huji.ac.il
                          <userid>@astro.huji.ac.il
WWW: http://www.fiz.huji.ac.il/
Founded: 1945
Staff: 40
Activities: cosmology * galaxies and large-scale structure * high-energy and relativistic astrophysics * relativity and gravitation * stellar structure and evolution
City Reference Coordinates: 035°13'00"E 31°46'00"N

Israel Academy of Sciences and Humanities
Albert Einstein Square
43 Jabotinsky Street
P.O. Box 4040
Jerusalem 91040
Telephone: (0)2-5676222
Telefax: (0)2-5666059
Electronic Mail: <userid>@academy.ac.il
WWW: http://www.academy.ac.il/
Founded: 1959
Membership: 70
Activities: cultivating and promoting scholarly and scientific endeavour * advising government * maintaining contact with similar bodies abroad * publishing
Periodicals: "Proceedings of the Israel Academy of Sciences and Humanities"
City Reference Coordinates: 035°13'00"E 31°46'00"N

Israel Institute of Technology, Asher Space Research Institute
Technion City
Haifa 32000
Telephone: (0)4-293020
Telefax: (0)4-230956
Electronic Mail: <userid>@technion.technion.ac.il
WWW: http://www.technion.ac.il/technion/trdf/research.html
Founded: 1983
Staff: 21
Activities: space technology * astrophysics
City Reference Coordinates: 035°00'00"E 32°49'00"N

Israel Meteorological Service
P.O. Box 25
Bet Dagan 50250
Telephone: (0)3-9682121
                   (0)3-9682116
Electronic Mail: meteo@serv.gov.il
City Reference Coordinates: 034°40'00"E 32°00'00"N

Israel Physical Society (IPS)
c/o G. Shaviv (President)
Physics Department
Israel Institute of Technology
Technion City
Haifa 32000
Telephone: (0)4-293020
Telefax: (0)4-230950
Electronic Mail: phr80gs@technion.bitnet
WWW: http://www.bin.ac.il/PH/IPS/
Founded: 1954
Membership: 250
City Reference Coordinates: 035°00'00"E 32°49'00"N

Israel Society of Aeronautics and Astronautics (ISAA)
P.O. Box 2956
Tel-Aviv 61028
Founded: 1951
Membership: 400
Activities: lectures * conventions
Periodicals: "BIAF"
City Reference Coordinates: 034°45'00"E 32°07'00"N

Lasky Planetarium
c/o Eretz Israel Museum
P.O. Box 17068
Tel-Aviv 61170
or :
2 Haim Lebanon
Ramat-Aviv
Tel-Aviv
Telephone: (0)3-6415244
Telefax: (0)3-6412408
Founded: 1967
Staff: 5
Activities: astronomy and space educational multimedia presentations
City Reference Coordinates: 034°45'00"E 32°07'00"N

Racah Institute of Physics
● See "Hebrew University, Racah Institute of Physics"

Raymond and Beverly Sackler Institute of Astronomy
● See "Tel Aviv University, Department of Physics and Astronomy"

Sackler Institute of Astronomy (Raymond and Beverly_)
● See "Tel Aviv University, Department of Physics and Astronomy"

Standards Institution of Israel (SII)
42 Chaim Levanon Street
Tel-Aviv 69977
Telephone: (0)3-6465154
Telefax: (0)3-6419683
Electronic Mail: standard@netvision.net.il
WWW: http://www.sii.org.il/
Founded: 1945 (as "Israeli Institute of Standards")
City Reference Coordinates: 034°45'00"E 32°07'00"N

Tel Aviv University, Department of Geophysics and Planetary Science
Ramat-Aviv
Tel-Aviv 69978
Telephone: (0)3-6408633
Telefax: (0)3-6409282
WWW: http://www.tau.ac.il/geophysics/
Founded: 1968
Staff: 20
Activities: research
City Reference Coordinates: 034°45'00"E 32°07'00"N

Tel Aviv University, School of Physics and Astronomy
Ramat-Aviv
Tel-Aviv 69978
Telephone: (0)3-6408208
Telefax: (0)3-6408179
Electronic Mail: <userid>@wise.tau.ac.il
WWW: http://www.tau.ac.il:81/~stupp/
        http://www.tau.ac.il/~shir/institute.html (Sackler Institute)
Founded: 1965
City Reference Coordinates: 034°45'00"E 32°07'00"N

Tel Aviv University, Florence and George Wise Observatory
Ramat-Aviv
Tel-Aviv 69978
or :
P.O. Box 90
Mitzpe Ramon 80650
Telephone: (0)3-6409279
        (0)3-6408729
        (0)7-6588133 (Mitzpe Ramon)

Telefax: (0)3-6408179
Electronic Mail: <userid>@wise.tau.ac.il
WWW: http://wise-obs.tau.ac.il/
Founded: 1971
Staff: 12
Activities: stars * galaxies * solar system * direct imaging * spectroscopy * photometry
Coordinates: 034°45'48"E 30°35'45"N H900m
City Reference Coordinates: 034°45'00"E 32°07'00"N (Tel-Aviv)

Weizmann Institute of Science, Department of Condensed Matter Physics
Rehovot 76100
Telephone: (0)8-343897
Electronic Mail: <userid>@wis-eyal.weizmann.ac.il
WWW: http://www.weizmann.ac.il/¿physics/cndnsd.html
Founded: 1933 (Institute)
City Reference Coordinates: 034°48'00"E 31°53'00"N

Weizmann Institute of Science, Department of Particle Physics
Rehovot 76100
Telephone: (0)8-343835
Telefax: (0)8-344106
Electronic Mail: <userid>@weizmann.weizmann.ac.il
WWW: http://www.weizmann.ac.il/physics/particles.html
Founded: 1995 (1954 as "Department of Nuclear Physics")
City Reference Coordinates: 034°48'00"E 31°53'00"N

Wise Observatory (Florence and George_)
• See "Tel Aviv University, Florence and George Wise Observatory"

Italy

Abdus Salam International Centre for Theoretical Physics (ICTP)
Strada Costiera 11
I-34014 Trieste
Telephone: (0)40-2240111
Telefax: (0)40-224163
Electronic Mail: sci_info@ictp.trieste.it
                              <userid>@ictp.trieste.it
WWW: http://www.ictp.trieste.it/
Founded: 1964
Staff: 130
Activities: theoretical physics and mathematics with applications
Periodicals: (1) "Annual Report" (ISSN 1020-7007, circ.: 300); (6) "Newsletter" (circ.: 2,000); "Preprints"
Awards: (1) "ICTP Prizes"; (2) "Dirac Medals"
● Formerly "International Centre for Theoretical Physics (ICTP)"
City Reference Coordinates: 013°46'00"E 45°40'00"N

Accademia delle Scienze di Torino
Via Maria Vittoria 3
I-10123 Torino
Telephone: (0)11-5620047
Telefax: (0)11-5620047
Founded: 1783
Membership: 250
Activities: conferences * publishing * lectures * library * archives
Periodicals: (1) "Annuario della Accademia delle Scienze di Torino" (circ.: 400) "Atti - Classe di Scienze Fisiche " (circ.: 600), "Atti - Classe di Scienze Morali" (circ.: 600), "Memorie - Classe di Scienze Fisiche" (circ.: 600) "Memorie - Classe di Scienze Morali" (circ.: 600)
Awards: (1) "Premio Gautieri", "Premio Gili-Agostinelli", "Premio Bonavera"; (1/2) "Premio Bressa", "Premio Martinetto"; (1/3) "Premio Italgas"; (1/4) "Premio Internazionale Maria Luisa Ferrari Soave e Dottore Luigi Soave" (biology); "Premio Internazionale Modesto Panetti" (applied mechanics), "Premio Internazionale Herlitzka" (physiology), "Premio Vallauri"; (1/5) "Premio Ravani-Pellati"; (1/10) "Premio Pollini"; "Premio Franco-Simone", "Premio Sergio Lupi"
City Reference Coordinates: 007°40'00"E 45°03'00"N

Accademia Nazionale dei Lincei
Palazzo Corsini
Via della Lungara 10
I-00165 Roma
Telephone: (0)6-6838831
Telefax: (0)6-6893616
WWW: http://t-and-m.com/lincei/
Founded: 1603
Activities: promoting culture and science
Periodicals: "Rendiconti" (ISSN 0001-4435)
City Reference Coordinates: 012°29'00"E 41°54'00"N

Agenzia Spaziale Italiana (ASI), Base di Lancio
S.S. 113 - n. 174 Contrada Milo
I-91100 Trapani
Telephone: (0)923-539928
              (0)923-539036
Telefax: (0)923-538493
Electronic Mail: <userid>@asitp.tp.asi.it
              http://www.asi.it/
Founded: 1975 (ASI: 1988)
Activities: stratospheric balloon launching site
City Reference Coordinates: 012°31'00"E 38°01'00"N

Agenzia Spaziale Italiana (ASI), Centro di Geodesia Spaziale "Giuseppe Colombo"
Localitá Terlecchia
Casella Postale 11
I-75100 Matera
Telephone: (0)835-3779
Telefax: (0)835-339005
Electronic Mail: <userid>@asimt0.mt.asi.it
WWW: http://www.asi.it/00HTL/asi/asicgs/cgs.html
              http://www.asi.it/
Founded: 1988 (ASI)
City Reference Coordinates: 016°37'00"E 40°40'00"N

Agenzia Spaziale Italiana (ASI), Sede
Via di Villa Patrizi 13

I-00161 Roma
Telephone: (0)6-85671
        (0)6-4405051
        (0)6-4405054
Telefax: (0)6-8567209
WWW: http://www.asi.it/
Founded: 1988
Staff: 150
Activities: Italian space agency
City Reference Coordinates: 012°29'00"E 41°54'00"N

Agenzia Spaziale Italiana (ASI), Uffici

Viale Regina Margherita 202
I-00198 Roma
Telephone: (0)6-85671
        (0)6-85679
Telefax: (0)6-8567267
WWW: http://www.asi.it/
Founded: 1988
City Reference Coordinates: 012°29'00"E 41°54'00"N

Alenia Spazio SpA

Via Saccomuro 24
I-00131 Roma
Telephone: (0)6-43681
        (0)6-41511
Telefax: (0)6-4190773
        (0)6-4190675
WWW: http://www.aleniaerospazio.com/
Staff: 2100
City Reference Coordinates: 012°29'00"E 41°54'00"N

Amateurastronomen Max Valier (AAMV)

Casella Postale 28
I-39100 Bolzano
or :
Neustifterweg 5
I-39100 Bozen
Telephone: (0)338-8106997
Electronic Mail: info@maxvalier.org
WWW: http://www.maxvalier.org/
Founded: 1985
Membership: 90
Activities: amateur astronomy * Meetings * lectures * trips
City Reference Coordinates: 011°22'00"E 46°30'00"N

Associazione Acquavivese Astrofili (AAA) "Hertzsprung-Russell"

Piazza Vittorio Emanuele 23
I-70021 Acquaviva
or :
Via Corso 41
I-70021 Acquaviva
Telephone: (0)80-761627
        (0)80-762854
Electronic Mail: giulani@teseo.it
Founded: 1975
Membership: 23
Activities: education * popularization * lectures * photography (planets, deep sky)
Coordinates: 016°47'48"E 40°51'14"N H441m (Curtomartino)
City Reference Coordinates: 012°25'00"E 43°57'00"N

Associazione Amici dei Planetari (AAP)

c/o Centro Studi e Ricerche Serafino Zani
Via Bosca 24
Casella Postale 104
I-25066 Lumezzane
Telephone: (0)30-871861
Telefax: (0)30-872545
WWW: http://www.cityline.it/cult/grup_sci/planeta.html
Founded: 1986
Membership: 74
Activities: promoting planetarium activities (publications, meetings, exhibitions, conferences) * census of Italian planetariums and material suppliers * organizing annual meetings and international days * national archive of planetaria * international exchanges
Periodicals: (8) "Circular"; (1) "Proceedings of Annual Meeting of Italian Planetariums"
City Reference Coordinates: 010°13'00"E 45°33'00"N (Brescia)

Associazione ASTRIS
● See "Associazione Astrofili della Società di Telecomunicazioni del Raggrupamento IRI-STET"

Associazione Astrofili Altair
Via Stefano Cansacchi 88
I-00056 Ostia Lido
Telephone: (0)6-56338743
          (0)6-5662688
Electronic Mail: astro.altair@usa.net
WWW: http://altair.freeweb.org
City Reference Coordinates: 012°17'00"E 41°43'00"N

Associazione Astrofili Aurunca
c/o Scuola Media F. De Sanctis
Via G. Bruno 1
I-81037 Sessa Aurunca
Telephone: (0)823-938627
Telefax: (0)823-937117
Electronic Mail: pago@sessa.peoples.it (Pascuale Ago, President)
WWW: http://utenti.tripod.it/astrofili/
Founded: 1997
Membership: 30
Activities: research * popularization
Periodicals: "Altair"
Awards: (1) "Giovanni Bruno"
City Reference Coordinates: 013°56'00"E 41°14'00"N

Associazione Astrofili Bolognesi (AAB)
Casella Postale 313
I-40100 Bologna
or :
Via Polese 13
I-40122 Bologna
Telephone: (0)51-306583
Telefax: (0)51-750360
Electronic Mail: astrofil@iperbole.bologna.it
WWW: http://www.bo.astro.it/aab/aabhome.html
Founded: 1967
Membership: 100
Activities: education * observing * popularization
Periodicals: (4) "Giornale dell'AAB" (ISSN 0392-3932)
Coordinates: 011°09'13"E 44°21'28"N H651m (Felsina)
City Reference Coordinates: 011°20'00"E 44°29'00"N

Associazione Astrofili del Basso Vicentino (AABV) "Edmund Halley"
c/o Municipio di Sossano
Piazza Mazzini 1
I-36040 Sossano
WWW: http://www.mclink.it/mclink/astro/ass/aabv/
        http://astrolink.mclink.it/ass/aabv/
Founded: 1992
City Reference Coordinates: 011°53'00"E 45°25'00"N (Asiago)

Associazione Astrofili della Società di Telecomunicazioni del Raggrupamento IRI-STET
(Associazione ASTRIS)
Via Sebino 11
I-00199 Roma
Telephone: (0)6-8414338
Telefax: (0)6-8414338
WWW: http://www.mclink.it/mclink/astro/ass/astris/
        http://astrolink.mclink.it/ass/astris/
Founded: 1990
Membership: 80
Activities: popularization
Periodicals: (4) "ASTRIS News"
City Reference Coordinates: 012°29'00"E 41°54'00"N

Associazione Astrofili di Piombino
c/o Meucci Stefano (President)
Via Cellini 18
I-57025 Piombino
Telephone: (0)565-220142
WWW: http://www.arcetri.astro.it/CDAs/homepages/home21.html
City Reference Coordinates: 010°32'00"E 42°56'00"N

Associazione Astrofili Fiorentini (AAF)
c/o Walter Benedetti
Piazetta Valdambra 9
I-50127 Firenze
Telephone: (0)55-410264
Telefax: (0)55-604854
Electronic Mail: aaf@geocities.com
                aafl@dada.it
                sanpolo@dada.it
WWW: http://www.geocities.com/~aaf
       http://www.geocities.com/CapeCanaveral/1229/
Founded: 1958
Membership: 35
Activities: photometry * astrometry
Coordinates: 011°10'26"E 43°43'41"N H140m (San Polo a Mosciano)
City Reference Coordinates: 011°15'00"E 43°46'00"N

Associazione Astrofili "Galileo Galilei" (AAGG)
Via Vecchia Fiorentina 424
I-56015 Riglione
Telephone: (0)50-542514
WWW: http://www.arcetri.astro.it/CDAs/homepages/G_galilei/
Founded: 1982
Membership: 100
City Reference Coordinates: 010°20'00"E 43°43'00"N (Pisa)

Associazione Astrofili Garfagnana (AAG)
Via della Centrale 4
I-55032 Castelnuovo di Garfagnana
WWW: http://www.arcetri.astro.it/CDAs/homepages/Aag/home27.html
City Reference Coordinates: 010°24'00"E 44°06'00"N

Associazione Astrofili "Geminiano Montanari"
Via Concordia 200
I-41032 Cavezzo
Telephone: (0)535-58755
Telefax: (0)535-58755
Electronic Mail: gmengoli@arcanet.it (Giorgio Mengoli)
WWW: http://www.arcanet.it/oss_astronomico/
Founded: 1972
Activities: observing
Coordinates: 011°00'20"E 44°51'50"N H18m
City Reference Coordinates: 011°02'00"E 44°50'00"N

Associazione Astrofili Imolesi (AAI)
Via Emilia 147
I-40026 Imola
or :
Via Comezzano 21
I-40026 Imola
Telephone: (0)542-684335
Electronic Mail: imoastro@freemail.it
WWW: http://www.angelfire.com/ct/imoastro
Founded: 1983
Membership: 100
Activities: popularization * photography (chemical and digital)
Periodicals: (1) "Almanacco Astronomico"
Coordinates: 011°38'15"E 44°20'16"N H250m (Osservatorio "A. Betti")
City Reference Coordinates: 011°43'00"E 44°22'00"N

Associazione Astrofili "Ionico Etnei"
c/o G. Catanzaro
Corso V. Bellini 84
I-95013 Fiumefreddo di Sicilia
Electronic Mail: gca@sunct.ct.astro.it
WWW: http://www.mclink.it/mclink/astro/ass/aaie/
Founded: 1989
Membership: 30
City Reference Coordinates: 015°13'00"E 37°47'00"N

Associazione Astrofili Mantovani (AAM)
c/o Luciano Luppi
Via Dugoni 24
I-46027 San Benedetto Po
Telephone: (0)376-615156

Electronic Mail: reald@tin.it
            assoc.astrofili@polirone.mn.it
WWW: http://www.freeweb.org/associazioni/aam/home.htm
Founded: 1992
Activities: popularization
City Reference Coordinates: 010°55'00"E 45°03'00"N

Associazione Astrofili Monti della Tolfa (AAMT)
Via degli Orti 30
I-00053 Civitavecchia
Telephone: (0)766-542936
Electronic Mail: bocci@roma2.infn.it (Valerio Bocci)
            giuseppe@etruria.net (Giuseppe Fusco)
WWW: http://www.mclink.it/mclink/astro/ass/aamt/
        http://astrolink.mclink.it/mclink/ass/aamt/
        http://www.mclink.it/mclink/astro/ass/ass0004.htm
Founded: 1985
Membership: 30
Activities: observing * photography * education
Coordinates: 011°47'00"E 42°06'00"N H10m (Osservatorio San Pio X)
City Reference Coordinates: 011°48'00"E 42°06'00"N

Associazione Astrofili Pegaso
c/o Antonio Citati
Via San Antonio 2
I-28021 Borgomanero
Telephone: (0)322-836224
Telefax: (0)322-860717
Electronic Mail: info@astropegaso.org
WWW: http://www.astropegaso.org/
City Reference Coordinates: 008°27'00"E 45°42'00"N

Associazione Astrofili Sardi (AAS)
c/o Marco Massa
Vico IV San Giacomo 1
I-09033 Decimomannu
or :
Stazione Astronomica
Loc. Poggio dei Pini
Strada 54
I-09012 Capoterra
Telephone: (0)70-655454
Electronic Mail: utzeri@mbox.vol.it (Andrea Utzeri)
WWW: http://www.ca.astro.it/astrofili/
Founded: 1977
Membership: 50
Activities: meetings * lectures * education * star parties * photography * CCD observing (deep sky, planets, minor planets)
City Reference Coordinates: 008°58'00"E 39°19'00"N (Decimomannu)
                008°58'00"E 39°11'00"N (Capoterra)

Associazione Astrofili Segusini (AAS)
c/o Andrea Ainardi (President)
Corso Couvert 5
I-10059 Susa
or :
Corso Trieste 15
I-10059 Susa
Telephone: (0)122-622766
Telefax: (0)122-32060
Electronic Mail: ainardi@tin.it
WWW: http://www.mclink.it/mclink/astro/ass/grange/
        http://astrolink.mclink.it/ass/grange/
Founded: 1973
Membership: 45
Activities: lectures * observing * astrometry * photography
Periodicals: "Circolare Interna"
Observatories: 1 (Osservatorio Astronomico Grange - see separate entry)
City Reference Coordinates: 007°03'00"E 45°08'00"N

Associazione Astrofili Spezzini (AAS)
Casella Postale 11
I-19100 La Spezia
or :
c/o Istituto Tecnico Statale Nautico "N. Sauro"
Viale Italia 88
I-19100 La Spezia

Telephone: (0)187-502240
Electronic Mail: mc7316@mclink.it
              scarfi@sp.itline.it
WWW: http://www.tamnet.interbusiness.it/htmlpages/tam_ita/aas/aas.html
Founded: 1978
Membership: 35
Activities: popularization * minor-planet astrometry * SN CCD survey * digital image processing * archaeoastronomy * sundials
Periodicals: (3) "Astronomica"; "Comunicando"
Coordinates: 009°47'00"E 44°08'26"N H350m (Monte Vissegi)
City Reference Coordinates: 009°50'00"E 44°07'00"N

Associazione Astrofili Teatini (AAT)

Via Crociferi 33
I-66100 Chieti
Telephone: (0)871-348921
Founded: 1974
Membership: 20
Activities: observing (planets) * education * public lectures * exhibitions
Coordinates: 014°10'00"E 42°21'02"N H330m
City Reference Coordinates: 014°10'00"E 42°21'00"N

Associazione Astrofili Tethys (AAT)

Via Indipendenza 14
I-27055 Rivanazzano
Telephone: (0)383-944160
Telefax: (0)131-887268
Electronic Mail: giorgia.c@aznet.it
WWW: http://www.aat.idp.it/
Founded: 1987
Membership: 90
Activities: telescope making * computational astronomy * education
Periodicals: (4) "Albireo"; (1) "Almanaco Astronomico"
City Reference Coordinates: 009°01'00"E 44°56'00"N

Associazione Astrofili Trentini (AAT)

c/o Museo Tridentino di Scienze Naturali
Via Calepina 14
I-38100 Trento
Telephone: (0)461-270311
Telefax: (0)461-233820
Electronic Mail: aat@mtsn.tn.it
WWW: http://www.mtsn.tn.it/astrofili/
Founded: 1976
Membership: 500
Activities: popularization
Periodicals: (4) "Notizario AAT"
City Reference Coordinates: 011°08'00"E 46°04'00"N

Associazione Astrofili Trevigiani

Borgo Covour 40
I-31100 Treviso
Telephone: (0)422-411725
Founded: 1974
Membership: 70
Activities: popularization * education * lectures * variable stars * meteorology * solar activity * planetarium (Collegio Pio X) * observing
Periodicals: (1) "Lezioni di Astronomia"
Coordinates: 012°13'00"E 45°40'00"N H25m
City Reference Coordinates: 012°13'00"E 45°40'00"N

Associazione Astrofili Valdinievole (AAV)

Casella Postale 156
I-51015 Monsummano Terme
or :
Via di Gragnano 349
I-51015 Monsummano Terme
Telephone: (0)572-51741
Telefax: (0)572-81192
Electronic Mail: aav@mercurio.iet.unipi.it
WWW: http://www.italway.it/associazioni/aav/
Founded: 1979
Membership: 40
Activities: photography * meetings * observing * lectures * software
Periodicals: (1) "Appunti di Astronomia"
Coordinates: 010°48'35"E 43°52'21"N H20m

City Reference Coordinates: 010°48'00"E 43°52'00"N

Associazione Astrofili Valtellinesi (AAV)
Casella Postale 52
I-23100 Sondrio
or :
Biblioteca Civica
I-23026 Ponte in Valtellina
Telephone: (0)342-219111
             (0)347-2461569
Electronic Mail: mcioccar@novanet.it
WWW: http://www.novanet.it/vvol/assoc/astrofili/
Founded: 1996
Membership: 47
Activities: observing * lectures * meetings
Coordinates: 009°57'00"E 46°10'00"N H485m
City Reference Coordinates: 009°52'00"E 46°11'00"N (Sondrio)

Associazione Astrofili Veneziani (AAV)
Casella Postale 433
I-30100 Venezia
or :
Convento San Nicolò
I-30126 Venezia Lido
Telephone: (0)41-770745 (Secretary)
WWW: http://www.mclink.it/mclink/astro/ass/venezia/
        http://astrolink.mclink.it/ass/venezia/
Founded: 1976
Activities: observing * popularization * planetarium (Venezia Lido)
Periodicals: (12) "Circolare Informativa"
Coordinates: 012°38'00"E 45°42'00"N H5m (Lido)
City Reference Coordinates: 012°38'00"E 45°42'00"N

Associazione Astronomica Cassino (AAC)
c/o Gianni Fardelli
Via Cavatelle 6
I-03043 Cassino
Telephone: (0)776-337761
Electronic Mail: giafar@officine.it
WWW: http://www.officine.it/citylife/astro/astronom.htm
        http://bertario.officine.it/citylife/aac/
Founded: 1995
City Reference Coordinates: 013°49'00"E 41°30'00"N

Associazione Astronomica Cortina (AAC)
Via Pecol 95
I-32043 Cortina d'Ampezzo
or :
Casella Postale 193
I-32043 Cortina d'Ampezzo
Electronic Mail: aac@sunrise.it
WWW: http://www.sunrise.it/associazioni/aac/
Founded: 1972
Membership: 100
Periodicals: "Cortina Astronomica"
Observatories: 1 (Col Drusciè - H1780m)
City Reference Coordinates: 012°08'00"E 46°32'00"N

Associazione Astronomica Feltrina "G.J. Rheticus"
Casella Postale 2
I-32032 Feltre
Telephone: (0)439-304366 (President)
WWW: http://www.mclink.it/mclink/astro/ass/ass0009.htm
        http://astrolink.mclink.it/ass/ass0009.htm
Founded: 1989 (1973 as "Gruppo Astrofili Feltrini")
Membership: 90
Activities: popularization * comets * sundials
Periodicals: (4) "Rheticus"
Coordinates: 011°54'43"E 46°03'17"N H462m (Vignui)
City Reference Coordinates: 011°54'00"E 46°01'00"N

Associazione Astronomica Frusinate (AAF)
Via Fosse Ardeatine 234
I-03100 Frosinone
Telephone: (0)775-833737

Telefax: (0)775-211238
WWW: http://www.rtmol.it/aocc/ (Campo Catino Observatory)
Founded: 1981
Observatories: 1 (Campo Catino - see separate entry)
City Reference Coordinates: 013°22'00"E 41°28'00"N

Associazione Astronomica Madonna di Campiglio
c/o Matteo Maturi
Via Cima Tosa 24
I-38084 Madonna di Campiglio
Telephone: (0)465-441010
Electronic Mail: hotzeled@well.it
WWW: http://www.geocities.com/CapeCanaveral/Lab/6914/
City Reference Coordinates: 010°49'00"E 46°13'00"N

Associazione Astronomica Milanese (AAM)
c/o Roberto Boccadoro
Viale Zara 118
I-20125 Milano
Telephone: (0)2-6686263
Electronic Mail: roberto_boccadoro@lotus.com
WWW: http://www.mclink.it/mclink/astro/ass/ass0003.htm
        http://astrolink.mclink.it/ass/ass0003.htm
Founded: 1991
Membership: 50
City Reference Coordinates: 009°12'00"E 45°28'00"N

Associazione Astronomica Quasar
c/o Centro di Scienze Naturali (CSN)
Via di Galceti 74
I-59100 Prato
Telephone: (0)574-38960
        (0)574-460503 (CSN)
Electronic Mail: quasar@comune.prato.it
WWW: http://www.comune.prato.it/csn/gen/htm/planet.htm (Planetarium)
City Reference Coordinates: 011°06'00"E 43°53'00"N

Associazione Cernuschese Astrofili (ACA)
Centro Cardinal Colombo
Piazza Matteotti 20
I-20063 Cernusco sul Naviglio
Telephone: (0)2-9231747
Electronic Mail: erusso@pointest.com (Emanuele Russo)
            astral@freenet.hut.fi (Gabriele Barletta)
WWW: http://lasvegas.pointest.com/astrofili/
Founded: 1988
Membership: 40
Activities: observing * education
City Reference Coordinates: 009°19'00"E 45°31'00"N

Associazione Friulana di Astronomia e Meteorologia (AFAM)
Casella Postale 179
I-33100 Udine
or :
Via San Stefano
I-33047 Remanzacco
Telephone: (0)432-668176
Electronic Mail: sostero@elettra.trieste.it
            afam@conecta.it
WWW: http://www.conecta.it/afam/dex.htm
Founded: 1969
Membership: 160
Activities: education * observing (visual, photographic, photometric) * variable stars * radioastronomy
Periodicals: (4) "L'Osservatorio" (circ.: 250)
Coordinates: 013°18'59"E 46°05'11"N H113m (Remanzacco)
City Reference Coordinates: 013°14'00"E 46°03'00"N (Udine)

Associazione IDRA
Piazzetta Arnella 4
I-80100 Napoli
or :
c/o Giuseppe Borrelli
Via Jannelli 18
I-80100 Napoli
Telephone: (0)81-5791228

Electronic Mail: mvivaldi@mbx.idn.it
WWW: http://astrolink.mclink.it/ass/idra/
City Reference Coordinates: 014°17'00"E 40°51'00"N

Associazione Ligure Astrofili Polaris
Via Galata 33/5
I-16121 Genova
Telephone: (0)10-5533045
Telefax: (0)10-5531775
Electronic Mail: bigatti@dima.unige.it (A.Bigatti)
WWW: http://www.mclink.it/mclink/astro/ass/polaris/
        http://astrolink.mclink.it/ass/polaris/
        http://members.xoom.com/astropolaris
Founded: 1994
Membership: 75
Activities: public lectures * observing * photography
Periodicals: "Notiziario"
City Reference Coordinates: 008°57'00"E 44°25'00"N

Associazione Ligure per lo Studio e la Divulgazione dell'Astronomia e dell'Astronautica
● See "Urania"

Associazione Marchigiana Astrofili (AMA)
c/o Massimo Morroni
Istituto Nautico A. Elia
Lungomare Vanvitelli 76
I-60100 Ancona
Telephone: (0)71-203444
Founded: 1969 (Planetarium: 1985)
Membership: 80
Activities: popularization * running a planetarium
Coordinates: 013°30'00"E 43°30'00"N H130m (Senigalliesi)
City Reference Coordinates: 013°30'00"E 43°38'00"N

Associazione Maremmana Studi Astronomici (AMSA)
Casella Postale 112
I-58100 Grosseto
Telephone: (0)564-455326
Electronic Mail: amsa@gol.grosseto.it
                emanfucc@gol.grosseto.it (Enrico Manfucci)
WWW: http://www.gol.grosseto.it/asso/amsa/amsa.htm
Founded: 1983
Membership: 60
Activities: public observing * photography * lectures
Periodicals: "Bollettino AMSA"
Coordinates: 011°10'15"E 42°48'44"N H100m
City Reference Coordinates: 011°08'00"E 42°46'00"N

Associazione Ogliastrina di Astronomia (AOA)
Casella Postale 87bis
I-08045 Lanusei
Founded: 1989
Membership: 60
Activities: popularization towards public and schools * Ferdinando Caliumi Observatory
Coordinates: 009°30'11"E 39°52'35"N H1,150m (Monte Armidda)
City Reference Coordinates: 009°34'00"E 39°52'00"N

Associazione per lo Studio e la Ricerca Astronomica (ASTRA)
c/o Osservatorio Astronomico "Galileo Galilei"
Via G. Carducci 172
I-73050 Salve
Telephone: (0)833-520426
        (0)832-302712
WWW: http://astrolink.mclink.it/ass/astra/
        http://astra.educations.net/
Founded: 1997
City Reference Coordinates: 018°11'00"E 40°23'00"N (Lecce)

Associazione Ravennate Astrofili Rheyta (ARAR)
c/o Planetario di Ravenna
Viale Santi Baldini 4A
I-48100 Ravenna
Telephone: (0)544-62534
Electronic Mail: arar@linknet.it
WWW: http://racine.ra.it/planet/

Founded: 1973
Membership: 33
Activities: popularization * planetarium * observing
Periodicals: (10) "Notiziario Rheyta"
Coordinates: 012°12'00"E 44°24'00"N
City Reference Coordinates: 012°12'00"E 44°24'00"N

Associazione Reggiana di Astronomia (ARA)
c/o Alessandro Geom. Guatteri
Via Cavagnola 1/1
I-42024 Castelnuovo di Sotto
or :
c/o Osservatorio Astronomico "Padre Angelo Secchi"
Via Prati Landi
I-42024 Castelnuovo di Sotto
Telephone: (0)522-682266
              (0)522-683641
Telefax: (0)522-682235
Electronic Mail: andzmb@tin.it
                     ara@coopsette.it
WWW: http://www.bo.astro.it/~pigi/Astrofili_Reggiani/
Founded: 1976
Membership: 12
Activities: Sun (photospheric activity, Wolf number) * planets (visual and photographic observing) * astrometry of comets and minor planets * photography * popularization
Periodicals: (4) "Cielo" (circ.: 400)
Coordinates: 010°33'46"E 44°48'12"N H28m
City Reference Coordinates: 010°34'00"E 44°48'00"N

Associazione Romana Astrofili (ARA)
Casella Postale 4011
I-00100 Roma
Telephone: (0)6-79960075
Electronic Mail: mc3325@mclink.it (Claudio Costa)
                     mc3383@mclink.it (Mario Farina)
WWW: http://www.mclink.it/mclink/astro/ass/ara/
         http://astrolink.mclink.it/ass/ara/
Founded: 1982
City Reference Coordinates: 012°29'00"E 41°54'00"N

Associazione Sabina Astrofili (ASA)
Via del Vivaio 5A
I-02045 Limiti Di Greccio
Telephone: (0)746-204960
              (0)746-274539
Electronic Mail: vascleri@geocities.com
WWW: http://www.geocities.com/CapeCanaveral/Hangar/9330/
Founded: 1995
Membership: 30
City Reference Coordinates: 012°51'00"E 42°24'00"N (Rieti)

Associazione Tuscolana di Astronomia (ATA)
c/o Istituto di Astrofisica Spaziale del CNR
Casella Postale 67
I-00044 Frascati
or :
Viale della Galassia 43
I-00040 Rocca Priora
Telephone: (0)6-9406339
Electronic Mail: atamail@hotmail.com
                     msitrm01.szxqc4@eds.com (Emilio Sassone Corsi)
WWW: http://www.mclink.it/mclink/astro/ass/ata/
         http://astrolink.mclink.it/ass/ata/
Founded: 1995
Membership: 40
City Reference Coordinates: 012°29'00"E 41°54'00"N (Frascati)
                                012°45'00"E 41°48'00"N (Rocca Priora)

Associazione Valdostana Scienze Astronomiche (AVSA)
Regione Crou 17
I-11100 Aosta
Telephone: (0)165-33853
              (0)165-555192
Electronic Mail: lravello@aostanet.com (Luciano Ravello, President)
WWW: http://www.aostanet.com/astro-radio/
Founded: 1986

Membership: 3
Activities: observing * education
City Reference Coordinates: 007°19'00"E 45°43'00"N

Associazione Vigevanese Divulgazione Astronomica (AVDA)

c/o Biblioteca Civica
Corso Cavour 82
I-27019 Vigevano
Telephone: (0)381-70149
Telefax: (0)381-70149
Electronic Mail: marcmoli@tin.it
WWW: http://www.vigevano.vol.it/cultura/avda/
            http://www.geocities.com/CapeCanaveral/Campus/6640/avda5.htm
Founded: 1996
Membership: 60
Activities: education * observing
City Reference Coordinates: 008°51'00"E 45°19'00"N

Astro Club Voyager (ACV)

Via Grande 2/A
I-31030 Castello di Godego
WWW: http://digilander.iol.it/acv
City Reference Coordinates: 012°15'00"E 45°40'00"N (Treviso)

astronomia (L'_)

● See "L'astronomia"

Astronomical Observatory of Campo Catino (AOCC)

Colle Pannunzio
I-03016 Guarcino
Telephone: (0)775-833737
            (0)775-435945
Telefax: (0)775-211238
Electronic Mail: oss.astronomico.campocatino@rtmol.stt.it
                mario.di.sora@rtmol.stt.it (Mario Di Sora, Director)
WWW: http://www.rtmol.it/aocc/
Founded: 1987
Membership: 7
Coordinates: 013°19'47"E 41°49'16"N
City Reference Coordinates: 013°19'00"E 41°48'00"N

Auriga Srl

Via Mario Fabio Quintiliano 30
I-20138 Milano
Telephone: (0)2-5097780
Telefax: (0)2-5097251
Electronic Mail: auriga@mbox.vol.it
WWW: http://www.auriga.it/
Founded: 1983
Staff: 13
Activities: distributing planetariums, telescopes, accessories and optical products
City Reference Coordinates: 009°12'00"E 45°28'00"N

BeppoSAX Scientific Data Center (SDC)

c/o Nuova Telespazio
Via Corcolle 19
I-00131 Roma
Telephone: (0)6-40796307
Telefax: (0)6-40796291
Electronic Mail: helpdesk@sax.sdc.asi.it
WWW: http://www.sdc.asi.it/
Founded: 1995
Staff: 11
City Reference Coordinates: 012°29'00"E 41°54'00"N

BPD Difesa & Spazio

Corso Garibaldi 22
I-00034 Colleferro
Telephone: (0)6-97291
Telefax: (0)6-97292299
Founded: 1913
Staff: 800
Activities: designing, manufacturing and testing propulsion systems for satellites and launchers * solid-propellant launchers for small satellites * advanced propulsion systems
City Reference Coordinates: 012°59'00"E 41°44'00"N

Centro Astrofili Bolzano (CAB)
Via M.Longon 3
I-39100 Bolzano
Telephone: (0)471-273165
Electronic Mail: mfrasc@em.parsec.it
WWW: http://www.parsec.it/cab/
Founded: 1990
City Reference Coordinates: 011°22'00"E 46°30'00"N

Centro Astronomico "Neil Armstrong" (CANA)
Via Canali 17
I-84121 Salerno
Telephone: (0)81-5750414
(0)89-790803
Electronic Mail: astro@salerno.infn.it
astro@vaxsa.csied.unisa.it
WWW: http://www.gisa.it/nucleo/cana/cana.htm
http://www.xcom.it/cana/
Founded: 1982
Activities: popularization * star parties * occultations * SN search * variable stars
City Reference Coordinates: 014°47'00"E 40°41'00"N

Centro Astronomico Orione (CAO)
Via 2 Giugno 31
I-00043 Ciampino
Telephone: (0)6-7912381
Electronic Mail: gbacaloni@pelagus.it (Rino Bacaloni)
Founded: 1994
Membership: 24
Activities: annual meetings * observing * public meetings
Periodicals: (4) "Bollettino Astronomico"
Coordinates: 012°35'42"E 41°48'00"N H185m (Stazione Astronomica Rigel)
City Reference Coordinates: 012°36'00"E 41°48'00"N

Centro di Cultura Scientifica Ettore Majorana (CCSEM)
(Ettore Majorana Centre for Scientific Culture - EMCSC)
Via Guarnotta 26
I-91016 erice
Telephone: (0)923-869133
Telefax: (0)923-869226
Electronic Mail: hq@emcsc.ccsem.infn.it
WWW: http://emcsc.ccsem.infn.it/
Founded: 1963
City Reference Coordinates: 012°36'00"E 38°02'00"N

Centro per l'Astronomia Infrarossa e lo Studio del Mezzo Interstellare (CAISMI)
● See "Consiglio Nazionale delle Ricerche (CNR), Centro per l'Astronomia Infrarossa e lo Studio del Mezzo Interstellare (CAISMI)"

Channel Srl
Il Girasole Palazzo 3/05 A
I-20084 Lacchiarella
Telephone: (0)2-90091773
Telefax: (0)2-90091787
WWW: http://www.channel.it/
Founded: 1984
Staff: 21
● Software and hardware distributor
City Reference Coordinates: 009°08'00"E 45°19'00"N

Circolo Astrofili Bergamaschi (CAB)
Via A. Maj 16/B
I-24121 Bergamo
Telephone: (0)35-223376
Telefax: (0)35-570641
Electronic Mail: cab@uninetcom.it
WWW: http://www.uninetcom.it/astro/
Founded: 1974
Membership: 80
Activities: popularization * SN search * photography * variable stars * sundials * archaeoastronomy
Coordinates: 009°45'10"E 45°45'20"N H1300m (Selvino)
City Reference Coordinates: 009°43'00"E 45°41'00"N

Circolo Astrofili di Mestre "Guido Ruggieri"
Via Padre Egidio Gelain 7

I-30175 Marghera
Telephone: (0)41-936177
              (0)41-951497
Electronic Mail: fasanta@tin.it
Founded: 1975
Membership: 30
Activities: variable stars * planets * popularization * observing
City Reference Coordinates: 012°15'00"E 45°29'00"N (Marghera)
                                012°14'00"E 45°30'00"N (Mestre)

Circolo Astrofili di Milano (CAM)
c/o Civico Planetario "Ulrico Hoepli"
Corso di Porta Venezia 57
I-20121 Milano
WWW: http://www.micronet.it/italian/astronomia/circolo/circolohome.html
Founded: 1932
City Reference Coordinates: 009°12'00"E 45°28'00"N

Circolo Astrofili Nord Sardegna
Via Mentana 34
I-07046 Porto Torres
Telephone: (0)79-510046
Founded: 1988
Membership: 10
Activities: popularization * education * photography
City Reference Coordinates: 008°24'00"E 40°51'00"N

Circolo Astrofili Talmassons
Via XXIV Maggio 10
I-33030 Talmassons
Telephone: (0)432-920670
Telefax: (0)432-920670
Electronic Mail: furlanetto@palmanet.it (Lucio Furlanetto, Secretary)
WWW: http://www.palmanet.it/cast/
Founded: 1993
Membership: 100
Activities: SN search * meteors * Moon * planets * photography * CCDs * education * lectures
Periodicals: (4) "Notiziario"
Coordinates: 013°06'14"E 45°55'21"N H28m
City Reference Coordinates: 013°07'00"E 45°56'00"N

Circolo Astrofili Veronesi (CAV) "Antonio Cagnoli"
Casella Postale 2016
I-37100 Verona
or :
Centro d'Incontro Circoscrizione
2 Largo Stazione Vecchia
Parona
I-37100 Verona
Telephone: (0)45-574345 (S. Moltomoli)
Electronic Mail: grejbear@tin.it
                     peanuts@rcvr.vr.it
                     circolo.astrofili.veronesi@rcvr.org
WWW: http://www.rcvr.org/assoc/astro/main.htm
Founded: 1977
Membership: 130
Activities: astronomy * observing * photography * CCDs * popularization * astronautics
Periodicals: (3) "CAV Notiziario"
Coordinates: 010°56'53"E 45°28'47"N H70m
City Reference Coordinates: 011°00'00"E 45°27'00"N

Circolo Astronomico Dorico "Paolo Andrenelli"
Casella Postale 70
I-60100 Ancona
Telephone: (0)71-200913
              (0)71-2862087
Telefax: (0)71-2862087
Electronic Mail: stefano@ascu.unian.it (Stefano Rosoni, President)
Founded: 1990
Membership: 20
Coordinates: 013°30'00"E 43°37'00"N
City Reference Coordinates: 013°30'00"E 43°37'00"N

Circolo Casolese Astrofili "Betelgeuse"
c/o Maurizio Cabibbo

Via delle Querce 33
I-53031 Casole d'Elsa
Telephone: (0)577-948284
           (0)577-948373
WWW: http://www.arcetri.astro.it/CDAs/homepages/casole/chisiamo.html
Founded: 1995
Membership: 16
Activities: public observing * popularization * education
Coordinates: 011°02'00"E 43°20'00"N H420m
City Reference Coordinates: 011°02'00"E 43°20'00"N

Circolo Culturale Astronomico di Farra d'Isonzo (CCAF)

Strada della Colombara 11
I-34070 Farra d'Isonzo
Telephone: (0)481-888540
Electronic Mail: ccaf@tmedia.it
WWW: http://www.btech.net/ccaf/
Founded: 1975
Activities: education * CCD astrometry of minor planets
Periodicals: (1) "Lunario" (circ.: 2,000)
Coordinates: 013°31'33"E 45°54'57"N H53m (Colombara)
City Reference Coordinates: 013°31'00"E 45°55'00"N

Circolo Pinerolese Astrofili Polaris

Via San Antonio 3
I-10060 Abbadia Alpina di Pinerolo
Telephone: (0)11-9070367
WWW: http://www.piw.it/val_lemina/osserva.htm
City Reference Coordinates: 007°19'00"E 44°53'00"N (Pinerolo)

Civico Planetario "Ulrico Hoepli"

Corso di Porta Venezia 57
I-20121 Milano
Telephone: (0)2-2895785
Telefax: (0)2-2047259
Electronic Mail: planet@imiucca.csi.unimi.it
WWW: http://www.brera.mi.astro.it/~planet/
Founded: 1929
Activities: education
City Reference Coordinates: 009°12'00"E 45°28'00"N

Coelum

Via Appia 18
I-30173 Venezia-Mestre
Telephone: (0)41-5321476
Telefax: (0)41-5327427
Electronic Mail: redaz@coelum.com
WWW: http://www.coelum.com/
● Periodical
City Reference Coordinates: 012°38'00"E 45°42'00"N (Venezia)
                                    012°14'00"E 45°30'00"N (Mestre)

Comitato Italiano per il Controllo delle Affermazioni sul Paranormale (CICAP)

Casella Postale 1117
I-35100 Padova
or :
Casella Postale 60
I-27058 Voghera
Telephone: (0)426-22013
Telefax: (0)426-22013
Electronic Mail: cicap@tin.it
WWW: http://www.valnet.it/cicap/
Founded: 1989
Membership: 19
Activities: lectures * investigations * media resource
Periodicals: (3) "Scienza & Paranormale" (circ.: 1,500)
City Reference Coordinates: 011°53'00"E 45°25'00"N (Padova)
                                    009°01'00"E 44°59'00"N (Voghera)

Consiglio Nazionale delle Ricerche (CNR)

Piazzale Aldo Moro 7
I-00185 Roma
Telephone: (0)6-49931
Telefax: (0)6-4461954
WWW: http://www.cnr.it/

Founded: 1923
City Reference Coordinates: 012°29'00"E 41°54'00"N

Consiglio Nazionale delle Ricerche (CNR), Centro per l'Astronomia Infrarossa e lo Studio del Mezzo Interstellare (CAISMI)

Largo Enrico Fermi 5
I-50125 Firenze
Telephone: (0)55-2752248
Telefax: (0)55-220039
Electronic Mail: <userid>@arcetri.astro.it
                    tirgo@arcetri.astro.it
WWW: http://www.arcetri.astro.it/irlab/tirgo/
        http://www.cnr.it/
Founded: 1981
Staff: 10
Activities: developing IR instrumentation * managing the TIRGO telescope located at the Gornergrat station in Switzerland * IR astronomical research
Coordinates: 007°47'04"E 45°59'04"N H3,130m (Gornergrat Station)
City Reference Coordinates: 011°15'00"E 43°46'00"N

Consiglio Nazionale delle Ricerche (CNR), Istituto di Astrofisica Spaziale (IAS)

Area di Ricerca di Tor Vergata
Via del Fosso del Cavaliere
I-00133 Roma
Telephone: (0)6-49934472
            (0)6-49934473
Telefax: (0)6-20660188
Electronic Mail: <userid>@saturn.ias.rm.cnr.it
WWW: http://www.ias.rm.cnr.it/
Founded: 1970
Staff: 65
Activities: galactic and stellar evolution * high-energy astrophysics * diffuse matter in space * planetology * X-ray, gamma-ray, IR and UV astronomy * instrumentation data analysis
Periodicals: "Rapporto Interno IAS"
City Reference Coordinates: 012°41'00"E 41°41'00"N

Consiglio Nazionale delle Ricerche (CNR), Istituto di Astrofisica Spaziale (IAS), Reparto di Planetologia

Area di Ricerca di Tor Vergata
Via del Fosso del Cavaliere
I-00133 Roma
Telephone: (0)6-49934472
            (0)6-49934473
Telefax: (0)6-20660188
Electronic Mail: <userid>@saturn.ias.rm.cnr.it
WWW: http://www.ias.rm.cnr.it/
Founded: 1974
Staff: 12
Activities: solar system studies * space missions
City Reference Coordinates: 012°29'00"E 41°54'00"N

Consiglio Nazionale delle Ricerche (CNR), Istituto di Cosmogeofisica

Corso Fiume 4
I-10133 Torino
Telephone: (0)11-658979
Telefax: (0)11-658972
WWW: http://www.cnr.it/
Founded: 1969
Staff: 12
Activities: cosmic rays * neutrinos * cosmogenesis * elementary particles * geophysics * climatology * oceanography * teledetection
City Reference Coordinates: 007°40'00"E 45°03'00"N

Consiglio Nazionale delle Ricerche (CNR), Istituto di Fisica Cosmica ed Applicazioni dell'Informatica (IFCAI)

Via Ugo La Malfa 153
I-90146 Palermo
Telephone: (0)91-6809690
            (0)91-6809577
Telefax: (0)91-6882258
Electronic Mail: <userid>@ifcai.pa.cnr.it
WWW: http://www.ifcai.pa.cnr.it/
Founded: 1981
Staff: 25
Activities: astrophysics * X-ray astronomy * gamma-ray astronomy * high-energy cosmic rays * space research * stratospheric balloons * data analysis methods

City Reference Coordinates: 013°21'00"E 38°07'00"N

Consiglio Nazionale delle Ricerche (CNR), Istituto di Fisica dell'Atmosfera
Area di Ricerca di Tor Vergata
Via del Fosso del Cavaliere
I-00133 Roma
Telephone: (0)6-49934296
Telefax: (0)6-49934292
Electronic Mail: <userid>@atmos.cnr.it
WWW: http://www.cnr.it/
Founded: 1962
Staff: 80
Activities: atmospheric physics * dynamics * radiation * meteorology * climatology * geomagnetism * aeronomy * energy * environment
Observatories: stations in Italy and Antarctica
City Reference Coordinates: 012°29'00"E 41°54'00"N

Consiglio Nazionale delle Ricerche (CNR), Istituto di Fisica dello Spazio Interplanetario (IFSI)
Via del Fosso del Cavaliere
I-00133 Roma
Telephone: (0)6-49934488
           (0)6-49934490
Telefax: (0)6-49934374
Electronic Mail: <userid>@hp.ifsi.fra.cnr.it
                candidi@ifsi.rm.cnr.it (M. Candidi, Director)
WWW: http://www.ifsi.rm.cnr.it/
Staff: 48
City Reference Coordinates: 012°41'00"E 41°41'00"N (Frascati)
                                 012°29'00"E 41°54'00"N (Roma)

Consiglio Nazionale delle Ricerche (CNR), Istituto di Metrologia "G. Colonnetti" (IMGC)
Strada delle Cacce 73
I-10135 Torino
Telephone: (0)11-39771
Telefax: (0)11-346761
Electronic Mail: <userid>@itoimgc.bitnet
WWW: http://www.cnr.it/
Founded: 1968
Staff: 105
Activities: metrology * measurements * instrument testing * thermodynamics * optics * optoelectronics * standards * fundamental constants * gravimetry
Periodicals: (1/2) "Report"; "Notizie di Metrologia"
City Reference Coordinates: 007°40'00"E 45°03'00"N

Consiglio Nazionale delle Ricerche (CNR), Istituto di Radioastronomia (IRA), Bologna
Via Gobetti 101
I-40129 Bologna
Telephone: (0)51-6399385
Telefax: (0)51-6399431
Electronic Mail: <userid>@ira.bo.cnr.it
WWW: http://www.ira.bo.cnr.it/
Founded: 1970
Staff: 46
Activities: radioastronomy * cosmology * VLBI * instrumentation
City Reference Coordinates: 011°20'00"E 44°29'00"N

Consiglio Nazionale delle Ricerche (CNR), Istituto di Radioastronomia (IRA), Stazione di Medicina
Località Aia Cavicchio
Via Fiorentina
I-40059 Medicina
Telephone: (0)51-6965041
           (0)51-6965001
           (0)51-6965122
Telefax: (0)51-6965105
WWW: http://trantor.ira.bo.cnr.it/HomePage.html
      http://www.ira.bo.cnr.it/
      http://www.cnr.it/
Founded: 1983
Staff: 20
Activities: radioastronomical observing
Coordinates: 011°38'43"E 44°31'14"N H42m
City Reference Coordinates: 011°38'00"E 44°28'00"N

Consiglio Nazionale delle Ricerche (CNR), Istituto di Radioastronomia (IRA), Stazione di Noto
Casella Postale 141

I-96017 Noto
or :
Contrada Renna Bassa
Località Casa di Mezzo
I-96017 Noto
Telephone: (0)931-822302
　　　　　　(0)931-835622
Telefax: (0)931-822359
　　　　　(0)931-573265
WWW: http://trantor.ira.bo.cnr.it/HomePage.html
　　　　http://www.ira.bo.cnr.it/
　　　　http://www.cnr.it/
Founded: 1988
Activities: radioastronomical observing
Coordinates: 014°59'19"E 36°52'36"N H100m
City Reference Coordinates: 015°05'00"E 36°53'00"N

Consiglio Nazionale delle Ricerche (CNR), Istituto di Ricerca sulle Onde Elettromagnetiche
Via Panciatichi 64
I-50127 Firenze
Telephone: (0)55-4378512
Telefax: (0)55-410893
WWW: http://www.cnr.it/
Founded: 1958
Staff: 106
Activities: teledetection * optical fibers * solid-state physics * computing * integrated optics
City Reference Coordinates: 011°15'00"E 43°46'00"N

Consiglio Nazionale delle Ricerche (CNR), Istituto di Tecnologie e Studio delle Radiazioni Extraterrestri (TESRE)
Via Gobetti 101
I-40129 Bologna
Telephone: (0)51-6398694
Telefax: (0)51-6398724
Electronic Mail: <userid>@botes1.bo.cnr.it
WWW: http://www.tesre.bo.cnr.it/
　　　　http://www.cnr.it/
Founded: 1969
Staff: 37
Activities: X-ray, gamma-ray, IR astronomy * cosmic-ray physics * balloon and space technologies
City Reference Coordinates: 011°20'00"E 44°29'00"N

Consiglio Nazionale delle Ricerche (CNR), Istituto per Ricerche in Fisica Cosmica e Tecnologie Relative (IFCTR)
Via Bassini 15
I-20133 Milano
Telephone: (0)2-23699302 (Secretary)
Telefax: (0)2-2362946
　　　　　(0)2-2666017
Electronic Mail: <userid>@ifctr.mi.cnr.it
WWW: http://www.ifctr.mi.cnr.it/
Founded: 1967
Staff: 34
Activities: observing (X-ray, gamma-ray, optical, UV) * instrumentation * data analysis * compact galactic sources * AGNs
* clusters of galaxies
City Reference Coordinates: 009°12'00"E 45°28'00"N

Consorzio Internazionale per l'Astrofisica Relativistica
● See "International Center for Relativistic Astrophysics (ICRA)"

D'Appolonia SpA
Via San Nazaro 19
I-16145 Genova
Telephone: (0)10-3628148
Telefax: (0)10-3621078
Electronic Mail: dappolonia@pn.itnet.it
WWW: http://www.dappolonia.it/
Founded: 1981
Activities: consultancy * technology transfer
City Reference Coordinates: 008°57'00"E 44°25'00"N

Denitron SnC
Via Milite Ignoto 81
I-21027 Ispra
Telephone: (0)332-782398

Telefax: (0)332-782398
Electronic Mail: ambrogio.nico@irc.it
Founded: 1988
Staff: 2
Activities: designing electronic instruments
City Reference Coordinates: 008°36'00"E 45°49'00"N

Ente Nazionale Italiano di Unificazione (UNI)
Via Battistotti Sassi 11
I-20123 Milano
Telephone: (0)2-700241
Telefax: (0)2-70106106
Electronic Mail: uni@uni.unicei.it
WWW: http://www.unicei.it/uni/
City Reference Coordinates: 009°12'00"E 45°28'00"N

ESRIN
● See "European Space Agency (ESA), ESRIN"

Ettore Majorana Centre for Scientific Culture (EMCSC)
● See "Centro di Cultura Scientifica Ettore Majorana (CCSEM)"

European Association for Research Managers and Administrators (EARMA)
c/o INFM
Corso Perrone 24
I-16152 Genova
WWW: http://www.cineca.it/earma/
Founded: 1995
City Reference Coordinates: 008°57'00"E 44°25'00"N

European Space Agency (ESA), ESRIN
Via Galileo Galilei
Casella Postale 64
I-00044 Frascati
Telephone: (0)6-941801
Telefax: (0)6-94180361
Electronic Mail: <userid>@esrin.esa.it
WWW: http://www.esrin.esa.it/
         http://www.esa.int/
Founded: 1964 (ESA)
Activities: ESA's information processing and distribution centre * bibliographic databases through the Information Retrieval System (IRS) * internal ESA administrative information * Earth observation data through LEDA and ERS-1 catalogues
Periodicals: "Space Information Systems Newsletter"
City Reference Coordinates: 012°41'00"E 41°41'00"N

Gruppo Amici del Cielo (GAC)
c/o Biblioteca Comunale di Barzago
Via Roma 1
I-23890 Barzago
Electronic Mail: amicicielo@geocities.com
WWW: http://www.geocities.com/CapeCanaveral/Launchpad/1331/
Founded: 1997
Activities: meetings * popularization * observing
City Reference Coordinates: 009°20'00"E 45°46'00"N

Gruppo Antares
Via Ronchi 78
I-20025 Legnano
Electronic Mail: betax6@tin.it
WWW: http://rcl.nemo.it/reteciv/associaz/culturali/antares/
Founded: 1975
Activities: popularization * lectures * star parties * photography
City Reference Coordinates: 008°54'00"E 45°36'00"N

Gruppo Astrofili "Arthur Eddington" (GAAE)
Via Pasquale Ferraro 18
I-73030 Diso
Telephone: (0)836-921228
Electronic Mail: astro_eddington@hotmail.com
Founded: 1997
Membership: 20
Activities: education * lectures * observing * photography
City Reference Coordinates: 012°24'00"E 40°01'00"N

Gruppo Astrofili Aretini
Agazzi 72
I-52100 Arezzo
Telephone: (0)575-27880
Telefax: (0)575-323359
Electronic Mail: progecol@ats.it
WWW: http://www.arcetri.astro.it/CDAs/homepages/home10.html
City Reference Coordinates: 011°53'00"E 43°28'00"N

Gruppo Astrofili Astigiani (GAA) "Beta Andromedae"
Via C. Battisti 58
I-14030 Penango
or :
Via Brovardi 40
I-14100 Asti
Telephone: (0)141-916473
            (0)141-210066
Electronic Mail: gaa@oasi.shiny.it
WWW: http://oasi.shiny.it/Homes/GAA/
City Reference Coordinates: 008°12'00"E 44°54'00"N (Asti)

Gruppo Astrofili Catanesi "Guido Ruggieri"
c/o Luigi Prestinenza
Via L. Capuana 125
I-95124 Catania
Telephone: (0)95-535387
WWW: http://members.tripod.com/~astrofilicatanesi/
Founded: 1977
Membership: 150
City Reference Coordinates: 015°06'00"E 37°30'00"N

Gruppo Astrofili Columbia
c/o Ferruccio Zanotti
Via Magoni 21
I-44100 Ferrara
or :
Via Baccanazza 13
I-42025 Santa Maria Maddalena
Telephone: (0)532-91489 (Ferruccio Zanotti)
            (0)338-5264372 (Massimiliano Di Giuseppe)
Electronic Mail: columbia@global.it
                 jester@global.it
WWW: http://www.ferrara.com/columbia/
Founded: 1985
City Reference Coordinates: 011°38'00"E 44°50'00"N (Ferrara)

Gruppo Astrofili del Dopolavoro Ferroviario di Rimini
Via Roma 70
I-47037 Rimini
Telephone: (0)541-703242
Electronic Mail: gadlf@rimini.com
WWW: http://www.rimini.com/arte/astro/
Founded: 1984
City Reference Coordinates: 012°34'00"E 44°04'00"N

Gruppo Astrofili di Cinisello Balsamo (GACB)
c/o Autoscuola Ricci
Via Piave 6
I-20092 Cinisello Balsamo
Telephone: (0)2-6184578
Telefax: (0)2-66011843
Electronic Mail: benatti@imiucca.csi.unimi.it
WWW: http://www.mclink.it/mclink/astro/ass/gacb/
        http://astrolink.mclink.it/ass/gacb/
Founded: 1983
Membership: 30
Activities: variable stars * photography * comets * deep sky
Periodicals: (4) "Il Bollettino"
Observatories: 1 (Presolana)
City Reference Coordinates: 009°13'00"E 45°33'00"N

Gruppo Astrofili di Padova (GAP)
Corso Garibaldi 41
I-35100 Padova

Telephone: (0)049-8073312
WWW: http://www.mripermedia.com/AltraPD/Astrofili/index.shtml
       http://scuolaworld.provincia.padova.it/astrofili/
Founded: 1965
Membership: 70
Activities: popularization * education * research * Planetario "Galileo Galilei"
Periodicals: (1) "Annuario"
Coordinates: 011°53'48"E 45°24'00"N (Osservatorio G. Colombo)
City Reference Coordinates: 011°53'00"E 45°25'00"N

Gruppo Astrofili di Palermo
Casella Postale 1210
I-90100 Palermo
Telephone: (0)91-670231
WWW: http://www.mclink.it/mclink/astro/ass/ass0007.htm
       http://astrolink.mclink.it/ass/gap/
Founded: 1993
City Reference Coordinates: 013°21'00"E 38°07'00"N

Gruppo Astrofili di Schio (GAS)
Via Tiziano Vecellio
Casella Postale 115
I-36015 Schio
WWW: http://dns.lead.it/AstroSchio/
       http://www.lead.it/AstroSchio/
Founded: 1976
Membership: 50
Activities: observing * photography * popularization
Coordinates: 011°18'41"E 45°46'05"N H1,506m (Monte Novegno)
City Reference Coordinates: 011°21'00"E 45°43'00"N

Gruppo Astrofili di Rozzano (GAR)
c/o Biblioteca Civica
Piazza G. Foglia
I-20089 Rozzano
Electronic Mail: mbini@micronet.it
WWW: http://utenti.micronet.it/mbini/gar.htm
Founded: 1982
Membership: 40
City Reference Coordinates: 009°12'00"E 45°28'00"N (Milano)

Gruppo Astrofili Frentani
c/o studio Mancinone
Via Parma 2
I-66034 Lanciano
Electronic Mail: gruppoastrofilifrentani@yahoo.com
WWW: http://members.xoom.com/gaf97/
Founded: 1997
City Reference Coordinates: 014°23'00"E 42°13'00"N

Gruppo Astrofili Genovesi (GAG)
Casella Postale 836
I-16100 Genova
or :
Via A. Doria 9
I-16127 Genova
Telephone: (0)10-3460065
Electronic Mail: astro@gsi.it
WWW: http://www.gsi.it/astronomia/gag/
Founded: 1997
Membership: 36
Activities: meetings * astrophotography
Periodicals: "Superba"
City Reference Coordinates: 008°57'00"E 44°25'00"N

Gruppo Astrofili "Giovanni e Angelo Bernasconi"
Via San Giuseppe 34
I-21047 Saronno
Telephone: (0)331-830704
Telefax: (0)331-830704
Electronic Mail: lec.inc@tread.it
WWW: http://www.pangea.va.it/bernasconi/
Founded: 1965
Membership: 47
Activities: comets * planets * variable stars * instrumentation * lectures * scientific expeditions * photography * education

Periodicals: "Nihil Sub Astris Novum"
City Reference Coordinates: 009°02'00" E 45°38'00" N

Gruppo Astrofili Hipparcos
c/o CCCDS
Via Nomentana 175
I-00161 Roma
Telephone: (0)6-44250561
Electronic Mail: hipparcos.cds@mclink.it
WWW: http://diamante.uniroma3.it/hipparcos/index.htm
City Reference Coordinates: 012°29'00" E 41°54'00" N

Gruppo Astrofili "Isaac Newton"
Via Francesca Sud 329
I-56020 Santa Maria a Monte
Telephone: (0)587-706694 (Mauro Bachini)
             (0)587-706813 (Marco Novi)
Electronic Mail: marco.novi@ipermedia.net
WWW: http://www.arcetri.astro.it/CDAs/homepages/home3.html
City Reference Coordinates: 010°20'00" E 43°43'00" N (Pisa)

Gruppo Astrofili "La Nuova Selene"
c/o Davide Salerno (President)
Via De Deo 21
I-72015 Fasano
Telephone: (0)330-701580
Telefax: (0)80-4392336
Electronic Mail: lanuovaselene@puglianet.it
WWW: http://www.puglianet.it/lanuovaselene/
Founded: 1997
Membership: 15
Activities: education
Coordinates: 017°19'00" E 40°45'00" N H450m
City Reference Coordinates: 017°21'00" E 40°50'00" N

Gruppo Astrofili Lariani (GAL)
c/o Circoscrizione 6
Via Grandi 21
I-22100 Como
Telephone: (0)31-272196
Electronic Mail: sminardi@ing.unico.it
Founded: 1974
Membership: 100
Activities: popularization * observing
Sections: Variable Stars * Planets
Periodicals: (4) "L'Astrofilo Lariano" (circ.: 300-500)
Observatories: 1 (Ramponio Verna H1,000m)
City Reference Coordinates: 009°05'00" E 45°47'00" N

Gruppo Astrofili Manfredonia (GAM)
c/o Antonio Rubino
Lungomare del Sole 30
I-71043 Manfredonia
Telephone: (0)884-542883
Founded: 1975
Membership: 25
Activities: observing * education * introduction to astronomy
Periodicals: (12) "Bollettino Astronomico"
City Reference Coordinates: 015°55'00" E 41°38'00" N

Gruppo Astrofili Massesi (GAM)
Via Godola 42
I-54100 Massa
Telephone: (0)585-790594
Telefax: (0)585-790594
WWW: http://www.zia.ms.it/gam/
City Reference Coordinates: 010°09'00" E 44°02'00" N

Gruppo Astrofili Menkalinan (GAM)
Casa delle Culture
I-87100 Cosenza
Electronic Mail: polaris@diemme.it
                orbiter@tin.it
WWW: http://www.geocities.com/Area51/Dimension/5189/gam.htm
Founded: 1997

City Reference Coordinates: 016°16'00"E 39°17'00"N

Gruppo Astrofili "N. Copernico"
Via Pulzona 1708
Santa Maria del Monte
I-47835 Saludecio
Telephone: (0)541-857026
Telefax: (0)541-21082
Electronic Mail: copernic@iper.net
WWW: http://www.iper.net/koppernick/
Founded: 1976
Membership: 12
Activities: education * observing
Periodicals: (2) "Notiziario di Astronomia"
Coordinates: 012°24'18"E 43°53'08"N H150m
City Reference Coordinates: 012°40'00"E 43°52'00"N

Gruppo Astrofili ORSA
• See "Organizzazione Ricerche e Studi di Astronomia (ORSA)"

Gruppo Astrofili Pavese (GAP)
c/o Davide Re
Via Torino 23
I-27100 Pavia
Telephone: (0)382-463030
Electronic Mail: calimero@venus.it (Michele Moroni)
WWW: http://www.asanet.it/ospiti/gap/
Founded: 1997
City Reference Coordinates: 009°10'00"E 45°10'00"N

Gruppo Astrofili Persicetani
c/o Romano Serra
Vicolo Baciadonne 1
I-40017 San Giovanni in Persiceto
Telephone: (0)51-827067
WWW: http://members.it.tripod.de/san_giovanni/San_giovanni/osservatorio.htm
          http://members.it.tripod.de/san_giovanni/San_giovanni/Planetario.htm
Membership: 25
City Reference Coordinates: 011°11'00"E 44°38'00"N

Gruppo Astrofili Pesarese (GAP)
c/o Maurizio Mucci (President)
Via Ferrari 7
I-61100 Pesaro
Telephone: (0)721-51935
Electronic Mail: maurmucc@tin.it
              acono@abanet.it (Fabio Arcidiacono)
WWW: http://www.comune.pesaro.ps.it/allegati/astrofili/cielo.htm
Founded: 1985
Membership: 20
City Reference Coordinates: 012°55'00"E 43°54'00"N

Gruppo Astrofili Piceni (GAP)
c/o Maurizio Morricone
Viale De Gasperi 101
I-63039 San Benedetto del Tronto
Telephone: (0)73583416
Electronic Mail: gap@insinet.it
WWW: http://www.insinet.it/gap/
City Reference Coordinates: 013°53'00"E 42°57'00"N

Gruppo Astrofili Reggini M 31 del Dopolavoro Ferroviario
c/o Gaetano De Benedetto (President)
Via Ciccarello Trav. IV 14
Casella Postale 148
I-89100 Reggio di Calabria
Telephone: (0)965-622239
Telefax: (0)965-622239
Founded: 1976
Membership: 30
Activities: popularization * observing (occultations, minor planets, variable stars)
Coordinates: 015°19'05"E 38°06'25"N H93m (Osservatorio A. Righi)
City Reference Coordinates: 015°39'00"E 38°07'00"N

Gruppo Astrofili Rigel
c/o Zuffi Valerio
Via Adua 31
I-20010 Arluno
WWW: http://www.geocities.com/CapeCanaveral/Galaxy/3204/index.htm
City Reference Coordinates: 009°12'00"E 45°28'00"N (Milano)

Gruppo Astrofili Romani (GAR)
c/o Parrocchia San Filippo Neri
Via Martino V 28
I-00167 Roma
Telephone: (0)6-39730250 (President)
Electronic Mail: gianluca.rossi@agippetroli.emi.it (Gianlucca Rossi, Presidente)
                memory.line@agora.stm.it (GAR News)
WWW: http://astrolink.mclink.it/ass/gar/
        http://www.freeweb.org/associazioni/GAR/
Founded: 1996
Membership: 26
Activities: education * observing * popularization * exhibitions * photography (deep sky, planets, Moon)
Periodicals: "GAR News"
City Reference Coordinates: 012°29'00"E 41°54'00"N

Gruppo Astrofili Savonesi (GAS)
Casella Postale 5
I-17100 Savona
Telephone: (0)19-822925 (Fabrizio Ciliberto, Secretary)
             (0)19-853110 (Roberto Bracco, President)
Electronic Mail: gas@ils.org
WWW: http://www.publinet.it/arte/gas/
Founded: 1969
Membership: 50
Activities: observing * popularization * education
Periodicals: (2) "Cielosservare" (circ.: 100)
Coordinates: 008°28'47"E 44°23'44"N H395m
City Reference Coordinates: 008°30'00"E 44°17'00"N

Gruppo Astrofili Soresinesi (GAS)
c/o Angelo Marchesini
Osservatorio Astronomico Pubblico
Via Matteotti 2
Casella Postale 21
I-26015 Soresina
Telephone: (0)374-43722 (Observatory)
             (0)374-70186 (President)
Telefax: (0)373-30901 (call for connection)
Founded: 1974
Membership: 100
Activities: popularization * image processing * software
City Reference Coordinates: 009°51'00"E 45°17'00"N

Gruppo Astrofili Vesuviano (GAV)
Via Mariotti 32
I-80047 San Giuseppe Vesuviano
Telephone: (0)81-5293941
Telefax: (0)81-5296195
Electronic Mail: astrofilivesuviani@yahoo.it
WWW: http://digilander.iol.it/gav/
Founded: 1998
Membership: 20
Activities: popularization * education * observing
City Reference Coordinates: 014°31'00"E 40°50'00"N

Gruppo Astrofili Vicentini (GAV) "Giorgio Abetti"
Centro Civico
Villa Lattes
Via Thaon di Revel
I-36100 Vicenza
Electronic Mail: gav@keycomm.it
WWW: http://www.keycomm.it/~gav/
Founded: 1986
Membership: 180
Activities: popularization
Coordinates: 011°32'09"E 45°29'50"N H190m
City Reference Coordinates: 011°33'00"E 45°33'00"N

Gruppo Astrofili "William Herschel"

Corso Monte Cucco 137
I-10141 Torino
Telephone: (0)11-3835279
          (0)11-354570
          (0)11-336844
Electronic Mail: piero@beatrix.cselt.stet.it
          a.bertoglio@agora.stm.it
WWW: http://www.valnet.it/WH/
City Reference Coordinates: 007°40'00"E 45°03'00"N

Gruppo Astronomia Digitale (GAD)

c/o Claudio Lopresti
Via Castellazzo 8/D
I-19125 La Spezia
Telephone: (0)187-715391
Telefax: (0)187-715391
Electronic Mail: iras@village.it
WWW: http://www.itsyn/astro
      http://www.luna.it/associazioni/gad/
Periodicals: "Notiziario GAD"
Founded: 1992
Membership: 228
Activities: SM * image processing * digital astronomy * computing
City Reference Coordinates: 009°50'00"E 44°07'00"N

Gruppo Astronomico Tradatese (GAT)

c/o Biblioteca Civica
Via Mameli 13
I-21049 Tradate
Telephone: (0)331-841820
Telefax: (0)331-820317
      (0)331-810117
Electronic Mail: gatrad@gwtradate.tread.it (Lorenzo Comolli)
WWW: http://gwtradate.tread.it/tradate/gat/Welcome.html
Founded: 1975
Membership: 350
Activities: planets * Sun * CCD photography * exhibitions * education
Periodicals: (6) "Lettera ai Soci"
Observatories: 1 (Monte San Martino, Valcuvia)
Awards: (1) "Premio Eros Benatti"
City Reference Coordinates: 008°54'00"E 45°43'00"N

Gruppo Astronomico Viareggio (GAV)

c/o Martellini Davide
Casella Postale 406
I-55049 Viareggio
Telephone: (0)584-395895
Electronic Mail: gav.it@usa.net
WWW: http://members.tripod.it/gav/
Founded: 1973
Membership: 30
Activities: popularization * research
Periodicals: (6) "Astronews"
City Reference Coordinates: 010°14'00"E 43°52'00"N

Gruppo "G.E.D. Alcock"

c/o Marco Vincenzi
Via Accademia dei Virtuosi 4
I-00147 Roma
Telephone: (0)6-6281880
Electronic Mail: mc9648@mclink.it (Giovanni Guerrieri)
          mc8255@mclink.it (Marco Vincenzi)
WWW: http://www.mclink.it/mclink/astro/ass/ass0002.htm
      http://astrolink.mclink.it/ass/ass0002.htm
City Reference Coordinates: 012°29'00"E 41°54'00"N

Gruppo Marsicano Astrofili (GMA) "F. Angelitti"

c/o Paolo Maria Ruscitti
Via Opi 5
I-67051 Avezzano
Telephone: (0)863-414799
Electronic Mail: zauri@aquila.infn.it (Renato Zauri)
WWW: http://moloch.univaq.it/~zauri/astro/
Founded: 1985
City Reference Coordinates: 013°26'00"E 42°02'00"N

IEI 83 Srl
Via Fratelli Dandolo 10
I-50135 Firenze
Telephone: (0)55-678811
Telefax: (0)55-678811
Electronic Mail: mc7977@mclink.it
Activities: manufacturing and distributing optical accessories
City Reference Coordinates: 011°15'00"E 43°46'00"N

International Center for Relativistic Astrophysics (ICRA)
(Consorzio Internazionale per l'Astrofisica Relativistica)
c/o Dipartimento di Fisica
Università degli Studi di Roma
Piazzale Aldo Moro 5
I-00185 Roma
Telephone: (0)6-49914254 (Secretary)
                    (0)6-49914304 (Director)
Telefax: (0)6-4454992
WWW: http://www.icra.it/
Founded: 1985
Activities: theoretical, experimental and observational astrophysics justifying international collaboration * managing instruments and equipment * appropriate methodologies and technologies
City Reference Coordinates: 012°29'00"E 41°54'00"N

International Centre for Theoretical Physics (ICTP)
● See now "Abdus Salam International Centre for Theoretical Physics"

International School for Advanced Studies (ISAS)
● See "Scuola Internazionale Superiore di Studi Avanzati (SISSA)"

International Supernovae Network (ISN)
c/o Stefano Pesci
Via Birolli 3
I-20125 Milano
Electronic Mail: peste@micronet.it
                       villi@mbox.queen.it
WWW: http://www.queen.it/web4you/noprofit/isn/isn.htm
          http://www.supernovae.net/isn.htm
Founded: 1996
Membership: 110
Activities: SN search and observation
City Reference Coordinates: 009°12'00"E 45°28'00"N

International Union of Amateur Astronomers (IUAA)
c/o Luigi Baldinelli
Casella Postale 1630
I-40100 Bologna
Telephone: (0)51-247784
Telefax: (0)51-247393
Electronic Mail: mc8070@mclink.it
Founded: 1969
Activities: coordinating activities of amateur astronomers throughout the world * promoting the study of astronomy in all its aspects
Periodicals: "Communications"
City Reference Coordinates: 011°20'00"E 44°29'00"N

International Union of Geological Sciences (IUGS)
c/o Attilio C. Boriani (Secretary General)
Dipartimento di Scienze della Terra
Università di Milano
I-20133 Milano
Telephone: (0)2-23698310
Telefax: (0)2-70638681
Electronic Mail: boriani@r10.terra.unimi.it
WWW: http://www.iugs.org/
Founded: 1961
Membership: 110 (nations)
Activities: organizing international geoscience meetings and cooperative scientific programs
Sections: (commissions) Comparative Planetology * Geological Documentation * Geology Teaching * Global Sedimentary Geology * History of Geological Sciences * Igneous and Metamorphic Petrogenesis * Marine Geology * Storage, Automatic Processing and Retrieval of Geological Data * Stratigraphy * Systematics in Petrology * Tectonics * Fossil Fuels * Geoscience for Environmental Planning
Periodicals: (4) "Episodes" (ISSN 0705-3797)
City Reference Coordinates: 009°12'00"E 45°28'00"N

Istituto di Astrofisica Spaziale (IAS)
● See "Consiglio Nazionale delle Ricerche (CNR), Istituto di Astrofisica Spaziale (IAS)"

Istituto di Fisica Cosmica ed Applicazioni dell'Informatica (IFCAI)
● See "Consiglio Nazionale delle Ricerche (CNR), Istituto di Fisica Cosmica ed Applicazioni dell'Informatica (IFCAI)"

Istituto di Fisica dello Spazio Interplanetario (IFSI)
● See "Consiglio Nazionale delle Ricerche (CNR), Istituto di Fisica dello Spazio Interplanetario"

Istituto di Metrologia G. Colonnetti (IMGC)
● See "Consiglio Nazionale delle Ricerche (CNR), Istituto di Metrologia G. Colonnetti (IMGC)"

Istituto di Radioastronomia (IRA)
● See "Consiglio Nazionale delle Ricerche (CNR), Istituto di Radioastronomia (IRA)"

Istituto di Tecnologie e Studio delle Radiazioni Extraterrestri (TESRE)
● See "Consiglio Nazionale delle Ricerche (CNR), Istituto di Tecnologie e Studio delle Radiazioni Extraterrestri (TESRE)"

Istituto e Museo di Storia della Scienza (IMSS), Planetario
Piazza dei Giudici 1
I-50122 Firenze
Telephone: (0)55-293493
            (0)55-2398876
Telefax: (0)55-288257
WWW: http://galileo.imss.firenze.it/indice.html (Italian)
        http://galileo.imss.firenze.it/general/ (English)
Founded: 1977 (Museum: 1930)
Staff: 12
Activities: education
Periodicals: (2) "Nuncius"
City Reference Coordinates: 011°15'00"E 43°46'00"N

Istituto Nazionale di Fisica Nucleare (INFN), Laboratori Nazionali del Gran Sasso (LNGS)
S.S. 17/bis - Km 18.910
I-67010 Assergi
Telephone: (0)862-4371
Telefax: (0)862-410795
        (0)862-437570
Electronic Mail: segreteria@lngs.infn.it
WWW: http://www.lngs.infn.it/
Founded: 1988
City Reference Coordinates: 013°42'00"E 42°27'00"N

Istituto Nazionale di Fisica Nucleare (INFN), Laboratori Nazionali di Frascati (LNF)
Casella Postale 13
I-00044 Frascati
Telephone: (0)6-94031
Telefax: (0)6-94032582
Electronic Mail: <userid>@lnf.infn.it
WWW: http://www.lnf.infn.it/
Founded: 1955
City Reference Coordinates: 012°41'00"E 41°41'00"N

Istituto Nazionale di Fisica Nucleare (INFN), Sezione di Napoli, Astrofisica Solare e Stellare
Mostra d'Oltremare
Pad. 20
I-80125 Napoli
Telephone: (0)81-7253111
Telefax: (0)81-2394508
Electronic Mail: <userid>@na.infn.it
WWW: http://www.na.infn.it/Astr/solar/report.html
City Reference Coordinates: 014°17'00"E 40°51'00"N

Istituto Nazionale di Geofisica (ING)
Via di Vigna Murata 605
I-00143 Roma
Telephone: (0)6-518601
Telefax: (0)6-5041181
Electronic Mail: <userid>@ingrm.it
WWW: http://www.ingrm.it/
Founded: 1936
Staff: 220
Activities: physics of the Earth * seismology * geomagnetism * aeronomy
Periodicals: (12) "Annali di Geofisica"

City Reference Coordinates: 012°29'00"E 41°54'00"N

Istituto per Ricerche in Fisica Cosmica e Tecnologie Relative (IFCTR)
● See "Consiglio Nazionale delle Ricerche (CNR), Istituto per Ricerche in Fisica Cosmica e Tecnologie Relative (IFCTR)"

Istituto Spezzino Ricerche Astronomiche (IRAS)
c/o Claudio Lopresti
Via Castellazzo 8/D
I-19125 La Spezia
Telephone: (0)187-715391
Telefax: (0)187-715391
Electronic Mail: iras@village.it
WWW: http://www.village.it/iras
        http://www.luna.it/associationi/iras
Founded: 1991
Membership: 78
Activities: popularization * photometry of variable stars * SN * image processing * digital astronomy * computing
Periodicals: "Notiziario IRAS"
Coordinates: 010°06'22"E 46°11'21"N H178m (Gragnola)
City Reference Coordinates: 009°50'00"E 44°07'00"N

Istituto Tecnico Nautico "Artiglio", Planetario
Via dei Pescatori
I-55049 Viareggio
Telephone: (0)584-390282
Telefax: (0)584-392090
Electronic Mail: info@nauticoartiglio.lu.it
WWW: http://www.nauticoartiglio.lu.it/
        http://www.nauticoartiglio.lu.it/planetario/planetar00.htm
City Reference Coordinates: 010°14'00"E 43°52'00"N

Istituto Tecnico Nautico Statale, Planetario
Via Mazzini 26
I-66026 Ortona
Telephone: (0)85-9063441
Telefax: (0)85-9063441
Electronic Mail: nautico.ortona@tin.it
WWW: http://www.nauticortona.it/
Founded: 1921
Staff: 6
Activities: education
Coordinates: 014°24'18"E 42°21'00"N H80m
City Reference Coordinates: 014°24'00"E 42°21'00"N

Istituto Universitario Navale, Cattedra di Astronomia Nautica
Via Acton 38
I-80100 Napoli
Telephone: (0)81-5475229
Telefax: (0)81-5513977
        (0)81-5519314
Electronic Mail: <userid>@naval.uninav.it
City Reference Coordinates: 014°17'00"E 40°51'00"N

Konus Italia Group Srl
Via Mirandola 45
I-37026 Settimo di Pescantina
Telephone: (0)45-6767670
Telefax: (0)45-6767671
Founded: 1979
● Dealer and distributor
City Reference Coordinates: 010°51'00"E 45°29'00"N (Pescantina)

Laben SpA
Strada Padana Superiore 290
I-20090 Vimodrone
Telephone: (0)2-250751
Telefax: (0)2-2505515
Electronic Mail: info@www.laben.i
WWW: http://www.laben.it/
Founded: 1958
Staff: 400
Activities: manufacturing satellite data equipments, checkout systems, electronic instrumentation * radioactivity control networks
City Reference Coordinates: 009°17'00"E 45°31'00"N

Laboratori Nazionali del Gran Sasso (LNGS)
• See "Istituto Nazionale di Fisica Nucleare (INFN), Laboratori Nazionali del Gran Sasso (LNGS)"

Laboratori Nazionali di Frascati (LNF)
• See "Istituto Nazionale di Fisica Nucleare (INFN), Laboratori Nazionali di Frascati (LNF)"

L'astronomia
Edizioni Media Presse Srl
Via Nino Bixio 30
I-20129 Milano
Telephone: (0)2-2043941
Telefax: (0)2-2046507
Founded: 1979
Staff: 6
• Journal (11) (circ.: 50,000)
City Reference Coordinates: 009°12'00"E 45°28'00"N

MSB Software
Via Romea Vecchia 67
I-48100 Classe
Telephone: (0)544-473589
Telefax: (0)544-473589
Electronic Mail: msbsoftware@tin.it
WWW: http://www.sira.it/msb
Founded: 1997
Staff: 2
• Software producer
Coordinates: 012°12'25"E 44°24'55"N
City Reference Coordinates: 012°12'00"E 44°25'00"N (Ravenna)

Museo Cassiniano
Piazza San Antonio
I-18030 Perinaldo
WWW: http://www.rosenet.it/perinaldo/museo.htm
Founded: 1990
Coordinates: 007°40'00"E 43°51'00"N H573m
City Reference Coordinates: 007°40'00"E 43°51'00"N

Museo Civico di Rovereto, Planetario
Borgo Santo Caterina 43
I-38068 Rovereto
Telephone: (0)464-439055
Telefax: (0)464-439487
Electronic Mail: museo@museocivico.rovereto.tn.it
WWW: http://www.museocivico.rovereto.tn.it
       http://www.museocivico.rovereto.trento.it/museo/planetario.htm
City Reference Coordinates: 011°03'00"E 45°53'00"N

Nuova Telespazio SpA
Via Tiburtina 965
I-00156 Roma
Telephone: (0)6-40791
City Reference Coordinates: 012°29'00"E 41°54'00"N

Organizzazione Ricerche e Studi d'Astronomia (ORSA)
Via Tramontana 28
I-90140 Palermo
or :
Via Zandonai 22
I-90144 Palermo
Telephone: (0)91-6703029
          (0)91-6813137
Electronic Mail: demaria@mbox.vol.it
WWW: http://www.vol.it/astrorsa/
       http://www.ifcai.pa.cnr.it/ORSA/
Activities: popularization * observing
City Reference Coordinates: 013°21'00"E 38°07'00"N

Osservatorio Astrofisico di Arcetri (OAA)
Largo Enrico Fermi 5
I-50125 Firenze
Telephone: (0)55-27521
Telefax: (0)55-220039
Electronic Mail: <userid>@arcetri.astro.it
WWW: http://www.arcetri.astro.it/

http://lbtwww.arcetri.astro.it/ (Large Binocular Telescope - LBT)
Founded: 1872
Staff: 75
Activities: solar physics * ISM * extragalactic astrophysics * high-energy astrophysics * instrumentation * adaptive optics * LBT project * TIRGO telescope
Periodicals: "Arcetri Astrophysical Preprints", "Arcetri Technical Reports"; (1) "Arcetri Astrophysical Observatory Annual Report"
Coordinates: 011°15'19"E 43°45'14"N H180m
007°47'30"E 45°59'04"N H3,135m (TIRGO Telescope, Zermatt, Switzerland)
City Reference Coordinates: 011°15'00"E 43°46'00"N

Osservatorio Astrofisico di Asiago
● See "Università di Padova, Dipartimento di Astronomia"

Osservatorio Astrofisico di Catania (OAC)
Città Universitaria
Viale Andrea Doria 6
I-95125 Catania
Telephone: (0)95-7332111
Telefax: (0)95-330592
Electronic Mail: <userid>@astrct.ct.astro.it
WWW: http://www.ct.astro.it/
Founded: 1880
Staff: 56
Activities: solar and stellar activity * interplanetary and interstellar matter * minor bodies of the solar system * relativity * cosmology * variable stars
Periodicals: "Bollettino Dati Solari" (ISSN 1120-8430)
Coordinates: 015°04'21"E 37°31'43"N H193m
014°58'24"E 37°41'30"N H1,735m (Serra la Nave, Etna)
City Reference Coordinates: 015°06'00"E 37°30'00"N

Osservatorio Astronomico di Bologna
Via Ranzani 1
I-40127 Bologna
Telephone: (0)51-6305727
Telefax: (0)51-6305700
Electronic Mail: <userid>@astbo3.bo.astro.it
WWW: http://www.bo.astro.it/
http://www.bo.astro.it/loiano/LoianoHome.html (Loiano Observatory)
Founded: 1985
Staff: 65
Activities: galactic and stellar astrophysics * large-scale structures and cosmology * numerical methods for astronomy * instrumentation
Periodicals: "Bologna Astrophysical Preprints", "Technical Reports"
Coordinates: 011°20'12"E 44°15'30"N H785m (Loiano)
City Reference Coordinates: 011°20'00"E 44°29'00"N

Osservatorio Astronomico di Brera (OAB), Merate
Via E. Bianchi 46
I-23807 Merate
Telephone: (0)39-999111
Telefax: (0)39-9991160
Electronic Mail: <userid>@merate.mi..astro.it
WWW: http://www.merate.mi.astro.it/
http://albinoni.brera.unimi.it/
http://www.mi.astro.it/
Founded: 1927
Staff: 35
Activities: stellar pulsations * observational cosmology * X-ray technology * celestial mechanics * meteorology * X-ray stellar sources * stellar evolution
Coordinates: 009°25'42"E 45°42'00"N H340m
City Reference Coordinates: 009°25'00"E 45°42'00"N (Merate)
009°12'00"E 45°28'00"N (Milano)

Osservatorio Astronomico di Brera (OAB), Milano
Via Brera 28
I-20121 Milano
Telephone: (0)2-723201
Telefax: (0)2-72001600
Electronic Mail: <userid>@brera.mi.astro.it
WWW: http://www.brera.mi.astro.it/
http://albinoni.brera.unimi.it/
Founded: 1760
Staff: 25
Activities: stellar pulsations * observational cosmology * X-ray technology * celestial mechanics * meteorology * X-ray stellar sources * stellar evolution

Coordinates: 009°11'30"E 45°28'00"N H146m
City Reference Coordinates: 009°12'00"E 45°28'00"N (Milano)

Osservatorio Astronomico di Cagliari (OAC)
(Stazione Astronomica di Cagliari)
Strada 54
Località Poggio dei Pini
I-09012 Capoterra
Telephone: (0)70-711801
Telefax: (0)70-71180222
Electronic Mail: scisec@ca.astro.it
          <userid>@ca.astro.it
WWW: http://www.ca.astro.it/
Founded: 1899
Staff: 28
Activities: geodynamics * astrophysics
Coordinates: 008°58'24"E 39°08'12"N (Capoterra)
          008°18'44"E 39°08'09"N (Carloforte)
          009°19'47"E 39°59'42"N (Bruncu Spina)
City Reference Coordinates: 009°06'00"E 39°13'00"N (Cagliari)

Osservatorio Astronomico di Capodimonte (OAC)
Via Moiariello 16
I-80131 Napoli
Telephone: (0)81-440101
Telefax: (0)81-456710
Electronic Mail: <userid>@oacosf.na.astro.it
WWW: http://oacosf.na.astro.it/
Founded: 1819
Staff: 50
Activities: stars * Sun * galaxies
Coordinates: 014°15'18"E 40°51'48"N H150m
City Reference Coordinates: 014°17'00"E 40°51'00"N

Osservatorio Astronomico di Colluriana-Teramo (OACT)
Via Mentore Maggini
I-64100 Teramo
Telephone: (0)861-210490
Telefax: (0)861-210492
Electronic Mail: <userid>@astrte.te.astro.it
WWW: http://terri1.te.astro.it/oact.home/home.html
Founded: 1892
Staff: 18
Activities: stellar evolution * minor planets * clusters * nucleosynthesis * RR Lyrae stars * Cepheids * dwarf galaxies * population synthesis
Coordinates: 013°44'00"E 42°39'27"N H398m (Colluriana)
          013°33'40"E 42°26'13"N H2,120m (Campo Imperatore)
City Reference Coordinates: 013°42'00"E 42°39'00"N (Teramo)

Osservatorio Astronomico di Cuneo
c/o Fulvio Romano
Liceo Scientifico "G. Peano"
Via Monte Zovetto 8
I-12100 Cuneo
Telephone: (0)171-694125
          (0)171-692906
Electronic Mail: fulvio.romano@pmn.it
Founded: 1990
Activities: solar activity * CCD
City Reference Coordinates: 007°32'00"E 44°23'00"N

Osservatorio Astronomico di Genova
● See "Università Popolare Sestrese, Osservatorio Astronomico di Genova"

Osservatorio Astronomico di Merate
● See "Osservatorio Astronomico di Brera (OAB), Merate"

Osservatorio Astronomico di Milano
● See "Osservatorio Astronomico di Brera (OAB), Milano"

Osservatorio Astronomico di Padova
Vicolo dell'Osservatorio 5
I-35122 Padova
Telephone: (0)49-8293411
          (0)424-462032 (Monte Ekar)

Telefax: (0)49-8759840
			(0)424-462884 (Monte Ekar)
Electronic Mail: <userid>@pd.astro.it
				<userid>@pd.astro.it (Monte Ekar)
WWW: http://www.pd.astro.it/
Founded: 1767
Staff: 70
Activities: solar system * stellar astrophysics * galactic and extragalactic astronomy * cosmology * high-energy astrophysics * space research * headquarters of the "Telescopio Nazionale Galileo (TNG - see separate entries in Italy and Spain)"
Coordinates: 011°52'18"E 45°24'00"N H38m
			011°34'17"E 45°50'36"N H1,350m (Monte Ekar)
City Reference Coordinates: 011°53'00"E 45°25'00"N

Osservatorio Astronomico di Palermo (OAP) "Giuseppe S. Vaiana"
Palazzo dei Normanni
I-90134 Palermo
Telephone: (0)91-6570451
Telefax: (0)91-488900
Electronic Mail: astropa@unipa.it
WWW: http://www.astropa.unipa.it/
Founded: 1790 (re-established independent: 1989)
Staff: 21
Activities: high-energy astronomy * X-rays * solar and stellar coronae * history of astronomy
Periodicals: (1) "Elementi Astronomici"
Coordinates: 013°21'22"E 38°06'44"N H72m
City Reference Coordinates: 013°21'00"E 38°07'00"N

Osservatorio Astronomico di Perugia
● See "Università di Perugia, Osservatorio Astronomico"

Osservatorio Astronomico di Roma (OAR)
Viale del Parco Mellini 84
I-00136 Roma
Telephone: (0)6-347056
Telefax: (0)6-3498236
WWW: http://oar.rm.astro.it/
		http://www.rm.astro.it/
		http://oar.rm.astro.it/home.html
Founded: 1938
Staff: 21
Activities: cosmology * stellar evolution * AGNs
Coordinates: 013°33'37"E 42°26'35"N H2,200m (Campo Imperatore)
			012°27'27"E 41°55'19"N H150m (Monte Mario)
City Reference Coordinates: 012°29'00"E 41°54'00"N

Osservatorio Astronomico di Roma (OAR), Monteporzio Catone
Via dell'Osservatorio 5
I-00040 Monteporzio
Telephone: (0)6-9449019
WWW: http://oar.rm.astro.it/
		http://www.rm.astro.it/
		http://www.mporzio.astro.it/oar-home.html
Founded: 1938 (OAR)
Activities: stellar spectroscopy * stellar evolution * clusters of galaxies
Coordinates: 012°42'00"E 41°48'00"N H350m
City Reference Coordinates: 012°42'00"E 41°48'00"N

Osservatorio Astronomico di Teramo
● See "Osservatorio Astronomico di Colluriana-Teramo"

Osservatorio Astronomico di Torino (OATo)
Strada Osservatorio 20
I-10025 Pino Torinese
Telephone: (0)11-4619000
Telefax: (0)11-4619030
Electronic Mail: <userid>@to.astro.it
WWW: http://otoxd2.to.astro.it/
		http://www.unito.it/
		http://astroserver.to.astro.it/ (Astrometric Group)
Founded: 1759
Staff: 52
Activities: astronomy * astrophysics * planetology * astrometry
Coordinates: 007°46'29"E 45°02'16"N H622m
City Reference Coordinates: 007°40'00"E 45°03'00"N

Osservatorio Astronomico di Trieste (OAT)
Casella Postale Succursale Trieste 5
Via G.B. Tiepolo 11
I-34131 Trieste
Telephone: (0)40-3199241 (Secretary)
Telefax: (0)40-309418
Electronic Mail: <userid>@oat.ts.astro.it
WWW: http://www.oat.ts.astro.it/
http://www.oat.ts.astro.it/oat-home.html
Founded: 1753
Staff: 60
Activities: solar physics * solar radioastronomy * cometary physics * stellar physics * stellar evolution * stellar atmospheres * ISM * extragalactic astrophysics * cosmology * observational archives * image processing * supercomputing * astronomy and space technologies
Periodicals: (4) "Osservazioni Solari"
Coordinates: 013°01'22"E 45°38'35"N H77m
City Reference Coordinates: 013°46'00"E 45°40'00"N

Osservatorio Astronomico Grange
Casella Postale 54
I-10053 Bussoleno
or :
Via M. d'Azeglio 34
I-10053 Bussoleno
Telephone: (0)122-640797
Telefax: (0)122-640797
Electronic Mail: grange@mclink.it
pognant@tin.it
mc2213@mclink.it
WWW: http://www.mclink.it/mclink/astro/ass/grange/grange1.htm
http://www.mclink.it/mclink/astro/ass/grange/english.htm
Founded: 1993
Staff: 2
Activities: astrometry of minor planets
Coordinates: 007°08'29"E 45°08'31"N H470m
City Reference Coordinates: 007°09'00"E 45°08'00"N

Osservatorio Astronomico Pubblico, Soresina
c/o Erinio Pini
Via Matteotti 4
I-26015 Soresina
Telephone: (0)374-43722
Founded: 1974
Staff: 15
City Reference Coordinates: 009°51'00"E 45°17'00"N

Osservatorio Astronomico "Serafino Zani"
c/o Centro Studi e Ricerche Serafino Zani
Via Bosca 24
Casella Postale 104
I-25066 Lumezzane
Telephone: (0)30-871861
Telefax: (0)30-872545
WWW: http://www.cityline.it/cult/zani/
Founded: 1993
Staff: 15
Activities: observing * photography * CCDs * education * projects * planetariums
Periodicals: (4) "Sagittario"
Coordinates: 010°14'25"E 45°39'53"N H830m
Awards: (1/2) "Shadows of Time"
City Reference Coordinates: 010°13'00"E 45°33'00"N (Brescia)

Osservatorio Astronomico Sharru
Via Giovanni XXIII 13
I-24050 Covo
Telephone: (0)363-93102
Telefax: (0)363-93102
Founded: 1975
Staff: 2
Activities: visual and photoelectric observing (novae, SN, cataclysmic variable stars)
Coordinates: 009°46'15"E 45°29'50"N H115m
City Reference Coordinates: 009°43'00"E 45°41'00"N (Bergamo)

Osservatorio San Giuseppe
Parr. San Donato a Livizzano
I-50056 Montelupo

Telephone: (0)571-671935
                (0)571-671106
Telefax: (0)571-675835
Electronic Mail: mirediz@logo.it (Claudio Allegri, Observatory Manager)
Founded: 1990
Staff: 8
Activities: popularization * deep-sky observing
Coordinates: 011°10'00"E 43°41'40"N H196m
City Reference Coordinates: 011°01'00"E 43°44'00"N

Osservatorio San Vittore
Via San Vittore 44
I-40136 Bologna
Founded: 1969
Staff: 6
Activities: private observatory * astrometry of minor planets and comets
Coordinates: 011°20'24"E 44°28'05"N H280m
City Reference Coordinates: 011°20'00"E 44°29'00"N

Pieri Astronomy Observatory
c/o Alessandro Pieri
Via Fratelli Bandiera 19
I-04100 Latina
Telephone: (0)773-695322
Electronic Mail: alex@uni.net
WWW: http://www.mclink.it/mclink/astro/ass/pao/
        http://astrolink.mclink.it/ass/pao/
● Private observatory
City Reference Coordinates: 012°52'00"E 41°28'00"N

Planetario Comunale di Modena
c/o M.U. Lugli
Viale Jacopo Barozzi 31
Casella Postale 34
Succursale 2
I-41100 Modena
Telephone: (0)59-224726
Telefax: (0)59-224726
WWW: http://www.isesitalia.it/SCUOLA/PV1000/ASTRO.HTM
City Reference Coordinates: 010°55'00"E 44°39'00"N

Planetario Comunale di Pisa
Via Mario Lalli 4
I-56127 Pisa
Telephone: (0)50-580456
Telefax: (0)50-580456
Electronic Mail: planetario@comune.pisa.it
WWW: http://www.comune.pisa.it/doc/istruzione/mep.htm
Founded: 1989
Staff: 2
City Reference Coordinates: 010°20'00"E 43°43'00"N

Planetario del Museo Provinciale di Storia Naturale
Via Roma 234
I-57127 Livorno
Telephone: (0)586-802294
Founded: 1978
Staff: 30
Activities: education * popularization * education * lectures * observing
City Reference Coordinates: 010°19'00"E 43°33'00"N

Planetario di Ravenna
Viale Santi Baldini 4A
I-48100 Ravenna
Telephone: (0)544-62534
Telefax: (0)544-22928
WWW: http://racine.ra.it/planet/homeplan.html
Founded: 1985
Staff: 3
Activities: education * popularization
Periodicals: "Almanaco Astronomico"
City Reference Coordinates: 012°12'00"E 44°25'00"N

San Gersolè Planetary Group
Via San Gersolè 2

I-50020 Monteoriolo
Telephone: (0)55-678811
Electronic Mail: iei83_g_quarra@mclink.it (Giovanni Quarra)
               d.sarocchi@mclink.it (Damiano Sarocchi)
WWW: http://www.chim1.unifi.it/group/education/caat/sgpg/
      http://blu.chim1.unifi.it/group/education/caat/sgpg/
Founded: 1988
City Reference Coordinates: 011°15'00"E 43°46'00"N (Firenze)

SAX Scientific Data Center (SDC)
● See "BeppoSAX Scientific Data Center (SDC)"

Scuola Internazionale Superiore di Studi Avanzati (SISSA)
(International School for Advanced Studies - ISAS)
Via Beirut 2-4
I-34014 Trieste
Telephone: (0)40-37871
Telefax: (0)40-3787528
Electronic Mail: <userid>@neumann.sissa.it
                <userid>@sissa.it
WWW: http://www.sissa.it/
      http://babbage.sissa.it/ (Preprint Server)
Founded: 1978
Staff: 46
Activities: physics * mathematics * biophysics * geometry * astrophysics * cognitive neurosciences
City Reference Coordinates: 013°46'00"E 45°40'00"N

Scuola Normale Superiore, Pisa
Piazza dei Cavalieri 7
I-56126 Pisa
Telephone: (0)50-509111
Telefax: (0)50-563513
Electronic Mail: <userid>@sns.it
Founded: 1810
Staff: 5
Activities: galactic dynamics * spiral and elliptical galaxies * cooling flows * dark matter * plasma astrophysics * n-body simulations * gravitational lensing * cluster of galaxies
City Reference Coordinates: 010°20'00"E 43°43'00"N

Società Astronomica "G.V. Schiaparelli"
c/o Salvatore Furia (President)
Via Andrea del Sarto 3
I-21100 Varese
or :
c/o Osservatorio Astronomico "G.V.Schiaparelli"
Campo dei Fiori
I-21100 Varese
Telephone: (0)332-235491
Telefax: (0)332-237143
Electronic Mail: astrogeo@astrogeo.va.it
WWW: http://www.astrogeo.va.it/astronom/astronom.htm
Founded: 1956
Membership: 400
Activities: popularization * astronomy * meteorology * botanics * seismology
Coordinates: 008°46'15"E 45°52'04"N
City Reference Coordinates: 008°49'00"E 45°49'00"N

Società Astronomica Italiana (SAIt)
c/o The Secretary
Osservatorio Astrofisico di Arcetri
Largo Enrico Fermi 5
I-50125 Firenze
Telephone: (0)55-435939
Telefax: (0)55-220039
WWW: http://www.sait.it/
Founded: 1920
Membership: 900
Periodicals: (4) "Memorie della Società Astronomica Italiana" (ISSN 0037-8720); 'Il Giornale di Astronomia" (ISSN 0390-1106)
City Reference Coordinates: 011°15'00"E 43°46'00"N

Società Astronomica Urania (SAU)
Strada Stradella 9
I-15067 Novi Ligure
Telephone: (0)143-329565

Electronic Mail: fabry@mediacomm.it
WWW: http://www.mediacomm.it/urania/
Founded: 1994
Activities: popularization * observing * photography
Coordinates: 008°48'30"E 44°47'51"N H347m
City Reference Coordinates: 008°47'00"E 44°46'00"N

Società Astronomica Versiliese (SAV)
Via San Agostino 1
I-55045 Pietrasanta
Telephone: (0)584-791122
Electronic Mail: bimbi@versilia.toscana.it (Marco Bimbi, President)
WWW: http://www.versilia.toscana.it/pietrasanta/sav/
Founded: 1976
Membership: 28
Activities: popularization * education
Coordinates: 010°13'14"E 43°57'31"N H22m (Osservatorio Spartaco Palla)
City Reference Coordinates: 010°14'00"E 43°57'00"N

Società Italiana di Fisica (SIF)
Via Castiglione 101
I-40136 Bologna
Telephone: (0)51-331554
Telefax: (0)51-581340
Electronic Mail: sif@bo.infn.it
WWW: http://www.sif.it/
Founded: 1897
Staff: 11
Membership: 6300
Activities: publishing * conventions * schools
Periodicals: "Il Nuovo Cimento" (ISSN 0392-6737), "La Rivista del Nuovo Cimento" (ISSN 0393-697x), "Giornale di Fisica",
"Il Nuovo Saggiatore"
Awards: (1) "Prize for graduates in physics"
City Reference Coordinates: 011°20'00"E 44°29'00"N

Società Italiana di Relatività Generale e Fisica della Gravitazione (SIGRAV)
Via Carlo Alberto 10
I-10123 Torino
Telephone: (0)11-5613096
Telefax: (0)11-5613096
Electronic Mail: sigrav@dm.unito.it
Founded: 1990
Activities: general relativity * relativistic astrophysics
Awards: (1/2) "SIGRAV Prize"
City Reference Coordinates: 007°40'00"E 45°03'00"N

Specola Astronomica Lodi
c/o Domenico Gellera
Via Gaetano Benaglia 5
I-26900 Lodi
Telephone: (0)371-35175
Founded: 1963
Staff: 1
Activities: double stars * astrometry * lunar and planetary cartography * daily local meteorological and seismographic reports
* restoring ancient astronomical instruments
Coordinates: 009°30'12"E 45°18'24"N H78m
City Reference Coordinates: 009°30'00"E 45°18'00"N

Stazione Astronomica di Cagliari
● See "Osservatorio Astronomico di Cagliari (OAC)"

Telescopio Nazionale Galileo (TNG)
Riviera Tiso Camposampiero 28
I-35122 Padova
Telephone: (0)49-8754343
Telefax: (0)49-8754345
Electronic Mail: barbieri@pd.astro.it (Cesare Barbieri, Director)
WWW: http://www.pd.astro.it/TNG/TNG.html
Founded: 1990
Periodicals: "Telescopio Nazionale Galileo Newsletter"
Coordinates: 017°52'35"W 28°45'34"N H2,373m
City Reference Coordinates: 011°53'00"E 45°25'00"N

Third World Academy of Sciences (TWAS)
c/o The Abdus Salam International Centre for Theoretical Physics

Casella Postale 586
Via Beirut 6
I-34014 Trieste
Telephone: (0)40-2240327
Telefax: (0)40-224559
Electronic Mail: twas@ictp.trieste.it
WWW: http://www.ictp.trieste.it/~twas/
Founded: 1983
Staff: 10
Membership: 480
Activities: prizes for scientists from developing countries * research grants * fellowships * donations * support for scientific meetings in the third world
Periodicals: (4) "TWAS Newsletter"; (1) "Year Book", "Annual Report"; "Proceedings"
Awards: (1) awards in biology, chemistry, physics, mathematics and basic medical sciences; (1/2) "Abdus Salam Medal in Science and Technology"; prizes for young scientists in developing countries - in collaboration with national academies and research councils (various frequencies)
City Reference Coordinates: 013°46'00"E 45°40'00"N

Third World Network of Scientific Organizations (TWNSO)

c/o The Abdus Salam International Centre for Theoretical Physics
Enrico Fermi Building
Room 112
Casella Postale 586
Via Beirut 6
I-34014 Trieste
Telephone: (0)40-2240386
Telefax: (0)40-224559
Electronic Mail: twas@ictp.trieste.it
WWW: http://www.ictp.trieste.it/TWAS/TWNSO.html
Founded: 1988
Membership: 147
City Reference Coordinates: 013°46'00"E 45°40'00"N

Unione Astrofili Bresciani (UAB)

Traversa XII 172
Villaggio Sereno
I-25125 Brescia
Telefax: (0)30-872525
WWW: http://www.cityline.it/cult/grup_sci/index.html#astrofili
Founded: 1975
Membership: 120
Activities: observing (variable stars, minor planets, comets) * photography * education * special interest groups * meetings * planetarium
Periodicals: (1) "Annuario Nautico"; "Circolare"
Coordinates: 010°14'25"E 45°39'53"N H830m (Serafino Zani Obs. - see separate entry)
City Reference Coordinates: 010°13'00"E 45°33'00"N

Unione Astrofili Italiani (UAI)

c/o Dipartimento di Astronomia
Università degli Studi di Padova
Vicolo dell'Osservatorio 5
I-35122 Padova
Telephone: (0)2-6686263 (Roberto Boccadoro)
Telefax: (0)2-8437382 (Roberto Boccadoro)
Electronic Mail: uai@mclink.it
WWW: http://www.mclink.it/mclink/astro/uai.htm
        http://astrolink.mclink.it/uai.htm
Founded: 1967
Membership: 1000
Staff: 7
Activities: coordinating and publishing research by Italian amateur astronomers * collaborating with national and international amateur and professional organizations * meetings * schools
Sections: Variable Stars * Planets * Lunar Occultations * Moon * Sun * Meteors * Minor Planets * Comets * Sundials * SN Search * Deep Sky * CCD Astronomy * Internet
Periodicals: (6) "Astronomia UAI" (ISSN 0392-2308, circ.: 1,200); (1) "Almanacco UAI"
City Reference Coordinates: 011°53'00"E 45°25'00"N

Unione Astrofili Cosentini

c/o Gianni Rende
Via Fellini 6
I-87100 Cosenza
Electronic Mail: astrocs@geocities.com
WWW: http://www.geocities.com/Area51/Shire/2346/
City Reference Coordinates: 016°16'00"E 39°17'00"N

Unione Astrofili Napoletani (UAN)
Via Guido de Ruggiero 37
I-80128 Napoli
Telephone: (0)81-5792234
                   (0)81-5575527
Electronic Mail: uan@oacosf.na.astro.it
WWW: http://oacosf.na.astro.it/uan/uanhome.htm
Founded: 1974
Membership: 150
Activities: observing (planets, variable stars) * photometry * photography * education * lectures
Periodicals: (12) "UANotizie"
City Reference Coordinates: 014°17'00"E 40°51'00"N

Unione Astrofili Senesi (UAS)
Centro Civico La Meridiana
Via G. di Vittorio
I-53100 Siena
Telephone: (0)577-46351
Electronic Mail: tutorl@tin.it
WWW: http://www.mclink.it/mclink/astro/ass/uas/
           http://astrolink.mclink.it/ass/uas/
           http://www.geocities.com/capecanaveral/hangar/6473.html
Founded: 1977
Membership: 50
Activities: popularization
Periodicals: (4) "Sky Watchers"
Coordinates: 011°19'22"E 43°18'54"N H300m
City Reference Coordinates: 011°20'00"E 43°19'00"N

Università degli Studi di Bologna
● See "Università di Bologna"

Università degli Studi di Catania
● See "Università di Catania"

Università degli Studi di Padova
● See "Università di Padova"

Università degli Studi di Trieste
● See "Università di Trieste"

Università di Bologna, Dipartimento di Astronomia
Via Ranzani 1
I-40127 Bologna
Telephone: (0)516305727
Telefax: (0)516305700
Electronic Mail: <userid>@bo.astro.it
WWW: http://www.bo.astro.it/dip/
Founded: 1725
Staff: 26
Activities: stellar astrophysics * extragalactic astrophysics * instrumentation
Periodicals: "Bologna Astrophysical Preprints", "Bologna Technical Reports"
Coordinates: 011°20'12"E 44°15'32"N H785m
                    011°20'04"E 44°15'23"N H795m
● Full university name: "Università degli Studi di Bologna"
City Reference Coordinates: 011°20'00"E 44°29'00"N

Università di Catania, Istituto di Astronomia
Città Universitaria
Viale Andrea Doria 6
I-95125 Catania
Telephone: (0)95-7332211
                   (0)95-7332210
Telefax: (0)95-330592
Electronic Mail: <userid>@alpha4.ct.astro.it
                           <userid>@astrct.ct.astro.it
WWW: http://www.ct.astro.it/
Founded: 1956
Staff: 15
Activities: solar and stellar activity * interplanetary and interstellar matter * solar-system minor bodies * relativity * cosmology * variable stars * nuclear astrophysics
Coordinates: 015°04'21"E 37°31'43"N H193m
● Full university name: "Università degli Studi di Catania"
City Reference Coordinates: 015°06'00"E 37°30'00"N

Università di Firenze, Dipartimento di Astronomia e Scienza dello Spazio
Largo Enrico Fermi 5
I-50125 Firenze
Telephone: (0)55-27521
Telefax: (0)55-224193
Electronic Mail: <userid>@arcetri.astro.it
WWW: http://www.arcetri.astro.it/Dipartimento/
Founded: 1967
Activities: far-IR and mm astronomy and spectroscopy * atmospheric physics * molecular spectroscopy * interstellar clouds * cosmic dust * UV observing
Periodicals: "Internal Reports"
City Reference Coordinates: 011°15'00"E 43°46'00"N

Università di Lecce, Dipartimento di Fisica, Gruppo di Astrofisica
Via Per Arnesano
Casella Postale 193
I-73100 Lecce
Telephone: (0)832-320501
Telefax: (0)832-320505
Electronic Mail: <userid>@le.infn.it
WWW: http://www.unile.it/ateneo/
Founded: 1967
Staff: 8
Activities: IR astronomy * ISM * galactic dynamics
City Reference Coordinates: 018°11'00"E 40°23'00"N

Università di Milano, Dipartimento di Fisica
Via Celoria 16
I-20133 Milano
Telephone: (0)2-2392272
                (0)2-2392206
Telefax: (0)2-70638413
                (0)2-2392205
Electronic Mail: <userid>@mi.infn.it
                      <userid>@uni.mi.astro.it
WWW: http://www.mi.infn.it/
Activities: plasma astrophysics * laboratory plasmas * stellar and galactic structure and evolution * cosmic microwave background and related technology * cosmic rays * astronomical databases * organizing workshops * education
Coordinates: 009°11'37"E 45°27'34"N
City Reference Coordinates: 009°12'00"E 45°28'00"N

Università di Padova, Dipartimento di Astronomia
Vicolo dell'Osservatorio 5
I-35122 Padova
Telephone: (0)49-8293411
Telefax: (0)49-8759840
Electronic Mail: <userid>@astrpd.pd.astro.it
                      dipastro@www.pd.astro.it
WWW: http://www.pd.astro.it/~dipastro/
          http://www.pd.astro.it/asiago (Osservatorio Astrofisico di Asiago)
Founded: 1986
Staff: 43
Activities: solar system * stellar astrophysics * galactic and extragalactic astronomy * cosmology * high-energy astrophysics * space research
Coordinates: 011°52'18"E 45°24'00"N H38m (Osservatorio Astrofisico di Asiago)
● Full university name: "Università degli Studi di Padova"
City Reference Coordinates: 011°53'00"E 45°25'00"N

Università di Pavia, Dipartimento di Fisica Nucleare e Teorica (DFNT)
Via A. Bassi 6
I-27100 Pavia
Telephone: (0)382-507436
Telefax: (0)382-507752
Electronic Mail: dfnt@pv.infn.it
                      <userid>@pv.infn.it
WWW: http://www.pv.infn.it/~dfntwww/dfnt.html
Founded: 1983
Staff: 100
Activities: theoretical and experimental nuclear and subnuclear physics * space physics * gravitation
City Reference Coordinates: 009°10'00"E 45°10'00"N

Università di Perugia, Osservatorio Astronomico
Via Bonfigli
I-06100 Perugia
Telephone: (0)75-20632
                (0)75-5853042

Electronic Mail: <userid>@vaxpg.pg.infn.it
WWW: http://wwwospg.pg.infn.it/
Founded: 1988
Activities: variable stars * stellar evolution * IR * history of astronomy
Periodicals: "Contributi Serie A", "Contributi Serie B"
Coordinates: 012°24'00"E 43°00'00"N H430m
City Reference Coordinates: 012°24'00"E 43°00'00"N

Università di Pisa, Dipartimento di Fisica, Sezione di Astronomia e Astrofisica
Piazza Torricelli 2
I-56100 Pisa
Telephone: (0)50-91111
Telefax: (0)50-48277
Electronic Mail: <userid>@astrpi.difi.unipi.it
WWW: http://astr17pi.difi.unipi.it/
Founded: 1730
Staff: 8
Activities: ISM * UV extinction * molecular clouds * OB stars * star formation * minor planets * galaxy evolution * stellar evolution * open and globular clusters
City Reference Coordinates: 010°20'00"E 43°43'00"N

Università di Pisa, Dipartimento di Matematica, Gruppo di Meccanica Spaziale
Via F. Buonarroti 2
I-56127 Pisa
Telephone: (0)50-599554
         (0)50-599552
Telefax: (0)50-599524
Electronic Mail: <userid>@icnucevm.bitnet
WWW: http://adams.dm.unipi.it/
Activities: celestial mechanics * planets * space geodesy * minor planets * small natural and artificial satellites
Periodicals: "Preprints"
City Reference Coordinates: 010°20'00"E 43°43'00"N

Università di Roma La Sapienza, Istituto Astronomico
Via G.M. Lancisi 29
I-00161 Roma
Telephone: (0)6-4403734
         (0)6-4403735
Electronic Mail: <userid>@astrm2.rm.astro.it
WWW: http://astrm2.rm.astro.it/home.html
       http://astro1.astro.uniroma1.it/home.html
Founded: 1981
Staff: 15
Activities: stellar physics * high-energy astrophysics * AGNs * clusters of galaxies * structure and dynamics of the Galaxy
Coordinates: 012°12'20"E 41°48'49"N H380m (Monteporzio)
City Reference Coordinates: 012°29'00"E 41°54'00"N

Università di Roma Tor Vergata, Dipartimento di Fisica, Astrofisica
Via della Ricerca Scientifica
I-00133 Roma
Telephone: (0)6-72594301
Telefax: (0)6-2023507
Electronic Mail: <userid>@roma2.infn.it
               <userid>@itovf2.roma2.infn.it
               vaxtov::<userid>
WWW: http://itovf2.roma2.infn.it/
       http://www.roma2.infn.it/
Founded: 1984
Staff: 12
Activities: cosmology * microwave background * AGNs * clusters of galaxies * solar physics * interplanetary plasma
City Reference Coordinates: 012°29'00"E 41°54'00"N

Università di Roma Tre, Dipartimento di Fisica
Via della Vasca Navale 84
I-00146 Roma
Telefax: (0)6-55179303
Electronic Mail: <userid>@fis.uniroma3.it
WWW: http://www.fis.uniroma3.it/new/
       http://193.204.162.110/
       http://193.204.162.110/~astro/
Activities: astrophysics * space physics * nuclear physics * particle physics * theoretical physics * environmental physics
City Reference Coordinates: 012°29'00"E 41°54'00"N

Università di Trieste, Dipartimento di Astronomia
Casella Postale Succursale Trieste 5

Via G.B. Tiepolo 11
I-34131 Trieste
Telephone: (0)40-3199255 (Secretary)
Telefax: (0)40-309418
Electronic Mail: <userid>@astrts.astro.it
WWW: http://www.oat.ts.astro.it/
Founded: 1985
Staff: 14
Activities: solar physics * stellar atmospheres * mass loss in stars * interacting binar.es * globular clusters * stellar evolution * classification of stellar spectra * AGNs * clusters of galaxies * data analysis * image processing * astronomical technologies including space
● Full university name: "Università degli Studi di Trieste"
City Reference Coordinates: 013°46'00"E 45°40'00"N

Università Popolare Sestrese, Osservatorio Astronomico di Genova
Piazzetta dell'Università Popolare 4
I-16154 Genova
Telephone: (0)10-678368
WWW: http://astrolink.mclink.it/oss_gen.htm
Founded: 1984 (1907 as association)
Membership: 34
Activities: education * popularization * minor planets * comets * astrometry * variaɔle stars * Sun
Periodicals: (4) "Circolare" (circ.: 300), "Contributo" (circ.: 100); (2) "Bollettino" (circ.: 150)
Coordinates: 008°50'13"E 44°26'03"N H124m
City Reference Coordinates: 008°57'00"E 44°25'00"N

Urania
(Associazione Ligure per lo Studio e la Divulgazione dell'Astronomia e dell'Astronautica)
c/o Museo Civico di Storia Naturale "G.Doria"
Via Brigata Liguria 9
I-16121 Genova
or :
c/o Centro Astronomico
Osservatorio di Rovegno
I-16028 Rovegno
Telephone: (0)10-8352882
Electronic Mail: associazione.urania@tiscalinet.it
WWW: http://www.icom.it/freeweb/urania/
Founded: 1951
Membership: 125
Activities: public lectures * education * photography * planetology * Sun * comets * observing
Periodicals: (6) "Circolare"
Coordinates: 009°18'06"E 44°34'31"N H918 (Alta Val Trebbia)
City Reference Coordinates: 008°57'00"E 44°25'00"N (Genova)
                                        009°17'00"E 44°35'00"N (Rovegno)

Japan

Aichi University of Education, Department of Physics and Astronomy
Hirosawa 1
Igaya-cho
Kariya 448
Telephone: (0)566-36-3111
Telefax: (0)566-36-4337
Electronic Mail: <userid>@auephyas.aichi-edu.ac.jp
WWW: http://www.physics.aichi-edu.ac.jp/physics.htm
Founded: 1988
Staff: 5
Activities: education * research * spiral galaxies * magnetic fields * MHD * black holes * AGNs
City Reference Coordinates: 136°59'00"E 34°59'00"N

Akeno Observatory
● See "University of Tokyo, Institute for Cosmic Ray Research (ICCR), Akeno Observatory"

Arianespace, Tokyo Office
Kasumigaseki Building
31st Floor
3-2-5 Kasumigaseki
Chiyoda-ku
Tokyo 100-6031
Telephone: (0)3-3592-2766
Telefax: (0)3-3592-2768
WWW: http://www.arianespace.com/
Founded: 1980
Activities: liaison office (see main entry in France) * satellite launching services
City Reference Coordinates: 139°46'00"E 35°42'00"N

Astronomical Data Analysis Center (ADAC)
● See "National Astronomical Observatory of Japan (NAOJ), Astronomical Data Analysis Center (ADAC)"

Astronomical Society of Japan (ASJ)
c/o National Astronomical Observatory of Japan
2-21-1 Osawa
Mitaka
Tokyo 181
Telephone: (0)422-31-1359
Telefax: (0)422-31-1359
Electronic Mail: <userid>@tenmon.or.jp
           pasj@tenmon.or.jp (PASJ)
WWW: http://www.tenmon.or.jp/
       http://www.tenmon.or.jp/pasj/ (PASJ)
Founded: 1908
Membership: 2800
Activities: developing and popularizing astronomy
Periodicals: (12) "The Astronomical Herald" (ISSN 0374-2466); (6) "Publications of the Astronomical Society of Japan (PASJ)" (ISSN 0004-6264)
City Reference Coordinates: 139°46'00"E 35°42'00"N

Astronomical Society of Wakabadai (ASW)
4-21-206 Wakabadai Asahiku
Yokohama 241
Electronic Mail: suchida@mxb.meshnet.or.jp (Shigemi Uchida)
WWW: http://www2a.meshnet.or.jp/~wakaba/
Founded: 1996
City Reference Coordinates: 139°39'00"E 35°27'00"N

Bisei Astronomical Observatory (BAO)
1723-70 Ohkura
Bisei
Oda-gun
Okayama 714-14
Telephone: (0)866-87-4222
Telefax: (0)866-87-4224
Electronic Mail: info@bao.go.jp
WWW: http://www.urban.ne.jp/home/bao/index-e.html
Founded: 1993
Coordinates: 133°34'27"E 34°40'36"N H516m
City Reference Coordinates: 133°55'00"E 34°49'00"N

Bisei Hydrographic Observatory
● See "Japan Hydrographic Department (JHD), Bisei Hydrographic Observatory"

Communications Research Laboratory (CRL), Hiraiso Solar-Terrestrial Research Center
3601 Isozaki-cho
Hitachinaka-shi
Ibaraki 311-12
Telephone: (0)292-65-7121
Telefax: (0)292-65-7209
Electronic Mail: <userid>@crl.go.jp
WWW: http://hiraiso.crl.go.jp/
Founded: 1915 (CRL: 1952)
Staff: 12
Activities: space weather forecasting * solar radio and optical observing * radio disturbance prediction * geomagnetic activities
Periodicals: (12) "Solar Radio Emission Hiraiso" (circ.: 40)
Coordinates: 140°37'30"E 36°21'54"N H26m
City Reference Coordinates: 140°26'00"E 36°17'00"N

Communications Research Laboratory (CRL), Inubo Radio Observatory
9961 Tennodai
Choshi
Chiba 288
Telephone: (0)479-22-0871
Telefax: (0)479-25-0675
Electronic Mail: <userid>@crl.go.jp
WWW: http://www.crl.go.jp/inb/overview.html
Founded: 1945 (CRL: 1952)
Coordinates: 140°51'30"E 35°42'12"N
City Reference Coordinates: 140°05'00"E 35°36'00"N

Communications Research Laboratory (CRL), Kansai Advanced Research Center (KARC)
588-2 Iwaoka
Nishi-ku
Kobe
Hyogo 651-24
Telephone: (0)78-969-2100
Telefax: (0)78-969-2200
Electronic Mail: <userid>@crl.go.jp
WWW: http://www-karc.crl.go.jp/
Founded: 1989 (CRL: 1952)
City Reference Coordinates: 135°00'00"E 35°00'00"N

Communications Research Laboratory (CRL), Kashima Space Research Center (KSRC)
893-1 Hirai
Kashima-machi
Ibaraki 314
Telephone: (0)299-82-1211
Telefax: (0)299-84-6860
Electronic Mail: <userid>@crl.go.jp
WWW: http://www.crl.go.jp/ka/
Founded: 1964 (CRL: 1952)
Staff: 12
Activities: VLBI * radio astronomy
Periodicals: (3) "Journal of the Communications Research Laboratory" (ISSN 0914-9260)
Coordinates: 140°39'46"E 35°57'15"N H79m (Kashima)
City Reference Coordinates: 140°26'00"E 36°17'00"N

Communications Research Laboratory (CRL), Okinawa Radio Observatory
829-3 Daigusukubaru
Aza-Kuba
Nakagusuku-son
Nakagami-gun
Okinawa 901-24
Telephone: (0)98-895-2045
Telefax: (0)98-895-4010
Electronic Mail: <userid>@crl.go.jp
WWW: http://www.crl.go.jp/okn/overview.html
Founded: 1972 (CRL: 1952)
Coordinates: 127°48'24"E 26°16'54"N
City Reference Coordinates: 127°59'00"E 26°31'00"N

Communications Research Laboratory (CRL), Tokyo
4-2-1 Nukuikitamachi
Koganei-shi
Tokyo 184

Telephone: (0)423-21-1211
Telefax: (0)423-27-7596
Electronic Mail: <userid>@crl.go.jp
WWW: http://www.crl.go.jp/
Founded: 1952
Staff: 425
Activities: space communications * VLBI * SLR * space environment research * communications technology * remote sensing * time & frequency standards
Periodicals: (12) "Ionospheric Data in Japan" (ISSN 0021-0382), "Standard Frequency and Time Service Bulletin" (ISSN 0387-5857); (3) "Journal of the Communications Research Laboratory" (ISSN 0914-9260); (2) "Ionospheric Data at Syowa Station (Antarctica)" (ISSN 0389-8237)
Coordinates: 139°29'18"E 35°42'24"N (Space Optical Comm. Research Center)
            140°40'00"E 35°57'12"N (Kashima - see separate entry)
            140°37'30"E 36°22'54"N H26m (Hiraiso - see separate entry)
City Reference Coordinates: 139°46'00"E 35°42'00"N

Communications Research Laboratory (CRL), Wakkanai Radio Observatory
3-20 Midori 2-chome
Wakkanai
Hokkaido 097
Telephone: (0)162-23-3386
Telefax: (0)162-24-3227
Electronic Mail: <userid>@crl.go.jp
WWW: http://www.crl.go.jp/wak/overview.html
Founded: 1946 (CRL: 1952)
        http://www.crl.go.jp/
Coordinates: 141°41'06"E 45°23'36"N
City Reference Coordinates: 141°40'00"E 45°25'00"N (Wakkanai)

Communications Research Laboratory (CRL), Yamagawa Radio Observatory
2719 Narikawa
Yamagawa-machi
Ibusuki-gun
Kagoshima 891-05
Telephone: (0)993-34-0077
Telefax: (0)993-35-2077
Electronic Mail: <userid>@crl.go.jp
WWW: http://www.crl.go.jp/yam/overview.html
Founded: 1946 (CRL: 1952)
Coordinates: 130°37'06"E 31°12'06"N
City Reference Coordinates: 130°33'00"E 31°36'00"N

Dodaira Observatory
● See "National Astronomical Observatory of Japan (NAOJ), Dodaira Observatory"

Fujitok Corp.
1-9-16 Kami-Jujo
Kita-ku
Tokyo 114-0034
Telephone: (0)3-3909-1791
Telefax: (0)3-3908-6450
WWW: http://www.barrassociates.com/
        http://www.fujitok.co.jp/
Activities: designing and manufacturing custom astronomical, space-based and other precision optical filters
City Reference Coordinates: 139°46'00"E 35°42'00"N

Fukuoka University of Education, Department of Earth Sciences and Astronomy
1-729 Akama
Munakata
Fukuoka 811-41
Telephone: (0)940-35-1375
Telefax: (0)940-33-7730
Electronic Mail: <userid>@fueipc.fukuoka-edu.ac.jp
WWW: http://www.fukuoka-edu.ac.jp/index-e.html
Founded: 1949
Staff: 4
Activities: stellar spectroscopy * radio astronomy
Periodicals: (1) "Bulletin of the Fukuoka University of Education - Part III: Mathematics, Natural Sciences and Technology"

Coordinates: 130°35'48"E 33°48'42"N H70m
City Reference Coordinates: 130°24'00"E 33°35'00"N

Gotoh Planetarium and Astronomical Museum
2-21-12 Shibuya
Shibuya-ku

Telephone: (0)3-3407-7409
WWW: http://www.f-space.co.jp/goto-planet/
City Reference Coordinates: 139°46'00"E 35°42'00"N (Tokyo)

GOTO Optical Manufacturing Co.
4-16 Yazaki-cho
Fuchu-shi
Tokyo 183-8530
Telephone: (0)42-362-5311
Telefax: (0)42-361-9571
Electronic Mail: info@goto.co.jp
                <userid>@goto.co.jp
WWW: http://www.goto.co.jp/
Founded: 1926
Staff: 185
Activities: manufacturing planetariums, telescopes, pano-hemispheric movie projectors
Coordinates: 138°18'27"E 35°53'25"N H1,035m (Yatsugakate)
City Reference Coordinates: 139°46'00"E 35°42'00"N

Gunma University, Department of Science Education
4-2 Aramaki-machi
Maebashi
Gunma 371
Telephone: (0)272-32-1611
Telefax: (0)272-33-9231
Electronic Mail: <userid>@storm.edu.gunma-u.ac.jp
WWW: http://waza.edu.gunma-u.ac.jp/
Founded: 1949
Activities: close binary systems
Periodicals: "Science Reports of the Faculty of Education of Gunma University" (ISSN 0017-5668)
City Reference Coordinates: 139°00'00"E 36°24'00"N

Hida Observatory
● See "Kyoto University, Kwasan and Hida Observatories, Hida Observatory"

Hiraiso Solar-Terrestrial Research Center
● See "Communications Research Laboratory (CRL), Hiraiso Solar-Terrestrial Research Center"

Hiroshima Planetarium
c/o Hiroshima Children's Museum
5-83 Motomachi
Nakaku
Hiroshima 730
Telephone: (0)82-222-5346
WWW: http://mothra.rerf.or.jp/ENG/Hiroshima-old/Art/Hiroshima-Children.html
        http://www.tourism.city.hiroshima.jp/english/level7/h040100007.html
Founded: 1980
Activities: shows * lectures
Periodicals: (4) "Planetarium Show"
Coordinates: 132°27'23"E 34°23'43"N H4m
City Reference Coordinates: 132°27'00"E 34°24'00"N

Hitotsubashi University, Laboratory of Astronomy and Geophysics
Naka 2-1
Kunitachi
Tokyo 186
Telephone: (0)425-72-1101
Telefax: (0)425-71-1893
Electronic Mail: nakajima@higashi.hit-u.ac.jp (Kaichi Nakajima)
Founded: 1984
Staff: 2
Activities: education (astronomy, computer science, information science) * astronomical data service
City Reference Coordinates: 139°46'00"E 35°42'00"N

Hokkaido University, Department of Physics, Astrophysics Laboratory
Kita-10 Nishi-8
Kita-ku
Sapporo 060
Telephone: (0)11-716-2111
Telefax: (0)11-727-3498
Electronic Mail: <userid>@astro1.phys.hokudai.ac.jp
                <userid>@phys.hokudai.ac.jp
WWW: http://avalon.phys.hokudai.ac.jp/avalon.html
        http://astro3.phys.hokudai.ac.jp/index_e.html
City Reference Coordinates: 141°21'00"E 43°03'00"N

Hoshinoko Yakata Observatory

1470-24 Aoyama
Himeji
Hyogo 671-2222
Telephone: (0)792-67-3050
Telefax: (0)792-67-3055
Electronic Mail: hosinoko@memenet.or.jp
WWW: http://www.obs.misato.wakayama.jp/~hosinoko
Founded: 1992
Staff: 3
Activities: astronomy education for children
Coordinates: 134°37'55"E 34°50'59"N H71m
City Reference Coordinates: 135°00'00"E 35°00'00"N

Hyogo College of Medicine, Department of Physics

Nishinomiya
Hyogo 663
Telephone: (0)798-45-6443
Telefax: (0)798-48-6261
Founded: 1972
Activities: solar granulation * Jovian decametric radiation
City Reference Coordinates: 135°00'00"E 35°00'00"N

Ibaraki University, Institute for Astrophysics and Planetary Science

2-1-1 Bunkyo
Mito 310
Telephone: (0)29-228-8333
Telefax: (0)29-228-8407
Electronic Mail: <userid>@mito.ipc.ibaraki.ac.jp
WWW: http://orion.sci.ibaraki.ac.jp/sci2a-e.html
City Reference Coordinates: 140°28'00"E 36°22'00"N

Institute of Space and Astronautical Science (ISAS)

3-1-1 Yoshinodai
Sagamihara
Kanagawa 229
Telephone: (0)427-51-3911 x2021
Telefax: (0)427-59-4251
Electronic Mail: isasero_mac01@isasmac1.newslan.isas.ac.jp
            astro@astro.isas.ac.jp
WWW: http://www.isas.ac.jp/index-e.html
        http://www.astro.isas.ac.jp/ (Astronomy Group)
Founded: 1981
Staff: 279
Activities: space science * space technology * scientific satellites * sounding rockets * balloons * launchers
Periodicals: (12) "ISAS News" (ISSN 0285-2861); "ISAS Report" (ISSN 0285-6808), "Uchuken Hokoku" (ISSN 0285-2853)

Coordinates: 131°04'45"E 31°15'00"N H210m (Kagoshima Space Center)
City Reference Coordinates: 139°38'00"E 35°28'00"N (Kanagawa)

International Latitude Observatory, Mizusawa

● See "National Astronomical Observatory of Japan (NAOJ), Mizusawa Astrogeodynamics Observatory"

Inubo Radio Observatory

● See "Communications Research Laboratory (CRL), Inubo Radio Observatory"

Itabashi Science and Education Center Planetarium

4-14-1 Tokiwadai
Itabashi-shi
Tokyo 174
Telephone: (0)3-559-6561
Telefax: (0)3-559-6000
Founded: 1988
Staff: 30
Activities: planetarium shows * science exhibitions
City Reference Coordinates: 139°46'00"E 35°42'00"N

Japan Academy (The_)

● See "The Japan Academy"

Japan Association for Information Processing in Astronomy (JAIPA)

c/o S. Ichikawa
National Astronomical Observatory of Japan
2-21-1 Osawa
Mitaka

Tokyo 181
Electronic Mail: jaipa@rl.mtk.nao.ac.jp
WWW: http://bandai.mtk.nao.ac.jp/jaipa/
Founded: 1990
Membership: 70
Activities: supporting ADAC/NAOJ (see separate entry) * developing astronomical software * exchanging information about computing, and so on
City Reference Coordinates: 139°46'00"E 35°42'00"N

Japanese Dark-Sky Association (JDA)

c/o Jun-ichi Watanabe
National Astronomical Observatory of Japan
2-21-1 Osawa
Mitaka
Tokyo 181
Telephone: (0)422-34-3644
Telefax: (0)422-34-3810
Electronic Mail: watanabe@sl9mtk.nao.ac.jp
Founded: 1993
Membership: 200
Activities: meetings * press releases * publishing
Periodicals: (2) "Kougai Bushin"
City Reference Coordinates: 139°46'00"E 35°42'00"N

Japanese Industrial Standards Committee (JISC)

c/o Standards Department
Agency of Industrial Science and Technology
Ministry of International Trade and Industry
1-3-1 Kasumigaseki
Chiyoda-ku
Tokyo 100
Telephone: (0)3-3501-9295
              (0)3-3501-9296
Telefax: (0)3-3580-1418
Electronic Mail: jisc_iso@jsa.or.jp
WWW: http://www.aist.go.jp/jisc/htm/jisc00.htm
City Reference Coordinates: 139°46'00"E 35°42'00"N

Japan Hydrographic Department (JHD), Bisei Hydrographic Observatory

Bisei-cho
Oda-gun
Okayama 714-1045
Telephone: (0)8668-7-3355
Electronic Mail: bho@oka.urban.ne.jp
WWW: http://www.jhd.go.jp/
        http://www.jhd.go.jp/jhd-E.html
Founded: 1949 (JHD: 1871)
Activities: lunar occultations * satellite geodesy
Coordinates: 133°34'27"E 34°40'36"N H516m
City Reference Coordinates: 133°55'00"E 34°39'00"N

Japan Hydrographic Department (JHD), Geodesy and Geophysics Division

5-3-1 Tsukiji
Chuo-ku
Tokyo 104
Telephone: (0)3-3541-3816
Telefax: (0)3-3545-2885
Electronic Mail: tenmon@ws12.cue.jhd.go.jp
WWW: http://www.jhd.go.jp/
        http://www.jhd.go.jp/jhd-E.html
Founded: 1871 (JHD)
Staff: 31
Activities: computation of Japanese Ephemeris * services for International Lunar Occultation Center (ILOC) * satellite geodesy * satellite laser ranging * gravimetry * magnetometry * lunar occultations
Periodicals: (1) "Japanese Ephemeris" (ISSN 0373-3696), "Report of Hydrographic Observations - Series of Astronomy and Geodesy" (ISSN 0287-2633), "Report of Hydrographic Observations - Series of Satellite Geodesy" (ISSN 0914-5753), "Report of Hydrographic Researches" (ISSN 0373-3602), "Report of Lunar Occultation Observations" (ISSN 1341-7282), "Report of International Lunar Occultation Center"
Observatories: 3 (Bisei, Simosato and Sirahama Hydrographic Obs. - see separate entries)
Coordinates: 139°46'11"E 35°39'41"N H41m
City Reference Coordinates: 139°46'00"E 35°42'00"N

Japan Hydrographic Department (JHD), Shimosato Hydrographic Observatory (SHO)

Shimosato
Nachikatsuura-cho
Higashimuro-gun

Wakayama 649-5142
Telephone: (0)7355-8-0084
Telefax: (0)7355-8-1535
Electronic Mail: simosato@po.iijnet.or.jp
                <userid>@shimosato.obs.jodc.ati.jp%sprint.com
WWW: http://www.jhd.go.jp/
       http://www.jhd.go.jp/jhd-E.html
Founded: 1954 (JHD: 1871)
Staff: 6
Activities: lunar occultations * satellite laser ranging * satellite geodesy
Periodicals: "Data Report of Hydrographic Observations - Series of Satellite Geodesy"
Coordinates: 135°56'23"E 33°34'27"N H63m
City Reference Coordinates: 135°11'00"E 34°13'00"N

Japan Hydrographic Department (JHD), Shirahama Hydrographic Observatory
Shirahama 3347
Simoda
Sizuoka 415-0012
Telephone: (0)5582-2-0865
Electronic Mail: sirasui@mail.wbs.ne.jp
WWW: http://www.jhd.go.jp/
       http://www.jhd.go.jp/jhd-E.html
Founded: 1942 (JHD: 1871)
Activities: lunar occultations * satellite geodesy
Coordinates: 138°59'20"E 34°42'47"N H172m
City Reference Coordinates: 138°23'00"E 34°58'00"N

Japan Information Center of Science and Technology (JICST)
● See now "Japan Science and Technology Corporation (JST)"

Japan Meteorological Agency
1-3-4 Otemachi
Chiyoda-ku
Tokyo 100
Telephone: (0)3-3212-8341
          (0)3-3211-4966
Telefax: (0)3-3211-3032
Electronic Mail: pro@hq.kishou.go.jp
WWW: http://www.kishou.go.jp/
Periodicals: "Geophysical Magazine" (ISSN 0016-8017), "Seismological Bulletin" (ISSN 0446-5059)
City Reference Coordinates: 139°46'00"E 35°42'00"N

Japan Physical Society (JPS)
Room 211
Kikai-Shinko Building
3-5-8 Shiba-Koen
Minato-ku
Tokyo 105
Telephone: (0)3-3434-2671
Telefax: (0)3-3432-0997
Electronic Mail: <userid>@jps.or.jp
WWW: http://wwwsoc.nacsis.ac.jp/jps/jps/
Founded: 1946
Staff: 13
Membership: 18000
Activities: holding scientific meetings * publishing journals and other documents * cooperation and exchange with other societies * documentation centre for members * sponsoring activities within the objectives of the society
Periodicals: (13) "Journal of the Physical Society of Japan" (ISSN 0031-9015, circ.: 3,000); (12) "Butsuri" (ISSN 0029-0181, circ.: 19,000)
City Reference Coordinates: 139°46'00"E 35°42'00"N

Japan Science and Technology Corporation (JST)
Kawaguchi Center Building
4-1-8 Honcho
Kawaguchi-shi
Saitama 332-0012
Telephone: (0)48-226-5628
Telefax: (0)48-226-5751
WWW: http://www.jst.go.jp/
Founded: 1996
Staff: 420
Activities: developing scientific and technological databases and disseminating them worldwide * developing machine translation system
Periodicals: (12) "Current Bibliography on Science and Technology (CBST)", "Journal of Information Processing and Management"; (1) "White Paper on Science and Technology"
● Formerly "Japan Information Center of Science and Technology (JICST)"

City Reference Coordinates: 139°30'00"E 36°00'00"N

Kagawa University, Department of Astronomy and Earth Sciences
Saiwaicho
Takamatsu-shi
Kagawa 760-8522
Telephone: (0)87-832-1466
Telefax: (0)87-832-1615
Electronic Mail: <userid>@ed.kagawa-u.ac.jp
Founded: 1966
Staff: 2
Activities: astronomy * geophysics
Periodicals: "Memoirs of the Faculty of Education - Kagawa University (Part II)" (ISSN 0389-3057)
City Reference Coordinates: 134°02'00"E 34°15'00"N

Kakuda Propulsion Center (KPC)
● See "National Space Development Agency of Japan (NASDA), Kakuda Propulsion Center (KPC)"

Kamioka Underground Observatory
● See "University of Tokyo, Institute for Cosmic Ray Research (ICCR), Kamioka Underground Observatory"

Kanagawa University, Department of Physics
3-27-1 Rokkakubashi
Kanagawa-ku
Yokohama 221
Telephone: (0)45-481-5661
Telefax: (0)45-791-4915
WWW: http://www.kanagawa-u.ac.jp/
Staff: 10
Activities: cosmic-ray astrophysics * solar-terrestrial physics * solid-state physics
City Reference Coordinates: 139°39'00"E 35°27'00"N

Kansai Advanced Research Center (KARC)
● See "Communications Research Laboratory (CRL), Kansai Advanced Research Center (KARC)"

Kashima Space Research Center (KSRC)
● See "Communications Research Laboratory (CRL), Kashima Space Research Center (KSRC)"

Katsushika City Museum, Planetarium
3-25-1 Shiratori
Katsushika-ku
Tokyo 125-0063
Telephone: (0)3-3838-1101
Telefax: (0)3-5680-0849
Electronic Mail: ldq05214@nifty.ne.jp (Tatsuyuki Arai, Astronomy Section)
Founded: 1991 (Museum)
WWW: http://www.obs.misato.wakayama.jp/~katusika/index-e.html
Activities: exhibitions * planetarium shows
City Reference Coordinates: 139°46'00"E 35°42'00"N

Katsuura Tracking and Data Acquisition Station
● See "National Space Development Agency of Japan (NASDA), Katsuura Tracking and Data Acquisition Station"

Kiso Observatory
● See "University of Tokyo, Institute of Astronomy, Kiso Observatory"

Kobe University, Department of Earth and Planetary Sciences
Rokkoudai-machi
Nada-ku
Kobe 657-8501
Telephone: (0)78-803-0574
         (0)78-803-0575
Telefax: (0)78-803-0490
Electronic Mail: <userid>@icluna.kobe-u.ac.jp
Founded: 1973
Staff: 4
Activities: accretion * computational fluid dynamics * cosmic-gas and dust dynamics * origin and evolution of solar system
City Reference Coordinates: 135°10'00"E 34°41'00"N

Kumamoto Civil Astronomical Observatory (KCAO)
Tsukawara
Johnan-machi
Simomasiki-gun
Kumamoto 861-4226

Telephone: (0)964-28-6060
Electronic Mail: tuyasima@kmt-technopolis.or.jp
WWW: http://192.47.99.5/KUMA/htm_e/home_e.html
        http://denouken.kmt-technopolis.or.jp/KUMA/htm_e/index_e.html
Founded: 1982
City Reference Coordinates: 130°43'00"E 32°48'00"N

Kumamoto University, Department of Physics

2-39-1 Kurokami
Kumamoto 860
Telephone: (0)96-344-2111
WWW: http://www.eecs.kumamoto-u.ac.jp/
Founded: 1949
Staff: 1
Activities: cosmology * black holes * accretion disks * nucleosynthesis * stellar evolution
Periodicals: (1/2) "Physics Reports" (ISSN 0303-4070)
City Reference Coordinates: 130°43'00"E 32°48'00"N

Kwasan Observatory

● See "Kyoto University, Kwasan and Hida Observatories, Kwasan Observatory"

Kyoto Sangyo University, Department of Physics

Kita-ku
Kamigamo
Kyoto 603
Telephone: (0)75-701-2151
Telefax: (0)75-705-1640
Electronic Mail: <userid>@cc.kyoto-su.ac.jp
WWW: http://www.kyoto-su.ac.jp/department/ph/index-j.html
Founded: 1965
Staff: 15
Activities: theoretical astrophysics * galaxy formation * distribution of matter in the universe * X-rays * celestial mechanics * mathematical physics
Periodicals: (4) "Acta Humanistica et Scientifica Universitatis Sangio Kyotiensis"
City Reference Coordinates: 135°45'00"E 35°00'00"N

Kyoto University, Department of Astronomy

Kitashirakawa
Sakyo-ku
Kyoto 606-8502
Telephone: (0)75-753-3890
Telefax: (0)75-753-3897
Electronic Mail: <userid>@kusastro.kyoto-u.ac.jp
WWW: http://www.kusastro.kyoto-u.ac.jp/
Founded: 1929
Staff: 9
Activities: accretion disks * circumstellar matter * early-type stars * eclipsing binaries * galaxies * ISM * stellar rotation * globular clusters * stellar spectroscopy * HII regions
Periodicals: "Contributions" (ISSN 0388-0230)
Coordinates: 135°47'00"E 35°01'48"N (Ouda Station)
City Reference Coordinates: 135°45'00"E 35°00'00"N

Kyoto University, Department of Astronomy, Ouda Station

Mochi
Ouda-cho
Uda
Nara 633-21
Telephone: (0)7458-3-3110
Electronic Mail: <userid>@kusastro.kyoto-u.ac.jp
WWW: http://www.kusastro.kyoto-u.ac.jp/
Founded: 1977
Activities: CCD imagery, photometry, and spectroscopy of galactic and extragalactic nebulae
Coordinates: 135°57'00"E 34°28'12"N H389m
City Reference Coordinates: 135°45'00"E 35°00'00"N (Kyoto)
                            135°50'00"E 34°41'00"N (Nara)

Kyoto University, Department of Geophysics

Sakyo-ku
Kyoto 606
Telephone: (0)75-753-3920
Telefax: (0)75-753-3717
Electronic Mail: <userid>@scphys.kyoto-u.ac.jp
WWW: http://www.kugi.kyoto-u.ac.jp/tosho/index.htm
Founded: 1920
Staff: 16

Activities: geodesy * physics of the solid Earth * geodynamics * seismology * applied geophysics * physical oceanography * meteorology * atmospheric physics * geomagnetism * space physics
Coordinates: 135°46'00"E 35°04'00"N H170m (Kamigamo Geophysical Observatory)
City Reference Coordinates: 135°45'00"E 35°00'00"N

Kyoto University, Department of Physics
Kitashirakawa
Sakyo
Kyoto 606-8502
WWW: http://www.scphys.kyoto-u.ac.jp/index-e.html
        http://www-tap.scphys.kyoto-u.ac.jp/index-e.html (Theoretical Astrophysics Group)
        http://www-cr.scphys.kyoto-u.ac.jp/ (Cosmic-Ray Group)
Founded: 1898
City Reference Coordinates: 135°45'00"E 35°00'00"N

Kyoto University, Kwasan and Hida Observatories, Hida Observatory
Kamitakara
Gifu 506-1314
Telephone: (0)578-6-2311
Telefax: (0)578-6-2118
Electronic Mail: <userid>@kwasan.kyoto-u.ac.jp
WWW: http://www.kuastro.kyoto-u.ac.jp/index-e.html
        http://www.kwasan.kyoto-u.ac.jp/
Founded: 1968
Staff: 10
Activities: solar active phenomena * solar atmosphere * planets * comets * solar system
Periodicals: "Contributions of the Kwasan and Hida Observatories, Kyoto University" (ISSN 0388-2349)
Coordinates: 137°18'30"E 36°14'56"N H1,276m
City Reference Coordinates: 136°45'00"E 35°25'00"N

Kyoto University, Kwasan and Hida Observatories, Kwasan Observatory
Yamashina
Kyoto 607
Telephone: (0)75-581-1235
Telefax: (0)75-593-9617
Electronic Mail: <userid>@kwasan.kyoto-u.ac.jp
WWW: http://www.kuastro.kyoto-u.ac.jp/index-e.html
        http://www.kwasan.kyoto-u.ac.jp/
Founded: 1929
Activities: solar and solar-system physics
Periodicals: "Contributions of the Kwasan and Hida Observatories, Kyoto University" (ISSN 0388-2349)
Coordinates: 135°47'46"E 34°59'26"N H221m
City Reference Coordinates: 135°45'00"E 35°00'00"N (Kyoto)

Kyoto University, Yukawa Institute for Theoretical Physics
Kitashirakawa-Owaike-cho
Sakyo-ku
Kyoto 606
Telephone: (0)75-753-7008
Telefax: (0)75-753-7010
Electronic Mail: <userid>@yukawa.kyoto-u.ac.jp
WWW: http://www.yukawa.kyoto-u.ac.jp/
Founded: 1953
Staff: 24
Activities: fundamental theoretical physics
Periodicals: "Progress of Theoretical Physics" (ISSN 0033-068x)
City Reference Coordinates: 135°45'00"E 35°00'00"N

Masuda Tracking and Data Acquisition Station
● See "National Space Development Agency of Japan (NASDA), Masuda Tracking and Data Acquisition Station"

Minolta Planetarium Co. Ltd.
2-30 Toyotsu-cho
Suita
Osaka 564-0051
Telephone: (0)6386-2050
Telefax: (0)6386-2027
Electronic Mail: mp-osk@mom.minolta.co.jp
WWW: http://www.minolta.com/japan/mo1/mo14/mo144.html
Founded: 1984
Staff: 76
Activities: manufacturing i.a. planetariums
City Reference Coordinates: 135°30'00"E 34°40'00"N

Misato Observatory
Misato
Wakayama 640-1366
Telephone: (0)73-498-0305
Telefax: (0)73-498-0306
Electronic Mail: kenkyu@obs.misato.wakayama.jp
WWW: http://www.obs.misato.wakayama.jp/index-e.html
Founded: 1995
Staff: 5
Activities: education * popularization * observing
Coordinates: 135°24'37"E 34°08'29"N H430m
City Reference Coordinates: 135°11'00"E 34°13'00"N

Mizusawa Astrogeodynamics Observatory
● See "National Astronomical Observatory of Japan (NAOJ), Mizusawa Astrogeodynamics Observatory"

Nagoya City Science Museum, Astronomy Section
17-1 Sakae 2-chome
Naka-ku
Nagoya 460
Telephone: (0)552-201-4486 x201
Telefax: (0)552-203-0788
Electronic Mail: astro@ncsm.tokai-ic.or.jp
WWW: http://www2.tokai-ic.or.jp/ncsm/index_E.html
Activities: exhibitions * planetarium shows
City Reference Coordinates: 136°55'00"E 35°10'00"N

Nagoya University, Department of Physics and Astrophysics
Furo-cho
Chikusa-ku
Nagoya 464-8601
Telephone: (0)52-789-2840
Telefax: (0)52-789-2845
Electronic Mail: <userid>@a.phys.nagoya-u.ac.jp
WWW: http://www.a.phys.nagoya-u.ac.jp/
Coordinates: 138°19'12"E 36°31'18"N H1,280m (Sugadaira Station)
138°36'42"E 35°26'36"N H1,000m (Fujigane Station)
138°58'24"E 35°08'54"N H75m (Radio Astronomy Laboratory)
City Reference Coordinates: 136°55'00"E 35°10'00"N

Nagoya University, Solar-Terrestrial Environment Laboratory
3-13 Honohara
Toyokawa
Aichi 442
Telephone: (0)5338-6-3154
Telefax: (0)5338-6-0811
Electronic Mail: <userid>@stelab.nagoya-u.ac.jp
WWW: http://www.stelab.nagoya-u.ac.jp/ste-www1/
Founded: 1990
Staff: 59
Activities: O3, NOx and aerosols observing * LIDAR * aurorae * ULF, VLF & LF-waves observing * interplanetary scintillation * cosmic rays * UHF telescopes * 3D MHD simulations
Coordinates: 130°43'00"E 31°29'00"N H20m (Kagoshima Observatory)
137°03'00"E 34°44'00"N H3m (Sakushima Observatory)
137°22'18"E 34°50'12"N H18m (Toyokawa Observatory)
137°32'00"E 35°35'00"N H330m (Sakashita Observatory)
137°38'00"E 35°48'00"N H1,100m (Kiso Station)
138°37'00"E 35°26'00"N H1,000m (Fuji Observatory)
138°19'00"E 36°31'00"N H1,300m (Sugadaira Station)
142°16'00"E 44°22'00"N H290m (Moshiri Observatory)
City Reference Coordinates: 137°15'00"E 35°00'00"N

Nagoya University, Solar-Terrestrial Environment Laboratory, Cosmic-Ray Section
Furo-cho
Chikusa-ku
Nagoya 464-8601
Telephone: (0)52-789-4314
Telefax: (0)52-789-4313
Electronic Mail: <userid>@stelab.nagoya-u.ac.jp
WWW: http://www.stelab.nagoya-u.ac.jp/omosaic/crle.html
Founded: 1947
Staff: 11
Activities: solar neutrons * cosmic rays
Periodicals: "Annual Report"
City Reference Coordinates: 136°55'00"E 35°10'00"N

National Astronomical Observatory of Japan (NAOJ)

2-21-1 Osawa
Mitaka
Tokyo 181
Telephone: (0)422-34-3600
Telefax: (0)422-34-3690
Electronic Mail: <userid>@mtk.nao.ac.jp
WWW: http://adac.mtk.nao.ac.jp/
         http://www.nao.ac.jp/
Founded: 1988
Staff: 255
Activities: almost all fields of ground-based, space, and theoretical astronomy
Periodicals: "Publications of the National Astronomical Observatory" (ISSN 0915-3640), "Report of the National Astronomical Observatory" (ISSN 0915-6321), "National Astronomical Observatory Reprint" (ISSN 0915-0021)
Coordinates: 139°32'29"E 35°40'21"N H59m (Mitaka)
              133°35'48"E 34°34'26"N H372m (Okayama - see separate entry)
              137°33'19"E 36°06'49"N H2,876m (Norikura - see separate entry)
              138°28'48"E 35°56'18"N H1,350m (Nobeyama - see separate entry)
              139°11'48"E 36°00'12"N H879m (Dodaira - see separate entry)
              141°07'52"E 39°08'03"N H61m (Mizusawa - see separate entry)
City Reference Coordinates: 139°46'00"E 35°42'00"N

National Astronomical Observatory of Japan (NAOJ), Astrometry and Celestial Mechanics Division

2-21-1 Osawa
Mitaka
Tokyo 181
Telephone: (0)422-34-3633
Telefax: (0)422-34-3793
Electronic Mail: <userid>@c1.mtk.nao.ac.jp
WWW: http://pluto.mtk.nao.ac.jp/
         http://adac.mtk.nao.ac.jp/
         http://www.nao.ac.jp/
Founded: 1988
Staff: 24
Activities: celestial mechanics * non-linear dynamics * position astronomy * ephemerides * time-keeping * gravity-wave detection * relativistic astrometry * stellar dynamics * solar-system dynamics
City Reference Coordinates: 139°46'00"E 35°42'00"N

National Astronomical Observatory of Japan (NAOJ), Astronomical Data Analysis Center (ADAC)

2-21-1 Osawa
Mitaka
Tokyo 181
Telephone: (0)422-34-3633
Telefax: (0)422-34-3793
Electronic Mail: <userid>@c1.mtk.nao.ac.jp
WWW: http://www.cc.nao.ac.jp/index-e.html
         http://adac.mtk.nao.ac.jp/
         http://www.nao.ac.jp/
Founded: 1988 (NAOJ)
City Reference Coordinates: 139°46'00"E 35°42'00"N

National Astronomical Observatory of Japan (NAOJ), Division of Earth Rotation

c/o Mizusawa Astrogeodynamics Observatory
2-12 Hoshigaoka-machi
Mizusawa-shi
Iwate 023
Telephone: (0)197-22-7111
Telefax: (0)197-22-7120
Electronic Mail: <userid>@sinet.ad.jp
WWW: http://adac.mtk.nao.ac.jp/
         http://www.nao.ac.jp/
Founded: 1988 (NAOJ)
Coordinates: 141°07'52"E 39°08'03"N H61m
City Reference Coordinates: 141°22'00"E 39°37'00"N

National Astronomical Observatory of Japan (NAOJ), Division of Theoretical Astrophysics

2-21-1 Osawa
Mitaka
Tokyo 181
Telephone: (0)422-41-3740
Telefax: (0)422-41-3746
Electronic Mail: <userid>@nova.mtk.nao.ac.jp
WWW: http://betha.mtk.nao.ac.jp/
         http://adac.mtk.nao.ac.jp/
         http://www.nao.ac.jp/
Founded: 1988

Staff: 8
Activities: cosmology * galaxy evolution * QSOs * star formation * planetary formation * ISM
City Reference Coordinates: 139°46'00"E 35°42'00"N

National Astronomical Observatory of Japan (NAOJ), Dodaira Observatory

Togikawa
Hiki
Saitama 355-03
Telephone: (0)493-67-0224
Telefax: (0)493-67-0824
WWW: http://www.nao.ac.jp/nao/doudaira-e.html
http://adac.mtk.nao.ac.jp/
http://www.nao.ac.jp/
Founded: 1962
Activities: variable stars * AGNs * photometry * polarimetry
Coordinates: 139°11'48"E 36°00'12"N H879m
City Reference Coordinates: 139°30'00"E 36°00'00"N

National Astronomical Observatory of Japan (NAOJ), Mizusawa Astrogeodynamics Observatory

2-12 Hoshigaoka-machi
Mizusawa-shi
Iwate 023
Telephone: (0)197-22-7111
Telefax: (0)197-22-7120
Electronic Mail: <userid>@sinet.ad.jp
WWW: http://www.nao.ac.jp/nao/mizusawa-e.html
http://adac.mtk.nao.ac.jp/
http://www.nao.ac.jp/
Founded: 1988
Staff: 29
Activities: astronomy * Earth rotation * geodesy * geophysics
Periodicals: (1) "Technical Reports of the Mizusawa Kansoku Center National Astronomical Observatory" (ISSN 0915-3780), "Annual Report of the Mizusawa Astrogeodynamics Observatory - Time Service and Geophysical Observations" (ISSN 0916-6343)
Coordinates: 141°07'52"E 39°08'03"N H61m
141°20'07"E 39°08'53"N H393m (Esashi Earth Tides Station)
City Reference Coordinates: 141°22'00"E 39°37'00"N

National Astronomical Observatory of Japan (NAOJ), Nobeyama Radio Observatory (NRO)

Nobeyama
Minamimaki-mura
Nagano 384-13
Telephone: (0)267-63-4300
Telefax: (0)267-98-2884
Electronic Mail: <userid>@nro.nao.ac.jp
WWW: http://solar.nro.nao.ac.jp/
http://www.nro.nao.ac.jp/
http://adac.mtk.nao.ac.jp/
http://www.nao.ac.jp/
Founded: 1982
Staff: 62
Activities: solar and cosmic radio observing * molecular lines * continuum * polarization * solar and cosmic fine structures * dark clouds * SNR * galactic activities
Periodicals: "NRO Report" (ISSN 0911-5501), "NRO Newsletter" (ISSN 0911-5870)
Coordinates: 138°28'48"E 35°56'18"N H1,350m
City Reference Coordinates: 138°11'00"E 36°39'00"N

National Astronomical Observatory of Japan (NAOJ), Norikura Solar Observatory

Mount Norikura
Azumi-mura
Nagano 390-1500
Telephone: (0)263-33-7455
WWW: http://www.nao.ac.jp/
Founded: 1949
Staff: 13
Activities: spectroscopic observing of solar corona, prominences, chromosphere and white-light flares * 5303Å emission corona
Coordinates: 137°33'19"E 36°06'49"N H2,876m
City Reference Coordinates: 138°11'00"E 36°39'00"N

National Astronomical Observatory of Japan (NAOJ), Okayama Astrophysical Observatory

Kamogata
Asakuchi
Okayama 719-02
Telephone: (0)8654-4-2155
Telefax: (0)8654-4-2360
WWW: http://www.cc.nao.ac.jp/oao/

http://adac.mtk.nao.ac.jp/
http://www.nao.ac.jp/
Founded: 1960
Activities: optical and near-IR observing of stars and galaxies * solar magnetic field
Coordinates: 133°35'48"E 34°34'26"N H372m (Mount Chikurinji)
City Reference Coordinates: 133°55'00"E 34°39'00"N

National Astronomical Observatory of Japan (NAOJ), Optical and Infrared Astronomy Division
2-21-1 Osawa
Mitaka
Tokyo 181
Telephone: (0)422-34-3650
Telefax: (0)422-34-3608
Electronic Mail: masa@optik.mtk.nao.ac.jp
          <userid>@optik.mtk.nao.ac.jp
WWW: http://masa.mtk.nao.ac.jp/
        http://sl9.mtk.nao.ac.jp/
        http://www.nao.ac.jp/
Founded: 1888
Staff: 250
Activities: observing * theoretical studies * instrumentation * observatory operation
Periodicals: (1) "Annual Report" (ISSN 0915-6410)
Observatories: 2 (Okayama, Dodaira - see separate entries)
City Reference Coordinates: 139°46'00"E 35°42'00"N

National Astronomical Observatory of Japan (NAOJ), Radio Astronomy Division
c/o Nobeyama Radio Observatory
Nobeyama
Minamimaki-mura
Nagano 384-13
Telephone: (0)267-98-2831
Telefax: (0)267-98-2884
WWW: http://adac.mtk.nao.ac.jp/
        http://www.nao.ac.jp/
Founded: 1988 (NAOJ)
City Reference Coordinates: 138°11'00"E 36°39'00"N

National Astronomical Observatory of Japan (NAOJ), Solar Physics Division
2-21-1 Osawa
Mitaka
Tokyo 181-8588
Telephone: (0)422-34-3716
Telefax: (0)422-41-3700
Electronic Mail: <userid>@solar.mtk.nao.ac.jp
WWW: http://solarwww.mtk.nao.ac.jp/
Founded: 1988
Staff: 24
Activities: solar atmosphere * solar activity (sunspots, flares, corona, magnetic fields)
City Reference Coordinates: 139°46'00"E 35°42'00"N

National Institute for Fusion Science (NIFS)
Oroshi-cho
Toki
Gifu 509-5292
Telephone: (0)572-58-2222
Telefax: (0)572-58-2601
        (0)572-58-2607
WWW: http://www.nits.ac.jp/
Founded: 1989
Staff: 300
Activities: nuclear fusion * plasma physics * atomic processes
Periodicals: (1) "Annual Report" (ISSN 0917-1185)
City Reference Coordinates: 137°10'00"E 35°20'00"N

National Institute of Polar Research (NIPR)
Kaga 1-9-10
Itabashi-ku
Tokyo 173
Telephone: (0)3-962-4711
WWW: http://www.nipr.ac.jp/Welcome.html
Founded: 1970
Staff: 115
Activities: organizing and carrying out Japanese antarctic research expeditions and arctic research programmes including atmospheric physics, meteorology, glaciology, Earth sciences and biology
Coordinates: 039°35'00"E 69°00'00"S (Syowa Station)
          044°20'00"E 70°42'00"S H2,230m (Mizuho Station)

024°08'00"E 71°32'00"S H930m (Asuka Camp)
City Reference Coordinates: 139°46'00"E 35°42'00"N

National Science Museum, Department of Physical Sciences
Ueno Park
Taito-ku
Tokyo 110-8718
Telephone: (0)3-3822-0111
Telefax: (0)3-5814-9899
Activities: meteorites * variable stars * eclipsing binaries * observing (sunspots) * education
Periodicals: "Bulletin of the National Science Museum, Series E (Physical Sciences and Engineering)"
Coordinates: 139°46'48"E 35°42'47"N
City Reference Coordinates: 139°46'00"E 35°42'00"N

National Space Development Agency of Japan (NASDA), Earth Observation Center (EOC)
1401 Numanoue
Ohashi
Hatoyama-machi
Hiki-gun
Saitama 350-03
Telephone: (0)492-96-1611
Telefax: (0)492-96-0217
WWW: http://www.nasda.go.jp/welcome_e.html
Founded: 1978
Staff: 34
Activities: data receiving station
Periodicals: (2) "EOC News" (circ.: 100)
Coordinates: 139°22'00"E 36°00'00"N
City Reference Coordinates: 139°22'00"E 36°00'00"N

National Space Development Agency of Japan (NASDA), Headquarters
World Trade Center Building
2-4-1 Hamamatsu-cho
Minato-ku
Tokyo 105
Telephone: (0)3-5470-4111
Telefax: (0)3-436-2928
WWW: http://www.nasda.go.jp/welcome_e.html
Founded: 1969
Staff: 973
Activities: implementing Japan's space activities
Periodicals: (4) "NASDA Report"
City Reference Coordinates: 139°46'00"E 35°42'00"N

National Space Development Agency of Japan (NASDA), Kakuda Propulsion Center (KPC)
1 Takakuzo
Jinjiro
Kakuda-shi
Miyagi 981-15
Telephone: (0)224-68-3211
Telefax: (0)224-67-1032
Electronic Mail: <userid>@rd.tksc.nasda.go.jp
WWW: http://www.nasda.go.jp/welcome_e.html
Founded: 1978
Activities: research and development of space transportation system and propulsion system
City Reference Coordinates: 140°52'00"E 38°22'00"N

National Space Development Agency of Japan (NASDA), Katsuura Tracking and Data Acquisition Station
1-14 Hanatateyama
Haga
Katsuura-chi
Chiba 299-53
Telephone: (0)470-73-0654
Telefax: (0)470-70-7001
WWW: http://yyy.tksc.nasda.go.jp/Home/Guide/Guide-e/katsuura_e.html
http://www.nasda.go.jp/welcome_e.html
Founded: 1969 (NASDA)
City Reference Coordinates: 140°07'00"E 35°36'00"N

National Space Development Agency of Japan (NASDA), Masuda Tracking and Data Acquisition Station
Nakatane-machi
Kumage-gun
Kagoshima 891-36

Telephone: (0)9972-7-1990
Telefax: (0)9972-4-2000
WWW: http://yyy.tksc.nasda.go.jp/Home/Guide/Guide-e/masuda_e.html
        http://www.nasda.go.jp/welcome_e.html
Founded: 1969 (NASDA)
City Reference Coordinates: 130°33'00"E 31°36'00"N

National Space Development Agency of Japan (NASDA), Okinawa Tracking and Data Acquisition Station

1712-1 Kinrabaru
Afuso - Onna-son
Kunigami-gun
Okinawa 904-04
Telephone: (0)98-967-8211
Telefax: (0)98-983-3001
WWW: http://yyy.tksc.nasda.go.jp/Home/Guide/Guide-e/okinawa_e.html
        http://www.nasda.go.jp/welcome_e.html
Founded: 1969 (NASDA)
City Reference Coordinates: 127°59'00"E 26°31'00"N

National Space Development Agency of Japan (NASDA), Ogasawara Downrange Station

Chichijima
Ogasawara-mura
Tokyo 100-21
WWW: http://www.nasda.go.jp/welcome_e.html
Founded: 1969 (NASDA)
City Reference Coordinates: 139°46'00"E 35°42'00"N

National Space Development Agency of Japan (NASDA), Tanegashima Space Center

Mazu - Kukinaga
Minamitane-cho
Kumage-gun
Kagoshima 891-3703
WWW: http://www.nasda.go.jp/welcome_e.html
Founded: 1969
Coordinates: 130°58'00"E 30°23'00"N
City Reference Coordinates: 130°33'00"E 31°36'00"N

National Space Development Agency of Japan (NASDA), Tsukuba Space Center

2-1-1 Sengen
Tsukuba-shi
Ibaraki 305
Telephone: (0)298-52-2211
Telefax: (0)298-52-2384
WWW: http://yyy.tksc.nasda.go.jp/Home/Guide/Guide-e/tacs1_e.html
        http://yyy.tksc.nasda.go.jp/Home/Guide/Guide-e/tacs_e.html
        http://www.nasda.go.jp/welcome_e.html
Founded: 1972
Staff: 250
Activities: research and development * space technology * spacecraft environment test * spacecraft tracking and control
City Reference Coordinates: 140°06'00"E 36°13'00"N

Nippon Meteor Society (NMS)

c/o Takatsugu Yoshida
110 Mukaiyama
Gyoyu-cho
Toyokawa-shi
Aichi 441-0211
Telephone: (0)533-88-6884
Telefax: (0)533-88-6884
Electronic Mail: lmj53851@biglobe.ne.jp
Founded: 1969
Membership: 350
Activities: observing (visual, telescopic, photographic, radio and video) and investigating meteors
Periodicals: "Astronomical Circular" (ISSN 0388-5852), "Ryusei-sokuho" (Meteor Advance Report), "Star Friends" (ISSN 0389-0341), "Journal of Meteor Observations"
Coordinates: 136°05'35"E 35°09'13"N H100m (Shiga)
City Reference Coordinates: 137°15'00"E 35°00'00"N

Nishi-Harima Astronomical Observatory (NHAO)

Sayo-cho
Hyogo 679-5313
Telephone: (0)790-82-3886
Telefax: (0)790-82-3514
Electronic Mail: harima@nhao.go.jp

WWW: http://www.nhao.go.jp/
Founded: 1990
Activities: popularization * education * observing * research
Periodicals: (1) "Annual Report of the Nishi-Harima Astronomical Observatory" (ISSN 0917-6926)
Coordinates: 134°20'19"E 35°01'21"N H446m
City Reference Coordinates: 135°00'00"E 35°00'00"N

Nobeyama Radio Observatory (NRO)
● See "National Astronomical Observatory of Japan (NAOJ), Nobeyama Radio Observatory (NRO)"

Norikura Solar Observatory
● See "National Astronomical Observatory of Japan (NAOJ), Norikura Solar Observatory"

Ogasawara Downrange Station
● See "National Space Development Agency of Japan (NASDA), Ogasawara Downrange Station"

Okayama Astrophysical Observatory
● See "National Astronomical Observatory of Japan (NAOJ), Okayama Astrophysical Observatory"

Okinawa Radio Observatory
● See "Communications Research Laboratory (CRL), Okinawa Radio Observatory"

Okinawa Tracking and Data Acquisition Station
● See "National Space Development Agency of Japan (NASDA), Okinawa Tracking and Data Acquisition Station"

Osaka Kyoiku University, Astronomical Institute
Asahigaoka
Kashiwara
Osaka 582
Telephone: (0)729-76-3211 x3124
Telefax: (0)729-76-3269
Electronic Mail: <userid>@cc.osaka-kyoiku.ac.jp
WWW: http://oak.sci.osaka-kyoiku.ac.jp/nsystem/sadakane/home.html
Founded: 1949
Staff: 3
Activities: accretion disks * AGNs * astrodynamics * astrophysical jets * galaxy * stars
Periodicals: (2) "Memoirs of Osaka Kyoiku University"
Coordinates: 135°39'07"E 34°32'53"N
City Reference Coordinates: 135°30'00"E 34°40'00"N

Osaka Prefectural Education Center
13-23 Karita 4-chome
Sumiyoshi-ku
Osaka 558-0011
Telephone: (0)6-692-1882
Telefax: (0)6-692-1898
Founded: 1962
Staff: 100
Activities: training of teachers * teaching material for astronomy * research
Periodicals: (1) "Osaka and Science Education" (in Japanese) (ISSN 0916-2917, circ.: 1,000)
Coordinates: 135°31'21"E 34°35'54"N H39m
City Reference Coordinates: 135°30'00"E 34°40'00"N

Ouda Station
● See "Kyoto University, Department of Astronomy, Ouda Station"

Preserving Starry-Sky Association (PSA)
c/o Shigemi Uchida (Section Organizer)
4-21-206 Wakabadai Asahiku
Yokohama 241
or :
c/o Satoru Ohotomo (Public Relations Officer)
3545 Kiyosato Takane-chu
Kitakoma-gun
Yamanashi 407-33
Telephone: (0)551-48-3822 (Shigemi Uchida)
              (0)45-921-2334 (Satoru Ohotomo)
Telefax: (0)551-48-3822 (Shigemi Uchida)
              (0)45-921-2334 (Satoru Ohotomo)
Electronic Mail: suchida@mxb.meshnet.or.jp
WWW: http://www2a.meshnet.or.jp/~wakaba/
Founded: 1996
City Reference Coordinates: 139°39'00"E 35°27'00"N (Yokohama)
                               138°40'00"E 35°40'00"N (Yamanashi)

Saji Astronomical Observatory
1071-1 Takayama
Saji-son
Yazu-gun
Tottiri 689-1312
Telephone: (0)858-89-1011
Telefax: (0)858-88-0103
Electronic Mail: sajinet@infosakyu.ne.jp
WWW: http://www.infosakyu.ne.jp/sajinet
Founded: 1994
City Reference Coordinates: 134°12'00"E 35°32'00"N

Science Council of Japan
22-34 Roppongi, 7-chome
Minato-ku
Tokyo 106-8555
Telephone: (0)3-3403-1949
Telefax: (0)3-3403-1942
Founded: 1949
Membership: 210
Activities: IAU contact through the "Japanese National Committee for Astronomy"
City Reference Coordinates: 139°46'00"E 35°42'00"N

Sendai Children's Space Center
Izumi Ward
Sendai 980
Telephone: (0)22-373-0999
WWW: http://english.itp.ne.jp/jtd/jinfo/sendai.html
Activities: exhibitions * planetarium shows
City Reference Coordinates: 140°52'00"E 38°16'00"N

Shibayama Scientific Co. Ltd.
3 Chome 11-8
Minami-Otsuka
Toshima-ku
Tokyo 170
Telephone: (0)3-3987-4151
Telefax: (0)3-3987-4155
Activities: manufacturer * trading company * consultant for planetariums and museums
City Reference Coordinates: 139°46'00"E 35°42'00"N

Shiga University, Department of Earth Science
2-5-1 Hiratsu
Otsu 520
Telephone: (0)775-37-7780
Telefax: (0)775-37-7839
WWW: http://www.sue.shiga-u.ac.jp/
Founded: 1949 (University)
Activities: solar activity (prominences, spicules, corona, oscillations)
Periodicals: (1) "Memoirs of the Faculty of Education, Shiga University"
Coordinates: 135°54'30"E 34°56'48"N
City Reference Coordinates: 135°52'00"E 35°00'00"N

Shimosato Hydrographic Observatory (SHO)
• See "Japan Hydrographic Department (JHD), Shimosato Hydrographic Observatory (SHO)"

Shirahama Hydrographic Observatory
• See "Japan Hydrographic Department (JHD), Shirahama Hydrographic Observatory"

Society of Geomagnetism and Earth, Planetary and Space Sciences (SGEPSS)
c/o Gakkai Jimu Center
5-16-9 Honkomagome
Bunkyo-ku
Tokyo 113-8622
or :
c/o Masaru Kono (President)
Department of Earth and Planetary Physics
University of Tokyo
7-3-1 Hongo
Tokyo 113-0033
Telephone: (0)3-3812-2111 x4310
Telefax: (0)3-3818-3247
Electronic Mail: sgepss@kurasc.kyoto-u.ac.jp
WWW: http://rasc5.kurasc.kyoto-u.ac.jp/sgepss/index-e.html
Founded: 1946

Membership: 763
Periodicals: "Earth, Planets and Space" (ISSN 0022-1392)
City Reference Coordinates: 139°46'00"E 35°42'00"N (Tokyo)

Sophia University, Department of Physics
7-1 Kioi-cho
Chiyoda-ku
Tokyo 102-8554
Telephone: (0)3-3238-3431
Telefax: (0)3-3238-3341
Electronic Mail: <userid>@hoffman.cc.sophia.ac.jp
WWW: http://www.sophia.ac.jp/
Founded: 1962
Staff: 24
Activities: education and research in physics (including astrophysics)
City Reference Coordinates: 139°46'00"E 35°42'00"N

Sunshine Planetarium
3-1-3 Higashi Ikebukuro
Toshima-ku
Telephone: (0)3-3989-3475
WWW: http://www.daichi.co.jp/tamate/plane/map/ikebue.html
    http://www.princehotels.co.jp/english/info1/05000100/08001008.html
City Reference Coordinates: 139°46'00"E 35°42'00"N

Tanegashima Space Center
● See "National Space Development Agency of Japan (NASDA), Tanegashima Space Center"

Terra Scientific Publishing Co.
2003 Sansei Jiyugaoka Haimu
5-27-19 Okusawa
Setagaya-ku
Tokyo 158
● Publisher
City Reference Coordinates: 139°46'00"E 35°42'00"N

The Japan Academy
7-32 Ueno Park
Taito-ku
Tokyo 110-0007
Telephone: (0)3-3822-2101
Telefax: (0)3-3822-2105
Founded: 1879
Membership: 133
Activities: prizes * publishing * international exchanges * public lectures
Periodicals: "Proceedings of the Japan Academy A: Mathematical Sciences" (ISSN 0386-2194), "Proceedings of the Japan Academy B: Physical and Biological Sciences" (ISSN 0386-2208)
City Reference Coordinates: 139°46'00"E 35°42'00"N
Awards: "Imperial Prize", "The Japan Academy Prize", "The Duke of Edinburgh Prize"

Tohoku University, Astronomical Institute
Aramaki Aza Aoba
Sendai
Aoba
Miyagi 980
Telephone: (0)22-222-1800 x3327
Telefax: (0)22-262-6609
Electronic Mail: <userid>@jpntohok.bitnet
WWW: http://www.astr.tohoku.ac.jp/
Founded: 1934
Activities: stellar structure and evolution * pulsating stars * stellar winds and magnetospheres * accretion discs * emission-line stars * ISM * structure and dynamics of galaxies * protogalaxy * radio & optical observing
Periodicals: (3) "The Scientific Reports of the Tohoku University, Eighth Series: Physics and Astronomy" (ISSN 0388-5607); "Sendai Astronomiaj Raportoj"
Coordinates: 140°50'33"E 38°15'26"N H153m
City Reference Coordinates: 140°52'00"E 38°16'00"N (Sendai)

Tohoku University, Geophysical Institute
Aramaki Aza Aoba
Sendai
Aoba
Miyagi 980-77
Telephone: (0)22-217-6493
WWW: http://www.geophys.tohoku.ac.jp/
Founded: 1949

Staff: 20
Activities: seismology * meteorology * geomagnetism * physical oceanography
Periodicals: (4) "Tohoku Geophysical Journal" (ISSN 0040-8794)
City Reference Coordinates: 140°52'00"E 38°16'00"N (Sendai)

Tokai University, Research Institute of Civilization
1117 Kitakaname
Hiratsuka-shi
Kanagawa 259-12
Telephone: (0)463-58-1211
Telefax: (0)463-35-2456
Electronic Mail: <userid>@keyaki.cc.u-tokai.ac.jp
                     <userid>@rh.u-tokai.ac.jp
WWW: http://www.u-tokai.ac.jp/English/
Founded: 1959
Staff: 36
Activities: stellar spectroscopy * SETI
Periodicals: "Bulletin of the Research Institute of Civilization" (ISSN 0285-0818)
City Reference Coordinates: 139°38'00"E 35°28'00"N

Tokushima Kainan Observatory
WWW: http://www-cc.ee.tokushima-u.ac.jp/kainan/
Founded: 1992
City Reference Coordinates: 134°34'00"E 34°03'00"N

Tokyo Astronomical Observatory
● See now "National Astronomical Observatory of Japan (NAOJ)"

Tokyo Gakugei University, Department of Astronomy and Earth Sciences
4-1-1 Nukuikitamachi
Koganei
Tokyo 184-8501
Telephone: (0)42-329-7111
Telefax: (0)42-329-7538
Electronic Mail: <userid>@u-gakugei.ac.jp
WWW: http://www.u-gakugei.ac.jp/~kouhou/div/e-index.html
Founded: 1949
Staff: 3 (astronomy group)
Activities: galaxies * clusters * variable stars * pre-main-sequence stars * ISM * education
Periodicals: (1) "Bulletin of the Tokyo Gakugei University - Section IV: Mathematics and Natural Sciences" (ISSN 0371-6813)

Coordinates: 139°30'00"E 35°42'00"N H95m
City Reference Coordinates: 139°30'00"E 35°42'00"N

Tokyo Metropolitan University (TMU), Department of Physics
Minamiohsawa 1-1
Hachioji
Tokyo 192-03
Telephone: (0)426-77-1111
Telefax: (0)426-77-2483
Electronic Mail: <userid>@phys.metro-u.ac.jp
WWW: http://www.phys.metro-u.ac.jp/physe.htm
Founded: 1949
Staff: 50
Activities: research * education * theoretical astrophysics * physics
City Reference Coordinates: 139°46'00"E 35°42'00"N

Toyama Astronomical Observatory (TAO)
San-no-kuma 49-4
Toyama 930-0155
Telephone: (0)76-434-9098
Telefax: (0)76-434-9228
Electronic Mail: watanabe@tsm.toyama.toyama.jp (Watanabe Makoto, Staff Member)
WWW: http://www.tsm.toyama.toyama.jp/tao/
Founded: 1956
Staff: 6
Activities: annex of Toyama Science Museum (TSM - see separate entry) * observing * education * exhibitions * research
Coordinates: 137°06'06"E 36°39'30"N
City Reference Coordinates: 137°13'00"E 36°41'00"N

Toyama Science Museum (TSM)
Nishinakano 1-8-31
Toyama 939-8084
Telephone: (0)76-491-2123
Telefax: (0)76-421-5950

Electronic Mail: watanabe@tsm.toyama.toyama.jp (Watanabe Makoto, Staff Member)
WWW: http://www.tsm.toyama.toyama.jp/
Founded: 1979
Staff: 27
Activities: exhibitions (astronomy, physics, chemistry, biology, geology, meteorology) * planetarium * public observatory (see separate entry)
Periodicals: "Bulletin of Toyama Science Museum", "Special Publication of Toyama Science Museum"
City Reference Coordinates: 137°13'00"E 36°41'00"N

Toyama University, Department of Mathematics
3190 Gofu-ku
Toyama 930-8555
Telephone: (0)764-45-6572
Telefax: (0)764-45-6573
Electronic Mail: <userid>@sci.toyama.ac.jp
WWW: http://www.sci.toyama-u.ac.jp/math/
Founded: 1989
Staff: 18
Activities: algebra * geomety * mathematical analysis * functional analysis * mathematical information theory * mathematical physics
Periodicals: (1) "Mathematical Journal of Toyama University" (ISSN 0916-6009)
City Reference Coordinates: 137°13'00"E 36°41'00"N

Toyama University, Department of Physics
3190 Gofu-ku
Toyama 930-8555
Telephone: (0)764-45-6590
Telefax: (0)764-45-6549
Electronic Mail: <userid>@sci.toyama-u.ac.jp
WWW: http://www.sci.toyama-u.ac.jp/sci/phys/
Founded: 1949
Staff: 4
Activities: physics of molecules * interstellar molecules * ionized regions * radio astronomy
City Reference Coordinates: 137°13'00"E 36°41'00"N

Toyo Corp.
P.O. Box 5014
Tokyo International 100-31
or :
1-2 Hongokucho 1-chome
Nihonbashi
Chuo-ku
Tokyo 103
Telephone: (0)3-3279-0771
Telefax: (0)3-3246-0645
Electronic Mail: <userid>@toyo.co.jp
City Reference Coordinates: 139°46'00"E 35°42'00"N

Tsukuba Space Center
● See "National Space Development Agency of Japan (NASDA), Tsukuba Space Center"

University of Tokyo, Department of Astronomy
11-16 Yagoi 2-chome
Bunkyo-ku
Tokyo 113
Telephone: (0)3-812-2111 x4251
Telefax: (0)3-3813-9439
Electronic Mail: <userid>@tansei.cc.u-tokyo.ac.jp
                              <userid>@astron.s.u-tokyo.ac.jp
WWW: http://www.astron.s.u-tokyo.ac.jp/
Founded: 1877
Periodicals: "Contributions" (ISSN 0563-8038)
City Reference Coordinates: 139°46'00"E 35°42'00"N

University of Tokyo, Department of Earth and Planetary Physics
7-3-1 Hongo
Tokyo 113-0033
Telephone: (0)3-3812-2111
Telefax: (0)3-3818-3247
WWW: http://www.geoph.s.u-tokyo.ac.jp/
City Reference Coordinates: 139°46'00"E 35°42'00"N (Tokyo)

University of Tokyo, Department of Earth Science and Astronomy
3-8-1 Komaba
Meguro-ku

Tokyo 153
Telephone: (0)3-3467-1171
Telefax: (0)3-3485-2904
Electronic Mail: <userid>@tansei.cc.u-tokyo.ac.jp
                <userid>@astron.s.u-tokyo.ac.jp
WWW: http://valis.c.u-tokyo.ac.jp/
Activities: N-body simulations * stellar dynamics * stellar structure * supernovae * rotating stars * relativistic astrophysics
City Reference Coordinates: 139°46'00"E 35°42'00"N

University of Tokyo, Department of Physics, Theoretical Astrophysics Group

Hongo 7-3-1
Bunkyo-ku
Tokyo 113-0033
Electronic Mail: <userid>@phys.s.u-tokyo.ac.jp
WWW: http://www-utap.phys.s.u-tokyo.ac.jp/index-e.html
        http://www.phys.s.u-tokyo.ac.jp/
City Reference Coordinates: 139°46'00"E 35°42'00"N

University of Tokyo, Institute for Cosmic Ray Research (ICCR)

3-2-1 Midori-cho
Tanashi
Tokyo 188
Telephone: (0)424-61-4131
Telefax: (0)424-68-1438
Electronic Mail: <userid>@icrr.u-tokyo.ac.jp
WWW: http://www.icrr.u-tokyo.ac.jp/
Founded: 1953
Staff: 26
Activities: cosmic-ray research * high-energy astrophysics * interplanetary dust * particle astrophysics * neutrino astronomy
Periodicals: "ICRR Report" (ISSN 0919-8296), "Annual Report"
Coordinates: 137°33'00"E 36°06'00"N H2,770m (Norikura)
              138°20'00"E 35°04'42"N H900m (Akeno - see separate entry)
              137°19'11"E 36°25'26"N H-1,000m (Kamioka - see separate entry)
              139°49'48"E 36°34'48"N H-80m (Ohya)
              138°44'00"E 35°22'00"N H3,750m (Mount Fuji)
              139°51'00"E 35°09'00"N H-180m (Nokogiriyama)
City Reference Coordinates: 139°46'00"E 35°42'00"N

University of Tokyo, Institute for Cosmic Ray Research (ICCR), Akeno Observatory

Akeno-mura
Kitakoma-gun
Yamanashi 407-02
Telephone: (0)551-25-2301
Telefax: (0)551-35-2303
WWW: http://www.icrr.u-tokyo.ac.jp/akeno/akeno.html
        http://www.icrr.u-tokyo.ac.jp/
Founded: 1953 (ICCR)
Activities: see main entry
Periodicals: see main entry
City Reference Coordinates: 138°40'00"E 35°40'00"N

University of Tokyo, Institute for Cosmic Ray Research (ICCR), Kamioka Underground Observatory

Higashi-mozumi
Kamioka-cho
Yoshiki-gun
Gifu 506-12
Telephone: (0)578-5-2169
Telefax: (0)578-5-2169
Electronic Mail: <userid>@suketto.icrr.u-tokyo.ac.jp
WWW: http://suketto.icrr.u-tokyo.ac.jp/
        http://www-sk.icrr.u-tokyo.ac.jp/
        http://www.icrr.u-tokyo.ac.jp/
Founded: 1981
Staff: 29
Activities: solar neutrinos
Periodicals: see main entry
Coordinates: 137°19'11"E 36°25'26"N H-1000m
City Reference Coordinates: 136°45'00"E 35°25'00"N

University of Tokyo, Institute for Cosmic Ray Research (ICCR), Nokogiriyama Station

3551 Kanaya
Futtsu 299-18
Telephone: (0)439-69-2758 (Laboratory)
          (0)439-69-8788 (Cottage)
WWW: http://www.icrr.u-tokyo.ac.jp/
Founded: 1953 (ICCR)

Activities: see main entry
Periodicals: see main entry
Coordinates: 139°49'00"E 35°19'00"N H-180m
City Reference Coordinates: 139°49'00"E 35°19'00"N

University of Tokyo, Institute of Astronomy, Kiso Observatory

Tarusawa
Mitake-mura
Kiso-gun
Nagano 397-0101
Telephone: (0)264-52-3360
Telefax: (0)264-52-3361
WWW: http://www.ioa.s.u-tokyo.ac.jp/kiso_obs/
Founded: 1974
Staff: 7
Activities: structure of galaxies * survey of UV-excess objects * image processing
Periodicals: (1) "Annual Report of the Kiso Observatory" (ISSN 0915-5392)
Coordinates: 137°37'42"E 35°47'39"N H1,130m
City Reference Coordinates: 138°11'00"E 36°39'00"N

University of Tokyo, Research Center for the Early Universe (RESCEU)

School of Science
7-3-1 Hongo
Bunkyo-ku
Tokyo 113-0033
Telefax: (0)3-5802-8691
Electronic Mail: <userid>@resceu.s.u-tokyo.ac.jp
WWW: http://www.resceu.s.u-tokyo.ac.jp/Welcome.html
Founded: 1995
Staff: 10
Activities: structure and evolution of early universe and galaxies
City Reference Coordinates: 139°46'00"E 35°42'00"N

University of Tokyo, Tokyo Astronomical Observatory

● See now "National Astronomical Observatory of Japan (NAOJ)"

Variable Star Observers League in Japan (VSOLJ)

c/o National Science Museum
Ueno Park
Taito-ku
Tokyo 110
Electronic Mail: nah01147@niftyserve.or.jp (Makoto Watanabe, Database Manager)
WWW: http://www.kusastro.kyoto-u.ac.jp/vsnet/VSOLJ/vsolj.html
Periodicals: "VSOLJ Variable Star Bulletin"
City Reference Coordinates: 139°46'00"E 35°42'00"N

Wakkanai Radio Observatory

● See "Communications Research Laboratory (CRL), Wakkanai Radio Observatory"

World Data Center C2 for Ionosphere

c/o Communications Research Laboratory
4-2-1 Nukuikitamachi
Koganei-shi
Tokyo 184
Telephone: (0)423-27-7478
Telefax: (0)423-27-7606
Electronic Mail: <userid>@crl.go.jp
WWW: http://wdc-c2.crl.go.jp/index_eng.html
Periodicals: (1) "Catalogue of Data in World Data Center C2 for Ionosphere" (ISSN 0389-8229)
City Reference Coordinates: 139°46'00"E 35°42'00"N

World Data Center for Solar Activity

c/o National Astronomical Observatory of Japan
2-21-1 Osawa
Mitaka
Tokyo 181-8588
Telephone: (0)422-34-3716
Telefax: (0)422-34-3700
Electronic Mail: sunspot@solar.mtk.nao.ac.jp
Founded: 1978
Staff: 6
Periodicals: (12) "Monthly Bulletin on Solar Phenomena"; (4) "Quarterly Bulletin on Solar Activity (QBSA)"
City Reference Coordinates: 139°46'00"E 35°42'00"N

Yamagawa Radio Observatory
● See "Communications Research Laboratory (CRL), Yamagawa Radio Observatory"

Yamaguchi Prefectural Museum
8-2 Kasuga-cho
Yamaguchi 753-0073
Telephone: (0)839-22-0294
Telefax: (0)839-22-0353
Electronic Mail: yamahaku@ymg.urban.ne.jp
WWW: http://www.pref.yamaguchi.jp/e2top1f.htm
Founded: 1912
Staff: 19
Activities: research and public education (astronomy, earth sciences, botany, zoology, engineering, Japanese history)
Periodicals: (1) "Bulletin of the Yamaguchi Museum" (ISSN 0288-4232, circ.: 500), "Yamaguchi-ken no shizen" (Nature Study of Yamaguchi-ken) (ISSN 0288-4240, circ.: 800), "Annual Report of the Yamaguchi Museum" (circ.: 1300)
Coordinates: 131°28'32"E 34°10'46"N H55m
City Reference Coordinates: 131°29'00"E 34°10'00"N

Yokohama Science Center (YSC), Astronomy Section
5-2-1 Yokodai
Isogo-ku
Yokohama 235-0045
Telephone: (0)45-832-1166
Telefax: (0)45-832-1161
Electronic Mail: <userid>@ysc.go.jp
WWW: http://www.city.yokohama.jp/yhspot/ysc/ysc/ysc.html
Founded: 1984
Staff: 8
Activities: increasing understanding and appreciation of science including astronomy * astronomy information service for the public and amateur astronomers * hot line of astronomical news * computer bulletin board * planetarium * public observing
Coordinates: 139°35'52"E 35°22'26"N H60m
City Reference Coordinates: 139°39'00"E 35°27'00"N

Yukawa Institute for Theoretical Physics
● See "Kyoto University, Yukawa Institute for Theoretical Physics"

Jordan

Jordanian Astronomical Society (JAS)
P.O. Box 141568
Amman 11814
Telephone: (0)6-5534754
Telefax: (0)6-5534826
Electronic Mail: odehjas@jas.org.jo (Mohammad Shawkat Odeh)
WWW: http://www.jas.org.jo/
Founded: 1987 (as former "Jordanian Amateur Astronomers Society")
Membership: 50
City Reference Coordinates: 035°56'00"E 31°77'00"N

Jordanian Institution for Standards and Metrology (JISM)
P.O. Box 941287
Amman 11194
Telephone: (0)6-680139
Telefax: (0)6-681099
City Reference Coordinates: 035°56'00"E 31°77'00"N

Jordan Meteorological Department
P.O. Box 341011
Amman
Telephone: (0)6-892408
             (0)6-892409
Telefax: (0)6-894409
City Reference Coordinates: 035°56'00"E 31°77'00"N

Kazakhstan

Astra Astronomical Club
c/o Junior Art Palace
459120 Rudnyj
Telephone: (0)36378
Founded: 1974
Membership: 20
Activities: observing * excursions * education * planetarium
Coordinates: 064°00'00"E 53°59'00"N H200m
City Reference Coordinates: 064°00'00"E 53°59'00"N

Fesenkov Astrophysical Institute (V.G._)
● See "National Academy of Sciences, V.G. Fesenkov Astrophysical Institute"

National Academy of Sciences, Space Research Institute (SRI)
Shevchenko Ul. 15
480021 Almaty
Telephone: (0)3272-615853
Telefax: (0)3272-494355
Electronic Mail: zak@kaziki.alma-ata.su
WWW: http://smis.iki.rssi.ru/inform/sri-kaz.htm
Founded: 1991
Staff: 125
City Reference Coordinates: 076°57'00"E 43°15'00"N

National Academy of Sciences, V.G. Fesenkov Astrophysical Institute
Kamenskoye Plato
480068 Almaty
Telephone: (0)3272-648311
               (0)3272-624040
               (0)3272-677018
               (0)3272-255425
Periodicals: "Trudy"
Coordinates: 076°57'24"E 43°11'18"N H1,450m
City Reference Coordinates: 076°57'00"E 43°15'00"N

National Academy of Sciences, V.G. Fesenkov Astrophysical Institute, Tien-Shan Observatory
Kamenskoye Plato
480068 Almaty
Telephone: (0)3272-658040
               (0)3272-252092
Electronic Mail: kurt@afi.academ.alma-ata.su (Kenes S. Kuratov, Chief)
City Reference Coordinates: 076°57'00"E 43°15'00"N

Republican Palace of School Children Observatory
Lenin Prospekt 114
480051 Almaty
Telephone: (0)3272-643866
Founded: 1983
Staff: 5
Activities: education * observing (Sun, comets, variable stars)
Coordinates: 075°40'00"E 43°16'00"N
City Reference Coordinates: 076°57'00"E 43°15'00"N

State Information Center for Standardization
Ul. Pushkina 166
473000 Astana
Telephone: (0)3172-752991
Telefax: (0)3172-320540
Electronic Mail: gic@asdc.kz
Founded: 1997
Staff: 22
Activities: national norms and standards
City Reference Coordinates: 076°57'00"E 43°15'00"N (Almaty)

Tien-Shan Observatory
● See "National Academy of Sciences, V.G. Fesenkov Astrophysical Institute, Tien-Shan Observatory"

V.G. Fesenkov Astrophysical Institute
● See "National Academy of Sciences, V.G. Fesenkov Astrophysical Institute"

Kenya

Kenya Bureau of Standards (KEBS)
Off Mombasa Road
Behind Belle Vue Cinema
P.O. Box 54974
Nairobi
Telephone: (0)2-502210
              (0)2-502219
Telefax: (0)2-503293
Electronic Mail: kebs@users.africaonline.co.ke
WWW: http://www.kebs.org/
City Reference Coordinates: 036°49'00"E 01°17'00"S

Kenya Meteorological Department
Dagoretti Corner
Ngong Road
P.O. Box 30259
Nairobi
Telephone: (0)2-567880
Telefax: (0)2-567889
Electronic Mail: director@lion.meteo.go.ke
WWW: http://www.meteo.go.ke/
City Reference Coordinates: 036°49'00"E 01°17'00"S

Kenya National Academy of Sciences (KNAS)
P.O. Box 39450
Nairobi
Telephone: (0)2-721345
Telefax: (0)2-721138
Electronic Mail: knas@ken.healthnet.org
Founded: 1986
Staff: 10
Activities: conference and scientific meetings * project development * scientific documentation and information * promotion, creation and guidance of scientific bodies
Periodicals: (1) "Kenya Journal of Sciences (KJS) - Series A: Physical Sciences", "Kenya Journal of Sciences (KJS) - Series B: Biological Sciences", "Kenya Journal of Sciences (KJS) - Series C: Humanities and Social Sciences"
Awards: "Scholastic Awards", "General Award", "Distinguished Professional Contribution Award"
City Reference Coordinates: 036°49'00"E 01°17'00"S

Survey of Kenya
Ardhi House
First Ngong Avenue
P.O. Box 30046
Nairobi
Telephone: (0)2-718050
Founded: 1903
Staff: 617
Activities: surveying * mapping * land information systems
City Reference Coordinates: 036°49'00"E 01°17'00"S

Korea (DPRK)

Academy of Sciences of the DPRK, Pyongyang Astronomical Observatory (PAO)
Taesong District
Pyongyang
Telephone: (0)521-4883
(0)521-9231
Staff: 100
Activities: astrophysics * astrometry * radio astronomy * ephemerides * artificial satellites * planetarium
City Reference Coordinates: 127°16'00"E 38°26'00"N

Committee for Standardization of the DPRK (CSK)
Zung Gu Yok Seungli-Street
Pyongyang
Telephone: (0)2-571576
City Reference Coordinates: 127°16'00"E 38°26'00"N

Pyongyang Astronomical Observatory (PAO)
● See "Academy of Sciences of the DPRK, Pyongyang Astronomical Observatory (PAO)"

Korea (ROK)

Bohyunsan Optical Astronomy Observatory (BOAO)
• See "Korea Astronomy Observatory (KAO), Bohyunsan Optical Astronomy Observatory (BOAO)"

Chungnam National University (CNU), Department of Astronomy and Space Science
Taejon 305-764
Telephone: (0)42-8215461
Telefax: (0)42-8228380
Electronic Mail: <userid>@astrol.chungnam.ac.kr
WWW: http://astrol.chungnam.ac.kr/intro.html
Founded: 1988
Staff: 5
Activities: radioastronomy * stellar dynamics * external galaxies * MHD * planetology
City Reference Coordinates: 127°26'00"E 36°20'00"N

Daeduk Radio Astronomy Observatory (DRAO)
• See now "Taeduk Radio Astronomy Observatory (TRAO)"

Korea Advanced Institute of Science and Technology (KAIST), Department of Physics, Space Science Laboratory
Taejon 305-701
Telephone: (0)42-869-2565
Telefax: (0)42-869-5525
Electronic Mail: <userid>@space.kaist.ac.kr
WWW: http://space.kaist.ac.kr/
Founded: 1987
City Reference Coordinates: 127°26'00"E 36°20'00"N

Korea Astronomy Observatory (KAO)
61-1 Whaam-dong
Yusong-gu
Taejon 305-348
Telephone: (0)42-8643230
Telefax: (0)42-8653201
     (0)42-8615610
Electronic Mail: <userid>@hanul.issa.re.kr
WWW: http://www.issa.re.kr/
     http://www.issa.re.kr/kaoe.html (English page)
Founded: 1974
Staff: 89
Activities: radio observing * Galaxy * clusters * eclipsing binaries * sunspots
Periodicals: "Publications of the Korea Astronomy Observatory"
Coordinates: 127°22'28"E 36°23'58"N H120m
City Reference Coordinates: 127°26'00"E 36°20'00"N

Korea Astronomy Observatory (KAO), Bohyunsan Optical Astronomy Observatory (BOAO)
P.O. Box 1
Jacheon Post Office
Young-Cheon
Kyungpook 770-820
Telephone: (0)563-3301000
Telefax: (0)563-3369450
Electronic Mail: <userid>@seeru.boao.re.kr
WWW: http://www.boao.re.kr/
Founded: 1995
Staff: 19
Activities: optical observing * Galaxy * clusters * eclipsing binaries * solar observing * sunspots flares
Periodicals: "Publications of the Korea Astronomy Observatory"
Coordinates: 128°58'36"E 36°09'53"N H1,162m
City Reference Coordinates: 128°59'00"E 36°10'00"N

Korea Astronomy Observatory (KAO), Sobaeksan Optical Astronomy Observatory (SOAO)
P.O.Box 132
Tanyang Post Office
Tanyang
Chungbuk 395-800
Telephone: (0)444-4221108
     (0)444-4234879
Telefax: (0)444-4221108
Electronic Mail: sobaek@hanul.issa.re.kr
WWW: http://www.issa.re.kr/~sobaek/sao.html

Founded: 1976
Staff: 7
Coordinates: 128°27'27"E 36°56'04"N H1,378m
City Reference Coordinates: 128°27'00"E 36°56'00"N

Korea Astronomy Observatory (KAO), Taeduk Radio Astronomy Observatory (TRAO)
San 36-1
Whaam-dong
Yuseong-gu
Taejon 305-348
Telephone: (0)42-8653282
Telefax: (0)42-8615610
Electronic Mail: <userid>@hanul.issa.re.kr
WWW: http://w3.trao.re.kr/trao/
         http://www.issa.re.kr/trao/
Founded: 1988
Staff: 8
Activities: radio observing * Galaxy * clusters * eclipsing binaries * sunspots
Periodicals: "Publications of the Korea Astronomy Observatory"
Coordinates: 127°22'19"E 36°23'53"N H120m
● Previously known as "Daeduk Radio Astronomy Observatory (DRAO)"
City Reference Coordinates: 127°26'00"E 36°20'00"N

Korean Astronomical Society (KAS)
c/o Korea Astronomy Observatory
36-1 Kwa-am Dong
Yusung Ku
Taejon 305-348
Telephone: (0)42-8653219
Telefax: (0)42-8615610
Electronic Mail: kas@hanul.issa.re.kr
WWW: http://astro.snu.ac.kr/kas/
Founded: 1965
Membership: 565
Activities: organizing conferences
City Reference Coordinates: 127°26'00"E 36°20'00"N

Korean Meteorological Administration
1 Songwol-dong
Chongno-gu
Seoul 110-101
Telephone: (0)2-7230011
           (0)2-7230017
           (0)2-7330565
Telefax: (0)2-7370325
         (0)2-7370368
         (0)2-7370369
         (0)2-7222049
WWW: http://www.kma.go.kr/
Founded: 1990 (1948 as "Central Meteorological Office")
City Reference Coordinates: 126°58'00"E 37°33'00"N

Korean National Institute of Technology and Quality (KNITQ)
2 Joongang-dong
Kwachon
Kyunggi-do 427-010
Telephone: (0)2-5074369
Telefax: (0)2-5037977
Electronic Mail: int_coop@mail.nitq.go.kr
WWW: http://www.nitq.go.kr
City Reference Coordinates: 127°15'00"E 37°30'00"N

Korean Physical Society (KPS)
Yuksam-Dong 635-4
Kangnam-Ku
Seoul 135-703
Telephone: (0)2-5564737
Telefax: (0)2-5541643
WWW: http://www.kps.or.kr/
City Reference Coordinates: 126°58'00"E 37°33'00"N

Korean Space Science Society (KSSS)
c/o Department of Astronomy and Space Science
Yonsei University
134 Shinchon

Seodaemun-ku
Seoul 120-749
Telephone: (0)2-3612693
Telefax: (0)2-3625135
Electronic Mail: ksss@csa.yonsei.ac.kr
WWW: http://csaweb.yonsei.ac.kr/~ksss/
Founded: 1984
Membership: 440
Staff: 2
Activities: promoting advancement of astronomy and space sciences * meetings
Periodicals: (2) "Journal of Astronomy and Space Sciences (JASS)" (ISSN 1225-052x), "Bulletin of the Korean Space Science Society"
City Reference Coordinates: 126°58'00"E 37°33'00"N

Kyung Hee University, Department of Astronomy and Space Science
Yong-In
Kyunggi-do 170-73
Telephone: (0)331-86131
WWW: http://photon.kyunghee.ac.kr/
Activities: radio studies of molecular clouds * binary stars
Coordinates: 127°04'20"E 37°14'30"N
City Reference Coordinates: 127°15'00"E 37°30'00"N

Kyungpook National University (KNU), Department of Astronomy and Atmospheric Sciences (DA-AS)
Taegu 702-701
Telephone: (0)53-950-6360
Telefax: (0)53-950-6359
Electronic Mail: <userid>@bh.kyungpook.ac.kr
WWW: http://sirius.kyungpook.ac.kr/
Founded: 1988
City Reference Coordinates: 128°35'00"E 35°52'00"N

Pusan National University, Department of Earth Sciences
Jangjeon-dong
Keum-ku
Pusan 609-735
Telephone: (0)51-5101626
Telefax: (0)51-5137495
Electronic Mail: <userid>@astrophys.pusan.ac.kr
WWW: http://mercury.es.pusan.ac.kr/
Staff: 3
Activities: education * research
Coordinates: 129°04'56"E 35°13'51"N
          129°04'39"E 35°13'44"N
City Reference Coordinates: 129°03'00"E 35°06'00"N

Sejong University, Daeyang Observatory
98 Koonja-dong
Sungdon-ku
Seoul 133-747
Telephone: (0)2-4600345
Telefax: (0)2-4600299
WWW: http://www.sejong.ac.kr/institutions/observatory/obserFram.html
City Reference Coordinates: 126°58'00"E 37°33'00"N

Seoul National University (SNU), Department of Astronomy
San 56-1
Shinrim-dong
Kwanak-ku
Seoul 151-742
Telephone: (0)2-8806621
Telefax: (0)2-8871435
Electronic Mail: <userid>@astro.snu.ac.kr
WWW: http://astro.snu.ac.kr/
Founded: 1958
Activities: education * research
City Reference Coordinates: 126°58'00"E 37°33'00"N

Sobaeksan Optical Astronomy Observatory (SOAO)
• See "Korea Astronomy Observatory (KAO), Sobaeksan Optical Astronomy Observatory (SOAO)"

Taeduk Radio Astronomy Observatory (TRAO)
• See "Korea Astronomy Observatory (KAO), Taeduk Radio Astronomy Observatory (TRAO)"

Yonsei University, Department of Astronomy and Space Science
134 Shinchon
Seodaemun-ku
Seoul 120-749
Telephone: (0)2-3612680
             (0)2-3613439 (Observatory)
Telefax: (0)2-3135033
Electronic Mail: <userid>@galaxy.yonsei.ac.kr
WWW: http://galaxy.yonsei.ac.kr/
Founded: 1967
Staff: 4
Activities: stellar evolution * globular clusters * elliptical galaxies * blue compact dwarf galaxies * eclipsing binaries * celestial mechanics * artificial satellites
Coordinates: 126°59'00"E 37°34'00"N H67m
               126°48'30"E 37°41'01"N (Ilsan Station)
City Reference Coordinates: 126°59'00"E 37°34'00"N

Latvia

Latvian Academy of Sciences, Radioastrophysical Observatory
● See now "University of Latvia, Institute of Astronomy"

Latvian Academy of Sciences, Ventspils International Radioastronomy Centre (VIRAC)
Akademijas Laukums 1
LV-1050 Riga
Telephone: (0)722-8321
Telefax: (0)722-1153
Electronic Mail: berv@acad.latnet.lv (Edgars Bervalds, Director)
WWW: http://www.lanet.lv/members/LU/astro/ast_ven.html
Founded: 1996
Staff: 13
Activities: conversion of former military antennas into radiotelescopes
Coordinates: 021°51'18"E 57°33'10"N H15m
City Reference Coordinates: 024°08'00"E 56°53'00"N

Latvian Astronomical Society
(Latvijas Astronomijas Biedriba - LAB)
P.O. Box 332
LV-1098 Riga
or :
Boulevard Rainis 19
LV-1586 Riga
Telephone: (0)7223637
Telefax: (0)7820180
Electronic Mail: astro@acad.latnet.lv
WWW: http://www.lanet.lv/members/LU/astro/ast_lab.html
Founded: 1947
Membership: 120
Activities: Summer camps * popularization
Periodicals: "Astronomical Calendar"
Coordinates: 024°51'00"E 57°09'00"N (Sigulda)
City Reference Coordinates: 024°08'00"E 56°53'00"N

Latvian Hydrometeorological Agency (HMP)
165 Maskavas Street
LV-1019 Riga
Telephone: (0)7-113274
Telefax: (0)7-145154
Electronic Mail: epoc@meteo.lv
                    <userid>@meteo.lv
WWW: http://www.meteo.lv/
City Reference Coordinates: 024°08'00"E 56°53'00"N

Latvian National Center for Standardization and Metrology (LVS)
Kr. Valdemara Street 157
LV-1013 Riga
Telephone: (0)2-378165
Telefax: (0)2-362805
Electronic Mail: lvs@mail.eunet.lv
City Reference Coordinates: 024°08'00"E 56°53'00"N

Latvian Physical Society
c/o Janis Berzins (President)
Nuclear Research Center
Miera Street 31
Salaspils 1
LV-2169 Riga
Telephone: (0)2-945835
Telefax: (0)2-945858
Electronic Mail: jbrz@mhd.iph.sal.lv
                    brzs@lanet.lv
Membership: 70
City Reference Coordinates: 024°08'00"E 56°53'00"N

University of Latvia, Institute of Astronomy
Boulevard Rainis 19
LV-1586 Riga
Telephone: (0)7223149
Telefax: (0)7820180
Electronic Mail: astra@latnet.lv

WWW: http://www.astr.lu.lv/
        http://www.lanet.lv/members/LU/astro
Founded: 1946 (1997 under its current name)
Staff: 30
Activities: astrometry * space geodesy (SLR, GPS) * education * instrumentation * carbon and related stars (photometry, spectroscopy, synthetic spectra, evolution) * solar radio emission from active regions
Periodicals: "Acta Universitatis Latviensis (Astronomy)", "Investigations of the Sun and Red Stars" (ISSN 0135-1303); (4) "Zvaigžnotá Debéss" (ISSN 0135-129x - see separate entry)
Coordinates: 024°07'00"E 56°57'08"N
           024°24'15"E 56°46'36"N H75m (Baldone)
City Reference Coordinates: 024°08'00"E 56°53'00"N

Ventspils International Radioastronomy Centre (VIRAC)
• See "Latvian Academy of Sciences, Ventspils International Radioastronomy Centre (VIRAC)"

Zvaigžnotá Debéss
(The Starry Sky)
Editorial Offices
Akademijas Laukumas 1
LV-1527 Riga
Telephone: (0)7-226796
Telefax: (0)7-228784
Electronic Mail: astra@acad.latnet.lv
WWW: http://www.lanet.lv/members/LU/astro/ast_zd.html
Founded: 1958
Staff: 8
• Periodical (4) (ISSN 0135-129x)
City Reference Coordinates: 024°08'00"E 56°53'00"N

Lebanon

Lebanese Astronomical Society (LAS)
Beirut
Electronic Mail: lebast@geocities.com
WWW: http://www.geocities.com/CapeCanaveral/Hall/6865/
Founded: 1999
Activities: monthly meetings * education * observing
City Reference Coordinates: 035°30'00"E 33°52'00"N
• Formerly know as "Lebanese Amateur Astronomical Society"

Lebanese Standards Institution
P.O. Box 55120
Sin El-Fil
Beirut
or :
Gedco Center 3
Bloc B - 10th floor
Mkalles Hayed Avenue
Sin El-Fil
Beirut
Telephone: (0)1-485927
              (0)1-485928
Telefax: (0)1-485929
City Reference Coordinates: 035°30'00"E 33°52'00"N

Lithuania

Institute of Physics, Stellar Systems Department
A. Goštauto 12
Vilnius 2600
Electronic Mail: wladas@ktl.mii.lt (Vladas Vansevicius, Head)
                <userid>@ktl.mii.lt
WWW: http://193.219.53.135/Directions/iv/010/010.htm
Founded: 1990
Coordinates: 025°16'48"E 54°41'23"N
City Reference Coordinates: 025°19'00"E 54°41'00"N

Institute of Theoretical Physics and Astronomy
A. Goštauto 12
Vilnius 2600
Telephone: (0)2-620668
Telefax: (0)2-225361
Electronic Mail: astro@itpa.fi.lt
                astro@itpa.lt
WWW: http://www.itpa.lt/
                http://www.itpa.lt/~astro/ba/ (Baltic Astronomy)
Founded: 1989
Staff: 140
Activities: theoretical physics * stellar astronomy * galactic structure * interstellar extinction * atomic and molecular spectra * comets
Periodicals: (4) "Baltic Astronomy" (ISSN 1392-0049)
Coordinates: 025°33'45"E 55°19'00"N H200m (Moletai - see separate entry)
                066°52'30"E 38°41'03"N H2,540m (Maidanak - Uzbekistan)
City Reference Coordinates: 025°19'00"E 54°41'00"N

Institute of Theoretical Physics and Astronomy, Moletai Astronomical Observatory
Astronomijos Observatorija
Moletai 4150
Telephone: (0)2-951728
                (0)2-945425
Telefax: (0)2-224694
Electronic Mail: astro@itpa.lt
WWW: http://www.itpa.lt/~astro/obs/mao/
Founded: 1969
Staff: 10
Activities: stellar photometry * interstellar extinction * astronomical instrumentation * variable stars
Coordinates: 025°33'45"E 55°19'00"N H200m
City Reference Coordinates: 025°25'00"E 55°14'00"N

Institute of Theoretical Physics and Astronomy, Planetarium
Ukmerges 12A
Vilnius 2005
Telephone: (0)2-724177
Telefax: (0)2-724177
Electronic Mail: gintaras@itpa.lt
                planet@astro.lt
WWW: http://www.astro.lt/planet/
Founded: 1962
Staff: 16
Activities: lectures * education * shows
City Reference Coordinates: 025°19'00"E 54°41'00"N

Lithuanian Academy of Sciences
Gedimino Pr. 3
Vilnius 2600
Telephone: (0)2-613651
Telefax: (0)2-618464
Electronic Mail: prezidum@ktl.mii.lt
WWW: http://neris.mii.lt/Akademija/inf.html
City Reference Coordinates: 025°19'00"E 54°41'00"N

Lithuanian Hydrometeorological Service (LHMS)
6 Rudnios Street
Vilnius 2600
Electronic Mail: lhmt@meteo.lt
WWW: http://www.meteo.lt/
Staff: 500
City Reference Coordinates: 025°19'00"E 54°41'00"N

Lithuanian Physical Society (LPS)
c/o G. Gaigalas (Secretary)
Institute of Theoretical Physics and Astronomy
A. Goštauto 12-346
Vilnius 2600
Telephone: (0)2-620668
Telefax: (0)2-224694
Electronic Mail: tmkc.plls@wllb.lt (Zenonas Rudzikas, President)
WWW: http://www.itpa.lt/~lfd/
Founded: 1963
Membership: 165
Activities: to contribute to physical sciences, studies and education * to unite all interested parties in physics and its applications
Periodicals: "Lithuanian Physics Journal"
City Reference Coordinates: 025°19'00"E 54°41'00"N

Lithuanian Standards Board (LST)
T. Kosciuškos g. 30
Vilnius 2600
Telephone: (0)2-226962
Telefax: (0)2-226252
Electronic Mail: lstboard@ktl.mii.lt
Founded: 1990
Staff: 46
Activities: standardization * accreditation * metrology
Periodicals: (12) "LST Bulletin" (circ.: 600)
City Reference Coordinates: 025°19'00"E 54°41'00"N

Moletai Astronomical Observatory
● See "Institute of Theoretical Physics and Astronomy, Moletai Astronomical Observatory"

Palace of Pupils, Technical Activities, Astronomical Laboratory
Zirmunu 1b
Vilnius 2012
Telephone: (0)122-773614
Founded: 1976
Staff: 3
Activities: education * expeditions * Summer camps
City Reference Coordinates: 025°19'00"E 54°41'00"N

Vilnius Planetarium
● See "Institute of Theoretical Physics and Astronomy, Planetarium"

Vilnius University, Astronomical Observatory
Čiurlionio 29
Vilnius 2009
Telephone: (0)2-633343
Telefax: (0)2-635648
Electronic Mail: <userid>@ff.vu.lt
WWW: http://www.vu.lt/menu/depart/physics/physics.html
Founded: 1753
Staff: 10
Activities: stellar photometry * variable stars * ISM * galactic structure and evolution * instrumentation * photometric classification
Coordinates: 025°17'12"E 54°41'00"N H122m
025°34'00"E 55°19'00"N H200m (Moletai - see separate entry)
066°54'00"E 38°41'03"N H2,600m (Maidanak - Uzbekistan)
City Reference Coordinates: 025°19'00"E 54°41'00"N

Luxembourg

Amateurs Astronomes du Luxembourg (AAL)
Boîte Postale 1711
L-1017 Luxembourg
Telephone: (0)395343
Telefax: (0)50501420
Electronic Mail: aal@aal.lu
WWW: http://www.aal.lu/
Founded: 1971
Membership: 150
Activities: public observing * lectures * education
City Reference Coordinates: 006°09'00"E 49°36'00"N

Service de la Météorologie et de l'Hydrologie
16, route d'Esch
Boîte Postale 1904
L-1019 Luxembourg
Telephone: (0)443232
Telefax: (0)443232
City Reference Coordinates: 006°09'00"E 49°36'00"N

Service de l'Énergie de l'État (SEE), Département de Normalisation
34, avenue de la Porte-Neuve
Boîte Postale 10
L-2010 Luxembourg
Telephone: (0)4697461
Telefax: (0)46974639
Electronic Mail: see.normalisation@.ed.etat.lu
WWW: http://www.etat.lu/SEE
Founded: 1967
Activities: certification * normalization * testing
City Reference Coordinates: 006°09'00"E 49°36'00"N

Macedonia

Hydrometeorological Institute of Macedonia
P.O. Box 213
91001 Skopje
WWW: http://www.meteo.gov.mk/
Staff: 230
City Reference Coordinates: 021°26'00"E 41°59'00"N

Society of Physicists of the Republic of Macedonia
c/o Viktor Urumov (President)
Faculty of Natural Sciences and Mathematics
Cyril and Methodius University
P.O. Box 162
91000 Skopje
Telephone: (0)91-261330
Telefax: (0)91-228141
Electronic Mail: urumov@innona.pmf.ukim.edu.mk
Membership: 300
City Reference Coordinates: 021°26'00"E 41°59'00"N

Zavod za Standardizacija i Metrologija (ZSM)
(Bureau of Standardization and Metrology)
Samoilova 10
91000 Skopje
Telephone: (0)91-131102
            (0)91-131160
Telefax: (0)91-110263
Electronic Mail: sofijak@lotus.mpt.com.mk
Founded: 1995
Staff: 22
Activities: standardization * metrology * related activities
Periodicals: "Bulletin"
City Reference Coordinates: 021°26'00"E 41°59'00"N

Malaysia

Bahagian Kajian Sains Angkasa (BAKSA)
● See "Space Science Studies Division"

Infinity Infocus
7 Jalan SS14/5E
Subang Jaya
47500 Petaling Jaya
Telephone: (0)3-7342519
Telefax: (0)3-7346308
Electronic Mail: infocus@pc.jaring.my
Founded: 1988
Staff: 3
Activities: distributing astronomical equipment and accessories * consultancy
Coordinates: 101°37'00"E 03°04'00"N H15m
City Reference Coordinates: 101°39'00"E 03°07'00"N

Malaysian Meteorological Service (MMS)
Jalan Sultan
46667 Petaling Jaya
Telephone: (0)3-7569422
Electronic Mail: mms@kjc.gov.my
WWW: http://www.kjc.gov.my/
City Reference Coordinates: 101°39'00"E 03°07'00"N

National Planetarium, Malaysia
53 Jalan Perdana
50480 Kuala Lumpur
Telephone: (0)3-2735484
           (0)3-2735485
           (0)3-2735486
Telefax: (0)3-2735488
Electronic Mail: baksa@po.jaring.my
WWW: http://www.mastic.gov.my/kstas/baksa/
Founded: 1989
Staff: 50
Activities: planetarium shows * observing * space science exhibits * education * research
City Reference Coordinates: 101°43'00"E 03°09'00"N

National University of Malaysia, Department of Physics
43600 Bangi
Selangor
Telephone: (0)3-8250001
Telefax: (0)3-8256484
WWW: http://www.ukm.my/
Founded: 1970
Activities: M-star survey of the Galaxy * education
City Reference Coordinates: 101°43'00"E 03°09'00"N

Sheikh Tahir Astronomical Centre
c/o Astronomy and Atmospheric Research Unit
University of Science Malaysia
11800 Penang
or :
Pantai Acheh
Balik Pulau
11010 Penang
Telephone: (0)4-6577888
Telefax: (0)4-6576155
WWW: http://www.usm.my/aaru/sheikh_tahir_centre.html
Founded: 1991
Staff: 2
Activities: observing * Moon * education
City Reference Coordinates: 100°19'00"E 05°24'00"N

Space Science Studies Division
(Bahagian Kajian Sains Angkasa - BAKSA)
National Planetarium Complex
53 Jalan Perdana
50480 Kuala Lumpur
Telephone: (0)3-22735484

Telefax: (0)3-22735484
Electronic Mail: baksa@baksa.gov.my
WWW: http://www.baksa.gov.my/
Founded: 1994
Staff: 50
Activities: planetarium shows * observing * astronomy and space science exhibits
Periodicals: (2) "Semesta" (ISSN 1394-8636)
City Reference Coordinates: 101°43'00"E 03°09'00"N

Standards and Industrial Research Institute of Malaysia (SIRIM)

P.O. Box 35
40700 Shah Alam
Telephone: (0)3-5592601
Telefax: (0)3-5508095
Electronic Mail: central@dsm4.gov.my
WWW: http://mastic.gov.my/kstas/dsm.htm
City Reference Coordinates: 101°31'00"E 03°02'00"N

University of Science Malaysia, Astronomy and Atmospheric Research Unit (AARU)

11800 Penang
Telephone: (0)4-6577888 x2115
         (0)4-6572859
Telefax: (0)4-6576155
WWW: http://www.cs.usm.my/aaru/
Founded: 1990
Activities: observing * data centre * Sun and Moon ephemerides * light flux measurements * calendar * time * atmospheric studies * radiation * ozone
Observatories: 2 (University Campus & Pantai Acheh)
● See also "Sheikh Tahir Astronomical Centre"
City Reference Coordinates: 100°19'00"E 05°24'00"N

Malta

Malta Astronomical Society (MAS)
P.O. Box 174
Valletta
or :
"Le Méridien"
Triq Is-Sedqa
Attard BZN 12
Telephone: (0)417401
Telefax: (0)683396
Electronic Mail: edwincam@hotmail.com (Edwin Camilleri, Treasurer)
WWW: http://www.cyberlobby.com/maltastro
       http://www.geocities.com/maltastro/
Founded: 1985
Membership: 100
Activities: observing (meteors, variable stars, comets, sunspots, lunar and solar eclipses, occultations) * lectures * slide and video shows * education * computing * radio and television programmes * camps * junior club * research on local astronomy * public forums
Periodicals: (12) "Newsletter"; (4) "Big Bang"; (1) "Sky Almanac for Maltese Islands"; "Sirius"
Awards: (1) "Astronomer of the Year Award"
City Reference Coordinates: 014°27'00"E 35°53'00"N

Malta Standardisation Authority (MSA)
Evans Building - Second Floor
Merchant Street
Valletta VLT 03
Telephone: (0)242420
Telefax: (0)242406
Electronic Mail: info@msa.com.mt
Founded: 1996
Staff: 6
City Reference Coordinates: 014°31'00"E 35°54'00"N

Meteorological Office
c/o Department of Civil Aviation
Airport
Luqa
Telephone: (0)249170 x332
       (0)249170 x308 (forecast)
Telefax: (0)246694
       (0)239278
Founded: 1922
Staff: 39
Activities: observing * statistics
City Reference Coordinates: 014°30'00"E 35°52'00"N

Mauritius

Mauritius Radio Telescope (MRT)
c/o Faculty of Science
University of Mauritius
Reduit
or :
Brad D'ean
Poste-De Flacq
Telephone: (0)4541041
Telefax: (0)4549642
Electronic Mail: mrt@dodo.uom.ac.mu
                    <userid>@uom.ac.mu
WWW: http://icarus.uom.ac/mu/
         http://www.uom.ac.mu/fos/fos.htm
Founded: 1992
Staff: 3
Activities: radio astronomy * surveys
Coordinates: 057°44'00"E 20°08'00"S
City Reference Coordinates: 057°44'00"E 20°08'00"S

Mauritius Standards Bureau (MSB)
Reduit
Telephone: (0)4541933
Telefax: (0)4641144
Electronic Mail: msb@intnet.mu
City Reference Coordinates: 057°44'00"E 20°08'00"S

Meteorological Department
Vacoas
Telephone: (0)6861031
         (0)6861032
         (0)6965626
Telefax: (0)6861033
City Reference Coordinates: 057°29'00"E 20°18'00"S

Mexico

Academia Mexicana de Ciencias (AMC)
Avenida San Jeronimo 260
Colonia Jardines del Pedregal
Apartado Postal 69-692
México, DF 04500
Telephone: (0)5503906
            (0)5507133
            (0)5504000
Telefax: (0)5501143
            (0)5500389
Electronic Mail: aic@servidor.unam.mx
WWW: http://www.unam.mx/academia/index.html
        http://serpiente.dgsca.unam.mx/academia/index_english.html
Founded: 1959
Activities: gathering distinguished Mexican scientists * promoting scientific research in Mexico * organizing meetings * exchanges of scientists
Periodicals: "Ciencia"; "Boletín de la Academia"
Awards: "Prizes of Scientific Research"
● Formerly known as "Academia de la Investigación Científica (AIC)"
City Reference Coordinates: 099°09'00"W 19°24'00"N

Centro Cultural Alfa
Apartado Postal 1177
Monterrey, NL 64000
or :
c/o Martha Cortinas
Roberto Garza Sada 1000
Fracc. Carrizalejo
Garza García Z.C.
Monterrey, NL 64254
Telephone: (0)83-565225
            (0)83-565285
Telefax: (0)83-565945
WWW: http://www.planetarioalfa.org.mx/ (Planetarium)
Founded: 1978
Activities: permanent exhibitions of astronomy, physics, technology and art * observing
City Reference Coordinates: 100°19'00"W 25°40'00"N

Centro de Ciencias de Sinaloa
Avenida de las Americas 2771 Norte
Culiacan, SIN 80100
Telephone: (0)67-122889
Telefax: (0)67-123061
Electronic Mail: planetario@computo.ccs.net.mx
WWW: http://www.ccs.net.mx/MuseoCientifico/Planetario.htm
City Reference Coordinates: 107°23'00"W 24°50'00"N

Centro de Convenciones y Exposiciones de Morelia, Planetario
Avenida Ventura Puente
Esquina Avenida Camelinas
Morelia, MICH 58070
WWW: http://michoacan.gob.mx/turismo/3036/cconvenciones.htm
City Reference Coordinates: 101°11'00"W 19°40'00"N

Centro de Investigaciones en Optica (CIO)
Apartado Postal 1-948
León, GTO 37000
or :
Loma del Bosque 115
Colonia Lomas del Campestre
León, GTO 37150
Telephone: (0)47-184425
            (0)47-731018
Telefax: (0)47-175000
Founded: 1980
Staff: 20
Activities: basic and applied research in optics, interferometry, holography, image processing and lasers * technological development in thin layers, optical design and instrumentation * building Newtonian telescopes
City Reference Coordinates: 101°42'00"W 21°10'00"N

Consejo Cultural Mundial (CCM)
Apartado Postal 11-823
México, DF 06101
or :
Medellín 201 - 3er Piso
Colonia Roma
México, DF 06700
Telephone: (0)5-5745039
Telefax: (0)5-5745039
Electronic Mail: worldcouncil@compuserve.com
Founded: 1982
Staff: 12
Activities: acknowledging through awards individuals and institutions who made outstanding achievements in science, education, and arts.
Awards: (1) "Albert Einstein World Award of Science"; (1/2) "Leonardo da Vinci World Award of Arts", "Jose Vasconcelos World Award of Education"
City Reference Coordinates: 099°09'00"W 19°24'00"N

Consejo Nacional de Ciencia y Tecnología (CONACYT)
Constituyentes 1046
Lomas Altas
México, DF 11950
Telephone: (0)2-3277626
Telefax: (0)2-3277532
Electronic Mail: <userid>@mailer.main.conacyt.mx
WWW: http://www.conacyt.mx/
Founded: 1971
Activities: funding research and development
City Reference Coordinates: 099°09'00"W 19°24'00"N

Dirección General de Normas (DGN)
Calle Puente de Tecamachalco 6
Lomas de Tecamachalco
Sección Fuentes
Naucalpán de Juárez
México, DF 53950
Telephone: (0)5-7299300
Telefax: (0)5-7299484
Electronic Mail: cidgn@secofi.gob.mx
WWW: http://www.secofi.gob.mx/normas/dgn1.shtml
City Reference Coordinates: 099°09'00"W 19°24'00"N

Instituto Nacional de Astrofísica, Optica y Electronica (INAOE), Observatorio Astrofísico Guillermo Haro (OAGH)
Avenida Sinaloa 25
Apartado Postal 125
Cananea, SON 84620
Telephone: (0)633-22915
Telefax: (0)633-22655
WWW: http://chiltin.inaoep.mx/~astrofi/cananea/
http://www.inaoep.mx/~astrofi/cananea/
http://www.inaoep.mx/
Founded: 1987
Staff: 17
Coordinates: 110°18'01"W 30°58'57"N H2,400m (Cerro de la Mariquita)
City Reference Coordinates: 110°18'00"W 30°57'00"N

Instituto Nacional de Astrofísica, Optica y Electronica (INAOE), Santa María Tonantzintla
Apartado Postal 51 y 216
Puebla, PUE 72000
Telephone: (0)22-472011
Telefax: (0)22-472231
(0)22-472580
Electronic Mail: <userid>@tonali.inaoep.mx
WWW: http://www.inaoep.mx/
http://www.inaoep.mx/~astrofi/ (Astrophysics Department)
Founded: 1942
Staff: 15
Activities: flare stars * star formation * open clusters * HII regions * variable stars * stellar atmospheres * supergiant stars
Coordinates: 098°18'50"W 19°01'58"N H2,174m
City Reference Coordinates: 097°50'00"W 18°45'00"N

Instituto Politécnico Nacional (IPN), Centro de Investigación y de Estudios Avanzados (CINVES-TAV)
Apartado Postal 14-740

México, DF 07000
or :
Avenida del Instituto Politécnico Nacional 2508
Colonia San Pedro Zacatenco
México, DF 07360
Telephone: (0)5-7540200
Telefax: (0)5-7548707
WWW: http://www.fis.cinvestav.mx/
Founded: 1961
Activities: biology * physics * engineering * mathematics * education
Periodicals: (6) "Avance y Perspectiva"
City Reference Coordinates: 099°09'00"W 19°24'00"N

Observatorio Astrofísico Guillermo Haro (OAGH)
● See "Instituto Nacional de Astrofísica, Optica y Electronica (INAOE), Observatorio Astrofísico Guillermo Haro (OAGH)"

Observatorio Astronómico Nacional (OAN), Mexico
● See "Universidad Nacional Autonóma de México (UNAM), Observatorio Astronómico Nacional (OAN)"

Observatorio de San Pedro Mártir
● See "Universidad Nacional Autonóma de México (UNAM), Observatorio Astronómico Nacional (OAN)"

Planetario de Cuidad Victoria
Boulevard Fidel Velazquez s/n
Colonia Horacio Teran
Apartado Postal 242
Ciudad Victoria, TAMPS 87000
Telephone: (0)131-51215
Telefax: (0)131-51215
Electronic Mail: joseba@spin.com.mx
WWW: http://www.uat.mx/Vinculos/planeta/
Founded: 1982
Staff: 15
Activities: scientific and technological popularization
City Reference Coordinates: 099°08'00"W 23°44'00"N

Servicio Meteorológico Nacional (SMN)
Avenida Observatorio 192
Colonia Observatorio
México, DF 11860
Telephone: (0)6-268650
Telefax: (0)6-268695
Electronic Mail: desparza@gsmn.cna.gob.mx
WWW: http://smn.cna.gob.mx/SMN.html
Founded: 1877
Staff: 1058
Activities: weather forecast * climatological data center
Periodicals: "Daily Forecast Bulletin"
Observatories: 72 observatories, 3,500 meteorological stations, 15 radiosonde stations and 12 radars
City Reference Coordinates: 099°09'00"W 19°24'00"N

Sociedad Astronómica Amateur de Sinaloa (SAASIN)
Angel Flores 119-1 Norte
Los Mochis, SIN 81200
Telephone: (0)68-182327
Telefax: (0)68-153097
Electronic Mail: info@saasin.org
WWW: http://www.saasin.org/
Founded: 1999
Membership: 10
Activities: observing * study * meetings
City Reference Coordinates: 109°00'00"W 25°48'00"N

Sociedad Astronómica de Aragón Ilhuícatl
Rosas Moreno 114-201
Colonia San Rafael
México, DF 06470
Telephone: (0)915-592-1553
         (0)915-623-0832
Electronic Mail: astro@hp-720.aragon.unam.mx
WWW: http://informatica.aragon.unam.mx/ilhuicatl/
Membership: 55
Activities: optics * telescope control systems * image processing * education * space technologies
Periodicals: (6) "Ilhuícatl"
City Reference Coordinates: 099°09'00"W 19°24'00"N

Sociedad Astronómica de la Faculdad de Ingenería (SAFIR)
Ciudad Universitaria
México, DF 04510
Telephone: (0)5-6220861
Telefax: (0)5-6162890
Electronic Mail: sidereus@servidor.unam.mx
WWW: http://odin.fi-b.unam.mx/sidereus/
Founded: 1997
Membership: 600
Activities: lectures * observing * links between astronomy and engineering * visits
Periodicals: "Sidereus Nuncius"
Coordinates: 099°11'03"W 19°19'50"N H2,280m
City Reference Coordinates: 099°09'00"W 19°24'00"N

Sociedad Astronómica de México (SAM) AC
Parque Felipe S. Xicoténcatl
Colonia Álamos
Apartado Postal M-9647
México, DF 03480
Telephone: (0)5-5194730
WWW: http://www.spin.com.mx/sam/sam.html
Founded: 1902
Activities: meetings * lectures * library * observing
Coordinates: 099°08'30"W 19°23'55"N H2,245m (Observatorio Luis G. León)
                 099°31'23"W 19°47'24"N H3,070m (Observatorio Cerro Las Ánimas)
City Reference Coordinates: 099°09'00"W 19°24'00"N

Sociedad Espacial Mexicana (SEM) AC
Apartado Postal 5-75
Guadalajara, JAL 45042
Telephone: (0)3-8104563
City Reference Coordinates: 103°20'00"W 20°40'00"N

Universidad Autónoma Metropolitana-Iztapalapa (UAM-I), Departamento de Física, Area Gravitación y Astrofísica
Avenida Michoacan y Purisima s/n
Colonia Vicentina
Apartado Postal 55-534
México, DF 09340
Telephone: (0)5-7244623
                 (0)5-6860322
Telefax: (0)5-6861717
WWW: http://www.iztapalapa.uam.mx/iztapala.www/division.cbi/fisica/progravi.htm
Founded: 1974
Staff: 8
Activities: astrophysics * classical and quantum cosmology * physics education
City Reference Coordinates: 099°09'00"W 19°24'00"N

Universidad de Guadalajara, Instituto de Astronomía y Meteorología
Avenida Vallarta 2602
Guadalajara, JAL 44140
Telephone: (0)3-6164937
Telefax: (0)3-6159829
WWW: http://www.udg.mx/udg/red/cu/cucei/iam/
Founded: 1903
Coordinates: 103°23'09"W 20°40'32"N H1,583m
City Reference Coordinates: 103°20'00"W 20°40'00"N

Universidad de Guanajuato, Departamento de Astronomía
Apartado 144
Guanajuato, Gto 36000
Telephone: (0)473-29548
                 (0)473-29607
Telefax: (0)473-20253
Electronic Mail: <userid>@astro.ugto.mx
WWW: http://www.astro.ugto.mx/
Founded: 1995
Staff: 12
Activities: variable stars * hot stars * binary stars * HI in galaxies * ISM * interactive galaxies * clusters of galaxies * radio sources * large-scale structure
City Reference Coordinates: 101°15'00"W 21°01'00"N

Universidad de Sonora, Centro de Investigación en Física, Área de Astronomía
Apartado Postal 5-088
Hermosillo, SON 83190

Telephone: (0)62-592156
Telefax: (0)62-126649
Electronic Mail: <userid>@cajeme.cifus.uson.mx
WWW: http://spin.cifus.uson.mx/unison/EOS/eos.htm
        http://cosmos.cifus.uson.mx/eosdata.htm (Estación de Observación Solar - EOS)
Founded: 1990
Activities: solar physics * observing * extragalactic astronomy
City Reference Coordinates: 110°58'00"W 29°04'00"N

Universidad Nacional Autónoma de México (UNAM), Instituto de Astronomía
Apartado Postal 70-264
México, DF 04510
or :
Circuito de la Investigación Científica
Ciudad Universitaria
México, DF 04510
Telephone: (0)5-6223906/11
Telefax: (0)5-6160653
Electronic Mail: <userid>@astroscu.unam.mx
WWW: http://www.astroscu.unam.mx/
Founded: 1863 (originally called "Observatorio Astronómico Nacional")
Staff: 90
Activities: ISM * HII regions * kinematics and dynamics * galaxies * cosmology * instrumentation * stars
Periodicals: (2) "Revista Mexicana de Astronomía y Astrofísica" (ISSN 0185-1101); "Revista Mexicana de Astronomía y Astrofísica - Serie de Conferencias" (ISSN 1485-2059)
Awards: (1/2) "Premio Harold Johnson"; "Premio Guillermo Haro Barraza"
City Reference Coordinates: 099°09'00"W 19°24'00"N

Universidad Nacional Autónoma de México (UNAM), Instituto de Astronomía, Unidad Morelia
J.J. Tablada 1006
Lomas de Santa María
Morelia, MICH 58090
Telephone: (0)43-236162
Telefax: (0)43-236165
Electronic Mail: <userid>@astrosmo.unam.mx
WWW: http://www.astrosmo.unam.mx/
Founded: 1995
Staff: 16
Activities: star formation * radio astronomy * ISM * HII regions * gas dynamics * MHD * radiative transfer * atmospheric turbulence * infrared astronomy
City Reference Coordinates: 101°11'00"W 19°40'00"N

Universidad Nacional Autónoma de México (UNAM), Instituto de Geofísica
Circuito Interior s/n
Ciudad Universitaria
Coyoacán
México, DF 04510
Telephone: (0)5-6224122
          (0)5-6162344
Telefax: (0)5-5502486
Electronic Mail: juf@tonatiuh.igeofcu.unam.mx (Jaime Urrutia Fucugauchi, Director)
WWW: http://www.igeofcu.unam.mx/
Founded: 1949
Staff: 110
Activities: impact craters * paleoclimatology * volcanology * paleomagnetism * geodynamics * hydrogeology * mathematical geophysics * remote sensing * solar-terrestrial relationships * cosmic rays * geomagnetism
Periodicals: (4) "Geofísica Internacional" (ISSN 0016-7169);
Awards: (1) "Julio Monges"
City Reference Coordinates: 099°09'00"W 19°24'00"N

Universidad Nacional Autónoma de México (UNAM), Observatorio Astronómico Nacional (OAN)
Apartado Postal 877
Ensenada, BC 22800
or :
Carretera de Tijuana Km. 106
Ensenada, BC 22860
Telephone: (0)617-44548
          (0)617-44580
          (0)617-44593
Telefax: (0)617-44607
Electronic Mail: oan@bufadora.astrosen.unam.mx
              <userid>@bufadora.astrosen.unam.mx
WWW: http://bufadora.astrosen.unam.mx/
Founded: 1868
Staff: 39
Activities: ISM * HII regions * dust * kinematics and dynamics * galaxies * PNs * instrumentation * stars

Coordinates: 115°27'49"W 31°02'40"N H2,890m (San Pedro Mártir)
                 098°19'00"W 19°02,00"N H2,130m (Tonantzintla)
● See also addresses in USA-CA
City Reference Coordinates: 116°37'00"W 31°52'00"N

Universum - Museo de las Ciencias
c/oJorge Flores Valdés
Cto. Mario de la Cueva s/n
Zona Cultural de la Ciudad Universitaria
Apartado Postal 70-487
México, DF 04510
Telephone: (0)5-6653761
Telefax: (0)5-6653769
Electronic Mail: universum@servidor.unam.mx
WWW: http://www.universum.unam.mx/
City Reference Coordinates: 099°09'00"W 19°24'00"N

Moldova

Department of Standards, Metrology and Technical Supervision
Coca 28
2064 Kishinev
Telephone: (0)2-748588
Telefax: (0)2-750581
Electronic Mail: moldovastandard@standart.mldnet.com
WWW: http://www.moldova.md/
City Reference Coordinates: 028°50'00"E 47°00'00"N

Integral Applied Studies Institute (IASI), Astrophysical Observatory
60 Mateevici Street
Building 4a
2009 Kishinev
Telephone: (0)2-577726
          (0)2-577459
Telefax: (0)2-762365
Electronic Mail: iasi@usm.md
Founded: 1973
Staff: 10
Activities: variable stars * stellar photometry * atmospheric sciences * remote sensing * environment * optics * education
Coordinates: 028°30'00"E 47°00'00"N H400m
City Reference Coordinates: 028°50'00"E 47°00'00"N

Mongolia

Mongolian Academy of Sciences, Centre of Astronomy and Geophysics
P.O. Box 788
Ulaanbataar 210613
Telephone: (0)1-50218
          (0)1-358849
Telefax: (0)1-358849
Electronic Mail: gnoon@magicnet.mn
              bechtur@csj.mn
Founded: 1997
Staff: 70
Activities: latitude * longitude * Earth rotation * microwave satellite observing * solar physics * radiation transfer * spectral-line formation * seismology * dynamic study of the Earth
Coordinates: 107°03'08"E 47°51'54"N (Khurel Togót Observatory)
City Reference Coordinates: 106°53'00"E 47°55'00"N

Mongolian National Institute for Standardisation and Metrology (MNISM)
Peace Street
Ulaanbataar 51
Telephone: (0)1-358032
          (0)1-53529
Telefax: (0)1-358032
Electronic Mail: mncsm@magicnet.mn
Founded: 1953
Staff: 120
Activities: standardization * metrology * quality assurance
Periodicals: (12) "Standards and Metrology"
City Reference Coordinates: 106°53'00"E 47°55'00"N

Morocco

Centre National de Coordination et de Planification de la Recherche Scientifique et Technique
Boîte Postale 8027
Rabat-Agdal
or :
52 Charia Omar Ibn Khattab
Agdal
Telephone: (0)7-72803
                (0)7-74215
Telefax: (0)7-771288
City Reference Coordinates: 006°51'00"W 34°02'00"N (Rabat)

Direction de la Météorologie Nationale (DMN)
Aéroport CASA/ANFA
Boîte Postale 8106 CASA/OASIS
Casablanca 20103
Telephone: (0)2-913297
                (0)2-913329
Telefax: (0)2-913554
                (0)2-913797
Founded: 1962
Staff: 1150
Activities: meteorology * climatology * applications * national network of stations
Periodicals: "Bulletin Météorologiques Quotidiens du Maroc", "Pluies Décadaires du Maroc", "Bulletin Agrométéorologique du Maroc", "Metagrhyd"
City Reference Coordinates: 007°35'00"W 33°39'00"N

Service de Normalisation Industrielle Marocaine (SNIMA)
Quartier Administratif
Rabat Chellah
or :
Tour Hassan
1, Place Sefrou
Rabat
Telephone: (0)7-721678
                (0)7-763733
Telefax: (0)7-760675
                (0)7-766296
Electronic Mail: snima@mcinet.gov.ma
WWW: http://www.mcinet.gov.ma/
City Reference Coordinates: 006°51'00"W 34°02'00"N (Rabat)

Université de Rabat, Laboratoire d'Astrophysique
Boîte Postale 8027 CP 10102
Agdal
Rabat
Telephone: (0)7-774215
                (0)7-770796
Telefax: (0)7-771288
Founded: 1985
Staff: 4
Activities: astrophysics * image processing * adaptive optics
Coordinates: 007°00'00"W 31°00'00"N (Oukaimeden)
City Reference Coordinates: 006°51'00"W 34°02'00"N

Netherlands

A.F. Philips Sterrenwacht (Dr._)
● See "Dr. A.F. Philips Sterrenwacht"

Amsterdamse Weer- en Sterrenkundige Kring (AWSK)
● See "Volkssterrenwacht Amsterdam (VSA)"

Artis Planetarium
Postbus 20164
NL-1000 HD Amsterdam
or :
Plantage Kerklaan 38-40
NL-1018 CZ Amsterdam
Telephone: (0)20-5233452
Telefax: (0)20-5233481
         (0)20-5233518
Electronic Mail: planetarium@artis.nl
WWW: http://www.artis.nl/
Founded: 1988
Staff: 20
Activities: shows * exhibitions * lectures * shop
City Reference Coordinates: 004°54'00"E 52°22'00"N

Astronomische Vereniging Wega
Gen. de Wetstraat 31
NL-5021 TK Tilburg
Telephone: (0)13-5422534
Electronic Mail: spaninks@freemail.nl
Founded: 1972
Membership: 60
Activities: observing * lectures
Periodicals: (3) "de Lier"
City Reference Coordinates: 005°05'00"E 51°34'00"N

Astronomy and Astrophysics (A&A), Main Journal
c/o H.J. Habing (Editor-in-Chief)
Sterrewacht Leiden
Postbus 9513
NL-2300 RA Leiden
or :
Niels Bohrweg 2
NL-2333 CA Leiden
Telephone: (0)71-5275916
Telefax: (0)71-5275803
Electronic Mail: aanda@strw.leidenuniv.nl
WWW: http://link.springer.de/link/service/journals/00230/
         http://cdsweb.u-strasbg.fr/abstract/Abstractlist.html (Abstracts at CDS)
Founded: 1969
● Journal (24) (ISSN 0004-6361)
City Reference Coordinates: 004°30'00"E 52°09'00"N

Baltzer Science Publishers
Postbus 221
NL-140 AE Bussum
or :
Asterweg 1A
NL-1031 HL Amsterdam
Telephone: (0)35-6954250
Telefax: (0)35-6954258
Electronic Mail: publish@baltzer.nl
WWW: http://www.baltzer.nl/
Founded: 1980
Periodicals: "Advances in Computational Mathematics" (ISSN 1019-7168), "Annals of Numerical Mathematics" (ISSN 1022-7091), "Approximation Theory and its Applications" (ISSN 1000-9221), "Computational Geosciences" (ISSN 1420-0597), "Numerical Algorithms" (ISSN 1017-1398)
● Publisher
City Reference Coordinates: 005°10'00"E 52°17'00"N (Bussum)
                        004°54'00"E 52°22'00"N (Amsterdam)

Beleidscommissie Remote Sensing (BCRS)
(Netherlands Remote Sensing Board)
Kanaalweg 4

Postbus 5023
NL-2600 GA Delft
Telephone: (0)15-2691111
Telefax: (0)15-2618962
Electronic Mail: p.b.bcrs@mdi.rws.minvenw.nl
WWW: http://www.minvenw.nl/rws/mdi/bcrs
Founded: 1986
Staff: 7
Activities: stimulating usage of remote sensing by subsidising project * national focal point for remote sensing
Periodicals: (6) "Remote Sensing Nieuwsbrief"
City Reference Coordinates: 004°21'00"E 52°00'00"N

Centaurus-A

c/o B.A. Gerritsen (Secretary)
Postbus 24
NL-6680 AA Bemmel
or :
Wardstraat 33
NL-6681 CG Bemmel
Telephone: (0)481-461368
Telefax: (0)481-465289
Founded: 1962
Activities: popularization * lectures * telescope making
City Reference Coordinates: 005°54'00"E 51°54'00"N

Centrum voor Constructie en Mechatronica (CCM)

Postbus 12
NL-5670 AA Nuenen
or :
De Pinehart 24
NL-5674 CC Nuenen
Telephone: (0)40-2834405
Telefax: (0)40-2837135
Founded: 1969
Staff: 85
Activities: technical research * development projects * product development * automation * signal processing * automatic instruments for biological experiments in space
City Reference Coordinates: 005°33'00"E 51°29'00"N

Centrum voor Hoge-Energie Astrofysica (CHEAF)

Kruislaan 403
NL-1098 SJ Amsterdam
Telephone: (0)20-5257491
                (0)20-5257492
Telefax: (0)20-5257484
Electronic Mail: <userid>@astro.uva.nl
Founded: 1988
Activities: X-ray astronomy * compact stars * neutrino astrophysics * cosmology * plasma astrophysics
• This center gathers scientists from the University of Amsterdam, the University of Utrecht and the National Institute for Nuclear and High-Energy Physics
City Reference Coordinates: 004°54'00"E 52°22'00"N

Centrum voor Wiskunde en Informatica (CWI)

Kruislaan 413
NL-1098 SJ Amsterdam
Telephone: (0)20-5929333
Telefax: (0)20-5924199
WWW: http://www.cwi.nl/
Founded: 1946
Staff: 202
Activities: research in mathematics and computer sciences
Periodicals: (4) "CWI Quarterly"
City Reference Coordinates: 004°54'00"E 52°22'00"N

Corona Borealis

c/o H.J.M. v. Rongen (Secretary)
Schuttersweg 143
NL-7314 LG Apeldoorn
or :
Eiberstraat 14
NL-6883 EJ Velp
Telephone: (0)55-3551044
Electronic Mail: bvrongen@tip.nl
WWW: http://www.doge.nl/~bveltman/corbor.html
Founded: 1977
Staff: 3

Activities: public observatory * lectures * working groups * radioastronomy *education
Periodicals: (4) "Corona Borealis"
Coordinates: 005°59'11"E 51°59'11"N H16m (Velp)
            003°11'44"E 43°42'29"N H270m (Lunas, France)
City Reference Coordinates: 005°58'00"E 52°13'00"N (Apeldoorn)
                        005°59'00"E 52°00'00"N (Velp)

Delft Institute for Earth-Oriented Space Research (DEOS)
● See "Technische Universiteit Delft, Delft Institute for Earth-Oriented Space Research (DEOS)"

Delft Instruments NV
Postbus 103
NL-2600 AC Delft
or :
Mercuriusweg 1
NL-2624 BC Delft
Telephone: (0)15-2601200
Telefax: (0)15-2602279
City Reference Coordinates: 004°21'00"E 52°00'00"N

Delft University of Technology
● See "Technische Universiteit Delft"

De Zonnewijzerkring
Van Gorkumlaan 39
NL-5641 WN Eindhoven
Telephone: (0)40-2817818
Electronic Mail: ferdv@iae.nl
WWW: http://www.iae.nl/users/ferdv
Founded: 1978
Membership: 160
Activities: sundials (registration, construction, restoration, research, design, advice)
Periodicals: (3) "Bulletin"
City Reference Coordinates: 005°28'00"E 51°26'00"N

Dinfa BV
Postbus 45
NL-2690 AA 's-Gravenzande
or :
Fultonstraat 11
NL-2691 HA 's-Gravenzande
Telephone: (0)174-414441
Telefax: (0)174-420100
WWW: http://www.multin.nl/dinfa/index.html
Founded: 1962
Staff: 50
Activities: precision engineering * subcontracting precision parts and equipment
City Reference Coordinates: 004°10'00"E 52°00'00"N

Dr. A.F. Philips Sterrenwacht
Burg. Mollaan 8
NL-5582 CK Waalre
Telephone: (0)40-2214355
Founded: 1936
Staff: 5
Periodicals: "De Sterrenkijker"
City Reference Coordinates: 005°26'00"E 51°24'00"N

Dutch Meteor Society (DMS)
c/o Hans Betlem
Lederkarper 4
NL-2318 NB Leiden
Telephone: (0)71-5223817
Telefax: (0)71-5223817
Electronic Mail: betlem@strw.leidenuniv.nl
WWW: http://www.dmsweb.org/
Founded: 1979
Membership: 80
Activities: meteor research * observing * spectrography * photography * all-sky network * statistics * photoelectric meteor registration
Periodicals: (6) "Radiant" (ISSN 0925-8566)
City Reference Coordinates: 004°30'00"E 52°09'00"N

Dutch Occultation Association (DOA)
Frescobaldistraat 21

NL-1323 BB Almere
WWW: http://web.inter.NL.net/hcc/elimburg.doa/
Founded: 1946
Membership: 65
City Reference Coordinates: 005°12'00"E 52°22'00"N

Eise Eisinga's Planetarium
Eise Eisingastraat 3
NL-8801 KE Franeker
Telephone: (0)517-393070
Founded: 1781
Staff: 1
Activities: museum * shows * exhibitions
City Reference Coordinates: 005°32'00"E 53°11'00"N

Elsevier Science BV
Sara Burgerhartstraat 25
Postbus 103
NL-1005 KV Amsterdam
Telephone: (0)20-4852532
Telefax: (0)20-4852580
Electronic Mail: nlinfo-f@elsevier.nl
                nhpdesked@elsevier.nl
                m.kolman@elsevier.nl (Michiel Kolman, Publishing Editor)
WWW: http://www.elsevier.nl/
Founded: 1950
Staff: 1600
Activities: publishing, among many titles, "Advances in Space Research" (ISSN 0273-1177), "Astroparticle Physics" (ISSN 0927-6505), "Chinese Astronomy and Astrophysics" (ISSN 0275-1062), "Computer Physics Communications (CPC)" (ISSN 0010-4655), "COSPAR Information Bulletin", "Earth and Planetary Science Letters" (ISSN 0012-821x), "New Astronomy" (ISSN 1384-1076), "New Astronomy Reviews" (ISSN 1387-6473), "Physics Reports" (ISSN 0303-4070), "Planetary and Space Science" (ISSN 0032-0633),
City Reference Coordinates: 004°54'00"E 52°22'00"N

ESTEC
● See "European Space Agency (ESA), ESTEC"

Euregio Publiekscentrum voor Sterrenkunde (EPS)
Postbus 93
NL-7590 AB Denekamp
or :
Frensdorferweg 22
NL-2635 NK Lattrop
Telephone: (0)541-229700
Electronic Mail: sterrenwacht@cnt.antenna.nl
WWW: http://www.introweb.nl/~eps/
Founded: 1960
Membership: 50
Activities: public observatory
City Reference Coordinates: 007°00'00"E 52°23'00"N (Denekamp)
                            006°59'00"E 52°26'00"N (Lattrop)

European Association for International Education (EAIE)
Van Diemenstraat 344
NL-1013 CR Amsterdam
Telephone: (0)20-6252727
Telefax: (0)20-6209406
Electronic Mail: eaie@eaie.nl
Founded: 1989
Membership: 1700
Activities: annual conference * publishing * training courses
Periodicals: (4) "EAIE Newsletter" (ISSN 0927-572x)
City Reference Coordinates: 004°54'00"E 52°22'00"N

European Association of Geoscientists and Engineers (EAGE)
Postbus 59
NL-3990 DB Houten
or :
Standerdmolen 10
NL-3995 AA Houten
Telephone: (0)30-6354055
Telefax: (0)30-6343524
Electronic Mail: eage@eage.nl
WWW: http://www.eage.nl/
Founded: 1995

Staff: 10
Activities: promoting exploration geophysics * fostering fellowship and cooperation among those working, studying or being otherwise interested in this field
Periodicals: "Geophysical Prospecting" (ISSN 0016-8025), "First Break", "Proceedings", "Petroleum Geoscience", "Extended Abstracts", "Yearbook" (ISSN 0531-2728)
Awards: (1) "Conrad Schlumberger Award", "Best Poster Award", "Best Paper Award", "Distinguished Lecturer Award"
City Reference Coordinates: 005°10'00"E 52°02'00"N

European Space Agency (ESA), ESTEC
(European Space Research and Technology Centre)
Postbus 2200
NL-2200 AG Noordwijk
or :
Keplerlaan 1
NL-2201 AZ Noordwijk
Telephone: (0)71-5656555
　　　　　(0)71-5656565
Telefax: (0)71-5657400
Electronic Mail: <userid>@estec.esa.nl
WWW: http://www.estec.esa.nl/
　　　http://www.esa.int/
Founded: 1963
Staff: 1500
Activities: European space scientific and technical centre
City Reference Coordinates: 004°26'00"E 52°14'00"N

European Space Agency (ESA), ESTEC, Astrophysics Division
Keplerlaan 1
Postbus 299
NL-2200 AG Noordwijk
Telephone: (0)71-5653557
Telefax: (0)71-5654690
Electronic Mail: astro@astro.estec.esa.nl
　　　　　　　<userid>@astro.estec.esa.nl
WWW: http://astro.estec.esa.nl/
　　　http://www.esa.int/
Founded: 1964
Staff: 30
Activities: space astrophysics * developing instruments * observational astronomy (space and ground based) in gamma rays, X-ray, UV, optical, IR and radio P * managing the ESA astrophysics archives
Periodicals: (3) "Astronews"; (1/2) "ISO Info"
City Reference Coordinates: 004°26'00"E 52°14'00"N

European Space Agency (ESA), ESTEC, ESA Publications Division (EPD)
Keplerlaan 1
Postbus 299
NL-2200 AG Noordwijk
Telephone: (0)71-5653408 (Secretary)
Telefax: (0)71-5655433
Electronic Mail: <userid>@estec.esa.nl
WWW: http://esapub.esrin.esa.it/publicat/epdi.htm
　　　http://www.esa.int/
Founded: 1964 (ESA)
Periodicals: (4) "Reaching for the Skies" (ISSN 1013-9044), "Earth Observation Quarterly" (ISSN 0256-596x), "Preparing for the Future"; (3) "Microgravity News from ESA"; (1) "Annual Report"; "ESA Bulletin" (ISSN 0376-4265), "ESA Special Publications" (ISSN 0379-6566), "ESA Brochures" (ISSN 0250-1589),
City Reference Coordinates: 004°26'00"E 52°14'00"N

European Space Research and Technology Centre (ESTEC)
● See "European Space Agency (ESA), ESTEC"

Fédération Internationale d'Information et de Documentation (FID)
(International Federation for Information and Documentation)
Postbus 90402
NL-2509 LK Den Haag
or :
Prins-Willem-Alexanderhof 5
NL-2595 BE Den Haag
Telephone: (0)70-3140671
Telefax: (0)70-3140667
Electronic Mail: secretariat@fid.nl
　　　　　　　fid@python.konbib.nl
WWW: http://fid.conicyt.cl:8000/
Founded: 1895 (as "Institut International de Bibliographie - IIB")
Staff: 5
Activities: information management * classification

Periodicals: (6) "FID Review"
City Reference Coordinates: 004°18'00"E 52°06'00"N

Fokker Space and Systems (FSS) BV
Postbus 32070
NL-2303 DB Leiden
or :
Newtonweg 1
NL-2333 CP Leiden
Telephone: (0)71-5245000
Telefax: (0)71-5245999
WWW: http://www.fokkerspace.nl/
Founded: 1968
Staff: 450
Activities: aerospace manufacturer
City Reference Coordinates: 004°30'00"E 52°09'00"N

FOM, Institute for Atomic and Molecular Physics (AMOLF)
Kruislaan 407
NL-1098 SJ Amsterdam
Telephone: (0)20-6081234
Telefax: (0)20-6684106
Electronic Mail: secr@amolf.nl
                  <userid>@amolf.nl
WWW: http://www.amolf.nl/
Founded: 1949
Staff: 180
City Reference Coordinates: 004°54'00"E 52°22'00"N

FOM, Institute for Plasma Physics
Postbus 1207
NL-3430 BE Nieuwegein
Telephone: (0)30-6096999
Telefax: (0)30-6031204
Electronic Mail: <userid>@rijnh.nl
Founded: 1959
Staff: 60
Activities: plasma physics
City Reference Coordinates: 005°08'00"E 52°05'00"N (Utrecht)

Galaxis, Vereniging voor Amateur Astronomie
Piet Slagerstraat 50
NL-5213 XT 's-Hertogenbosch
Telephone: (0)73-6410257 (C.V. Huijkelom)
Founded: 1977
Membership: 58
Activities: monthly meetings * lectures * experience sharing * education
Periodicals: (4) "Galaxis"
City Reference Coordinates: 005°19'00"E 51°41'00"N

Gebiedsbestuur Aard- en Levenswetenschappen (ALW)
Postbus 93120
NL-2509 AC Den Haag
or :
Laan van Nieuw Oost Indië 131
NL-2593 BM Den Haag
Telephone: (0)70-3440780
            (0)70-3440619
Telefax: (0)70-3832173
          (0)70-3819033
Electronic Mail: alw@nwo.nl
WWW: http://www.nwo.nl/alw/
Founded: 1999
Staff: 21
Activities: funding agency for fundamental and strategic research regarding geosciences and life sciences
Periodicals: (12) "KNMG/ALW Nieuwsbrief"
City Reference Coordinates: 004°18'00"E 52°06'00"N

Interdepartmental Committee on Space
c/o Ministry of Economic Affairs
Postbus 20201
NL-2500 EC Den Haag
Telephone: (0)70-3797523
            (0)70-3796157
Telefax: (0)70-3796508

WWW: http://info.minez.nl/
City Reference Coordinates: 004°18'00"E 52°06'00"N

International Association for Statistical Computing (IASC)
428 Prinses Beatrixlaan
Postbus 950
NL-2270 AZ Voorburg
Telephone: (0)70-3375737
Telefax: (0)70-3860025
Electronic Mail: isi@cbs.nl
WWW: http://www.stat.unipg.it/iasc/
Founded: 1977
Membership: 900
Activities: fostering world-wide interest in effective statistical computing and exchange of technical knowledge * organising COMPSTAT meetings
Periodicals: (10) "Computational Statistics & Data Analysis" (ISSN 0167-9473); "Statistical Software Newsletter"
City Reference Coordinates: 004°23'00"E 52°05'00"N

International Federation for Information and Documentation
● See "Fédération Internationale d'Information et de Documentation (FID)"

International Federation of Library Associations and Institutions (IFLA)
Postbus 95312
NL-2509 CH Den Haag
Telephone: (0)70-3140884
Telefax: (0)70-3834827
Electronic Mail: ifla.hq@ifla.nl
WWW: http://www.nlc-bnc.ca/ifla
      http://ifla.inist.fr/
Founded: 1927
Periodicals: (4) "IFLA Journal" (ISSN 0340-0352); (3) "Bibliothek Forschung und Praxis" (ISSN 0341-4183); (1) "IFLA Directory" (ISSN 0074-6002)
Awards: "Gustav Hofmann Grant", "Hans-Peter Geh Grant", "Guust van Wesemael Literacy Prize"
City Reference Coordinates: 004°18'00"E 52°06'00"N

International Institute for Aerospace Survey and Earth Science (ITC)
Hengelosestraat 99
Postbus 6
NL-7500 AA Enschede
Telephone: (0)53-4874444
Telefax: (0)53-4874400
Electronic Mail: pr@itc.nl
WWW: http://www.itc.nl/
Founded: 1950
Staff: 350
Activities: education, research and consulting in the collection, handling, presentation and use of geoinformation
Periodicals: (4) "ITC Journal" (ISSN 0303-2434); (1) "Annual Report"
Awards: (1) "ITC Research Award", "ITC Education Award"
City Reference Coordinates: 006°53'00"E 52°12'00"N

International Soil Reference and Information Centre (ISRIC)
Postbus 353
NL-6700 AJ Wageningen
or :
Duivendaal 9
NL-6701 AR Wageningen
Telephone: (0)317-471711
Telefax: (0)317-471700
Electronic Mail: soil@isric.nl
Founded: 1966
Staff: 30
Activities: documentation centre on world soils * improvement of research methodologies * project implementations
City Reference Coordinates: 005°40'00"E 51°58'00"N

International Statistical Institute (ISI)
Postbus 950
428 Prinses Beatrixlaan
NL-2270 AZ Voorburg
Telephone: (0)70-3375737
Telefax: (0)70-3860025
Electronic Mail: isi@cbs.nl
WWW: http://www.cbs.nl/isi/
Founded: 1885
Staff: 10
Membership: 2000 (plus 3000 members from ISI sections)

Activities: developing and improving statistical methods and their application through the promotion of international activity and co-operation * conferences * study committees * special programmes
Sections: International Association for Statistical Education * Bernouilli Society for Mathematical Statistics and Probability * International Association of Survey Statisticians * International Association for Statistical Computing * International Association for Official Statistics
Periodicals: "Bulletin of the International Statistical Institute", "International Statistical Review", "Statistical Theory and Method Abstracts", "Short Book Reviews", "ISI Newsletter" "ISI Directories"
City Reference Coordinates: 004°23'00"E 52°05'00"N

Jan Paagman Sterrenwacht (JPS)
Postbus 203
NL-5720 AE Asten
or :
Ostaderstraat 23
NL-5721 WC Asten
Telephone: (0)493-691865
Electronic Mail: jpsasten@iaehv.nl
WWW: http://www.iaehv.nl/users/jpsasten/
Founded: 1980
Membership: 120
Activities: popularization * open days * education * Pieterse Planetarium
Periodicals: (12) "Interkomeet" (circ.: 120)
Coordinates: 005°44'04"E 51°24'17"N H25m
City Reference Coordinates: 005°45'00"E 51°24'00"N

Joint Institute for VLBI in Europe (JIVE)
Postbus 2
NL-7990 AA Dwingeloo
or :
Oude Hoogeveensedijk 4
NL-7991 PD Dwingeloo
Telephone: (0)521-595100
Telefax: (0)521-597332
Electronic Mail: jive@nfra.nl
WWW: http://www.nfra.nl/jive/
        http://www.nfra.nl/jive/evn/evn.html (European VLBI Network)
Founded: 1993
Staff: 17
Activities: designing and constructing a 16-station VLBI data processor * supporting operations of the European VLBI Network (EVN) * supporting users of EVN
Periodicals: "EVN Newsletter"; (1) "Annual Report"
City Reference Coordinates: 006°22'00"E 52°49'00"N

Kapteyn Sterrenkundig Instituut
● See "Universiteit Groningen, Kapteyn Sterrenkundig Instituut"

Kapteyn Sterrenwacht - StarTel BV
Mensingheweg 20
NL-9301 KA Roden
Telephone: (0)50-5028888
Telefax: (0)50-5028800
Electronic Mail: wjhverkaik@startel.nl
WWW: http://www.startel.nl/
Founded: 1992
Staff: 15
Activities: education * image control software
Coordinates: 006°26'37"E 53°07'44"N H12m
City Reference Coordinates: 006°26'00"E 53°07'00"N

Katholieke Universiteit Nijmegen
● See "Universiteit Nijmegen"

Kluwer Academic Publishers (KAP)
Spuiboulevard 50
Postbus 17
NL-3300 AA Dordrecht
Telephone: (0)78-6392315
Telefax: (0)78-6392254
Electronic Mail: harry.blom@wkap.nl (Harry [J.J.] Blom)
WWW: http://www.wkap.nl/
Founded: 1978
Staff: 50
Periodicals: "Astrophysics and Space Science Library", "Astrophysics and Space Science" (ISSN 0004-640x), "Celestial Mechanics and Dynamical Astronomy" (ISSN 0923-2958), "Earth, Moon and Planets" (ISSN 0167-9295), "Experimental Astronomy" (ISSN 0922-6435), "Solar Physics" (ISSN 0038-0938), "Space Science Reviews" (ISSN 0038-6308), "Journal of Reducing Space Mission Cost" (ISSN 1385-7479)

• Publisher
City Reference Coordinates: 004°40'00"E 51°49'00"N

Koninklijke Nederlandse Akademie van Wetenschappen (KNAW)
Postbus 19121
Kloveniersburgwal 29
NL-1000 GC Amsterdam
Telephone: (0)20-5510700
Telefax: (0)20-6204941
Electronic Mail: knaw@bureau.knaw.nl
WWW: http://www.knaw.nl/
Founded: 1808
Membership: 400
Activities: advisory body to the Dutch government in all fields of science * promoting science, scientific communication, exchanges and collaboration between Dutch and foreign members * managing institutes of fundamental research
Periodicals: "Academy News"
Awards: "Academy Medal"
City Reference Coordinates: 004°54'00"E 52°22'00"N

Koninklijke Nederlands Meteorologisch Instituut (KNMI)
Wilhelminalaan 10
Postbus 201
NL-3730 AE De Bilt
Telephone: (0)30-2206911
Telefax: (0)30-2210407
Electronic Mail: <userid>@knmi.nl
WWW: http://www.knmi.nl/
Founded: 1854
Staff: 530
Activities: weather forecast * climatic research * ocean-climate interactions * seismology * network of about 15 observing stations
Periodicals: (daily) "Daily Weather Bulletin"; (12) "Monthly Weather Bulletin", "Monthly Rain Bulletin", "Seismological Bulletin"
City Reference Coordinates: 005°10'00"E 52°06'00"N

Landelijk Samenwerkende Volkssterrenwachten (LSV)
Zonnenburg 2
NL-3512 NL Utrecht
Telephone: (0)30-2321456
Telefax: (0)30-2300483
Founded: 1977
Membership: 19 (observatories)
Staff: 5
Activities: stimulating member observatories * setting up programs * raising funds * popularization
City Reference Coordinates: 005°08'00"E 52°05'00"N

Meteo Consult BV
Postbus 617
NL-6700 AP Wageningen
Telephone: (0)317-423300
Telefax: (0)317-423164
Electronic Mail: <userid>@meteocon.nl
WWW: http://www.meteocon.nl/
City Reference Coordinates: 005°40'00"E 51°58'00"N

Museum Boerhaave
• See "Rijksmuseum voor de Geschiedenis van de Natuurwetenschappen en van de Geneeskunde"

Nationaal Instituut voor Kernfysica en Hoge-Energie Fysica (NIKHEF)
Postbus 41882
NL-1009 DB Amsterdam
or :
Kruislaan 409
NL-1098 SJ Amsterdam
Telephone: (0)20-5922000
Telefax: (0)20-5922165
         (0)20-5925155
Electronic Mail: <userid>@nikhef.nl
WWW: http://www.nikhef.nl/
Founded: 1946
Staff: 270
Activities: sub-atomic physics
Coordinates: 004°57'05"E 52°22'24"N H-5m
City Reference Coordinates: 004°54'00"E 52°22'00"N

Nationaal Lucht- en Ruimtevaartlaboratorium (NLR)
(National Aerospace Laboratory - Netherlands)
Postbus 90502
NL-1006 BM Amsterdam
or :
Anthony Fokkerweg 2
NL-1059 CM Amsterdam
Telephone: (0)20-5113113
Telefax: (0)20-5113210
Electronic Mail: info@nlr.nl
WWW: http://www.nlr.nl/
Founded: 1937 (1919 as "Government Service for Aeronautical Studies")
Staff: 900
Activities: research and development * fluid dynamics * flight testing * simulation * structures and materials * remote sensing * space user support * information technology * electronics
Periodicals: "NLR News"
City Reference Coordinates: 004°54'00"E 52°22'00"N

Nederlands Comité Astronomie (NCA)
c/o J. Lub
Sterrewacht Leiden
Postbus 9513
NL-2300 RA Leiden
or :
Niels Bohrweg 2
NL-2333 CA Leiden
Telephone: (0)71-5275835
Telefax: (0)71-5275819
Electronic Mail: lub@strw.leidenuniv.nl
City Reference Coordinates: 004°30'00"E 52°09'00"N

Nederlandse Astronomenclub (NAC)
c/o E.R. Deul
Sterrewacht Leiden
Postbus 9513
NL-2300 RA Leiden
or :
c/o J.B.G.M. Bloemen (Secretary)
Laboratory for Space Research
Sorbonnelaan 2
NL-3584 CA Utrecht
Telephone: (0)71-5275827 (Deul)
           (0)30-2535600 (Bloemen)
Telefax: (0)71-5275829 (Deul)
        (0)30-2540860 (Bloemen)
Electronic Mail: bloemen@sron.rug.nl
               deul@strw.leidenuniv.nl
Founded: 1918
Membership: 600
Activities: meetings with lectures on astronomical topics * annual Dutch astronomers conference
City Reference Coordinates: 004°30'00"E 52°09'00"N (Leiden)
                          005°08'00"E 52°05'00"N (Utrecht)

Nederlandse Natuurkundige Vereniging (NNV)
Postbus 302
NL-1170 AH Badhoevedorp
Telephone: (0)30-352516 (President)
           (0)30-285626 (Secretary)
Telefax: (0)30-518689 (President)
        (0)30-251381 (Secretary)
Electronic Mail: brussaar@fys.ruu.nl (P.J. Brussaard, President)
                hoogenb@fys.ruu.nl (A.M. Hoogenboom, Secretary)
WWW: http://www.nat.vu.nl/~claud
      http://www.nikhef.nl/www/pub/eps/europa/nederl.html
Founded: 1921
Membership: 3600
Periodicals: "Nederlandse Tijdschrift voor Natuurkunde"
City Reference Coordinates: 004°46'00"E 52°21'00"N

Nederlandse Organisatie voor Toegepast-Natuurwetenschappelijk Onderzoek (TNO)
Schoemakerstraat 97
Postbus 6070
NL-2600 JA Delft
Telephone: (0)15-2696900
Telefax: (0)15-2612403
WWW: http://www.tno.nl/

Founded: 1932
Staff: 5000
Activities: research in industrial technology, space, energy, environment, nutrition and food, health, defence, building and infrastructure
City Reference Coordinates: 004°21'00"E 52°00'00"N

Nederlandse Organisatie voor Wetenschappelijk Onderzoek (NWO)
Postbus 93138
NL-2509 AC Den Haag
or :
Laan van Nieuw Oost Indië 131
NL-2593 BM Den Haag
Telephone: (0)70-3440640
Telefax: (0)70-3850971
Electronic Mail: nwo@nwo.nl
                    <userid>@nwo.nl
Founded: 1950
Staff: 4500
Activities: initiation, stimulation and coordination of scientific research in the Netherlands by funding and advisory work
Periodicals: (10) "Onderzoekberichten" (ISSN 0928-6640); (6) "Research Reports from the Netherlands" (ISSN 0927-880x), "Forschungsnachrichten aus den Niederlanden" (ISSN 0928-6403), "Bulletin de Recherche des Pays-Bas" (ISSN 0928-6411); (4) "Hypothese" (ISSN 1381-5652); (1) "Jaarboek" (ISSN 0167-6792); "NWO FActs"
City Reference Coordinates: 004°18'00"E 52°06'00"N

Nederlandse Vereniging voor Ruimtevaart
c/o Stichting De Koepel
Zonnenburg 2
NL-3512 NL Utrecht
Telephone: (0)30-2311360
Electronic Mail: nvr@dataweb.nl
WWW: http://www.dataweb.net/~nvr/
Founded: 1951
Membership: 800
Activities: meetings on space sciences * collaboration with Stichting De Koepel
Periodicals: (6) "Ruimtevaart"
City Reference Coordinates: 005°08'00"E 52°05'00"N

Nederlandse Vereniging voor Weer- en Sterrenkunde (NVWS)
c/o Stichting De Koepel
Zonnenburg 2
NL-3512 NL Utrecht
Telephone: (0)30-2311360
WWW: http://stkwww.fys.ruu.nl:8000/~mathlenr/Zenit.html (Zenit)
        http://www.fys.ruu.nl/~wwwzenit/Zenit.html
Founded: 1901
Membership: 3000
Activities: meetings * excursions
Sections: Meteors * Star Occultations * Sun * Variable Stars * Comets * Satellites * Meteorology * Instrumentation * Special Group for Teenagers * Photography * Moon and Planets
Periodicals: (11) "Zenit" (ISSN 0165-0211); (1) "Sterrengids" (in collaboration with Stichting de Koepel)
City Reference Coordinates: 005°08'00"E 52°05'00"N

Nederlands Instituut voor Vliegtuigontwikkeling en Ruimtevaart (NIVR)
Postbus 35
NL-2600 AA Delft
or :
Kluyverweg 4
NL-2629 HT Delft
Telephone: (0)15-2788025
Telefax: (0)15-2623096
Electronic Mail: nivr@lr.tudelft.nl
WWW: http://www.satserv.nl/NIVR
Founded: 1946
Staff: 24
Activities: promoting industrial aerospace activities primarily through R&D funding
City Reference Coordinates: 004°21'00"E 52°00'00"N

Nederlands Normalisatie-Instituut (NNI)
Kalfjeslaan 2
Postbus 5059
NL-2600 GB Delft
Telephone: (0)15-2690390
Telefax: (0)15-2690190
Electronic Mail: info@nni.nl
WWW: http://www.nni.nl/
Founded: 1916

Staff: 180
Activities: standardization * publishing * documentation services
Periodicals: (1) "Jaarverslag"; "Newsletter", "Catalogues", "CD-ROMs"
City Reference Coordinates: 004°21'00"E 52°00'00"N

Netherlands Foundation for Research in Astronomy (NFRA)
● See "Stichting Astronomisch Onderzoek in Nederland (ASTRON)"

Netherlands Foundation for Radio Astronomy (NFRA)
● Now merged with Stichting Astronomisch Onderzoek in Nederland (ASTRON) (see separate entry)

Netherlands Industrial Space Organisation (NISO)
Postbus 32070
NL-2303 DB Leiden
or :
Newtonweg 1
NL-2333 CP Leiden
Telephone: (0)71-5245124
Telefax: (0)71-5245125
Founded: 1989
Membership: 23 (Dutch companies)
Activities: association of Dutch industries and research institutes * defining common interests * developing and advocating common strategies * harmonizing Dutch governmental policies and industrial strategies * stimulating use and application of space facilities
City Reference Coordinates: 004°30'00"E 52°09'00"N

Netherlands Remote Sensing Board
● See "Beleidscommissie Remote Sensing (BCRS)"

Nonius BV
Postbus 811
NL-2600 AV Delft
or :
Röntgenweg 1
NL-2624 BD Delft
Telephone: (0)15-2698300
Telefax: (0)15-2627401
Electronic Mail: info@nonius.nl
WWW: http://www.enraf-nonius.delftny.com/
Founded: 1925
Staff: 90
Activities: developing, manufacturing, marketing and supporting X-ray diffraction instruments
City Reference Coordinates: 004°21'00"E 52°00'00"N

Noordwijk Space Expo (NSE)
Postbus 277
NL-2200 AG Noordwijk
or :
Keplerlaan 3
NL-2201 AZ Noordwijk
Telephone: (0)71-3646446
Telefax: (0)71-3646453
Founded: 1990
Staff: 9
Activities: permanent space exhibition * education * information * group visits * films * guided tours * press conferences * meetings * visitors' centre for ESA/ESTEC
Periodicals: (1) "Annual Report"; "SpaceTalk"
City Reference Coordinates: 004°26'00"E 52°14'00"N

Nutssterrenwacht Ommen
Chevalleraustraat 8
NL-7731 EE Ommen
or :
Molener 75
NL-7731 BV Ommen
or :
(after Summer 2000)
Koperwiekstraat 4
NL-7731 ZH Ommen
Telephone: (0)529-451849
Founded: 1992
Staff: 4
Activities: public education on astronomy, meteorology and astronautics
Periodicals: "Stand van Zaken"
Coordinates: 006°25'00"E 50°32'00"N H8m

City Reference Coordinates: 006°25'00"E 50°32'00"N

Omniversum Space Theater and Digital Planetarium
President Kennedylaan 5
NL-2517 JK Den Haag
Telephone: (0)70-3547479
         (0)70-3545454
Telefax: (0)70-3524280
WWW: http://www.omniversum.nl/
Founded: 1984
Staff: 17
Activities: popularization of science in general and astronomy more particularly
City Reference Coordinates: 004°18'00"E 52°06'00"N

Paagman Sterrenwacht (JPS) (Jan_)
● See "Jan Paagman Sterrenwacht (JPS)"

Philips Sterrenwacht (Dr. A.F._)
● See "Dr. A.F. Philips Sterrenwacht"

Phoenix Public Observatory
Vordenseweg 6
NL-7241 SB Lochem
Telephone: (0)573-254310
WWW: http://www.nerdnet.nl/~angelo/phoenix/
City Reference Coordinates: 006°25'00"E 52°10'00"N

Planetron
Drift 11b
NL-7911 AA Dwingeloo
Telephone: (0)521-593535
Telefax: (0)521-593541
Electronic Mail: info@elders.nl
WWW: http://www.elders.nl/attract/planetr.htm
City Reference Coordinates: 006°22'00"E 52°49'00"N

Radiosterrenwacht Dwingeloo
Postbus 2
NL-7990 AA Dwingeloo
or :
Oude Hoogeveensedijk 4
NL-7991 PD Dwingeloo
Telephone: (0)521-595100
Telefax: (0)521-597332
Electronic Mail: secretary@nfra.nl
                <userid>@nfra.nl
WWW: http://www.nfra.nl/
Founded: 1956
Staff: 90
Activities: observing * instrument building * data processing
Coordinates: 006°23'49"E 52°48'47"N H25m
City Reference Coordinates: 006°22'00"E 52°49'00"N

Radiosterrenwacht Westerbork
Schattenberg 1
NL-9433 TA Zwiggelte
Telephone: (0)593-592421
Telefax: (0)593-592486
Electronic Mail: <userid>@nfra.nl
WWW: http://www.nfra.nl/wsrt/wsrtpage.htm
Founded: 1968
Staff: 4
Activities: radio astronomical data acquisition * operating the Westerbork Synthesis Radio Telescope (WSRT)
Coordinates: 006°36'15"E 52°55'00"N H16m
City Reference Coordinates: 006°35'00"E 52°52'00"N

Rijksmuseum voor de Geschiedenis van de Natuurwetenschappen en van de Geneeskunde
(Museum Boerhaave)
Postbus 11280
NL-2301 EG Leiden
or :
Lange Sint Agnietenstraat 10
NL-2312 WC Leiden
Telephone: (0)71-5214224
Telefax: (0)71-5120344

WWW: http://www.museumboerhaave.nl/
Founded: 1928
Staff: 40
Activities: collection, exposition and scientific research of historical scientific instruments
City Reference Coordinates: 004°30'00"E 52°09'00"N

Rijksuniversiteit Groningen
● See "Universiteit Groningen"

Rijksuniversiteit Leiden
● See "Universiteit Leiden"

Rijksuniversiteit Utrecht
● See "Universiteit Utrecht"

Simon Stevin Public Astronomical Observatory
Bovenstraat 89
NL-4741 SK Hoeven
Telephone: (0)165-502439
                    (0)165-502493
Telefax: (0)165-504959
Electronic Mail: vsw@signet.nl
WWW: http://www.signet.nl/vsw/
Founded: 1961
Staff: 3
Activities: popularization * optical and radioastronomical facilities * education * exhibitions * meteorology * excursions of astronomical interest * planetarium
Coordinates: 004°33'48"E 51°34'00"N H9m
Awards: (1) "Simon Stevin Award"
City Reference Coordinates: 004°35'00"E 51°35'00"N

Space Research Organization Netherlands (SRON), Groningen
Postbus 800
NL-9700 AV Groningen
or :
Landleven 12
NL-9747 AD Groningen
Telephone: (0)50-3634074
Telefax: (0)50-3634033
Electronic Mail: <userid>@sron.nl
                    secr@sron.nl
WWW: http://www.sron.nl/
Founded: 1984 (1966 as KNAW branch)
Staff: 40
Activities: designing, manufacturing and operating space instrumentation for atmospheric and astronomical photometry and spectroscopy in (far) IR and sub-mm wavelengths
Periodicals: (1) "Annual Report"
City Reference Coordinates: 006°33'00"E 53°13'00"N

Space Research Organization Netherlands (SRON), Utrecht
Sorbonnelaan 2
NL-3584 CA Utrecht
Telephone: (0)30-2535600
Telefax: (0)30-2540860
Electronic Mail: <userid>@sron.nl
                    secr@sron.nl
WWW: http://www.sron.nl/
Founded: 1983
Activities: designing, developing and building space instruments * processing and interpreting satellite data in cooperation with university groups * X rays * IR * Earth observation
City Reference Coordinates: 005°08'00"E 52°05'00"N

StarTel BV
● See "Kapteyn Sterrenwacht - StarTel BV"

Sterrenkundig Instituut Anton Pannekoek
● See "Universiteit Amsterdam, Sterrenkundig Instituut Anton Pannekoek"

Sterrenwacht De Tiendesprong
Gen. de Wetstraat 31
NL-5021 TK Tilburg
Telephone: (0)13-5422534
Electronic Mail: spaninks@freemail.nl
Founded: 1988
Staff: 1

Activities: observing (Sun, Moon, planets) * education * popularization * lectures * shows
Coordinates: 005°04'48"E 51°32'49"N
City Reference Coordinates: 005°05'00"E 51°34'00"N

Sterrenwacht Halley
Halleyweg 1
Postbus 110
NL-5384 ZJ Heesch
Telephone: (0)412-452383
Founded: 1985
Membership: 250
Activities: popularization * observing
Periodicals: (4) "Halley Periodiek"
Coordinates: 005°29'16"E 51°42'15"N
City Reference Coordinates: 005°32'00"E 51°44'00"N

Sterrewacht Leiden
● See "Universiteit Leiden, Sterrewacht Leiden"

Sterrenwacht Schrieversheide
Schaapskooiweg 95
NL-6414 EL Heerlen
Telephone: (0)455-225543
Telefax: (0)455-224562
Electronic Mail: hercunet@cuci.nl
WWW: http://www.cuci.nl/~hercunet/
Founded: 1976
Staff: 3
Activities: observing * exhibitions * holography * meteorology * multimedia productions
Periodicals: (11) "Hercules"
Observatories: 2
Coordinates: 005°59'00"E 50°54'00"N
● Formerly: Stichting Volkssterrewacht Hercules
City Reference Coordinates: 005°59'00"E 50°54'00"N

Stichting Astronomisch Onderzoek in Nederland (ASTRON)
(Netherlands Foundation for Research in Astronomy - NFRA)
Postbus 2
NL-7990 AA Dwingeloo
or :
Oude Hoogeveensedijk 4
NL-7991 PD Dwingeloo
Telephone: (0)521-595100
Telefax: (0)521-597332
Electronic Mail: secretary@nfra.nl
                    <userid>@nfra.nl
WWW: http://www.nfra.nl/
         http://www.nfra.nl/nfra/NFRA_home.html
         http://www.nfra.nl/jive/ (Joint Institute for VLBI in Europe/JIVE - see separate entry)
Founded: 1949
Staff: 125
Activities: designing and commissioning instrumentation and software * operating Radiosterrenwacht Dwingeloo and Radiosterrenwacht Westerbork (see separate entries) * providing research grants for university research in astronomy
Periodicals: (2) "ASTRON/NFRA Newsletter"; (1) "Annual Report"
City Reference Coordinates: 006°22'00"E 52°49'00"N

Stichting De Koepel
c/o Sterrenwacht Sonnenborgh
Zonnenburg 2
NL-3512 NL Utrecht
Telephone: (0)30-2311360
Telefax: (0)30-2342852
Electronic Mail: sonborgh@fys.ruu.nl (Sterrenwacht Sonnenborgh)
WWW: http://stkwww.fys.ruu.nl:8000/~mathlenr/Zenit.html (Zenit)
         http://www.dru.nl/onderwijs/sonnenborgh/sonnenborgh.html (Sterrenwacht Sonnenborgh)
         http://www.fys.ruu.nl/~sonborgh/ (Sterrenwacht Sonnenborgh)
Founded: 1973
Membership: 6000
Staff: 2
Activities: coordinating amateur activities in Holland * publishing
Periodicals: (11) "Zenit" (ISSN 0165-0211); (10) "Informatieblad"; (1) "Sterrengids"
Coordinates: 005°07'48"E 52°05'12"N H14m (Sonnenborgh)
City Reference Coordinates: 005°08'00"E 52°05'00"N

Stichting Geologisch, Oceanografisch en Atmosferisch Onderzoek (GOA)
● See now "Gebiedsbestuur Aard- en Levenswetenschappen (ALW)"

Stichting Radiostraling van Zon en Melkweg (SRZM)
● Now merged with Stichting Astronomisch Onderzoek in Nederland (ASTRON) (see separate entry)

Stichting Skepsis
Postbus 2657
NL-3500 GR Utrecht
Telephone: (0)50-3129893
Electronic Mail: skepsis@wxs.n
WWW: http://www.skepsis.org/
Activities: promoting science and critical thinking * exchange and dissemination of ideas and research results * parascience debunking * education
City Reference Coordinates: 005°08'00"E 52°05'00"N

Stichting Volkssterrenwacht Saturnus
Frans Halsstraat 4
NL-1701 JL Heerhugowaard
or :
Veenweg 98
NL-1701 HH Heerhugowaard
Telephone: (0)72-5745057
(0)72-5745323
Telefax: (0)72-5745323
Electronic Mail: saturnus@redernet.nl
Founded: 1976
Membership: 136
Activities: meetings * public observatory activities * telescope making * education * microscopy * weather satellite ground station * photography * electronics * computing
Periodicals: (12) "Titan"
City Reference Coordinates: 004°50'00"E 52°40'00"N

Stichting Volkssterrewacht Hercules
● See now "Sterrenwacht Schrieversheide"

Stork Product Engineering (SPE) BV
Czaar Peterstraat 229
Postbus 379
NL-1000 AJ Amsterdam
Telephone: (0)20-5563444
Telefax: (0)20-5563556
Electronic Mail: <userid>@spe.stork.nl
Founded: 1987
Staff: 70
Activities: carbon fibre technology * mechanical parts * ignition and gas generation systems * propulsion technology * electronic control units * electronic test and simulation equipment
City Reference Coordinates: 004°54'00"E 52°22'00"N

Swets & Zeitlinger BV
Postbus 830
NL-2160 SZ Lisse
or :
Heereweg 347 B
NL-2161 CA Lisse
Telephone: (0)252-435111
Telefax: (0)252-415888
Electronic Mail: infoho@swets.nl
WWW: http://www.swets.nl/
Staff: 750 (worldwide)
Activities: publishing * distributing books and journals * subscription agent
City Reference Coordinates: 004°33'00"E 52°15'00"N

Technische Universiteit Delft, Delft Institute for Earth-Oriented Space Research (DEOS)
Postbus 5058
NL-2600 GB Delft
or :
Kluyverweg 1
NL-2629 HS Delft
Telephone: (0)15-2782072
Telefax: (0)15-2785322
Electronic Mail: ssr&t-secr@lr.tudelft.nl
WWW: http://deos.lr.tudelft.nl/
Founded: 1945 (Faculty of Engineering)

Staff: 8
Activities: orbit mechanics * geophysics * oceanic physics * satellite systems * earth observation systems * remote sensing * tracking and satellite communications
City Reference Coordinates: 004°21'00"E 52°00'00"N

Technische Universiteit Delft, Vakgroep Telecommunicatie- en Verkeersbegeleidingssystemen (TVS)

Postbus 5031
NL-2600 GA Delft
Telephone: (0)15-2786193
Telefax: (0)15-2781774
Electronic Mail: <userid>@et.tudelft.nl
WWW: http://dutetvg.et.tudelft.nl/
Founded: 1842
Staff: 35
Activities: telecommunications * teleinformation techniques * communication networks * space-based and terrestrial integrated radionavigation systems
City Reference Coordinates: 004°21'00"E 52°00'00"N

Twin Press

Oosterkamp 9
NL-8381 BV Vledder
Telefax: (0)521-383610
Electronic Mail: g.kiers@twinpress.nl (Gert Kiers)
WWW: http://www.twinpress.nl/
Founded: 1993
• Publisher
City Reference Coordinates: 006°12'00"E 52°52'00"N

Universiteit Amsterdam, Sterrenkundig Instituut Anton Pannekoek

Kruislaan 403
NL-1098 SJ Amsterdam
Telephone: (0)20-5257491
          (0)20-5257492
Telefax: (0)20-5257484
Electronic Mail: secr-astro@astro.uva.nl
                <userid>@astro.uva.nl
WWW: http://www.astro.uva.nl/
Founded: 1922
Staff: 28
Activities: stellar evolution * neutron stars * black holes * white dwarfs * binaries * accretion * X-ray and optical astronomy * mass loss * circumstellar matter * ISM * IR emissions * galaxies * star formation * stellar populations
City Reference Coordinates: 004°54'00"E 52°22'00"N
• Full university name: "Universiteit van Amsterdam"

Universiteit Amsterdam, Faculteit der Natuurkunde en Sterrenkunde

De Boelelaan 1081
NL-1081 HV Amsterdam
Telephone: (0)20-4447956
          (0)20-4447957
Telefax: (0)20-4447899
Electronic Mail: <userid>@nat.vu.nl
                secretariaat@nat.vu.nl
WWW: http://www.nat.vu.nl/
          http://www.nat.vu.nl/vakgroepen/ster/english/
Staff: 11
Activities: light scattering in the solar system * laboratory astrophysics using lasers * light scattering in ISM
• Full university name: "Vrije Universiteit Amsterdam"
City Reference Coordinates: 004°54'00"E 52°22'00"N

Universiteit Groningen, Kapteyn Sterrenkundig Instituut

Landleven 12
NL-9747 AD Groningen
Telephone: (0)50-3634073
Telefax: (0)50-3636100
Electronic Mail: secr@astro.rug.nl
WWW: http://kapteyn.astro.rug.nl/
          http://www.astro.rug.nl/
Founded: 1885
Staff: 70
Activities: galaxies * cosmic evolution * ISM and interaction with stars
Periodicals: (1) "Annual Report"
• Full university name: "Rijksuniversiteit Groningen"
City Reference Coordinates: 006°33'00"E 53°13'00"N

Universiteit Leiden, Laboratorium Astrofysica, Huygens Laboratorium
Niels Bohrweg 2
Postbus 9504
NL-2300 RA Leiden
Telephone: (0)71-5275894
    (0)71-5275917
Telefax: (0)71-5275819
Electronic Mail: <userid>@rulgm5.leidenuniv.nl
WWW: http://rulgm5.leidenuniv.nl/
Staff: 5
Activities: education * interstellar and circumstellar matter * comets * origin of life
Coordinates: 004°29'00"E 52°09'20"N H12m
● Full university name: "Rijksuniversiteit Leiden"
City Reference Coordinates: 004°30'00"E 52°09'00"N

Universiteit Leiden, Sterrewacht Leiden
Postbus 9513
NL-2300 RA Leiden
or :
Niels Bohrweg 2
NL-2333 CA Leiden
Telephone: (0)71-5275835
Telefax: (0)71-5275819
Electronic Mail: strw@strw.leidenuniv.nl
WWW: http://www.strw.leidenuniv.nl/
Founded: 1633
Staff: 50
Activities: ISM * galactic structure * active galaxies * cosmology
● Full university name: "Rijksuniversiteit Leiden"
City Reference Coordinates: 004°30'00"E 52°09'00"N

Universiteit Nijmegen, Sterrenkunde
Toernooiveld
NL-6525 ED Nijmegen
Telephone: (0)24-3652080
Telefax: (0)24-3652191
Electronic Mail: kuijpers@hef.kun.nl (Jan Kuijpers)
WWW: http://lahore.hef.kun.nl/
Coordinates: 005°52'06"E 51°49'31"N H62m
● Full university name: "Katholieke Universiteit Nijmegen"
City Reference Coordinates: 005°52'00"E 51°50'00"N

Universiteit Utrecht, Instituut voor Marien en Atmosferisch Onderzoek Utrecht (IMAU)
Postbus 80.005
NL-3508 TA Utrecht
or :
Princetonplein 5
NL-3584 CC Utrecht
Telephone: (0)30-2533275
Telefax: (0)30-2543163
Electronic Mail: imau@phys.uu.nl
WWW: http://www.phys.uu.nl/~wwwimau/
Founded: 1966
Staff: 20
Activities: meteorology * glaciology * oceanography * atmospheric chemistry * coastal dynamics
● Full university name: "Rijksuniversiteit Utrecht"
City Reference Coordinates: 005°08'00"E 52°05'00"N

Universiteit Utrecht, Sterrekundig Instituut
Princetonplein 5
Postbus 80000
NL-3508 TA Utrecht
Telephone: (0)30-2535200
Telefax: (0)30-2531601
Electronic Mail: <userid>@fys.ruu.nl
    minnaert@fys.ruu.nl
WWW: http://stkwww.fys.ruu.nl:8000/
   http://ruunfs.fys.ruu.nl/
   http://www.fys.ruu.nl/~wwwstk/
Founded: 1642
Staff: 14
Activities: research in astrophysics
Periodicals: (1) "Annual Report"
Coordinates: 005°07'48"E 52°05'12"N H14m (Sterrenwacht Sonnenborgh)
● Full university name: "Rijksuniversiteit Utrecht"

City Reference Coordinates: 005°08'00"E 52°05'00"N

Vereniging voor Weer- en Sterrenkunde Noord Drenthe
c/o G. van den Braak
Leenakkersweg 46
NL-9471 CL Zuidlaren
Telephone: (0)50-4093909
Electronic Mail: gudbraak@wxs.nl
WWW: http://www.astro.rug.nl/~nvws/ndrenthe
Founded: 1948
Membership: 52
Activities: lectures * meetings * photography
Periodicals: "Scoopium"
Coordinates: 006°40'50"E 53°05'51"N H2m (Zuidlaren)
             006°34'20"E 53°03'59"N H6m (Vries)
             006°37'07"E 53°52'47"N H4m (Elp)
City Reference Coordinates: 006°41'00"E 53°06'00"N

Vesta Volkssterrenwacht
Zuideinde 195-197
NL-1511 GD Oostzaan
Telephone: (0)75-6843148
           (0)75-6845039
Founded: 1975
Membership: 80
Activities: lectures * stargazing * photography * telescope making * computing * Meteosat image reception * seismological station
Periodicals: "Vesta"
Coordinates: 004°53'13"E 52°25'34"N
City Reference Coordinates: 004°52'00"E 52°26'00"N

Volkssterrenwacht Amsterdam (VSA)
Nieuwe Teertuinen 17
NL-1013 LV Amsterdam
Telephone: (0)20-6238717
Electronic Mail: nobelw@logica.com (Wim Nobel)
WWW: http://vsa.intouch.net/
Founded: 1991 (AWSK: 1918)
Membership: 20
Activities: meetings * lectures * excursions * group observing * instrumentation
Periodicals: (4) "De Cluster" (circ.: 500)
Coordinates: 004°53'11"E 52°23'12"N H10m
             005°06'24"E 52°19'13"N H0m (Muiderberg Observatory)
● Includes also the "Amsterdamse Weer- en Sterrenkundige Kring (AWSK)"
City Reference Coordinates: 004°54'00"E 52°22'00"N

Volkssterrenwacht Bussloo (VSB)
Bussloselaan 4
NL-7383 RP Bussloo-Voorst
Telephone: (0)313-652716
Founded: 1975
Membership: 15
Activities: public observing * lectures * meteor astronomy * planetarium
Periodicals: (4) "Astrovisie"
Coordinates: 006°07'13"E 52°11'53"N H7m
City Reference Coordinates: 006°10'00"E 52°10'00"N (Voorst)

Volkssterrenwacht Philippus Lansbergen
Herengracht 52
NL-4331 PX Middelburg
Telephone: (0)118-638018
Telefax: (0)118-633825
Electronic Mail: philpp3@mygale.org
WWW: http://www.mygale.org/o3/philipp3
Founded: 1969
Membership: 100
Activities: observing * photography * slide shows * lectures * exhibitions * education
Periodicals: (5) "Observator et Emergo" (circ.: 250)
Coordinates: 003°37'00"E 51°30'00"N H-1m
City Reference Coordinates: 003°37'00"E 51°30'00"N

Volkssterrenwacht Stichting Copernicus
Postbus 2507
NL-2002 RA Haarlem
or :

Observatory
Vergierdeweg 269
NL-2026 BJ Haarlem
Telephone: (0)23-5716705 (Secretary)
Founded: 1974
Membership: 117
Activities: popularization
Periodicals: (9) "Contacten"
Coordinates: 004°39'33" E 52°25'23" N
City Reference Coordinates: 004°38'00" E 52°23'00" N

Vrije Universiteit Amsterdam
● See "Universiteit Amsterdam"

Weer- en Sterrenkundige Kring Eemsmond
c/o A.A.M. Heeres (Secretary)
Kleine Belt 119
NL-9933 RD Delfzijl
Telephone: (0)596-611715
Electronic Mail: winzanstra@freemail.nl (W.T. Zanstra, President)
Founded: 1978
Membership: 62
Activities: observing * lectures * excursions * telescope making * exhibitions
Periodicals: (8) "De Vangspiegel"
City Reference Coordinates: 006°46'00" E 53°19'00" N

Zonnewijzerkring (De_)
● See "De Zonnewijzerkring"

New Zealand

Astronautics Association of New Zealand (AANZ) Inc.
P.O. Box 11-734
Wellington
Telephone: (0)4-4773632 (Home)
　　　　　　(0)4-5765158 (Office)
Telefax: (0)4-4773632 (Home)
Founded: 1979
Membership: 30
Activities: promoting awareness of space exploration
Periodicals: (6) "AANZ Journal" (ISSN 1170-7372)
City Reference Coordinates: 174°46'00"E 41°18'00"S

Astronomical Pocket Diary (APD)
c/o Norbert Haley
Poste Restante
P.O. Wellesley Street
Auckland
Electronic Mail: norb@kcbbs.gen.nz
　　　　　　　　norb@geocities.com
WWW: http://members.tripod.com/~apd2/apd.htm
Founded: 1989
Staff: 1
Activities: publishing pocket-size almanac
City Reference Coordinates: 174°46'00"E 36°55'00"S

Auckland Astronomical Society, Inc.
P.O. Box 24-187
Royal Oak
Auckland 6
Telephone: (0)9-656945
Telefax: (0)9-6250019
Electronic Mail: christie@iconz.co.nz (Grant Watson Christie, Secretary-Treasurer)
WWW: http://www.astronomy.org.nz/
Founded: 1922
Membership: 230
Activities: observing * research * lectures * education
Periodicals: (12) "Journal of Auckland Astronomical Society"
Coordinates: 174°46'00"E 36°55'00"S (Auckland Observatory)
City Reference Coordinates: 174°46'00"E 36°55'00"S

Auckland Observatory and Planetarium Trust
P.O. Box 24-180
Auckland 6
or :
One Tree Hill Domain
Auckland 3
Telephone: (0)9-624-1246
　　　　　　(0)9-624-3744
Telefax: (0)9-625-2394
Electronic Mail: auckobs@iconz.co.nz
WWW: http://www.stardome.org.nz/
Founded: 1967
Staff: 6
Activities: lectures * photoelectric photometry * eclipsing binaries * Miras * dwarf novae
Coordinates: 174°46'42"E 36°54'24"S H80m
City Reference Coordinates: 174°46'00"E 36°55'00"S

Blaxall and Steven Ltd.
163 Saint Asaph Street
P.O. Box 25095
Christchurch
Telephone: (0)3-366-2828
Telefax: (0)3-365-2072
Electronic Mail: blaxall@voyager.co.nz (Michael T. Blaxall, General Manager)
Founded: 1954
Staff: 6
Activities: importing optical material
City Reference Coordinates: 172°38'00"E 43°42'00"S

Canterbury Astronomical Society (CAS)
P.O. Box 25-137

Victoria Street
Christchurch
Telephone: (0)3-374-9549
Telefax: (0)3-325-3865
Electronic Mail: l.hussey@lincoln.ac.nz
WWW: http://www.phys.canterbury.ac.nz/cas/
Founded: 1957
Membership: 150
Activities: observing (grazing occultations, minor-planet occultations, variable stars) * photometry * monthly meetings * public nights * public education
Periodicals: (11) "CASMAG"
Coordinates: 172°20'58"E 43°30'02"S H118m (Joyce Memorial Observatory)
City Reference Coordinates: 172°38'00"E 43°42'00"S

Canterbury Astronomical Society (CAS), Ashburton Branch
c/o Ken Lucas (President)
11 Queens Drive
Ashburton
Telephone: (0)3-308-9203
WWW: http://www.phys.canterbury.ac.nz/cas/ (CAS)
Founded: 1970
Membership: 12
Activities: amateur observing * monthly meetings with guest speakers * public nights * public education * seminars
Coordinates: 171°45'03"E 43°53'38"S
City Reference Coordinates: 171°46'00"E 43°54'00"S

Carter Observatory
P.O. Box 2909
Wellington 1
or :
40 Salamanca Road
Kelburn
Wellington
Telephone: (0)4-472-8167
Telefax: (0)4-472-8320
Electronic Mail: astronomy@clear.net.nz
                    wayne.orchiston@vuw.ac.nz (Wayne Orchiston, Executive Director)
WWW: http://www.carterobs.ac.nz/
Founded: 1941
Staff: 13
Activities: National Observatory of New Zealand * astronomical research (dark matter, extra-solar planets, supernovae, open clusters, globular clusters, Vilnius southern standards, variable stars, asteroids, meteorites, history of astronomy, astronomical optics) * Golden Bay Planetarium * national astronomical heritage collection
Sections: research * public astronomy * education * astronomical heritage
Periodicals: (12) "Carter Observatory Newsletter" (ISSN 1173-8812); (1) "Annual Report" (ISSN 1173-5392), "Astronomical Handbook" (ISSN 0114-2216); "Journal of Astronomical History and Heritage"
City Reference Coordinates: 174°46'00"E 41°18'00"S

Gisborne Astronomical Society (GAS)
P.O. Box 678
Gisborne
Telephone: (0)6-867-7901
Electronic Mail: hac.gas@xtra.co.nz
WWW: http://mysite.xtra.co.nz/~AstroGisborne/
City Reference Coordinates: 178°02'00"E 38°41'00"S

Golden Bay Planetarium
● See "Carter Observatory"

Hamilton Astronomical Society (HAS)
P.O. Box 9153
Hamilton 2001
or :
Next to 200 Brymer Road
Hamilton
Telephone: (0)7-849-8522
Electronic Mail: secretary@has.org.nz
WWW: http://www.has.org.nz/
Founded: 1933
Membership: 50
Activities: meetings * lectures * education * open public nights
Periodicals: (11) "Hamilton Astronomical Society Bulletin"
Coordinates: 175°13'40"E 37°46'32"S H100m
City Reference Coordinates: 175°15'00"E 37°45'00"S

Institute of Geological and Nuclear Sciences (IGNS) Inc.
69B Gracefield Road
P.O. Box 30-368
Lower Hutt
Telephone: (0)4-5701444
Telefax: (0)4-5704600
WWW: http://www.gns.cri.nz/
Founded: 1865
Staff: 220
Activities: Earth sciences research (including seismology, volcanology, paleontology, isotopes, hydrocarbons)
Periodicals: "Monographs IGNS", "Geological Maps IGNS"
City Reference Coordinates: 174°46'00"E 41°18'00"S

Meteorological Service of New Zealand Ltd.
30 Salamanca Road
P.O. Box 722
Wellington 6015
Telephone: (0)4-472-9379
Telefax: (0)4-499-1942
Electronic Mail: <userid>@met.co.nz
             service@met.co.n
WWW: http://metcon.met.co.nz/
       http://www.met.co.nz/
Founded: 1992
Staff: 177
Activities: weather forecasting * managing about 100 observing sites
City Reference Coordinates: 174°46'00"E 41°18'00"S

Mount John University Observatory (MJUO)
● See "University of Canterbury, Department of Physics and Astronomy"

New Zealand Institute of Physics (NZIP)
c/o IRL
P.O. Box 31310
Lower Hutt
Electronic Mail: nzip@irl.cri.nz
WWW: http://nzip.rsnz.govt.nz/
City Reference Coordinates: 174°46'00"E 41°18'00"S

New Zealand Spaceflight Association (NZSA) Inc.
P.O. Box 5829
Wellesley Street
Auckland 1
Telephone: (0)9-480-7900
Founded: 1977
Membership: 200
Activities: studying and promoting manned spaceflight and robotic craft
Periodicals: (6) "Liftoff"
City Reference Coordinates: 174°46'00"E 36°55'00"S

Pallasite Press
P.O. Box 33-1218
Takapuna
Auckland
Telephone: (0)9-486-2428
Telefax: (0)9-489-6750
Electronic Mail: j.schiff@auckland.ac.nz
WWW: http://meteor.co.nz/
Founded: 1994
Staff: 6
Periodicals: (4) "Meteorite!" (ISSN 1173-2245)
City Reference Coordinates: 174°46'00"E 36°55'00"S
● Publisher

Royal Astronomical Society of New Zealand (RASNZ) Inc.
P.O. Box 3181
Wellington
Electronic Mail: rasnz@rasnz.org.nz
WWW: http://www.rasnz.org.nz/
Founded: 1920
Membership: 220
Activities: publications * annual conference * observing
Sections: Aurorae and Sun * Comets and Minor Planets * Education * Occultations * Photometry * Variable Stars
Periodicals: (10) "RASNZ Newsletter" (ISSN 1173-8138); (4) "Southern Stars" (ISSN 0049-1640)

Awards: (1) "Murray Geddes Prize"
City Reference Coordinates: 174°46'00"E 41°18'00"S

Royal Society of New Zealand (RSNZ)
P.O. Box 598
Wellington
or :
4 Halswell Street
Thorndon
Wellington
Telephone: (0)4-472-7421
Telefax: (0)4-473-1841
Electronic Mail: ceo@rsnz.govt.nz
WWW: http://www.rsnz.govt.nz/
Founded: 1867
Membership: 20000
Staff: 31
Activities: statutory body incorporating the "New Zealand National Committee for Astronomy" and the "New Zealand Standing Committee on the Astronomical Sciences" (astronomy, space science, radio astronomy and astrophysics)
Periodicals: "Journal of The Royal Society of New Zealand", "New Zealand Journal of Agricultural Research", "New Zealand Journal of Botany", "New Zealand Journal of Crop and Horticultural Science", "New Zealand Journal of Geology and Geophysics", "New Zealand Journal of Marine and Freshwater Research", "New Zealand Journal of Zoology"
City Reference Coordinates: 174°46'00"E 41°18'00"S

Sky Laboratories International Ltd.
163 Saint Asaph Street
P.O. Box 25095
Christchurch
Telephone: (0)3-366-2827
Telefax: (0)3-365-2072
Electronic Mail: optics.orders@voyager.co.nz
WWW: http://www.skylab.co.nz/
Founded: 1979
Staff: 2
Activities: retailing optical material
City Reference Coordinates: 172°38'00"E 43°42'00"S

Southland Astronomical Society (SAS)
Southland Museum
P.O. Box 1012
Invercargill
Electronic Mail: esler@southnet.co.nz (Secretary)
WWW: http://www.es.co.nz/~bevans/sas.htm
City Reference Coordinates: 168°21'00"E 46°26'00"S

Standards New Zealand (SNZ)
Private Bag 2439
Wellington 6020
or :
Standards House
155 The Terrace
Wellington 6001
Telephone: (0)4-498-5990
Telefax: (0)4-498-5994
Electronic Mail: snz@standards.synet.net.nz
WWW: http://www.standards.co.nz/
Founded: 1932
Staff: 60
Activities: national standards authority
Periodicals: (11) "Standards New Zealand" (ISSN 0110-4667)
City Reference Coordinates: 174°46'00"E 41°18'00"S

The Phoenix Astronomical Society (TPAS)
P.O. Box 2217
Wellington
Electronic Mail: sacha.hall@dear.net.nz
WWW: http://www.phoenix.org.nz/
Founded: 1997
City Reference Coordinates: 174°46'00"E 41°18'00"S

University of Canterbury, Department of Physics and Astronomy
Private Bag 4800
Christchurch 1
Telephone: (0)3-366-7001
Telefax: (0)3-364-2469

Electronic Mail: <userid>@csc.canterbury.ac.nz
        psi%0530130000034::<userid>
WWW: http://www.phys.canterbury.ac.nz/
        http://www.canterbury.ac.nz/
Founded: 1872
Staff: 12
Activities: variable-star photometry * stellar spectroscopy * CCD photometry * minor-planet and comet astrometry
Observatories: 1 (Mount John University Observatory - see separate entry)
City Reference Coordinates: 172°38'00"E 43°42'00"S

University of Canterbury, Department of Physics and Astronomy, Mount John University Observatory (MJUO)

P.O. Box 56
Lake Tekapo 8770
Telephone: (0)3-680-6000
Electronic Mail: mjuo@phys.canterbury.ac.nz
WWW: http://www.phys.canterbury.ac.nz/
        http://www.mjuo.canterbury.ac.nz/mjuo/
Founded: 1965
Staff: 8
Activities: variable-star photometry * stellar spectroscopy * solar-system astrometry * CCD photometry
Coordinates: 170°27'54"E 43°59'12"S H1,029m
City Reference Coordinates: 170°29'00"E 44°02'00"S

University of Waikato, Mathematics Department

Private Bag 3105
Hamilton
Telephone: (0)7-856-2889
Telefax: (0)7-838-4666
Electronic Mail: <userid>@waikato.ac.nz
WWW: http://hoiho.math.waikato.ac.nz/
Founded: 1964
Staff: 8
Activities: theoretical astrophysics * MHD
City Reference Coordinates: 175°17'00"E 37°47'00"S

Wairarapa Astronomical Society

c/o Alison Adams (Secretary)
5 McKenzie Terrace
Carterton
or :
c/o B. Harvey
186d Colombo Road
Masterton
Telephone: (0)6-378-7865
Founded: 1985
Membership: 16
Activities: variable stars * occultations * comets * minor planets * aurorae * solar flares * photography * visits to schools * mobile observatory
City Reference Coordinates: 175°39'00"E 40°57'00"S (Masterton)

Wanganui Astronomical Society

17 Rata Street
Wanganui
b
c/o David Calder
12 Rees Street
Wanganui
Telephone: (0)6-345-6113
Founded: 1947
Membership: 100
Activities: public education * observing
City Reference Coordinates: 175°02'00"E 39°56'00"S

Wellington Astronomical Society (WAS)

P.O. Box 3126
Wellington
Telephone: (0)4-528-3468 (Mike Clear)
Electronic Mail: clearm@agriquality.co.nz
WWW: http://astronomy.wellington.net.nz/
Periodicals: (12) "Newsletter"
City Reference Coordinates: 174°46'00"E 41°18'00"S

Whangarei Astronomical Society Inc.

P.O. Box 436

Whangarei
Telephone: (0)4389630
Founded: 1963
Membership: 32
Activities: lectures * discussions * observing * grazes * talks
Periodicals: "Octantis"
City Reference Coordinates: 174°20'00"E 35°43'00"S

Nigeria

Meteorological Department
Oshodi
Lagos
or :
Private Mail Bag 12542
Lagos
Telephone: (0)1-631792
            (0)1-631717
            (0)1-2633371
Telefax: (0)1-2634489
          (0)1-2636097
City Reference Coordinates: 003°24'00"E 06°27'00"N

Nigerian Academy of Sciences (NAS)
P.M.B. 1004
University of Lagos Post Office
Akoka
Yaba
Lagos
Telephone: (0)1-863874
Telefax: (0)1-863874
Founded: 1977
Staff: 5
Membership: 85
Activities: public lectures * conferences * symposia
Periodicals: "Proceedings" (ISSN 0794-7976)
Awards: (1) "MAN National Science Prize"
City Reference Coordinates: 003°24'00"E 06°27'00"N

Standards Organization of Nigeria (SON)
Federal Secretariat
Phase 1 - 9th Floor
Ikoyi
Lagos
Telephone: (0)1-2696178
Telefax: (0)1-2696176
City Reference Coordinates: 003°24'00"E 06°27'00"N

University of Nigeria, Department of Physics and Astronomy
Nsukka
Telephone: (0)42-770500
Telefax: (0)42-770644
Founded: 1960
Staff: 30
Activities: radio astronomy * high-energy astrophysics * solar energy * solid-state physics * optical astronomy * cosmology
Coordinates: 007°27'13"E 06°52'07"N
City Reference Coordinates: 007°29'00"E 06°51'00"N

Norway

Andøya Rocket Range (ARR)
● See "Norwegian Space Centre (NSC), Andøya Rocket Range (ARR)"

European Incoherent Scatter Facility (EISCAT), Ionospheric Heating Facility
Ramfjordmoen
N-9027 Ramfjordbotn
Telephone: 77692171
Telefax: 77692360
　　　 77692380
Electronic Mail: mike.rietveld@eiscat.uit.no (Mike Rietveld, Senior Scientist)
WWW: http://www.eiscat.uit.no/heating/
Founded: 1975 (EISCAT)
Staff: 2
Coordinates: 019°13'00"E 69°35'00"N
City Reference Coordinates: 019°00'00"E 69°40'00"N

European Incoherent Scatter Facility (EISCAT), Tromsø Station
Ramfjordmoen
N-9027 Ramfjordbotn
Telephone: 77692166
Telefax: 77692360
　　　 77692380
Electronic Mail: eiscat@eiscat.uit.no
　　　　　 <userid>@eiscat.uit.no
WWW: http://www.eiscat.uit.no/
Founded: 1975 (EISCAT)
Staff: 15
Activities: mesospheric, ionospheric and magnetospheric research using incoherent scatter radars (transmitter and receiver site)
Coordinates: 019°13'00"E 69°35'00"N H30m
City Reference Coordinates: 019°13'00"E 69°35'00"N

European Incoherent Scatter Facility (EISCAT), Svalbard Radar
P.O. Box 432
N-9170 Longyearbyen
Telephone: 79021236
Telefax: 79021751
Electronic Mail: esr@esr.eiscat.no
　　　　　 <userid>@esr.eiscat.no
WWW: http://www.esr.eiscat.no/
Founded: 1975 (EISCAT)
Staff: 4
City Reference Coordinates: 015°40'00"E 78°12'00"N

Maidanak Foundation (The_)
● See "The Maidanak Foundation"

Nordlysplanetariet Tromsø as
● See "Northern Lights Planetarium"

Norsk Astronautisk Forening (NAF)
(Norwegian Astronautical Association)
Postboks 52
Blindern
N-0313 Oslo
Telephone: 88003130
Founded: 1951
Membership: 350
Periodicals: (8) "Smånytt om Romfart"; (4) "Nytt om Romfart" (ISSN 0332-5962)
City Reference Coordinates: 010°45'00"E 59°55'00"N

Norsk Astronomisk Selskap (NAS)
(Norwegian Astronomical Society - NAS)
Postboks 1029
Blindern
N-0315 Oslo
Telephone: 22456513
Telefax: 22856505
WWW: http://www.ifi.uio.no/~mikkels/nas/nas.html
　　　 http://www.ifi.uio.no/~mikkels/nas/InterNAS.html

http://www.astro.uio.no/nas/
Founded: 1938
Membership: 800
Activities: observing groups * publishing * lectures * excursions
Sections: Meteors * Sun * Variable Stars * Supernova Search * Occultations * Comets
Periodicals: (4) "Astronomi", "Astronomisk Tidsskrift" (ISSN 0004-6345)
City Reference Coordinates: 010°45'00"E 59°55'00"N

Norske Videnskaps-Akademi

(Norwegian Academy of Science and Letters)
Postboks 7585
Skillebekk
N-0205 Oslo
or :
Drammensveien 78
N-0271 Oslo
Telephone: 2444296
Telefax: 2562656
WWW: http://hotell.nextel.no/dnva/
Founded: 1857
Staff: 5
Membership: 610
Activities: promoting science
Periodicals: (1) "Academy's Yearbook" (ISSN 0332-6909), "Kristian Birkeland Lecture"
Awards: "Fridtjof Nansen Medal"
City Reference Coordinates: 010°45'00"E 59°55'00"N

Norsk Teknisk Museum (Teknoteket), Saint-Exupéry Planetarium

Kjelsåsvn 143
N-0491 Oslo
Telephone: 22796000
Founded: 1985
Staff: 4
City Reference Coordinates: 010°45'00"E 59°55'00"N

Northern Lights Planetarium

(Nordlysplanetariet Tromsø as)
Universitetsområdet
Breivika
N-9037 Tromsø
Telephone: 77676000
Telefax: 77675700
Electronic Mail: nptweb@geronimo.uit.no
                    nptweb@npt.uit.n
WWW: http://www.uit.no/npt/homepage-npt.en.html
         http://www.uit.no/planetariet
Founded: 1989
Activities: shows
City Reference Coordinates: 018°58'00"E 69°40'00"N

Norwegian Academy of Science and Letters

● See "Norske Videnskaps-Akademi"

Norwegian Astronautical Association

● See "Norsk Astronautisk Forening (NAF)"

Norwegian Astronomical Society (NAS)

● See "Norsk Astronomisk Selskap (NAS)"

Norwegian Defence Research Establishment (NDRE), Division for Electronics

(Forsvarets Forskningsinstitutt - FFI)
Postboks 25
N-2007 Kjeller
Telephone: 63807000
Telefax: 63807212
Electronic Mail: <userid>@ffi.no
WWW: http://wwwspace.ffi.no/
Founded: 1946
Staff: 200
Activities: mainly defence-related research * ionospheric and thermospheric research * theoretical plasma physics * positioning * geodesy
Periodicals: "FFI Publications"
City Reference Coordinates: 009°26'00"E 56°17'00"N

Norwegian Geotechnical Institute (NGI)
Sognsveien 72
Postboks 3930
Ullevaal Stadion
N-0806 Oslo
Telephone: 22023000
Telefax: 22230448
Electronic Mail: ngi@ngi.no
WWW: http://www.ngi.no/
Founded: 1953
Staff: 136
Activities: geotechnical engineering * arctic engineering * construction control * cross hole seismic * drainage design * earthquake engineering * engineering geology * environmental geotechnology * erosion control * feasibility studies * field instrumentation * foundation engineering * geochemistry * geodynamics * geological mapping * grouting design * in situ testing * model testing * numerical analysis * performance observing * permeability testing * permafrost investigations * petroleum reservoir technology * pollution * risk analysis * rock blast design * rock mechanics * sampling * seepage control * site investigation * snow mechanics * soil dynamics * soil mechanics * stability analysis * vibration control
Periodicals: "NGI Publications Series" (ISSN 0078-1193)
City Reference Coordinates: 010°45'00"E 59°55'00"N

Norwegian Meteorological Institute (DNMI)
Niels Henrik Aabelsvei 40
Postboks 43
Blindern
N-0313 Oslo
Telephone: 22963000
Telefax: 22963050
WWW: http://www.dnmi.no/
Founded: 1866
Staff: 450
Activities: national weather forecasting centre * meteorological research * NOAA and Meteosat receiving and processing
City Reference Coordinates: 010°45'00"E 59°55'00"N

Norwegian Physical Society
c/o T. Henriksen (President)
Department of Physics
University of Oslo
Postboks 1048
Blindern
N-0316 Oslo
Telephone: 2855648
Telefax: 2855671
Electronic Mail: henrik@master.uio.no
WWW: http://www.fys.uio.no/~tovesv/nfs.htm
Founded: 1953
Membership: 850
City Reference Coordinates: 010°45'00"E 59°55'00"N

Norwegian Space Centre (NSC)
Drammensveien 105
Postboks 113
Skoyen
N-0212 Oslo
Telephone: 22511800
Telefax: 22511801
Electronic Mail: spacecentre@spacecentre.no
　　　　　　　　<userid>@spacecentre.no
WWW: http://www.spacecentre.no/
Founded: 1987
Staff: 75
Activities: national agency coordinating Norwegian space activities * operating the "Andøya Rocket Range (ARR)" and the "Tromsø Satellite Station (TSS)" (see separate entries)
Periodicals: (1) "Space Research in Norway", "Space Technology and Industries in Norway", "Annual Report", "National Long-Term Plan for Space Activities"
City Reference Coordinates: 010°45'00"E 59°55'00"N

Norwegian Space Centre (NSC), Andøya Rocket Range (ARR)
Postboks 54
N-8480 Andenes
Telephone: 76141644
Telefax: 76141857
Electronic Mail: info@rocketrange.no
WWW: http://www.rocketrange.no/
　　　　http://www.spacecentre.no/
Founded: 1962

Staff: 31
Activities: launching sounding rockets and scientific balloons
Coordinates: 016°01'00"E 69°17'00"N
City Reference Coordinates: 016°08'00"E 69°16'00"N

Norwegian Space Centre (NSC), Tromsø Satellite Station (TSS)
● See "Tromsø Satellite Station (TSS)"

Norwegian Standards Association
(Norges Standardiseringsforbund - NSF)
Drammensveien 145
Postboks 353
Skøyen
N-0212 Oslo
or :
Hegdehaugsveien 31
Postboks 7020
Homansbyen
N-0306 Olso
Telephone: 22466094
          22049200
Telefax: 22464457
          22049211
Electronic Mail: firmapost@nsf.telemax.no
WWW: http://www.standard.no/nsf
Founded: 1923
City Reference Coordinates: 010°45'00"E 59°55'00"N

Norwegian University of Science and Technology, Institute of Physics
Haakon Magnussons gate 3
N-7055 Dragvoll
Telephone: 73591850
Telefax: 73591852
Electronic Mail: <userid>@avh.unit.no
Founded: 1922
Staff: 15
Activities: theoretical physics * astrophysics (AGNs, compact stars, cosmology, gravitation) * elementary particles * field theory
City Reference Coordinates: 010°25'00"E 63°25'00"N (Trondheim)

Research Council of Norway
Postboks 2700
St. Hanshaugen
N-0131 Oslo
Telephone: 22037000
Telefax: 22037001
Electronic Mail: <userid>@nfr.no
WWW: http://www.sn.no/
Founded: 1993
Activities: bioproduction and processing * industry and energy * culture and society * medicine and health * environment and development * natural sciences and technology
City Reference Coordinates: 010°45'00"E 59°55'00"N

Saint-Exupéry Planetarium
● See "Norsk Teknisk Museum (Teknoteket), Saint-Exupéry Planetarium"

Scandinavian University Press
Postboks 2959
Tøyen
N-0608 Oslo
Telephone: 22575400
Telefax: 22575454
Electronic Mail: <userid>@scup.no
WWW: http://www.scup.no/
Staff: 180
● Publisher
City Reference Coordinates: 010°45'00"E 59°55'00"N

Skibotn Astronomical Observatory
N-9048 Skibotn
Telephone: 77715820
Coordinates: 020°22'00"E 69°21'00"N H155m
City Reference Coordinates: 020°22'00"E 69°21'00"N

Stavanger Astronomical Association
Postboks 453
N-4001 Stavanger
or :
Byhaugen Kafe
Carl Sundt Hansensgatan 15
N-4024 Stavanger
Telephone: 51566813 (Wednesdays 19:00-23:00)
Electronic Mail: safstyret@bsda.hsr.no
WWW: http://www.ux.his.no/saf/
Founded: 1969
Membership: 63
Activities: variable stars * meteorites * deep sky * planets
Periodicals: (3) "Astro"
Coordinates: 005°42'08"E 58°58'26"N
City Reference Coordinates: 005°45'00"E 58°58'00"N

Svalsat
● See "Tromsø Satellite Station (TSS), SvalSat"

Teknoteket
● See "Norsk Teknisk Museum (Teknoteket)"

The Maidanak Foundation
Welhavens Gate 61
N-5006 Bergen
Telefax: 55962663
Electronic Mail: henrik.nilsen@fi.uib.no (Henrik E. Nilsen, Chairman)
WWW: http://www.maidanak.org/
Founded: 1999
Staff: 6
Activities: supporting scientific teams running Mount Maidanak Observatory (Uzbekistan) * providing funding for scientific equipment
City Reference Coordinates: 005°20'00"E 60°23'00"N

Tromsø Satellite Station (TSS)
Prestvannsveien 38
N-9291 Tromsø
Telephone: 4777600250
Telefax: 4777600299
Electronic Mail: tss@tss.no
WWW: http://www.tss.no/
Founded: 1967
Staff: 40
Activities: reception and distribution of data from polar orbiting satellites * search and rescue via satellites * monitoring services based on Earth observation data
● Jointly and equally owned by the "Norwegian Space Centre (NSC)" and the "Swedish Space Corp. (SSC), Earth Observation Division"
Coordinates: 018°56'00"E 69°39'00"N
City Reference Coordinates: 018°56'00"E 69°39'00"N

Tromsø Satellite Station (TSS), SvalSat
P.O. Box 458
N-9171 Longyearbyen
Telephone: 4779022550
Telefax: 4779023784
Electronic Mail: svalsat@tss.no
WWW: http://www.svalsat.com/
Founded: 1997
Activities: tracking and command of polar orbiting satellites
City Reference Coordinates: 015°40'00"E 78°12'00"N

University of Bergen, Department of Physics
Allégaten 55
N-5007 Bergen
Telephone: 55213050
Telefax: 55317229
           55318334
Electronic Mail: <userid>@fi.uib.no
WWW: http://www.fi.uib.no/
Founded: 1948
Staff: 78
Activities: space physics * magnetospheric research
City Reference Coordinates: 005°20'00"E 60°23'00"N

University of Bergen, Institute of Solid Earth Physics (ISEP)
Allégaten 41
N-5007 Bergen
Telephone: 55583420
Telefax: 55589699
WWW: http://www.ifjf.uib.no/ifjf2.html
Founded: 1990
City Reference Coordinates: 005°20'00"E 60°23'00"N

University of Oslo, Department of Physics
Postboks 1048
Blindern
N-0316 Oslo
Telephone: 22856428
Telefax: 22856422
Electronic Mail: admin@fys.uio.no
                 <userid>@fys.uio.no
WWW: http://www.fys.uio.no/english.html
Founded: 1964 (University: 1811)
Activities: general physics * plasma and space physics * high energy * solid state * electronics * nuclear physics * biophysics * materials * theoretical physics
City Reference Coordinates: 010°45'00"E 59°55'00"N

University of Oslo, Institute of Theoretical Astrophysics
Postboks 1029
Blindern
N-0315 Oslo
Telephone: 22856501
         22856502
Telefax: 22856505
Electronic Mail: mats.carlsson@astro.uio.no (Mats Carlsson, Director)
WWW: http://www.astro.uio.no/
Founded: 1934 (University: 1811)
Staff: 13
Activities: solar transition zone * prominences * sunspots * cool stars * solar wind * plasma physics * planetary systems * cosmology * gravitational lenses
Periodicals: "Institute of Theoretical Astrophysics, Reports" (ISSN 0078-6780)
City Reference Coordinates: 010°45'00"E 59°55'00"N

University of Tromsø, Auroral Observatory
Nordlysobservatoriet
Prestvannsveien 40
N-9037 Tromsø
Telephone: 77644000
         77644001
         77645150
Telefax: 77648850
Electronic Mail: <userid>@phys.uit.no
WWW: http://geo.phys.uit.no/
Founded: 1928
Staff: 4
Activities: geomagnetic observations
Periodicals: "Journal of Magnetic Observations" (ISSN 0373-4854)
City Reference Coordinates: 018°58'00"E 69°40'00"N

University of Tromsø, Department of Physics
Nordlysobservatoriet
Prestvannsveien 40
N-9037 Tromsø
Telephone: 77645150
Telefax: 77645580
Electronic Mail: <userid>@phys.uit.no
WWW: http://www.phys.uit.no/
Founded: 1976
Staff: 25
Activities: stellar activity * degenerate objects * interplanetary dust * cosmic geophysics * plasma physics * applied physics * molecular quantum mechanics * Northern Lights Planetarium (see separate entry)
Coordinates: 020°21'54"E 69°20'54"N (Skibotn Astronomical Observatory)
City Reference Coordinates: 018°58'00"E 69°40'00"N

University of Tromsø, Department of Physics, Astrophysics Group
Nordlysobservatoriet
Prestvannsveien 40
N-9037 Tromsø
Telephone: 77645191
         77715820 (Skibotn Astronomical Observatory)

Telefax: 77645595
Electronic Mail: <userid>@phys.uit.no
WWW: http://www.phys.uit.no/fysikk/astro/index-e.html
Founded: 1972
Staff: 1
Activities: degenerate objects * Northern Lights Planetarium (see separate entry)
Coordinates: 020°21'54"E 69°20'54"N H157m (Skibotn Astronomical Observatory)
City Reference Coordinates: 018°58'00"E 69°40'00"N

University of Trondheim, Group for Theoretical Physics
N-7034 Trondheim
Telephone: 73593646
Electronic Mail: <userid>@phys.unit.no
WWW: http://www.ntnu.no/ntnu/old/ten_minutes/avh.html
Periodicals: "Arkiv for det Fysiske Seminar i Trondheim" (ISSN 0365-2459)
City Reference Coordinates: 010°25'00"E 63°25'00"N

Pakistan

Lahore Astronomical Society Pakistan (LAST)
36 Tariq Block
New Garden Town
Lahore
Telephone: (0)5831727
Electronic Mail: pulsar@brain.net.pk
WWW: http://www.geocities.com/Broadway/8701/
Founded: 1976
City Reference Coordinates: 074°18'00"E 31°35'00"N

Pakistan International Airlines (PIA), Planetarium, Lahore
University Ground
Lake Road
Lahore 54000
Telephone: (0)42-210810
Founded: 1987
Staff: 19
Activities: shows for public and groups of students and teachers
City Reference Coordinates: 074°18'00"E 31°35'00"N

Pakistan Meteorological Department (PMD)
P.O. Box 8454
University Road
Karachi 75270
Telephone: (0)21-8112223
Telefax: (0)21-8112885
Electronic Mail: pmd@paknet3.ptc.pk
WWW: http://met.gov.pk/
Founded: 1947
Staff: 2000
Activities: meteorology * climatology * hydrology * seismology * geophysics * atmospheric physics * environmental pollution
Periodicals: (12) "Monthly Climatic Summary of Pakistan"; "Agromet Bulletin of Pakistan"
City Reference Coordinates: 067°03'00"E 24°52'00"N

Pakistan Space & Upper Atmosphere Research Commission (SUPARCO)
Sector 28, Gulzar-e-Hijri
Off University Road
P.O. Box 8402
Karachi 75270
Telephone: (0)21-472630
           (0)21-470158
Telefax: (0)21-4960553
Founded: 1961
Staff: 220
Activities: national space agency for the promotion of peaceful uses and applications of space sciences and technology * remote sensing * satellite communications * ionospheric research
Periodicals: (4) "Space Horizons" (ISSN 0259-9163)
City Reference Coordinates: 067°03'00"E 24°52'00"N

Pakistan Standards Institution (PSI)
39 Garden Road
Saddar
Karachi 74400
Telephone: (0)21-7729527
Telefax: (0)21-7728124
Electronic Mail: pakqltyk@super.net.pk
Founded: 1951
Staff: 152
Activities: drawing up standards and promoting their adoption
City Reference Coordinates: 067°03'00"E 24°52'00"N

Paraguay

Club de Astrofísica del Paraguay (CAP)
14 de Mayo 1565
Casilla de Correo 1978
Asunción
Telephone: (0)71541
Telefax: (0)606297
Founded: 1985
Membership: 50
Activities: education * meetings * observing
Periodicals: (6) "Astrocosas"
Coordinates: 057°24'44"W 25°17'48"S
City Reference Coordinates: 057°40'00"W 25°15'00"S

Dirección Nacional de Aeronáutica Civil (DINAC), Dirección de Meteorología e Hidrología (DMH)
Asunción
Electronic Mail: dmh@highway.com.py
WWW: http://www.highway.com.py/dinac/tiempo.html
City Reference Coordinates: 057°40'00"W 25°15'00"S

Instituto Nacional de Tecnología y Normalización (INTN)
Casilla de Correo 967
Asunción
Telephone: (0)2-1290160
Telefax: (0)2-1290873
City Reference Coordinates: 057°40'00"W 25°15'00"S

Peru

Consejo Nacional de Ciencia y Tecnología (CONCYTEC)
Apartado 1984
Lima 100
or :
Avenida Paseo de la República 3505
6° Piso
Lima 27
or :
Edificio del Ministerio de Educación
Piso 18
Parque Universitario s/n
Lima 11
WWW: http://www.uni.edu.pe/RPCyT/concytec.htm
Activities: funding research * dissemination of knowledge
City Reference Coordinates: 077°03'00"W 12°03'00"S

Instituto Geofísico del Perú (IGP)
Apartado 3747
Lima 100
or :
Calatrava 216
Urbanización Camino Real
La Molina
Lima
Telephone: (0)14-368437
Telefax: (0)14-370258
Electronic Mail: <userid>@igpcam.pe
WWW: http://www.ee.cornell.edu/~spp/radar/jro/jicamarca.html (Jicamarca)
Founded: 1947
Staff: 70
Activities: geomagnetism * seismology * volcanology * upper atmosphere physics * solar physics * meteorology * climatology
Coordinates: 075°19'22"W 12°02'18"S H3,312m (Huancayo)
          077°09'00"W 11°46'12"S H40m (Ancón)
          076°52'25"W 11°56'55"S H500m (Jicamarca)
City Reference Coordinates: 077°03'00"W 12°03'00"S

Instituto Nacional de Defensa de la Competencia y de la Protección de la Propiedad Intelectual (INDECOPI)
Prolongación Avenida Guardia Civil 400
San Borja
Lima 41
Telephone: (0)14-711777
Telefax: (0)14-711617
Electronic Mail: <userid>@indecopi.gob.pe
WWW: http://www.indecopi.gob.pe/
Founded: 1992
Staff: 120
Activities: defending free competition * protecting intellectual property * offering consumer protection * acting against dumping and unfair practices * normalization and quality control
City Reference Coordinates: 077°03'00"W 12°03'00"S

Jicamarca Radio Observatory (JRO)
● See "Instituto Geofísico del Perú (IGP)"

Servicio Nacional de Meteorología e Hidrología (SENAMHI)
Casilla Postal 1308
Lima 11
or :
Jr. Cahuide 785
Jesus Maria
Lima 11
Telephone: (0)4-717287
          (0)4-714565
          (0)4-704863
Telefax: (0)4-717287
Electronic Mail: senamhi@senamhi.gob.pe
WWW: http://www.senamhi.gob.pe/
City Reference Coordinates: 077°03'00"W 12°03'00"S

Universidad Nacional Mayor de San Marcos (UNMSM), Seminario Pemanente de Astronomía y Ciencias Espaciales (SPACE)
Apartado 20077
Succ. 51 Colmena
Lima 1
or :
Pabellón de Física
Ciudad Universitaria UNMSM
Avenida Venezuela s/n
Lima 1
Telephone: (0)1-4243961
            (0)1-4521343
Telefax: (0)1-4521343
Electronic Mail: d220002@unmsm.edu.pe
                A3_user2@unmsm.edu.pe (Maria Luisa Aguilar, Chairman)
                jorge@iagusp.usp.br (Jorge Melendez Moreno)
WWW: http://www.iagusp.usp.br/~jorge/saa.html
Founded: 1982
Staff: 58
Activities: research * education * popularization
Periodicals: (3) "Boletín Informativo SAA"; (1) "Revista Peruana de Astronomía y Astrofísica"
Coordinates: 075°05'00"W 12°04'00"S H40m
City Reference Coordinates: 077°03'00"W 12°03'00"S

Philippines

Bureau of Product Standards (BPS)
Trade and Industry Building
361 Sen. Gil J. Puyat Avenue
P.O. Box 3328 MCPO
Makati 1200
Telephone: (0)2-8904965
Telefax: (0)2-8904926
Electronic Mail: dtibpsrp@mnl.sequel.net
WWW: http://www.dti.gov.ph/bps
Founded: 1964 (as former "Division of Standards")
Staff: 84
Activities: standards development * product certification and testing * information dissemination and training * laboratory accreditation * accreditation of conformity assessment bodies * registration of quality assessors
Periodicals: (4) "BPS Directions" (ISSN 0118-3648)
City Reference Coordinates: 121°01'00"E 14°34'00"N

Manila Observatory (MO)
P.O. Box 122
UP Post Office
1101 Quezon City
or :
Ateneo de Manila Campus
Loyola Heights
Quezon City
Telephone: (0)2-4265921
              (0)2-4265922
              (0)2-4265923
Telefax: (0)2-4266141
Electronic Mail: mo@admu.edu.ph
WWW: http://www.admu.edu.ph/
Founded: 1865
Staff: 36
Activities: crustal deformation * inventory of greenhouse gases * climate change * air quality monitoring * upper atmosphere * focal mechanism of earthquakes
Periodicals: "Solar Patrol", "Ionospheric Data", "Solar Maps and Activity"
Coordinates: 121°04'35"E 14°38'10"N H58m
City Reference Coordinates: 121°02'00"E 14°39'00"N

National Museum Planetarium
P. Burgos Street
Rizal Park
P.O. Box 2659
Manila 2801
Telephone: (0)5271830
Founded: 1975
Staff: 11
Activities: planetarium shows * lectures * demonstrations
City Reference Coordinates: 120°59'00"E 14°35'00"N

Philippine Astronomical Society (PAS)
P.O. Box 122
UP Post Office 1101
Quezon City
Telephone: (0)2-9241751
Telefax: (0)2-9241751
WWW: http://www.pworld.net.ph/user/jkty/pas.html
Founded: 1971
Membership: 120
Activities: meetings * observing * instrumentation * lectures * consulting * astronomical clearinghouse
Periodicals: (12) "The Appulse"
Awards: (1) "Padre Faura Astronomy Award"
Coordinates: 121°04'35"E 14°38'10"N H58m
City Reference Coordinates: 121°02'00"E 14°39'00"N

Philippine Atmospheric, Geophysical and Astronomical Services Administration (PAGASA), Astronomy Research and Development Section
P.O. Box 2277
Manila
or :
Asia Trust Building
1424 Quezon Avenue

Quezon City
Metro Manila
Telephone: (0)9228416
Telefax: (0)9221872
Electronic Mail: pagasa@mail.ph.net
WWW: http://max.ph.net/~pagasa/pagimg.htm
        http://www.pagasa.dost.ph/
Founded: 1954
Staff: 33
Activities: astronomical observing * disseminating data/space science information * recording and collecting solar data * keeping and disseminating time * research in astronomy and related fields * planetarium
Periodicals: (1) "Philippines Astronomical Handbook", "Almanac for Geodetic Engineers", "Moonrise, Moonset, Sunrise, Sunset and Twilight Tables", "Bulletin of Astronomical Observations", "Calendar Data", "Annual Report"
Coordinates: 121°04'18"E 14°39'12"N H70m
City Reference Coordinates: 120°59'00"E 14°35'00"N (Manila)
                                  121°00'00"E 14°38'00"N (Quezon City)

Poland

Acta Astronomica
Al. Ujazdowskie 4
PL-00-478 Warszawa
Telephone: (0)22-6295346
Telefax: (0)22-6294967
Electronic Mail: acta@sirius.astrouw.edu.pl
Founded: 1925
Staff: 3
● Journal (4) (ISSN 0001-5237)
City Reference Coordinates: 021°00'00"E 52°15'00"N

Copernicus Astronomical Center (NCAC) (N._)
● See "Polish Academy of Sciences, N. Copernicus Astronomical Center (NCAC)"

Copernicus Museum (Nicolas_)
● See "Nicolas Copernicus Museum"

Copernicus Planetarium and Observatory in Chorzów (Nicolas_)
● See "Nicolas Copernicus Planetarium and Observatory in Chorzów"

Frombork Planetarium and Observatory
Ul. Elblaska 2
P.O. Box 6
PL-14-530 Frombork
Telephone: (0)506-7392
WWW: http://www.frombork.art.pl/Pol08.htm
Founded: 1983
Membership: 67
Activities: Summer astronomical camps * assistance in operating the planetarium and observatory in Nicolas Copernicus Museum (see separate entry) * observing (meteors, comets)
Periodicals: (4) "Wiadomości Pulsara"
City Reference Coordinates: 019°41'00"E 54°22'00"N

Gdansk University, Institute of Theoretical Physics and Astrophysics
Ul. Wita Stwosza 57
PL-80-952 Gdansk
Telephone: (0)415241 x188
           (0)415241 x210
           (0)415241 x243
Electronic Mail: <userid>@halina.univ.gda.pl
WWW: http://www.univ.gda.pl/
Founded: 1981
Staff: 22
Activities: stellar spectroscopy * stellar atmospheres (observing and modelling) * late-type stars * RS CVn binaries * IUE spectra
City Reference Coordinates: 018°40'00"E 54°23'00"N

Institute of Geodesy and Cartography
Ul. Jasna 2/4
PL-00-950 Warszawa
Telephone: (0)22-270328
Telefax: (0)22-270328
Electronic Mail: jnstgeo@plearn.bitnet
WWW: http://www.gik.pw.edu.pl/
Founded: 1930
Staff: 11
Activities: geodetic astronomy * time service * satellite geodesy
Periodicals: (1) "Astronomical Yearbook"
Coordinates: 021°02'06"E 52°28'36"N H110m (Borowa Gòra)
City Reference Coordinates: 021°00'00"E 52°15'00"N

Institute of Meteorology and Water Management
Ul. Podleśna 61
PL-01-673 Warszawa
Telephone: (0)22-341651
Telefax: (0)22-345466
Electronic Mail: <userid>@imgw.pl
WWW: http://www.imgw.pl/
Founded: 1919
Staff: 1500

Activities: meteorology * hydrology * oceanology * water management * hydraulics * water engineering * water resources * water quality * network of stations around the country
Periodicals: (6) "Journal of the Institute of Meteorology and Water Management Observer" (ISSN 0208-4325, circ.: 4,000); (4) "Reports of the Institute of Meteorology and Water Management" (ISSN 0208-6263, circ.: 480); "Research Papers": "Hydrology and Oceanology" (ISSN 0239-6297, circ.: 250), "Meteorology" (ISSN 0239-6262, circ.: 300), "Water Engineering" (ISSN 0239-6254, circ.: 150), "Water Management and Water Protection" (ISSN 0239-6238, circ.: 250); "Bibliography of Hydrology and Oceanology" (ISSN 0239-6246, circ.: 150), "Bibliography of Meteorology" (ISSN 0239-6270, circ.: 150), "Bibliography of Water Management and Engineering" (ISSN 0239-622x, circ.: 150)
Coordinates: 020°58'00"E 52°17'00"N H99m
City Reference Coordinates: 021°00'00"E 52°15'00"N

Jagiellonian University
● See "Kraków Jagiellonian University"

Józefoslaw University of Technology
● See "Warsaw Józefoslaw University of Technology"

Kalisz Observatory
Ul. Ułanska 9c
PL-62-800 Kalisz
Telephone: (0)62-75443
Founded: 1964
Membership: 15
Activities: observing (stars * planets) * bibliographical service
Coordinates: 018°08'00"E 51°46'00"N H116m
City Reference Coordinates: 018°08'00"E 51°46'00"N

Kraków Jagiellonian University, Astronomical Observatory
Ul. Orla 171
PL-30-244 Kraków
Telephone: (0)12-4251294
Telefax: (0)12-4251318
Electronic Mail: <userid>@oa.uj.edu.pl
WWW: http://www.oa.uj.edu.pl/
Founded: 1792
Staff: 24
Activities: comets * variable stars * solar physics * AGNs * extragalactic radiosources * ISM * intergalactic matter * magnetic fields in galaxies * cosmology * cosmic rays
Periodicals: (12) "Monthly Report on Solar Radio Emission"; (1) "Rocznik Astronomiczny Obserwatorium Krakowskiego" (Ephemeris of eclipsing variable stars)
Coordinates: 019°49'38"E 50°03'16"N H348m (Fort Skala Station)
                 022°15'00"E 49°10'00"N H840m (Bieszczady Mountains Station)
City Reference Coordinates: 019°58'00"E 50°03'00"N

Kraków Pedagogical University, Department of Physics, Mount Suhora Observatory
Ul. Podchorazych 2
PL-30-084 Kraków
Telephone: (0)12-6379747
              (0)90-330126 (Observatory Dome)
Telefax: (0)12-6372243
Electronic Mail: <userid>@cyf-kr.edu.pl
WWW: http://www.as.wsp.krakow.pl/
Founded: 1987
Staff: 7
Activities: photometry of eclipsing binaries * cataclysmic variable stars * spot stars
Coordinates: 020°05'24"E 49°34'48"N H1,000m
City Reference Coordinates: 019°58'00"E 50°03'00"N

Maria Curie-Sklodowska University, Institute of Physics, Astrophysics and Didactics, Astronomy Laboratory
Maria Curie-Sklodowska Square 1
PL-20-031 Lublin
Telephone: (0)81-5376186
Telefax: (0)81-5376191
Electronic Mail: lgladysz@tytan.umcs.lublin.pl (Longin Gładyszewski, Head)
Founded: 1944
Staff: 4
Activities: education * meteorites * solar physics * solar radioastronomy
Periodicals: "Annales UMCS", "Annual Report"
Coordinates: 022°32'37"E 51°14'47"N H210m
City Reference Coordinates: 022°35'00"E 51°15'00"N

Mlodziezowe Obserwatorium Astronomiczne
c/o Aleksander Trebacz
Ul. M. Kopernika 2

PL-32-005 Niepolomicexi
Telephone: (0)12-811561
Telefax: (0)12-812131
Electronic Mail: moa@pkpf.if.uj.edu.pl
WWW: http://www.cyf_kr.edu.pl/~ufjochym/MOA/
City Reference Coordinates: 020°13'00"E 50°02'00"N

Mount Suhora Observatory

● See "Kraków Pedagogical University, Department of Physics, Mount Suhora Observatory"

N. Copernicus Astronomical Center (NCAC)

● See "Polish Academy of Sciences, N. Copernicus Astronomical Center (NCAC)"

Nicolas Copernicus Museum

Ul. Katedralna 8
PL-14-530 Frombork
Telephone: (0)55-2437218
Telefax: (0)55-2437218
Electronic Mail: frombork@softel.elblag.pl
Founded: 1948
Staff: 35
Activities: exhibitions * planetarium shows * observing
Periodicals: "Komentarze Fromborskie"
Coordinates: 019°40'59"E 54°20'31"N H47m
City Reference Coordinates: 019°40'00"E 54°21'00"N

Nicolas Copernicus Planetarium and Observatory in Chorzów

P.O. Box 10
PL-41-501 Chorzów
Telephone: (0)546330
Telefax: (0)413296
WWW: http://www.man.katowice.pl/katowice/informator/tekst/english/s08.shtml
Founded: 1955
Staff: 11
Activities: popularization * astrometry (comets, minor planets) * meteorological and seismological station * software production
Coordinates: 018°59'30"E 50°17'33"N H325m
City Reference Coordinates: 018°56'00"E 50°19'00"N

Opole Pedagogical University, Department of Astrophysics

Ul. Oleska 48
PL-45-951 Opole
Telephone: (0)77-35841
Telefax: (0)77-38387
Electronic Mail: <userid>@sparc-1.wsp.opole.pl
WWW: http://www.uni.opole.pl/
Founded: 1950
Staff: 4
Activities: atomic spectroscopy * line profiles * stellar spectra * atomic partition functions * white dwarfs
City Reference Coordinates: 017°55'00"E 50°41'00"N

Planetarium and Astronomical Observatory, Grudziądz

Ul. Hoffmanna 1/7
PL-86-300 Grudziądz
Telephone: (0)51-22794
Founded: 1972
Staff: 1
Activities: popularization * seminars
Coordinates: 018°45'16"E 53°28'45"N H72m
City Reference Coordinates: 018°45'00"E 53°29'00"N

Planetarium and Astronomical Observatory, Olsztyn

Al. Marszałka Piłsudskiego 38
PL-10-450 Olsztyn
Telephone: (0)334951
　　　　　(0)335178
WWW: http://www.planetarium.olsztyn.pl/
Founded: 1973
Staff: 6
Activities: programs for public and schools * lectures * exhibitions * education for young people * public observing
Coordinates: 020°29'25"E 53°46'25"N H150m
City Reference Coordinates: 020°29'00"E 53°48'00"N

Polish Academy of Sciences, Astrogeodynamic Observatory

Borowiec

Ul. Drapalka 4
PL-62-035 Kórnik
Telephone: (0)61-8170187
Telefax: (0)61-8170219
Electronic Mail: sch@cbk.poznan.pl (Stanislaw Schillak, Head of Observatory)
WWW: http://www.cbk.poznan.pl/
Founded: 1953
Staff: 15
Activities: satellite geodesy * space research * geodynamics * Earth rotation * GPS * SLR * time
Periodicals: (1) "Borowiec Laser Station Operational Report"
Coordinates: 017°04'30"E 52°16'36"N H80m
● Formerly known as "Astronomical Latitude Observatory"
City Reference Coordinates: 017°04'00"E 52°17'00"N

Polish Academy of Sciences, N. Copernicus Astronomical Center (NCAC)
Ul. Bartycka 18
PL-00-716 Warszawa
Telephone: (0)22-411086
Telefax: (0)22-410046
Electronic Mail: <userid>@camk.edu.pl
WWW: http://www.camk.edu.pl/
Founded: 1974
Staff: 48
Activities: binary stars * stellar oscillations * stellar atmospheres * circumstellar matter * neutron stars * high-energy astrophysics * cosmology * n-body dynamics * ionosphere plasma * elementary particles
City Reference Coordinates: 021°00'00"E 52°15'00"N

Polish Academy of Sciences, N. Copernicus Astronomical Center, Department of Astrophysics I
Ul. Rabianska 8
PL-87-100 Torun
Telephone: (0)56-6219249
            (0)56-6219319
            (0)56-6219341
Telefax: (0)56-6219381
Electronic Mail: <userid>@ncac.torun.pl
WWW: http://www.ncac.torun.pl/
Founded: 1974 (NCAC)
Staff: 9
City Reference Coordinates: 018°35'00"E 53°02'00"N

Polish Academy of Sciences, Space Research Center
Ordona 21
PL-01-237 Warszawa
Telephone: (0)22-362885
Telefax: (0)22-368961
Electronic Mail: cbk@cbk.waw.pl
WWW: http://www.cbk.waw.pl/
Founded: 1977
Staff: 55
Activities: space physics * physics and dynamics of planets and comets * heliophysics * astrometry * satellite geodesy
Periodicals: "Artificial Satellites" (two series "Space Physics" and "Planetary Geodesy") (ISSN 0571-205x)
Coordinates: 017°04'30"E 52°16'36"N H85m (Borowiec)
City Reference Coordinates: 021°00'00"E 52°15'00"N

Polish Academy of Sciences, Space Research Center, Solar Physics Division
Ul. Kopernika 11
PL-51-622 Wrocław
Telephone: (0)71-483238
Telefax: (0)71-729372
Electronic Mail: js@cbk.pan.wroc.pl
WWW: http://cbk.pan.wroc.pl/
Founded: 1969
Staff: 16
Activities: solar physics * X-rays * space instrumentation * data analysis
City Reference Coordinates: 017°00'00"E 51°06'00"N

Polish Amateur Astronomical Society
Ul. Sw. Tomasza 30/8
PL-31-027 Kraków
Telephone: (0)12-223892
Electronic Mail: ptma@oa.uj.edu.pl
WWW: http://www.oa.uj.edu.pl/~ptma/
Founded: 1919
Membership: 2040
Activities: popularization * observing (occultations, variable stars, sunspots, comets, planets, meteors, eclipses) * telescope making

Periodicals: (12) "Urania" (ISSN 0042-0794); "Biblioteka Uranii"
Awards: (1/3) "Gold Award", "Silver Award"; "Honorable Membership"
City Reference Coordinates: 019°58'00"E 50°03'00"N

Polish Astronautical Society

Ul. Krasińskiego 54
PL-01-755 Warszawa
Telephone: (0)22-6852703
Founded: 1954
Membership: 386
Activities: research * education * popularizing astronautics
Periodicals: (6) "Astronautyka" (ISSN 0004-623x); (4) "Postepy Astronautyki" (ISSN 0373-5982)
City Reference Coordinates: 021°00'00"E 52°15'00"N

Polish Astronomical Society

Ul. Bartycka 18
PL-00-716 Warszawa
Telephone: (0)410041 x76
Telefax: (0)410828
Electronic Mail: library@camk.edu.pl
Founded: 1923
Membership: 213
Activities: developing astronomical sciences * education and popularization of astronomy * organizing symposia, meetings and Summer schools
Periodicals: (4) "Postepy Astronomii"
Awards: (1/2) "W. Zonn Medal"; (1) "Prize for Young Astronomers"
City Reference Coordinates: 021°00'00"E 52°15'00"N

Polish Committee for Standardization

(Polski Komitet Normalizacyjny - PKN)
Ul. Elektoralna 2
PL-00-139 Warszawa
Telephone: (0)22-6205434
Telefax: (0)22-6200741
Electronic Mail: polknor@atos.warman.com.pl
Founded: 1994
Staff: 306
Periodicals: (12) "Normalizacja" (ISSN 0029-179x); "Biuletyn Informacyjny" (ISSN 1230-8242)
City Reference Coordinates: 021°00'00"E 52°15'00"N

Polish Geophysical Society

Ul. Podleśna 61
PL-01-673 Warszawa
Telephone: (0)341651 x528
              (0)341651 x321
Telefax: (0)345466
Founded: 1947
Membership: 420
Activities: meteorology * hydrology * geophysics
Periodicals: "Review of Geophysics" (ISSN 0033-2135)
Awards: (1/2) "Marian Molga Scientific Award", "Award of the Polish Geophysical Society"
City Reference Coordinates: 021°00'00"E 52°15'00"N

Polish Physical Society

Ul. Hoza 69
PL-00-681 Warszawa
Telephone: (0)2-6212668
Telefax: (0)2-6212668
Electronic Mail: ptf@fuw.edu.pl
WWW: http://www.fuw.edu.pl/~ptf/
Founded: 1920
Membership: 1800
City Reference Coordinates: 021°00'00"E 52°15'00"N

Poznań University A. Mickiewicza, Astronomical Observatory

Ul. Słoneczna 36
PL-60-286 Poznań
Telephone: (0)61-8679670
Telefax: (0)61-8686511
Electronic Mail: secretary@mail.astro.amu.edu.pl
WWW: http://www.astro.amu.edu.pl/welcome.html
Founded: 1919
Staff: 24
Activities: celestial mechanics * fundamental astrometry * theory of refraction * dynamics of artificial satellites * small bodies of the solar system * minor planets * stellar astrophysics

Coordinates: 016°52'50"E 52°23'53"N H85m
City Reference Coordinates: 016°53'00"E 52°25'00"N

Société Européenne pour l'Astronomie dans la Culture (SEAC)

c/o S. Iwaniszewski (Secretary)
State Archaeological Museum
P.O. Box 69
PL-00-950 Warszawa
Telephone: (0)22-313221
Telefax: (0)22-315195
Founded: 1993
Membership: 80
Activities: promotion of research on astronomical practice in its cultural context
Periodicals: "Newsletter"
City Reference Coordinates: 021°00'00"E 52°15'00"N

Torun Planetarium

Ul. Franciszka 19
PL-87-100 Torun
Telephone: (0)56-25066
WWW: http://www.man.torun.pl/Planetarium/
City Reference Coordinates: 018°35'00"E 53°02'00"N

Torun Radio Astronomy Observatory (TRAO)

• See "Torun University Nicolaus Copernicus, Radio Astronomy Observatory"

Torun University Nicolaus Copernicus, Institute of Astronomy

Ul. Chopina 12/18
PL-87-100 Torun
Telephone: (0)56-26018 (Institute)
            (0)56-11655 (Observatory)
Telefax: (0)56-24602 (University)
Electronic Mail: <userid>@astri.uni.torun.pl
WWW: http://www.astri.uni.torun.pl/
       file://copernicus.astro.torun.edu.pl/pub/
Founded: 1947
Staff: 25
Activities: physics of peculiar stars * local galactic structure * physics of planets and comets * motion of natural and artificial satellites
Periodicals: "Preprints"
Coordinates: 018°33'15"E 53°05'48"N H90m
City Reference Coordinates: 018°35'00"E 53°02'00"N

Torun University Nicolaus Copernicus, Radio Astronomy Observatory

Ul. Gagarina 11
PL-87-100 Torun
Telephone: (0)56-783327
            (0)56-11651
Telefax: (0)56-11651
WWW: http://www.astro.uni.torun.pl/
Founded: 1979
Staff: 32
Activities: VLBI study of extragalactic sources * solar observing (127 MHz) * developing new instrumentation and data analysis methods
Coordinates: 018°33'42"E 53°05'44"N H80m (Piwnice Observatory)
City Reference Coordinates: 018°35'00"E 53°02'00"N

Warsaw Józefoslaw University of Technology, Institute of Geodesy and Geodetic Astronomy

Pl. Politechniki 1
PL-00-661 Warszawa
Telephone: (0)22-6228515
            (0)22-6607754
Telefax: (0)22-6210052
Electronic Mail: <userid>@gik.pw.edu.pl
Founded: 1949
Activities: GPS observing for IGS and EUREF * GPS evaluation and analysis centre * gravity measurements (absolute and tidal) * meteorology * VZT latitude observing
Periodicals: "Latitude Circulars", "Scientific Publications of Józefoslaw Warsaw University of Technology - Series: Geodesy", "Reports on Geodesy"
Coordinates: 021°02'00"E 52°05'50"N
City Reference Coordinates: 021°00'00"E 52°15'00"N

Warsaw University, Astronomical Observatory

Al. Ujazdowskie 4
PL-00-478 Warszawa

Telephone: (0)22-294011
Telefax: (0)22-294967
Electronic Mail: <userid>@sirius.astrouw.edu.pl
WWW: http://www.astrouw.edu.pl/
Founded: 1825
Staff: 15
Activities: variable stars (observing and theory) * stellar atmospheres * ISM dynamics * statistics of extragalactic objects
Coordinates: 021°25'12"E 52°05'23"N H138m (Ostrowik)
City Reference Coordinates: 021°00'00"E 52°15'00"N

Warsaw University of Technology
● See "Warsaw Józefoslaw University of Technology"

Wrocław University, Astronomical Institute
Ul. Kopernika 11
PL-51-622 Wrocław
Telephone: (0)71-482434
         (0)71-729373
         (0)71-729374
Telefax: (0)71-729378
Electronic Mail: <userid>@astro.uni.wroc.pl
WWW: http://astro1.astro.uni.wroc.pl/
      http://www.ift.uni.wroc.pl/wroc_uni/faculties/MP/astronomy.html
Founded: 1791
Periodicals: "Contributions", "Reprints"
Coordinates: 017°05'18"E 51°06'42"N H117m
City Reference Coordinates: 017°00'00"E 51°06'00"N

Portugal

Associação Portuguesa de Astronomos Amadores (APAA)
Rua Alexandre Herculano 57 - 4° Dto
P-1250 Lisboa
Telefax: (0)21-7784153
WWW: http://www.terravista.pt/FerNoronha/1475/
City Reference Coordinates: 009°08'00"W 38°43'00"N

Associação Portuguesa para o Ensino da Astronomia
Apartado 52503
Amial
P-4202-301 Porto
Electronic Mail: astroportugal@ip.pt
City Reference Coordinates: 008°36'00"W 41°11'00"N (Porto)

Centro de Astrofísica da Universidade do Porto (CAUP)
● See "Universidade do Porto, Centro de Astrofísica"

Centro de Observação Astronômica do Algarve (COAA)
Poio
P-8500-149 Portimão
Telephone: 282471180
Telefax: 282471516
Electronic Mail: coaa@mail.telepac.pt
WWW: http://www.ip.pt/coaa/
Founded: 1987
Staff: 3
Activities: providing facilities for visiting astronomers
Periodicals: "COAA News";(1) "Annual Report"
Coordinates: 008°35'37"W 37°11'29"N H65m
City Reference Coordinates: 008°32'00"W 37°08'00"N

Fundação para a Ciência e a Tecnologia (FCT)
Avenida D. Carlos I 126
P-1200 Lisboa
Telephone: 213924300
Telefax: 213907481
Electronic Mail: presidencia@fct.mct.pt
WWW: http://www.fct.mct.pt/
Founded: 1997
Staff: 150
Activities: promoting, following up and funding science and technology activities in Portugal
City Reference Coordinates: 009°08'00"W 38°43'00"N

Instituto de Meteorologia
Rua C
Aeroporto de Lisboa
P-1700 Lisboa
Telephone: 218472880
          218472890
      218402370
Electronic Mail: im@meteo.pt
          <userid>@meteo.pt
WWW: http://www.meteo.pt/
Founded: 1916
City Reference Coordinates: 009°08'00"W 38°43'00"N

Instituto para a Cooperação Científica e Tecnológica International (ICCTI)
Rua Castillo 5
4th Floor
P-1250-066 Lisboa
or :
Avenida Don Carlos I 126
P-1200 Lisboa Codex
Telephone: 213924300
Telefax: 213975144
Electronic Mail: iccti@mail.telepac.pt
WWW: http://www.iccti.mct.pt/
Founded: 1997
Staff: 40
Activities: supporting, guiding and coordinating all international collaborations in the fields of science and technology

City Reference Coordinates: 009°08'00"W 38°43'00"N

Instituto Português da Qualidade (IPQ)
Rua C à Avenida dos Três Vales
P-2825 Monte da Caparica
Telephone: 212948100
Telefax: 212948101
Electronic Mail: lcmds@correio.ipq.gtw-ms.mailpac.pt
              ipqmail@mail.ipq.pt
WWW: http://www.ipq.pt/
Founded: 1986
Staff: 210
Activities: standardization * metrology * certification * accreditation
Periodicals: "OPÇÃO" (ISSN 0872-2884), "Qualirama","Metrologia", "Catálogo IPQ"
Awards: "Prémio Excelência"
City Reference Coordinates: 009°08'00"W 38°43'00"N (Lisboa)

Observatório Astronômico de Lisboa (OAL)
Tapada da Ajuda
P-1300 Lisboa
Telephone: 213637351
Telefax: 213621722
WWW: http://astro.cc.fc.ul.pt/oal.html
Founded: 1861
Activities: astrometry * meridian astronomy * time and latitude * lunar occultations * double and multiple stars * astronomical information to the national community
Periodicals: (1) "Dados Astronômicos para os Almanaques"; "Bulletin of the Astronomical Observatory of Lisbon"
Coordinates: 009°11'10"W 38°42'31"N H111m
City Reference Coordinates: 009°08'00"W 38°43'00"N

Observatório Astronômico, Coimbra
● See "Universidade de Coimbra, Observatório Astronômico"

Observatório Astronômico Prof. Manuel de Barros
● See "Universidade do Porto, Observatório Astronômico Prof. Manuel de Barros"

Observatório das Ciências e das Tecnologias (OCT)
Rua das Praças 13-B
P-1200-765 Lisboa
Telephone: 213926000
Telefax: 213950979
Electronic Mail: geral@oct.mct.pt
WWW: http://www.oct.mct.pt/
Founded: 1996
Staff: 40
Activities: collection, processing, analysis and publication of information on R&D and on the national S&T system
City Reference Coordinates: 009°08'00"W 38°43'00"N

Planetário Calouste Gulbenkian (PCG)
Praça do Império
P-1400-206 Lisboa
Telephone: 213610508
           213610123
Telefax: 213636005
Electronic Mail: planetario@mail.telepac.pt
WWW: http://www.marinha.pt/planetario
Founded: 1965
Staff: 14
Activities: shows for children, students and public
City Reference Coordinates: 009°08'00"W 38°43'00"N

Rede Nacional de Observação Astronómica (RNOA)
Trv. 1° de Maio 16
P-2430 Marinha Grande
Telephone: 244566986 (Carlos Reis, Presidente)
           244503820 (João Clérigo, Coordinator)
Electronic Mail: rnoa@mail.telepac.pt
                info@rnoa.rcts.pt
WWW: http://www.anoa.pt/
     http://www.rnoa.rcts.pt/
Founded: 1995
Periodicals: "Boletim RNOA"
City Reference Coordinates: 008°45'00"W 39°45'00"N

Secção Portuguesa das Uniões Internacionais Astronômica e Geodesica e Geofisica (SPUIAGG)
Rua Artlharia Um. 107
P-1099-052 Lisboa
Telephone: 213819600
Telefax: 213819699
Founded: 1923
Staff: 9
Activities: coordinating astronomical, geodetical and geophysical activities * Portuguese representation at IAU and IGGU * seminars * conferences
City Reference Coordinates: 009°08'00"W 38°43'00"N

Sociedade Portuguesa de Física (SPF)
c/o J. Bessa Sousa (President)
Laboratório da Física
Universidade do Porto
Avenida da República 37-4°
P-1000 Lisboa
or :
Praça Gomes Teixeira
P-4000 Porto
Telephone: 222001653
           217973251
Telefax: 217952349
Electronic Mail: spf@nautilus.fis.uc.pt
              jbsousa@fcl.fc.up.pt
WWW: http://nautilus.fis.uc.pt/~spf/
        http://atlas.cii.fc.ul.pt/spf/
Membership: 800
City Reference Coordinates: 009°08'00"W 38°43'00"N (Lisboa)
                                008°36'00"W 41°11'00"N (Porto)

Universidade de Coimbra, Observatório Astronômico
Caixa Postal 147
P-3002 Coimbra Codex
Telephone: 239814947
WWW: http://www.fis.uc.pt/
        http://www.fis.uc.pt/old/df_home.html
Founded: 1772
Activities: spectroscopy of K line * astrolabe astrometry
Periodicals: (1) "Astronomical Ephemeris", "Annais" (ISSN 0870-2856); "Communications"
Coordinates: 008°26'37"W 40°11'53"N H99m
City Reference Coordinates: 008°25'00"W 40°12'00"N

Universidade de Évora, Departamento de Física, Grupo de Astronomia
Rua Romão Ramalho 59
P-7000 Évora
Telephone: 266744616/8
Telefax: 266744546
Electronic Mail: <userid>@astro.ce.uevora.pt
WWW: http://www.lca.uevora.pt/
Founded: 1994
Activities: galactic physics * pulsars * planet formation and evolution * interstellar jets * gas dynamics * ISM * camputational astrophysics
City Reference Coordinates: 007°54'00"W 38°34'00"N

Universidade de Lisboa, Departamento de Física, Centro de Astronomia e Astrofísica
Observatório Astronômico de Lisboa (OAL)
Tapada da Ajuda
P-1349-018 Lisboa
Telephone: 213616739
Telefax: 213616752
Electronic Mail: <userid>@oal.ul.pt
WWW: http://www.oal.ul.pt/
Founded: 1992
Staff: 10
Activities: education * research * star formation * galactic structure and evolution * active galaxies * planetary astronomy
Coordinates: 009°09'20"W 38°45'28"N
City Reference Coordinates: 009°08'00"W 38°43'00"N

Universidade do Porto, Centro de Astrofísica
Rua das Estrelas
P-4150-762 Porto
Telephone: 226089830
Telefax: 226089839

Electronic Mail: www@astro.up.pt
WWW: http://www.astro.up.pt/
          http://www.astro.up.pt/planetario/ (Planetarium)
Founded: 1989
Staff: 22
Activities: research * education * popularization
City Reference Coordinates: 008°36'00"W 41°11'00"N

Universidade do Porto, Observatório Astronômico Prof. Manuel de Barros

Monte da Virgem
P-4430-146 Vila Nova de Gaia
Telephone: 227820404
Telefax: 227827253
Electronic Mail: observatorio@oa.fc.up.pt
WWW: http://www.fc.up.pt/oa
Founded: 1948
Staff: 14
Activities: astrophysics * solar radioastronomy * geodesy * remote sensing
Periodicals: "Publicações do Observatório Astronômico Prof. Manuel de Barros" (ISSN 0871-1542), "Informações do Observatório Astronômico Prof. Manuel de Barros" (ISSN 0871-1550)
Coordinates: 008°35'15"W 41°06'28"N H232m
City Reference Coordinates: 008°34'00"W 41°45'00"N

Puerto Rico

Arecibo Observatory
● See "National Astronomy and Ionosphere Center (NAIC), Arecibo Observatory"

National Astronomy and Ionosphere Center (NAIC), Arecibo Observatory
P.O. Box 995
Arecibo, PR 00613-0995
Telephone: 878-2612
Telefax: 878-1861
         878-0662
Electronic Mail: <userid>@naic.edu
WWW: http://aosun.naic.edu/
Founded: 1963
Staff: 14
Activities: radioastronomy * planetary radar * atmospheric physics
Periodicals: "NAIC/Arecibo Observatory Newsletter"
Coordinates: 066°45'11"W 18°20'37"N H497m
City Reference Coordinates: 066°44'00"W 18°29'00"N

Sociedad de Astronomia de Puerto Rico
c/o Ernesto E. Santiago-Jordan
34th Street P-2
Urbanización Bairoa
Caguas, PR 00725
WWW: http://fast.to/Astronomia
City Reference Coordinates: 066°04'00"W 18°14'00"N

University of Puerto Rico (UPR), Humacao, Observatory
Department of Physics
Humacao, PR 00791
Telephone: 850-9381
         850-9336
Telefax: 852-4638
         850-9471
Electronic Mail: <userid>@upr.clu.edu
               rj_muller@cuhac.upr.clu.edu (Rafael J. Muller, Observatory Director)
WWW: http://cuhwww.upr.clu.edu/fisica/observatorio/fiobs.html (Observatory)
        http://cuhwww.upr.clu.edu/~observ/fiobs.htm (Observatory)
        http://cuhwww.upr.clu.edu/
Founded: 1985
Staff: 2
Activities: observing * CCDs * photometry
Coordinates: 065°50'00"W 18°08'00"N H0m
City Reference Coordinates: 065°50'00"W 18°08'00"N

University of Puerto Rico (UPR), Rio Piedras, Department of Physics
P.O. Box 23343
Rio Piedras, PR 00931
Telephone: 764-0620
Telefax: 764-4063
Electronic Mail: <userid>@upr1.upr.clu.edu
WWW: http://physd.upr.clu.edu/
Founded: 1903
City Reference Coordinates: 066°03'00"W 18°24'00"N

Qatar

Department of of Civil Aviation and Meteorology (DCAM)
P.O. Box 3000
Doha
Telephone: (0)426262
             (0)429070
Electronic Mail: aliali96@qatar.net.qa
Founded: 1975
Staff: 110
Activities: weather forecasting
Periodicals: (12) "Monthly Weather and Climate Report"; (1) "Annual Weather and Climate Report"
City Reference Coordinates: 051°32'00"E 25°17'00"N

Department of Standards, Measurements and Consumer Protection
c/o Ministry of Finance, Economy and Commerce
P.O. Box 1968
Doha
Telephone: (0)408555
Telefax: (0)478849
Staff: 110
Activities: establishing standards * compliance testing of local and imported products * granting certificates of metrology, conformity and quality marks * calibration of taximeters
City Reference Coordinates: 051°32'00"E 25°17'00"N

University of Qatar, Scientific and Applied Research Centre (SARC)
P.O. Box 2713
Doha
Telephone: (0)874961
Telefax: (0)860680
Electronic Mail: sarc@qu.edu.qa
Founded: 1981
Staff: 50
Activities: scientific and applied research * education * training * consultancy
City Reference Coordinates: 051°32'00"E 25°17'00"N

Romania

Cluj-Napoca Astronomical Observatory
• See "Romanian Academy, Astronomical Institute, Cluj-Napoca Astronomical Observatory"

National Institute of Meteorology and Hydrology
Soseaua Bucureşti-Ploieşti 97
RO-71552 Bucureşti 18
Telephone: (0)1-2303116
Telefax: (0)1-2303143
Electronic Mail: info@meteo.inmh.ro
WWW: http://www.inmh.ro/
Founded: 1884
Activities: synoptic meteorology * dynamic meteorology * numerical forecasting * aeronautical meteorology * climatology * atmospheric physics * aerology * agrometeorology * air pollution * operational hydrology * underground hydrology
Periodicals: "Romanian Journal of Meteorology" (ISSN 1223-1118), "Romanian Journal of Hydrology and Water Resources" (ISSN 1223-1126), "Studii si Cercetari de Meteorologie", "Studii si Cercetari de Hidrologie"
City Reference Coordinates: 026°06'00"E 44°26'00"N

Romanian Academy, Astronomical Institute
Cuţitul de Argint 5
RO-75212 Bucureşti 28
Telephone: (0)1-3356892
　　　　　 (0)1-3358010
Telefax: (0)1-3373389
Electronic Mail: <userid>@roastro.astro.ro
WWW: http://roastro.astro.ro/
Founded: 1908
Staff: 52
Activities: variable stars * solar physics * photographic astrometry * reference systems * Earth rotation * cosmology * celestial mechanics * upper terrestrial atmosphere * history of astronomy
Periodicals: (2) "Romanian Astronomical Journal" (ISSN 1210-5168, circ.: 300); "Observations Solaires" (circ.: 230), "Anuarul Astronomic" (circ.: 300)
Coordinates: 026°05'47"E 44°24'50"N H80m (Bucureşti)
　　　　　　 023°35'53"E 46°42'48"N H730m (Cluj-Napoca)
　　　　　　 021°13'45"E 45°44'15"N H300m (Timişoara - see separate entry)
City Reference Coordinates: 026°06'00"E 44°26'00"N

Romanian Academy, Astronomical Institute, Cluj-Napoca Astronomical Observatory
Calea Cireşilor 19
RO-3400 Cluj-Napoca
Telephone: (0)64-194592
Telefax: (0)64-192820
Electronic Mail: academy1@mail.soroscj.ro
Founded: 1920
Staff: 14
Activities: celestial mechanics * variable stars
Coordinates: 023°35'52"E 46°42'48"N H750m (Faget Station)
City Reference Coordinates: 023°36'00"E 46°47'00"N

Romanian Academy, Astronomical Institute, Timişoara Astronomical Observatory
Piata Axente Sever 1
RO-1900 Timişoara
Telephone: (0)56-162838
Telefax: (0)56-162838
Electronic Mail: astro@astrotm.sorostm.ro
WWW: http://www.sorostm.ro/astro/
Founded: 1959
Coordinates: 021°24'55"E 45°44'15"N H88m
　　　　　　 021°13'45"E 45°44'15"N H300m
City Reference Coordinates: 021°13'00"E 45°45'00"N

Romanian Academy, Commission on Astronautics
Calea Victoriei 125
RO-71102 Bucureşti
Telephone: (0)1-6507193
Telefax: (0)1-3120209
Founded: 1961
Staff: 90
City Reference Coordinates: 026°06'00"E 44°26'00"N

Romanian Institute of Normalization (IRS)
Jean-Louis Calderon 13

RO-70201 Bucureşti
Telephone: (0)1-6114043
          (0)1-2113296
Telefax: (0)1-3120823
          (0)1-2110833
City Reference Coordinates: 026°06'00"E 44°26'00"N

Romanian Space Agency (ROSA)

21-25 Mendeleev Street
RO-70168 Bucureşti 1
Telephone: (0)1-6504222
Telefax: (0)1-3128804
Electronic Mail: <userid>@rosa.ro
WWW: http://www.rosa.ro/
Founded: 1995
City Reference Coordinates: 026°06'00"E 44°26'00"N

Societatea Română de Fizică

(Romanian Physical Society)
c/o Alexandru Calboreanu (General Secretary)
Tandem Laboratory
Institute of Atomic Physics
P.O. Box MG-6
RO-70000 Bucureşti
Telephone: (0)1-7807040
Telefax: (0)1-4231650
Electronic Mail: calbo@roifa.ifa.ro
Founded: 1890 (re-founded in 1990)
Membership: 1200
Activities: organizing national conferences * consulting * international representation
Periodicals: "Curierul de Fizică"
City Reference Coordinates: 026°06'00"E 44°26'00"N

Russia

A.F. Ioffe Physical Technical Institute
● See "Russian Academy of Sciences, A.F. Ioffe Physical Technical Institute"

Astronomical Club Parsec
Ul. Mendeleeva 13
Dvorec Unior
456780 Ozersk
Telephone: (0)351-7179018
Founded: 1979
Membership: 17
Activities: Sun * Moon * planets * meteors * comets * photography
Coordinates: 060°29'51"E 55°40'48"N H281m
City Reference Coordinates: 060°30'00"E 55°41'00"N

Astronomical Council
● See "Russian Academy of Sciences, Astronomical Council"

Astronomical-Geodetical Society
24 Sadovaja-Kudrinskaja Ul.
103001 Moscow
Telephone: (0)095-2915896
Founded: 1932
Membership: 8000 (of which 1500 juniors)
Activities: astronomical and geodetical education and popularization * national development projects for astronomy, geodesy and cartography * publishing * meteors * meteorites * NLC * comets * variable stars * telescope designing
Periodicals: (6) "Zemlja y Vsselennaja" (ISSN 0044-3948), "Astronomicheskij Vestnik" (ISSN 0320-930x); (3-5) "Circulars" (ISSN 0201-7342), "Soobtshenija" (ISSN 0235-3431)
City Reference Coordinates: 038°50'00"E 55°05'00"N

Astro Space Center (ASC)
● See "Russian Academy of Sciences, P.N. Lebedev Physics Institute, Astro Space Center (ASC)"

Baksan Neutrino Observatory (BNO)
● See "Russian Academy of Sciences, Baksan Neutrino Observatory (BNO)"

Center for Astronomical Data (CAD)
● See "Russian Academy of Sciences, Institute for Astronomy, Center for Astronomical Data (CAD)"

Committee for Standardization, Metrology and Certification (GOST)
Lenin Avenue 9
117049 Moscow
Telephone: (0)095-2364044
Telefax: (0)095-2376032
Electronic Mail: info@gost.ru
WWW: http://www.gost.ru/
City Reference Coordinates: 038°50'00"E 55°05'00"N

Complex Independent Tunguska Expedition
c/o Astronomical Observatory
P.O. Box 1106
634010 Tomsk
Telephone: (0)3822-212466
           (0)3822-909721
Electronic Mail: niipmm@urania.tomsk.su
           ok@siberia-ltd.tomsk.su
Founded: 1958
Membership: 100
Activities: investigating the Tunguska meteorite fall * Siberian meteorites * cosmic dust * ecological investigation
Coordinates: 101°53'00"E 60°53'00"N (Tunguska Meteorite Reserve)
City Reference Coordinates: 085°05'00"E 56°30'00"N

Engelhardt Astronomical Observatory
● See "Kazan State University, Engelhardt Astronomical Observatory"

Euro-Asian Astronomical Society (EAAS)
c/o Sternberg State Astronomical Institute
University Avenue 13
119899 Moscow
Telephone: (0)095-9328844
Telefax: (0)095-9328844

(0)095-9328841
Electronic Mail: boch@astronomy.msk.su
WWW: http://www.issp.ac.ru/univer/astro/eaas_e.html
Founded: 1990
Membership: 750
Activities: maintaining the development of astronomy * reinforcing communication between national and foreign astronomers * organizing scientific conferences * supporting astronomical education
Periodicals: "Astronomical and Astrophysical Transactions" (ISSN 1055-6796), "Solar Data", "EAAS Bulletin", "Astrocourier", "Bulletin of Association of Planetaria of Russia", "Universe and Ourselves", "Zvezdociot"
City Reference Coordinates: 038°50'00"E 55°05'00"N

Federal Service for Hydrometeorology and Monitoring of Environment
2 Novovagan'kovsky Ul.
123242 Moscow
Telephone: (0)095-2520808
            (0)095-2552493
Telefax: (0)095-2521158
City Reference Coordinates: 038°50'00"E 55°05'00"N

Hydrometeorological Research Centre of the Russian Federation (RHMC)
Bolshoi Predtechensky Per. 9-11
123242 Moscow
Telephone: (0)095-2555026
Telefax: (0)095-2551582
Electronic Mail: rusgmc@glas.apc.org
WWW: http://www.mecom.ru/
Founded: 1929
Staff: 470
Activities: weather forecast * hydrometeorology and related activities
Periodicals: (12) "Meteorology and Hydrology"
City Reference Coordinates: 038°50'00"E 55°05'00"N

Institute for Physical-Technical and Radio-Technical Measurements (VNIIFTRI)
Mendeleevo
141570 Moscow
Telephone: (0)095-5358490
            (0)095-5352401
Telefax: (0)095-5357386
Electronic Mail: root@ftri.extech.msk.su
WWW: http://www.extech.msk.su:8082/src_eng/catalog/40/40e-gnz.htm
Founded: 1956
Activities: Earth rotation parameters * data acquisition, processing and publishing
Periodicals: "Universal Time and Pole Coordinates": "Circular A" (ISSN 0135-2415), "Circular E" (ISSN 0234-1069)
City Reference Coordinates: 038°50'00"E 55°05'00"N

INTERSPUTNIK
2 Smolensky Per. 1/4
121099 Moscow
Telephone: (0)095-2440333
Telefax: (0)095-2539906
Electronic Mail: dir@intersputnik.com
WWW: http://intersputnik.com/
Founded: 1971
Staff: 15
Activities: international organization of space communications
City Reference Coordinates: 038°50'00"E 55°05'00"N

Ioffe Physical Technical Institute (A.F._)
• See "Russian Academy of Sciences, A.F. Ioffe Physical Technical Institute"

Irkutsk Society of Young Amateur Astronomers
Goncharov Ul. 3
P.O. Box 3512
664054 Irkutsk
Telephone: (0)395-2461148
Electronic Mail: root@sitmis.irkutsk.su (c/o Nefedyev)
Founded: 1981
Activities: instrumentation * observing * expeditions * education
Observatories: 2 (at Baikal Lake: Listvyanka and Mondy)
City Reference Coordinates: 104°20'00"E 52°16'00"N

Irkutsk State University, Astronomical Observatory
Sovetskaya 119A
664009 Irkutsk
Telephone: (0)3952-270294

Electronic Mail: uuastra@astra.isu.runnet.ru
Founded: 1931
Staff: 12
Activities: Earth rotation parameters * NLC * upper atmosphere dynamics * education
Coordinates: 104°20'42"E 52°36'44"N H468m
City Reference Coordinates: 104°20'00"E 52°16'00"N

IZMIRAN
● See "Russian Academy of Sciences, Institute of Terrestrial Magnetism, Ionosphere and Radio Wave Propagation (IZMIRAN)"

Joint Scientific and Educational Center (JSEC)
Politechnicheskaya Street 21
194021 Saint Petersburg
Telephone: (0)812-2472223
Telefax: (0)812-2475062
Electronic Mail: <userid>@brown.nord.nw.ru
WWW: http://brown.nord.nw.ru/
City Reference Coordinates: 030°15'00"E 59°55'00"N

Kazan State University, Department of Astronomy
Ul. Lenina 18
420008 Kazan
Telephone: (0)323641
WWW: http://astro.ksu.ru/
Founded: 1810
Staff: 30
Activities: stellar atmospheres * double stars * astrometry * Moon * comets * meteors * celestial mechanics * photometry * geodesy
Periodicals: "Trudy" (ISSN 0371-8247)
Coordinates: 049°07'18"E 55°47'24"N H79m
City Reference Coordinates: 049°08'00"E 55°45'00"N

Kazan State University, Engelhardt Astronomical Observatory
Observatoria Station
422526 Kazan
Telephone: (0)324827
WWW: http://astro.ksu.ru/
Founded: 1901
Staff: 40
Activities: astrometry * meteors
Periodicals: (1) "Izvestiya" (ISSN 0321-1762)
Coordinates: 048°48'54"E 55°50'18"N H98m
             041°26'36"E 43°39'10"N H2,040m
City Reference Coordinates: 049°08'00"E 55°45'00"N

Keldysh Institute of Applied Mathematics
● See "Russian Academy of Sciences, Keldysh Institute of Applied Mathematics"

Kirov Planetarium
Ul. Kosmonavta Volkova 6
School 27
610021 Kirov
Telephone: (0)8833-26361
           (0)8833-291965
           (0)8833-291980
Founded: 1960
Staff: 51
Activities: lectures * readings * meetings * observing * popularization
Coordinates: 049°37'00"E 58°39'00"N
City Reference Coordinates: 049°37'00"E 58°39'00"N

Kislovodsk Astronomical Station
● See "Russian Academy of Sciences, Pulkovo Observatory, Kislovodsk Astronomical Station"

Lebedev Physics Institute (P.N._)
● See "Russian Academy of Sciences, P.N. Lebedev Physics Institute"

Lomonosov Moscow State University (M.V._)
● See M.V. Lomonosov Moscow State University

Meteoritic and Cosmic Dust Siberian Commission
P.O. Box 1106
634010 Tomsk

Telephone: (0)3822-909721
(0)3822-212466
Electronic Mail: niipmm@urania.tomsk.su
Founded: 1960
Membership: 25
Activities: meteorites * comets * minor planets * Tunguska 1908 explosion
Coordinates: 101°53'00"E 60°53'00"N (Tunguska Meteorite Reserve)
City Reference Coordinates: 085°05'00"E 56°30'00"N

Mir
2 Pervy Rizhsky Per.
129820 Moscow
Telephone: (0)095-2861783
(0)095-2889522
Electronic Mail: <userid>@mir.msk.su
Founded: 1946
Staff: 165
● Publisher
City Reference Coordinates: 038°50'00"E 55°05'00"N

Moscow Physical-Technical Institute
8 Onezhskaya
125438 Moscow
Telephone: (0)095-4564608
Electronic Mail: alex@post.mipt.rssi.ru hahttp://www.mipt.rssi.ru/
Founded: 1951
Staff: 23
Activities: space power engineering
City Reference Coordinates: 038°50'00"E 55°05'00"N

Moscow Planetarium
Sadovaja-Kudrinskaya Ul. 5
123242 Moscow
Telephone: (0)095-2520217
Telefax: (0)095-2003227
WWW: http://www.planetarium.ru/english/english.html
Founded: 1929
Staff: 50
Activities: shows * observing * lectures * excursions
City Reference Coordinates: 038°50'00"E 55°05'00"N

Moscow State University (M.V. Lomonosov_)
● See M.V. Lomonosov Moscow State University

Moscow State University of Geodesy and Cartography
Gorokovsky Per. 4
103064 Moscow
Telephone: (0)095-2613152
Telefax: (0)095-2674681
Electronic Mail: amport@msugc.msk.ru
Founded: 1779
Staff: 800
Activities: education * photogrammetry * space geodesy * remote sensing
Periodicals: (6) "Geodesy and Air Photo Surveying" (in Russian) (ISSN 0536-101x)
City Reference Coordinates: 038°50'00"E 55°05'00"N

M.V. Lomonosov Moscow State University, Institute of Nuclear Physics
119899 Moscow
Telephone: (0)095-939-5097
Telefax: (0)095-939-0896
Electronic Mail: <userid>@taspd.npi.msu.su
WWW: http://www.npi.msu.su/
Founded: 1946
Staff: 1000
City Reference Coordinates: 038°50'00"E 55°05'00"N

National Committee of Russian Physicists
Leninsky Prospect 32A
117993 Moscow
Telephone: (0)095-9381695
Telefax: (0)095-9381714
Electronic Mail: keldysh@oofa.msk.su
Founded: 1991
Staff: 5
City Reference Coordinates: 038°50'00"E 55°05'00"N

Novokuzneck Planetarium
Metallurgy Avenue 18-A
654000 Novokuzneck
Telephone: (0)3843-445114
Founded: 1959
Staff: 20
Activities: lectures in astronomy, cosmonautics, geography and biology for school children and adults
City Reference Coordinates: 087°06'00"E 53°45'00"N

Orlyonok All-Russia Children's Centre Observatory
352842 Tuapse
Telephone: (0)91-549
Telefax: (0)91-567
Founded: 1975
Staff: 4
Activities: lectures * workshops * SETI * observing
Coordinates: 038°49'00"E 44°15'15"N H60m
City Reference Coordinates: 039°05'00"E 44°06'00"N

P.N. Lebedev Physics Institute
● See "Russian Academy of Sciences, P.N. Lebedev Physics Institute"

Pulkovo Observatory
● See "Russian Academy of Sciences, Pulkovo Observatory"

Pushchino Radioastronomy Observatory (PRAO)
● See "Russian Academy of Sciences, P.N. Lebedev Physics Institute, Pushchino Radioastronomy Observatory (PRAO)"

Rostov State University (RSU), Department of Space Research
Zorge Ul. 5
344090 Rostov-na-Donu
Telephone: (0)8632-220858
Telefax: (0)8632-285044
Electronic Mail: physdep@rsu.rnd.runnet.ru
WWW: http://www.unird.ac.ru/
Founded: 1970
Staff: 10
Activities: stars * stellar atmospheres * structure, evolution, dynamics and chemical composition of galaxies * ISM * nucleosynthesis * large-scale structure * background radiation
City Reference Coordinates: 039°25'00"E 57°11'00"N

Rostov State University (RSU), Institute of Physics
Zorge Ul. 5
344090 Rostov-na-Donu
Telephone: (0)8632-220857
Telefax: (0)8632-220884
Electronic Mail: <userid>@phys.rsu.ru
WWW: http://www.phys.rsu.ru/
Founded: 1967
Staff: 14
Activities: galactic structure and evolution * ISM * abundances * cosmic background radiation
City Reference Coordinates: 039°25'00"E 57°11'00"N

Russian Academy of Sciences, A.F. Ioffe Physical Technical Institute, Department of Theoretical Astrophysics
Politekhnicheskaya 26
194021 Saint Petersburg
Telephone: (0)812-2479326
Telefax: (0)812-5504890
Electronic Mail: <userid>@astro.ioffe.rssi.ru
WWW: http://www.ioffe.rssi.ru/pti_ppap.html
　　　　http://www.ras.ru/
Founded: 1963
Staff: 22
Activities: QSOs * cosmology * ISM * gamma rays * star formation regions * neutron stars * education
Periodicals: (1) "Annual Report"
City Reference Coordinates: 030°15'00"E 59°55'00"N

Russian Academy of Sciences, Astronomical Council
c/o Institute for Astronomy
Ul. Pyatnitskaya 48
109017 Moscow
Telephone: (0)095-2315461
Telefax: (0)095-2302081
Electronic Mail: <userid>@airas.msk.su

City Reference Coordinates: 038°50'00"E 55°05'00"N

Russian Academy of Sciences, Baksan Neutrino Observatory (BNO)
P/O Neutrino
Tyrnyauz
361609 Kabarkian-Balkar
Telephone: (0)866-22-77475
Telefax: (0)866-22-77475
Electronic Mail: alexeev@neutr.novoch.ru
WWW: http://www.ras.ru/
Founded: 1967
Staff: 60
Activities: anisotropy and chemical composition of primary cosmic rays * collapse * dark matter * double beta decay * gamma astronomy * solar neutrino flux registration
Coordinates: 043°32'00"E 43°40'40"N
City Reference Coordinates: 043°32'00"E 43°41'00"N

Russian Academy of Sciences, Committee on Meteorites
c/o Yuri A. Shukolyukov (Chairman)
Vernadsky Institute of Geochemistry and Analytical Chemistry
Kosygin Street 19
117975 Moscow
Telephone: (0)095-1374270
Telefax: (0)095-9382054
WWW: http://www.ras.ru/
Founded: 1939
Membership: 30
Activities: meteorites * meteors * meteoritic craters
Periodicals: "Meteoritika"
City Reference Coordinates: 038°50'00"E 55°05'00"N

Russian Academy of Sciences, Department of General Physics and Astronomy (DGPA)
Lenin Avenue 32a
117993 Moscow
Telephone: (0)095-9381695
         (0)095-9385500
Telefax: (0)095-9381714
Electronic Mail: <userid>@oofa.msk.su
               <userid>@gpad.ac.ru
WWW: http://www.ras.ru/RAS/oofa.html
      http://www.gpad.ac.ru/
Staff: 120
City Reference Coordinates: 038°50'00"E 55°05'00"N

Russian Academy of Sciences, Geophysical Center
Ul. Molodezhnaya 3
117296 Moscow
Telephone: (0)095-9300546
Telefax: (0)095-9305509
Electronic Mail: <userid>@wdeb.rssi.ru
WWW: http://www.wdeb.rssi.ru/
City Reference Coordinates: 038°50'00"E 55°05'00"N

Russian Academy of Sciences, Institute for Astronomy (INASAN)
Ul. Pyatnitskaya 48
109017 Moscow
Telephone: (0)095-9515461
Telefax: (0)095-2302081
Electronic Mail: <userid>@inasan.rssi.ru
WWW: http://www.inasan.rssi.ru/
Founded: 1936
Staff: 100
Activities: physics and evolution of stars * stellar spectroscopy * variable stars * astronomical catalogues * dynamics of stellar and planetary systems * satellite geodesy * geophysics * geodynamics * space research
Periodicals: "Peremennye Zvezdy" (ISSN 0373-7683) (circ.: about 700)
Coordinates: 036°46'26"E 55°41'40"N H173m (Zvenigorod)
City Reference Coordinates: 038°50'00"E 55°05'00"N

Russian Academy of Sciences, Institute for Astronomy, Center for Astronomical Data (CAD)
Ul. Pyatnitskaya 48
109017 Moscow
Telephone: (0)095-2315461
Telefax: (0)095-2302081
Electronic Mail: olgad@inasan.rssi.ru
WWW: http://www.inasan.rssi.ru/inasan/CAD.html

http://www.ras.ru/
Staff: 20
City Reference Coordinates: 038°50'00"E 55°05'00"N

Russian Academy of Sciences, Institute for Theoretical Astronomy (ITA)

Ul. Naberezhnaya Kutuzova 10
191187 Saint Petersburg
Telephone: (0)812-2788809
          (0)812-2788810
          (0)812-2790667
Telefax: (0)812-2797968
Electronic Mail: <userid>@iipah.spb.ru
WWW: http://www.whitenights.com/ita.htm
       http://www.ras.ru/
Founded: 1919
Staff: 60
Activities: celestial mechanics * ephemeris * minor planets * natural satellites * comet dynamics * satellite geodesy
Periodicals: "Trudy ITA" (ISSN 0568-6016); (2) "Bulletin ITA" (ISSN 0002-3302); (1) "Astronomical Yearbook", "Ephemeris of Minor Planets"
City Reference Coordinates: 030°15'00"E 59°55'00"N

Russian Academy of Sciences, Institute of Applied Astronomy

Zdanovskaya Ul. 8
197110 Saint Petersburg
Telephone: (0)812-2307414
Telefax: (0)812-2307413
Electronic Mail: iparan@ipa.rssi.ru
WWW: http://www.ipa.rssi.ru/
Founded: 1988
Staff: 232
Activities: radioastronomy * VLBI * astrometry * geodesy * geophysics
Periodicals: "Communications of the Institute of Applied Astronomy"
Coordinates: 029°46'54"E 60°32'00"N H80m (Svetloe - see separate entry)
City Reference Coordinates: 030°15'00"E 59°55'00"N

Russian Academy of Sciences, Institute of Applied Astronomy, Irkutsk Department

Lermontova Ul. 297-B-45
664033 Irkutsk
Telephone: (0)3952-460822
WWW: http://www.ipa.rssi.ru/
Founded: 1988
Staff: 8
Activities: radioastronomy * VLBI * astrometry * geodesy * geophysics
City Reference Coordinates: 104°20'00"E 52°16'00"N

Russian Academy of Sciences, Institute of Applied Astronomy, Svetloe Observatory

188833 Svetloe
Telephone: (0)812-3123628
Electronic Mail: tarasov@light.ipa.rssi.ru
WWW: http://www.ipa.rssi.ru/
Founded: 1997
Staff: 29
Activities: radioastronomy * VLBI * astrometry * geodesy * geophysics
Coordinates: 029°46'54"E 60°32'00"N H80m
City Reference Coordinates: 030°15'00"E 59°55'00"N

Russian Academy of Sciences, Institute of Applied Astronomy, Zelenchuk Department

Kalinin Ul. 1a
357140 Zelenchukskaya
Telephone: (0)87878-43402
Electronic Mail: aps@ipa.rssi.ru
WWW: http://www.ipa.rssi.ru/
Founded: 1988
Staff: 28
City Reference Coordinates: 041°36'00"E 43°53'00"N

Russian Academy of Sciences, Institute of Applied Physics, Millimeter Astronomy Group

Uljanov Ul. 46
603600 Nizhny Novgorod
Telephone: (0)8312-367253
Telefax: (0)8312-362061
Electronic Mail: zin@appl.sci-nnov.ru
WWW: http://zin.appl.sci-nnov.ru/mm-astro/
Founded: 1977 (Institute)
Staff: 10

Activities: ISM * star formation * dense cores * mm-wave radio astronomy
City Reference Coordinates: 044°00'00"E 56°20'00"N

Russian Academy of Sciences, Institute of Cosmophysical Research and Aeronomy (IKFIA)
Lenin Avenue 31
677891 Yakutsk
Telephone: (0)411-2222551
Telefax: (0)95-2302919 (for IKFIA)
Electronic Mail: centr@ikfia.yacc.yakutia.su
WWW: http://mx.iki.rssi.ru/IKFIA.html
        http://www.ras.ru/
Founded: 1962
Staff: 300
Activities: cosmic rays * solar wind * ionosphere * magnetosphere * magnetic-field radiations
Coordinates: 129°40'00"E 62°02'00"N (Yakutsk)
             128°45'00"E 71°40'00"N (Tixie Bay)
City Reference Coordinates: 129°40'00"E 62°02'00"N

Russian Academy of Sciences, Institute of Spectroscopy
142092 Troitsk
Telephone: (0)095-3340579
Telefax: (0)095-3340886
Electronic Mail: <userid>@isan.msk.su
WWW: http://www.isan.troitsk.ru/
        http://www.ras.ru/
Founded: 1968
Staff: 270
Activities: optical spectroscopy
City Reference Coordinates: 037°19'00"E 55°28'00"N

Russian Academy of Sciences, Institute of Terrestrial Magnetism, Ionosphere and Radio Wave Propagation (IZMIRAN)
142092 Troitsk
Telephone: (0)095-3340120
Telefax: (0)095-3340124
Electronic Mail: kvd@sci.izmiran.troitsk.su
                 kvd@charley.izmiran.rssi.ru
                 <userid>@izmiran.rssi.ru
WWW: http://www.izmiran.rssi.ru/
        http://mx.iki.rssi.ru/IZMIRAN.html
        http://www.ras.ru/
Founded: 1940
Staff: 730
Activities: solar physics * STP * radiophysics * ionosphere * geophysics * geomagnetism * space
Coordinates: 037°19'00"E 55°28'00"N H190m
City Reference Coordinates: 037°19'00"E 55°28'00"N

Russian Academy of Sciences, Institute of Terrestrial Magnetism, Ionosphere and Radio Wave Propagation (IZMIRAN), Arkhangelsk Complex Magnetic Ionospheric Observatory (ACMIO)
Primorskii
164420 Arkhangelsk
Telephone: (0)430593
Electronic Mail: <userid>@izmiran.rssi.ru
WWW: http://www.izmiran.rssi.ru/
        http://mx.iki.rssi.ru/IZMIRAN.html
        http://www.ras.ru/
Founded: 1967
Staff: 27
Activities: geomagnetic, ionospheric, and auroral data * cosmic-noise absorption
Coordinates: 040°19'30"E 64°34'20"N H2m
             044°16'00"E 65°50'00"N (Mezen)
             044°24'00"E 64°00'00"N (Karpogory)
City Reference Coordinates: 040°32'00"E 64°34'00"N

Russian Academy of Sciences, Institute of Terrestrial Magnetism, Ionosphere and Radio Wave Propagation (IZMIRAN), Kaliningrad Magnetic Ionospheric Observatory (KMIO)
Pr. Pobedy 41
236017 Kaliningrad
Telephone: (0)11-215606
Electronic Mail: <userid>@magion.izmiran.koenig.su
                 <userid>@izmiran.rssi.ru
WWW: http://www.izmiran.rssi.ru/
        http://mx.iki.rssi.ru/IZMIRAN.html
        http://www.ras.ru/
Founded: 1965
Staff: 40

Activities: magnetic and ionospheric data * radiophysical research * theoretical atmospheric modelling
Coordinates: 020°37'00"E 54°42'00"N H15m
City Reference Coordinates: 020°37'00"E 54°42'00"N

Russian Academy of Sciences, Institute of Terrestrial Magnetism, Ionosphere and Radio Wave Propagation (IZMIRAN), Saint Petersburg Branch
Muchnoi per. 2
P.O. Box 188
191023 Saint Petersburg
Telephone: (0)812-3105232
Telefax: (0)821-3105035
Electronic Mail: <userid>@izmiran.rssi.ru
WWW: http://www.izmiran.rssi.ru/
    http://mx.iki.rssi.ru/IZMIRAN.html
    http://www.ras.ru/
Founded: 1938
Staff: 126
Activities: geomagnetism * ionosphere * solar-terrestrial physics * cosmic rays
Coordinates: 030°42'00"E 59°57'00"N (Voeikovo)
City Reference Coordinates: 030°15'00"E 59°55'00"N

Russian Academy of Sciences, Institute of Theoretical and Experimental Physics
Ul. B. Cheremushkinskaya 25
117259 Moscow
Telephone: (0)095-1230292
        (0)095-1236584
WWW: http://www.itep.ru/
Activities: SN * stellar structure * low-energy neutrinos
City Reference Coordinates: 038°50'00"E 55°05'00"N

Russian Academy of Sciences, Keldysh Institute of Applied Mathematics
Miusskaja 4
125047 Moscow
Telephone: (0)095-2581314
WWW: http://www.keldysh.ru/
Staff: 30
Activities: planets * celestial mechanics * cosmology * relativistic astrophysics
City Reference Coordinates: 038°50'00"E 55°05'00"N

Russian Academy of Sciences, National Geophysical Committee
Molodezhnaya 3
117296 Moscow
Telephone: (0)095-9305629
Telefax: (0)095-9305509
Electronic Mail: <userid>@wdeb.rssi.ru
WWW: http://www.wdeb.rssi.ru/NGC/
Founded: 1955
Membership: 100
Periodicals: (20) "Results of Researches on the International Geophysical Projects"
City Reference Coordinates: 038°50'00"E 55°05'00"N

Russian Academy of Sciences, Nuclear Physics Institute
Gatchina
188350 Saint Petersburg
Telephone: (0)812-2949132
        (0)812-2949146
        (0)812-2983538
Telefax: (0)812-2980257
        (0)812-7137196
Electronic Mail: ¡usierd¿@pnpi.ru
WWW: http://www.pnpi.spb.ru/
    http://www.ras.ru/
City Reference Coordinates: 030°15'00"E 59°55'00"N

Russian Academy of Sciences, P.N. Lebedev Physics Institute
Lenin Avenue 53
117924 Moscow
Telephone: (0)095-1357980
        (0)095-1357980
        (0)095-1326573
Telefax: (0)095-1357880
        (0)095-9382251
Electronic Mail: <userid>@sci.fian.msk.su
WWW: http://www.lpi.msk.su/
    http://www.lpi.msk.su/LPI/NPAD/NPAD.html (Nuclear Physics and Astrophysics Division)

http://www.ras.ru/
City Reference Coordinates: 038°50'00"E 55°05'00"N

Russian Academy of Sciences, P.N. Lebedev Physics Institute, Astro Space Center (ASC)
Ul. Profsojuznaya 84/32
117810 Moscow
Telephone: (0)095-3332378
Telefax: (0)095-3332378
Electronic Mail: <userid>@asc.rssi.ru
WWW: http://www.asc.rssi.ru/
Activities: space physics * interferometry * solar atmosphere * UV * cosmic rays * upper atmosphere
City Reference Coordinates: 038°50'00"E 55°05'00"N

Russian Academy of Sciences, P.N. Lebedev Physics Institute, Pushchino Radioastronomy Observatory (PRAO)
142292 Pushchino
Telephone: (0)967-732780
Telefax: (0)967-732482
Electronic Mail: <userid>@prao.psn.ru
WWW: http://www.prao.psn.ru/
       http://psun32.prao.psn.ru/ (Pulsar Astrometry)
City Reference Coordinates: 037°52'00"E 56°01'00"N

Russian Academy of Sciences, P.N. Lebedev Physics Institute, Nuclear Physics and Astrophysics
Lenin Avenue 53
117924 Moscow
Telephone: (0)095-1354264
Telefax: (0)095-1357880
       (0)095-9382251
Electronic Mail: <userid>@lpi.msk.su
WWW: http://www.lpi.msk.su/
Founded: 1956
Staff: 33
City Reference Coordinates: 038°50'00"E 55°05'00"N

Russian Academy of Sciences, Pulkovo Observatory
65 Pulkovo
196140 Saint Petersburg
Telephone: (0)812-2982242
Electronic Mail: <userid>@gao.spb.su
WWW: http://www.gao.spb.ru/
Founded: 1839
Periodicals: "Izvestiya" (ISSN 0367-8466), "Solar Data Bulletin"
Coordinates: 030°19'36"E 59°46'24"N H75m
City Reference Coordinates: 030°15'00"E 59°55'00"N

Russian Academy of Sciences, Pulkovo Observatory, Kislovodsk Astronomical Station
357700 Kislovodsk
Telephone: (0)86537-33088
       (0)812-1234096
Telefax: (0)812-1231922
       (0)812-3143360
Electronic Mail: <userid>@gaosun.spb.su
WWW: http://www.gao.spb.ru/english/staff/dsp/mas.html
Founded: 1948
Staff: 34
Activities: solar physics * astrometry
Periodicals: "Solnechnye Dannye"
Coordinates: 042°40'12"E 43°44'47"N H2,130m
City Reference Coordinates: 042°44'00"E 42°55'00"N

Russian Academy of Sciences, Siberian Division, Institute of Solar-Terrestrial Physics
Ul. Lermontov 126
P.O. Box 4026
664033 Irkutsk
Telephone: (0)3952-462365
       (0)3952-460265
Telefax: (0)3952-462557
Electronic Mail: root@sitmis.irkutsk.su
       <userid>@sitmis.irkutsk.su
WWW: http://www.iszf.irk.ru/
       http://www.geocities.com/CapeCanaveral/Lab/6261/ssrt.html (SSRT)
       http://www.ras.ru/
Founded: 1960
Staff: 635

Activities: Sun * STP * instrumentation * Siberian Solar Radio Telescope (SSRT)
Periodicals: (4) "Issledovania po Geomagnetismu, Aeronomii i Fisike Solntsa"
Coordinates: 100°55'30"E 51°37'19"N H2,012m (Sayan Solar Observatory)
                 104°55'00"E 51°50'45"N H600m (Baikal Astrophysical Observatory)
                 102°12'30"E 51°45'27"N H832m (Sayan Mountains Radiophysical Observatory)
City Reference Coordinates: 104°20'00"E 52°16'00"N

Russian Academy of Sciences, Space Research Institute (IKI)

Ul. Profsojuznaya 84/32
117810 Moscow
Telephone: (0)095-3333122
           (0)095-3335212
Telefax: (0)095-3107023
        (0)095-3335178
        (0)095-9133040
Electronic Mail: <userid>@sovamsu.sovusa.com
WWW: http://www.iki.rssi.ru/
        http://www.ras.ru/
Founded: 1967
Staff: 1500
Activities: space physics * space plasma * astrophysics * solar system
Awards: "Zel'dovich Award"
City Reference Coordinates: 038°50'00"E 55°05'00"N

Russian Academy of Sciences, Special Astrophysical Observatory (SAO)

Nizhnij Arkhyz
357147 Karachai
Telephone: (0)901-4982931
Telefax: (0)901-4982931
Electronic Mail: <userid>@sao.ru
WWW: http://www.sao.ru/
Founded: 1966
Staff: 120
Activities: observing (optical and radio) * cosmology * extragalactic astronomy * instrument building
Periodicals: "Bulletin of the Special Astrophysical Observatory" (ISSN 0320-9318)
Coordinates: 041°26'30"E 43°39'12"N H2,100m (6m altazimutal reflector)
                 041°35'25"E 43°49'32"N H973m (600m radiotelescope)
City Reference Coordinates: 041°57'00"E 43°44'00"N

Russian Academy of Sciences, Ussurijsk Astrophysical Observatory (UAO)

P/O Gornotayozhnoye
692533 Ussurijsk
Telephone: 42341-91121
Telefax: 42341-91121
Electronic Mail: baranov@ml.ussurijsk.ru
WWW: http://www.ras.ru/
Founded: 1954
Staff: 34
Activities: Sun * solar activity * comets
Periodicals: (1) "Solar Activity and its Influence on the Earth" (in Russian)
Coordinates: 132°10'00"E 43°40'00"N H250m
City Reference Coordinates: 131°99'00"E 43°88'00"N

Russian Space Agency

Shukin Street
Building 42
129857 Moscow
Electronic Mail: <userid>@rka.ru
WWW: http://www.rka.ru/
        http://www.rka.ru/english/eindex.htm
City Reference Coordinates: 038°50'00"E 55°05'00"N

Saint Petersburg University, Astronomical Institute

Bibliotechnaja Pl. 2
Petrodvorets
198904 Saint Petersburg
Telephone: (0)812-4287129
           (0)812-4284265
Telefax: (0)812-4284259
        (0)812-4287129
Electronic Mail: ai@astro.spbu.ru
                <userid>@astro.spbu.ru
WWW: http://www.astro.spbu.ru/
Founded: 1881
Staff: 90

Activities: stellar structure * radiation transfer * active galaxies * cosmic-gas dynamics * celestial dynamics * stellar dynamics * astrometry * Earth rotation parameters * fundamental stellar catalogues * solar radio astronomy
Periodicals: (1/2) "Transactions of the Saint Petersburg Astronomical Institute" (ISSN 0136-8109 & 0136-8141)
Coordinates: 030°17'45"E 59°56'32"N H0m (Saint Petersburg)
                        044°17'30"E 40°20'07"N H1,500m (Byurakan, Armenia)
City Reference Coordinates: 030°15'00"E 59°55'00"N

Saint Petersburg University, Department of Applied Mathematics and Control Processes
Fakultet PM-PU
Bibliotechnaya Pl. 2
198904 Saint Petersburg
Telephone: (0)812-4287159
Telefax: (0)812-4287159
Electronic Mail: lapetr@robot.apmath.spb.su
Founded: 1969
Staff: 24
Activities: control in problems of celestial mechanics * Moon motion * N-body problem * stellar dynamics * star clusters * data processing
Periodicals: (4) "Vestnik Sankt-Peterburskogo Universiteta - Series of Mathematics, Mechanics, and Astronomy" (ISSN 0132-4624)
City Reference Coordinates: 030°15'00"E 59°55'00"N

Scientific and Research Centre on Space Hydrometeorology "Planeta"
7 Bolshoy Predtechensky Per.
123242 Moscow
Telephone: (0)95-2551263
                        (0)95-2523717
Telefax: (0)95-2004210
Electronic Mail: asmus@ns.planeta.rssi.ru (Vassily Asmus, Director)
WWW: http://sputnik.infospace.ru/
Founded: 1974
Staff: 300
Activities: research * remote sensing * environment * hydrometeorology * satellite operations * receiving, processing, archiving and disseminating data
City Reference Coordinates: 038°50'00"E 55°05'00"N

Special Astrophysical Observatory (SAO)
● See "Russian Academy of Sciences, Special Astrophysical Observatory (SAO)"

Sternberg State Astronomical Institute (SAI)
University Avenue 13
119899 Moscow
Telephone: (0)095-9392858
Telefax: (0)095-9398841
Electronic Mail: <userid>@sai.msu.su
WWW: http://www.sai.msu.su/
                http://comet.sai.msu.su/ (Radio Astronomy Department)
                http://xray.sai.msu.su/ (X-Ray Group)
                http://crydee.sai.msu.ru/ (Heliophysics)
Founded: 1931
Staff: 400
Activities: Earth * stars * Sun * galaxies * X-ray sources * planets * ISM * radio sources
Periodicals: "Trudy" (ISSN 0371-6791)
Coordinates: 037°32'40"E 55°42'00"N H195m (Moscow)
                        034°01'30"E 44°33'42"N (Crimea, Ukraine)
City Reference Coordinates: 038°50'00"E 55°05'00"N

Tomsk Planetarium
Batenkova Street 3
634069 Tomsk
Telephone: (0)3822-909721
                        (0)3822-232780
Electronic Mail: niipmm@urania.tomsk.su
Founded: 1948
Staff: 12
Activities: lectures * observing * meetings * expeditions
City Reference Coordinates: 085°05'00"E 56°30'00"N

Tomsk State University, Astronomical Observatory
P.O. Box 1106
634010 Tomsk
Telephone: (0)3822-410576
Electronic Mail: niipmm@urania.tomsk.su
WWW: http://www.tsu.tomsk.su/
Founded: 1922

Staff: 15
Activities: small bodies (satellites, minor planets, comets, meteorites, meteoroids) * celestial mechanics * astrometry * geodesy
Periodicals: (1) "Astronomy and Geodesy" (ISSN 0201-5099)
Coordinates: 084°56'47"E 56°28'07"N H130m
City Reference Coordinates: 085°05'00"E 56°30'00"N

Udmurt State University, Department of Astronomy and Mechanics

Universitetskaya Street 1
426034 Izhevsk
Telephone: (0)3412-761794
Electronic Mail: <userid>@uni.udm.ru
WWW: http://www.uni.udm.ru/
Founded: 1991
Staff: 7
Activities: stellat dynamics * celestial mechanics * cataclysmic variables * potential theory * planets
Periodicals: "Vestnik Udmurtskogo Universiteta"
Coordinates: 053°10'48"E 56°51'12"N
City Reference Coordinates: 053°11'00"E 56°49'00"N

Ufa Planetarium

Pl. Lenina 3
450075 Ufa
Telephone: (0)3472-357023
Telefax: (0)3472-528653 (c/o Planetarium)
Electronic Mail: postmaster@uctlgf.bashkiria.su (c/o Planetarium)
Founded: 1964
Staff: 15
Activities: education * shows
City Reference Coordinates: 055°56'00"E 54°44'00"N

Ural State University, Department of Astronomy and Geodesy

Pr. Lenina 51
620083 Ekaterinburg
Telephone: (0)223386
              (0)220729
WWW: http://www.usu.ru/eng/usu/faculty/astron.htm
        http://www.usu.ru/eng/usu/subdivisions/observ.htm (Kourov)
Founded: 1965
Staff: 50
Activities: stellar astronomy * solar physics * satellites
Coordinates: 059°32'00"E 57°02'00"N H500m
City Reference Coordinates: 059°32'00"E 57°02'00"N

Ussurijsk Astrophysical Observatory (UAO)

● See "Russian Academy of Sciences, Ussurijsk Astrophysical Observatory (UAO)"

VNIIFTRI

● See "Institute for Physical Technical and Radio-Technical Measurements (VNIIFTRI)"

World Data Center B for Solar-Terrestrial Physics

c/o Geophysical Institute
Russian Academy of Sciences
Ul. Molodezhnaya 3
117296 Moscow
Telephone: (0)095-9305619
Telefax: (0)095-9305509
Electronic Mail: <userid>@wdcb.ru
WWW: http://www.wdcb.rssi.ru/WDCB/wdcb_stp.shtml
Founded: 1957
Staff: 83
Activities: collecting, storing and exchanging solar activity data * distributing data upon request
Periodicals: (10) "Materials of the World Data Center B"
City Reference Coordinates: 038°50'00"E 55°05'00"N

Zabaikalsky State Pedagogical University

Babushkin Ul. 129
672045 Chita
Telephone: (0)302-267317
              (0)302-234874
Telefax: (0)302-267935
Founded: 1938
Staff: 443 (one astronomer)
Activities: teacher training * physics * astronomy * mathematics * computing * chemistry
Coordinates: 113°29'00"E 52°03'00"N
City Reference Coordinates: 113°29'00"E 52°03'00"N

Saudi Arabia

King Abdulaziz City for Science and Technology (KACST)
P.O. Box 6086
Riyadh 11442
Telephone: (0)1-4813547
Telefax: (0)1-4813523
Electronic Mail: <userid>@kacst.edu.sa
WWW: http://www.kacst.edu.sa/
Founded: 1977
Staff: 25
Activities: formulating national science and technology policies * setting priorities for funding and coordinating research activities * establishing research institutes relevant to national development
Periodicals: "Science and Technology" (in Arabic)
Coordinates: 039°37'48"E 21°22'12"N H333m (Makka)
036°04'12"E 29°10'12"N H330m (H. Ammar)
036°22'48"E 26°27'00"N H100m (Wajh)
040°54'36"E 27°15'00"N H1,000m (Hail)
046°24'00"E 23°33'36"N H853m (Hariq)
042°36'00"E 19°13'48"N H2,377m (Namass)
City Reference Coordinates: 046°43'00"E 24°38'00"N

King Abdulaziz University, Astronomy and Geophysics Research Institute (AGRI)
P.O. Box 9028
Jeddah 21413
Telephone: (0)2-6952284
WWW: http://www.kacst.edu.sa/
Founded: 1979
Activities: PN * radioastronomy * artificial satellites * sunspots * celestial mechanics * lunar occultations * stellar photometry * solar spectroscopy
Coordinates: 039°49'00"E 21°28'00"N
City Reference Coordinates: 039°49'00"E 21°28'00"N

King Saud University (KSU), College of Science, Astronomy Department
P.O. Box 2455
Riyadh 11451
Telephone: (0)1-4676324
Telefax: (0)1-4674253
WWW: http://www.mohe.gov.sa/ksu/
Founded: 1986
Staff: 5
Activities: education * research * astrophysics * external galaxies * history of astronomy
Coordinates: 046°43'02"E 24°37'58"N
City Reference Coordinates: 046°43'00"E 24°38'00"N

Meteorology and Environmental Protection Administration (MEPA)
P.O. Box 1358
Jeddah 21431
Telephone: (0)2-6512312
(0)2-6710448
(0)2-6710439
(0)2-6713238
Telefax: (0)2-6511424
Electronic Mail: sidc@mepa.org.sa
WWW: http://www.mepa.org.sa/
City Reference Coordinates: 039°49'00"E 21°28'00"N

Saudi Arabian Standards Organization (SASO)
P.O. Box 3437
Riyadh 11471
Telephone: (0)1-4520000
(0)1-4520132
Telefax: (0)1-4520086
(0)1-4520133
Electronic Mail: sasoinfo@saso.org
WWW: http://www.sao.org/
Founded: 1972
Staff: 351
Activities: developing national standards * promoting and distributing them * conformity certification * quality marking * metrology * testing * technical information
City Reference Coordinates: 046°43'00"E 24°38'00"N

Singapore

Astronomical Society of Singapore (TASOS) (The_)
● See "The Astronomical Society of Singapore (TASOS)"

Astro Scientific Centre Pte. Ltd. (ASCPL)
c/o Singapore Science Centre
Level 2
Science Centre Road
Singapore 609081
Telephone: (0)567-4163
Telefax: (0)567-4826
Electronic Mail: astro@astro.com.sg
                    sales@astro.com.sg
WWW: http://astro.com.sg/
Founded: 1986
Staff: 7
Activities: distributing and servicing telescopes, binoculars, ...
City Reference Coordinates: 103°48'00"E 01°22'00"N

Meteorological Service Singapore (MSS)
P.O. Box 8
Changi Airport
Singapore 918141
Telephone: (0)545-7193
          (0)545-7190
          (0)545-7191
Telefax: (0)545-7192
Electronic Mail: mss_operations@mss.gov.sg
WWW: http://www.gov.sg/metsin/
Staff: 125
City Reference Coordinates: 103°48'00"E 01°22'00"N

National University of Singapore (NUS), Department of Mathematics
5 Lower Kent Ridge Road
Singapore 119260
Telephone: (0)874-2738
Telefax: (0)779-5452
Founded: 1929
WWW: http://www.math.nus.edu.sg/
Staff: 2
Activities: education * research * radiative transfer in atmospheres
City Reference Coordinates: 103°48'00"E 01°22'00"N

Singapore Institute of Standards and Industrial Research (SISIR)
Kent Ridge
1 Science Park Drive
Singapore 118221
Telephone: (0)7787777
          (0)2786666
Telefax: (0)7780086
          (0)7781280
WWW: http://www.psb.gov.sg/
Founded: 1973
Staff: 500
Activities: materials technology advising * electronic and computer applications * consultancy * product and process development and design * standards and certification authority
Periodicals: (4) "SISIR News" (ISSN 0129-0908); (1) "Annual Report"
City Reference Coordinates: 103°48'00"E 01°22'00"N

Singapore Productivity and Standards Board
PSB Building
2 Bukit Merah Central
Singapore 159835
Telephone: (0)2786666
Telefax: (0)2786665
          (0)2786667
Electronic Mail: queries@psb.gov.sg
WWW: http://www.psb.gov.sg/
City Reference Coordinates: 103°48'00"E 01°22'00"N

Singapore Science Centre, Omnitheatre and Observatory
Science Centre Road
Off Jurong Town Hall Road
Singapore 609081
Telephone: (0)560-3316
                (0)425-2500
Telefax: (0)565-9533
Electronic Mail: <userid>@sci-ctr.edu.sg
WWW: http://www.sci-ctr.edu.sg/
Founded: 1977
Activities: omnimax movies * planetarium programmes * public lectures * observing sessions
Periodicals: (12) "Skytrack"
Coordinates: 103°44'15"E 01°20'03"N
City Reference Coordinates: 103°48'00"E 01°22'00"N

The Astronomical Society of Singapore (TASOS)
c/o Astro Scientific Centre Pte. Ltd.
Level 2
Science Centre Road
Singapore 609081
Telephone: (0)567-4163
Telefax: (0)567-4826
Electronic Mail: tasos@astro.com.sg
WWW: http://astro.com.sg/tasos.html
Founded: 1992
Membership: 200
Activities: observing * photography * lectures
Periodicals: (4) "Moonstarer"
City Reference Coordinates: 103°48'00"E 01°22'00"N

World Scientific Publishing Co. (WSPC) Pte. Ltd.
Farrer Road
P.O. Box 128
Singapore 9128
or :
1022 Tai Seng Avenue
Tai Seng Industrial Estate
Singapore 534415
Telephone: (0)382-5663
Telefax: (0)382-5919
Electronic Mail: wspc@wspc.com.sg
WWW: http://www.wspc.com.sg/
Founded: 1981
Staff: 120
• Publisher
City Reference Coordinates: 103°48'00"E 01°22'00"N

Slovak Republic

Banská Bystrica Astronomical Observatory
P.O. Box 86
975 90 Banská Bystrica
Electronic Mail: hvezdar@isternet.sk
Telephone: (0)88-4115015
Founded: 1961
Staff: 7
Activities: popularization * observing * meteors * solar photosphere * occultations
Coordinates: 019°09'19"E 48°43'06"N H568m (Vartovka)
City Reference Coordinates: 019°10'00"E 48°44'00"N

Comenius University, Institute of Astronomy
Mlynská Dolina
842 15 Bratislava
Telephone: (0)7-65424000
Telefax: (0)7-65425882
Electronic Mail: <userid>@fmph.uniba.sk
Founded: 1998
Staff: 16
Activities: Sun * interplanetary matter * solar system * minor planets * comets * meteors
Coordinates: 017°07'12"E 48°09'18"N H530m (Modra - see separate entry)
City Reference Coordinates: 017°10'00"E 48°10'00"N

Comenius University, Institute of Astronomy, Astronomical and Gecphysical Observatory
P.O. Box 4
900 01 Modra
Telephone: (0)704-475261
Electronic Mail: ago@fmph.uniba.sk
WWW: http://center.fmph.uniba.sk/~ago
Founded: 1993
Staff: 7
Activities: interplanetary matter * minor planets * comets * meteors * geomagnetic measurments * seismic monitoring * Schumann resonance
Coordinates: 017°16'34"E 48°22'23"N H530m
City Reference Coordinates: 017°17'00"E 48°21'00"N

Kozmos
Konventna 19
811 03 Bratislava
Telephone: (0)7-54414133
Telefax: (0)7-54414133
Electronic Mail: kozmos@netlab.sk
WWW: http://www.ta3.sk/kozmos/kozmos.html
Founded: 1970
Staff: 2
● Periodical
City Reference Coordinates: 017°10'00"E 48°10'00"N

Michalovce Observatory
Hrádok 1
071 01 Michalovce
Telephone: (0)946-6443259
         (0)946-6443260
Telefax: (0)946-6443260
Electronic Mail: hvezmice@in4.sk
WWW: http://www.astro.sk/hvezdarne/michalovce/
        http://www.home.sk/www/astromice/
Founded: 1981
Staff: 4
Activities: solar photosphere * lunar occultations * meteors * symbiotic stars
Coordinates: 021°53'34"E 48°45'59"N H115m
City Reference Coordinates: 021°55'00"E 48°45'00"N

Nitra Regional Cultural Centre, Astronomical Cabinet
Fatranská 3
949 01 Nitra
Telephone: (0)87-531544
Founded: 1973
Staff: 1
Activities: lectures * observing * education
Coordinates: 018°05'00"E 48°19'00"N H190m

City Reference Coordinates: 018°05'00"E 48°19'00"N

Partizánske Observatory
P.O. Box 59
958 01 Partizánske
or :
Dolinky
Malé Bielice
Telephone: (0)815-497108
Telefax: (0)815-497108
Electronic Mail: 0petrik@center.dnet.fmph.uniba.sk
                 astropet@auriga.ta3.sk
WWW: http://www.coseco.sk/hvezdaren/
Founded: 1988
Staff: 4
Activities: popularization * Sun * photography * symbiotic and cataclysmic variable stars * minor planets * occultations
Coordinates: 018°20'25"E 48°37'42"N H325m
City Reference Coordinates: 018°23'00"E 48°39'00"N (Partizánske)

Prešov Observatory and Planetarium
Dilongova 17
080 01 Prešov
Telephone: (0)91-722065
Telefax: (0)91-722065
Electronic Mail: hvezdapl@vadium.sk
WWW: http://www.hap.vadium.sk/
Founded: 1948
Staff: 19
Activities: observing (solar photosphere) * lectures * planetarium shows
Periodicals: (1) "Bulletin of Solar Photospheric Observations in Slovakia"
Coordinates: 021°15'26"E 48°58'57"N H304m
City Reference Coordinates: 021°10'00"E 49°00'00"N

Rimavská Sobota Observatory
Tomasovska 63
979 01 Rimavská Sobota
Telephone: (0)866-5624709
Telefax: (0)866-5624709
Electronic Mail: astrors@bb.telecom.sk
Founded: 1975
Staff: 5
Activities: Sun * occultations * meteors * variable stars * comets
Coordinates: 020°00'25"E 48°22'29"N H228m
City Reference Coordinates: 020°00'00"E 48°22'00"N

Roztoky Observatory
090 11 Roztoky
Telephone: (0)937-7592320
Telefax: (0)937-7592320
WWW: http://www.ta3.sk/hvezdaren/roztoky/
Founded: 1980
Staff: 5
Activities: education * public observing * variable stars
Coordinates: 021°29'00"E 49°24'00"N
City Reference Coordinates: 021°29'00"E 49°24'00"N

Slovak Academy of Sciences, Astronomical Institute, Department of Interplanetary Matter
842 28 Bratislava
Telephone: (0)7-375157
Founded: 1953
Electronic Mail: <userid>@ta3.sk
WWW: http://www.ta3.sk/
Activities: comets * minor planets * meteor streams
Coordinates: 020°14'42"E 49°11'20"N H1,783m (Skalnaté Pleso Observatory)
City Reference Coordinates: 017°10'00"E 48°10'00"N

Slovak Academy of Sciences, Astronomical Institute, Department of Solar Physics
059 60 Tatranská Lomnica
Telephone: (0)969-467866
         (0)92-730709 (Lomnický Štít Coronal Observatory)
Telefax: (0)969-467656
Electronic Mail: <userid>@ta3.sk
WWW: http://www.ta3.sk/
Founded: 1943
Staff: 13

Activities: solar physics and activity * STP * solar corona * eclipses * space-born coronagraph
Periodicals: "Contributions"
Coordinates: 020°17'27"E 49°09'19"N H805m (Stará Lesná Observatory)
           020°13'12"E 49°11'48"N H2,632m (Lomnický Štít Coronal Observatory)
City Reference Coordinates: 020°17'00"E 49°09'00"N

Slovak Academy of Sciences, Astronomical Institute, Stellar Department

059 60 Tatranská Lomnica
Telephone: (0)969-467866
           (0)92-730762 (Skalnaté Pleso Observatory)
Telefax: (0)969-467656
Electronic Mail: <userid>@ta3.sk
WWW: http://www.ta3.sk/
Founded: 1953
Staff: 11
Activities: stellar photometry * stellar spectroscopy * interacting binaries * peculiar stars * eclipsing variable stars * symbiotic stars
Periodicals: (1) "Contributions of the Astronomical Observatory Skalnaté Pleso"
Coordinates: 020°14'42"E 49°11'20"N H1,783m (Skalnaté Pleso Observatory)
           020°17'27"E 49°09'19"N H805m (Stará Lesná Observatory)
City Reference Coordinates: 020°17'00"E 49°09'00"N

Slovak Academy of Sciences, Astronomical Institute, Tatranská Lomnica

059 60 Tatranská Lomnica
Telephone: (0)969-467866
           (0)92-730709 (Lomnický Štít Coronal Observatory)
           (0)92-730762 (Skalnaté Pleso Observatory)
Telefax: (0)969-467656
Electronic Mail: astrinst@ta3.sk
WWW: http://www.ta3.sk/
Founded: 1953
Staff: 18
Activities: solar physics * solar activity * STP * comets * minor planets * meteors * stellar photometry * interacting binaries * peculiar stars
Periodicals: (1) "Contributions of the Astronomical Observatory Skalnaté Pleso"
Coordinates: 020°14'42"E 49°11'20"N H1,783m (Skalnaté Pleso Observatory)
           020°17'27"E 49°09'19"N H805m (Stará Lesná Observatory)
           020°13'12"E 49°11'48"N H2,632m (Lomnický Štít Coronal Observatory)
City Reference Coordinates: 020°17'00"E 49°09'00"N

Slovak Academy of Sciences, Geophysical Institute

Dúbravská cesta 9
842 28 Bratislava
Telephone: (0)7-59410601
Telefax: (0)7-59410626
Electronic Mail: geoflabi@savba.sk
WWW: http://gpi.savba.sk/
Founded: 1953
Staff: 73
Activities: geophysics (seismology, geomagnetism, geothermics, gravity field) * STP * meteorology
Periodicals: "Contributions of the Geophysical Institute of the Slovak Academy of Sciences", "Bulletin of the Slovak Seismological Stations", "Results of Geomagnetic Observations at the Hurbanovo Geomagnetic Observatory"
Observatories: 5 for geophysics (Hurbanovo, Sroborova, Vyhne, Bratislava & Skalnaté Pleso) and 3 for meteorology (Mlynany, Stará Lesná & Skalnaté Pleso)
Coordinates: 020°14'00"E 49°11'00"N H1,778m (Skalnaté Pleso)
           020°17'00"E 49°09'00"N H810m (Stará Lesná)
           018°20'00"E 48°19'00"N H195m (Mlynany)
City Reference Coordinates: 017°10'00"E 48°10'00"N

Slovak Astronomical Society (SAS)

059 60 Tatranská Lomnica
Telephone: (0)969-467866
Telefax: (0)967-467656
Electronic Mail: sas@ta3.sk
WWW: http://www.ta3.sk/sas/sas.html
Founded: 1959
Membership: 253
Activities: observing (Sun, comets, meteors, variable stars) * education * lectures * amateur-professional collaborations
City Reference Coordinates: 020°15'00"E 49°11'00"N

Slovak Central Observatory (SCO)

Komárnanská 134
947 01 Hurbanovo
Telephone: (0)818-7602484
Telefax: (0)818-7602487

Electronic Mail: suh@kemar.sk
WWW: http://www.geomag.sk/
Founded: 1871
Membership: 29
Activities: observing (Sun, variable stars, meteors, occultations) * education * planetarium shows * lectures * public observing

Periodicals: (6) "Kozmos" (see separate entry); (1) "Astronomical Yearbook", "Astronomical Calendar"
Coordinates: 018°11'22"E 47°52'27"N H112m
• Formerly "Slovak Centre of Amateur Astronomy"
City Reference Coordinates: 018°10'00"E 47°53'00"N

Slovak Hydrometeorological Institute
Jeséniova 17
P.O. Box 37
833 15 Bratislava
Telephone: (0)7-371247
(0)7-3785226
Telefax: (0)7-374593
(0)7-375670
(0)7-376197
Electronic Mail: <userid>@shmuvax.shmu.sk
WWW: http://www.shmu.sk/
Founded: 1969
Staff: 700
Activities: hydrology * meteorology * climatology * environment * water quality and quantity * air pollution
Periodicals: "Bulletin of Air Pollution", "Collection of Papers", "Hydrological Yearbook"
Observatories: meteorological, hydrological, climatological and air pollution monitoring network of stations
City Reference Coordinates: 017°10'00"E 48°10'00"N

Slovak Office of Standards, Metrology and Testing (UNMS)
P.O. Box 75
810 05 Bratislava
or :
Štefanovičova 3
814 39 Bratislava
Telephone: (0)7-391085
Telefax: (0)7-391050
Electronic Mail: <userid>@normoff.gov.sk
WWW: http://www.normoff.gov.sk/unms_sr/
Founded: 1993
Staff: 56
Activities: national body for standardization, testing, certification and metrology
Periodicals: "Bulletin of the Slovak Office of Standards, Metrology and Testing", "Standardization", "Metrology and Testing"

City Reference Coordinates: 017°10'00"E 48°10'00"N

Slovak Physical Society
c/o Dalibor Krupa (President)
Institute of Physics
Dúbravská cesta 9
842 28 Bratislava
Telephone: (0)7-395676
Telefax: (0)7-395676
Electronic Mail: krupa@savba.sk
sfs@savba.sk
WWW: http://www.savba.sk/~fyzisfs
http://www.sfs.sk/
Founded: 1993
Membership: 280
Activities: promoting advancement of physics research, teaching and training
Periodicals: "Gradient"
Awards: (1) "Young Physicists Competition"
City Reference Coordinates: 017°10'00"E 48°10'00"N

Slovak Union of Amateur Astronomers
c/o Rimavská Sobota Observatory
Tomasovska 63
979 01 Rimavská Sobota
Telephone: (0)866-5624709
Telefax: (0)866-5624709
Electronic Mail: astrors@bb.telecom.sk
Founded: 1970
Membership: 350
Activities: photography * popularization * history of astronomy * observing (Sun, meteors, variable stars, eclipsing binaries, comets)
Periodicals: (6) "Kozmos" (see separate entry)

Coordinates: 020°00'25"E 48°22'29"N H228m (Rimavská Sobota - see separate entry)
City Reference Coordinates: 020°00'00"E 48°22'00"N

Slovak University of Technology, Observatory
Radlinského 11
813 68 Bratislava
Telephone: (0)7-52498047
Telefax: (0)7-52925476
Electronic Mail: <userid>@cvt.stuba.sk
                 hefty@cvt.stuba.sk (Ján Hefty, Director)
WWW: http://www.stuba.sk/
Founded: 1956
Staff: 4
Activities: Earth rotation * astrometry * geodynamics * GPS
Coordinates: 017°07'11"E 48°09'18"N H170m
City Reference Coordinates: 017°10'00"E 48°10'00"N

Žilina Observatory
Horný Val 20/41
P.O. Box B 153
012 42 Žilina
Telephone: (0)89-43780
Electronic Mail: rksza@za.sanet.sk
Founded: 1961
Staff: 5
Activities: popularization * observing (Sun, comets, variable stars, occultations, eclipses, meteors)
Coordinates: 018°45'13"E 49°12'22"N H404m (Malý Diel)
City Reference Coordinates: 018°40'00"E 49°14'00"N

Slovenia

Hydrometeorological Institute of Slovenia
Vojkova 1b
P.O. Box 2549
1001 Ljubljana
Telephone: (0)61-1784255
Telefax: (0)61-1331396
WWW: http://www.rzs-hm.si/
Founded: 1947
Staff: 210
Activities: weather forecast * climatology * observing * air pollution * hydrology * hydrometry * water quality * ground water * agrometeorology
Coordinates: 013°51'00"E 46°23'00"N H2,514m (Kredarica)
City Reference Coordinates: 014°31'00"E 46°03'00"N

Javornik Astronomical Society
Tavcarjeva 2
P.O. Box 504
61000 Ljubljana
or :
c/o Ivo Babarovic
Pot na Fuzine 53
1000 Ljubljana
Telephone: (0)61-315198
        (0)61-446311
Electronic Mail: ivo.babarovic@guest.arnes.si
            dintinjana@ean.uni-mb.ac.mail.yu
WWW: http://www2.arnes.si/~ljastronom1/adj.html
Founded: 1979
Membership: 120
Activities: popularization * education * Summer camp * workshops * observing
Periodicals: (2) "Astronom"; "Circular of ADJ"
Coordinates: 014°02'00"E 45°50'00"N H1,150m
City Reference Coordinates: 014°31'00"E 46°03'00"N

Slovenian Academy of Sciences and Arts (SAZU)
Novi trg 3
61000 Ljubljana
Telephone: (0)61-1256-068
Telefax: (0)61-1253423
Electronic Mail: sazu@sazu.si
WWW: http://www.sazu.si/
Founded: 1938
Membership: 171
City Reference Coordinates: 014°31'00"E 46°03'00"N

Society of Mathematicians, Physicists and Astronomers of Slovenia
c/o M. Mankoc-Borstnik (President)
P.O. Box 64
61111 Ljubljana
Telephone: (0)61-265061
Telefax: (0)61-217281
Electronic Mail: norma.mankoc@ijs.si
City Reference Coordinates: 014°31'00"E 46°03'00"N

Standards and Metrology Institute of Slovenia (SMIS)
Kotnikova 6
61000 Ljubljana
Telephone: (0)61-1312322
        (0)61-1783000
Telefax: (0)61-314882
        (0)61-1783196
Electronic Mail: ic@usm.mzt.si
            smis@usm.mzt.si
WWW: http://www.usm.mzt.si/
Founded: 1991
Staff: 70
Activities: national standards organization * legal metrology * national accreditation service
Periodicals: (12) "Sporočila/Messages"
City Reference Coordinates: 014°31'00"E 46°03'00"N

University of Ljubljana, Department of Physics, Astronomical and Geophysical Observatory
Pot na Golovec 25
1000 Ljubljana
Telephone: (0)61-1401353
Telefax: (0)61-1405370
Electronic Mail: <userid>@uni-lj.si
WWW: http://www.fiz.uni-lj.si/astro/
        http://david.fiz.uni-lj.si/astro/
Founded: 1948
Staff: 4
Activities: education * photometry * spectroscopy * general relativity * pulsars
Periodicals: (1) "Astronomical Ephemeris" (ISSN 1318-0614)
Coordinates: 014°31'38"E 46°02'37"N H401m
City Reference Coordinates: 014°31'00"E 46°03'00"N

South Africa

Astronomical Society of Southern Africa (ASSA)
P.O. Box 9
7935 Observatory
Telephone: (0)21-4470025
Electronic Mail: basbs@bremner.uct.ac.za (Brian Skinner, Honorary Secretary)
                     mgs@maties.sun.ac.za
                     mnassa@saao.ac.za (MNASSA)
WWW: http://www.saao.ac.za/sky/freq.html#assa
Founded: 1912
Membership: 1000
Activities: meetings * discussions * observations and searches
Sections: variable stars * comets and meteors * minor planets occultations * grazing occultations * nova search * computing * ordinary occultations * solar observing * double stars * imaging
Periodicals: (6) "Monthly Notes of the Astronomical Society of Southern Africa (MNASSA)" (ISSN 0024-8266); (1) "ASSA Handbook" (ISSN 0571-7191)
Awards: (1) "Gill Medal"
City Reference Coordinates: 018°22'00"E 33°55'00"S

Astronomical Society of Southern Africa (ASSA), Cape Centre
P.O. Box 13018
7705 Mowbray
Founded: 1912
Activities: lectures * observing * instrumentation * education
Periodicals: (12) "Newsletter"; (4) "The Cape Observer"
Coordinates: 018°32'00"E 33°55'00"S
City Reference Coordinates: 018°32'00"E 33°55'00"S

Astronomical Society of Southern Africa (ASSA), Garden Route Centre
c/o Hans Daehne (Secretary)
9A Ironside Street
6529 George
Telephone: (0)44-8745902
Telefax: (0)44-3431736
Electronic Mail: janhers@pixie.co.za (Jan Hers, Chairman)
Founded: 1998
Membership: 35
Activities: monthly meetings * observing
City Reference Coordinates: 022°28'00"E 33°57'00"S

Astronomical Society of Southern Africa (ASSA), Johannesburg Centre
P.O. Box 93145
2143 Yeoville
or :
18A Gill Street
Observatory
Johannesburg
Telephone: (0)11-7163199
Telefax: (0)11-3392926
Electronic Mail: assa_jhb@aqua.co.za
WWW: http://www.aqua.co.za/assa_jhb.htm
Founded: 1922
Membership: 144
Activities: amateur astronomy * popularization * public observing
City Reference Coordinates: 028°00'00"E 26°15'00"S

Astronomical Society of Southern Africa (ASSA), Natal Centre
P.O. Box 5330
4001 Durban
                     sthomson@mweb.co.za
WWW: http://www.astronomical.lia.net/
City Reference Coordinates: 031°00'00"E 29°53'00"S

Astronomical Society of Southern Africa (ASSA), Natal Midlands Centre (Pietermaritzburg)
c/o The Secretary
P.O. Box 2106
3200 Pietermaritzburg
Telephone: (0)33-3433646
WWW: http://www.botany.unp.ac.za/nmc/nmc.htm
Founded: 1974
Membership: 50
Activities: variable stars * observing * occultations * education

Periodicals: (12) "Stardust"
Observatories: 1 (World's View Observatory)
City Reference Coordinates: 030°16'00"E 29°37'00"S

Astronomical Society of Southern Africa (ASSA), Pretoria Centre
c/o N.P. Young (Secretary)
201 Kritzinger street
Meyers Park
0184 Pretoria
or :
c/o Membership Secretary
P O Box 11151
Queerswood 0121
Telephone: (0)12-833765
Telefax: (0)12-3257534
WWW: http://mafadi.aero.csir.co.za/assa/
Founded: 1968
Membership: 51
Activities: amateur astronomy * observing * data gathering
Periodicals: (12) "Newsletter"
Coordinates: 028°16'03"E 25°44'11"S
City Reference Coordinates: 028°12'00"E 25°45'00"S (Pretoria)

Cederberg Observatory
P.O. Box 5203
8000 Cape Town
Electronic Mail: zasah52@iafrica.com (Bill Hollenbach)
WWW: http://www.cs.uct.ac.za/~iwebb/obs/
Founded: 1986
Coordinates: 019°25'00"E 33°29'30"S
City Reference Coordinates: 018°22'00"E 33°55'00"S

Council for Scientific and Industrial Research (CSIR)
P.O. Box 395
0001 Pretoria
Telephone: (0)12-8412911
Telefax: (0)12-3491153
WWW: http://www.csir.co.za/
Founded: 1945
Staff: 3000
City Reference Coordinates: 028°10'00"E 25°45'00"S

Durban Natural Science Museum, Astronomy Interest Group
P.O. Box 4085
4000 Durban
or :
City Hall - 1st Floor
Smith Street
Durban
Telephone: (0)31-3006212
Telefax: (0)31-3006308
Electronic Mail: mario@dimaggio.org (Mario Di Maggio, Education Officer)
WWW: http://www.kwazuzulwazi.co.za/AIG-default.htm
Founded: 1998
Membership: 106
Activities: multimedia presentations * demonstrations * observing
Periodicals: (12) "Newsletter" (circ.: 450)
City Reference Coordinates: 030°56'00"E 29°55'00"S

Foundation for Research Development (FRD)
P.O. Box 2600
0001 Pretoria
Telephone: (0)12-8412879
Telefax: (0)12-8413688
      (0)12-8042679
Electronic Mail: info@frd.ac.za
             <userid>@frd.ac.za
WWW: http://www.frd.ac.za/
City Reference Coordinates: 028°10'00"E 25°45'00"S

Foundation for Research Development (FRD), Hartebeesthoek Radio Astronomy Observatory (HartRAO)
P.O. Box 443
1740 Krugersdorp
Telephone: (0)11-6424692

Telefax: (0)11-6422424
Electronic Mail: <userid>@bootes.hartrao.ac.za
WWW: http://www.hartrao.ac.za/
        http://apies.frd.ac.za/hartrao.html
Founded: 1975
Staff: 33
Activities: national facility for radioastronomy * VLBI * spectroscopy * pulsars * masers * variable radio sources * surveys
Periodicals: (1) "Annual Report"
Coordinates: 027°41'07"E 25°53'23"S H1,416m
City Reference Coordinates: 027°35'00"E 26°05'00"S

Hartebeesthoek Radio Astronomy Observatory (HartRAO)
● See "Foundation for Research Development (FRD), Hartebeesthoek Radio Astronomy Observatory (HartRAO)"

Potchefstroom University for Christian Higher Education, Department of Physics, Space Research Unit (SRU)
2520 Potchefstroom
Telephone: (0)18-2992423
           (0)18-2992403
Telefax: (0)18-2992421
Electronic Mail: <userid>@puknet.puk.ac.za
WWW: http://www.puk.ac.za/fskdocs/ern_topE.html
        http://www.puk.ac.za/fskdocs/phys_topE.html
Founded: 1961
Staff: 11
Activities: cosmic rays * magnetospheric and heliospheric physics * observing (solar flares, galactic cosmic-rays) * gamma-ray astronomy * pulsars * X-ray binaries * cataclysmic variable stars
Observatories: 4 (Potchefstroom, Hermanus, Tsumeb - Namibia and Sanae - Antarctica)
Coordinates: 026°54'09"E 27°10'32"S H1,408m (Potchefstroom)
City Reference Coordinates: 027°06'00"E 27°42'00"S

Rhodes University, Astronomy and Ham Radio Society
Grahamstown
Electronic Mail: queries@astrosoc.soc.ru.ac.za
               barry@rucus.ru.ac.za (Barry Burdis)
               keith@rucus.ru.ac.za (Keith Burdis)
WWW: http://astrosoc.soc.ru.ac.za/
Founded: 1991
City Reference Coordinates: 026°31'00"E 33°19'00"S

Rhodes University, Department of Physics and Electronics
P.O. Box 94
6140 Grahamstown
Telephone: (0)461-318111
Telefax: (0)461-25049
Electronic Mail: <userid>@hippo.ru.ac.za
WWW: http://phlinux.ru.ac.za/physics/
Founded: 1957
Staff: 9
Activities: 2.3 GHz radio-continuum survey of the Southern sky * radio continuum mapping * ionospheric model for Southern African region
Coordinates: 026°30'00"E 33°32'00"S
City Reference Coordinates: 026°31'00"E 33°19'00"S

Sedgefield Observatory
P.O. Box 48
6573 Sedgefield
Telephone: (0)44-3431736
Telefax: (0)44-3431736
Electronic Mail: janhers@pixie.co.za (Jan Hers)
Founded: 1979
Staff: 1
Activities: variable stars (visual and photoelectric measures)
Coordinates: 022°47'46"E 34°00'44"S
City Reference Coordinates: 022°48'00"E 34°01'00"S

South African Association of Science and Technology Centres (SAASTEC)
c/o Rudi Horak
Gold Fields Exploratorium
Faculty of Science
University of Pretoria
0002 Pretoria
Telephone: (0)12-4202865
Telefax: (0)12-4203874
Electronic Mail: rudi@gold.up.ac.za

WWW: http://www.saastec.co.za/
Founded: 1996
Membership: 20 (institutions) and 19 (individuals)
Activities: stimulating interest in and establishing more science and technology centres * education * popularization * workshops
City Reference Coordinates: 028°10'00"E 25°45'00"S

South African Astronomical Observatory (SAAO)
P.O. Box 9
7935 Observatory
Telephone: (0)21-470025
Telefax: (0)21-473639
Electronic Mail: <userid>@saao.ac.za
WWW: http://www.saao.ac.za/
        http://da.saao.ac.za/~salt/ (Southern African Large Telescope - SALT)
Founded: 1820
Staff: 42
Activities: optical region and IR photometry * spectroscopy * CCD photometry
Periodicals: (1) "SAAO Annual Report" (ISSN 0250-0671); "SAAO Circulars", "SAAO Newsletter"
Coordinates: 018°28'07"E 33°56'03"S H18m (Cape Town)
             020°48'07"E 32°22'07"S H1,771m (Sutherland - see separate entry)
City Reference Coordinates: 018°22'00"E 33°55'00"S (Cape Town)

South African Astronomical Observatory (SAAO), Sutherland Observing Station
c/o SAAO
P.O. Box 9
7935 Observatory
Telephone: (0)23-5711205
Telefax: (0)23-5714135
Electronic Mail: sutherland@saao.ac.za
WWW: http://www.saao.ac.za/
Founded: 1972
Staff: 18
Coordinates: 020°48'07"E 32°22'07"S H1,771m
City Reference Coordinates: 018°22'00"E 33°55'00"S

South African Bureau of Standards (SABS)
Private Bag X 191
0001 Pretoria
or :
1 Dr. Lategan Road
Groenkloof
Telephone: (0)12-4287911
Telefax: (0)12-3441568
Electronic Mail: info@sabs.co.za
WWW: http://www.sabs.co.za/
City Reference Coordinates: 028°10'00"E 25°45'00"S (Pretoria)

South African Institute of Physics (SAIP)
c/o Physics Department
University of Zululand
3886 Kwa Dlangezwa
Telephone: (0)24-93911
Telefax: (0)24-93571
Electronic Mail: <userid>@unizul1.uzulu.ac.za
                 bspoelst@pan.uzulu.ac.za (B. Spoelstra, Honorary Secretary)
WWW: http://www.sun.ac.za/local/academic/fisika/saip/saip.html
        http://www.sun.ac.za/local/academic/natural/fisika/saip/saip.html
Founded: 1955
Membership: 550
Awards: (1/2) "Gold Medal for Outstanding Achievement", "Silver Medal for Outstanding Achievement of Physicist under 35yrs"
City Reference Coordinates: 030°56'00"E 29°55'00"S (Durban)

South African Museum, Planetarium
P.O. Box 61
8000 Cape Town
or :
25 Queen Victoria Street
8001 Cape Town
Telephone: (0)21-4243330
Telefax: (0)21-4246716
Electronic Mail: t.ferreira@samuseum.ac.za (Theo Ferreira)
WWW: http://www.museums.org.za/sam/planet/planetar.htm
Founded: 1987
Staff: 10

Activities: public and school shows * education * concerts * special events * show kits
Periodicals: (1) "Annual Report"
City Reference Coordinates: 018°22'00"E 33°55'00"S

South African Weather Bureau (SAWB)
Private Bag X-097
0001 Pretoria
or :
Forum Building
Struben Street
0001 Pretoria
Telephone: (0)12-3093120
Telefax: (0)12-3093127
Electronic Mail: <userid>@cirrus.sawb.gov.za
WWW: http://cirrus.sawb.gov.za/
Founded: 1860
Staff: 310
Activities: meteorology * surface and upper air observations * climatology * forecasting * research * instrumentation
Periodicals: (12) "SAWB Daily Weather Bulletin" (ISSN 0011-5517), "Climate Summary of Southern Africa"; "SAWB Technical Papers", "SAWB Technical Reports"; (1) " SAWB Yearly Weather Report"
City Reference Coordinates: 028°10'00"E 25°45'00"S

Spreeufontein Observatory
c/o Albert G. Jansen
Markstraat 3
6930 Prince Albert
Telephone: (0)23-5411871
Telefax: (0)23-5411871
Electronic Mail: agjansen@ilink.nis.za
WWW: http://www.nis.za/~agjansen/spreeu.htm
Founded: 1995
Staff: 1
Activities: visual, photographic and CCD observing
Coordinates: 022°11'45"E 32°56'44"S H720m
City Reference Coordinates: 022°02'00"E 32°13'00"S

Sutherland Observing Station
● See "South African Astronomical Observatory (SAAO), Sutherland Observing Station"

University of Cape Town (UCT), Department of Applied Mathematics
Private Bag
7700 Rondebosch
Telephone: (0)21-6502340
Telefax: (0)21-6502334
Electronic Mail: <userid>@maths.uct.ac.uk
WWW: http://shiva.mth.uct.ac.za/ (Cosmology Group)
Founded: 1974
Staff: 4
Activities: theoretical research * cosmology * relativity
City Reference Coordinates: 018°22'00"E 33°55'00"S (Cape Town)

University of Cape Town (UCT), Department of Astronomy
Private Bag
7700 Rondebosch
Telephone: (0)21-6502391
Telefax: (0)21-6503726
Electronic Mail: astro.uctvax@f4.n494.zs.fidonet.org
                     ucthpx!<userid>@uunet.uu.net.uucp
WWW: http://www.uct.ac.za/depts/astronomy/
Founded: 1970
Activities: high-speed photometry * variable stars (cataclysmic, rapid) * AGNs
Periodicals: "Publications of the Department of Astronomy of the University of Cape Town"
Coordinates: 018°28'36"E 33°56'03"S
City Reference Coordinates: 018°22'00"E 33°55'00"S (Cape Town)

University of Cape Town (UCT), Department of Physics
Private Bag
7700 Rondebosch
Telephone: (0)21-6503331
         (0)21-6503326
Telefax: (0)21-6503342
Electronic Mail: <userid>@physci.uct.ac.za
WWW: http://www.uct.ac.za/depts/physics/
Founded: 1893
City Reference Coordinates: 018°22'00"E 33°55'00"S (Cape Town)

University of Natal, Space Physics Research Institute (SPRI)
King George V Avenue
4001 Durban
Telephone: (0)31-8162775
Telefax: (0)31-2616550
Electronic Mail: rash.undphp@f4.n494.z5.fidonet.org
WWW: http://swift.ph.und.ac.za:8080/spri/homepage.html
Founded: 1987
Staff: 6
Activities: aurorae * VLF * whistlers * pulsations * ozone
Coordinates: 002°24'00"W 70°18'00"S H100m (Sanae, Antarctica)
City Reference Coordinates: 030°56'00"E 29°55'00"S

University of South Africa (UNISA), Department of Mathematics, Applied Mathematics and Astronomy
P.O. Box 392
0003 UNISA
Telephone: (0)12-4296345
                 (0)12-4296202 (Secretary)
                 (0)12-4293381 (Observatory)
Telefax: (0)12-4296064
Electronic Mail: dps@astro.unisa.ac.za
WWW: http://www.unisa.ac.za/dept/wis/
           http://astro.unisa.ac.za/~uniobs (Observatory)
Founded: 1962 (Department) (Observatory: 1992)
Staff: 2 (Astronomy)
Activities: education * mathematics and applied mathematics * dust in galaxies * OH masers
Coordinates: 028°11'58"E 25°46'06"S H1,448m
City Reference Coordinates: 028°10'00"E 25°45'00"S (Pretoria)

University of the Witwatersrand, Department of Computational and Applied Mathematics
P.O. Wits
2050 Johannesburg
Johannesburg
Telephone: (0)11-7163808
Telefax: (0)11-3397620
Electronic Mail: <userid>@gauss.cam.wits.ac.za
                         block@gauss.cam.wits.ac.za (David L. Block, Astronomy)
WWW: http://www.cam.wits.ac.za/
Founded: 1922 (University)
Staff: 24
City Reference Coordinates: 028°00'00"E 26°15'00"S

University of the Witwatersrand, Department of Physics
P.O. Wits
2050 Johannesburg
Johannesburg
Telephone: (0)11-7162439
Telefax: (0)11-3397965
             (0)11-3398262
Electronic Mail: <userid>@physnet.phys.wits.ac.za
WWW: http://www.wits.ac.za/wits/fac/science/physics/physics.html
Founded: 1922 (University)
Staff: 50
City Reference Coordinates: 028°00'00"E 26°15'00"S

Spain

Agrupación Astronáutica Española (AAE)
C/ Copérnico 94 Entlo 3°
E-08006 Barcelona
Telephone: 932097339
        932017208
Telefax: 932097339
Founded: 1950
Activities: education
City Reference Coordinates: 002°11'00"E 41°23'00"N

Agrupación Astronómica Aragonesa (AAA)
Avenida Navarra 54
Aula 2
E-50010 Zaragoza
Electronic Mail: komet@arrakis.es
WWW: http://www.astrored.org/aaa
Founded: 1989
Membership: 150
Activities: education * popularization * observing
Periodicals: (6) "Boletin"
Coordinates: 000°52'48"W 41°39'24"N
City Reference Coordinates: 000°53'00"W 41°38'00"N

Agrupación Astronómica Cántabra (AAC)
Apartado 573
E-39080 Santander
Electronic Mail: aacantabra@pagina.de
WWW: http://matsun1.matesco.unican.es/~gcampos/aac.html
       http://pagina.de/AACantabra
Founded: 1982
Membership: 30
Activities: variable stars * solar physics * occultations * planets
Periodicals: (6) "Estela"
City Reference Coordinates: 003°48'00"W 43°28'00"N

Agrupación Astronómica Complutense (AAC)
Apartado 199
E-28880 Alcalá de Henares
or :
Casa de la Juventud
Paseo del Val 2
E-28800 Alcalá de Henares
Telephone: 918896612
Electronic Mail: felixm@wanadoo.es (Félix L. Moreno, Secretary)
Membership: 150
Activities: popularization * talks * meetings
Periodicals: (4) "AAC Actividades"
Coordinates: 003°22'05"W 40°28'53"N
City Reference Coordinates: 003°22'00"W 40°28'00"N

Agrupación Astronómica de Barcelona (ASTER)
Passeig de Gràcia 71 (àtic)
E-08008 Barcelona
Telephone: 932151531
Electronic Mail: aster@encomix.es
WWW: http://www.encomix.es/~aster/
Founded: 1948
Membership: 200
Activities: observing (deep sky, Sun, Moon, meteorites, planets, variable stars) * photography
Sections: Astrophotography * Computing * Sun * Planets * Meteorites * Meteorology * Variable Stars * Library
Periodicals: (2) "ASTER" (ISSN 0212-6036); (4) "Ephemerides"
Coordinates: 000°02'10"E 41°23'35"N H69m
City Reference Coordinates: 002°11'00"E 41°23'00"N

Agrupación Astronómica de Cáceres
c/o Gabino Muriel Brillo
Apartado 153
E-10080 Cáceres
WWW: http://www.geocities.com/CapeCanaveral/Hall/5175/
       http://www.arrakis.es/~aacc
City Reference Coordinates: 006°23'00"W 39°29'00"N

Agrupación Astronómica de Córdoba (AAC)
C/ Huerto de San Pedro El Real 1
Apartado 701
E-14080 Córdoba
Telephone: 957456077
Electronic Mail: i52cacaj@uco.es
WWW: http://www.uco.es/~i52cacaj
Founded: 1982
Membership: 40
Activities: education for beginners * observing camps * visits to observatories * lectures * library * Summer camps
Periodicals: (4) "Boletín Informativo de la AAC"
Observatories: 1 (Santa Maria de Trassierra)
City Reference Coordinates: 004°46'00"W 37°53'00"N

Agrupación Astronómica de Fuerteventura (AAF)
C/ Los Coroneles 16
E-35640 La Oliva
Fuerteventura
Islas Canarias
Electronic Mail: cvera@arrakis.es (Carlos Vera)
WWW: http://www.arrakis.es/~cvera/aaf
Founded: 1995
Membership: 12
Activities: observing * photography * popularization *education
Coordinates: 013°56'00"W 28°37'00"N H200m
City Reference Coordinates: 013°53'00"W 28°36'00"N

Agrupación Astronómica de Gran Canaria (AAGC)
Apartado 4240
E-35080 Las Palmas de Gran Canaria
Islas Canarias
Telephone: 928247128
Electronic Mail: aagc@edi.ulpgc.es
WWW: http://ccdis.dis.ulpgc.es:8086/AAGC/aagc.html
Founded: 1990
Periodicals: "SPACEX 90 - Nuevos Horizontes"
City Reference Coordinates: 015°24'00"W 28°06'00"N

Agrupación Astronómica de Huesca (AAH)
Apartado 61
E-22080 Huesca
or :
Costanilla de Oteiza 1
E-22000 Huesca
Telephone: 974223255
WWW: http://dftuz.unizar.es/aah/aah.html
         http://dftuz.unizar.es/externo/aah/aah.html
Founded: 1994
Membership: 100
City Reference Coordinates: 000°25'00"W 42°08'00"N

Agrupación Astronómica de la Autónoma
Universidad Autónoma de Barcelona
Aula del Cosmos
Paseo Mas Roig 57 Bajos
E-08190 Sant Cugat del Vallès
WWW: http://tau.uab.es/~a3/
City Reference Coordinates: 002°05'00"E 41°28'00"N

Agrupación Astronómica de la Palma (AAP)
Apartado 449
E-38700 Santa Cruz de la Palma
Islas Canarias
Telephone: 922411974
         607592175
Telefax: 922412203
Electronic Mail: jafami@santandersupernet.com (José Antonio Fernandez Arozena, Secretary)
WWW: http://www.ing.iac.es/AAP/AAP.html
Founded: 1986
Membership: 7
Activities: meteors * variable stars * photoelectric photometry
Periodicals: "Circulars"
Coordinates: 017°52'06"W 28°45'10"N H2,400m (Roque de los Muchachos)
City Reference Coordinates: 017°46'00"W 28°41'00"N

Agrupación Astronómica de la Rioja
c/o Victor Lanchares (President)
Departamento de Matemática y Computación
Universidad de la Rioja
E-26004 Logroño
Electronic Mail: astrorioja@knet.es
WWW: http://astrorioja.knet.es/
City Reference Coordinates: 002°26'00"W 42°28'00"N

Agrupación Astronómica de los Dolores (AAD)
Centro Cultural de Los Dolores
Plaza del Mortero s/n
E-30570 Los Dolores
Telephone: 988259323
Electronic Mail: galileo@arrakis.es
WWW: http://www.Geocities.com/CapeCanaveral/Galaxy/9337
        http://www.arrakis.es/~galileo/
City Reference Coordinates: 001°01'00"W 37°39'00"N

Agrupación Astronómica de Madrid (AAM)
Argumosa 39
Apartado 1039
E-28080 Madrid
Telephone: 914671268
Electronic Mail: aam@sicyd.es
                aam@eucmos.sim.ucm.es
WWW: http://www.iac.es/AA/AAM/AAM.html
Founded: 1975
Membership: 600
Activities: variable stars * comets * photography * occultations * planets
Periodicals: (4) "Boletín Informativo" (circ.: 800)
Coordinates: 004°58'00"W 40°32'00"N H1,190m (Hija de Dios, Avila)
City Reference Coordinates: 003°41'00"W 40°24'00"N

Agrupación Astronómica de Manresa (AAM)
Apartado 233
E-08240 Manresa
Telephone: 938730636
          938720948
WWW: http://www2.ninorisa.es/aam
Founded: 1984
Membership: 61
Activities: popularization * photography * solar physics * telescope making * Messier objects * observing camps * deep-sky objects * variable stars * computing * CCD
Periodicals: (6) "Bulleti Informatiu"
Coordinates: 001°50'29"E 41°43'06"N H290m (La Culla)
City Reference Coordinates: 001°50'00"E 41°44'00"N

Agrupación Astronómica de Sabadell (AAS)
(Agrupació Astronòmica de Sabadell - AAS)
Apartado 50
E-08200 Sabadell
or :
Carrer Prat de la Riba, s/n
Parc Catalunya
E-08200 Sabadell
Telephone: 937255373
Telefax: 937272941
Electronic Mail: secretaria@astrosabadell.org
WWW: http://www.astrosabadell.org/
Founded: 1960
Membership: 724
Activities: education * observing * meetings * library * advising * CCD imagery
Periodicals: (6) "Astrum" (ISSN 0210-4105)
Coordinates: 002°05'29"E 41°33'03"N H231m (Parc Catalunya)
City Reference Coordinates: 002°07'00"E 41°33'00"N

Agrupación Astronómica de Santa Pola
C/ Prudencia 3-1
E-03130 Santa Pola
WWW: http://www.geocities.com/CapeCanaveral/Launchpad/1627/
City Reference Coordinates: 000°32'00"W 38°12'00"N

Agrupación Astronómica de Setenil "Stellae"
Finca "El Alambique"

Carretera de Alcalá s/n
E-11692 Setenil de las Bodegas
Founded: 1986
Membership: 14
Activities: amateur astronomy
Periodicals: (4) "Stellae"
Coordinates: 005°10'08"W 36°51'50"N H550m
City Reference Coordinates: 005°10'00"W 36°52'00"N

Agrupación Astronómica de Tenerife (AAT)

Apartado 10644
E-38080 Santa Cruz de Tenerife
Islas Canarias
Telephone: 922617879 (President)
Telefax: 922563295
Electronic Mail: aatwww@arrakis.es
WWW: http://www.iac.es/AA/AAT/
Founded: 1987
Membership: 250
Activities: photography * comets * meteors * deep sky * planets * minor planets * popularization
Periodicals: (3) "Nova 87-a" (circ.: 100)
City Reference Coordinates: 016°14'00"W 28°27'00"N

Agrupación Astronómica d'Osona (AAO)

Carrer del Pare Xifré 1-3 3°
E-08500 Vic
Telephone: 938864154
Electronic Mail: aao@rpodos.fcr.es
WWW: http://www.infomet.fcr.es/clima/osona/aao.htm
Founded: 1986
Coordinates: 002°15'30"E 41°55'50"N H512m
City Reference Coordinates: 002°16'00"E 41°56'00"N

Agrupación Astronómica Tamix

C/ Geranio 10
E-29650 Mijas Costa
Telephone: 952463530
Electronic Mail: mateoperez@nexo.es
Founded: 1991
Membership: 25
Periodicals: (4) "Tamix Journal"
Coordinates: 004°42'57"W 36°33'36"N
City Reference Coordinates: 004°38'00"W 36°36'00"N (Mijas)

Agrupación Astronómica Universitaria de Alicante (AAUA)

Apartado 99
E-03080 Alicante
Electronic Mail: cancillo@ua.es
WWW: http://www.psi.ua.es/aaua.htm
Founded: 1996
Membership: 20
City Reference Coordinates: 000°29'00"W 38°21'00"N

Alternativa Racional a las Pseudociencias (ARP)

Apartado 310
E-08860 Castelldefels
or :
Apartado 1516
E-50080 Zaragoza
Telephone: 976591671
Electronic Mail: arp@kepler.unizar.es
                   arp_sapc@yahoo.com
WWW: http://www.encomix.es/~ceus/arp.htm
           http://zar.unizar.es/~arp
Founded: 1986
Membership: 200
Activities: conferences * lectures * publishing
Periodicals: (4) "La Alternativa Racional (LAR)" (ISSN 1130-491x)
City Reference Coordinates: 001°57'00"E 41°17'00"N (Castelldefels)
                                   000°53'00"W 41°38'00"N (Zaragoza)

Animaciencia

Apartado 1081
E-02080 Albacete
Telephone: 619775960

Founded: 1998
Membership: 100
Activities: scientific camps * education * workshops * trips * meetings
City Reference Coordinates: 001°52'00"W 39°00'00"N

Antares, Ciencia y Ediciones SA
Apartado 5314
E-08080 Barcelona
or :
Gran Vía 646-4° 29
E-08007 Barcelona
Telephone: 933011717
Telefax: 933011765
Electronic Mail: universo@antares.es
WWW: http://universo.home.ml.org/
Founded: 1995
Staff: 10
Activities: publishing * conferences * courses * popularization
Periodicals: (12) "Universo, Astronomía y Astronaútica" (ISSN 1135-2876, circ.: 10,000)
City Reference Coordinates: 002°11'00"E 41°23'00"N

Asesores Astronómicos Cacereños (AAC)
Apartado 409
E-10080 Cáceres
Telephone: 927232363
Telefax: 927236486
Electronic Mail: pacoviolat@redestb.es (Francisco A. Violat Bordonau, President)
Founded: 1992
Membership: 35
Activities: CCD astronomy * variable stars * planets
Periodicals: (6) "Alcor"
Coordinates: 006°23'25"W 39°27'16"N H468m
City Reference Coordinates: 006°23'00"W 39°29'00"N

Asociación Española de Normalización y Certificación (AENOR)
C/ Génova 6
E-28004 Madrid
Telephone: 914326000
Telefax: 913104976
Electronic Mail: aenor@aenor.es
WWW: http://www.aenor.es/
Founded: 1986
Periodicals: "Informe Anual"
City Reference Coordinates: 003°41'00"W 40°24'00"N

Asociación Valenciana de Astronomía (AVA)
Apartado 2069
E-46080 Valencia
or :
Marià de Cavia 45 - 34
E-46014 Valencia
Telephone: 963579951
Founded: 1972
Membership: 300
Activities: popularization * observing (Sun, Moon, planets, variable stars, comets, deep sky) * computing
Periodicals: (4) "Revista"; "Efemérides", "Circulares de Actividades Periódicas"
City Reference Coordinates: 000°22'00"W 39°28'00"N

Calar Alto Observatory
• See "Centro Astronómico Hispano-Alemán, Observatorio de Calar Alto"
• See also "Observatorio Astronómico Nacional (OAN), Observatorio de Calar Alto"

Casa de las Ciencias, Planetario
Parque de Santa Margarita s/n
E-15005 La Coruña
Telephone: 981279144
Telefax: 981277777
Electronic Mail: domus@casaciencias.org
WWW: http://www.casaciencias.org/
Founded: 1985
Staff: 50 (Casa)
Activities: lectures * education * shows * science fairs
City Reference Coordinates: 008°23'00"W 43°22'00"N

Catalana de Telescopios
C/ John Lennon 18
E-08800 Vilanova i la Geltrú
Telephone: 938150153
Telefax: 938150153
Founded: 1989
Staff: 1
Activities: manufacturing and distributing telescopes * domes * import * maintenance * Foucaults's pendulums * interactive modules for science museums * small and medium planetariums
City Reference Coordinates: 001°43'00"E 41°13'00"N

Centro Astronómico de Yebes (CAY)
● See "Observatorio Astronómico Nacional (OAN), Centro Astronómico de Yebes (CAY)"

Centro Astronómico Hispano-Alemán, Observatorio de Calar Alto
Jesús Durbán Remón 2 - 2°
Apartado 511
E-04080 Almería
Telephone: 950225576
Telefax: 950225562
Electronic Mail: <userid>@caha.es
WWW: http://www.mpia-hd.mpg.de/CAHA/
          http://caserv.caha.es/
Founded: 1972
Staff: 46
Activities: young stellar objects * active galaxies * early universe * cosmology
Coordinates: 002°32'47"W 37°13'24"N H2,168m
● Jointly operated by the "Max-Planck-Institut für Astronomie", Heidelberg (Germany) and the "Comisión Nacional de Astronomía", Spain
City Reference Coordinates: 002°27'00"W 36°50'00"N

Ciutat de les Arts i les Ciències de València
Autovía del Saler
Esquina C/Instituto Obrero s/n
E-46013 Valencia
Telephone: 902100031
Electronic Mail: buzon@cac.es
WWW: http://www.cac.es/paginac.html a*museum of arts and sciences * planetarium (L'Hemisfèric)
City Reference Coordinates: 000°22'00"W 39°28'00"N

Consejo Superior de Investigaciones Científicas (CSIC)
Serrano 117
E-28006 Madrid
Telephone: 915855001
Telefax: 914113077
Electronic Mail: <userid>@cti.csic.es
WWW: http://www.csic.es/
Founded: 1939
City Reference Coordinates: 003°41'00"W 40°24'00"N

Consejo Superior de Investigaciones Científicas (CSIC), Instituto de Astronomía y Geodesia (IAG)
● See "Instituto de Astronomía y Geodesia (IAG)"

Consejo Superior de Investigaciones Científicas (CSIC), Instituto de Astrofísica de Andalucia (IAA)
Apartado 3004
E-18080 Granada
or :
C/ Sancho Panza s/n
E-18008 Granada
Telephone: 958121311
Telefax: 958814530
Electronic Mail: <userid>@iaa.es
WWW: http://www.iaa.es/
Founded: 1975
Staff: 80
Activities: star formation * VLBI * clusters of galaxies * active galaxies * airglow * planetary atmospheres * binary stars * open clusters * pulsating stars
Coordinates: 000°13'38"W 37°03'50"N H3,000m (Sierra Nevada)
            000°10'12"E 37°13'27"N H2,200m (Calar Alto)
City Reference Coordinates: 003°41'00"W 37°13'00"N

Consejo Superior de Investigaciones Científicas (CSIC), Instituto de Física de Cantabria
● See "Instituto de Física de Cantabria"

Consejo Superior de Investigaciones Científicas (CSIC), Observatorio del Ebro
● See "Universidad Ramón Llull (URL), Observatorio del Ebro"

Equipo Sirius
C/ Desengaño 12-4° 1
E-28004 Madrid
Telephone: 915216008
Telefax: 915319032
Electronic Mail: astronomia@mad.servicom.es
WWW: http://www.laeff.esa.es/~tribuna/sirius/sirius.html
Founded: 1985
Staff: 10
Activities: publishing astronomy books and journals
Periodicals: (12) "Tribuna de Astronomía" (ISSN 0213-5892, circ.: 8,500)
City Reference Coordinates: 003°41'00"W 40°24'00"N

European Space Agency (ESA), Villafranca Satellite Tracking Station (VILSPA), Infrared Space Observatory (ISO)
Apartado 50727
E-28080 Madrid
Telephone: 918131100
Telefax: 918131172
Electronic Mail: <userid>@vilspa.esa.es
WWW: http://isowww.estec.esa.nl/
          http://www.iso.vilspa.esa.es/
          http://www.ipac.caltech.edu/iso (ISO US Support Center)
          http://www.esa.int/
Founded: 1977
Activities: operations of the "Infrared Space Observatory" * astrophysics
Periodicals: "ISO Info"
City Reference Coordinates: 003°41'00"W 40°24'00"N

Grup d'Estudis Astronomics (GEA)
(Grupo de Estudios Astronómicos - GEA)
Apartado 9481
E-08080 Barcelona
Telephone: 935930225
Telefax: 935930331
          935702148
Electronic Mail: gea@astro.gea.cesca.es
                info@astro.gea.cesca.es
                gea@montsant.cesca.es
WWW: http://astro.gea.cesca.es/texte.html
Founded: 1984
Membership: 40
Activities: photoelectric photometry (solar system, variable stars) * photography
Periodicals: (2) "Circular GEA"
Coordinates: 001°46'34"E 41°32'24"N H375m (Hostalets)
            002°12'29"E 41°32'24"N H70m (Mollet)
            000°22'44"W 41°38'31"N H465m (Monegrillo)
            001°45'18"E 41°31'23"N H324m (Piera)
            001°49'19"E 42°22'12"N H1,070m (Sampsor)
            002°53'53"E 39°38'49"N H200m (Sencelles)
● Original name: "Grup d'Estudis Astronòmics (GEA)"
City Reference Coordinates: 002°11'00"E 41°23'00"N

Grupo Astronómico Silos (GAS)
Apartado 200
E-50080 Zaragoza
Electronic Mail: jltrisan@lander.es (Jose Luis Trisan, Public Relations & Treasurer)
Founded: 1983
Membership: 40
Activities: Sun * Moon * Jupiter * occultations * Messier objects * planets * variable stars * photography * CCDs * instrumentation * meteorology
Periodicals: "Boletín Informativo del GAS" (ISSN 1135-223x), "Planetary Circular"
Coordinates: 000°52'42"W 41°39'28"N
City Reference Coordinates: 000°53'00"W 41°38'00"N

Grupo Canario de Estrellas Variables (GCEV)
Las Palmas de Gran Canaria
WWW: http://aagc.dis.ulpgc.es/gcev.html
Founded: 1994
City Reference Coordinates: 015°24'00"W 28°06'00"N

Grupo de Estudios Astronómicos (GEA)
● See "Grup d'Estudis Astronomics (GEA)"

Grupo de Estudios Astronómicos de Puertollano(GEAP)
Travesma Alta 13
E-13500 Puertollano
Telephone: 926411806
Electronic Mail: geap@teleline.es
WWW: http://teleline.terra.es/personal/jaduque/
Founded: 1998
City Reference Coordinates: 004°07'00"W 38°41'00"N

Grupo Europeo de Observaciones Estelares (GEOS)
c/o Luis Rivas
Calle Colón 9 - 12°
E-46016 Tabernes Blanques
Telephone: 961850948
WWW: http://www.upv.es/geos/
Founded: 1978
Membership: 12
Activities: photoelectric and visual observing (variable stars) * participating in European observational campaigns * computing periods and orbits * searching new variable stars * holding a yearly international congress, a photoelectric photometry school and Summer camp * international meetings
Periodicals: "GEOS Circulares", "Notas Circulares", "Fichas Técnicas"
Coordinates: 000°49'54"W 38°15'37"N H275m (Observatorio CEMA, Crevillente)
City Reference Coordinates: 000°16'00"W 39°04'00"N

Grupo Joven de Astronomía Scorpius Elche
C/ Capitan Gaspar Ortiz 106, 2°-2
E-03201 Elche
Telephone: 970340219
WWW: http://www.geocities.com/CapeCanaveral/Launchpad/1627/scorpius.htm
City Reference Coordinates: 000°41'00"W 38°16'00"N

Infrared Space Observatory (ISO)
● See "European Space Agency (ESA), Villafranca Satellite Tracking Station (VILSPA) Infrared Space Observatory (ISO)"

Instituto de Astrofísica de Andalucia (IAA)
● See "Consejo Superior de Investigaciones Científicas (CSIC), Instituto de Astrofísica de Andalucia (IAA)"

Instituto de Astrofísica de Canarias (IAC)
C/ Vía Láctea s/n
E-38200 La Laguna
Islas Canarias
Telephone: 922605200
          922329100 (Teide Observatory)
          922405500 (Roque de los Muchachos Observatory - see this entry)
Telefax: 922605201
          922329117 (Teide Observatory)
Electronic Mail: <userid>@iac.es
WWW: http://www.iac.es/home.html
Founded: 1975
Staff: 164
Activities: Spanish research organization managing two international observatories * structure of the universe * cosmology * structure and evolution of galaxies * star formation and evolution * planetary systems * Sun * ISM * instrumentation
Periodicals: "Preprint Series", "Publicaciones del IAC", "Noticias del IAC" (ISSN 0213-893x), "Memoria del IAC"
Coordinates: 016°30'25"W 28°17'56"N H2,398m (Teide)
          017°52'34"W 28°45'34"N H2,400m (Roque de los Much. - see this entry)
City Reference Coordinates: 016°19'00"W 28°29'00"N

Instituto de Astrofísica de Canarias (IAC), Observatorio del Roque de los Muchachos (ORM)
Apartado 303
E-38700 Santa Cruz de la Palma
Islas Canarias
Telephone: 922405500
Telefax: 922405501
WWW: http://www.iac.es/
Founded: 1975 (IAC)
Coordinates: 017°52'34"W 28°45'34"N H2,400m
City Reference Coordinates: 017°46'00"W 28°41'00"N

Instituto de Astronomía y Geodesia (IAG)
c/o Facultad de Ciencias Matemáticas
Cuidad Universitaria

E-28040 Madrid
Telephone: 913944585
Telefax: 913944607
Founded: 1947
Staff: 20
Activities: Earth tides * geodynamics * gravimetry * Earth rotation * spatial geodesy * geoid
Periodicals: "Publicaciones" (ISSN 0213-6198)
● Jointly operated by the "Consejo Superior de Investigaciones Científicas (CSIC)" and the "Universidad Complutense"
City Reference Coordinates: 003°41'00"W 40°24'00"N

Instituto de Física de Cantabria
c/o Facultad de Ciencias
Avenida de los Castros s/n
E-39005 Santander
Telephone: 942201461
Telefax: 942201402
Electronic Mail: <userid>@astro.unican.es
WWW: http://www.ifca.unican.es/
Founded: 1995
Staff: 6
Activities: extragalactic astronomy * observational cosmology
● Jointly operated by the "Consejo Superior de Investigaciones Científicas (CSIC)" and the "Universidad de Cantabria"
City Reference Coordinates: 003°48'00"W 43°28'00"N

Instituto de Radioastronomía Milimétrica (IRAM)
Núcleo Central
Avenida Divina Pastora 7
E-18012 Granada
Telephone: 958228899
  958226696
  958482002 (30m telescope)
Telefax: 958222363
  958481148 (30m telescope)
Electronic Mail: <userid>@iram.es
WWW: http://www.iram.es/irame_home.html
  http://iram.fr/
Founded: 1979
Staff: 31
Activities: radio observing at mm-wavelengths
Coordinates: 003°23'35"W 37°03'59"N H2,916m (Observatorio Pico de Veleta)
City Reference Coordinates: 003°41'00"W 37°13'00"N

Instituto Nacional de Meteorología (INM)
Apartado 285
E-28040 Madrid
or :
Camino de las Morenas s/n
E-28040 Madrid
Telephone: 915819630
  915819810 (Meteorological information)
Telefax: 915819845
Electronic Mail: <userid>@inm.es
WWW: http://www.inm.es/
City Reference Coordinates: 003°41'00"W 40°24'00"N

Instituto Nacional de Tecnica Aerospacial (INTA), División de Investigaciones Espaciales
Carretera de Ajalvir Km. 4
E-28850 Torrejón de Ardóz
Telephone: 916270999
Telefax: 916270945
Electronic Mail: <userid>@inta.es
WWW: http://www.inta.es/homepage.html
Founded: 1991
Staff: 55
Activities: space science research * astrophysics * fundamental physics * remote sensing * Earth atmosphere
City Reference Coordinates: 003°29'00"W 40°27'00"N

Instituto Nacional de Tecnica Aerospacial (INTA), Laboratorio de Astrofísica Espacial y Física Fundamental (LAEFF)
Apartado 50727
E-28080 Madrid
or :
c/o VILSPA
E-28691 - Villanueva de la Cañada
Telephone: 918131161
Telefax: 918131160

Electronic Mail: <userid>@laeff.esa.es
WWW: http://www.laeff.esa.es/
Founded: 1990
Staff: 12
Activities: space astrophysics * stellar evolution * binary stars * galactic clusters * high-energy astrophysics * UV astronomy * fundamental physics * cosmology * astroparticle physics * gravity * elementary particles * radio astronomy
City Reference Coordinates: 003°41'00"W 40°24'00"N (Madrid)

Instituto Politécnico Maritimo Pesquero de Pasajes, Planetario
Marinos 2
Apartado 5
E-20110 Pasajes
Telephone: 943399340
Telefax: 943390756
Founded: 1970
Staff: 1
Activities: maritime astronomy * education
City Reference Coordinates: 001°55'00"W 43°20'00"N

Isaac Newton Group (ING), Observatorio del Roque de los Muchachos (ORM)
Apartado 321
E-38780 Santa Cruz de la Palma
Islas Canarias
Telephone: 922425410
            922405655
Telefax: 922425401
           922405646
Electronic Mail: <userid>@ing.iac.es
WWW: http://ing.iac.es/
        http://www.ast.cam.ac.uk/ING/
Founded: 1984
Staff: 75
Activities: operating the Isaac Newton Group of telescopes: 4.2m William Herschel Telescope (WHT), 2.5m Isaac Newton Telescope (INT) and 1.0m Jacobus Kapteyn Telescope (JKY)
Periodicals: "Spectrum"
Coordinates: 017°52'41"W 28°45'38"N H2,372m
City Reference Coordinates: 017°46'00"W 28°41'00"N

Museo de la Ciencia y el Cosmos (MCC)
C/ Vía Láctea s/n
E-38200 La Laguna
Islas Canarias
Telephone: 922263454
            922315080
            922315265
Telefax: 922263295
Electronic Mail: museo@cosmos.mcc.rcanaria.es
WWW: http://www.mcc.rcanaria.es/
        http://www.mcc.rcanaria.es/inform/inform.htm#Planetario
Founded: 1993
Staff: 15
Activities: popularization * education * shows * workshops
Coordinates: 016°18'00"W 28°30'00"N H700m
City Reference Coordinates: 016°19'00"W 28°29'00"N

Museu de la Ciència
Teodor Roviralta 55
E-08022 Barcelona
Telephone: 932126050
Telefax: 934170381
Electronic Mail: musciencia.fundacio@lacaixa.es
                info.fundacio@lacaixa.es
WWW: http://www.fundacio.lacaixa.es/eng/equips/museu.htm
Founded: 1981
Activities: science museum * planetarium
City Reference Coordinates: 002°11'00"E 41°23'00"N

Nordic Optical Telescope (NOT)
c/o Observatorio del Roque de los Muchachos
Apartado 474
E-38700 Santa Cruz de la Palma
Islas Canarias
Telephone: 922425473
Telefax: 922425475
Electronic Mail: <userid>@not.iac.es
WWW: http://www.not.iac.es/

Founded: 1984
Coordinates: 017°52'48"W 28°45'30"N H2,372m
City Reference Coordinates: 017°46'00"W 28°41'00"N

Nuevas Tecnologías Observacionales (NTO)
Candido Soto 12
E-28223 Pozuelo de Alarcón
WWW: http://www.nto.org/
City Reference Coordinates: 003°49'00"W 40°26'00"N

Observatorio Astronómico del Garraf (OAG)
(Observatori Astronòmic del Garraf - OAG)
C/ Tetuan 6
E-08800 Vilanova i la Geltrú
Telephone: 938143189
Electronic Mail: rho@arrakis.es
WWW: http://www.arrakis.es/~jaumeplan/
Founded: 1990
Staff: 5
Activities: double stars * education
Coordinates: 001°43'59"E 41°13'29"N
City Reference Coordinates: 001°43'00"E 41°13'00"N

Observatorio Astronómico de Mallorca (OAM)
Camino de Observatorio s/n
E-07144 Costitx
Telephone: 971876019
Telefax: 971892380
Electronic Mail: astroam@bitel.es
WWW: http://www.oam.es/
Founded: 1989
Membership: 20
Activities: planetary atmospheres * comets * minor planets * astrometry * photometry
Periodicals: (3) "Astrosplai i Natura"
Coordinates: 002°57'05"E 39°38'38"N
City Reference Coordinates: 002°57'00"E 39°39'00"N

Observatorio Astronómico de Valencia
● See "Universidad de Valencia, Observatorio Astronómico"

Observatorio Astronómico Nacional (OAN), Spain
Campus Universitario de Alcalá de Henares
Apartado 1143
E-28800 Alcalá de Henares
Telephone: 918855060
                918855061
Telefax: 918855062
Electronic Mail: <userid>@oan.es
WWW: http://www.oan.es/
Founded: 1790
Staff: 56
Activities: astronomy (observations, theory, technical developments) * time signal distribution * museum * popularization
Periodicals: (1) "Anuario del Observatorio Astronómico Nacional"
Coordinates: 003°41'17"W 40°24'30"N H670m
                003°05'21"W 40°31'24"N H930m (Centro Astronómico de Yebes - CAY)
                002°32'54"W 37°13'27"N H2,165m (Observatorio de Calar Alto)
City Reference Coordinates: 003°22'00"W 40°29'00"N

Observatorio Astronómico Nacional (OAN), Centro Astronómico de Yebes (CAY)
Apartado 148
E-19080 Guadalajara
or :
Cerro de la Palera s/n
E-19041 Yebes
Telephone: 949290311
Telefax: 949290063
Electronic Mail: <userid>@cay.oan.es
WWW: http://www.oan.es/cay/
City Reference Coordinates: 003°10'00"W 40°37'00"N (Guadalajara)
                              003°07'00"W 40°31'00"N (Yebes)

Observatorio Astronómico Nacional (OAN), Observatorio de Calar Alto
Apartado 793
E-04080 Almería
or :

c/o Delegación del Instituto Geográfico Nacional
Edificio de Servicios Múltiples
C/ Hermanos Machado 4, 6° planta
E-04004 Almería
Telephone: 951225576 x256
951234507
950521034 (Dome)
950632596 (Dome)
Telefax: 950632585 (Dome)
Electronic Mail: eoca@oan.es
<userid>@caserv.caha.es
WWW: http://caserv.caha.es/
http://www.oan.es/calar/
Founded: 1790 (OAN)
Activities: spectroscopy * photometry
Coordinates: 002°32'54"W 37°13'27"N H2,160m
City Reference Coordinates: 002°27'00"W 36°50'00"N (Almaría)

Observatorio Astronómico Ramón María Aller
● See "Universidad de Santiago de Compostela, Observatorio Astronómico Ramón María Aller"

Observatorio de Calar Alto
● See "Centro Astronómico Hispano-Alemán, Observatorio de Calar Alto"
● See also "Observatorio Astronómico Nacional (OAN), Observatorio de Calar Alto"

Observatorio del Ebro
● See Universidad Ramón Llull (URL), Observatorio del Ebro"

Observatorio del Roque de los Muchachos (ORM)
● See "Instituto de Astrofísica de Canarias (IAC), Observatorio del Roque de los Muchachos (ORM)"
● See also "Isaac Newton Group (ING), Observatorio del Roque de los Muchachos (ORM)"
● See also "Nordic Optical Telescope (NOT)"
● See also "Royal Swedish Academy of Sciences, Research Station for Astrophysics"
● See also "Telescopio Nazionale Galileo (TNG)"

Observatorio del Teide
● See "Instituto de Astrofísica de Canarias (IAC)"

Observatorio Pico de Veleta
● See "Instituto de Radioastronomía Milimétrica (IRAM)"

Parque de las Ciencias
Avenida del Mediterraneo s/n
E-18006 Granada
Telephone: 958131900
Telefax: 958 133582
Electronic Mail: cpciencias@parqueciencias.com
WWW: http://www.parqueciencias.com/
http://www.parqueciencias.com/planet.html (Planetarium)
City Reference Coordinates: 003°41'00"W 37°13'00"N

Planetario de Madrid
Avenida Planetario 16
E-28045 Madrid
Telephone: 914673461
914673578
914673898 (answering machine)
Telefax: 914681154
Electronic Mail: buzon@planetmad.es
WWW: http://www.planetmad.es/
Founded: 1986
Staff: 19
Activities: shows * scientific exhibitions
City Reference Coordinates: 003°41'00"W 40°24'00"N

Planetario de Pamplona
Sancho Ramirez s/n
E-31008 Pamplona
Telephone: 948260004
948260056
Telefax: 948261919
Electronic Mail: armentia_javier@euskom.spritel.es
planetario@cin.es
WWW: http://www.ucm.es/OTROS/Astrof/pamplona/pp-casa.html
http://www.ucm.es/info/Astrof/pamplona/pp-casa.html

Founded: 1993
Staff: 13
Activities: education * popularization * science centre * museum * exhibitions * lectures * conferences
City Reference Coordinates: 001°38'00"W 42°49'00"N

Real Instituto y Observatorio de la Armada (ROA)
Cecilio Pujazon s/n
E-11110 San Fernando
Telephone: 956599365
Telefax: 956599366
Electronic Mail: astro@roa.es
Founded: 1753
Staff: 17
Activities: ephemerides * astrometry * time and frequency * magnetism * sismology * artificial satellites * lasers * GPS
Periodicals: "Almanaque Náutico" (ISSN 0210-735x), "Efemérides Astronómicas" (ISSN 0080-5971), "Fenomenos Astronó-
micos" (ISSN 0210-8127), "Anales ROA"
Coordinates: 006°12'19"W 36°27'42"N H29m
City Reference Coordinates: 006°12'00"W 36°28'00"N

Real Sociedad Española de Física (RSEF)
c/o Facultad de Ciencias Físicas
Universidad Complutense
Ciudad Universitaria s/n
E-28040 Madrid
Telephone: 913944359
Telefax: 915433879
Electronic Mail: rsef@fis.ucm.es
WWW: http://www.ucm.es/info/rsef
Founded: 1903
Membership: 1400
Periodicals: (5) "Revista Española de Física" (ISSN 0213-862x); "Anales de Física"
Awards: (1) "Medalla de la Real Sociedad Española de Física"; (3) "Premio a Investigadores Noveles de la Real Sociedad
Española de Física"
City Reference Coordinates: 003°41'00"W 40°24'00"N

Royal Swedish Academy of Sciences, Research Station for Astrophysics
Grupo Sueco
Apartado 66
E-38700 Santa Cruz de la Palma
Islas Canarias
Telephone: 922405590
Telefax: 922405592
Electronic Mail: paco@not.iac.es (Paco Armas, Administrator)
WWW: http://www.astro.su.se/groups/solar/solar.html
Founded: 1979
Activities: high spatial resolution studies of the dynamic evolution of magnetic fields and velocity fields on the Sun
Coordinates: 017°52'48"W 28°45'30"N H2,372m
City Reference Coordinates: 017°46'00"W 28°41'00"N

Sociedad Astronómica de España y América (SADEYA)
Avenida Diagonal 377, 2°
E-08008 Barcelona
Telephone: 934160478
Electronic Mail: sadeya@sadeya.cesca.es
WWW: http://www.sadeya.cesca.es/
Founded: 1911
Membership: 750
Activities: popularizing astronomy and related sciences * lectures * education
Sections: comets * computing * observatory * optics * radioastronomy
Periodicals: (6) "Astronomía, Astrofotografía y Astronautica" (ISSN 0213-621x), "SADEYA"; (1) "Almanaque Astronómico";
"Circulars"
Coordinates: 000°56'54"E 42°22'00"N H1,400m (Antist)
City Reference Coordinates: 002°11'00"E 41°23'00"N

Sociedad Astronómica Granadina (SAG)
Apartado 195
E-18080 Granada
or :
Avenida Barcelona 20-10 C
E-18006 Granada
Telephone: 958136363
Electronic Mail: xsag@fedro.ugr.es
Founded: 1981
Membership: 340
Activities: popularization * observing * education

Periodicals: (2) "Halley"
Coordinates: 003°35'00"W 37°10'00"N H680m (Cubillas)
City Reference Coordinates: 003°41'00"W 37°13'00"N

Sociedad Astronómica SYRMA
Apartado 5380
E-47080 Valladolid
Telephone: 983339210
Electronic Mail: gua@uva.es
WWW: http://www.gvi.uva.es/~gua
Founded: 1981
Membership: 83
Activities: popularization * education * observing * library * conferences
Periodicals: "Astrea"
Coordinates: 004°21'00"W 41°39'00"N H885m (Cogeces del Monte)
City Reference Coordinates: 004°43'00"W 41°39'00"N

Sociedad de Astronomía Balear (SAB)
Apartado 1206
E-07080 Palma de Mallorca
Telephone: 971284390
Founded: 1979
Membership: 100
Activities: popularization * observing
Periodicals: "Circular"
City Reference Coordinates: 002°39'00"E 39°34'00"N

Sociedad de Ciencias Aranzadi, Sección Astronomía
c/o Museo de San Telmo
Plaza de Ignacio Zuloaga s/n
E-20003 San Sebastian
Telephone: 943422945
Telefax: 943421316
Electronic Mail: astroaranzadi@redestb.es
Founded: 1977
Membership: 110
Activities: solar physics * deep sky * planets * comets * occultations * lectures * ephemerides * photography * meteor showers * education
Periodicals: (3) "Boletín de Astronomía" (ISSN 1132-2306)
Coordinates: 001°59'03"W 43°17'46"N H105m (Torres Arbide)
City Reference Coordinates: 001°59'00"W 43°19'00"N

Sociedad de Observadores de Meteoros y Cometas de España (SOMYCE)
c/o Luis Ramón Bellot Rubio
Instituto de Astrofísica de Canarias
C/ Vía Láctea s/n
E-38200 La Laguna
Islas Canarias
Telephone: 922605369
Telefax: 922605210
Electronic Mail: lbellot@iac.es
            preyes@afrodita.fcu.um.es
            csantana@cic.teleco.ulpgc.es
            msolano@ext.step.es
WWW: http://sopa.dis.ulpgc.es/~rvr/somyce.html
       http://www.iac.es/AA/SOMYCE/somyce.html
Founded: 1988
Membership: 100
Activities: observing (comets, meteors) * analyzing meteor and comet data * orbit computations * visual, radio and photographic work
Periodicals: (6) "Meteoros" (circ.: 100)
City Reference Coordinates: 016°19'00"W 28°29'00"N

Sociedad Española de Astronomía (SEA)
c/o Facultad de Física
Universidad de Barcelona
Avenida Martí Franquès 1
E-08028 Barcelona
Telephone: 934021125
Telefax: 934021133
Electronic Mail: secretaria@sea.am.ub.es
WWW: http://sea.am.ub.es/
Founded: 1993
Membership: 300
Activities: promoting the development of astronomical activities in Spain
Periodicals: (2) "Boletin Informativo"

City Reference Coordinates: 002°11'00"E 41°23'00"N

Sociedad Española de Optica (SEDO)
Serrano 121
E-28006 Madrid
Telephone: 915616070
Telefax: 915645557
Founded: 1968
Membership: 320
Activities: promoting optical research and development in Spain
Periodicals: "Optica Pura y Aplicada" (ISSN 0030-3917)
City Reference Coordinates: 003°41'00"W 40°24'00"N

Sociedad Malagueña de Astronomía (SMA)
Apartado de Correos 6072.
E-29080 Málaga
or :
Centro Cultural José María Gutiérrez Romero
C/ República Argentina 9
Urbanización El Limonar
E-29016 Málaga
Electronic Mail: sma@laeff.esa.es
WWW: http://www.laeff.esa.es/~sma/
City Reference Coordinates: 004°25'00"W 36°43'00"N

Telescopio Nazionale Galileo (TNG)
Apartado 565
E-38700 Santa Cruz de la Palma
Islas Canarias
Telephone: 922425043
Telefax: 922420508
Electronic Mail: wwwstaff@tng.iac.es
WWW: http://www.tng.iac.es/
Founded: 1990
Coordinates: 017°52'35"W 28°45'34"N H2,373m
City Reference Coordinates: 017°46'00"W 28°41'00"N

Universidad Autonoma de Madrid (UAM), Grupo de Astrofísica
Departamento de Física Teórica
Modulo C-XI
Campus de Cantoblanco
E-28049 Madrid
Telephone: 913974880 (Office)
              913978605 (Observatory)
Telefax: 913973936
Electronic Mail: <userid>@uam.es
WWW: http://pollux.ft.uam.es/astro/
Founded: 1987
Staff: 15
Activities: active galaxies * chemical evolution * extragalactic astrophysics * HII regions * Galaxy formation * numerical cosmology * star formation
City Reference Coordinates: 003°41'00"W 40°24'00"N

Universidad Complutense, Departamento de Astrofísica
c/o Facultad de Físicas
Ciudad Universitaria
E-28040 Madrid
Telephone: 913944577
              913944592
Telefax: 913945195
Electronic Mail: <userid>@astrax.fis.ucm.es
WWW: http://www.ucm.es/info/Astrof/
Staff: 9
Activities: stellar activity (RS CVn stars) * stellar outer atmosphere * eclipsing binaries * stellar photometry * Ba stars * compact galaxies * abundance and stellar content of galaxies
City Reference Coordinates: 003°41'00"W 40°24'00"N

Universidad Complutense, Instituto de Astronomía y Geodesia (IAG)
● See "Instituto de Astronomía y Geodesia (IAG)"

Universidad de Barcelona, Departamento de Astronomía y Meteorología
Avenida Diagonal 647
E-08028 Barcelona
Telephone: 934021125
Telefax: 934021133

Electronic Mail: <userid>@mizar.am.ub.es
                 <userid>@alcor.am.ub.es
                 secre@mizar.am.ub.es
WWW: http://www.am.ub.es/
        http://mizar.am.ub.es/
Founded: 1970
Staff: 20
Activities: stellar kinematics and dynamics * SN * gravitational lenses * stellar formation * molecular clouds * HII regions * interplanetary particle events * galaxy clusters * Earth rotation * large-scale structures * nucleosynthesis * structure and evolution of galaxies * stellar clusters
City Reference Coordinates: 002°11'00"E 41°23'00"N

Universidad de Cantabria, Instituto de Física de Cantabria
● See "Instituto de Física de Cantabria"

Universidad de Granada, Departamento de Ciencias de la Computación e Inteligencia Artificial (DECSAI)
Apartado 590
E-18080 Granada
or :
c/o Facultad de Ciencias
Avenida Fuentenueva s/n
E-18071 Granada
Telephone: 958243317
Telefax: 958243317
Electronic Mail: <userid>@decsai.ugr.es
WWW: http://decsai.ugr.es/
Founded: 1986
Staff: 26
Activities: approximate reasoning * computer vision * models for general information
City Reference Coordinates: 003°41'00"W 37°13'00"N

Universidad de Granada, Departamento de Física Téorica y del Cosmos
c/o Facultad de Ciencias
Avenida Fuentenueva s/n
E-18002 Granada
Telephone: 958243305
Telefax: 958248529
Electronic Mail: <userid>@ugr.es
WWW: http://deneb.ugr.es/
Founded: 1982
Staff: 4
Activities: MHD in spiral galaxies * magnetic fields in intergalactic space * education * carbon stars * SN
City Reference Coordinates: 003°41'00"W 37°13'00"N

Universidad de las Islas Baleares (UIB), Departamento de Física
E-07071 Palma de Mallorca
Telephone: 971173228
Telefax: 971173426
Electronic Mail: <userid>@eps.uib.es
WWW: http://www.uib.es/depart/dfs/
Founded: 1973
Staff: 25
Activities: education * research * solar physics * relativity * material sciences * condensed matter * nuclear physics * geophysical fluid dynamics
City Reference Coordinates: 002°39'00"E 39°34'00"N

Universidad de Santiago de Compostela, Observatorio Astronómico Ramón María Aller
Avenida de las Ciencias s/n
Apartado 197
E-15706 Santiago de Compostela
Telephone: 981592747
Telefax: 981597054
Electronic Mail: <userid>@usc.es
WWW: http://www.usc.es/astro/
Founded: 1943
Staff: 3
Activities: celestial mechanics * double stars
Periodicals: (4) "Publicaciones del Observatorio Astronómico Ramón Maria Aller" (ISSN 1130-0892)
Coordinates: 008°33'33"W 42°52'32"N H240m
City Reference Coordinates: 008°33'00"W 42°53'00"N

Universidad de Valencia, Departamento de Astronomía y Astrofísica
Dr. Moliner 50
E-46100 Burjassot

Electronic Mail: <userid>@vlbi.matapl.uv.es
WWW: http://vlbi.uv.es/
Founded: 1995
City Reference Coordinates: 000°22'00"W 39°28'00"N

Universidad de Valencia, Faculdad de Matemáticas, Grupo de Astronomía y Ciencias del Espacio (GACE)
E-46100 Burjassot
Telephone: 963864573
Telefax: 963864364
Electronic Mail: <userid>@pollux.uv.es
WWW: http://pollux.uv.es/
Founded: 1990
City Reference Coordinates: 000°22'00"W 39°28'00"N

Universidad de Valencia, Observatorio Astronómico
Avenida Blasco Ibañez 13
E-46010 Valencia
Telephone: 963864773
Telefax: 963864773
Electronic Mail: alvaro.lopez@uv.es
Founded: 1909
Staff: 4
Activities: astrometry of minor planets * PZT * celestial mechanics
Coordinates: 000°22'00"W 39°28'41"N H30m
City Reference Coordinates: 000°22'00"W 39°28'00"N

Universidad de Zaragoza, Grupo de Mecanica Espacial
Edificio Matemáticas
E-50009 Zaragoza
Telephone: 976761000
Telefax: 976761140
Electronic Mail: <userid>@posta.unizar.es
WWW: http://gme.unizar.es/
Founded: 1986
Staff: 8
Activities: celestial mechanics * binary stars * artificial-satellite theory * algebra computation
City Reference Coordinates: 000°53'00"W 41°38'00"N

Universidad Politécnica de Cataluña, Departamento de Física Aplicada
C/ Jordi Girona Salgado 1-3
Módulo B5
Campus Nord
E-08034 Barcelona
Telephone: 934016802
Telefax: 934016090
Electronic Mail: <userid>@etseccpb.upc.es
WWW: http://www-fa.upc.es/
Activities: white dwarfs * cooling * stars * evolution * SN * convection
City Reference Coordinates: 002°11'00"E 41°23'00"N

Universidad Politécnica de Madrid, Departamento de Física
Ronda de Valencia 3
E-28012 Madrid
Telephone: 913367686
Telefax: 915309244
WWW: http://www.upm.es/
Activities: double stars * eclipsing binaries * Algol-type stars
City Reference Coordinates: 003°41'00"W 40°24'00"N

Universidad Ramón Llull (URL), Observatorio del Ebro
E-43520 Roquetes
Telephone: 977500511
Telefax: 977504660
Electronic Mail: <userid>@peb.wil.es
WWW: http://www.url.es/
Founded: 1904
Staff: 11
Activities: STP * solar physics * geomagnetism * ionosphere * meteorology * seismology
Periodicals: "Boletín del Observatorio del Ebro - Ionosfera" (ISSN 0211-5166), "Publicaciones del Observatorio del Ebro - Miscelánea" (ISSN 0211-4534), "Publicaciones del Observatorio del Ebro - Memoria"
Coordinates: 000°29'30"E 40°49'12"N H50m
● Associated with the "Consejo Superior de Investigaciones Científicas (CSIC)"
City Reference Coordinates: 000°30'00"E 40°50'00"N

Villafranca Satellite Tracking Station (VILSPA)
• See "European Space Agency (ESA), Villafranca Satellite Tracking Station (VILSPA)"

VILSPA
• See "European Space Agency (ESA), Villafranca Satellite Tracking Station (VILSPA)"

Sri Lanka

Department of Meteorology, Sri Lanka
383 Bauddhaloka Mawatha
Colombo 7
Telephone: (0)1-694846
Telefax: (0)1-691443
Founded: 1907
Activities: weather forecast * limited astronomical service * seismological service * supplying meteorological and climatological data
Periodicals: (1) "Annual Report"
Coordinates: 079°52'00"E 06°54'00"N H7m
City Reference Coordinates: 079°52'00"E 06°54'00"N

Institute of Fundamental Studies (IFS)
Hantana Road
Kandy
Telephone: (0)8-232002
Telefax: (0)8-232131
Electronic Mail: ifs@ifs.ac.lk
Founded: 1981
Staff: 150
Activities: fundamental research * science dissemination
Periodicals: (12) "Pragna"
City Reference Coordinates: 080°38'00"E 07°18'00"N

National Science Foundation (NSF)
47/5 Maitland Place
Colombo 7
Telephone: (0)1-696771
(0)1-696772
(0)1-696773
Telefax: (0)1-691691
Electronic Mail: info@nsf.ac.lk
WWW: http://www.nsf.ac.lk/
Founded: 1968
Staff: 98
Activities: research funding * S&T information * S&T policies
Periodicals: (2) "Journal of the National Science Foundation of Sri Lanka" (ISSN 0300-9254), "Sri Lanka Journal of Social Science" (ISSN 0258-9710); (1) "Annual Report"
Awards: (1/2) "Merit Awards for Scientific Research"
City Reference Coordinates: 079°51'00"E 06°56'00"N

Sri Lanka Standards Institution (SLSI)
53 Dharmapala Mawatha
P.O. Box 17
Colombo 3
Telephone: (0)1-326051
Telefax: (0)1-446018
Electronic Mail: slsi@slt.lk
WWW: http://www.naresa.ac.lk/slsi/
City Reference Coordinates: 079°51'00"E 06°56'00"N

Sweden

Aeronautical Research Institute of Sweden
Ranhammarsvägen 12-14
Ulvsunda
Box 11021
SE-161 11 Bromma
Telephone: (0)8-6341000
Telefax: (0)8-253481
Electronic Mail: <userid>@ffa.se
WWW: http://www.ffa.se/
Founded: 1940
Staff: 200
Activities: aeronautics and aerospace research * computational fluid dynamics * wind tunnel testing * development of experimental techniques, from subsonic to hypersonic speed * system analysis * flight dynamics * man-machine interface * flight safety * computational mechanics * composite structures * fatigue and fracture * aviation and environment * noise and vibration
City Reference Coordinates: 015°55'00"E 59°21'00"N

Astromedia AB
Box 7170
SE-402 33 Göteborg
Telephone: (0)31-694500
Telefax: (0)31-694508
Electronic Mail: info@astromedia.se
WWW: http://www.astromedia.se/
Founded: 1988
Activities: manufacturing and distributing telescopes and accessories, books, and star catalogues * exclusive dealer for Celestron, Vixen, and TeleVue in Sweden, and Vixen in Finland
Periodicals: (12) "TeleScoop"
City Reference Coordinates: 012°00'00"E 57°45'00"N

Astronomiska Sällskapet Tycho Brahe (ASTB)
● "Tycho Brahe Astronomical Society"

Broman Planetarium AB
Kärnvedsgatan 11
SE-416 80 Göteborg
Telephone: (0)31256475
Telefax: (0)31257477
Electronic Mail: pbr@planetarium.euromail.se (Per Broman)
                 lbr@planetarium.euromail.se (Lars Broman)
WWW: http://www.dalnet.se/~stella/broman/
Founded: 1985 (Incorporated: 1990)
City Reference Coordinates: 012°00'00"E 57°45'00"N

Chalmers Technical University
● See "Göteborg University, Chalmers Technical University"

Computer Solutions Europe, Comsol AB
Björnnäsvägen 21
SE-113 47 Stockholm
Telephone: (0)8-153022
Telefax: (0)8-157635
Electronic Mail: info@comsol.se
WWW: http://www.comsol.se/
Founded: 1986
Staff: 14
● Software manufacturer and distributor
City Reference Coordinates: 018°03'00"E 59°20'00"N

CosmoNorr
c/o Christer Strand
Poppelvägen 11
SE-832 54 Frösön
Telephone: (0)63-43785
Telefax: (0)63-105181
Electronic Mail: christer.strandh@micro.se
WWW: http://www.jamtnet.se/cosmonorr/
Coordinates: 014°30'00"E 63°06'00"N (Valla Observatory)
City Reference Coordinates: 014°32'00"E 63°11'00"N

Cosmonova
c/o Naturhistoriska Riksmuseet
Frescativägen 40
Box 50007
SE-104 05 Stockholm
Telephone: (0)8-51955101
Telefax: (0)8-51955100
Electronic Mail: tom.callen@nrm.se (Tom Callen)
WWW: http://www.nrm.se/cosmonova/
Founded: 1992
Staff: 11
City Reference Coordinates: 018°03'00"E 59°20'00"N

Erna and Victor Hasselblad Foundation
Ekmansgatan 8
SE-412 56 Göteborg
Telephone: (0)31-7781990
Telefax: (0)31-7784640
Electronic Mail: info@hasselbladfoundation.o.se
WWW: http://www2.hasselbladfoundation.o.se/
Founded: 1979
Staff: 2
Activities: promoting research and academic teaching in the natural sciences and photography
Awards: (1) "Hasselblad International Award in Photography"
City Reference Coordinates: 012°00'00"E 57°45'00"N

Esrange
● See "Swedish Space Corp. (SSC), Esrange"

European Incoherent Scatter Facility (EISCAT), Headquarters
Box 812
SE-981 28 Kiruna
Telephone: (0)980-79153
Telefax: (0)980-79161
Electronic Mail: eiscat@eiscathq.irf.se
                  <userid>@eiscathq.irf.se
WWW: http://snowflake.irf.se/
          http://www.eiscat.uit.no/eiscat.html (EISCAT Scientific Association)
Founded: 1975
Staff: 7 (all sites: 35)
Activities: incoherent scatter radar system site for atmospheric, ionospheric and magnetospheric research
Periodicals: "Annual Report" (ISSN 0349-2710), "Technical Reports"
Coordinates: 020°26'00"E 67°52'00"N H412m (Kiruna)
                019°14'00"E 69°35'00"N H30m (Tromsø, Norway - see separate entry)
                026°38'00"E 67°22'00"N H198m (Sodankylä, Finland - see separate entry)
                016°03'00"E 78°09'00"N (Longyearbyen, Norway - see seperate entry)
City Reference Coordinates: 020°27'00"E 67°52'00"N

European Space Agency (ESA), Satellite Station, Kiruna
Box 815
SE-981 28 Kiruna
Telephone: (0)980-76000
Telefax: (0)980-17121
WWW: http://www.esrin.esa.it/
          http://www.esa.int/
Founded: 1962 (ESA)
City Reference Coordinates: 020°27'00"E 67°52'00"N

Föreningen för Astronomi och Astronautik (FAA)
● See "Society of Astronomy and Astronautics"

Framtidsmuseet
● See "Future's Museum"

Future's Museum
(Framtidsmuseet)
Jussi Björlingväg 25
SE-784 32 Borlänge
Telephone: (0)243-793000
          (0)243-793900
Telefax: (0)243-226977
Electronic Mail: tnc@framtidsmuseet.se
WWW: http://www.framtidsmuseet.se/
Founded: 1986

Staff: 4
Activities: lectures * Kosmorama Rymdteater * multimedia shows
City Reference Coordinates: 015°25'00"E 60°29'00"N

Gislaved Astronomiska Sällskap (GAS) Orion

c/o Bo Ekström
Högåsstigen 9
SE-332 33 Gislaved
Telephone: (0)371-14213
Electronic Mail: bo.ekstrom@ebox.tninet.se
Founded: 1983
Membership: 40
Activities: lectures * observing
City Reference Coordinates: 013°30'00"E 57°19'00"N

Göteborg Astronomical Club

(Göteborgs Astronomiska Klubb - GAK)
c/o Naturhistoriska Museet
Box 7283
SE-402 05 Göteborg
Telephone: (0)31-144250
          (0)31-885189
WWW: http://www.tripnet.se/gak/
Founded: 1955
Membership: 200
Activities: observing * lectures
Periodicals: (4) "Aurora" (ISSN 1101-1718, circ.: 200)
Coordinates: 012°06'43"E 57°38'23"N H124m (Lahall Observatory)
City Reference Coordinates: 012°00'00"E 57°45'00"N

Göteborgs Astronomiska Klubb (GAK)

● See "Göteborg Astronomical Club"

Göteborg University, Chalmers Technical University, Department of Astronomy and Astrophysics

SE-412 96 Göteborg
Telephone: (0)31-7723135
          (0)31-7723136
          (0)31-7723137
          (0)31-7723138
          (0)31-7723139
          (0)31-7723140
          (0)31-7723141
Telefax: (0)31-7723204
Electronic Mail: <userid>@fy.chalmers.se
WWW: http://fy.chalmers.se/~marek/Department/astro.html
       http://www.chalmers.se/
Founded: 1985
Activities: interstellar abundances * red giant stars * molecular clouds * star formation processes * galaxy dynamics * accretion disks
City Reference Coordinates: 012°00'00"E 57°45'00"N

Göteborg University, Chalmers Technical University, Department of Physics

SE-412 96 Göteborg
Telephone: (0)31-7721000
Telefax: (0)31-165176
WWW: http://www.fy.chalmers.se/
       http://www.chalmers.se/
City Reference Coordinates: 012°00'00"E 57°45'00"N

Göteborg University, Chalmers Technical University, Onsala Space Observatory (OSO)

SE-439 92 Onsala
Telephone: (0)31-7725500
Telefax: (0)31-7725590
Electronic Mail: <userid>@oso.chalmers.se
WWW: http://www.oso.chalmers.se/
Founded: 1949
Staff: 50
Activities: radioastronomy * mm-wave spectroscopy * ISM * circumstellar shells * galaxies * VLBI * QSOs * plate tectonics * GPS * aeronomy * geodesy
Coordinates: 011°53'35"E 57°23'45"N H59m
         070°43'56"W 29°15'48"S H2,410m (La Silla, Chile)
City Reference Coordinates: 012°00'00"E 57°25'00"N

GrafiTex-Data

Storgatan 11

SE-590 40 Kisa
Telephone: (0)494-10077
Telefax: (0)494-10074
WWW: http://www.grafitex.se/
Founded: 1988
Staff: 3
Activities: distributing mathematical-assistant and desktop-publishing packages
City Reference Coordinates: 015°37'00"E 57°59'00"N

Halmstad Astronomical Association
(Halmstads Astronomiska Sällskap - HAS)
c/o T. Nilsson
Norra Vägen 9B
SE-302 31 Halmstad
Founded: 1958
Membership: 60
Activities: observing * education
City Reference Coordinates: 012°55'00"E 56°41'00"N

Halmstads Astronomiska Sällskap (HAS)
• See "Halmstad Astronomical Association"

Hasselblad Foundation (Erna and Victor _)
• See "Erna and Victor Hasselblad Foundation"

House of Technology
• See "Teknikens Hus"

Institutet för Rymdfysik (IRF)
• See "Swedish Institute of Space Physics"

International Meteorological Institute in Stockholm (MISU)
• See "Stockholm University, Department of Meteorology"

Karlskrona Astronomi Förening (KAF)
• See "Karlskrona Astronomy Association"

Karlskrona Astronomy Association
(Karlskrona Astronomi Förening - KAF)
c/o Bernth Svensson
N. Möllebacksgränd 2
SE-371 34 Karlskrona
Telephone: (0)455-17340
Telefax: (0)455-54444
Founded: 1975
Membership: 20
Activities: observing
Periodicals: (6) "KAF-NYTT"
City Reference Coordinates: 015°35'00"E 56°10'00"N

Kvistaberg Station
• See "Uppsala University, Astronomical Observatory, Kvistaberg Station"

Lund Observatory
• See "Lund University, Lund Observatory"

Lund Planetarium
• See "Lund University, Lund Observatory, Planetarium"

Lund University, Department of Physics
Box 118
SE-221 00 Lund
or :
Professorsgatan 1
SE-223 62 Lund
Telephone: (0)46-126097
Telefax: (0)46-2224709
Electronic Mail: <userid>@fysik.lu.se
WWW: http://ferrum.fysik.lu.se/
Founded: 1950
Activities: basic and applied research in atomic spectroscopy * laboratory astrophysics * fusion research * stellar spectroscopy
City Reference Coordinates: 013°11'00"E 55°42'00"N

Lund University, Lund Observatory
Svanegatan 9
Box 43
SE-221 00 Lund
Telephone: (0)46-2227300
Telefax: (0)46-2224614
Electronic Mail: <userid>@astro.lu.se
WWW: http://www.astro.lu.se/
Founded: 1672
Staff: 30
Activities: solar physics * meteors * stellar physics * stellar evolution * magnetic phenomena * rapid and high-energy processes * star formation * double and multiple stars * stellar clusters * stellar dynamics * galactic structure * ISM * galactic evolution * astrometry * telescopes * ancillary instrumentation * planetary nebulae * radial velocities
Periodicals: "Lund Observatory Reports" (ISSN 0349-4217)
Coordinates: 013°11'12"E 55°41'54"N H34m
013°26'00"E 55°37'24"N H145m (Jävan Station)
City Reference Coordinates: 013°11'00"E 55°42'00"N

Lund University, Lund Observatory, Planetarium
Svanegatan 9
Box 43
SE-221 00 Lund
Telephone: (0)46-2227302
(0)46-2220153 (Eva Mezey, Planetarium Director)
Telefax: (0)46-2224614
Electronic Mail: planetariet@astro.lu.se
WWW: http://www.astro.lu.se/~planetariet/
Founded: 1978
Staff: 1
Activities: public shows * education * exhibitions * information service
City Reference Coordinates: 013°11'00"E 55°42'00"N

Malmö Astronomi & Rymfarts Sällskap (MARS)
● "Malmö Astronomy and Space Society"

Malmö Astronomy and Space Society
(Malmö Astronomi & Rymfarts Sällskap - MARS)
Box 5017
SE-200 71 Malmö
Telephone: (0)40-547012
WWW: http://www.ludat.lth.se/~dat93jso/mars.html
http://www.mars.m.se/default.htm
http://www.mars.m.se/mars.htm
Founded: 1962
Membership: 69
Activities: observing * photography * lectures * travels * deep sky
Periodicals: (4) "MARS-Bulletinen" (ISSN 0284-6667)
Observatories: 1 (Tycho Brahe Observatory, Malmö)
City Reference Coordinates: 013°00'00"E 55°36'00"N

Mariestads Astronomiska Klubb (MAK)
c/o Rune Fogelquist (President)
Borgmästaregatan 7
SE-542 33 Mariestad
Telephone: (0)501-18167
Electronic Mail: makastro@algonet.se
WWW: http://www.algonet.se/~makastro
Founded: 1978
Membership: 150
Periodicals: (5) "Asterisken"
City Reference Coordinates: 013°51'00"E 58°43'00"N

Naturvetenskapliga Forskningsrådet (NFR)
● See "Swedish Natural Science Research Council"

Nobel Foundation
Box 5232
SE-102 45 Stockholm
or :
Sturegatan 14
Stockholm
Telephone: (0)86630920
Telefax: (0)86603847
WWW: http://www.nobel.se/

Founded: 1900
City Reference Coordinates: 018°03'00"E 59°20'00"N

Nordic Planetarium Association (NPA)
c/o Mariana Back (Secretary)
Framtidsmuseet
Jussi Björlingsväg 25
SE-781 50 Borlänge
Telephone: (0)243-80185
Telefax: (0)243-226977
WWW: http://www2.nrm.se/cosmonova/tc-wnpa.html
Founded: 1984
Activities: conferences * meetings
Periodicals: "NP Newsletter"
City Reference Coordinates: 015°25'00"E 60°29'00"N

Norrköpings Astronomiska Klubb (NAK)
c/o Hekan Norin
Sankt Persgatan 41
SE-602 33 Norrköping
Telephone: (0)11-160906
WWW: http://hem2.passagen.se/nrkastro/
Founded: 1978
City Reference Coordinates: 016°11'00"E 58°36'00"N

Observatoriemuseet
Drottningatan 120
SE-113 60 Stockholm
Telephone: (0)8-315810
Telefax: (0)8-315810
Electronic Mail: observatoriemuseet@swipnet.se
WWW: http://www.kva.se/sve/pg/museer/
Founded: 1991
Staff: 11
Activities: guided tours
City Reference Coordinates: 018°03'00"E 59°20'00"N

Observatoriet i Slottsskogen
• See "Slottsskogen Observatory"

Onsala Space Observatory (OSO)
• See "Göteborg University, Chalmers Technical University, Onsala Space Observatory (OSO)"

Östergötland Astronomical Society
(Östergötlands Astronomiska Sällskap - ÖAS)
c/o Ulf Sandberg (Secretary)
Hargs Gård
SE-590 40 Kisa
Telephone: (0)494-71670
Founded: 1978
Membership: 100
Activities: lectures * discussions * observing * study trips
Periodicals: (4) "Cygnus"
City Reference Coordinates: 015°37'00"E 57°59'00"N

Östergötlands Astronomiska Sällskap (ÖAS)
• See "Östergötland Astronomical Society"

Quantum Image Systems
Tjalmargatan 6
SE-831 45 Ostersund
Telephone: (0)63-181612
Telefax: (0)63-105181
Electronic Mail: info@quantimage.com
WWW: http://www.quantimage.com/
Founded: 1996
Staff: 3
• Software producer
City Reference Coordinates: 014°40'00"E 63°10'00"N

Royal Swedish Academy of Sciences
Lilla Frescativägen 4
Box 50005
SE-104 05 Stockholm

Telephone: (0)8-6739500
Telefax: (0)8-155670
WWW: http://www.kva.se/
Founded: 1739
Activities: promoting mathematics and the natural sciences
Periodicals: "Documenta" (ISSN 0347-5791), "Physica Scripta", "Ambio", "Zoolcgica Scripta", "Acta Zoologica", "Acta Mathematica", "Arkiv för Matematik"
Awards: (1) "Nobel Prizes" (Physics, Chemistry, Economic Sciences), "Crafoord Prize"; numerous grants
City Reference Coordinates: 018°03'00"E 59°20'00"N

Royal Swedish Academy of Sciences, Research Station for Astrophysics, Canary Islands
• See Spain

Saab Ericsson Space AB
SE-405 15 Göteborg
Telephone: (0)31-3350000
Telefax: (0)31-3359520
WWW: http://www.space.se/
Founded: 1992
Staff: 360
Activities: manufacturing on-board systems and equipment for launchers and satellites (computers, data handlers, microwave electronics, guidance and separation systems)
City Reference Coordinates: 012°00'00"E 57°45'00"N

Satellitbild
• See "Swedish Space Corp. (SSC), Satellitbild"

Slottsskogen Observatory
(Observatoriet i Slottsskogen)
c/o Naturhistoriska Museet
Box 7283
SE-402 05 Göteborg
Telephone: (0)31-126300
        (0)31-129807
WWW: http://www.tripnet.se/gak/slottsskogen
Founded: 1929 (present building: 1985)
Staff: 5
Activities: observing (Sun, night sky) for public and schoolchildren * lectures * exhibits * planetarium shows
Coordinates: 011°56'27"E 57°41'26"N H83m
City Reference Coordinates: 012°00'00"E 57°45'00"N

Society of Astronomy and Astronautics
(Föreningen för Astronomi och Astronautik - FAA)
Norra Kungsgatan 15
SE-803 20 Gävle
Telephone: (0)26-164112
Electronic Mail: grabowski_radek@hotmail.com (Radoslaw Grabowski, Secretary)
WWW: http://www.vasa.gavle.se/elevforening/faa/faahms3.htm
Founded: 1960
Membership: 15
Activities: observing * meetings * organization of the annual "Naturvetarkvällarna" (science evenings) with lectures, exhibitions, demonstrations, observing * visits
Coordinates: 017°08'40"E 60°40'43"N H30m
        017°15'47"E 60°35'41"N H45m
City Reference Coordinates: 017°10'00"E 60°41'00"N

Standardiseringen i Sverige (SIS)
• See "Swedish Standards Institution (SIS)"

Stella Nova Planetarium
c/o Falun Science Center
Östra Hamngatan 1
SE-791 71 Falun
Telephone: (0)23-25552
        (0)23-30166
Telefax: (0)23-10137
Electronic Mail: stella@dalnet.se
WWW: http://www.dalnet.se/~stella/nova/
Founded: 1992
City Reference Coordinates: 015°38'00"E 60°36'00"N

Stockholm Amateur Astronomers
(Stockholms Amatörastronomer - STAR)
c/o Gamla Observatoriet
Drottninggade 120

SE-113 60 Stockholm
WWW: http://www.astro.su.se/STAR/
Founded: 1988
Membership: 250
City Reference Coordinates: 018°03'00"E 59°20'00"N

Stockholm Observatory
● See "Stockholm University, Stockholm Observatory"

Stockholms Amatörastronomer (STAR)
● See "Stockholm Amateur Astronomers"

Stockholm University, Department of Meteorology
c/o Arrhenius Laboratory
SE-106 91 Stockholm
Telephone: (0)8-162000
Telefax: (0)8-157185
Electronic Mail: <userid>@misu.su.se
WWW: http://www.misu.su.se/
Founded: 1947
Staff: 50
Activities: dynamic meteorology * chemical meteorology * upper atmospheric physics * aeronomy * aerosols * hosting the International Meteorological Institute in Stockholm (IMI)
Periodicals: (1) "Annual Report" (ISSN 0349-0068); "Atmospheric Aerosol Science" (ISSN 1108-4818), "Chemical Meteorology" (ISSN 0028-445x), "Dynamical Meteorology" (ISSN 0349-0467), "Atmospheric Physics" (ISSN 0280-4441)

City Reference Coordinates: 018°03'00"E 59°20'00"N

Stockholm University, Stockholm Observatory
SE-133 36 Saltsjöbaden
Telephone: (0)8-164445
Telefax: (0)8-7174719
Electronic Mail: <userid>@astro.su.se
WWW: http://www.astro.su.se/home.html
           http://erling.astro.su.se/home.html
Founded: 1753
Staff: 40
Activities: stellar structure and evolution * ISM and star formation * radiative transfers * galactic structure and dynamics * AGNs * IR astronomy
Coordinates: 018°18'30"E 59°16'18"N H60m
City Reference Coordinates: 018°03'00"E 59°20'00"N

Sundsvalls Astronomiska Förening
c/o Johannes Nordvall
Juniskärsvägen 570
SE-862 91 Kvissleby
Telephone: (0)60-562351
Electronic Mail: johannes.nordvall@mailbox.swipnet.se
WWW: http://www.kuai.se/~jono/
           http://home1.swipnet/~w-14449
Founded: 1986
Coordinates: 017°25'24"E 62°17'47"N H35m
City Reference Coordinates: 017°22'00"E 62°22'00"N (Sundsvall)

Svensk Amatör Astronomisk Förening (SAAF)
● See "Swedish Amateur Astronomical Society"

Svenska Astronomiska Sällskapet (SAS)
● See "Swedish Astronomical Society"

Swedish Amateur Astronomical Society
(Svensk Amatör Astronomisk Förening - SAAF)
c/o Jan Persson
Eklanda Hage 31
SE-431 49 Mölndal
Telephone: (0)31-277820
Electronic Mail: andersl@saaf.se
WWW: http://www.astro.uu.se/popast/SAAF/astroweb.html
Founded: 1972
Membership: 600
Activities: photography * comets * Sun * meteors * deep sky * optics * telescope making * variable stars
Sections: Deep Sky * Comets * Optics and Telescope Manufacturing * Planets * Spaceflight * Sun * Variable Stars
Periodicals: (4) "Astro" (ISSN 0280-7173)
City Reference Coordinates: 012°01'00"E 57°39'00"N

Swedish Astronomical Society

(Svenska Astronomiska Sällskapet - SAS)
c/o Stockholm Observatory
SE-133 36 Saltsjöbaden
Telephone: (0)8-164483
(0)8-164477
Telefax: (0)8-7174719
Electronic Mail: sas@astro.su.se
dan@astro.su.se
WWW: http://www.astro.su.se/sas/sas.html
http://www.astro.su.se/~anders/sas/sas.html
Founded: 1919
Membership: 1300
Activities: lectures * colloquia
Periodicals: (4) "Astronomisk Tidsskrift" (ISSN 0004-6345)
City Reference Coordinates: 018°18'00"E 59°17'00"N

Swedish Institute of Space Physics

(Institutet för Rymdfysik - IRF)
Box 812
SE-981 28 Kiruna
Telephone: (0)980-79000
Telefax: (0)980-79050
Electronic Mail: irf@irf.se
WWW: http://www.irf.se/
Founded: 1957
Staff: 120
Activities: space physics research * observatory measurements * space physics education * hot-plasma instrumentation
Periodicals: (1) "Annual Report" (ISSN 0284-169x); "Kiruna Geophysical Data" (ISSN 0453-9478), "Ionospheric Data Sweden", "Scientific Reports"
Coordinates: 020°24'00"E 67°50'00"N H420m (Kiruna)
018°48'00"E 64°42'00"N (Lycksele)
017°36'00"E 59°48'00"N (Uppsala)
City Reference Coordinates: 020°24'00"E 67°50'00"N

Swedish Institute of Space Physics, Kiruna Division

(Institutet för Rymdfysik - IRF, Kiruna)
Box 812
SE-981 28 Kiruna
Telephone: (0)980-79000
Telefax: (0)980-79050
Electronic Mail: <userid>@irf.se
WWW: http://www.irf.se/irfK.html
http://www.irf.se/
Founded: 1957 (IRF)
City Reference Coordinates: 020°27'00"E 67°52'00"N

Swedish Institute of Space Physics, Laboratory of Mechanical Waves

(Institutet för Rymdfysik - IRF, Laboratoriet för Mekaniska Vågor - LMV)
Sörfors 634
SE-905 88 Umeå
Telephone: (0)90-30297
Telefax: (0)90-30468
Electronic Mail: <userid>@irf.se
WWW: http://www.irf.se/LMV/
http://www.irf.se/
Founded: 1957 (IRF)
City Reference Coordinates: 020°15'00"E 63°50'00"N

Swedish Institute of Space Physics, Solar-Terrestral Physics Division

(Institutet för Rymdfysik - IRF, Lund)
Solar-Terrestrial Physics Division
Scheelevägen 17
SE-223 70 Lund
Electronic Mail: <userid>@irf.se
WWW: http://www.irfl.lu.se/HeliosHome/irflund.html
http://www.irf.se/
Founded: 1957 (IRF)
City Reference Coordinates: 013°11'00"E 55°42'00"N

Swedish Institute of Space Physics, Umeå Division

(Institutet för Rymdfysik - IRF, Umeå)
c/o University of Umeå
SE-901 87 Umeå
Telephone: (0)90-130505
Telefax: (0)90-166673

Electronic Mail: <userid>@tp.umu.se
                    <userid>@physics.umu.se
WWW: http://www.tp.umu.se/Space/
         http://www.irf.se/
Founded: 1957 (IRF)
Staff: 6
Activities: space plasma theory
City Reference Coordinates: 020°15'00"E 63°50'00"N

Swedish Institute of Space Physics, Uppsala Division

(Institutet för Rymdfysik - IRF, Uppsala)
SE-755 91 Uppsala
Telephone: (0)18-303600
Telefax: (0)18-403100
Electronic Mail: <userid>@irfu.se
WWW: http://www.irfu.se/
         http://www.irf.se/
Founded: 1952 (IRF: 1957)
Staff: 31
Activities: magnetospheric and ionospheric physics * space plasma physics
Periodicals: "Annual Report"
City Reference Coordinates: 017°38'00"E 59°52'00"N

Swedish Meteorological and Hydrological Institute (SMHI)

Folkborgsvägen 1
SE-601 76 Norrköping
Telephone: (0)11-158000
Telefax: (0)11-170207
         (0)11-170208
Electronic Mail: <userid>@smhi.se
WWW: http://www.smhi.se/
Founded: 1873
Staff: 680
Activities: meteorological, hydrographical and oceanographical service
Periodicals: "Meteorology Report", "Hydrology Report", "Oceanography Report"
City Reference Coordinates: 016°11'00"E 58°36'00"N

Swedish National Space Board (SNSB)

Box 4006
SE-171 04 Solna
or :
Albygatan 107
Solna
Telephone: (0)8-6276480
Telefax: (0)8-6275014
Electronic Mail: rymdstyrelsen@snsb.se
                    <userid>@snsb.se
WWW: http://nos.snsb.se/
Founded: 1972
Staff: 10
Membership: 7
Activities: initiating R&D * coordinating Swedish activities within the fields of space technology, research and remote sensing
City Reference Coordinates: 018°01'00"E 59°22'00"N

Swedish Natural Science Research Council

(Naturvetenskapliga Forskningsrådet - NFR)
Box 7142
SE-113 87 Stockholm
or :
Regeringsgatan 56
3rd floor
SE-113 87 Stockholm
Telephone: (0)8-4544200
Telefax: (0)8-4544250
Electronic Mail: nfr@nfr.se
WWW: http://www.nfr.se/
Founded: 1977
Staff: 38
Activities: allocating research grants * creating research positions * supporting international collaborations * research information
Periodicals: (4) "carpe scientiam" (ISSN 1403-3542); (1) "NFR Yearbook"
City Reference Coordinates: 018°03'00"E 59°20'00"N

Swedish Physical Society

c/o Lotten Hägg (Secretary)
Manne Siegbahn Laboratory

Stockholm University
Frescativ. 24
SE-104 05 Stockholm
Telephone: (0)8-161021
Telefax: (0)8-158674
Electronic Mail: sfs@atom.msi.se
WWW: http://sfs.msi.se/
Founded: 1920
Membership: 950
City Reference Coordinates: 018°03'00"E 59°20'00"N

Swedish Space Corp. (SSC), Esrange
Box 802
SE-981 28 Kiruna
Telephone: (0)980-72000
Telefax: (0)980-12890
Electronic Mail: <userid>@ssc.se
WWW: http://www.ssc.se/
Founded: 1966
Activities: launching sounding rockets * releasing scientific balloons * satellite receiving and control stations
Coordinates: 021°04'00"E 67°56'00"N
City Reference Coordinates: 021°04'00"E 67°56'00"N

Swedish Space Corp. (SSC), Satellitbild
Box 816
SE-981 28 Kiruna
or :
Rymdhuset Österleden 15
SE-981 28 Kiruna
Telephone: (0)980-67100
Telefax: (0)980-16044
Electronic Mail: <userid>@ssc.se
                    custsupp@ssc.se
WWW: http://www.ssc.se/sb
Founded: 1982
Staff: 55
City Reference Coordinates: 020°27'00"E 67°52'00"N

Swedish Space Corp. (SSC), Solna
Albygatan 107
Box 4207
SE-171 04 Solna
Telephone: (0)8-6276200
Telefax: (0)8-987069
Electronic Mail: <userid>@ssc.se
WWW: http://www.ssc.se/
Founded: 1972
Staff: 320
Activities: design and development (satellites, equipment for sounding rockets, balloons, satellite navigation, ...) * launches (sounding rockets, balloons) * operation and control (satellites) * telecommunication services * data collection, processing, archiving and dissemination * value-added products * mapping services * remote sensing * methodologies * airborne maritime surveillance systems
City Reference Coordinates: 018°01'00"E 59°22'00"N

Swedish Standards Institution (SIS)
(Standardiseringen i Sverige - SIS)
St Eriksgatan 115
Box 6455
SE-113 82 Stockholm
or :
Box 3295
SE-103 66 Stockholm
Telephone: (0)8-6135200
            (0)8-6103000
Telefax: (0)8-4117035
          (0)8-307757
Electronic Mail: info@sis.se
WWW: http://www.sis.se/
Founded: 1922
Staff: 75
Activities: standardization
Periodicals: "Månadens Standard", "Teknik & Standard"
City Reference Coordinates: 018°03'00"E 59°20'00"N

Teknikens Hus
(House of Technology)

r,.Museums!Technology!Luleå
Högskoleområdet
SE-971 87 Luleå
Telephone: (0)920-72205
Telefax: (0)920-72202
Electronic Mail: age@teknikens-hus.se (Ann-Gerd Eriksson, Science Educator)
WWW: http://www.luth.se/th/
Founded: 1988
Staff: 16
Activities: science education * planetariums
City Reference Coordinates: 022°10'00"E 65°34'00"N

Tierps Astronomiska Klubb (TAK)
c/o Daniel Söderström
Sjukarby 2509
SE-815 92 Tierp
Telephone: (0)293-12254
Electronic Mail: astro.tak@swipnet.se
WWW: http://home1.swipnet.se/~w-12155
Founded: 1994
Membership: 20
City Reference Coordinates: 017°30'00"E 60°20'00"N

Tom Tits Experiment
Storgatan 33
SE-151 36 Södertälje
Telephone: (0)8-52252500
Telefax: (0)8-52252510
Electronic Mail: info@tomtit.se
WWW: http://www.tomtit.se/Planetary/Planetary_uk.html
Founded: 1987
City Reference Coordinates: 017°39'00"E 59°11'00"N

Tycho Brahe Astronomical Society
(Astronomiska Sällskapet Tycho Brahe - ASTB)
c/o Peter Linde
Box 43
SE-221 00 Lund
Telephone: (0)40-21 23 40
                (0)418-19946 (Bengt Rosengren, Secretary)
Electronic Mail: peter@astro.lu.se
Founded: 1937
WWW: http://www.astro.lu.se/~tycho/
City Reference Coordinates: 013°11'00"E 55°42'00"N

Tycho Brahe Observatory (TBO)
Box 2
SE-238 21 Oxie
Telephone: (0)40-547012
Electronic Mail: jan@astro.lu.se (Jan Sonnvik)
WWW: http://www.astro.lu.se/~jan/tbobs.html
                http://www.tbobs.lu.se/
Coordinates: 013°05'12"E 55°32'34"N
City Reference Coordinates: 013°04'00"E 55°33'00"N

Uppsala Amatörastronomer (UAA)
c/o Johan Warell (President)
Wennerbergsgatan 11
SE-754 21 Uppsala
Telephone: (0)18241342
Electronic Mail: johwar@astro.uu.se
WWW: http://www.astro.uu.se/uaa/
Founded: 1980
Membership: 60
Activities: observing * meetings * instrumentation * photography
Observatories: 1 (Sandvreten)
City Reference Coordinates: 017°38'00"E 59°52'00"N

Uppsala University, Astronomical Observatory
Box 515
SE-751 20 Uppsala
Telephone: (0)18-530265
Telefax: (0)18-527583
Electronic Mail: astro@astro.uu.se
                <userid>@astro.uu.se

WWW: http://www.astro.uu.se/
Founded: 1650
Staff: 37
Activities: extragalactic research * galactic structure * stellar atmospheres * solar system * observational astrophysics
Periodicals: "Uppsala Astronomical Observatory Reports", "Uppsala Preprints in Astronomy"
Coordinates: 017°37'30"E 59°51'30"N H21m
                017°36'24"E 59°30'06"N H35m (Kvistaberg Station - see separate entry)
                149°04'00"E 31°16'36"S H1,164m (Uppsala Southern Station)
City Reference Coordinates: 017°38'00"E 59°52'00"N

Uppsala University, Astronomical Observatory, Kvistaberg Station

SE-197 91 Bro
Telephone: (0)8-58240157
Telefax: (0)8-58240157
WWW: http://www.astro.uu.se/
Staff: 1
Founded: 1944
Activities: Schmidt observing * photoelectric photometry * CCDs * minor planets
Coordinates: 017°36'24"E 59°30'06"N H35m
City Reference Coordinates: 017°38'00"E 59°31'00"N

Uppsala University, Department of Geophysics

Villavägen 16
SE-752 36 Uppsala
Telephone: (0)18182370
Telefax: (0)18501110
Electronic Mail: <userid>@geophys.uu.se
WWW: http://www.geofys.uu.se/
Founded: 1961
Activities: space geodesy * planetary physics * geodynamics
City Reference Coordinates: 017°38'00"E 59°52'00"N

Västerås Astronomi och Rymdforsknings Förening

c/o Sven-Erik Persson
Dragverksgatan 29
SE-724 74 Västerås
Telephone: (0)21-356783
Electronic Mail: vart@vasteras.mail.teia.com
WWW: http://w1.213.telia.com/~u21304183
Founded: 1989
Membership: 33
Activities: popularization
Periodicals: "A&R-bladet"
City Reference Coordinates: 016°32'00"E 59°36'00"N

Victor Hasselblad Foundation (Erna and_)

● See "Erna and Victor Hasselblad Foundation"

Switzerland

Académie Suisse des Sciences Naturelles (ASSN)
● See "Schweizerische Akademie der Naturwissenschaften (SANW)"

Accademia Svizzera di Scienze Naturali (ASSN)
● See "Schweizerische Akademie der Naturwissenschaften (SANW)"

Association des Amis de l'Observatoire des Creusets à Arbaz (AOCA)
c/o Lycée-Collège des Creusets
Rue St-Guérin 34
CH-1950 Sion
Telephone: (0)27-3222930
Telefax: (0)27-3237920
WWW: http://www.cobweb.ch/obs-creusets/
Founded: 1995
City Reference Coordinates: 007°22'00"E 46°14'00"N (Sion)

Astroclub Solaris Aarau (ASA)
c/o Alte Kantonsschule
Bahnhofstrasse 91
CH-5001 Aarau
Telephone: (0)62-8271177
Electronic Mail: christian.roth@ibpv.unil.ch (Christian Roth, Treasurer)
WWW: http://homer.span.ch/~spaw2173/ASA/welcome.html
        http://homer.span.ch/~spaw2173/ASA/nuetziweid/Nuetziweid.html (Volkssternwarte Nütziweid)
Founded: 1979
Membership: 113
Activities: education * trips * observing
Coordinates: 008°03'00"E 47°24'00"N H707m (Volkssternwarte Nütziweid)
City Reference Coordinates: 008°03'00"E 47°24'00"N

Astronomie-Verein Olten (AVO)
c/o Marcel Lips
Allmendstrasse 40
CH-4658 Däniken
Telephone: (0)62-653259
          (0)62-2913259
Founded: 1977
Membership: 35
Activities: observing * meetings
Periodicals: "Newsletter"
Coordinates: 007°53'00"E 47°23'00"N H785m (Froburg)
City Reference Coordinates: 007°53'00"E 47°23'00"N

Astronomische Gesellschaft Baden (AGB)
c/o Jean-Marc Schweizer
Sooremattstrasse 6
CH-5212 Hausen bei Brugg
Telephone: (0)56-416703
WWW: http://www.astroinfo.ch/clubs/agb/
Founded: 1951
Membership: 70
Activities: lectures * displays * trips * discussions
Coordinates: 008°06'45"E 47°31'40"N H665m (Cheisacker)
City Reference Coordinates: 008°19'00"E 47°28'00"N (Baden)
                                008°13'00"E 47°29'00"N (Brugg)

Astronomische Gesellschaft Bern (AGB)
Hangweg
CH-3148 Lanzenhäusern
Telephone: (0)31-7311604
Founded: 1923
Membership: 242
Activities: lectures * discussions * observing
Coordinates: 007°25'42"E 46°57'12"N H550m (Muesmatt)
City Reference Coordinates: 007°26'00"E 46°57'00"N (Bern)

Astronomische Gesellschaft Oberwallis (AGO)
c/o Rudolf Arnold
Wierystrasse 101
CH-3902 Brig-Glis

Telephone: (0)27-9241387
Founded: 1982
Membership: 62
Activities: observing * lectures
Periodicals: "AGO-Mitteilungen"
City Reference Coordinates: 008°00'00"E 46°19'00"N

Astronomische Gesellschaft Solothurn (AGS)

c/o Fred Nicolet
Jupiterstrasse 6
CH-4500 Solothurn
Telephone: (0)32-6223020
Founded: 1954
Membership: 30
City Reference Coordinates: 007°32'00"E 47°13'00"N

Astronomische Gesellschaft Zürcher Oberland (AGZO)

c/o Walter Brändli
Oberer Hömel 32
CH-8636 Wald
Telephone: (0)55-954212
           (0)55-2461763
Telefax: (0)55-955162
Electronic Mail: astro@pax.eunet.ch
                  christoph.bosshard@astroinfo.ch (Christoph Bosshard)
WWW: http://www.astroinfo.ch/clubs/agzo/
Founded: 1967
Membership: 55
Activities: observing * education
Coordinates: 008°54'40"E 47°15'44"N H753m
City Reference Coordinates: 008°33'00"E 47°23'00"N

Astronomische Gesellschaft Zürcher Unterland (AGZU)

c/o Urs Stich
Gerstmattstrasse 41
CH-8172 Niederglatt
Telephone: (0)1-8506319
Telefax: (0)1-8506319
Electronic Mail: agzu@astronomie.ch
WWW: http://agzu.astronomie.ch/
Founded: 1970
Membership: 200
Activities: observing * lectures * excursions * special activities for young members
Periodicals: (6) "Himmelsbeobachter"
Observatories: 1 (Schul- und Volkssternwarte Bülach - see separate entry)
City Reference Coordinates: 008°31'00"E 47°29'00"N (Niederglatt)
                                  008°33'00"E 47°23'00"N (Zürich)

Astronomische Gruppe des Kantons Glarus

c/o Paul Zimmermann
Rufistrasse 4
CH-8762 Schwanden
Telephone: (0)58-617301
           (0)55-6442614
Telefax: (0)58-617301
Founded: 1961
Membership: 22
Activities: observing * education
City Reference Coordinates: 009°04'00"E 47°00'00"N

Astronomische Vereinigung Frauenfeld (AVF)

Thundorferstraße 139
CH-8500 Frauenfeld
Electronic Mail: p_guhl@eudoramail.com
WWW: http://www.astroinfo.ch/clubs/avf/
Founded: 1995
Observatories: 1 (Oberherten)
City Reference Coordinates: 008°54'00"E 47°34'00"N

Astronomische Vereinigung Kreuzlingen (AVK)

c/o Robert Testa
Waldheimstrasse 1
CH-8280 Kreuzlingen
Telephone: (0)71-6725301
           (0)71-6725855 (Observatory)

Telefax: (0)71-6725301
Founded: 1972
Membership: 250
Periodicals: (1) "Jahresbericht"
Coordinates: 009°09'41"E 47°38'33"N H481m
City Reference Coordinates: 009°11'00"E 47°39'00"N

Astronomische Vereinigung Sankt Gallen
c/o Rolf Burgstaller
Grünaustrasse 5
CH-9053 Teufen
Telephone: (0)71-3331374
Founded: 1955
Membership: 92
Activities: popularization * observing
City Reference Coordinates: 009°23'00"E 47°23'00"N

Astronomische Vereinigung Zürich (AVZ)
c/o Andreas Inderbitzin (President)
Winterthurerstrasse 420
CH-8051 Zürich
or :
c/o Dieter Späni
Bachmattstrasse 9
CH-8618 Oetwill-am-See
Telephone: (0)1-3067523
          (0)1-9291127
Electronic Mail: inderbitzin.a@bluewin.ch
WWW: http://www.astroinfo.ch/clubs/avz/
Founded: 1949
Membership: 190
Activities: observing
Periodicals: (4) "AVZ-Mitteilung"
City Reference Coordinates: 008°33'00"E 47°23'00"N (Zürich)

Astronomy Software for PCs
c/o Christian Nuesch
Haldenstrasse 12
CH-8320 Fehraltorf
Telephone: (0)1-955-13-93
Electronic Mail: nuesch@active.ch
WWW: http://www.astronomy.ch/
City Reference Coordinates: 008°45'00"E 47°23'00"N

Astrooptik Kohler (AOK)
Emmenweidstrasse
Bau 607/M4
CH-6020 Emmenbrücke
Telephone: (0)41-2601677
Telefax: (0)41-2601677
Electronic Mail: aokswiss@access.ch
WWW: http://www.interconnect.ch/customers/aokswiss/menu.html
Founded: 1988
Staff: 1
Activities: developing and producing astronomical telescope systems * distributing telescopes and optical accessories
City Reference Coordinates: 008°17'00"E 47°04'00"N

Bielser Observatorien
Hauptstrasse 22
CH-4132 Muttenz
Telephone: (0)79-6590414
Telefax: (0)61-4618177
Electronic Mail: bielser.gerold@datacomm.ch
WWW: http://www.astroinfo.org/bielser/
Activities: manufacturing domes
City Reference Coordinates: 007°39'00"E 47°31'00"N

Birkhäuser Verlag AG
Postfach 133
CH-4010 Basel
or :
Klosterberg 23
CH-4051 Basel
Telephone: (0)61-2050707
Telefax: (0)61-2050799

Electronic Mail: sales@birkhauser.ch
WWW: http://www.birkhauser.ch/
Founded: 1879
Staff: 65
• Publisher
City Reference Coordinates: 007°35'00"E 47°33'00"N

Centre Européen pour la Recherche Nucléaire (CERN)
(European Organization for Nuclear Research)
Route de Meyrin
CH-1211 Genève 23
Telephone: (0)22-7676111 (central switchboard)
           (0)22-7672210 (reception desk)
Telefax: (0)22-7677555 (central fax)
Electronic Mail: libdesk@cernvm.cern.ch (Scientific Information Service)
WWW: http://www.cern.ch/
         http://info.cern.ch/
Founded: 1954
Staff: 10000
Activities: theoretical and experimental research in elementary-particle physics * accelerator and detector design and development
Periodicals: (6) "CERN Courier" (ISSN 0304-288x), "Courrier CERN" (ISSN 0374-2288), "Liste des Publications Scientifiques" (ISSN 0304-2871), "Annual Report" (ISSN 0304-2901), "Rapport Annuel" (ISSN 0304-291x), "CERN-HERA Reports" (ISSN 0366-5690), "CERN Reports" (ISSN 0007-8328), "CERN School of Physics Proceedings" (ISSN 0531-4283)
City Reference Coordinates: 006°09'00"E 46°12'00"N

Club d'Astronomes Amateurs de Bienne (CAAB)
c/o Claudio Cerini
Route d'Aegerten 31
CH-2503 Bienne
Telephone: (0)32-3658074
Founded: 1981
Membership: 20
Activities: popularization * observing * photography
City Reference Coordinates: 007°12'00"E 47°10'00"N

Community of European Solar Radio Astronomers (CESRA)
c/o A.O. Benz
Institut für Astronomie
Eidgenössische Technische Hochschule
ETH-Zentrum
CH-8092 Zürich
Telephone: (0)1-6324223
Telefax: (0)1-6321205
Electronic Mail: benz@astro.phys.ethz.ch
Founded: 1972
Membership: 180
Activities: workshops * conferences * information exchange
City Reference Coordinates: 008°33'00"E 47°23'00"N

Comsol AG
Technopark
Morgenstrasse 129
CH-3018 Bern
Telephone: (0)31-9984411
Telefax: (0)31-9984418
Electronic Mail: info@comsol.ch
WWW: http://www.comsol.ch/
Founded: 1990
Staff: 8
Activities: specialized software for R&D * consultancy
City Reference Coordinates: 007°26'00"E 46°57'00"N

Consiglio Nazionale delle Ricerche (CNR), Centro per l'Astronomia Infrarossa e lo Studio del Mezzo Interstellare (CAISMI), Gornergrat Stazione
CH-3920 Zermatt
Telephone: (0)27-9671219
Telefax: (0)27-9674850
WWW: http://www.arcetri.astro.it/irlab/tirgo/
Founded: 1981 (CAISMI)
Activities: IR observing
Coordinates: 007°47'30"E 45°59'04"N H3,135m
City Reference Coordinates: 007°45'00"E 46°01'00"N

Cryophysics SA
Rue Rotschild 39
CH-1202 Genève
Telephone: (0)22-7329520
Telefax: (0)22-7385246
Electronic Mail: 100763.3270@compuserve.com
WWW: http://ourworld.compuserve.com/homepages/cryophysicsch
Founded: 1967
Staff: 80
Activities: manufacturing and distributing cryogenic equipment
Periodicals: "Cryophysics Newsletter"
City Reference Coordinates: 006°09'00"E 46°12'00"N

École Polytechnique Fédérale
● See "Eidgenössische Technische Hochschule (ETH)"

Eidgenössische Technische Hochschule (ETH), Institut für Astronomie
ETH-Zentrum
CH-8092 Zürich
Telephone: (0)1-6323813
Telefax: (0)1-6321205
Electronic Mail: <userid>@astro.phys.ethz.ch
WWW: http://www.astro.phys.ethz.ch/
http://www.ethz.ch/
Staff: 26
Activities: solar and stellar physics * radio astronomy * instrumentation
Coordinates: 009°40'06"E 46°47'00"N H2,050m
City Reference Coordinates: 008°33'00"E 47°23'00"N

Eidgenössische Technische Hochschule (ETH), Institut für Geophysik
Hönggerberg
CH-8093 Zürich
Telephone: (0)1-6332605
Telefax: (0)1-6331065
Electronic Mail: <userid>@ifg.ethz.ch
<userid>@seismo.ig.erdw.ethz.ch
WWW: http://www.geophys.ethz.ch/
http://www.ethz.ch/
City Reference Coordinates: 008°33'00"E 47°23'00"N

Eidgenössische Technische Hochschule (ETH), Laboratorium für Atmosphärenphysik
Hönggerberg HPP
CH-8093 Zürich
Telephone: (0)1-6332755
Telefax: (0)1-6331864
Electronic Mail: sekretariat@atmos.umnw.ethz.ch
WWW: http://www.umnw.ethz.ch/LAPETH/
http://www.ethz.ch/
Founded: 1963
Staff: 24
Activities: atmospheric physics * atmospheric modelling * radar meteorology * cloud physics * climate change * ozone research
City Reference Coordinates: 008°33'00"E 47°23'00"N

Etel SA
Zône Industrielle
CH-2112 Môtiers
Telephone: (0)38-611858
Telefax: (0)38-612419
Electronic Mail: etel@etel.ch
WWW: http://www.etel.ch/
Founded: 1974
Staff: 33
Activities: designing and manufacturing large torque motors for i.a. telescopes
City Reference Coordinates: 006°37'00"E 46°55'00"N

European Astronomical Society (EAS)
Case Postale 82
CH-1213 Petit-Lancy 2
Founded: 1991
WWW: http://www.iap.fr/eas/
● See main entry in Czech Republic
City Reference Coordinates: 006°09'00"E 46°12'00"N (Genève)

European Physical Society (EPS)

Chemin de la Vendée 27
Case Postale 69
CH-1213 Petit-Lancy 2
Telephone: (0)22-7931130
Telefax: (0)22-7931317
WWW: http://www.nikhef.nl/www/pub/eps/eps.html
       http://epswww.epfl.ch/
Founded: 1968
Staff: 4
Membership: 70000
Activities: coordinating conferences * publishing * education
Periodicals: "Europhysics News" (ISSN 0531-7479), "Europhysics Letters" (ISSN 0295-5075); "European Journal of Physics", "Europhysics Conference Abstracts"
Awards: (1) "Hewlett-Packard Europhysics Prize"; (1/2) "High-Energy and Particle Physics Prize"
City Reference Coordinates: 006°09'00"E 46°12'00"N (Genève)

Ferrovia Monte Generoso (FMG) SA

CH-6825 Capolago
Telephone: (0)91-6481105
Telefax: (0)91-6481107
Electronic Mail: info@montegeneroso.ch
WWW: http://www.montegeneroso.ch/
Founded: 1890
Staff: 20
Activities: touristic rack railway * sporting and recreational activities * excursions * restaurant * public observatory
Coordinates: 008°59'00"E 45°55'00"N H1870m
City Reference Coordinates: 008°59'00"E 45°55'00"N

Fondation de l'Observatoire François-Xavier Bagnoud

CH-3961 Saint-Luc
Telephone: (0)27-655808
Telefax: (0)27-652237
Electronic Mail: breguetj@aletsch.esis.vsnet.ch (Jacques Bréguet)
WWW: http://www.icare.ch/OFXB
       http://www.icare.ch/OFXB/welcomef.htm
Founded: 1993
Staff: 1
Activities: observing (Sun, night sky) for tourists and schools * instrumentation available for good amateurs
Coordinates: 007°36'36"E 46°13'51"N H2,200m
City Reference Coordinates: 007°37'00"E 46°13'00"N

Fondation Jungfraujoch-Gornergrat

● See "Internationale Stiftung Hochalpine Forschungsstationen Jungfraujoch und Gornergrat (HFSJG)"

Fondation Robert-A. Naef

Route du Petit-Épendes 45
CH-1731 Épendes
Telephone: (0)26-4131099
Electronic Mail: info@observatoire-naef.ch
WWW: http://www.observatoire-naef.ch/
Coordinates: 007°08'00"E 46°45'00"N H700m
City Reference Coordinates: 006°37'00"E 46°46'00"N

Fondo Nazionale Svizzero per la Ricerca Scientifica

● See "Schweizerischer Nationalfonds zur Förderung der Wissenschaftlichen Forschung"

Fonds National Suisse de la Recherche Scientifique

● See "Schweizerischer Nationalfonds zur Förderung der Wissenschaftlichen Forschung"

Forschungsstationen Jungfraujoch und Gornergrat (HFSJG) (Internationale Stiftung Hochalpine_)

● See "Internationale Stiftung Hochalpine Forschungsstationen Jungfraujoch und Gornergrat (HFSJG)"

Gesellschaft der Freunde der Urania-Sternwarte (GFUS)

c/o Volkshochschule des Kantons Zürich
Splügenstraße 10
CH-8002 Zürich
Telephone: (0)1-2058484
Telefax: (0)1-2059495
Electronic Mail: vhszh@access.ch
Founded: 1936
Activities: supporting and managing Urania Observatory
Coordinates: 008°32'26"E 47°22'32"N H408m
City Reference Coordinates: 008°33'00"E 47°23'00"N

Gornergrat Station
● See "Consiglio Nazionale delle Ricerche (CNR), Centro per l'Astronomia Infrarossa e lo Studio del Mezzo Interstellare (CAISMI), Gornergrat Stazione"
● See also "Internationale Stiftung Hochalpine Forschungsstationen Jungfraujoch und Gornergrat (HFSJG)"
● See also "Internationale Stiftung Hochalpine Forschungsstationen Jungfraujoch und Gornergrat (HFSJG), Gornergrat Station"

Institut Suisse de Météorologie (ISM)
● See "Istituto Svizzero di Meteorologia (ISM)"
● See also "Schweizerische Meteorologische Anstalt (SMA)"

Internationale Stiftung Hochalpine Forschungsstationen Jungfraujoch und Gornergrat (HFSJG)
Sidlerstrasse 5
CH-3012 Bern
Telephone: (0)31-6314052
        (0)22-7552611 (Bernard Nicolet)
Telefax: (0)31-6314405
     (0)22-7553983 (Bernard Nicolet)
Electronic Mail: debrunner@phim.unibe.ch
            nicolet@scsun.unige.ch (Bernard Nicolet)
Founded: 1931
Activities: running the Jungfraujoch and Gornergrat scientific stations (see also these entries)
City Reference Coordinates: 007°26'00"E 46°57'00"N

Internationale Stiftung Hochalpine Forschungsstationen Jungfraujoch und Gornergrat (HFSJG), Gornergrat Station
CH-3920 Zermatt
Telephone: (0)27-9671219 (Northern Tower)
        (0)27-9672715 (Southern Tower)
Telefax: (0)27-9674850 (Northern Tower)
Activities: IR and sub-mm spectroscopy and photometry
Founded: 1931 (HFSJG)
Coordinates: 007°47'04"E 45°59'04"N H3,130m
City Reference Coordinates: 007°45'00"E 46°01'00"N

Internationale Stiftung Hochalpine Forschungsstationen Jungfraujoch und Gornergrat (HFSJG), Jungfraujoch Station
CH-3801 Jungfraujoch
Telephone: (0)33-8568950 (Research Station)
        (0)31-6314052 (Administration)
Telefax: (0)31-6314405 (Administration)
Founded: 1930
Activities: environmental sciences * meteorology * glaciology * astronomy * medicine and physiology
Coordinates: 007°59'02"E 46°32'53"N H3,450m (Scientific Station) H3,580m (Observatory)
City Reference Coordinates: 007°59'00"E 46°33'00"N

International Organization for Standardization (ISO)
Rue de Varembe 1
Case Postale 56
CH-1211 Genève 20
Telephone: (0)22-7490111
Telefax: (0)22-7333430
Electronic Mail: central@iso.ch
WWW: http://www.iso.ch/
Founded: 1947
Membership: 129 (countries)
Staff: 167
Activities: international standardization in all fields (except electrical and electronic engineering) through technical committees (see list below)
Periodicals: (12) "ISO Bulletin" (ISSN 0303-805x); (1) "ISO Memento", "ISO Catalogue"; "ISO 9000 News"
Sections: (technical committees) 1. Screw Threads * 2. Fasteners * 4. Rolling Bearings * 5. Ferrous Metal Pipes and Metallic Fittings * 6. Paper, Board and Pulps * 8. Ships and Marine Technology * 10. Technical Drawings, Product Definition and Related Documentation * 11. Boilers and Pressure Vessels * 12. Quantities, Units, Symbols, Conversion Factors * 14. Shafts for Machinery and Accessories * 17. Steel * 18. Zinc and Zinc Alloys * 20. Aircraft and Space Vehicles * 21. Equipment for Fire Protection and Fire Fighting * 22. Road Vehicles * 23. Tractors and Machinery for Agriculture and Forestry * 24. Sieves, Sieving and Other Sizing Methods * 25. Cast Iron and Pig Iron * 26. Copper and Copper Alloys * 27. Solid Mineral Fuels * 28. Petroleum Products and Lubricants * 29. Small Tools * 30. Measurement of Fluid Flow in Closed Conduits * 31. Tyres, Rims and Valves * 33. Refractories * 34. Agricultural Food Products * 35. Paints and Varnishes * 36. Cinematography * 37. Terminology (Principles and Coordination) * 38. Textiles * 39. Machine Tools * 41. Pulleys and Belts (including Veebelts) * 42. Photography * 43. Acoustics * 44. Welding and Allied Processes * 45. Rubber and Rubber Products * 46. Information and Documentation * 47. Chemistry * 48. Laboratory Glassware and Related Apparatus * 51. Pallets for Unit Load Method of Materials Handling * 52. Light Gauge Metal Containers * 54. Essential Oils * 55. Sawn Timber and Sawlogs * 58. Gas Cylinders * 59. Building Construction * 60. Gears * 61. Plastics * 65. Manganese and Chromium Ores * 67. Materials, Equipment and Offshore Structures for Petroleum and Natural Gas Industries * 68. Banking, Securities and Other Financial Services * 69. Applications of Statistical Methods * 70. Internal Combustion Engines * 71. Concrete, Reinforced Concrete and Pre-stressed Concrete * 72. Textile Machinery and Machinery for Dry-Cleaning and Industrial Laundering * 74. Cement

and Lime * 76. Transfusion, Infusion and Injection Equipment for Medical Use * 77. Products in Fibre-Reinforced Cement * 79. Light Metals and their Alloys * 81. Common Names for Pesticides and Other Agrochemicals * 82. Mining * 83. Sports and Recreational Equipment * 84. Medical Devices for Injections * 85. Nuclear Energy * 86. Refrigeration * 87. Cork * 89. Wood-Based Panels * 91. Surface Active Agents * 92. Fire Safety * 93. Starch (including Derivatives and By-Products) * 94. Personal Safety - Protective Clothing and Equipment * 96. Cranes * 98. Bases for Design of Structures * 99. Semi-Manufactures of Timber * 100. Chains and Chain Wheels for Power Transmission and Conveyors * 101. Continuous Mechanical Handling Equipment * 102. Iron Ores * 104. Freight Containers * 105. Steel Wire Ropes * 106. Dentistry * 107. Metallic and Other Inorganic Coatings * 108. Mechanical Vibration and Shock * 110. Industrial Trucks * 111. Round Steel Link Chains, Chain Slings, Components and Accessories * 112. Vacuum Technology * 113. Hydrometric Determinations * 114. Horology * 115. Pumps * 116. Space Heating Appliances * 117. Industrial Fans * 118. Compressors, Pneumatic Tools and Pneumatic Machines * 119. Powder Metallurgy * 120. Leather * 121. Anaesthetic and Respiratory Equipment * 122. Packaging * 123. Plain Bearings * 126. Tobacco and Tobacco Products * 127. Earth-Moving Machinery * 130. Graphic Technology * 131. Fluid Power Systems * 132. Ferroalloys * 134. Fertilizers and Soil Conditioners * 135. Non-Destructive Testing * 136. Furniture * 137. Sizing System, Designations and Marking for Boots and Shoes * 138. Plastics Pipes, Fittings and Valves for the Transport of Fluids * 145. Graphical Symbols * 146. Air Quality * 147. Water Quality * 148. Sewing Machines * 149. Cycles * 150. Implants for Surgery * 153. Valves * 154. Documents and Data Elements in Administration, Commerce and Industry * 155. Nickel and Nickel Alloys * 156. Corrosion of Metals and Alloys * 157. Mechanical Contraceptives * 158. Analysis of Gases * 159. Ergonomics * 160. Glass in Building * 161. Control and Safety Devices for Non-Industrial Gas-Fired Appliances and Systems * 162. Doors and Windows * 163. Thermal Insulation * 164. Mechanical Testing of Metals * 165. Timber Structures * 166. Ceramic Ware, Glassware and Glass Ceramic Ware in Contact with Food * 167. Steel and Aluminium Structures * 168. Prosthetics and Orthotics * 170. Surgical Instruments * 171. Document Imaging Applications * 172. Optics and Optical Instruments * 173. Technical Systems and Aids for Disabled or Handicapped Persons * 174. Jewellery * 175. Fluorspar * 176. Quality Management and Quality Assurance * 177. Caravans * 178. Lifts, Escalators, Passenger Conveyors * 179. Masonry * 180. Solar Energy * 181. Safety of Toys * 182. Geotechnics * 183. Copper, Lead and Zinc Ores and Concentrates * 184. Industrial Automation Systems and Integration * 185. Safety Devices for Protection Against Excessive Pressure * 186. Cutlery and Table and Decorative Metal Hollow-Ware * 187. Colour Notations * 188 Small Craft * 189. Ceramic Tile * 190. Soil Quality * 191. Animal (Mammal) Traps * 192. Gas Turbines * 193. Natural Gas * 194. Biological Evaluation of Medical Devices * 195. Building Construction Machinery and Equipment * 196. Natural Stone * 197. Hydrogen Technologies * 198. Sterilization of Health Care Products * 199. Safety of Machinery * 201. Surface Chemical Analysis * 202. Microbeam Analysis * 203. Technical Energy Systems * 204. Transport Information and Control Systems * 205. Building Environment Design * 206. Fine Ceramics * 207. Environmental Management * 208. Thermal Turbines for Industrial Application (Steam Turbines, Gas Expansion Turbines) * 209. Cleanrooms and Associated Controlled Environments * 210. Quality Management and Corresponding General Aspects for Medical Devices * 211. Geographic Information/Geometrics * 212. Clinical Laboratory Testing and In Vitro Diagnostic Test Systems * 213 Dimensional and Geometrical Product Specifications and Verification * 214 Elevating Work Platforms * 215 Health Informatics * 216 Footwear * 217 Cosmetics
City Reference Coordinates: 006°09'00"E 46°12'00"N

International Space Science Institute (ISSI)
Hallerstrasse 6
CH-3012 Bern
Telephone: (0)31-6314896
Telefax: (0)31-6314897
Electronic Mail: <userid>@issi.unibe.ch
WWW: http://ubeclu.unibe.ch/issi/
Founded: 1995
Staff: 12
Activities: contributing to the achievement of a deeper understanding of the results from space-research missions, adding value to those results through multi-disciplinary research in an atmosphere of international cooperation
City Reference Coordinates: 007°26'00"E 46°57'00"N

International Telecommunication Union (ITU)
(Union Internationale des Télécommunications - UIT)
Place des Nations
CH-1211 Genève 20
Telephone: (0)22-7305111
Telefax: (0)22-7337256
Electronic Mail: itumail@itu.ch
WWW: http://www.itu.ch/
Founded: 1934 (1865 as former "International Telegraph Union")
Membership: 185 (countries)
Staff: 800
Activities: maintaining and extending international cooperation between all members * promoting and offering technical assistance to developing countries in the field of telecommunications * promoting the development of technical facilities and their most efficient operation * harmonizing national actions to these ends * coordinating efforts to eliminate harmful interferences * improving the use made of the radio frequency spectrum * coordinating efforts to harmonizing the development of telecommunication space techniques
Periodicals: (12) "ITU News" (ISSN 0497-137x)
City Reference Coordinates: 006°09'00"E 46°12'00"N

International Union of Amateur Astronomers (IUAA), European Section
c/o Rinaldo Roggero
Via R. Simen 3
CH-6600 Locarno
Telephone: (0)91-7515857
Telefax: (0)91-7515857
Electronic Mail: ragan@bluewin.ch

Founded: 1989
Membership: 5
Activities: promoting cooperation and exchange of information between astronomical societies within Europe
Coordinates: 008°47'26"E 46°10'02"N H215m
City Reference Coordinates: 008°48'00"E 46°10'00"N

Istituto Ricerche Solari Locarno (IRSOL)
Via Patocchi
CH-6605 Locarno-Monti
Telephone: (0)91-7434226
Telefax: (0)91-7434226
Electronic Mail: mbianda@ccsc.ch (Michele Bianda)
Founded: 1960
Staff: 2
Activities: solar research (visible spectrum)
Coordinates: 008°47'22"E 46°10'39"N H500m
City Reference Coordinates: 008°48'00"E 46°10'00"N

Istituto Svizzero di Meteorologia (ISM), Osservatorio Ticinese di Locarno-Monti
CH-6605 Locarno-Monti
Telephone: (0)93-312773
Telefax: (0)93-317838
WWW: http://www.sma.ch/
Founded: 1935
Staff: 14
Activities: weather forecast * regional climatology * Meteosat image processing * radar meteorology
City Reference Coordinates: 008°48'00"E 46°10'00"N

Jungfraujoch Station
● See "Internationale Stiftung Hochalpine Forschungsstationen Jungfraujoch und Gornergrat (HFSJG)"
● See also "Internationale Stiftung Hochalpine Forschungsstationen Jungfraujoch und Gornergrat (HFSJG), Jungfraujoch Station"

Kölner Observatorium für Submillimeter und Millimeter Astronomie (KOSMA)
● See "Universität Köln, Erstes Physikalisches Institut, Kölner Observatorium für Submillimeter und Millimeter Astronomie (KOSMA)"

LeCroy sa
Rue du Pré-de-la-Fontaine 2
Case Postale 341
CH-1217 Meyrin 1
Telephone: (0)22-7192111
Telefax: (0)22-7823915
WWW: http://www.lecroy.com/
Founded: 1972
Staff: 108
Activities: developing and producing digital oscilloscopes
City Reference Coordinates: 006°05'00"E 46°14'00"N

Leica, Geosystems AG
Heinrich-Wild-Strasse
CH-9435 Heerbrugg
Telephone: (0)71-7273131
Telefax: (0)71-7274674
WWW: http://www.leica.com/
Founded: 1925 (Leica)
● Manufacturer
City Reference Coordinates: 009°37'00"E 47°25'00"N

Les Pléiades - Société d'Astronomie de Saint-Imier
Passage de la Reine-Berthe 5
CH-2610 Saint-Imier
WWW: http://www.eisi.ch/pleiades/
City Reference Coordinates: 007°00'00"E 47°09'00"N

Murmel Spielwerkstatt und Verlag
Hermann Greulich-Strasse 60
CH-8004 Zürich
Telephone: (0)1-4015156
        (0)1-2421718
Telefax: (0)1-4015158
Founded: 1983
Activities: designing, producing and distributing games, including astronomical ones
City Reference Coordinates: 008°33'00"E 47°23'00"N

Musée Suisse des Transports
• See "Verkehrshaus der Schweiz"

Observatoire Cantonal de Neuchâtel
Rue de l'Observatoire 58
CH-2000 Neuchâtel
Telephone: (0)32-8896870
Telefax: (0)32-8896281
Electronic Mail: observatoire.cantonal@ne.ch
WWW: http://www.ne.ch/adm/dep/ocne/
Founded: 1858
Staff: 30
Activities: time * frequencies * atomic clocks * meteorology * geophysics
Periodicals: (1) "Rapport Annuel"
Coordinates: 006°57'30"E 46°59'54"N H488m
City Reference Coordinates: 006°55'00"E 46°59'00"N

Observatoire de Genève
Chemin des Maillettes 51
CH-1290 Sauverny
Telephone: (0)22-7552611
Telefax: (0)22-7553983
Electronic Mail: <userid>@obs.unige.ch
WWW: http://obswww.unige.ch/
Founded: 1772
Staff: 100
Activities: UV and multicolour photometry * stellar evolution * stellar dynamics * high-energy astrophysics * AGNs * radial velocities
Observatories: Jungfraujoch (Switzerland), Saint-Michel-l'Observatoire (France) and La Silla (Chile)
Coordinates: 006°04'00"E 46°18'24"N H455m
City Reference Coordinates: 006°09'00"E 46°12'00"N

Observatoire des Creusets
• See "Association des Amis de l'Observatoire des Creusets à Arbaz (AOCA)"

Observatoire François-Xavier Bagnoud (OFXB)
• See "Fondation de l'Observatoire François-Xavier Bagnoud"

Osservatorio Ticinese di Locarno-Monti
• See "Istituto Svizzero di Meteorologia, Osservatorio Ticinese di Locarno-Monti"

Organisation Météorologique Mondiale (OMM)
• See "World Meteorological Organization (WMO)"

Organisation Mondiale de la Propriété Intellectuelle (OMPI)
• See "World Intellectual Property Organization (WIPO)"

Physikalisch-Meteorologisches Observatorium Davos (PMOD), Weltstrahlungszentrum
(World Radiation Center - WRC)
Dorfstrasse 33
CH-7260 Davos-Dorf
Telephone: (0)81-4175111
Telefax: (0)81-4175100
Electronic Mail: <userid>@pmodwrc.ch
WWW: http://www.pmodwrc.ch/
Founded: 1907
Staff: 18
Activities: solar irradiance * solar variability * helioseismology * atmospheric physics
City Reference Coordinates: 009°50'00"E 46°48'00"N

Planetarium Zürich
Haldenstrasse 138
CH-8055 Zürich
Telephone: (0)1-4625500
Telefax: (0)1-4625501
Electronic Mail: plani@dial.eunet.ch
           planizh@zep.ch
WWW: http://www.echo.ch/-planetarium-zh/
Founded: 1990
Staff: 8
Activities: shows (movable planetarium) * development of instrumentation
City Reference Coordinates: 008°33'00"E 47°23'00"N

P. Wyss Photo-Video
Postfach
CH-8034 Zürich
or :
Dufourstrasse 124
CH-8008 Zürich
Telephone: (0)1-3830108
Telefax: (0)1-3830094
Electronic Mail: wyssproastro@access.ch
Founded: 1982
Staff: 6
• Dealer
City Reference Coordinates: 008°33'00"E 47°23'00"N

Rudolf Wolf Gesellschaft
c/o H.U. Keller (Secretary)
Kolbenhofstrasse 33
CH-8045 Zürich
Telephone: (0)1-4616814
Founded: 1992
Membership: 50
Activities: securing the continuation of Wolf's sunspot observations with Wolf's original telescope
Periodicals: (2) "Mitteilungen der Rudolf Wolf Gesellschaft" (ISSN 1021-8823)
City Reference Coordinates: 008°33'00"E 47°23'00"N

Schul- und Volkssternwarte Bülach
Postfach 282
CH-8180 Bülach
Telephone: (0)1-8608448 (Observatory)
             (0)1-8601221 (Bookings)
Telefax: (0)1-8604954
Electronic Mail: buelach@astronomie.ch
WWW: http://buelach.astronomie.ch/
Founded: 1983
Activities: observing * popularization
Coordinates: 008°34'22"E 47°31'13"N H550m
City Reference Coordinates: 008°32'00"E 47°32'00"N

Schweizerische Akademie der Naturwissenschaften (SANW)
(Académie Suisse des Sciences Naturelles - ASSN)
(Accademia Svizzera di Scienze Naturali - ASSN)
(Swiss Academy of Sciences - SAS)
Bärenplatz 2
CH-3011 Bern
Telephone: (0)31-3123375
Telefax: (0)31-3123291
Electronic Mail: schindler@sanw.unibe.ch
Founded: 1815
Activities: research fostering
Periodicals: (1) "Jahrbuch/Annuaire"; "Mitteilungsblatt/Bulletin d'Information", "Interne Mitteilungen/Communications Internes"
City Reference Coordinates: 007°26'00"E 46°57'00"N

Schweizerische Astronomische Gesellschaft (SAG)
(Société Astronomique de Suisse - SAS)
c/o Sue Kernen (Secretary)
Gristenbühl
CH-9315 Neukirch
Telephone: (0)71-4771743
Electronic Mail: astro_mod_4@ezinfo.vmsmail.ethz.ch
                noel.cramer@obs.unige.ch (Noël Cramer, Orion Editor-in-Chief)
WWW: http://ezinfo.ethz.ch/ezinfo/astro/kontakt/SAS.html
        http://ezinfo.ethz.ch/ezinfo/astro/kontakt/kon2_0.html
Founded: 1938
Membership: 3000
Activities: education * popularization * promoting cooperation between amateur and professional astronomers, as well as between astronomical groups
Periodicals: (6) "Orion" (ISSN 0030-557x, circ.: 2,800)
City Reference Coordinates: 009°23'00"E 47°32'00"N

Schweizerische Gesellschaft für Astrophysik und Astronomie (SGAA)
• See "Société Suisse d'Astrophysique et d'Astronomie (SSAA)"

Schweizerische Meteorologische Anstalt (SMA)
Krähbühlstrasse 58
CH-8044 Zürich
Telephone: (0)1-2569111
Telefax: (0)1-2569278
WWW: http://www.sma.ch/
Founded: 1881
Staff: 201
Activities: meteorology * climatology
● See also "Institut Suisse de Météorologie" and "Istituto Svizzero di Meteorologia"
City Reference Coordinates: 008°33'00"E 47°23'00"N

Schweizerische Normen-Vereinigung (SNV)
(Swiss Association for Standardization)
Mühlebachstraße 54
CH-8008 Zürich
Telephone: (0)1-2545454
Telefax: (0)1-2545474
Electronic Mail: info@snv.ch
WWW: http://www.snv.ch/
Founded: 1919
Membership: 600
Staff: 30
Activities: standardization (national, European, international)
Periodicals: (12) "Switec Information", (11) "SNV Bulletin"
City Reference Coordinates: 008°33'00"E 47°23'00"N

Schweizerische Physikalische Gesellschaft (SPG)
● See "Société Suisse de Physique (SSP)"

Schweizerischer Nationalfonds zur Förderung der Wissenschaftlichen Forschung
(Fonds National Suisse de la Recherche Scientifique)
(Fondo Nazionale Svizzero per la Ricerca Scientifica)
(Swiss National Science Foundation)
Wildhainweg 20
Postfach 8232
CH-3001 Bern
Telephone: (0)31-3082222
Telefax: (0)31-3013009
Electronic Mail: pri@snf.ch
WWW: http://www.snf.ch/
Founded: 1952
Staff: 93
Periodicals: (4) "Horizonte/Horizons"
City Reference Coordinates: 007°26'00"E 46°57'00"N

Società Astronomica Ticinese
c/o Specola Solare Ticinese
CH-6605 Locarno-Monti
Telephone: (0)91-7526376
　　　　　　　 (0)91-7562376
Telefax: (0)91-7526310
Founded: 1961
Membership: 150
Activities: observing (Sun, meteors, variable stars, planets)
Periodicals: (6) "Meridiana"
City Reference Coordinates: 008°48'00"E 46°10'00"N

Société Astronomique de Genève
Terreaux-du-Temple 6
CH-1211 Genève 1
Founded: 1923
Membership: 190
Activities: meetings * observing * popularization
Periodicals: "L'Observateur"
Observatories: 2 (Genève, Saint-Cergue)
City Reference Coordinates: 006°09'00"E 46°12'00"N

Société Astronomique de Suisse (SAS)
● See "Schweizerische Astronomische Gesellschaft (SAG)"

Société d'Astronomie de Saint-Imier
● See "Les Pléiades - Société d'Astronomie de Saint-Imier"

Société d'Astronomie du Haut-Léman (SAHL)
c/o Alain Bollschweiler
Rue du Conseil 13
CH-1800 Vevey
or :
c/o Observatoire
Avenue E. Bieler
CH-1800 Vevey
Telephone: (0)21-9215523 (Observatory)
          (0)21-9227995 (Secretary)
Telefax: (0)21-9227995 (Secretary)
Founded: 1970
Membership: 60
Activities: observing * photography * spectrography
Coordinates: 006°51'04"E 46°28'00"N H461m
City Reference Coordinates: 006°51'00"E 46°28'00"N

Société d'Astronomie du Valais Romand (SAVAR)
c/o Alain Kohler
Route de Vissigen 88
CH-1950 Sion
Telephone: (0)27-2031786
Founded: 1994
Membership: 70
Activities: observing * photography * CCDs
Periodicals: (6) "Bulletin de la Société d'Astronomie du Valais Romand"
City Reference Coordinates: 007°21'00"E 46°14'00"N

Société de Physique et d'Histoire Naturelle (SPHN)
c/o Muséum d'Histoire Naturelle
Route de Malagnon 1
Case Postale 6434
CH-1211 Genève 6
Telephone: (0)22-4186300
Telefax: (0)22-4183445
Founded: 1790
Staff: 12
Membership: 240
Activities: publishing * excursions * interdisciplinary forums * popularization
Periodicals: (3) "Archives des Sciences"; "Mémoires de la Société de Physique et d'Histoire Naturelle"
Awards: "Prix Augustin-Pyramus de Caudolle", "Prix Marc-Auguste Pictet"
City Reference Coordinates: 006°09'00"E 46°12'00"N

Société Jurassienne d'Astronomie (SJA)
c/o Michel Ory
Rue du Béridier 30
CH-2800 Délémont
Telephone: (0)32-4356562
          (0)32-4233156
Electronic Mail: michelory@pingnet.ch (Michel Ory, President)
Founded: 1980
Membership: 120
Activities: observing * photography * CCD
Coordinates: 007°25'20"E 47°21'20"N H505m (Vignes)
City Reference Coordinates: 007°21'00"E 47°22'00"N

Société Neuchâteloise d'Astronomie (SNA)
c/o Raoul Behrend
Bonne-Fontaine 6
CH-2304 La Chaux-de-Fonds
WWW: http://obswww.unige.ch/~behrend/page_sna.html
Founded: 1979
Membership: 100
City Reference Coordinates: 006°50'00"E 47°06'00"N

Société Suisse d'Astrophysique et d'Astronomie (SSAA)
(Schweizerische Gesellschaft für Astrophysik und Astronomie - SGAA)
c/o Roland Buser (President)
Astronomisches Institut
Universität Basel
Venusstrasse 7
CH-4102 Binningen
Telephone: (0)61-2055416
Telefax: (0)61-2055455
Electronic Mail: buser1@ubaclu.unibas.ch
Founded: 1969

Membership: 140
Activities: national society of professional astronomers
City Reference Coordinates: 007°35'00"E 47°33'00"N

Société Suisse de Physique (SSP)
(Schweizerische Physikalische Gesellschaft - SPG)
(Swiss Physical Society - SPS)
c/o Institut für Physik
Universität Basel
Klingelbergstrasse 82
CH-4056 Basel
Telephone: (0)61-2673715
Telefax: (0)61-2673716
Electronic Mail: sps@ubaclu.unibas.ch
WWW: http://www.sps.ch/sps/
Founded: 1908
Membership: 1400
City Reference Coordinates: 007°35'00"E 47°33'00"N

Société Valaisanne des Sciences Naturelles "La Murithienne"
Case Postale 2251
CH-1950 Sion 2 Nord
Telephone: (0)27-6064731
Telefax: (0)27-6064734
Founded: 1861
Membership: 650
Activities: excursions * lectures
Periodicals: (1) "Bulletin de La Murithienne" (ISSN 0374-6402)
City Reference Coordinates: 007°21'00"E 46°14'00"N

Société Vaudoise d'Astronomie (SVA)
Chemin des Grandes-Roches 8
Case Postale 190
CH-1018 Lausanne
Telephone: (0)22-3614030
Electronic Mail: abizzini@ulchi4.unil.ch
Founded: 1942
Membership: 285
Activities: photography * computing * education * CCD * popularization
Periodicals: (6) "Galaxie", "Bulletin SVA"
Coordinates: 006°37'28"E 46°32'07"N H595m
City Reference Coordinates: 006°38'00"E 46°31'00"N

Specola Solare Ticinese
CH-6605 Locarno-Monti
Telephone: (0)91-7562376
Telefax: (0)91-7562310
Electronic Mail: cortesi@webshuttle.ch (Sergio Cortesi, Director)
Founded: 1957
Staff: 3
Activities: observing (Sun, variable stars)
Coordinates: 008°47'21"E 46°10'23"N H367m
City Reference Coordinates: 008°48'00"E 46°10'00"N (Locarno)

Sternwarte Eschenberg
c/o Markus Griesser
Breitenstrasse 2
CH-8542 Wiesendangen
Telephone: (0)52-3372848
Telefax: (0)52-3372848
Electronic Mail: griesser@spectraweb.ch
Staff: 1
WWW: http://www.spectraweb.ch/~griess/Sternwarte/
Founded: 1979
Coordinates: 008°44'38"E 47°28'33"N H542m
City Reference Coordinates: 008°48'00"E 47°32'00"N

Sternwarte Rümlang
● See "Verein Sternwarte Rotgrueb Rümlang (VSRR)"

Sternwarte Sursee
Berufsschulhaus Kotten
CH-6210 Sursee/Lu
Electronic Mail: ens@ens.ch (Peter Ens)
WWW: http://www.ens.ch/sternwarte/

City Reference Coordinates: 008°07'00"E 47°11'00"N

Sternwarte Urania Burgdorf
Frommgutweg 6
CH-3400 Burgdorf
Telephone: (0)34-4227612
Electronic Mail: urania@urania.ch
markus.buetikofer@spectraweb.ch
WWW: http://www.urania.ch/
Founded: 1920
City Reference Coordinates: 007°37'00"E 47°04'00"N

Stiftung Sternwarte Uitikon
c/o Andreas Inderbitzin
Winterthurerstrasse 420
CH-8051 Zürich
Telephone: (0)1-3228736 (Private)
(0)1-3067523 (Office)
Electronic Mail: inderbitzin.a@bluewin.ch
WWW: http://www.uitikon.ch/sternw.htm
City Reference Coordinates: 008°33'00"E 47°23'00"N (Zürich)

Swiss Association for Standardization
● See "Schweizerische Normen-Vereinigung (SNV)"

Swiss National Science Foundation
● See "Schweizerischer Nationalfonds zur Förderung der Wissenschaftlichen Forschung"

Swiss Transport Museum
● See "Verkehrshaus der Schweiz"

Union Internationale des Télécommunications (UIT)
● See "International Telecommunication Union (ITU)"

Universität Basel, Astronomisches Institut
Venusstrasse 7
CH-4102 Binningen
Telephone: (0)61-2055454
Telefax: (0)61-2055455
Electronic Mail: <userid>@unibas.ch
WWW: http://www.astro.unibas.ch/
Founded: 1890
Staff: 20
Activities: stellar astronomy * stellar photometry * galactic structure * SN * extragalactic astronomy * cosmology
Periodicals: "Preprints"
Coordinates: 007°35'00"E 47°32'30"N H318m (Basel)
007°28'48"E 47°28'30"N H557m (Metzerlen)
City Reference Coordinates: 007°35'00"E 47°33'00"N

Universität Basel, Mathematisches Institut
Rheinsprung 21
CH-4051 Basel
Telephone: (0)61-2672690
Telefax: (0)61-2672695
Electronic Mail: <userid>@math.unibas.ch
WWW: http://www.math.unibas.ch/
City Reference Coordinates: 007°35'00"E 47°33'00"N

Universität Basel, Theoretische Kern/Teilchen- und Astrophysik
Institut für Physik
Klingelbergstrasse 82
CH-4056 Basel
Telephone: (0)61-2673111
(0)61-2673691
Telefax: (0)61-2673784
Electronic Mail: <userid>@quasar.physik.unibas.ch
WWW: http://quasar.physik.unibas.ch/
Staff: 24
Activities: nuclear astrophysics * nucleosynthesis * stellar evolution * novae * SN * hydrodynamics * neutron stars * gamma and X-ray bursts
City Reference Coordinates: 007°35'00"E 47°33'00"N

Universität Bern, Astronomisches Institut
Sidlerstrasse 5

CH-3012 Bern
Telephone: (0)31-6318591
Telefax: (0)31-6313869
Electronic Mail: <userid>@aiub.unibe.ch
WWW: http://www.cx.unibe.ch/aiub/
Founded: 1921
Staff: 25
Activities: fundamental astronomy * satellite geodesy
Periodicals: (1) "Mitteilungen der Satellitenbeobachtungsstation Zimmerwald" (circ.: 100)
Coordinates: 007°27'54"E 46°52'36"N H929m (Zimmerwald Station)
City Reference Coordinates: 007°26'00"E 46°57'00"N

Universität Bern, Institut für Angewandte Physik

Sidlerstrasse 5
CH-3012 Bern
Telephone: (0)31-6318911
Telefax: (0)31-6313765
Electronic Mail: <userid>@sun.iap.unibe.ch
                 iapemail@iap.unibe.ch
WWW: http://www.cx.unibe.ch/iap/
Founded: 1967
Activities: solar flares
City Reference Coordinates: 007°26'00"E 46°57'00"N

Universität Köln, Erstes Physikalisches Institut, Kölner Observatorium für Submillimeter und Millimeter Astronomie (KOSMA)

Observatorium Gornergrat-Süd
Gornergrat
CH-3920 Zermatt
Telephone: (0)279-672715
Telefax: (0)279-672715
Electronic Mail: observer@gg-KOSMA.unibe.ch
WWW: http://www.ph1.uni-koeln.de/kosma.html
        http://www.ph1.uni-koeln.de/kosma_observatory.html
Founded: 1980 (I. Phys. Inst.)
Coordinates: 007°47'04"E 45°59'04"N H3,130m
City Reference Coordinates: 007°45'00"E 46°01'00"N

Université de Lausanne, Institut d'Astronomie

Bâtiment des Sciences Physiques
Dorigny
CH-1015 Lausanne
Telephone: (0)21-6922373
Electronic Mail: <userid>@obs.unige.ch
WWW: http://obswww.unige.ch/
Founded: 1818
Staff: 10
City Reference Coordinates: 006°38'00"E 46°31'00"N

Université de Lausanne, Institut d'Astronomie, Observatoire

CH-1290 Chavannes-des-Bois
Telephone: (0)22-7552611
Telefax: (0)22-7553983
Electronic Mail: <userid>@obs.unige.ch
WWW: http://obswww.unige.ch/
        http://obswww.unige.ch/~mermio/ia/
Activities: stellar photometry (calibration, data base, CP stars, stellar clusters)
Coordinates: 006°08'12"E 46°18'24"N H455m
City Reference Coordinates: 006°08'00"E 46°18'00"N

Urania-Sternwarte

Uraniastrasse 9
CH-8000 Zürich
Electronic Mail: urania@ezinfo.ethz.ch
WWW: http://www.astroinfo.ch/obs/urania/
Founded: 1907
Staff: 10
Activities: education * public observing
City Reference Coordinates: 008°33'00"E 47°23'00"N

Verein Sternwarte Rotgrueb Rümlang (VSRR)

c/o Walter Bersinger (President)
Obermattenstrasse 9
CH-8153 Rümlang
Telephone: (0)1-8172813

Electronic Mail: bersingerw@bluewin.ch
WWW: http://ruemlang.astronomie.ch/
Periodicals: (6) "VSRR-Infoblatt" (circ.: 200)
Coordinates: 008°31'28"E 47°26'22"N H495m
City Reference Coordinates: 008°32'00"E 47°27'00"N

Verkehrshaus der Schweiz

(Musée Suisse des Transports)
(Swiss Transport Museum)
Lidostrasse 5
CH-6006 Luzern
Telephone: (0)41-3704444
Telefax: (0)41-3706168
WWW: http://www.verkehrshaus.ch/planet.htm
Founded: 1959 (Planetarium: 1969)
Activities: museum of transport and communication * planetarium * IMAX theatre
City Reference Coordinates: 008°18'00"E 47°03'00"N

Volkssternwarte Schanfigg Arosa (VSA)

Postfach 13
CH-7029 Peist
Telephone: (0)61-6927146
Telefax: (0)61-2673012
Electronic Mail: moehle@iet.mavt.ethz.ch
                     sibylle.moehle@ao-asif.ch
WWW: http://www.lkt.iet.ethz.ch/vsa/
City Reference Coordinates: 009°41'00"E 46°47'00"N

Wolf Gesellschaft (Rudolf-)

● See "Rudolf Wolf Gesellschaft"

World Intellectual Property Organization (WIPO)

(Organisation Mondiale de la Propriété Intellectuelle - OMPI)
Chemin des Colombettes 34
CH-1211 Genève 20
Telephone: (0)22-7309111
Telefax: (0)22-7335428
Electronic Mail: wipo.mail@wipo.int
WWW: http://www.wipo.int/
Founded: 1970
Staff: 750
Membership: 173 (states)
Activities: United Nations specialized agency responsible for the promotion of the protection of intellectual property throughout the world through cooperation among States and for the administration of various multilateral treaties dealing with the legal and administrative aspects of intellectual property
City Reference Coordinates: 006°09'00"E 46°12'00"N

World Meteorological Organization (WMO)

(Organisation Météorologique Mondiale - OMM)
Avenue Giuseppe-Motta 41
Case Postale 2300
CH-1211 Genève 2
Telephone: (0)22-7308111
Telefax: (0)22-7342326
Electronic Mail: <userid>@www.wmo.ch
                     ipa@www.wmo.ch (Information and Public Affairs)
WWW: http://www.wmo.ch/
Founded: 1950
Membership: 181 (states & territories)
Activities: meteorology * operational hydrology
Periodicals: (4) "World Meteorological Organization Bulletin" (ISSN 0042-9767); (1) "World Meteorological Organization Annual Report"
Awards: (1) "IMO Prize", "Vilho Väisälä Award", "Norbert Gerbier-Mumm International Award", "WMO Research Award for Young Scientists"
City Reference Coordinates: 006°09'00"E 46°12'00"N

World Radiation Center

● See "Physikalisch-Meteorologisches Observatorium Davos (PMOD), Weltstrahlungszentrum"

Wyss Photo-Video (P.-)

● See "P. Wyss Photo-Video"

Syria

Syrian Arab Organization for Standardization and Metrology (SASMO)
P.O. Box 11836
Damascus
Telephone: (0)11-450538
          (0)11-5128213
Telefax: (0)11-5128214
City Reference Coordinates: 036°18'00"E 33°30'00"N

Syrian Cosmological Society (SCS)
P.O. Box 13187
Damascus
Telephone: (0)11-7776729
          (0)11-5113004
Telefax: (0)11-3311609
Founded: 1980
Membership: 100
Activities: lectures * observing * seminars * training * theoretical research * several observatories (Damascus, Alleppo, ...)
City Reference Coordinates: 036°18'00"E 33°30'00"N

Syrian Meteorological Department
P.O. Box 4211
Damascus
Telephone: (0)11-6620554
Telefax: (0)11-6620553
Founded: 1952
Staff: 600
Activities: meteorology * climatology * air pollution
Periodicals: (12) & (1) "Climatological Data"
Observatories: 21 synoptic, 98 climatological and 290 precipitation stations
City Reference Coordinates: 036°18'00"E 33°30'00"N

Taiwan (ROC)

Academia Sinica Institute of Astronomy and Astrophysics (ASIAA), Preparatory Office
P.O. Box 1-87
Taipei 11529
or :
128 Yen Chiou Yuan Road
Section 2
Nankang
Taipei 11529
Telephone: (0)2-26522020
Telefax: (0)2-27881106
Electronic Mail: asiaa@asiaa.sinica.edu.tw
                     <userid>@asiaa.sinica.edu.tw
WWW: http://www.asiaa.sinica.edu.tw/
Founded: 1993
Staff: 40
Activities: radio-astronomy * optical observing * star and planetary system formation * nuclear astrophysics * cosmology * interacting galaxies * galactic nuclei * galactic dynamics * instrumentation
Coordinates: 121°37'00"E 25°03'00"N
City Reference Coordinates: 121°32'00"E 25°05'00"N

Central Weather Bureau (CWB), Astronomical Observatory
64, Kung Yuan Road
Taipei 10039
Telephone: (0)2-3491096
Electronic Mail: <userid>@cwb.gov.tw
WWW: http://www.cwb.gov.tw/
Founded: 1945
Staff: 5
Activities: solar observing
Periodicals: (2) "Report on Sunspot Observations"
Coordinates: 121°30'24"E 25°02'23"N
City Reference Coordinates: 121°32'00"E 25°05'00"N

Directorate General of Telecommunications (DGT), Lunping Observatory
180 Lunping
Kuanyin
Taoyuan 32814
Telephone: (0)3-4263467
Telefax: (0)3-4987444
Electronic Mail: lunping@dgt.gov.tw
Founded: 1965
Activities: sunspot * geomagnetism * ionosphere
Periodicals: (1) "Report of the Lunping Observatory - Sunspots" (ISSN 0254-3796), "Report of the Lunping Observatory - Geomagnetism" (ISSN 1019-8695)
Coordinates: 120°10'00"E 25°00'00"N H100m
City Reference Coordinates: 120°10'00"E 25°00'00"N

Lunping Observatory
● See "Directorate General of Telecommunications (DGT), Lunping Observatory"

National Central University (NCU), Institute of Astronomy
Chung-li 32054
Telephone: (0)3-4262302
Telefax: (0)3-4262304
Electronic Mail: astronomy@phyast.phy.ncu.edu.tw
WWW: http://www.phy.ncu.edu.tw/
Founded: 1992
Staff: 8
Activities: research in astronomy and astrophysics
Coordinates: 121°11'12"E 24°58'12"N H152m
City Reference Coordinates: 121°08'00"E 24°55'00"N

National Tsing Hua University, Department of Physics
Hsinchu 300
Telephone: (0)35-719037
Telefax: (0)35-723052
Electronic Mail: <userid>@phys.nthu.edu.tw
WWW: http://www.phys.nthu.edu.tw/
Founded: 1965
Staff: 70

City Reference Coordinates: 120°58'00"E 24°48'00"N

Nick Enterprise Co. Ltd.
198, 12F-3, Sec. 2
Roosevelt Road
Taipei
Telephone: (0)2-3655790
Telefax: (0)2-3687854
Founded: 1955
Staff: 25
Activities: importing and distributing scientific equipment and astronomical instrumentation * agent of GOTO Optical Manufacturing Co.
City Reference Coordinates: 121°32'00"E 25°05'00"N

Space Theater
c/o National Museum of Natural Science
1 Kuan Chien Road
P.O. Box 1412
Taichung
Telephone: (0)4-3226940
Telefax: (0)4-3222290
WWW: http://www.nmns.edu.tw/
Founded: 1986
Staff: 13
Activities: planetarium and OmniMax film shows
City Reference Coordinates: 120°41'00"E 24°09'00"N

Taipei Observatory
5 Sec. 4
Chung-shan N. Road
Yuan Shan
Taipei 10452
Telephone: (0)2-5926827
Telefax: (0)2-5910763
Founded: 1938
Staff: 23
Activities: planetarium shows * solar observations * orbit calculations
Periodicals: "Astronomy Newsletter", "Taipei Astro-Circular", "Sunspot Observations", "Almanac" (in Chinese)
Coordinates: 121°31'38"E 25°04'40"N H31m
City Reference Coordinates: 121°32'00"E 25°05'00"N

Tajikistan

Tajik Academy of Sciences, Institute of Astrophysics
Bukhoro Sir. 22
734042 Dushanbe
Telephone: (0)3772-274614
(0)3770-231432
(0)3770-275483 (Mount Sanglok Observatory)
Electronic Mail: pulat@astro.td.silk.org (P.B. Babadzhanov, Director)
WWW: http://www.glasnet.ru/~ermdtaj/projects/iastro.htm
Founded: 1932
Staff: 11
Activities: physics and dynamics of small solar-system bodies * astrometry * variable stars * structure and dynamics of galaxies * Earth ionopshere
Periodicals: "Komety y Meteory" (ISSN 0568-6199), "Bulletin" (ISSN 0568-6865)
Coordinates: 068°38'00"E 38°29'00"N H730m (Hisar Astronomical Observatory)
069°00'00"E 38°16'00"N H2300m (Mount Sanglok Observatory)
074°00'00"E 38°00'00"N H4350m (Pamir Branch)
City Reference Coordinates: 068°48'00"E 38°35'00"N

Thailand

Chiang Mai University, Department of Physics, Observatory
Chiang Mai 50002
Telephone: (0)53-222699 x3367
Telefax: (0)53-222268
WWW: http://www.chiangmai.ac.th/CMU/sccmu/dph.html
Founded: 1977
Staff: 6
Activities: photoelectric and CCD photometry of eclipsing binaries
Coordinates: 098°38'00"E 18°36'00"N H784m
City Reference Coordinates: 098°58'00"E 18°46'00"N

Chulalongkorn University, Department of Physics, Astronomy Unit
Bangkok 10330
Telephone: (0)2-2527985
                (0)2-2185280
Telefax: (0)2-2531150
            (0)2-2531130
Founded: 1917
Staff: 7
Activities: education
Coordinates: 100°32'00"E 13°45'00"N H15m
City Reference Coordinates: 100°32'00"E 13°45'00"N

Thai Astronomical Society (TAS)
c/o Bangkok Planetarium
928 Sukhumvit Road
Bangkok 10110
Telephone: (0)3-3902301
                (0)3-3902305
Founded: 1981
Membership: 3000
Activities: star parties * quiz contests * astronomical projects contests * lectures
Periodicals: (6) "Tang Chang Pueg" (ISSN 0858-2637)
Awards: (1) contest awards
Coordinates: 100°30'00"E 13°45'00"N
City Reference Coordinates: 100°30'00"E 13°45'00"N

Thai Industrial Standards Institute (TISI)
Rama VI Street
Bangkok 10400
Telephone: (0)2-2457802
Telefax: (0)2-2478741
Electronic Mail: thaistan@tisi.go.th
WWW: http://www.tisi.go.th/
Founded: 1968
Staff: 530
Activities: standardization * product and quality system certification * testing * laboratory accreditation
Periodicals: (4) "TISI Newsletter" (ISSN 0858-4648)
City Reference Coordinates: 100°32'00"E 13°45'00"N

Thai Meteorological Department
4353 Sukhumvit Road
Bang-Na
Phaakanong
Bangkok 10260
Telephone: (0)2-3989886
                (0)2-2580437
                (0)2-2587057
Telefax: (0)2-3989886
            (0)2-3994014
            (0)2-3984972
            (0)2-3989816
WWW: http://www.ncrt.go.th/htmlpages/met/wattana.html
        http://www.thaimet.tmd.go.th/
Founded: 1942
Staff: 1030
Activities: weather forecast * warning centre for severe weather conditions and floods
City Reference Coordinates: 100°32'00"E 13°45'00"N

Trinidad & Tobago

Caribbean Meteorological Organization (CMO)
P.O. Box 461
Port of Spain
Trinidad
Telephone: 624-4481
Telefax: 623-3634
Founded: 1951
Activities: meteorology * operational hydrology
City Reference Coordinates: 061°31'00"W 10°39'00"N

Meteorological Office, Crown Point
International Airport
Crown Point
Tobago
Telephone: 639-8780
                639-8715
Telefax: 639-7455
Founded: 1967
Staff: 18
Activities: weather radar observing * equipment maintenance
Coordinates: 060°50'00"W 11°09'00"N
City Reference Coordinates: 060°50'00"W 11°09'00"N

Meteorological Office, Piarco
International Airport
Piarco
Trinidad
Telephone: 664-5465
Telefax: 664-4727 (Forecast Office)
        664-4737 (Operations)
        664-4009 (Director)
Founded: 1946
Staff: 69
Activities: weather forecast * meteorological activities for aviation, public and agriculture * climatology * tropical cyclone monitoring * upper air * equipment maintenance
Periodicals: (4) "Climate Summary"; (1) "Climate Summary"
Coordinates: 061°21'00"W 10°35'00"N
City Reference Coordinates: 061°21'00"W 10°35'00"N

Trinidad and Tobago Bureau of Standards (TTBS)
P.O. Box 467
Port of Spain
or :
2 Century Drive
Trincity Industrial Estate
Macoya
Tunapuna
Telephone: 662-8827
Telefax: 663-4335
Electronic Mail: ttbs@opus.co.tt
WWW: http://www.opus.co.tt/ttbs/
Founded: 1974
Staff: 60
Activities: developing and monitoring national standards for non-food items * certification of products * quality assurance training
Periodicals: (4) "The Standard"
City Reference Coordinates: 061°31'00"W 10°39'00"N (Port of Spain)
                    061°23'00"W 10°38'00"N (Tunapuna)

University of West Indies (UWI), Trinidad, Department of Physics
Saint Augustine
Trinidad
Telephone: 663-1369
Telefax: 663-9686
Electronic Mail: <userid>@uwimona.edu.jm
Founded: 1964
Staff: 10
Activities: education * research
City Reference Coordinates: 061°31'00"W 10°39'00"N (Port of Spain)

Turkey

Ankara University, Department of Astronomy and Space Science
Degol Cad.
Tandogan
06100 Ankara
Telephone: (0)312-2126720
　　　　　(0)312-2121364
Telefax: (0)312-2232395
WWW: http://www.ankara.edu.tr/~astro
Founded: 1943
Staff: 17
Activities: close binary stars * stellar spectroscopy * photoelectric photometry * long-period variables
Coordinates: 032°46'48"E 39°50'37"N H1,257m
City Reference Coordinates: 032°52'00"E 39°56'00"N

Bilkent University, Department of Physics
Bilkent
06533 Ankara
Telephone: (0)312-2664397
Telefax: (0)312-2664579
Electronic Mail: astro@fen.bilkent.edu.tr
WWW: http://www.fen.bilkent.edu.tr/~physics/
　　　　http://www.fen.bilkent.edu.tr/~astro/
City Reference Coordinates: 032°52'00"E 39°56'00"N

Bogaziçi University, Kandilli Observatory and Earthquake Research Institute
Cengelköy
81220 Istanbul
Telephone: (0)216-3080514
Telefax: (0)216-3321711
Electronic Mail: kandil@boun.edu.tr
WWW: http://www.boun.edu.tr/
　　　　http://www.koeri.boun.edu.tr/astronomy/astronomy.html
Founded: 1911
Staff: 70
Activities: solar activity * flares * astrometry * history of astronomy * earthquakes and related topics * geodesy * geophysics
Coordinates: 029°03'39"E 41°03'49"N H120m
City Reference Coordinates: 028°58'00"E 41°01'00"N

Cukurova University, Department of Physics, Center for Space Sciences
Balcali
01330 Adana
Telephone: (0)71-326941
　　　　　(0)71-326942 x2480
　　　　　(0)71-326942 x2691
Telefax: (0)71-326945
　　　　(0)71-326070
WWW: http://www.cc.cu.edu.tr/
Founded: 1990
Staff: 6
Activities: radioastronomy * X- and gamma-ray astronomy * remote sensing applications * data analysis and image processing * telescope site selection * photometry
City Reference Coordinates: 035°18'00"E 37°01'00"N

Ege University, Department of Astronomy and Space Sciences
35100 Bornova
Telephone: (0)232-3880110 x2332
Electronic Mail: <userid>@ege.edu.tr
WWW: http://bornova.ege.edu.tr/
　　　　http://bornova.ege.edu.tr/~fenfak/astronomy/bolum.html
　　　　http://www.sci.ege.edu.tr/~euobs/
　　　　http://astronomy.sci.ege.edu.tr/
Founded: 1965
Staff: 20
Activities: photoelectric photometry * instrinsic and eclipsing variable stars * education * cosmology
Periodicals: "Publications of the Ege University Observatory"
Coordinates: 027°16'30"E 38°23'54"N H795m
　　　　　027°10'03"E 38°23'36"N H20m
City Reference Coordinates: 027°15'00"E 38°28'00"N

Inönü University, Physics Department
44069 Malatya

Telephone: (0)422-3410010
Telefax: (0)422-3410037
Founded: 1986
Activities: astrometry * stellar photometry * binary stars * variable stars
City Reference Coordinates: 038°19'00"E 38°21'00"N

Istanbul University, Astronomy and Space Sciences Department
34452 University
Telephone: (0)212-5223597
        (0)212-5283847
Telefax: (0)212-5190834
      (0)212-5283847
Electronic Mail: astronomy@istanbul.edu.tr
WWW: http://www.istanbul.edu.tr/fen/astronomy/
Founded: 1933
Staff: 21
Activities: stellar astrophysics * solar physics * galactic and extragalactic astronomy
Periodicals: (1) "Publications of the Istanbul University Observatory"
Coordinates: 028°57'52"E 41°00'40"N H65m
City Reference Coordinates: 028°58'00"E 41°01'00"N (Istanbul)

Kandilli Observatory and Earthquake Research Institute
● See "Bogaziçi University, Kandilli Observatory and Earthquake Research Institute"

Marmara Research Center
● See "Scientific and Technical Research Council of Turkey (TIBITAK), Marmara Research Center"

Marmara University, Department of Physics
81040 Fikirtepe
Telephone: (0)1-3459090 x145
Telefax: (0)1-3306022
WWW: http://www.marun.edu.tr/faculties/arts_sciences/physics/
City Reference Coordinates: 027°34'00"E 40°36'00"N (Marmara)

Middle East Technical University, Physics Department, Astrophysics Division
Inönü Bulvari
06531 Ankara
Telephone: (0)312-2101000 x3252
        (0)312-2101000 x3253
Telefax: (0)312-2101281
Electronic Mail: <userid>@newton.physics.metu.edu.tr
WWW: http://astroa.physics.metu.edu.tr/
Founded: 1960
Staff: 20
Activities: stellar astronomy * stellar structure and evolution * X-ray astronomy * neutron stars * pulsars * photometry * spectroscopy * data analysis * pulsating stars * rotating stars
Coordinates: 032°46'28"E 39°54'10"N H954m
City Reference Coordinates: 032°52'00"E 39°56'00"N

Scientific and Technical Research Council of Turkey (TUBITAK)
Ataturk Boulevard 221
06100 Karaklidere
Telephone: (0)4-1685300
Telefax: (0)4-1277489
WWW: http://www.tubitak.gov.tr/eng/index_eng.html
Founded: 1963
Activities: promoting, organising and coordinating research and development
Periodicals: (12) "Bilimve Teknik", "Içindekiler"
City Reference Coordinates: 032°52'00"E 39°56'00"N (Ankara)

Scientific and Technical Research Council of Turkey (TUBITAK), Marmara Research Center, Space Sciences Department
P.O. Box 21
Gebze
41470 Kocaeli
Telephone: (0)262-6412300 x3300, x2165
Telefax: (0)262-6412309
Founded: 1991
Staff: 26
Activities: radio astronomy * remote sensing * astrophysics * radio physics * instrumentation
Periodicals: (1) "Annual Report"
Observatories: 1 (Marmara Radio Telescope Observatory, Gebze)
City Reference Coordinates: 029°55'00"E 40°46'00"N

TUBITAK
● See "Scientific and Technical Research Council of Turkey (TUBITAK)"

TUBITAK Ulusal Gozlemevi (TUG)
● See "Turkish National Observatory"

Turkish National Observatory
(TUBITAK Ulusal Gozlemevi - TUG)
Akdeniz Üniversitesi Kampüsü
Dumlupinar Bulvari
07058 Antalya
Telephone: (0)242-2278401
Telefax: (0)242-2278400
Electronic Mail: tug@tug.tubitak.gov.tr
WWW: http://www.tug.tubitak.gov.tr/
Founded: 1997
Staff: 29
Activities: stellar photometry * spectroscopy * astrometry
Coordinates: 030°20'08"E 36°49'30"N H2,550m
City Reference Coordinates: 030°42'00"E 36°53'00"N (Antalya)

Turkish Physical Society
c/o N. Enduran (General Secretary)
Physics Department
Istanbul University
Vezneciler
34459 Istanbul
Electronic Mail: erduran@istanbul.edu.tr
Founded: 1950
Membership: 50
City Reference Coordinates: 028°58'00"E 41°01'00"N

Turkish Standards Institution (TSE)
Necatibey Cad. 112
Bakanliklar
Ankara
Telephone: (0)312-4178330
Telefax: (0)312-4254399
Electronic Mail: didb@tse.org.tr
Founded: 1954
Staff: 1000
Activities: preparing standards * certification * quality control * metrology * education * promotion * testing * applied industrial research * publishing
Periodicals: (12) "Standard"; (1) "Catalogue of Turkish Standards"
City Reference Coordinates: 032°52'00"E 39°56'00"N

Turkish State Meteorological Service
P.O. Box 401
Ankara
Telephone: (0)312-3141616
Telefax: (0)312-3593430
WWW: http://www.meteor.gov.tr/
Founded: 1937
Staff: 3500
Activities: meteorological data collecting and archiving * forecasting * 273 climatic, 91 synoptic, 30 marine, 34 aeronautic and 7 Rawinsonde stations
Periodicals: (12) "Meteorological Bulletin"
City Reference Coordinates: 032°52'00"E 39°56'00"N

Ukraine

Chernigovka Amateur Astronomical Society
c/o School n° 1
Babenko Ul. 9
71200 Chernigovka
Telephone: (0)61-4092911
Founded: 1972
Membership: 15
Activities: comets * instrumentation * eclipses
Coordinates: 036°14'58"E 47°11'14"N H120m
City Reference Coordinates: 036°12'00"E 47°11'00"N

Crimean Amateur Astronomical Society (CAAS)
Gogol'a 26
P.O. Box 52
333000 Simferopol
Telephone: (0)652-252558
Telefax: (0)652-270829
Electronic Mail: <userid>@ssaa.cris.crimea.ua
WWW: http://www.cris.net/~astro
Founded: 1946
Membership: 200
Activities: observing (meteors, Sun) * education
Coordinates: 034°07'00"E 44°58'00"N
            034°59'55"E 44°55'00"N (Sidak)
City Reference Coordinates: 034°07'00"E 44°58'00"N

Crimean Astrophysical Observatory (CrAO)
Nauchny
Bahchisarai
334413 Crimea
Telephone: (0)6554-71161
            (0)655-471166
Telefax: (0)6554-40704
            (0)655-471754
Electronic Mail: <userid>@crao.crimea.ua
WWW: http://www.sai.crimea.ua/
Founded: 1945
Staff: 380
Activities: galaxies * stars * gas and dust nebulae * Sun * solar system * solar and stellar activity * gamma-ray astronomy
Periodicals: "Izvestja" (ISSN 0367-8466); "Bulletin" (ISSN 0190-2717)
Coordinates: 034°01'00"E 44°43'42"N H550m (Nauchny)
City Reference Coordinates: 033°59'00"E 44°24'00"N

Crimean Astrophysical Observatory (CrAO), Radioastronomical Department, Katsiveli
334247 Crimea
Telephone: (0)654-727952
Telefax: (0)654-727961
Electronic Mail: <userid>@nad.crimea.ua
WWW: http://www.sai.crimea.ua/
Founded: 1945 (CrAO)
Coordinates: 033°58'47"E 44°23'53"N H25m
City Reference Coordinates: 033°59'00"E 44°24'00"N

Crimean Astrophysical Observatory (CrAO), Radioastronomical Department, Simeiz
Mount Koshka
334242 Crimea
Telephone: (0)654-771079
Telefax: (0)654-727961
Electronic Mail: <userid>@cat.crimea.ua
WWW: http://www.sai.crimea.ua/
Founded: 1945 (CrAO)
Coordinates: 033°59'48"E 44°24'12"N H346m
City Reference Coordinates: 033°59'00"E 44°24'00"N

Donetsk Planetarium
Artema Ul. 165
340048 Donetsk
Telephone: (0)622-552508
Telefax: (0)622-557462
Electronic Mail: filippov@kinetic.ac.donetsk.ua (Irina Filippova, Director)
Founded: 1962

Staff: 8
Activities: shows * education * lectures
City Reference Coordinates: 037°48'00"E 48°00'00"N

International Centre of Astronomical, Medical and Ecological Research
Golosiiv
252022 Kyiv
Telephone: (0)44-2662286
Telefax: (0)44-2662147
Electronic Mail: <userid>@mao.kiev.ua
Founded: 1992
Staff: 80
Activities: kinematical and physical characteristics of celestial objects * developing and improving astronomical instrumentation
Coordinates: 042°29'59"E 43°16'34"N H3,100m (Terskol)
City Reference Coordinates: 030°30'00"E 50°25'00"N

Kalinenkov Astronomical Observatory
● See "Mykolaiv State Pedagogical Institute, Kalinenkov Astronomical Observatory"

Kharkiv Astronomical Club
c/o Vadim Kaydach (Club Leader)
House of Culture
Pr. Traktorostroiteley 55
310153 Kharkiv
Founded: 1993
Electronic Mail: prool@infocom.kharkov.ua (with "for astronomical Club" in subject line)
WWW: http://www.infocom.kharkov.ua/misc/astro.html
City Reference Coordinates: 036°15'00"E 50°00'00"N

Kharkiv State University, Astronomical Observatory
Sumskaja Ul. 35
310022 Kharkiv
Telephone: (0)572-432428
Electronic Mail: <userid>@astron.kharkov.ua
WWW: http://khassm.virtualave.net/ (Multiwave Station of Solar Monitoring)
Founded: 1883
Staff: 50
Activities: minor planets * planets * Moon * Sun * astrometry * Earth rotation * speckle interferometry * atmospheric seeing * statistics of stellar systems * QSO statistics
Coordinates: 036°13'54"E 50°00'12"N H138m
                036°56'12"E 49°38'34"N (Chuguev)
City Reference Coordinates: 036°15'00"E 50°00'00"N

Kharkiv Y.A. Gagarin Planetarium
Per. Kravtsova 15
310003 Kharkiv
Telephone: (0)572-439190
Telefax: (0)572-433368
Founded: 1957
Staff: 20
Activities: education * observing
Coordinates: 036°10'00"E 50°00'00"N H130m
City Reference Coordinates: 036°15'00"E 50°00'00"N

Kyiv University, Astronomical Observatory
Observatorna 3
254053 Kyiv
Telephone: (0)44-2162691
Electronic Mail: <userid>@aoku.freenet.kiev.ua
Founded: 1845
Staff: 60
Activities: astrometry * Sun (activity, MHD) * relativistic astrophysics * comets * meteors
Periodicals: "Visnyk Kyivskoho Universitetu. Astronomia", "Comet Circular"
Coordinates: 030°30'08"E 50°27'12"N H184m
                030°31'35"E 50°17'53"N H131m (Lisnyky)
                029°55'00"E 50°35'00"N H140m (Pylypovychi)
● Full university name: "Kyiv Taras Shevchenko University"
City Reference Coordinates: 030°30'00"E 50°25'00"N

Kyiv University, Astronomy and Space Physics Department
pr. Glushkova 6
252002 Kyiv
Telephone: (0)44-2664457
Telefax: (0)44-2664507

Electronic Mail: astdept@astrophys.ups.kiev.ua
WWW: http://space.ups.kiev.ua/
● Full university name: "Kyiv Taras Shevchenko University"
City Reference Coordinates: 030°30'00"E 50°25'00"N

Kyiv University, Department of Astronomy
64 Volodymyrska Street
252017 Kyiv
Telephone: (0)44-2664507
Telefax: (0)44-2664507
Electronic Mail: <userid>@astron.univ.kiev.ua
WWW: http://www.univ.kiev.ua/
City Reference Coordinates: 030°30'00"E 50°25'00"N

Kyiv Youth Palace Astronomical Observatory
Janvarskogo Vostanija Ul. 13
252010 Kyiv
Telephone: (0)44-2904998
Founded: 1966
Staff: 53
Activities: popularization * observing * solar physics * astrometry
City Reference Coordinates: 030°30'00"E 50°25'00"N

L'viv Astronomical Society
c/o Chair of Astrophysics
L'viv University
Kyrylo & Mephody 8
290005 L'viv
Telephone: (0)322-729088
Telefax: (0)322-729088
Electronic Mail: vita@astro.lviv.ua
Founded: 1997
Membership: 28
Activities: popularization * assistance to professional and amateur astronomers
City Reference Coordinates: 024°00'00"E 49°50'00"N

L'viv Polytechnical State University, Department of Geodesy and Astronomy
12 Bandera Street
290646 L'viv
Telephone: (0)322-398804
Telefax: (0)322-744300
Electronic Mail: <userid>@polynet.lviv.ua
WWW: http://www.polynet.lviv.ua/
City Reference Coordinates: 024°00'00"E 49°50'00"N

L'viv University, Astronomical Observatory
Kyrylo & Mephody 8
290005 L'viv
Telephone: (0)322-729088
Telefax: (0)322-729088
Electronic Mail: director@astro.lviv.ua
                 star@astro.lviv.ua
Founded: 1769
Staff: 30
Activities: solar physics and activity * variable stars * PNs * star clusters * cosmology * satellites
Coordinates: 023°57'11"E 49°55'04"N H325m
City Reference Coordinates: 024°00'00"E 49°50'00"N

Main Astronomical Observatory
● See "Ukrainian National Academy of Sciences, Main Astronomical Observatory"

Mykolaiv Astronomical Observatory
1 Observatorna Street
327030 Mykolaiv
Telephone: (0)512-375714
          (0)512-375206
Telefax: (0)512-362420
         (0)512-362426
Electronic Mail: nao@ascor.nikolaev.ua
                 root@mao.nikolaev.ua
WWW: http://www.comcent.nikolaev.ua/project/mao
Founded: 1821
Staff: 79
Activities: astrometry * instrumentation * time and frequency service
Coordinates: 031°58'26"E 46°58'21"N H52m

City Reference Coordinates: 032°00'00"E 46°57'00"N

Mykolaiv State Pedagogical Institute, Kalinenkov Astronomical Observatory
24 Nikolska Street
327000 Mykolaiv
Telephone: (0)512-352213
Electronic Mail: <userid>@comcent.nikolaev.ua
City Reference Coordinates: 032°00'00"E 46°57'00"N

National Centre of the Ukrainian Youth Airspace Education
26 Gagarina Street
P.O. Box 5595
3200005 Dniepropetrovsk
Telephone: (0)562-466032
Telefax: (0)562-466171
Electronic Mail: sirius@unaec.dts.iskra.dp.ua
Founded: 1996
Staff: 40
Activities: developing and conducting airspace science and technology education
City Reference Coordinates: 035°00'00"E 48°29'00"N

Nikolaev Astronomical Observatory (NAO)
● See now "Mykolaiv Astronomical Observatory"

Odessa State University, Astronomical Observatory
T.G. Shevchenko Park
270014 Odessa
Telephone: (0)482-228442
Telefax: (0)482-228442
Electronic Mail: astro@paco.odessa.ua
Founded: 1871
Staff: 150
Activities: variable stars * close binaries * cool stars * models of atmospheres and envelopes * comets * meteors * dust components * spectrophotometry * polarimetry * astrometry * instrumentation * artificial satellites * photometry * time series analysis * sky patrol
Periodicals: (1) "Odessa Astronomical Publications"
Coordinates: 030°16'15"E 46°23'50"N H19m (Mayaki)
            030°48'00"E 46°33'42"N H40m (Kryzhanovska)
            030°45'30"E 46°28'36"N H60m (Odessa)
City Reference Coordinates: 030°44'00"E 46°28'00"N

Poltava Gravimetrical Observatory
● See "Ukrainian National Academy of Sciences, Poltava Gravimetrical Observatory"

State Committee of Ukraine for Standardization, Metrology and Certification (DSTU)
174 Gorkiy Ul.
252650 Kyiv
Telephone: (0)44-2262971
Telefax: (0)44-2262970
Electronic Mail: iso@dstu1.kiev.ua
City Reference Coordinates: 030°30'00"E 50°25'00"N

Sudak Astronomical Club
c/o Nick P. Rogov
Sudak Town College
Ul. Majakovskogo 2a
334 882 Sudak
Telephone: (0)6566-23605
Telefax: (0)6522-76236
Electronic Mail: yudjin@cris.crimea.ua
Founded: 1965
Membership: 38
Activities: observing meteors
Observatories: 4 (Sudak, Simferopol, Alushta, Karadag)
City Reference Coordinates: 034°59'00"E 44°52'00"N

Ukrainian Astronomical Association (UAA)
3 Observatorna Street
254053 Kyiv
Telephone: (0)44-2162691
Telefax: (0)44-2246387
Electronic Mail: uaa@aoku.freenet.kiev.ua
WWW: http://www.mao.kiev.ua/uaa
Founded: 1991
Membership: 54

Staff: 30
Activities: supporting astronomy and space research in Ukraine * representing in Ukraine the International Astronomical Union (IAU) and other organizations * organizing conferences * coordinating activities from Ukrainian astronomical institutions
Periodicals: (3) "Ukrainian Astronomical Association Information Bulletin"
Awards: (1/2) "Outstanding Contribution to Development of Astronomy in Ukraine"
City Reference Coordinates: 030°30'00"E 50°25'00"N

Ukrainian National Academy of Sciences, Institute of Radio Astronomy
4 Chervonopraporna Street
310002 Kharkiv
Telephone: (0)572-451009
Telefax: (0)572-476506
Electronic Mail: rai@ira.kharkov.ua
WWW: http://www.ira.kharkov.ua/
Founded: 1985
Staff: 300
Activities: decameter radioastronomy * space radiophysics * mm radioastronomical equipment * ionospheric and space plasma research * radio wave propagation * antennae
Periodicals: (4) "Radio Physics and Radio Astronomy"
Coordinates: 036°54'00"E 49°42'00"N H150m (Grakovo)
City Reference Coordinates: 036°15'00"E 50°00'00"N

Ukrainian National Academy of Sciences, Main Astronomical Observatory (MAO)
Golosiiv
252650 Kyiv 22
Telephone: (0)44-2663110
Telefax: (0)44-2662147
Electronic Mail: maouas@mao.kiev.ua
director@mao.kiev.ua
<userid>@mao.kiev.ua
WWW: http://www.mao.kiev.ua/
Founded: 1944
Staff: 200
Activities: positional astronomy * Earth rotation * planetary atmospheric physics * physics and evolution of stars and galaxies * solar physics * space plasma physics
Periodicals: "Kinematics and Physics of Celestial Bodies", "Space Science and Technology", "Astronomical Almanac"
Coordinates: 030°30'30"E 50°21'54"N H188m
042°31'00"E 43°16'00"N H3,100m (High Altitude Station Terskol)
City Reference Coordinates: 030°30'00"E 50°25'00"N

Ukrainian National Academy of Sciences, Poltava Gravimetrical Observatory
27/29 Myasoedova Street
314029 Poltava
Telephone: (0)5322-72039
Electronic Mail: geo@geo.kot.poltava.ua
City Reference Coordinates: 034°35'00"E 49°35'00"N

Ukrainian Physical Society
c/o V.O. Andreev (Executive Secretary)
Kyiv University
Academician Glushkov Avenue 6
252122 Kyiv
Telephone: (0)44-2664477
Telefax: (0)44-2208258
Electronic Mail: andreev@ipu.univ.kiev.ua
andreev@office.ups.kiev.ua
City Reference Coordinates: 030°30'00"E 50°25'00"N

Ukrainian Youth Aerospace Association "Suzirya"
Head Office
P.O. Box 4406
49000 Dniepropetrovsk
Telefax: (0)562-440608
Electronic Mail: suzirya@creator.dp.ua
WWW: http://www.suzir.org.ua/
Founded: 1991
Membership: 14000
Activities: popularization of space sciences and technologies
Periodicals: "Suzirya"
City Reference Coordinates: 035°00'00"E 48°29'00"N + e-mail (0000504)

Uzhgorod State University, Laboratory for Space Research
2A Daleka Street
294000 Uzhgorod
Telephone: (0)3122-36065

Electronic Mail: space@univ.uzhgorod.ua
City Reference Coordinates: 022°15'00"E 48°38'00"N

Zaporozhye Astronomical Club Altair

c/o Viktor N. Gladkij
Mayakovskogo Prospekt 14
330035 Zaporozhye
Telephone: (0)333126
Founded: 1975
Staff: 25
Activities: observing (meteors, Moon, variable stars) * instrumentation
Coordinates: 035°08'00"E 47°48'00"N
City Reference Coordinates: 035°08'00"E 47°48'00"N

Zaporozhye Palace of Youth Work, Planetarium

Leninskaya Prospekt 1
330006 Zaporozhye
Telephone: (0)25762
Founded: 1985
Membership: 30
Activities: education * shows
City Reference Coordinates: 035°08'00"E 47°48'00"N

United Kingdom

Aberdeen College, Planetarium
Gallowgate Centre
Aberdeen AB25 1BN
Scotland
Telephone: (0)1224-612000
Telefax: (0)1224-612001
Electronic Mail: g.russell@abcol.ac.uk (G.A. Russell, Planetarium Development Officer)
WWW: http://www.abcol.ac.uk/planetarium/
Founded: 1978
Staff: 2
Activities: education * entertainment * theatre
City Reference Coordinates: 002°04'00"W 57°10'00"N

Academia Europaea
31 Old Burlington Street
London W1X 1LB
Telephone: (0)207-7345402
Telefax: (0)207-2875115
Electronic Mail: acadeuro@compuserve.com
WWW: http://academia.darmstadt.gmd.de/
Founded: 1988
Staff: 4
Membership: 1900
Activities: annual meeting * study groups * conferences
Periodicals: (4) "European Review" (ISSN 1062-7987); (2) "Academia Europaea Newsletter" (ISSN 0968-3356); (1) "Academia Europaea Directory" (ISSN 1462-3854)
Awards: (1) "Erasmus Medal", "Gold Medal"
City Reference Coordinates: 000°10'00"W 51°30'00"N

Academic Press
24-28 Oval Road
London NW1 7DX
Telephone: (0)171-4822893
Telefax: (0)171-2674752
Electronic Mail: app@apuk.co.uk
WWW: http://www.europe.apnet.com/
          http://www.apnet.com/
Founded: 1959
● Publisher
City Reference Coordinates: 000°10'00"W 51°30'00"N

Altrincham and District Astronomical Society (ADAS)
c/o Derek McComiskey (Secretary)
10 Dale Road
Marple
Stockport SK6 6HA
Telephone: (0)161-4270350
Electronic Mail: adas@mccomiskey.u-net.com
WWW: http://www.u-net.com/ph/nwgas/altrinch/altrinch.htm
Founded: 1963
Membership: 25
City Reference Coordinates: 002°21'00"W 53°24'00"N (Altrincham)
                                002°10'00"W 53°25'00"N (Stockport)

Amateur Astronomy Centre (AAC)
Swallows Barn
Clough Bank
Bacup Road
Todmorden OL13 7HW
Telephone: (0)1706-816964
WWW: http://www.ravenfield.com/aac.htm
Founded: 1982
Membership: 2450
Activities: sponsoring a practical astronomical facility * building a 1m telescope * Planet:Earth Planetarium * education * observing
Periodicals: (3) "AAC Newsletter"
Coordinates: 002°00'00"W 53°00'00"N H350m
City Reference Coordinates: 002°00'00"W 53°00'00"N

Armagh Observatory
College Hill

Armagh BT61 9DG
Northern Ireland
Telephone: (0)1861-522928
Telefax: (0)1861-527174
Electronic Mail: <userid>@star.arm.ac.uk
WWW: http://star.arm.ac.uk/
Founded: 1790
Staff: 15
Activities: solar and flare-star studies in optical, radio, UV and X ray * early-type hypergiants
Periodicals: "Armagh Observatory Preprints"; "Irish Astronomical Journal" (jointly with Dunsink Observatory - see separate entry)
Coordinates: 006°38'54"W 54°21'12"N H64m
City Reference Coordinates: 006°39'00"W 54°21'00"N

Armagh Planetarium
College Hill
Armagh BT61 9DB
Northern Ireland
Telephone: (0)1861-524725
Telefax: (0)1861-526187
WWW: http://www.armagh-planetarium.co.uk/
Founded: 1968
Staff: 12
Activities: education * interactive star shows * exhibitions * school visits * public observing * mail order department
Coordinates: 006°38'54"W 54°21'11"N H30m
City Reference Coordinates: 006°39'00"W 54°21'00"N

Ashgate Publishing Ltd.
Gower House
Croft Road
Aldershot GU11 3HR
Telephone: (0)1252-331551
Telefax: (0)1252-344405
Electronic Mail: info@ashgatepub.demon.co.uk
Founded: 1967
● Publisher
● Formerly "Variorum"
City Reference Coordinates: 000°47'00"W 51°15'00"N

Association for Astronomy Education (AAE)
c/o Royal Astronomical Society
Burlington House
Piccadilly
London W1V 0NL
WWW: http://www.star.ucl.ac.uk/~aae/aaehomep.htm
Founded: 1981
Membership: 230
Activities: annual meetings * advising and supporting teachers
Periodicals: "Gnomon" (ISSN 0952-326x)
City Reference Coordinates: 000°10'00"W 51°30'00"N

Association for Computing Machinery (ACM), European Service Centre
108 Cowley Road
Oxford OX4 1JF
Telephone: (0)1865-382338
Telefax: (0)1865-381338
Electronic Mail: acm_europe@acm.org
WWW: http://www.acm.org/
Founded: 1947 (as former "Society of the Computing Community")
● See main entry in USA-NY
City Reference Coordinates: 001°15'00"W 51°46'00"N

Association for Information Management (ASLIB)
Information House
20-24 Old Street
London EC1V 9AP
Telephone: (0)171-2534488
Telefax: (0)171-4300514
Electronic Mail: internet@aslib.co.uk
WWW: http://www.aslib.co.uk/aslib/
Founded: 1924
Membership: 2000
City Reference Coordinates: 000°10'00"W 51°30'00"N

Association in Scotland to Research into Astronautics (ASTRA) Ltd.
c/o Duncan Lunan (Treasurer)
Flat 65 - Dalriada House
56 Blythswood Court
Glasgow G2 7PE
Scotland
Telephone: (0)141-2217658
Electronic Mail: info@astra.org.uk
WWW: http://www.astra.org.uk/
Founded: 1953
Membership: 85
Activities: lectures * exhibitions * publishing * technical projects * amateur astronomy * amateur rocketry * education
Periodicals: (12) "Spacereport"; "ASGARD"
Awards: (1) "Oscar Schwiglhofer Trophy"
Coordinates: 003°57'00"W 55°53'00"N H0m (Airdrie Observatory)
City Reference Coordinates: 004°15'00"W 55°53'00"N

Association of Learned and Professional Society Publishers (ALPSP)
c/o Sally Morris (Secretary-General)
South House
The Street
Clapham
Worthing BN13 3UU
Telephone: (0)1903-871686
Telefax: (0)1903-871457
Electronic Mail: alpsp@morris-assocs.demon.co.uk
                    alpsp@storrie.demon.co.uk
WWW: http://www.alpsp.org.uk/
Founded: 1972
Membership: 130
Periodicals: (4) "Learned Publishing" (ISSN 0260-9428)
City Reference Coordinates: 000°23'00"W 50°48'00"N

AstroArt
99 Southam Road
Hall Green
Birmingham B28 0AB
Telephone: (0)121-7771802
Telefax: (0)121-7772792
Electronic Mail: dave@hardyart.demon.co.uk (David A. Hardy)
WWW: http://www.hardyart.demon.co.uk/
Founded: 1970
Staff: 2
Activities: astronomical art * illustrations * books * slides * prints * computer graphics * picture library
City Reference Coordinates: 001°50'00"W 52°30'00"N

Astrodome
c/o Peter J. Golding
Gillingham ME7 2LP
Telephone: (0)1634-853541
Telefax: (0)1634-853541
Electronic Mail: astrodome@clara.co.net
WWW: http://www.astrodome.clara.co.uk/
Activities: mobile planetarium
City Reference Coordinates: 000°33'00"E 51°24'00"N

Astronomer (The_)
● See "The Astronomer"

Astronomical Science Group of Ireland (ASGI)
c/o Armagh Observatory
College Hill
Armagh BT61 9DG
Northern Ireland
Telephone: (0)1861-522928
Telefax: (0)1861-527174
Electronic Mail: csj@star.arm.ac.uk
WWW: http://star.arm.ac.uk/~csj/asgi
Founded: 1975
Membership: 120
Activities: one-day meeting held twice annually
Periodicals: "Meeting Proceedings" (published in the "Irish Astronomical Journal" - see separate entry)
City Reference Coordinates: 006°39'00"W 54°21'00"N

Astronomical Society of Edinburgh (ASE)
c/o Graham Rule (Secretary)
105/19 Causewayside
Edinburgh EH9 1QG
Scotland
or :
City Observatory
Calton Hill
Edinburgh EH7 5AA
Scotland
Telephone: (0)131-6670647
          (0)131-5564365
Electronic Mail: grahamrule@ed.ac.uk
WWW: http://www.roe.ac.uk/asewww/
Founded: 1924
Membership: 123
Activities: monthly meetings * lectures * observing
Periodicals: (12) "Bulletin"; (2) "Journal"
Awards: "Lorimer Medal"
City Reference Coordinates: 003°13'00"W 55°57'00"N

Astronomical Society of Glasgow (ASG)
Apartment 8/4
75 Plean Street
Glasgow G14 0YW
Scotland
Telephone: (0)141-9547063
Founded: 1894
Membership: 100
Activities: monthly lectures * education * observing * Summer trips
Periodicals: (12) "Newsletter"
City Reference Coordinates: 004°15'00"W 55°53'00"N

Astronomical Society of Haringey (ASH)
c/o Ken Creamer
23 Hemington Avenue
London N11 3LR
Telephone: (0)181-3680888
Founded: 1972
Membership: 70
Activities: observing * space and astronomical lectures * exhibitions
Periodicals: (12) "2002"
City Reference Coordinates: 000°10'00"W 51°30'00"N (London)

Astronomy Bookshop (The_)
● See "The Astronomy Bookshop"

Astronomy Now
P.O. Box 175
Tonbridge TN10 4ZY
Telephone: (0)1732-367542
Telefax: (0)1732-356230
Electronic Mail: syoung@astronow.compulink.co.uk
          pam@astronow.demon.co.uk
          editorial@astronow.demon.co.uk
WWW: http://www.astronomynow.com/
Founded: 1987
● Journal
City Reference Coordinates: 000°16'00"E 51°12'00"N

Astronomy Roadshow (The_)
● See "The Astronomy Roadshow"

Astronomy Technology Centre (ATC)
Royal Observatory
Blackford Hill
Edinburgh EH9 3HJ
Scotland
Telephone: (0)131-668-8100
Telefax: (0)131-668-8264
WWW: http://www.roe.ac.uk/atc/
Founded: 1998
City Reference Coordinates: 003°13'00"W 55°57'00"N

AstroSoc
● See "University of Bath, Astronomy Society"
● See also "University of Saint Andrews, Astronomy Society"

Automatic Systems Laboratories (ASL) Ltd.
28 Blundells Road
Bradville
Milton Keynes MK13 7HF
Telephone: (0)1908-320666
Telefax: (0)1908-322564
Electronic Mail: sales@aslco.com
WWW: http://www.aslco.com/
Founded: 1967
Staff: 40
Activities: manufacturing contact, non-contact, and optical displacement measurement transducers and systems * precision temperature measurement
City Reference Coordinates: 000°42'00"W 52°02'00"N

AWR Technology
The Old Bakehouse
Albert Road
Deal CT14 9DZ
Telephone: (0)1304-365918
Telefax: (0)1304-369737
Electronic Mail: awr.tech@dial.pipex.com
WWW: http://dspace.dial.pipex.com/awr.tech/
Founded: 1984
Staff: 5
Activities: designing, manufacturing and distributing telescope drive systems and sidereal clocks
City Reference Coordinates: 001°24'00"E 51°14'00"N

Aylesbury Astronomical Society
c/o A. Smith
182 Marley Fields
Leighton Buzzard LU7 8WN
or :
c/o Peter Biswell
14 The Pastures
Aylesbury HP20 1XL
or :
c/o Colin Hunt Observatory
Upper Winchendon
Aylesbury
Telephone: (0)1296-85125
              (0)152-5374258
Founded: 1962
Membership: 50
Activities: building an observatory * promoting local interest in astronomy
Periodicals: (12) "Aylesbury Astronomical Society Bulletin"
Coordinates: 000°55'46"W 51°48'42"N H133m (Colin Hunt Observatory)
City Reference Coordinates: 000°50'00"W 51°50'00"N (Aylesbury)

Ayrshire Astronomical Society (AAS)
1 Lova Street
Kilwinning KA13 7LQ
Scotland
Telephone: (0)1294-552250
Telefax: (0)1294-552250
Electronic Mail: tmcewan@kersland.u-net.com (Tom McEwan)
WWW: http://www.kersland.u-net.com/aas/aashome.htm
Founded: 1992
City Reference Coordinates: 004°38'00"W 55°28'00"N (Ayr)
                                   004°42'00"W 55°40'00"N (Kilwinning)

Barr Associates Ltd.
3 & 4 Home Farm Business Units
Yattendon
Newbury RG18 0XT
Telephone: (0)1635-201317
Telefax: (0)1635-202030
Electronic Mail: barruk@compuserve.com
WWW: http://www.barr-associates-uk.com/
Founded: 1972 (USA)
Activities: designing and manufacturing custom astronomical, space-based and other precision optical filters
City Reference Coordinates: 001°20'00"W 51°25'00"N

Beacon Hill Telescopes
112 Mill Road
Cleethorpes DN35 8JD
Telephone: (0)1472-692959
Telefax: (0)1472-506108
Founded: 1985
Staff: 5
Activities: manufacturing telescopes and scientific instruments
City Reference Coordinates: 000°02'00"W 53°34'00"N

Bedford Astronomical Society (BAS)
24 Swallowfield
Wyboston MK44 3AE
Telephone: (0)1480-406350
Electronic Mail: dave@bas.powernet.co.uk (Dave Eagle)
Founded: 1987
Membership: 40
Activities: observing * meetings
Periodicals: (12) "BAS Newsletter"
City Reference Coordinates: 000°29'00"W 52°08'00"N (Bedford)

BHC Aerovox Ltd.
20-21 Granby Industrial Estate
Weymouth DT4 9TE
Telephone: (0)1305-782871
Telefax: (0)1305-760670
Electronic Mail: sales@bhc.co.uk
WWW: http://www.bhc.co.uk/
Founded: 1968
Activities: aluminium electrolytic capacitors for AC and DC applications * aluminium electrolytic energy discharge capacitors * microwave oven capacitor
City Reference Coordinates: 002°28'00"W 50°36'00"N

Birmingham Astronomical Society (BAS)
c/o Peter Bolas (Membership Secretary)
4 Moat Bank
Burton-upon-Trent DE15 0QJ
Telephone: (0)1283-530786
Electronic Mail: pbolas@aol.com
                     jane@carty.demon.co.uk
WWW: http://www.carty.demon.co.uk/
Founded: 1950
Membership: 100
Activities: lectures * library * instrumentation
Periodicals: "Journal", "Communiqué"
Coordinates: 001°50'00"W 52°30'00"N
City Reference Coordinates: 001°50'00"W 52°30'00"N (Birmingham)
                                001°36'00"W 52°48'00"N (Burton-upon-Trent)

Blackpool and District Astronomical Society (BADAS)
c/o Terry Devon
30 Victory Road
Blackpool FY1 3JT
Telephone: (0)1253-625975
WWW: http://www.u-net.com/ph/nwgas/blackpl/blackpl.htm
Membership: 16
City Reference Coordinates: 003°03'00"W 53°50'00"N

Blackwell Science Ltd.
Osney Mead
Oxford OX2 0EL
Telephone: (0)1865-206206
Telefax: (0)1865-721205
Electronic Mail: <userid>@blacksci.co.uk
WWW: http://www.blacksci.co.uk/
Founded: 1939
Staff: 180
● Publisher
City Reference Coordinates: 001°15'00"W 51°46'00"N

Bolton Astronomical Society (BAS)
c/o Peter Miskiw
9 Hedley Street
Bolton BL1 3LF
Telephone: (0)1204-491568

WWW: http://www.u-net.com/ph/nwgas/bolton/bolton.htm
       http://www.netcomuk.co.uk/~sfield/bas/
Founded: 1979
Membership: 24
Activities: lectures * discussions * observing * trips
Periodicals: (4) "Newsletter"
City Reference Coordinates: 002°26'00"W 53°35'00"N

Border Astronomical Society
14 Shap Grove
Carlisle CA2 5QR
Telephone: (0)1228-532724
Electronic Mail: david.pettitt@tesco.net (David Pettitt, Honorary Secretary)
Founded: 1971
Membership: 20
Activities: photography * instrumentation * aurorae and magnetometry * observing (meteors, Sun, lunar occultations, planets)
Coordinates: 002°56'30"W 54°54'42"N H24m
City Reference Coordinates: 002°55'00"W 54°54'00"N

Bradford Astronomical Society (BAS)
c/o David Cooper
36 Pollard Lane
Undercliffe BD2 4RN
Electronic Mail: bas@akili.demon.co.uk
WWW: http://www.andybat.demon.co.uk/bas/
Founded: 1973
Membership: 40
City Reference Coordinates: 001°45'00"W 53°48'00"N (Bradford)

Braintree Astronomical Society (BAS)
c/o Iain Manning
11 Walnut Drive
Witham CM8 2ST
Telephone: (0)1376-521897
Electronic Mail: iain@manning.netlineuk.net
WWW: http://www.btinternet.com/~braintree.astro/
Founded: 1980
Membership: 45
Activities: monthly meetings
Periodicals: (12) "Braintree Astronomical Society Newsletter"
City Reference Coordinates: 000°32'00"E 51°53'00"N (Braintree)
                               000°38'00"E 51°48'00"N (Witham)

Bridgend Astronomical Society
c/o Clive Down (Secretary)
10 Glan-y-Llyn
North Cornelly
Mid Glamorgan CF33 4EF
Wales
Telephone: (0)1656-740754
Electronic Mail: clivedown@delphi.com
WWW: http://users.aol.com/astrowales/bridgend.html
Founded: 1981
Membership: 55
Activities: observing * lectures
Periodicals: (4) "Bridgend Astronomical Society Newsletter" (circ.: 200)
City Reference Coordinates: 003°35'00"W 51°31'00"N (Bridgend)

Bridgwater Astronomical Society (BAS)
c/o G. MacKenzie
Dennathorne
Spurwells
Ilton
Ilminster TA19 9HP
Telephone: (0)1278-423571
Founded: 1969
Membership: 15
Activities: photography * lectures * observing * history of astronomy
Observatories: 1 (Charterhouse Observatory)
City Reference Coordinates: 003°00'00"W 51°08'00"N (Bridgwater)
                               002°55'00"W 50°56'00"N (Ilminster)

Bridport Astronomy Society (BAS)
3 The Green
Walditch

Bridport DT6 4LB
Telephone: (0)1305-456250
Founded: 1985
Membership: 15
Activities: meetings * lectures * discussions * visits
City Reference Coordinates: 002°46'00"W 50°44'00"N

Bristol Astronomical Society (BAS)
c/o Membership Secretary
23 Chiltern Close
Whitchurch BS14 9RH
WWW: http://wkweb4.cableinet.co.uk/adg.avcomp/bashome.htm
Membership: 100
City Reference Coordinates: 002°35'00"W 51°27'00"N (Bristol)

British Association of Planetaria (BAP)
c/o London Planetarium
Marylebone Road
London NW1 5LR
Telephone: (0)171-4870200 (Administration)
Telefax: (0)171-4650862
Founded: 1976
Membership: 44
City Reference Coordinates: 000°10'00"W 51°30'00"N

British Astronomical Association (BAA)
Burlington House
Piccadilly
London W1V 9AG
Telephone: (0)171-7344145
Electronic Mail: 100257.735@compuserve.com (Hazel McGee, Journal Editor)
WWW: http://www.emoticon.com/emoticon/astro/baamain.html
http://www.ukindex.co.uk/ukastro/baamain.html
http://www.ast.cam.ac.uk/~baa/
http://www.u-net.com/ph/cfds/ (BAA Campaign for Dark Skies - CfDS)
http://www.star.ucl.ac.uk/~hwm (Journal of the British Astronomical Association)
Founded: 1890
Membership: 4000
Activities: observing (planets, Sun, Moon, meteors, comets, aurorae) * photography * artificial satellites * meetings * library
Periodicals: (6) "Journal of the British Astronomical Association" (ISSN 0007-0297); (1) "Handbook of the British Astronomical Association" (ISSN 0068-130x); "Circulars of the British Astronomical Association" (ISSN 0264-4185), "BAA Variable Star Section Circulars" (ISSN 0267-9272)
City Reference Coordinates: 000°10'00"W 51°30'00"N

British Geological Survey (BGS)
Kingsley Dunham Centre
Keyworth
Nottingham NG12 5GG
Telephone: (0)115-9363100
(0)115-9363578 (Public Relations and Media)
(0)115-9363241 (Sales Desk)
Telefax: (0)115-9363200
Electronic Mail: <userid>@nkw.ac.uk
WWW: http://www.nkw.ac.uk/bgs
Founded: 1835
Staff: 1000
Activities: providing relevant and up-to-date geoscience information and advice for the UK, onshore, offshore and internationally
City Reference Coordinates: 001°01'00"W 52°58'00"N

British Interplanetary Society (BIS)
27/29 South Lambeth Road
London SW8 1SZ
Telephone: (0)171-7353160
Telefax: (0)171-8201504
Electronic Mail: bis@cix.compulink.co.uk
WWW: http://freespace.virgin.net/bis.bis/Bis.htm
Founded: 1933
Membership: 4500
Activities: promoting and advancing knowledge relative to space research, technology and applications * reviewing national and international space activities and formulating forward-looking proposals for the advancement of space exploration and utilization
Periodicals: (12) "Spaceflight" (ISSN 0038-6340), "Journal of the British Interplanetary Society" (ISSN 0007-084x)
Awards: "Space Achievement Medals", "Patrick Moore Medals"
City Reference Coordinates: 000°10'00"W 51°30'00"N

British National Space Centre (BNSC)
151 Buckingham Palace Road
London SW1W 9SS
Telephone: (0)171-2150806
              (0)171-2150807
              (0)171-2150808
Telefax: (0)171-2150936
Electronic Mail: information@bnsc-hq.ccmail.compuserve.com
                 <userid>@bnsc-hq.ccmail.compuserve.com
WWW: http://www.open.gov.uk/bnsc/bnschome.htm
Founded: 1985
Staff: 50
Activities: coordinating UK interests and activities in space
Periodicals: (2) "Space News"; (1) "UK Space Index", "UK Space Activities"
City Reference Coordinates: 000°10'00"W 51°30'00"N

British Standards Institution (BSI)
389 Chiswick High Road
London W4 4AL
Telephone: (0)181-9969000
Telefax: (0)181-9967400
Electronic Mail: info@bsi.org.uk
WWW: http://www.bsi.org.uk/
City Reference Coordinates: 000°10'00"W 51°30'00"N

British Sundial Society (BSS)
112 Whitehull Road
London E4 6DW
or :
c/o Robert. B. Sylvester (Membership Secretary)
Barncroft
Grizebeck
Kirkby-in-Furness LA17 7XJ
Telephone: (0)181-5294880
              (0)1229-889716
WWW: http://www.sundials.co.uk/bsshome.htm
Founded: 1989
Membership: 540
Periodicals: (3) "Newsletter of the British Sundial Society"
City Reference Coordinates: 000°10'00"W 51°30'00"N (London)
                            002°33'00"W 54°38'00"N (Kirkby-in-Furness)

British UFO Research Association (BUFORA) Ltd.
BM Bufora
London WC1N 3XX
or :
16 Southway
Burgess Hill RH15 9ST
Telephone: (0)1444-236738
Telefax: (0)0924-444049
Electronic Mail: bufora@stairway.co.uk
WWW: http://www.bufora.org.uk/
Founded: 1964
Staff: 1
Membership: 1000
Activities: scientific investigation of the UFO phenomenon * monthly lectures * conferences
Periodicals: (4) "UFO Times" (2) "Journal of Transient Aerial Phenomena"
City Reference Coordinates: 000°07'00"W 50°57'00"N

Broadhurst Clarkson & Fuller Ltd.
Telescope House
63 Farringdon Road
London EC1M 3JB
Telephone: (0)171-4052156
              (0)171-4050448
Telefax: (0)171-4303471
Founded: 1750
Staff: 6
Activities: importing, renovating and distributing telescopes and accessories
City Reference Coordinates: 000°10'00"W 51°30'00"N

Butterworth-Heinemann Ltd.
Linacre House
Jordan Hill
Oxford OX2 8DP
Telephone: (0)1865-310366

Telefax: (0)1865-310898
WWW: http://www.bh.com/
● Publisher
City Reference Coordinates: 001°15'00"W 51°46'00"N

Cambridge Astronomical Association (CAA)

c/o Steve Jones (Membership Secretary)
17 Acrefield Drive
Cambridge CB4 1JW
or :
c/o Frank Murphy
98 High Street
Harlton
Cambridge CB3 7ES
Telephone: (0)1223-362530 (Jones)
            (0)1223-262421 (Murphy)
Electronic Mail: davidh@sdl.mdcbbs.com
WWW: http://www.mrc-dunn.cam.ac.uk/caa/caahome.html
Founded: 1959
Membership: 60
Activities: monthly lectures * education * observing * instrument making
Periodicals: (6) "Capella"
Coordinates: 000°02'00"W 52°09'30"N H20m
City Reference Coordinates: 000°08'00"E 52°13'00"N

Cambridge University Astronomical Society (CUAS)

c/o Institute of Astronomy
Madingley Road
Cambridge CB3 0HA
Telephone: (0)1223-337548
Telefax: (0)1223-337523
Electronic Mail: jwk21@cam.ac.uk (James Keen)
              jds1@phx.cam.ac.uk
              jrk14@phx.cam.ac.uk
WWW: http://www.chu.cam.ac.uk/home/jwk21/cuas.html
Founded: 1942
Membership: 450
Activities: weekly lectures in full term * observing facilities available for members * weekly social gatherings * visits to places of astronomical interest * quizes against other astronomical societies
Periodicals: (6) "Neptune"
Coordinates: 000°05'46"E 52°12'47"N H30m
City Reference Coordinates: 000°08'00"E 52°13'00"N

Cambridge University Press (CUP)

Edinburgh Building
Shaftesbury Road
Cambridge CB2 2RU
Telephone: (0)1223-325760
Telefax: (0)1223-315052
Electronic Mail: smitton@cup.cam.ac.uk
WWW: http://www.cup.cam.ac.uk/
Founded: 1534
● Publisher
City Reference Coordinates: 000°08'00"E 52°13'00"N

Cardiff Astronomical Society (CAS)

c/o Jackie Rudd (Membership Secretary)
19 Norris Close
Penarth
Vale of Glamorgan CF6 1QW
Wales
or :
c/o David Powell (Secretary)
1 Tal-y-Bont Road
Ely
Cardiff
Wales
Telephone: (0)1222-551704
Electronic Mail: cas@ilddat.demon.co.uk
WWW: http://www.astro.cf.ac.uk/cas/cas_home.html
        http://users.aol.com/astrocas/
        http://users.aol.com/astrocas/private/cas_home.html
        http://www.astro.cf.ac.uk/cas/cas_hom.html
Founded: 1975
Membership: 200
Activities: weekly lectures * observing * promoting public awareness

Periodicals: (5) "News Letter"
Awards: "Bill Sutherland Award"
City Reference Coordinates: 003°13'00" W 51°29'00" N (Cardiff)

Castle Point Astronomy Club (CPAC)
c/o Andrew P. Turner (Secretary)
3 Canewdon Hall close
Canewdon
Rochford SS4 3PY
Telephone: (0)1702-258640
Electronic Mail: neil@badlands.demon.co.uk (Neil Pellinacci)
                100436,2022@compuserve.com (Ted Rodway)
WWW: http://www.netlink.co.uk/users/badlands/cpag/
         http://www.cpac.freeserve.co.uk/
Founded: 1969
Periodicals: "Star News"
City Reference Coordinates: 000°43'00" E 51°36'00" N

Centronic Ltd.
Centronic House
King Henry's Drive
New Addington
Croydon CR9 0BG
Telephone: (0)1689-842121
Telefax: (0)1689-843053
Electronic Mail: rdsales@centronic.co.u
WWW: http://www.centronic.co.uk/
Founded: 1947
Staff: 150
Activities: manufacturing silicon photodiodes, Sun sensors, star mappers, remote sensors and detectors (UV, visible, IR, nuclear)
City Reference Coordinates: 000°10'00" W 51°30'00" N (London)

Chester Astronomical Society (CAS)
c/o Frank White
23 Cedar Grove
Hoole
Chester CH2 3LQ
Telephone: (0)1244-314265
WWW: http://www.u-net.com/ph/nwgas/chester/chester.htm
Founded: 1970
Membership: 60
Activities: lectures * observing * trips
Periodicals: (10) "CAS Newsletter"
City Reference Coordinates: 002°54'00" W 53°12'00" N

Chilbolton Observatory
● See "Council for the Central Laboratory of the Research Councils (CCLRC), Rutherford Appleton Laboratory (RAL), Chilbolton Observatory"

Cleethorpes and District Astronomical Society (CDAS)
112 Mill Road
Cleethorpes DN35 8JD
Telephone: (0)1472-692959
Founded: 1969
Membership: 40
Activities: education * observing
Coordinates: 000°02'00" W 53°33'00" N (Beacon Hill Observatory)
City Reference Coordinates: 000°02'00" W 53°33'00" N

Cleveland and Darlington Astronomical Society (CaDAS)
c/o J. McCue
40 Bradbury Road
Norton
Stockton-on-Tees TS20 1LE
Telephone: (0)1642-554725
Electronic Mail: dr@johnast.demon.co.uk
WWW: http://www.stocktonsfc.ac.uk/mccue/caseden.htm
Founded: 1979
Membership: 45
Activities: monthly meetings * lectures * observing (Castle Eden Public Observatory)
City Reference Coordinates: 001°34'00" W 54°31'00" N (Darlington)
                                 001°19'00" W 54°34'00" N (Stockton-on-Tees)

Cockermouth Astronomical Society (CAS)
c/o Stuart Atkinson (Secretary)
2 Horsman Street
Cockermouth CA13 0HE
Electronic Mail: stuartatk@aol.com
WWW: http://subnet.virtual-pc.com/gr537450/
Founded: 1992
Membership: 30
City Reference Coordinates: 003°21'00"W 54°40'00"N

Computer Journal (The_)
● See "The Computer Journal"

Cotswold Astronomical Society (CAS)
c/o J.A. Daniel
103 The Bassetts
Cashes Green
Stroud GL5 4SL
Telephone: (0)1453-757026
Electronic Mail: cotswoldas@xoommail.com
WWW: http://members.xoom.com/CotswoldAS
Founded: 1982
Membership: 35
Activities: photography * SN patrol
Periodicals: (6) "Mercury"
City Reference Coordinates: 002°12'00"W 51°45'00"N

Council for the Central Laboratory of the Research Councils (CCLRC)
c/o Tony Buckley (Head of Public Relations)
Daresbury Laboratory
Warrington WA4 4AD
Telephone: (0)1925-603000
Telefax: (0)1925-603100
Electronic Mail: a.g.buckley@daresbury.ac.uk
WWW: http://www.ccl.ac.uk/
  http://www.cclrc.ac.uk/
Periodicals: (1) "Annula Report" (ISSN 1359-5865)
City Reference Coordinates: 002°37'00"W 53°24'00"N

Council for the Central Laboratory of the Research Councils (CCLRC), Daresbury Laboratory
Warrington WA4 4AD
Telephone: (0)1925-603000
Telefax: (0)1925-603100
Electronic Mail: <userid>@daresbury.ac.uk
WWW: http://www.ccl.ac.uk/
  http://www.cclrc.ac.uk/
City Reference Coordinates: 002°37'00"W 53°24'00"N

Council for the Central Laboratory of the Research Councils (CCLRC), Rutherford Appleton Laboratory (RAL)
Chilton
Didcot OX11 0QX
Telephone: (0)1235-821900
  (0)1235-445789
Telefax: (0)1235-445808
  (0)1235-446665
Electronic Mail: <userid>@rl.ac.uk
WWW: http://www.clrc.ac.uk/
  http://star-www.rl.ac.uk/ssd.html
Founded: 1965
Staff: 1400
Activities: research in astrophysics, space science, remote sensing, particle physics, lasers, computing * providing facilities for universities' research
Periodicals: (1) "Annual Report" (ISSN 0263-8355); "Rutherford Appleton Laboratory Report Series"
Coordinates: 001°18'41"W 51°34'19"N
City Reference Coordinates: 001°15'00"W 51°37'00"N

Council for the Central Laboratory of the Research Councils (CCLRC), Rutherford Appleton Laboratory (RAL), Chilbolton Observatory
Stockbridge SO20 6BJ
Telephone: (0)1264-860391
Telefax: (0)1264-860142
WWW: http://www.rl.ac.uk/home.html
  http://www.rl.ac.uk/rutherford.html

Founded: 1965 (RAL)
Coordinates: 001°26'13"W 51°08'40"N H85m
City Reference Coordinates: 001°29'00"W 51°07'00"N

Council for the Central Laboratory of the Research Councils (CCLRC), Rutherford Appleton Laboratory (RAL), Space Science Department (SSD)

Chilton
Didcot OX11 0QX
Telephone: (0)1235-821900
Telefax: (0)1235-445848
Electronic Mail: <userid>@rl.ac.uk
WWW: http://star-www.rl.ac.uk/ssd.html
        http://www.ssd.rl.ac.uk/
        http://ast.star.rl.ac.uk/astro_home_page.html (Astrophysics Division)
Founded: 1965 (RAL)
Staff: 200
Activities: collaborating with universities in astronomy and geophysics research programmes on satellite instrument design, project management and other support activities * providing national facilities in data processing, engineering, test and satellite operations * collaborating with international agencies including ESA, NASA, and Russian and Japanese organisations * liaising technically with British National Space Centre (BNSC) partners
City Reference Coordinates: 001°15'00"W 51°37'00"N

Council for the Central Laboratory of the Research Councils (CCLRC), Rutherford Appleton Laboratory (RAL), STARLINK Project

Chilton
Didcot OX11 0QX
Telephone: (0)1235-821900
Telefax: (0)1235-445848
Electronic Mail: ussc@star.rl.ac.uk
WWW: http://www.starlink.rl.ac.uk/
Founded: 1979 (RAL: 1965)
Staff: 13
Activities: data reduction and analysis
Periodicals: "Starlink Documentation Series", "Starlink Software Collection", "Starlink Bulletin"
City Reference Coordinates: 001°15'00"W 51°37'00"N

CPC Program Library

c/o Department of Applied Mathematics and Theoretical Physics
Queen's University of Belfast
Belfast BT7 1NN
Northern Ireland
Telephone: (0)28-90273222
Telefax: (0)29-90239182
Electronic Mail: cpc@qub.ac.uk
WWW: http://cpc.cs.qub.ac.uk/
Founded: 1969
Staff: 3
Activities: storing and distributing all computer programs described and published in "Computer Physics Communications"
City Reference Coordinates: 005°55'00"W 54°35'00"N

Crayford Manor House Astronomical Society (CMHAS)

c/o R.D. Pickard
28 Appletons
Hadlow TN11 0DT
or :
The Manor House
Mayplace Road East
Crayford DA1 4HB
Telephone: (0)1732-850663
Electronic Mail: rdp@star.ukc.ac.uk
WWW: http://www.astronomy.freeserve.co.uk/
Founded: 1961
Membership: 40
Activities: variable stars * photoelectric photometry * occultations * innovative large amateur telescopes * collaborating with professionals
City Reference Coordinates: 000°20'00"E 51°14'00"N (Hadlow)
                            000°11'00"E 51°27'00"N (Crayford)

Croydon Astronomical Society (CAS)

c/o John Murrell (Secretary)
17 Dalmeny Road
Carshalton SM5 4PW
Telephone: (0)181-6475490
Electronic Mail: johnmurell@compuserve.com
WWW: http://www.users.dircon.co.uk/~tangel2/
Founded: 1956

Membership: 150
Activities: fortnightly meetings * observing * informal practical meetings * education * social activities
Periodicals: (6) "Altair" (circ.: 150)
Coordinates: 000°06'09"W 51°18'15"N H168m (Kenley Observatory)
City Reference Coordinates: 000°10'00"W 51°22'00"N

Culture and Cosmos
P.O. Box 1071
Bristol BS99 1HE
Telephone: (0)117-9291303
Electronic Mail: cuture@caol.demon.co.uk
Founded: 1997
Staff: 4
● Periodical (2) (ISSN 1368-6534)
City Reference Coordinates: 002°35'00"W 51°27'00"N

Daresbury Laboratory
● See "Council for the Central Laboratory of the Research Councils (CCLRC), Daresbury Laboratory"

Darlington Astronomical Society (CaDAS) (Cleveland and_)
● See "Cleveland and Darlington Astronomical Society (CaDAS)"

DERA Astronomy Society
c/o Paul Alner (Chairman)
Defence Evaluation and Research Agency
Ively Road
Farnborough GU14 0LX
Electronic Mail: pdalner@mail.dera.gov.uk (Phil Alner, Chairman)
WWW: http://www.ASTRONOMY.DERA.gov.uk/
Founded: 1998
Activities: meetings * lectures * observing
City Reference Coordinates: 000°46'00"W 51°17'00"N

Devon Astronomical Association (DAA)
c/o Lawrence Harris
5 Burnham Park Road
Peverell PL3 5QB
Electronic Mail: lawrenceh@peverell.demon.co.uk
WWW: http://www.peverell.demon.co.uk/daa.html
Founded: 1974
City Reference Coordinates: 004°10'00"W 50°23'00"N (Plymouth)

Dumfries Museum and Camera Obscura
The Observatory
Dumfries DG2 7SW
Scotland
Telephone: (0)1387-253374
Electronic Mail: info@dumfriesmuseum.demon.co.uk
WWW: http://www.dumfriesmuseum.demon.co.uk/
Founded: 1836
Activities: museum exhibition on the history of the Observatory and related activities
City Reference Coordinates: 003°37'00"W 55°04'00"N

Dundee Astronomical Society (DAS)
37 Polepark Road
Dundee DD1 5QT
Scotland
Telephone: (0)1382-667138
Founded: 1956
Membership: 40
Activities: lectures * debates * observing * education
Periodicals: (1) "Newsletter"
City Reference Coordinates: 003°00'00"W 56°28'00"N

Eastbourne Astronomical Society (EAS)
c/o Peter B.J. Gill
18 Selwyn House
Selwyn Road
Eastbourne BN21 2LF
Telephone: (0)1323-646853
Founded: 1960
Membership: 70
Activities: monthly lectures * film/video shows * observing
Periodicals: (10) "Orbit"
Coordinates: 000°16'00"E 50°45'25"N

City Reference Coordinates: 000°17'00"E 50°46'00"N

East Riding Astronomical Society (HERAS) (Hull and_)
● See "Hull and East Riding Astronomical Society (HERAS)"

Edinburgh Mathematical Society (EMS)
c/o The Honorary Secretary
James Clerk Maxwell Building
Mayfield Road
Edinburgh EH9 3JZ
Scotland
Electronic Mail: edmathsoc@maths.ed.ac.uk
WWW: http://www.maths.ed.ac.uk/~edmathsoc/
Founded: 1883
Periodicals: "Newsletter", "Proceedings"
City Reference Coordinates: 003°13'00"W 55°57'00"N

EM Electronics
The Rise
Brockenhurst SO42 7SJ
Telephone: (0)1590-622934
Telefax: (0)1590-622192
WWW: http://www.emelectronics.co.uk/
Founded: 1979
Staff: 3
Activities: designing and manufacturing scientific instruments, particularly ultra-low level DC measuring equipment
City Reference Coordinates: 001°34'00"W 50°49'00"N

Engineering and Physical Sciences Research Council (EPSRC)
Polaris House
North Star Avenue
P.O. Box 18
Swindon SN2 1ET
Telephone: (0)1793-411000
Telefax: (0)1793-411400
Electronic Mail: <userid>@epsrc.ac.uk/
WWW: http://www.epsrc.ac.uk/
City Reference Coordinates: 001°47'00"W 51°34'00"N

ERA Technology Ltd.
Cleeve Road
Leatherhead KT22 7SA
Telephone: (0)1372-367000
Telefax: (0)1372-367099
Electronic Mail: info@era.co.uk
WWW: http://www.era.co.uk/
Founded: 1920
Staff: 480
Activities: contract research organization * research and development * testing * communications engineering * radiofrequency technology * electronic engineering * electrical engineering
City Reference Coordinates: 000°20'00"W 51°18'00"N

Esk Valley College Planetarium (Jewell and_)
● See "Jewell and Esk Valley College Planetarium"

European Centre for Medium-Range Weather Forecasts (ECMWF)
Shinfield Park
Reading RG2 9AX
Telephone: (0)118-9499000
Telefax: (0)118-9869450
Electronic Mail: ecmwf-director@ecmwf.int
WWW: http://www.ecmwf.int/
Founded: 1975
Staff: 140
City Reference Coordinates: 000°59'00"W 51°28'00"N

Ewell Astronomical Society (EAS)
46 Stanton Close
Epsom KT19 9NP
or :
Bourne Hall
Spring Street
Ewell KT17 1UD
Telephone: (0)181-3970385
          (0)1737-353608

WWW: http://www.shine.demon.co.uk/EAS/group.htm
Founded: 1966
Membership: 107
Activities: monthly meetings * observing * instruments for loan * exhibitions * visits * excursions * books, magazines, videos for loan
Periodicals: (6) "Janus" (circ.: 110)
City Reference Coordinates: 000°15'00"W 51°21'00"N (Ewell)
　　　　　　　　　　　　　　　　　　000°16'00"W 51°20'00"N (Epsom)

Exeter Astronomical Society (EAS)
Penn Hill Cottage
Dunchideock
Exeter EX6 7YE
Telephone: (0)1392 832311
Electronic Mail: jruddy@cix.co.uk (John Ruddy, Chairman)
WWW: http://www.compulink.co.uk/~jruddy/eas.html
Membership: 30
Periodicals: "Helios"
City Reference Coordinates: 003°31'00"W 50°43'00"N

Explorers Tours
223 Coppermill Road
Wraysbury TW19 5NW
Telephone: (0)1753-680237
Telefax: (0)1753-682660
Electronic Mail: astro@explorers.co.uk
WWW: http://www.explorers.co.uk/
Founded: 1981
Staff: 10
Activities: organizing inclusive tours, including for observing total solar eclipses, meteor showers, aurorae and foreign observatories
City Reference Coordinates: 000°33'00"W 51°27'00"N

Federation of Astronomical Societies (FAS)
c/o Clive Down
10 Glan-y-Llyn
North Cornelly
Mid Glamorgan CF33 4EF
Wales
Telephone: (0)1656-740754
Electronic Mail: clivedown@btinternet.com
WWW: http://www.emoticon.com/emoticon/astro/fasmain.html
　　　　　http://www.ukindex.co.uk/ukastro/fasmain.html
　　　　　http://www.fedastro.demon.co.uk/
Founded: 1974
Membership: 104 (societies)
Activities: linking societies together * information service
Periodicals: (5) "Newsletter"; (1) "Handbook", "Astrocalendar"
City Reference Coordinates: 003°10'00"W 51°30'00"N (Cardiff)

Fieldview Astronomy Centre
West Barsham Road
East Barsham NR21 0AR
Telephone: (0)1328-820083
Electronic Mail: fieldview@csi.com
WWW: http://ourworld.compuserve.com/homepages/fieldview/
　　　　　http://www.earthandsky.co.uk/ (Earth and Sky)
Founded: 1984
Staff: 2
Activities: guest house for astronomers, birdwatchers, cyclists, naturalists * retailing books and slides
City Reference Coordinates: 001°25'00"W 53°15'00"N (Chesterfield)

Furness Astronomical Society
c/o Allen Hartley (Secretary)
78 Coronation Drive
Dalton-in-Furness LA15 8QH
Telephone: (0)1229-464917
WWW: http://www.u-net.com/ph/nwgas/furness/furness.htm
Founded: 1973
Membership: 22
Activities: promoting astronomy * public education
Coordinates: 003°10'40"W 54°08'40"N H65m (Newton-in-Furness)
City Reference Coordinates: 003°11'00"W 54°09'00"N

Galvoptics Ltd., Telescope Division
Harvey Road
Burnt Mills Industrial Estate
Basildon SS13 1ES
Telephone: (0)1268-728077
Telefax: (0)1268-590445
Founded: 1968
Staff: 29
Activities: manufacturing lenses, prisms, filters, mirrors * optical coatings on glass, low-expansion material, quartz, saphire, fused silica, and other optical materials * removal and recoat of mirrors
City Reference Coordinates: 000°25'00"E 51°35'00"N

George Pick Aerotours
83A London Road
Leicester LE2 0PF
Telephone: (0)116-2540588
Telefax: (0)116-2470834
WWW: http://www.gptravel.co.uk/gptg-welcome.html
Founded: 1984
Staff: 6
Activities: tour operators * educational visits to scientific institutions, museums, planetariums, and other astronomy and space sciences related fields
City Reference Coordinates: 001°05'00"W 52°38'00"N

Glasgow College of Nautical Studies (GCNS), Planetarium
21 Thistle Street
Glasgow G5 9XB
Scotland
Telephone: (0)141-4293201
         (0)141-5652700
Telefax: (0)141-4201690
         (0)141-5652599
Electronic Mail: maritime@glasgow-nautical.ac.uk
WWW: http://www.glasgow-nautical.ac.uk/
Founded: 1969
Activities: demonstrations to clubs, societies, schools and public
City Reference Coordinates: 004°15'00"W 55°53'00"N

Gordon and Breach Science Publishers
P.O. Box 90
Reading RG1 8JL
Telephone: (0)1734-560080
Telefax: (0)1734-568211
Electronic Mail: info@gbhap.com
WWW: http://www.gbhap.com/
Founded: 1961
● Publisher - see main entry in USA-PA
City Reference Coordinates: 000°59'00"W 51°28'00"N

Greenwich Observatory
● See "Old Royal Observatory (ORO)"

Gwynedd Astronomical Society
Ael-y-Bryn
Newborough
Llanfairpwll
Gwynedd, Wales
or :
18 Lon y Gamfa
Menai Bridge
Gwynedd LL59 5QJ
Telephone: (0)1248-440395
Electronic Mail: oss041@bangor.clss1
Founded: 1974
Membership: 15
Activities: lectures * observing
City Reference Coordinates: 004°08'00"W 53°13'00"N (Bangor)

Hampshire Astronomical Group
1 Conifer Close
Waterlooville PO8 8AF
Telephone: (0)1705-646875
WWW: http://www.hantsastro.demon.co.uk/
Founded: 1960

Membership: 80
Activities: observing * education * visits * weekly meetings * monthly lectures
Periodicals: (4) "The Hampshire Observer"
Awards: "Honorary Membership"
Coordinates: 001°01'05"W 50°56'14"N H161m (Clanfield Observatory)
City Reference Coordinates: 001°02'00"W 50°53'00"N

Harrogate Astronomical Society (HAS)
1 York Lane
Knaresborough HG5 0AJ
Telephone: (0)1423-862002
WWW: http://www.andybat.demon.co.uk/has.htm
Founded: 1986
Membership: 35
Activities: monthly meetings * observing * talks to local groups
Periodicals: (4) "Astro News" (circ.: 30)
Observatories: 1 (Harlow Hill Observatory, Harrogate)
City Reference Coordinates: 001°33'00"W 54°00'00"N (Harrogate)

Havering Astronomical Society
c/o Frances Ridgley
133 Severn Drive
Upminster, Essex
Telephone: (0)1708-227397
Electronic Mail: 100304.2143@compuserve.com
WWW: http://ourworld.compuserve.com/homepages/Nik_Szymanek/astro-so.htm
Founded: 1994
Membership: 50
City Reference Coordinates: 000°15'00"E 51°33'00"N

Hawksley Education Ltd. (PHEL) (Paton_)
● See "Paton Hawksley Education Ltd. (PHEL)"

Heart of England Astronomical Society (HOEAS)
c/o Keith Middleton (Secretary)
33 Digby Drive
Marston Green
Solihull B37 7DU
Electronic Mail: hoeas@yahoo.com
WWW: http://www.woden.com/~hoeas/
        http://wemfas.future.easyspace.com/hoeas.html
City Reference Coordinates: 001°45'00"W 52°25'00"N

Hebden Bridge Literary and Scientific Society, Astronomy Section
c/o P.G.H. Jackson
Royal Nook
44 Gilstead Lane
Gilstead
Bingley BD16 3NP
Founded: 1985
Activities: monthly meetings * lectures
City Reference Coordinates: 002°00'00"W 53°45'00"N (Hebden Bridge)

Heinemann Ltd. (Butterworth-_)
● See "Butterworth-Heinemann Ltd."

Heffers Booksellers
20 Trinity Street
Cambridge CB2 1TY
Telephone: (0)1223-568568
Telefax: (0)1223-568591
Electronic Mail: orders@heffers.co.uk
WWW: http://www.heffers.co.uk/
Founded: 1876
City Reference Coordinates: 000°08'00"E 52°13'00"N

Helius Designs
The White House
Aldington
Evesham WR11 5UB
Telephone: (0)1386-830083
Telefax: (0)1386-830083
Electronic Mail: 100674.431@compuserve.com
Founded: 1980
Staff: 4

Activities: designing and manufacturing computer-controlled telescopes and CCD imaging systems
City Reference Coordinates: 001°56'00"W 52°06'00"N

Hencoup Enterprises
Collins Cottage
Lower Road
Loosley Row HP27 0PF
Telephone: (0)1844-347493
Telefax: (0)1844-274782
Electronic Mail: hencoup@hencoup.demon.co.uk
Founded: 1985
Activities: promoting astronomy through TV, radio, newspapers, magazines, bulletin boards, videos, etc. * consultancy work
for astronomical institutions
City Reference Coordinates: 000°50'00"W 51°50'00"N (Aylesbury)

Herschel Museum (William_)
● See "William Herschel Society"

Herschel Society (William_)
● See "William Herschel Society"

H.H. Wills Physics Laboratory
● See "University of Bristol, H.H. Wills Physics Laboratory"

Highlands Astronomical Society (HAS)
c/o James F. Dick
8 Holm Park
Inverness IV2 4XT
Scotland
WWW: http://www.stoob.demon.co.uk/has.htm
City Reference Coordinates: 004°38'00"W 57°13'00"N

Huddersfield Astronomical and Philosophical Society (HAPS)
c/o Paul D. Harper (Treasurer)
45 Lydgate
Lepton HD8 0LT
Telephone: (0)1484-606832
Founded: 1968
Membership: 40
Activities: lectures * visits to other astronomical societies and places of interest * telescope making * elementary astronomy
classes * observing * slide shows * videos * quizes * planetarium
Periodicals: (4) "Omega"
Awards: (1) photographic cup
Coordinates: 001°50'00"W 53°38'00"N H300m (Crosland Hill)
City Reference Coordinates: 001°47'00"W 53°39'00"N (Huddersfield)
                                001°50'00"W 53°38'00"N (Lepton)

Hull and East Riding Astronomical Society (HERAS)
c/o A.G. Scaife
15 Beech Road
Elloughton
Brough HU15 1JX
Telephone: (0)1482-668665 (Helen Marshall, Secretary)
Founded: 1953
Membership: 45
Activities: monthly educational meetings
City Reference Coordinates: 002°19'00"W 54°32'00"N

IGG Component Technology Ltd.
Grove Road
Cosham
Portsmouth PO6 1LX
Telephone: (0)1705-210051
Telefax: (0)1705-210058
Electronic Mail: info@igg.co.uk
WWW: http://www.igg.co.uk/
Founded: 1978
Activities: providing procurement, parts engineering expertise, testing and qualification services for space-quality electronic,
electrical and electro-mechanical components
City Reference Coordinates: 001°05'00"W 50°48'00"N

**Imperial College of Science, Technology and Medicine (ICSTM), Physics Department, Astrophysics
Group**
Prince Consort Road
London SW7 2BZ

Telephone: (0)171-5895111
Telefax: (0)171-5899463
Electronic Mail: astro@ic.ac.uk
WWW: http://icstar5.ph.ic.ac.uk/
Founded: 1984
Staff: 24
Activities: astro-particle physics * satellite IR and X-ray astronomy * extragalactic astrophysics * cosmology * SNs * stellar winds
City Reference Coordinates: 000°10'00"W 51°30'00"N

Imperial College of Science, Technology and Medicine (ICSTM), Space and Atmospheric Physics Group

Blackett Laboratory
Prince Consort Road
London SW7 2BW
Telephone: (0)207-5947770
Telefax: (0)207-5947772
Electronic Mail: <userid>@ic.ac.uk
WWW: http://www.sp.ph.ic.ac.uk/
Founded: 1986
Staff: 35
Activities: space plasma physics * atmospheric physics * instrumentation for space and atmospheric missions
City Reference Coordinates: 000°10'00"W 51°30'00"N

Imperial College Press (ICP)

57 Shelton Street
Covent Garden
London WC2H 9HE
Telephone: (0)171-8363954
Telefax: (0)171-8362002
Electronic Mail: edit@icpress.co.uk
WWW: http://www.icpress.co.uk/
City Reference Coordinates: 000°10'00"W 51°30'00"N
Founded: 1995
Staff: 8
● Publisher

INSPEC

c/o Institution of Electrical Engineers
Michael Faraday House
Six Hills Way
Stevenage SG1 2AY
Telephone: (0)1438-313311
Telefax: (0)1438-742840
Electronic Mail: inspec@iee.org.uk
WWW: http://www.iee.org.uk/publish/inspec/
Founded: 1898 (as former "Science Abstracts")
Staff: 140
Activities: producing INSPEC database (over 5M records) covering literature in physics (including astronomy and astrophysics), electronics and computing * service available internationally online, as abstracts journals and on CD-ROMs
Periodicals: "Physics Abstracts" (ISSN 0036-8091), "Current Papers in Physics" (ISSN 0011-3786), "Computer and Control Abstracts", "Electrical and Electronics Abstracts", "Current Papers on Electrical and Electronics Engineering", "Current Papers on Computers and Control", plus other titles in Key Abstracts
City Reference Coordinates: 000°14'00"W 51°55'00"N

Institute for Scientific Information (ISI), Europe

Brunel Science Park
Uxbridge UB8 3PQ
Telephone: (0)1895-270016
Telefax: (0)1895-256710
Electronic Mail: eurohelp@isinet.com
uksales@isinet.com
WWW: http://www.isinet.com/
Founded: 1958
Staff: 29
Activities: publishing citation and table-of-contents databases linking the research community to scientific, technical and medical peer-reviewed journals
Periodicals: (52) "Current Contents (Physical, Chemical and Earth Sciences)" (ISSN 0163-2574)
City Reference Coordinates: 000°29'00"W 51°33'00"N

Institute of Physics (IOP)

76 Portland Place
London W1N 3DH
Telephone: (0)171-4704800
Telefax: (0)171-4704848
Electronic Mail: physics@iop.org

WWW: http://www.iop.org/
Founded: 1874
Membership: 20000
Activities: publishing books and journals * organising meetings and conferences * education
Periodicals: (50) "Journal of Physics C: Condensed Matter" (ISSN 0953-8984); (40) "Public Understanding of Science" (ISSN 0963-6625); (24) "Journal of Physics A: Mathematical and General" (ISSN 0305-4470), "Journal of Physics B: Atomic, Molecular and Optical Physics" (ISSN 0953-4075); (12) "Classical and Quantum Gravity" (ISSN 0264-9381), "Journal of Physics D: Applied Physics" (ISSN 0022-3727), "Journal of Physics G: Nuclear Physics" (ISSN 0954-3889), "Measurement Science and Technology" (ISSN 0957-0233), "Physics in Medicine and Biology" (ISSN 0031-9155), "Physics World" (ISSN 0953-8585), "Plasma Physics and Controlled Fusion" (ISSN 0741-3335), "Reports on Progress in Physics" (ISSN 0034-4885), "Semiconductor Science and Technology" (ISSN 0268-1242), "Superconductor Science and Technology" (ISSN 0953-2048); (6) "European Journal of Physics" (ISSN 0143-0807), "Inverse Problems" (ISSN 0266-5611), "Physics Education" (ISSN 0031-9120), "Nonlinearity" (ISSN 0951-7715), "OLE - Opto and Laser Europe" (ISSN 0966-9809), "Pure and Applied Optics" (ISSN 0963-9659), "Quantum Optics" (ISSN 0954-8998); (5) "Modelling and Simulation in Materials Science and Engineering" (ISSN 0965-0393); (4) "Distributed Systems Engineering" (ISSN 0967-1846), "High-Performance Polymers" (ISSN 0954-0083), "Journal of Hard Materials" (ISSN 0954-027x), "Journal of Micromechanics and Microengineering" (ISSN 0960-1317), "Journal of Radiological Protection" (ISSN 0952-4746), "Nanotechnology" (ISSN 0957-4484), "Network: Computation in Neural Systems" (ISSN 0954-898x), "Physiological Measurements" (ISSN 0967-3334), "Plasma Sources Science and Technology" (ISSN 0936-0252), "Smart Materials and Structures" (ISSN 0964-1726), "Waves in Random Media" (ISSN 0959-1726)
City Reference Coordinates: $000°10'00"$W $51°30'00"$N

Institute of Physics (IOP) Publishing Ltd.

Dirac House
Temple Back
Bristol BS1 6BE
Telephone: (0)117-9297481
Telefax: (0)117-9294318
Electronic Mail: custserv@ioppublishing.co.uk
                 <userid>@ioppublishing.co.uk
                 listproc@listserver.ioppublishing.com
WWW: http://www.ioppublishing.com/
Founded: 1874 (IOP)
Staff: 100
• Publisher (see list of journals under "Institute of Physics")
City Reference Coordinates: $002°35'00"$W $51°27'00"$N

Institution of Electrical Engineers (IEE)

Savoy Place
London WC2E 0BL
Telephone: (0)171-2401871
Telefax: (0)171-2407735
Electronic Mail: <userid>@iee.org.uk
WWW: http://www.iee.org.uk/Welcome.html
       http://www.iee.org.uk/Industry/Aerospace
Founded: 1871
Membership: 140000
Staff: 450
Activities: promoting advancement of electrical, manufacturing and information engineering * facilitating exchange of knowledge and ideas * publishing * bibliographical information service (see INSPEC)
Periodicals: "Computer and Control Abstracts" (ISSN 0036-8113), "Computing and Control Engineering Journal" (ISSN 0956-3385), "Electrical and Electronics Abstracts" (ISSN 0036-8105), "Electronics and Communication Engineering Journal" (ISSN 0954-0695), "Electronics Education" (ISSN 0265-0096), "Electronics Letters" (ISSN 0013-5194), "Engineering Management Journal" (ISSN 0960-7919), "Engineering Science and Education Journal" (ISSN 0963-7346), "IEE News" (ISSN 0308-0684), "IEE Proceedings - Circuits, Devices and Systems" (ISSN 1350-2409), "IEE Proceedings - Communications" (ISSN 1350-2425), "IEE Proceedings - Computers and Digital Techniques" (ISSN 1350-2387), "IEE Proceedings - Control Theory and Applications" (ISSN 1350-2379), "IEE Proceedings - Electric Power Applications" (ISSN 1350-2352), "IEE Proceedings - Generation, Transmission and Distribution" (ISSN 1350-2360), "IEE Proceedings - Microwaves, Antennas and Propagation" (ISSN 1350-2417), "IEE Proceedings - Optoelectronics" (ISSN 1350-2433), "IEE Proceedings - Radar, Sonar and Navigation" (ISSN 1350-2395), "IEE Proceedings - Science, Measurement and Technology" (ISSN 1350-2344), "IEE Proceedings - Software Engineering" (ISSN 0268-6961), "IEE Proceedings - Vision, Image and Signal Processing" (ISSN 1350-245x) "IEE Review" (ISSN 0013-5127), "Manufacturing Engineer" (ISSN 0956-9944), "Medical and Biological Engineering and Computing" (ISSN 0140-0118), "Microwaves, Antennas and Propagation" (ISSN 1350-2417), "Optoelectronics" (ISSN 1350-2433), "Physics Abstracts" (ISSN 0036-8091), "Power Engineering Journal" (ISSN 0950-3366), "Radar, Sonar and Navigation" (ISSN 1350-2395)
City Reference Coordinates: $000°10'00"$W $51°30'00"$N

International Association of Astronomical Artists (IAAA), European Office

c/o David A. Hardy (President)
99 Southam Road
Hall Green
Birmingham B28 0AB
Telephone: (0)121-7771802
Telefax: (0)121-7772792
Electronic Mail: iaaa@iaaa.org
               dave@hardyart.demon.co.uk
WWW: http://www.iaaa.org/
Founded: 1982 (USA)

Membership: 7
City Reference Coordinates: 001°50'00"W 52°30'00"N

International Association of Geomagnetism and Aeronomy (IAGA)
c/o Michael Gadsden
Physics Unit
Fraser Noble Building
Aberdeen University
Aberdeen AB9 2UE
Scotland
Telephone: (0)1224-574585
Telefax: (0)1224-584776
WWW: http://www.ngdc.noaa.gov/IAGA/iagahome.html
Founded: 1954 (1919 as "IGGU Section on Terrestrial Magnetism and Electricity")
Membership: 76 (countries)
Activities: promoting studies of magnetism, aeronomy, and space plasma physics of the Earth, planets and the interplanetary medium of the solar system
Periodicals: (1) "IAGA News"; "IAGA Bulletin"
Awards: "Long-Service Award"
City Reference Coordinates: 002°04'00"W 57°10'00"N

International Association of Physics Students (IAPS)
c/o Trinity College
Cambridge CB2 1TQ
Telephone: (0)1223-425894
Telefax: (0)1223-337918
Electronic Mail: iaps@nikhef.nl
WWW: http://www.nikhef.nl/pub/iaps/
Founded: 1987
Activities: organising the "International Conference of Physics Students - ICPS"
Periodicals: "Journal of the International Association of Physics Students (JIAPS)"
City Reference Coordinates: 000°08'00"E 52°13'00"N

International Information Services for the Physics and Engineering Communities
● See "INSPEC"

International Maritime Organization (IMO)
4 Albert Embankment
London SE1 7SR
Telephone: (0)20-77357611
Telefax: (0)20-75873210
Electronic Mail: info@imo.org
WWW: http://www.imo.org/
Founded: 1948
Staff: 300
Membership: 153 (states)
Periodicals: (4) "IMO News"
Sections: (committees) Maritime Safety * Maritime Environment Protection * Legal * Technical Cooperation * Facilitation
● Until 1982, known as Inter-Governmental Maritime Consultative Organization (IMCO)
City Reference Coordinates: 000°10'00"W 51°30'00"N

IOP Publishing Ltd.
● See "Institute of Physics (IOP) Publishing Ltd."

Ipswich School Astronomical Society (ISAS)
Henley Road
Ipswich IP1 3SG
Telephone: (0)1473-408300
Telefax: (0)1473-400058
Electronic Mail: isas@dial.pipex.com
WWW: http://dspace.dial.pipex.com/town/parade/lg53/
Founded: 1998
Activities: weekly meetings * observing * instrumentation
City Reference Coordinates: 001°10'00"E 52°04'00"N

Irish Astronomical Association (IAA)
4 Ailsa Park
Bangor BT19 1EA
Northern Ireland
WWW: http://star.pst.qub.ac.uk/iaa/
Founded: 1946
Membership: 220
Activities: meetings * observing * education
Periodicals: (4) "Stardust"
City Reference Coordinates: 005°40'00"W 54°40'00"N

Irish Astronomical Journal (IAJ)
c/o Editorial Office
12 Heather Lea Avenue
Dore
Sheffield S17 3DJ
Electronic Mail: iaj@star.arm.ac.uk
WWW: http://star.arm.ac.uk/iaj/
Founded: 1950
Staff: 2
● Journal (2) (ISSN 0021-1052, circ.: 300)
City Reference Coordinates: 001°30'00"W 53°23'00"N

Isaac Newton Institute for Mathematical Sciences
20 Clarkson Road
Cambridge CB3 0EH
Telephone: (0)1223-335999
Telefax: (0)1223-330508
Electronic Mail: info@newton.cam.ac.uk
WWW: http://www.newton.cam.ac.uk/
Founded: 1992
Staff: 50
Activities: research linked to mathematical sciences
City Reference Coordinates: 000°08'00"E 52°13'00"N

Island Planetarium (The_)
● See "The Island Planetarium"

Isle of Man Astronomical Society
c/o James W. Martin
Ballaterson Farm
Peel
Isle of Man 1M5 3AB
Telephone: (0)1624-842954
Electronic Mail: tremott@enterprise.net (Mike Kelly, Chairman)
                              gary.kewin@bigfoot.com (Gary Kewin, Vice Chairman)
WWW: http://www.netcomuk.co.uk/~fangorn/iomastro/iomastro.html
Founded: 1989
Membership: 70
Activities: meetings * star parties
Coordinates: 004°38'00"W 54°10'00"N (Foxdale)
City Reference Coordinates: 004°30'00"W 54°15'00"N

James Lockyer Planetarium
● See "Norman Lockyer Observatory and James Lockyer Planetarium"

Jane's Information Group (JIG)
Sentinel House
163 Brighton Road
Coulsdon CR5 2YH
Telephone: (0)181-7003700
Telefax: (0)181-7631006
Electronic Mail: info@janes.co.uk
WWW: http://www.janes.com/
Founded: 1898
Staff: 350
Activities: defence, aerospace, geopolitical and transport information provider
Periodicals: "Jane's Defense Weekly", "Jane's International Defense Review", "Jane's Navy International", "Jane's Defence Upgrades", "Jane's Missiles and Rockets", "Jane's Space Directory"
City Reference Coordinates: 000°08'00"W 51°19'00"N

Jewel and Esk Valley College Planetarium
24 Milton Road East
Edinburgh EH15 2PP
Scotland
Telephone: (0)131-6698461 x294
WWW: http://members.aol.com/DMaccon125/Leonardo/jewel.htm
Founded: 1978
Activities: shows for public and societies * school visits * elementary education
City Reference Coordinates: 003°13'00"W 55°57'00"N

Jodrell Bank Observatory
● See "University of Manchester, Department of Physics and Astronomy, Jodrell Bank Observatory"

Jodrell Bank Science Centre, Planetarium and Arboretum
Macclesfield SK11 9DL

Telephone: (0)1477-571339
Telefax: (0)1477-571695
Electronic Mail: <userid>@jb.man.ac.uk
WWW: http://www.jb.man.ac.uk/sc.html
Founded: 1966
Staff: 15
Activities: visitor attraction for general public and groups * education
City Reference Coordinates: 002°07'00"W 53°16'00"N

John Wiley and Sons Ltd., Head Office
Baffins Lane
Chichester PO19 1UD
Telephone: (0)1243-779777
Telefax: (0)1243-775878
Electronic Mail: compbks@jwiley.com
                 customer@wiley.co.uk
                 journals@wiley.co.uk
WWW: http://www.wiley.co.uk/
       http://www.awa.com/wiley/
Founded: 1807
Staff: 300
• Publisher
City Reference Coordinates: 000°48'00"W 50°50'00"N

Journal of Modern Optics
c/o P.L. Knight (Editor)
Optics Section
Imperial College
Blackett Laboratory
Prince Consort Road
London SW7 2BZ
Telephone: (0)171-5947727
Telefax: (0)171-8238376
WWW: http://www.tandfdc.com/JNLS/mop.htm
Founded: 1953
• Journal (ISSN 0950-0340) (formerly Optica Acta)
City Reference Coordinates: 000°10'00"W 51°30'00"N

JRA Aerospace & Technology Ltd.
CED House
Taylors Close
Marlow SL7 1PR
Telephone: (0)1628-891105
Telefax: (0)1628-890519
Electronic Mail: mail@jratech.co.uk
WWW: http://www.jratech.co.uk/
Founded: 1988
Activities: technology transfer consultancy (technology audit, assessment, bid support and marketing services)
City Reference Coordinates: 000°48'00"W 51°35'00"N

Kendall Hyde Ltd.
Kingsland Industrial Park
Stroudley Road
Basingstoke RG24 0UG
Telephone: (0)1256-840830
Telefax: (0)1256-840443
Founded: 1973
Staff: 20
Activities: optical coating engineering
City Reference Coordinates: 001°05'00"W 51°15'00"N

Kettering and District Amateur Astronomers Society (KDAAS)
• See now "Northants Amateur Astronomers (NAA)"

Krieger Publishing Co. (Eurospan)
3 Henrietta Street
Covent Garden
London WC2E 8LU
Telephone: (0)171-2400856
          (0)0800-526830 (UK only)
Electronic Mail: info@krieger-pub.com
WWW: http://web4u.com/krieger-publishing/
Telefax: (0)171-3790609
Founded: 1969 (USA)
• Publisher

City Reference Coordinates: 000°10'00"W 51°30'00"N

Lancaster and Morecambe Astronomical Society
4 Bedford Place
Scotforth
Lancaster LA1 4EB
Telephone: (0)1524-36211
             (0)1524-412726
Electronic Mail: ehelenerob@btinternet.com (E. Robinson, Secretary)
Founded: 1978
Membership: 30
Activities: monthly meetings * lectures
City Reference Coordinates: 002°48'00"W 54°03'00"N (Lancaster)
                            002°53'00"W 54°04'00"N (Morecambe)

Lancaster University, Environmental Science Department, Planetary Science Research Group (PS-RG)
Environmental Science Department
Lancaster LA1 4YQ
Telephone: (0)1524-593889
Telefax: (0)1524-593985
WWW: http://www.es.lancs.ac.uk/es/research/psrg/psrg.htm
Founded: 1996
Staff: 14
Activities: planetary geology, especially volcanology and tectonics
City Reference Coordinates: 002°48'00"W 54°03'00"N

Leeds Astronomical Society (LAS)
4 Belvedere Grove
Leeds LS17 8BP
Electronic Mail: r.emery@westview.demon.co.uk (Ray Emery, President)
                 astro@westview.demon.co.uk
WWW: http://www.dsellers.demon.co.uk/las/
      http://www.westview.demon.co.uk/las/
Founded: 1859
Membership: 45
Activities: meetings * visits * observing
Periodicals: (4) "Nebula"
City Reference Coordinates: 001°35'00"W 53°50'00"N

Leicester Astronomical Society (LAS)
c/o Ann Bonell (Vice-President)
53 Wardens Walk
Leicester Forest East LE3 3GG
WWW: http://www.leicastrosoc.freeserve.co.uk/
Founded: 1952
Membership: 60
Activities: meetings * observing
City Reference Coordinates: 001°05'00"W 52°38'00"N

Leicester University Astronomy Society
c/o Department of Physics and Astronomy
University of Leicester
University Road
Leicester LE1 7RH
Electronic Mail: klp5@le.ac.uk (President)
WWW: http://www.le.ac.uk/astrosoc/
Membership: 100
Activities: observing evenings * social events
City Reference Coordinates: 001°05'00"W 52°38'00"N

Letchworth and District Astronomical Society (LDAS)
c/o Colin Maynard (Secretary)
4 Churchfields
Stanstead Mountfitchet CM24 8RJ
WWW: http://www.thetrainingpost.co.uk/ldas/home.htm
Founded: 1991
Membership: 70
Activities: meetings * observing
City Reference Coordinates: 000°14'00"W 51°58'00"N (Letchworth)
                            000°12'00"E 51°54'00"N (Stanstead Mountfitchet)

Liverpool Astronomical Society (LAS)
31 Sandymount Drive
Wallasey L45 0LG

Telephone: (0)151-7945356
Electronic Mail: ggastro@liverpool.ac.uk (Gerard Gilligan)
WWW: http://www.liv.ac.uk/~ggastro/home.html
Founded: 1881
Activities: public star parties * meetings * education
Periodicals: (1) "Liverpool's Monthly Sky Diary", "Astronomical Events for the Liverpool Area"
Awards: "Silver Medal"
City Reference Coordinates: 002°55'00"W 53°25'00"N (Liverpool)
002°03'00"W 53°26'00"N (Wallasey)

Liverpool John Moores University, Astrophysics Research Institute
Twelve Quays House
Egerton Wharf
Birkenhead L41 1LD
Telephone: (0)151-2312337
Telefax: (0)151-2312475
Electronic Mail: <userid>@astro.livjm.ac.uk
WWW: http://www.livjm.ac.uk/astro/
Founded: 1992
Staff: 30
Activities: observational cosmology * galaxies * brown dwarfs * hot star environments * novae * star-forming regions * ISM * education
● See also "University of Liverpool"
City Reference Coordinates: 003°02'00"W 53°24'00"N

Liverpool Museum Planetarium
c/o Department of Earth and Physical Sciences
National Museums and Galleries on Merseyside
William Brown Street
Liverpool L3 8EN
Telephone: (0)151-2070001
Telefax: (0)151-4784390
WWW: http://www.itl.net/vc/europe/Liverpool/Tourism/museums.html
Founded: 1969
Staff: 5
Activities: displays * planetarium shows * instrument collection * education * research
City Reference Coordinates: 002°55'00"W 53°25'00"N

Lockyer Observatory (Norman_)
● See "Norman Lockyer Observatory and James Lockyer Planetarium"

Lockyer Planetarium (James_)
● See "Norman Lockyer Observatory and James Lockyer Planetarium"

London Planetarium
Marylebone Road
London NW1 5LR
Telephone: (0)171-4870200 (Administration)
Telefax: (0)171-4650862
WWW: http://www.kidsnet.co.uk/places/planetar.shtml
Founded: 1958
Membership: 100
Activities: planetarium shows for public and schools * exhibitions
City Reference Coordinates: 000°10'00"W 51°30'00"N

Longman Group UK Ltd.
Longman House
Burnt Mill
Harlow CM20 2JE
Telephone: (0)1279-426721
Telefax: (0)1279-431059
WWW: http://www.webserve.co.uk/clients/ivca/malphas/longman.htm
● Publisher
City Reference Coordinates: 000°08'00"W 51°47'00"N

LOT-Oriel Ltd.
1 Mole Business Park
Leatherhead KT22 7AU
Telephone: (0)1372-378822
Telefax: (0)1372-375353
Electronic Mail: info@lotoriel.co.uk
WWW: http://www.lotoriel.co.uk/
Founded: 1976
Staff: 12
Activities: distributing optical and electro-optical instruments

City Reference Coordinates: 000°20'00"W 51°18'00"N

Lumonics Ltd.
Cosford Lane
Swift Valley
Rugby CV21 1QN
Telephone: (0)1788-70321
Telefax: (0)1788-79824
WWW: http://www.lumonics.com/
Founded: 1970
Activities: manufacturing lasers and laser-based systems
City Reference Coordinates: 001°15'00"W 52°23'00"N

Lyons Instruments (LI) Ltd.
Brook Road
Waltham Cross EN8 7LR
or :
Ware Road
Hoddesdon EN11 9DX
Telephone: (0)1992-467161
          (0)1992-768888
Telefax: (0)1992-444513
          (0)1992-788000
Electronic Mail: li@lyons.demon.co.uk
WWW: http://www.lyons.demon.co.uk/li.html
Founded: 1968
Staff: 250
Activities: electrical metrology and standards * frequency and signal sources * technical software * test and measurement devices
Periodicals: (4) "LI News" (circ.: 5,000)
City Reference Coordinates: 000°02'00"W 51°42'00"N (Waltham Cross)
                          000°01'00"W 51°46'00"N (Hoddesdon)

Macclesfield Astronomical Society
c/o Cherry Moss
164A Chester Road
Macclesfield SK11 8PT
Electronic Mail: chris@motorlaw.co.uk (Christopher J. Rose, Vice-President)
WWW: http://www.employees.org/~macas/macc-as.htm
          http://www.g0-evp.demon.co.uk/
Founded: 1990
Membership: 150
City Reference Coordinates: 002°07'00"W 53°16'00"N

Maidenhead Astronomical Society
c/o T.V. Haymes
Hill Rise
Knowl Hill Common
Knowl Hill
Reading RG10 9YD
Telephone: (0)1628-822442
Electronic Mail: 101360.700@compuserve.com
Founded: 1957
Membership: 25
Activities: meetings * lectures * observing * exhibitions
City Reference Coordinates: 000°59'00"W 51°28'00"N (Reading)
                          000°44'00"W 51°32'00"N (Maidenhead)

Maidstone Astronomical Society
c/o Peter Mount
6 Orchard Bank
Chart Sutton
Maidstone ME17 3SG
Telephone: (0)1622-843904
Electronic Mail: pmount@maidast.demon.co.uk
WWW: http://www.demon.co.uk/finder/mas/
Founded: 1984
City Reference Coordinates: 000°32'00"E 51°17'00"N

Manchester Astronomical Society
c/o J.H.W. Davidson (Honorary Secretary)
The Godlee Observatory
Institute of Science and Technology
University of Manchester
Sackville Street

Manchester M60 1QD
Telephone: (0)161-2004977 (answering machine)
Telefax: (0)161-2287040
Electronic Mail: mike@ph.u-net.com
WWW: http://www.u-net.com/ph/mas/
Founded: 1903
Membership: 75
Activities: lectures * conventions * photography * instrumentation * Sun * Moon * meteors * occultations * popularization * advising
Periodicals: (2) "Current Notes"
Coordinates: 002°13'57"W 53°28'34"N H85m
City Reference Coordinates: 002°15'00"W 53°30'00"N

Mansfield and Sutton Astronomical Society (MSAS)
c/o Sherwood Observatory
Coxmoor Road
Sutton-in-Ashfield NG17 5LF
Telephone: (0)1623-552276
Electronic Mail: shepherd@innotts.co.uk (G.W. Shepherd, Secretary)
WWW: http://www.innotts.co.uk/~shepherd/
            http://ourworld.compuserve.com/homepages/Laurie_Cochrane/msas1.htm
Founded: 1970
Membership: 40
Activities: fostering interest in astronomy
Periodicals: (1) "Annual Reports and Accounts"
Coordinates: 001°18'21"W 53°06'50"N H188m
City Reference Coordinates: 001°11'00"W 53°09'00"N (Mansfield)
                                                001°15'00"W 54°14'00"N (Sutton)

Matra Marconi Space
Anchorage Road
Portsmouth PO3 5PU
Telephone: (0)1705-664966
Telefax: (0)1705-690455
WWW: http://www.matra-marconi-space.com/
Founded: 1962
Activities: manufacturing space equipment
City Reference Coordinates: 001°05'00"W 50°48'00"N

McGraw-Hill Book Co. Europe
Shoppenhangers Road
Maidenhead SL6 2QL
Telephone: (0)1628-23432
Telefax: (0)1628-788232
Electronic Mail: customer.service@mcgraw-hill.com
WWW: http://www.mcgraw-hill.co.uk/
            http://www.mcgraw-hill.com/
● Publisher
City Reference Coordinates: 000°44'00"W 51°32'00"N

Meteoritical Society
c/o Monica Grady (Secretary)
Natural History Museum
Cromwell Road
London SW7 5BD
WWW: http://www.uark.edu/studorg/metsoc/
            http://www.uark.edu/meteor/ (Meteoritics and Planetary Science)
Founded: 1931
Membership: 800
Activities: scientific investigations of meteorites, of related materials, and cf impact structures
Periodicals: (4) "Meteoritics and Planetary Science" (ISSN 0026-1114); "Geochimica et Cosmochimica Acta" (co-sponsored)
Awards: "Leonard Medal", "Barringer Award", "Nier Prize"
City Reference Coordinates: 000°10'00"W 51°30'00"N

Meteorological Office
London Road
Bracknell RG12 2SZ
Telephone: (0)1344-420242
                (0)1344-854260
Telefax: (0)1344-854412
                (0)1344-854948
Electronic Mail: <userid>@meto.govt.uk
WWW: http://www.meto.govt.uk/
Founded: 1854
Staff: 2300
Activities: meteorology * climatology * atmospheric sciences * national and international weather services

Periodicals: (12) "Monthly Weather Report"; (1) "Annual Report"; "Marine Observer"
City Reference Coordinates: 000°45'00"W 51°26'00"N

Mexborough and Swinton Astronomical Society (MSAS)
c/o The Secretary
14 Sandalwood Drive
Swinton S64 8PN
WWW: http://www.msasuk.force9.co.uk/
Founded: 1978
City Reference Coordinates: 001°17'00"W 53°30'00"N (Mexborough)
                                          002°21'00"W 53°31'00"N (Swinton)

Mid-Kent Astronomical Society
167 Shakespeare Road
Gillingham, Kent
Electronic Mail: 101752.3725@compuserve.com (Douglas Hull, Treasurer)
WWW: http://ourworld.compuserve.com/homepages/DouglasHull/
City Reference Coordinates: 000°33'00"E 51°24'00"N

Midlands Spaceflight Society (MSS)
c/o Andy Salmon (Secretary)
Olympus Mons
13 Jacmar Crescent
Smethwick B67 7LF
or :
c/o Mike Bryce
16 Yellowhammer Court
Kidderminster DY10 4RR
Telephone: (0)121-5654845
Electronic Mail: andy_salmon@compuserve.com
WWW: http://wemfas.future.easyspace.com/mss/mss.html
Founded: 1990
Membership: 53
Periodicals: (6) "CapCom"
City Reference Coordinates: 001°58'00"W 52°30'00"N (Smethwick)
                                          002°14'00"W 52°23'00"N (Kidderminster)

Mills Observatory
Balgay Park
Glamis Road
Dundee DD2 2UB
Scotland
Telephone: (0)1382-667138
                   (0)1382-435846
Telefax: (0)1382-435962
                   growe@mic.dundee.ac.uk
WWW: http://orange.mcs.dundee.ac.uk:8080/Test/astronomy/mills/home.htm
          http://alife.mic.dundee.ac.uk/Mills/
Founded: 1935
Staff: 2
Activities: popularization * education * observing (Moon, planets, Sun, meteors) aurorae) * information service * planetariums

Periodicals: (12) "The Sky this Month"
Coordinates: 003°00'40"W 56°27'54"N H152m
City Reference Coordinates: 003°00'00"W 56°28'00"N

Monthly Notices of the Royal Astronomical Society (MNRAS)
c/o The Executive Secretary
Royal Astronomical Society
Burlington House
Piccadilly
London W1V 0NL
Telephone: (0)171-7344582
Telefax: (0)171-4940166
Electronic Mail: jr@ras.org.uk (John Randall, Senior Editorial Assistant)
WWW: http://www.blacksci.co.uk/products/journals/mnras.htm
Founded: 1827
● Journal (24) (ISSN 0035-8711, circ.: 1,250)
City Reference Coordinates: 000°10'00"W 51°30'00"N

Mullard Radio Astronomy Observatory (MRAO)
Cavendish Laboratory
Madingley Road
Cambridge CB3 0HE
Telephone: (0)1223-337294

Telefax: (0)1223-354599
Electronic Mail: <userid>@mrao.cam.ac.uk
WWW: http://www.mrao.cam.ac.uk/
         http://www.phy.cam.ac.uk/www/research/ra/mrao.home.html
Founded: 1946
Staff: 40
Activities: radio astronomy * radio galaxies * QSOs * cosmology * radio and optical interferometry * molecular clouds * ISM * interplanetary medium * SNR * mm-wavelength astronomy
Coordinates: 000°02'48"E 52°10'12"N H17m
City Reference Coordinates: 000°08'00"E 52°13'00"N

Mullard Space Science Laboratory (MSSL)
● See "University College London (UCL), Mullard Space Science Laboratory (MSSL)"

Multi Resolutions Ltd.
23 Cranmore Park
Belfast BT9 6JF
Northern Ireland
Telephone: (0)28-90662277
Telefax: (0)28-90662277
Electronic Mail: info@multiresolution.com
WWW: http://www.multiresolution.com/
Founded: 1997
Staff: 2
Activities: software and harware solutions * image and signal processing * high-performance computing * formal methods * speech and text information retrieval * data mining * consultancy
City Reference Coordinates: 005°55'00"W 54°35'00"N

National Museum of Science and Industry (NMSI), Science Museum, Astronomy Section
Exhibition Road
London SW7 2DD
Telephone: (0)171-9388053
           (0)171-9388056
Telefax: (0)171-9388118
Electronic Mail: j.wess@nmsi.ac.uk (Jane Wess, Curator)
                 k.johnson@nmsi.ac.uk (Kevin Johnson, Associate Curator)
WWW: http://www.nmsi.ac.uk/
Founded: 1857
Staff: 400 (Museum)
Activities: acquisition and display of astronomical instrumentation * history and education (ground-based and space astronomy)
Awards: "Science Book Prize" (set up by the Committee on the Public Understanding of Science - COPUS - and sponsored by Science Museum)
City Reference Coordinates: 000°10'00"W 51°30'00"N

National Remote Sensing Centre (NRSC) Ltd.
Delta House
Southwood Crescent
Southwood
Farnborough GU14 0NL
Telephone: (0)1252-541464
Telefax: (0)1252-375016
WWW: http://www.nrsc.co.uk/
Founded: 1989
Staff: 120
Activities: supplying data, products and services based on information derived from Earth observation satellites and aerial photography, including GIS and software engineering capabilities
Periodicals: "Albedo"
City Reference Coordinates: 000°46'00"W 51°17'00"N

National Space Science Centre (NSSC)
Mansion House
41 Guildhall Lane
Leicester LE1 5FQ
Telephone: (0)116-2530811
Telefax: (0)116-2616800
Electronic Mail: info@nssc.co.uk
                 spacecentre@nssc.co.uk
WWW: http://www.nssc.co.uk/
Founded: 1997
Staff: 12
Activities: public understanding of space and science * developing visitor centre (due to open in 2001)
City Reference Coordinates: 001°05'00"W 52°38'00"N

Natural Environment Research Council (NERC)
Polaris House
North Star Avenue
Swindon SN2 1EU
Telephone: (0)1793-411500
Telefax: (0)1793-411501
Electronic Mail: <userid>@nerc.ac.uk
WWW: http://www.nerc.ac.uk/
Founded: 1965
City Reference Coordinates: 001°47'00"W 51°34'00"N

Natural History Museum, Meteoritic Research
Cromwell Road
London SW7 5BD
Telephone: (0)171-9389123
        (0)171-9389445
Telefax: (0)171-9389268
Electronic Mail: mmg@nhm.ac.uk (Monica Grady)
WWW: http://www.nhm.ac.uk/mineralogy/intro/project4/
        http://www.nhm.ac.uk/
City Reference Coordinates: 000°10'00"W 51°30'00"N

Nature
c/o Editorial Office
Porters South
4 Crinan Street
London N1 9XW
Telephone: (0)171-8334000
Telefax: (0)171-8434596
        (0)171-8434597
WWW: http://www.nature.com/
Founded: 1869
● Journal (51) (ISSN 0028-0836)
City Reference Coordinates: 000°10'00"W 51°30'00"N

Newcastle-upon-Tyne Astronomical Society (NAS)
c/o A. Petty
7 Elmfield Park
Gosforth
Newcastle upon Tyne NE3 4UX
Telephone: (0)191-2851706
Founded: 1904
Staff: 50
Activities: lectures * observing * star parties
Periodicals: (3) "Newsletter"
Coordinates: 001°18'06"W 54°59'18"N H55m (Wylam)
City Reference Coordinates: 001°35'00"W 54°59'00"N

Newton Institute for Mathematical Sciences (Isaac_)
● See "Isaac Newton Institute for Mathematical Sciences"

Norman Lockyer Observatory and James Lockyer Planetarium
Salcombe Hill
Sidmouth EX10 0BS
Telephone: (0)1395-579941
Electronic Mail: g.e.white@exeter.ac.uk (Gerald White)
WWW: http://www.azure.com/nlo/
Founded: 1912
Membership: 268
Activities: education * astronomy * radio astronomy * meteorology * amateur radio
Coordinates: 003°13'07"W 50°41'16"N
City Reference Coordinates: 003°15'00"W 50°41'00"N

Northamptonshire Natural History Society (NNHS), Astronomy Section
The Humfrey Rooms
10 Castilian Terrace
Northampton NN1 1LD
Electronic Mail: ram@hamal.demon.co.uk (Bob Marriott, Secretary)
WWW: http://www.teg1.demon.co.uk/nnhs/
Founded: 1957 (NNHS: 1876)
Activities: meetings * observing
City Reference Coordinates: 000°54'00"W 52°14'00"N

Northants Amateur Astronomers (NAA)
124 Senwick Drive
Wellingborough NN8 1RU
Telephone: (0)1933-272628
Electronic Mail: naastronomers@geocities.com
WWW: http://www.geocities.com/CapeCanaveral/Galaxy/6057/
Founded: 1983
Membership: 35
Activities: meetings
Periodicals: (12) "In Focus"
• Formerly known as "Kettering and District Amateur Astronomers Society (KDAAS)"
City Reference Coordinates: 000°42'00"W 52°19'00"N

North Devon Astronomical Society (NDAS)
c/o P. G. Vickery
12 Brd. Park Crescent
Ilfracombe EX34 8DX
Telephone: (0)1271-863224
Electronic Mail: adavies915@aol.com
WWW: http://members.aol.com/NDASTRO/
Founded: 1973
City Reference Coordinates: 004°08'00"W 51°13'00"N

North East London Astronomical Society (NELAS)
c/o B. Beeston
38 Abbey Road
Bush Hill Park
Enfield EN1 2QN
or :
Wanstead House (Meetings)
21 The Green
Wanstead
London E11 2NT
Telephone: (0)181-3635696
Founded: 1956
Membership: 15
Activities: meetings * talks * slide shows * films * videos * observing
Periodicals: (11) "Newsletter"
City Reference Coordinates: 000°10'00"W 51°30'00"N (London)

North Gwent Astronomical Society
12 Merthyr Road
Abergavenny NP7 0AA
Wales
Telephone: (0)1873-831293
Founded: 1989
Membership: 12
Activities: observing * debates * field trips * lectures
City Reference Coordinates: 003°00'00"W 51°50'00"N

North West Group of Astronomical Societies (NWGAS)
c/o Kevin Kilburn
66 Henshall Road
Bollington SK10 5DN
Telephone: (0)1625-572453
WWW: http://www.u-net.com/ph/nwgas
Founded: 1994
Membership: 12 (societies)
Activities: amateur astronomy
City Reference Coordinates: 002°06'00"W 53°18'00"N

Norton and Co. Ltd. (W.W._)
• See "W.W. Norton and Co. Ltd."

Norwich Astronomical Society (NAS)
c/o Malcolm Jones (Secretary)
Tabor House
Norwich Road
Mulbarton NR14 8JT
Telephone: (0)1508-578392
Telefax: (0)1508-570986
Electronic Mail: 100257.1434@compuserve.com
WWW: http://nas.theso.co.uk/nas/nashome.htm
        http://nas.hmso.gov.uk/nas/nashome.htm

http://nas.gurney.org.uk/
Founded: 1945
Membership: 120
Periodicals: (4) "Cygnus"
Coordinates: 001°14'00"E 52°37'00"N H40m (Seething Astronomical Observatory)
City Reference Coordinates: 001°14'00"E 52°37'00"N (norwich)

Nottingham Astronomical Society (NAS)
40 Swindon Close
The Vale
Giltbrook
Nottingham NG16 2WD
Telephone: (0)115-8548687
Electronic Mail: carl.stella@virgin.net (Carl Brennan, Secretary)
Founded: 1946
Activities: amateur astronomy * lectures * observing
Periodicals: (12) "Journal of the Nottingham Astronomical Society"
City Reference Coordinates: 001°01'00"W 52°58'00"N

Nuffield Radio Astronomy Laboratories (NRAL)
● See now "University of Manchester, Department of Physics and Astronomy, Jodrell Bank Observatory"

Numerical Algorithms Group (NAG) Ltd.
Wilkinson House
Jordan Hill Road
Oxford OX2 8DR
Telephone: (0)1865-311744
Telefax: (0)1865-311755
Electronic Mail: infodesk@nag.co.uk
                    <userid>@nag.co.uk
WWW: http://www.nag.co.uk/
Founded: 1971
Staff: 100
Periodicals: "NAG Newsletter" (ISSN 0269-0780)
● Software producer
City Reference Coordinates: 001°15'00"W 51°46'00"N

Observatory Magazine
c/o The Editors
Rutherford Appleton Laboratory
Building R25
Chilton
Didcot OX11 0QX
Telephone: (0)1235-446523
Telefax: (0)1235-445848
Electronic Mail: obs@ast.star.rl.ac.uk
WWW: http://www.ulo.ucl.ac.uk/obsmag/
Founded: 1877
Staff: 4 (Editors)
● Journal (3) (ISSN 0029-7704, circ.: 1,000)
City Reference Coordinates: 001°15'00"W 51°37'00"N

Old Royal Observatory (ORO)
c/o National Maritime Museum
Romney Road
Greenwich
London SE10 9NF
Telephone: (0)181-3126757
            (0)181-3126764
Telefax: (0)181-3126771
WWW: http://www.nmm.ac.uk/tm/oro.html
Founded: 1675 (Museum: 1956)
Activities: history of astronomy * school programmes * scientific instruments * Caird Planetarium
Coordinates: 000°00'00"E 51°28'38"N N47m
City Reference Coordinates: 000°10'00"W 51°30'00"N

Open University (OU), Department of Physics, Astronomy Research Group
Walton Hall
Milton Keynes MK7 6AA
Telephone: (0)1908-653229
Telefax: (0)1908-654192
Electronic Mail: <userid>@open.ac.uk
WWW: http://yan.open.ac.uk/ (Department of Physics)
        http://yan.open.ac.uk/research/astro/ (Astronomy Research Group)
Founded: 1997 (Department: 1969)

Staff: 30
Activities: interacting binaries * compact objects * accretion * Galactic chemical evolution * stellar abundances * plasma astrophysics * orbital dynamics * solar eclipses
City Reference Coordinates: 000°42'00"W 52°02'00"N

Open University (OU), Planetary Sciences Research Institute (PSRI)
Walton Hall
Milton Keynes MK7 6AA
Telephone: (0)1908-655169
Telefax: (0)1908-655910
Electronic Mail: psri@open.ac.uk
WWW: http://psri.open.ac.uk/
Founded: 1997
Staff: 30
Activities: instrumentation of space missions * extraterrestrial sample analysis * impact cratering research * meteorites
City Reference Coordinates: 000°42'00"W 52°02'00"N

Optometrics (UK) Ltd.
Unit 6
Cross Green Garth
Leeds LS9 0SF
Telephone: (0)113-2496973
Telefax: (0)113-2350420
Electronic Mail: optouk@aol.com
WWW: http://www.optometrics.com/
Activities: manufacturing filters
City Reference Coordinates: 001°35'00"W 53°50'00"N

Oriel Ltd. (LOT-_)
● See "LOT-Oriel Ltd."

Orpington Astronomical Society (OAS)
c/o Ian R. Carstairs (Membership Secretary)
28B Thicket Road
Pence
London SE20 8DD
Telephone: (0)181-6591096
Electronic Mail: oas@chocky.demon.uk
WWW: http://www.chocky.demon.co.uk/oas/
Founded: 1980
Membership: 100
Activities: lectures * observing * outings * visits
Periodicals: (5) "The Orpington Astronomical Society Times (TOAST)"
City Reference Coordinates: 000°10'00"W 51°30'00"N (London)
000°05'00"W 51°23'00"N (Orpington)

Orwell Astronomical Society Ipswich (OASI)
c/o R. Gooding (Secretary)
168 Ashcroft Road
Ipswich IP1 6AE
Electronic Mail: mike.harlow@bt-sys.bt.co.uk (Mike Harlow)
ipswich@ast.cam.ac.uk
WWW: http://www.ast.cam.ac.uk/~ipswich/
Founded: 1967
Membership: 70
Activities: observing
Periodicals: "Newsletter"
Coordinates: 001°13'57"E 52°00'33"N
City Reference Coordinates: 001°10'00"E 52°04'00"N

Oxford Instruments, Research Instruments
Tubney Woods
Abingdon
Oxon OX13 5QX
Telephone: (0)1865-393200
Telefax: (0)1865-393333
Electronic Mail: info.ri@oxinst.co.uk
WWW: http://www.oxford-instruments.com/
Founded: 1959
Staff: 300
Activities: manufacturing scientific equipment
Periodicals: "Research Matters"
City Reference Coordinates: 001°29'00"W 51°48'00"N

Oxford University, Mathematical Institute
24-29 Saint Giles'
Oxford OX1 3LB
Telephone: (0)1865-273525
Telefax: (0)1865-273583
Electronic Mail: maths@vax.ox.uk
WWW: http://www.maths.ox.ac.uk/
Activities: education * research
City Reference Coordinates: 001°15'00"W 51°46'00"N

Oxford University Space and Astronomical Society (OUSAS)
c/o Richard Smith
Balliol College
Oxford
Electronic Mail: space@sable.ox.ac.uk
WWW: http://users.ox.ac.uk/~space/
Founded: 1960
Membership: 100
Activities: promoting an interest in astronomy and space-related sciences among the students of Oxford University
City Reference Coordinates: 001°15'00"W 51°46'00"N

Oxford University Press (OUP)
Great Clarendon Street
Oxford OX2 6DP
Telephone: (0)1865-56767
Telefax: (0)1865-56646
Electronic Mail: enquiry@oup.co.uk
WWW: http://www.oup.co.uk/
          http://www.comlab.ox.ac.uk/archive/publishers/oup.html
Founded: 1478
Staff: 800
● Publisher
City Reference Coordinates: 001°15'00"W 51°46'00"N

Oxford University, UK Gemini Support Group (UKGSG)
Keble Road
Oxford OX1 3RH
Telephone: (0)1865-273294
Telefax: (0)1865-273390
Electronic Mail: caa@astro.ox.ac.uk (Colin Aspin)
WWW: http://www-astro.physics.ox.ac.uk/gemini/
Founded: 1998
Staff: 4
Activities: supporting Gemini Observatory in the UK
City Reference Coordinates: 001°15'00"W 51°46'00"N

Particle Physics and Astronomy Research Council (PPARC)
Polaris House
North Star Avenue
Swindon SN2 1SZ
Telephone: (0)1793-442000
Telefax: (0)1793-442002
Electronic Mail: <userid>@pparc.ac.uk
WWW: http://www.pparc.ac.uk/
Staff: 100
Periodicals: "Frontiers"
City Reference Coordinates: 001°47'00"W 51°34'00"N

Particle Physics and Astronomy Research Council (PPARC), Royal Observatory Edinburgh (ROE)
Blackford Hill
Edinburgh EH9 3HJ
Scotland
Telephone: (0)131-6688100
Telefax: (0)131-6688264
Electronic Mail: <userid>@roe.ac.uk
WWW: http://www.roe.ac.uk/
          http://www.roe.ac.uk/ukstu/ukst.html (UK Schmidt Telescope)
Founded: 1822
Staff: 115
Periodicals: "Edinburgh Astronomy Preprints", "ROE Computing Newsletter", "Annual Report" (ISSN 0309-0108), "The JCMT-UKIRT Newsletter", "Spectrum" (ISSN 0963-2700)
Coordinates: 003°11'00"W 55°55'30"N H146m
          155°28'18"W 19°49'54"N H4,200m (Mauna Kea, USA-HI)
City Reference Coordinates: 003°13'00"W 55°57'00"N

Paton Hawksley Education Ltd. (PHEL)
Rockhill Laboratories
59 Wellsway
Keynsham BS18 1PG
Telephone: (0)1117-9862364
Telefax: (0)1117-9868285
Founded: 1953
Staff: 13
Activities: manufacturing diffraction gratings and spectroscopes
City Reference Coordinates: 002°30'00"W 51°26'00"N

Pendle Astronomical Society
Electronic Mail: u5i30@cc.keele.ac.uk (Paul Brown)
WWW: http://members.tripod.com/~tf/pendle.html
Founded: 1996
Membership: 10
Periodicals: "A Drop in the Universe"
City Reference Coordinates: 002°09'00"W 53°52'00"N (Colne)

Penny & Giles Controls Ltd.
6 Nine Mile Point Industrial Estate
Cwmfelinfach
Gwent NP11 7HZ
Wales
Telephone: (0)1495-202000
Telefax: (0)1495-202006
Electronic Mail: sales@pgcontrols.com
WWW: http://www.penny-giles-controls.co.uk/
Founded: 1955
Activities: aerospace, industrial and audio/video products * rotary position sensors * joystick controllers * audio/video controllers
City Reference Coordinates: 003°00'00"W 51°30'00"N

Pick Aerospace Tours (George_)
● See "George Pick Aerospace Tours"

Planet:Earth Planetarium
● See "Amateur Astronomy Centre (AAC)"

Plymouth Astronomical Society (PAS)
c/o Lawrence D.J. Harris (Chairman)
5 Burnham Park Road
Peverell
Plymouth PL3 5QB
Telephone: (0)1752-775148
Telefax: (0)1752-775148
Electronic Mail: lawrenceh@peverell.demon.co.uk
WWW: http://www.itchycoo-park.freeserve.co.uk/pas.html
Founded: 1965
Membership: 35
Activities: promoting and publicizing astronomy
City Reference Coordinates: 004°10'00"W 50°23'00"N

Polyhedron Software Ltd.
Linden House
93 High Street
Standlake
Witney OX8 7RH
Telephone: (0)1865-300579
Telefax: (0)1865-300232
Electronic Mail: sales@polyhedron.com
WWW: http://www.polyhedron.com/
Founded: 1986
Staff: 8
● Software producer and distributor
City Reference Coordinates: 001°29'00"W 51°48'00"N

Powys County Observatory
Llanshay Lane
Knighton LD7 1LW
Powys Co., Wales
Telephone: (0)1547-520247
Telefax: (0)1547-520247
Electronic Mail: starlink@eidosnet.co.uk
WWW: http://pages.eidosnet.co.uk/~starlink

Founded: 1995
Staff: 2
Activities: guided tours * observing * meteorological station * planetarium * seismology * camera obscura * school visits
Coordinates: 003°01'05"W 52°19'34"N H417m
City Reference Coordinates: 003°03'00"W 52°21'00"N

Preston and District Astronomical Society (PADAS)
c/o Jeremiah Horrocks Observatory
Preston PR1 2HE
Telephone: (0)1772-257181 (Keith Robinson)
Electronic Mail: padas@btinternet.com
WWW: http://www.btinternet.com/~roy.jackson/
Founded: 1979
Coordinates: 002°42'12"W 53°46'30"N H33m
City Reference Coordinates: 002°42'00"W 53°46'00"N

Queen Mary and Westfield College (QMWC)
● See "University of London, Queen Mary and Westfield College (QMWC)"

Queensgate Instruments Ltd.
Queensgate House
Waterside Park
Bracknell RG12 1RB
Telephone: (0)1344-484111
Telefax: (0)1344-484115
Electronic Mail: <userid>@queensgate.co.uk
                 space@queensgate.co.uk
                 qi@queensgate.com
WWW: http://www.queensgate.com/
Founded: 1979
Staff: 30
Activities: designing and manufacturing optical interferometers and micropositioning devices * space-qualified systems
City Reference Coordinates: 000°45'00"W 51°26'00"N

Queen's University of Belfast (QUB), Department of Applied Mathematics and Theoretical Physics
David Bates Building
Belfast BT7 1NN
Northern Ireland
Telephone: (0)1232-273177
Telefax: (0)1232-239182
Electronic Mail: <userid>@vi.am.qub.ac.uk
WWW: http://www.am.qub.ac.uk/
Staff: 46
Activities: calculation and use of atomic data in the interpretation of spectra from the Sun and stars
City Reference Coordinates: 005°55'00"W 54°35'00"N

Queen's University of Belfast (QUB), Department of Pure and Applied Physics, Astrophysics and Planetary Science Division
Belfast BT7 1NN
Northern Ireland
Telephone: (0)1232-245133 x3554
Telefax: (0)1232-438918
Electronic Mail: <userid>@qub.ac.uk
WWW: http://star.pst.qub.ac.uk/
Founded: 1967
Staff: 9
Activities: optical and UV spectroscopy of stars * ISM spectroscopy * diagnostics for solar and laboratory plasmas * spectroscopy and photometry of comets and minor planets
City Reference Coordinates: 005°55'00"W 54°35'00"N

Reading Astronomical Society (RAS)
c/o Ruth Sumner (Secretary)
22 Anson Crescent
Shinfield
Reading RG2 8JT
WWW: http://www.users.zetnet.co.uk/astro-reading/
Founded: 1972
Membership: 130
Activities: Meetings * visits
Periodicals: (4) "RASTAR"
City Reference Coordinates: 000°59'00"W 51°28'00"N

Remote Sensing Society (RSS)
c/o School of Geography
University of Nottingham

Nottingham NG7 2RD
Telephone: (0)115-9515435
Telefax: (0)115-9515249
Electronic Mail: rss@nottingham.ac.uk
WWW: http://www.the-rss.org/
Founded: 1974
Membership: 800
Activities: promoting remote-sensing activities and training
Periodicals: (4) "Remote Sensing Society Newsletter"; (1) "Annual Report"; "International Journal of Remote Sensing (IJRS)"

Awards: "Gold Medal", "Len Curtis European Award", "Student Award", "Taylor & Francis Best Letter Award"
City Reference Coordinates: 001°01'00"W 52°58'00"N

Richmond and Kew Astronomical Society (RKAS)
41A Bruce Road
Mitcham CR4 2BJ
Founded: 1986
Membership: 60
Activities: observing * lectures * visits
Periodicals: (4) "Newsletter"
City Reference Coordinates: 000°18'00"W 51°29'00"N (Kew)
000°09'00"W 51°24'00"N (Mitcham)
000°19'00"W 51°28'00"N (Richmond)

Riding Astronomical Society (Hull and East_)
● See "Hull and East Riding Astronomical Society"

Roditi International Corp. Ltd. (RICL)
Carrington House
130 Regent Street
London W1R 6BR
Telephone: (0)171-4394390
Telefax: (0)171-4340896
WWW: http://www.roditi.co.uk/
Founded: 1935
Staff: 12
Activities: distributing optical components, filters and laser crystals
City Reference Coordinates: 000°10'00"W 51°30'00"N

Roper Scientific
P.O. Box 1192
43 High Street
Marlow SL7 1GB UK
Telephone: (0)1628-890858
Telefax: (0)1628-898381
Electronic Mail: info@roperscientific.co.uk
WWW: http://www.roperscientific.co.uk/
Founded: 1970
Staff: 96
Activities: designing, developing and manufacturing CCD digital imaging systems
City Reference Coordinates: 000°48'00"W 51°35'00"N

Royal Astronomical Society (RAS)
Burlington House
Piccadilly
London W1V 0NL
Telephone: (0)20-77343307
Telefax: (0)20-74940166
Electronic Mail: info@ras.org.uk
WWW: http://www.ras.org.uk/ras/
Founded: 1820
Membership: 2800
Staff: 12
Activities: encouragement and promotion of astronomy and geophysics by publishing the results of research * maintaining a library * holding meetings
Periodicals: (24) "Monthly Notices of the Royal Astronomical Society (MNRAS)" (see separate entry); (12) "Geophysical Journal International" (ISSN 0955-419x); (4) "Astronomy & Geophysics" (ISSN 1366-8781, circ.: 3,500)
Awards: (1/3) "Eddington Medal", "Chapman Medal", "Herschel Medal"; (1) "Gold Medal"; "Michael Penston Astronomy Prize"
City Reference Coordinates: 000°10'00"W 51°30'00"N

Royal Holloway and Bedford New College (RHBNC), Department of Physics
Egham Hill
Egham TW20 0EX
Telephone: (0)1784-443448

Telefax: (0)1784-472794
Electronic Mail: physics-department@rhbnc.ac.uk
                    <userid>@rhbnc.ac.uk
WWW: http://www.ph.rhbnc.ac.uk/
Founded: 1884
Staff: 20
Activities: particle physics * 4D string theory * higher-dimensional cosmology
City Reference Coordinates: 000°34'00"W 51°26'00"N

Royal Meteorological Society (RMS)
104 Oxford Road
Reading RG1 7LL
Telephone: (0)1189-568500
Telefax: (0)1189-568571
Electronic Mail: execsec@royal-met-soc.org.uk
                    <userid>@royal-met-soc.org.uk
WWW: http://itu.rdg.ac.uk/rms/rms.html
Founded: 1850
Membership: 3650
Staff: 9
Activities: meetings * education * awards * professional accreditation
Periodicals: (12) "International Journal of Climatology", "Weather" (ISSN 0043-1656); (4) "Quarterly Journal", "Meteorological Applications"
Awards: 3 annual, 5 biennial
City Reference Coordinates: 000°59'00"W 51°28'00"N

Royal Observatory Edinburgh (ROE)
● See "Particle Physics and Astronomy Research Council (PPARC), Royal Observatory Edinburgh (ROE)"

Royal Society (RS)
6 Carlton House Terrace
London SW1Y 5AG
Telephone: (0)207-8395561
Telefax: (0)207-9302170
Electronic Mail: info@royalsoc.ac.uk
WWW: http://www.royalsoc.ac.uk/
Founded: 1660
Membership: 1200
Activities: promoting the exchange and development of scientific ideas and knowledge * recognizing excellence in scientific scholarship and research * encouraging scientific research and promoting its application * promoting and facilitating international scientific relations * providing a source of independent advice on scientific matters * initiating public debate on matters of public importance * helping focus and represent the views and interests of the scientific community * providing science education, awareness and understanding * providing information on and support for the history of scientific endeavour * providing research appointments and grants * organizing meetings and lectures * library and information services
Periodicals: (1) "Year Book", "Annual Report"; "Philosophical Transactions A - Mathematical and Physical Sciences" (ISSN 0080-4614), "Philosophical Transactions B - Biological Sciences", "Proceedings A - Mathematical and Physical Sciences", "Proceedings B - Biological Sciences", "Biographical Memoirs of Fellows of the Royal Society", "Notes and Records", "Science and Public Affairs", "Reports"
Awards: (1) "Copley Medal", "Rumford Medal", "Royal Medals", "Davy Medal", "Darwin Medal", "Buchanan Medal", "Sylvester Medal", "Hughes Medal", "Leverhulme Medal", "Gabor Medal", "Mullard Award", "Esso Energy Award", "Wellcome Foundation Prize and Lecture", "Armourers & Brasiers' Award", "Michael Faraday Award"
City Reference Coordinates: 000°10'00"W 51°30'00"N

Royal Society of Chemistry (RSC)
Burlington House
Piccadilly
London W1V 0BN
Telephone: (0)171-4378656
Telefax: (0)171-4378883
Electronic Mail: info@rsc.org
WWW: http://www.rsc.org/
Founded: 1980
Membership: 46000
Staff: 300
Activities: result of the unification of the "Chemical Society" and the "Royal Institute of Chemistry" * advancement of chemistry and its applications * maintenance of high standards of competence and integrity among practising chemists * databases
Periodicals: (52) "Chemical Business Bulletins"; "Journal of the Chemical Society" subdivided into (24) "Chemical Communications" (ISSN 0022-4936), (24) "Dalton Transactions" (ISSN 0300-9246), (24) "Perkin Transactions I" (ISSN 0300-922x), (12) "Perkin Transactions II" (ISSN 0300-9580), and (24) "Faraday Transactions" (ISSN 0956-5000); (12) "Journal of Chemical Research" consisting of "Part S (Synopsis)" (ISSN 0308-2342), "Part M (Microfiche)" (ISSN 0308-2350), "Part M (Miniprint)" (ISSN 0000-037x), and "Vouchers"; (12) "Chemistry in Britain" (ISSN 0009-3106), "The Analyst" (ISSN 0003-2654), "Analytical Communications" (ISSN 1359-7337), "Chemical Hazards in Industry" (ISSN 0265-5721), "Laboratory Hazards Bulletin" (ISSN 0261-2917), "Methods in Organic Synthesis" (ISSN 0265-4245), "Natural Products Updates" (ISSN 0950-1711), "Current Biotechnology Abstracts" (ISSN 0264-3391), "Chemical Engineering Abstracts" (ISSN 0262-6438), "Mass Spectrometry Bulletin" (ISSN 0025-4738), "Chemical Business Update", "Journal of Materials Chemistry" (ISSN 0959-9428); (8) "Journal of

Analytical Atomic Spectrometry"; (6) "Natural Products Report" (ISSN 0265-0568), "Education in Chemistry" (ISSN 0013-1350), "Theoretical Chemical Engineering Abstracts" (ISSN 0040-5787); (4) "Chemical Society Reviews" (ISSN 0306-0012); (2) "Faraday Discussions"; (1) "Annual Reports on the Progress of Chemistry" consisting of "Section A: Inorganic Chemistry" (ISSN 0260-1818), "Section B: Organic Chemistry" (ISSN 0060-3030), and "Section C: Physical Chemistry" (ISSN 0670-1826); "Analytical Abstracts" (ISSN 0003-2689)
City Reference Coordinates: 000°10'00"W 51°30'00"N

Royal Society of Edinburgh (RSE)
22-24 George Street
Edinburgh EH2 2PQ
Scotland
Telephone: (0)131-2405000
Telefax: (0)131-2405024
Electronic Mail: rse@rse.org.uk
WWW: http://www.ma.hw.ac.uk/RSE/
Founded: 1783
Periodicals: (4) "RSE News" (ISSN 1352-3325)
City Reference Coordinates: 003°13'00"W 55°57'00"N

Royal Statistical Society (RSS)
12 Errol Street
London EC1Y 8LX
Telephone: (0)171-6388998
Telefax: (0)171-2567598
Electronic Mail: rss@rss.org.uk
WWW: http://www.rss.org.uk/
Founded: 1834
Membership: 6000
Periodicals: (12) "RSS News"; "Journal of the Royal Statistical Society - Series A, B, C and D", "Annual Report"
City Reference Coordinates: 000°10'00"W 51°30'00"N

Rutherford Appleton Laboratory (RAL)
• See "Council for the Central Laboratory of the Research Councils (CCLRC), Rutherford Appleton Laboratory (RAL)"

Safelight
Unit 889
Cowley Road
London W3 7YD
Telephone: (0)181-7401516
Telefax: (0)181-7401011
Electronic Mail: service@safelight.com
WWW: http://www.safelight.com/
Activities: multimedia presentations
City Reference Coordinates: 000°10'00"W 51°30'00"N

Salford Astronomical Society (SAS)
59 Vancouver Quay
Salford Quays
Manchester M5 2TU
Telephone: (0)1204-531411 x3476
Telefax: (0)1204-528768
Electronic Mail: salfordac@ast.man.ac.uk
WWW: http://www.salfordastro.org.uk/
Founded: 1967
City Reference Coordinates: 002°15'00"W 53°30'00"N (Manchester)
                                     002°16'00"W 53°30'00"N (Salford)

Salford Software Ltd.
Adephi House
Adelphi Street
Salford M3 6EN
Telephone: (0)161-8342454
Telefax: (0)161-8342148
Electronic Mail: sales@salford-software.com
                      support@salford-software.com
WWW: http://www.salford.ac.uk/ssl/
Founded: 1988
Staff: 12
• Software producer
City Reference Coordinates: 002°16'00"W 53°30'00"N

Salisbury Astronomical Society (SAS)
c/o Rita Collins
Mountains
3 Fairview Road

Salisbury SP1 1JX
Telephone: (0)1722-332892
Electronic Mail: ritacollins@compuserve.com
Founded: 1965
Membership: 25
Activities: observing (planets, Moon, occultations, deep sky) * computing
Periodicals: (12) "The Stargazer"
Coordinates: 001°55'00"W 51°02'00"N (Bishopstone)
001°50'00"W 51°10'00"N (Stonehenge)
001°39'00"W 51°05'00"N (Winterslow)
City Reference Coordinates: 001°55'00"W 51°02'00"N

Science and Technology Regional Organization (SATRO) - North Scotland
Marischal College
Broad Street
Aberdeen AB10 1YS
Scotland
Telephone: (0)1224-273161
Telefax: (0)1224-273160
Electronic Mail: info@setpoint.org.uk
WWW: http://www.setpoint.org.uk/
Founded: 1986
Membership: 12
Activities: creating an awareness in the community of the role of science and technology * promoting and enhancing science and technology education in schools and colleges
City Reference Coordinates: 002°04'00"W 57°10'00"N

Science History Publications Ltd.
c/o M.A. Hoskin (Director)
16 Rutherford Road
Cambridge CB2 2HH
Telephone: (0)1223-565532
Telefax: (0)1223-565532
Electronic Mail: shpltd@aol.com
Founded: 1969
Staff: 2
Periodicals: (5) "Journal for the History of Astronomy (JHA)" (ISSN 0021-8286, circ.: 750); (2) "Archaeoastronomy" (ISSN 0142-7253, circ.: 750)
City Reference Coordinates: 000°08'00"E 52°13'00"N
● Publisher

Science Museum
● See "National Museum of Science and Industry (NMSI), Science Museum"

Scottish Astronomers Group (SAG)
c/o Brian Kelly
Mills Observatory
Balgay Park
Glamis Road
Dundee DD2 2UB
Scotland
Telephone: (0)1382-667138
WWW: http://star-www.st-and.ac.uk/~fv/sag/sag.html
Founded: 1970
Membership: 50 + 11 affiliated societies
Activities: inter-society communication * lecture meetings * observing
Periodicals: (4) "SAG Newsletter"
City Reference Coordinates: 003°00'00"W 56°28'00"N

Sheffield Astronomical Society (SAS)
c/o Ken Eastburn (Treasurer)
5 Spinneyfield
Rotherham S60 3WH
Telephone: (0)14-2219601
Electronic Mail: arg@freenet.co.uk
WWW: http://www.saqqara.demon.co.uk/sas/sashome.htm
Founded: 1934
Membership: 45
Activities: promoting amateur astronomy
City Reference Coordinates: 001°20'00"W 53°26'00"N (Rotherham)
001°30'00"W 53°23'00"N (Sheffield)

Shropshire Astronomical Society (SAS)
c/o David Woodward
20 Station Road

Condover
Shrewsbury SY5 7BQ
Wales
Telephone: (0)1743-872991 (David Woodward, Chairman)
Electronic Mail: d.woodward@virgin.net
                 g.privett@astro.cf.ac.uk
WWW: http://www.astro.cf.ac.uk/sas/sasmain.html
Founded: 1996
Membership: 50
Activities: amateur astronomy
Periodicals: "Hermes"
City Reference Coordinates: 002°45'00"W 52°43'00"N

Sira Ltd., Electro-Optics Division
South Hill
Chislehurst BR7 5EH
Telephone: (0)181-4672636
Telefax: (0)181-4676515
Electronic Mail: info@sira.co.uk
                 marketing@siraeo.co.uk
WWW: http://www.sira.co.uk/
Founded: 1918
Staff: 220
Activities: R&D in instrumentation and control, optics, electronics, expert systems, engineering
Periodicals: (2) "Spotlight"
City Reference Coordinates: 000°04'00"E 51°25'00"N

Skeptic (The_)
● See "The Skeptic"

Skymap Software
9 Severn Road
Culcheth WA3 5ED
Electronic Mail: support@skymap.com
WWW: http://www.skymap.com/
Founded: 1990
● Software producer
City Reference Coordinates: 002°32'00"W 53°27'00"N

SOADAS Observatory Trust
c/o J.T. Harrison
92 Cottage Lane
Ormskirk L39 3NJ
Telephone: (0)1695-574516
Founded: 1980
Activities: promoting astronomy education
City Reference Coordinates: 002°54'00"W 53°35'00"N

Society for Interdisciplinary Studies (SIS)
10 Witley Green
Darley Heights
Stopsley LU2 8TR
Telephone: (0)113-2623865 (Editor)
Telefax: (0)113-2243557 (Editor)
Electronic Mail: iantresman@easynet.co.uk
WWW: http://www.knowledge.co.uk/xxx/cat/sis/
Founded: 1974
City Reference Coordinates: 000°29'00"W 52°08'00"N (Bedford)

Society for Popular Astronomy (SPA)
36 Fairway
Keyworth NG12 5DU
Electronic Mail: spastronomy@aol.com
WWW: http://www.popastro.com/
Founded: 1953
Membership: 2600
Staff: 14
Activities: quarterly meetings * outings * organisation for beginners of all ages
Sections: Aurora * Comet * Deep Sky * Lunar * Meteor * Occultation * Planetary * Solar * Variable Star
Periodicals: (4) "Popular Astronomy" (ISSN 0261-0892); (6) "News Circulars"
Awards: "Fred Best Award"
● Formerly "Junior Astronomical Society (JAS)"
City Reference Coordinates: 001°06'00"W 52°52'00"N

Society of British Aerospace Companies (SBAC) Ltd.
Duxbury House
60 Petty France
Victoria
London SW1H 9EU
Telephone: (0)171-2271000
Telefax: (0)171-2271067
Electronic Mail: post@sbac.co.uk
WWW: http://www.sbac.co.uk/
Membership: 350 (companies)
Periodicals: (1) "Annual Report"
City Reference Coordinates: 000°10'00"W 51°30'00"N

Solent Amateur Astronomers Astronomical Society
c/o Ken Medway
443 Burgess Road
Swaythling SO16 3BL
Telephone: (0)1703-582204
Founded: 1972
Membership: 75
Activities: practical astronomy (equipment, observatory construction) * occultations * monthly meetings
Periodicals: (12) "Newsletter"
Observatories: 2 (Toothill, Itchen)
City Reference Coordinates: 001°25'00"W 50°55'00"N (Southampton)

Southampton Astronomical Society (SAS)
124 Winchester Road
Shirley SO16 6US
Telephone: (0)1703-327952
Electronic Mail: hollies@tcp.co.uk (Michael Hobbs)
WWW: http://home.clara.net/lmhobbs/sas.html
Founded: 1924
Membership: 68
Activities: fostering interest in astronomy and related subjects * lectures * discussions * shows * visits * observing * telescope making
Periodicals: (6) "Newsletter"; (2) "Journal"
City Reference Coordinates: 001°48'00"W 52°24'00"N (Shirley)
                                001°25'00"W 50°55'00"N (Southampton)

South Coast Telescopes
3 Beacon Mount
Park Gate
Southampton SO31 7GN
Telephone: (0)1489-584192
             (0)976690928
Telefax: (0)1489-603318
Founded: 1996
Staff: 2
Activities: manufacturing mirror making kits
City Reference Coordinates: 001°25'00"W 50°55'00"N

South Downs Astronomical Society (SDAS)
c/o J. Green
46 Central Avenue
Bognor Regis PO21 5HH
Telephone: (0)1243-829868
WWW: http://www.port.ac.uk/sdas/sdas.htm
        http://www.port.ac.uk/sdas/sdpt.htm (South Downs Planetarium Trust)
Founded: 1971
Membership: 100
Activities: observing * popularization * circulation of literature
Periodicals: (12) "Newsletter"; (1) "Supernova"
City Reference Coordinates: 000°41'00"W 50°47'00"N

South East Kent Astronomical Society (SEKAS)
c/o Andrew McCarthy
25 St. Pauls Way
Sandgate
Folkestone CT20 3NT
Telephone: (0)1303-241933 (Andrew McCarthy)
             (0)1304-363719 (Chris Moore)
Electronic Mail: rperry@dial.pipex.com
WWW: http://dspace.dial.pipex.com/town/square/fd13/sekas1.htm
Founded: 1972
Membership: 110

Activities: public observing sessions * meetings
Periodicals: (4) "Eclipse"
City Reference Coordinates: 001°11'00"E 51°05'00"N

Southend Planetarium
c/o Central Museum
Victoria Avenue
Southend-on-Sea, Essex
Telephone: (0)1702-330214
WWW: http://www.southend.gov.uk/leisure/attract/planet.htm
Founded: 1994
Staff: 2
City Reference Coordinates: 000°43'00"E 51°33'00"N

Southern Area Group of Astronomical Societies (SAGAS)
31 Paignton Avenue
Copnor
Portsmouth PO3 6LL
Telephone: (0)2-392618113
Telefax: (0)2-392352460
Electronic Mail: pete@nightlife.demon.co.uk
WWW: http://www.port.ac.uk/sagas/
Founded: 1973
Membership: 15 (societies)
Activities: liaison between member societies
Periodicals: "SAGAS News Round-up"
City Reference Coordinates: 001°05'00"W 50°48'00"N

Southport Astronomical Society
c/o Patrick Brannon (Secretary)
Willow Cottage
90 Jacksmere Lane
Scarisbrick L40 9RS
Telephone: (0)1704-880027
Electronic Mail: brannonpm@aol.com
WWW: http://www.u-net.com/ph/nwgas/southpt/southpt.htm
Founded: 1986
Membership: 50
Activities: monthly meetings * lectures * observing
Periodicals: (4) "Sirius"
City Reference Coordinates: 002°56'00"W 53°37'00"N (Scarisbrick)
                                        003°01'00"W 53°39'00"N (Southport)

South Tyneside College, Planetarium and Observatory
Saint George's Avenue
South Shields NE34 6ET
Telephone: (0)191-4273589
Founded: 1964
Staff: 1
Activities: education * astronomy * navigation * school and public shows * public observing * teacher in service training * astronomy resource centre
Coordinates: 001°25'09"W 54°59'10"N (South Tyneside College Observatory)
City Reference Coordinates: 001°25'00"W 55°00'00"N (South Shields)

Space, Physics and the Advancement of Cosmic Exploration (SPACE)
c/o Physics Department
University of Kent at Canterbury
Canterbury CT2 7NZ
Electronic Mail: spacesoc@ukc.ac.uk
WWW: http://alethea.ukc.ac.uk/SU/Societies/SPACE/
         http://www.su.ukc.ac.uk/societies/space/
City Reference Coordinates: 001°05'00"E 51°17'00"N

Speedibrews
54 Lovelace Drive
Pyrford
Woking GU22 8QY
Telephone: (0)1932-346942
Telefax: (0)1932-346942
Electronic Mail: maunder@speedibrews.free-online.co.uk (Michael Maunder, Proprietor)
WWW: http://www.speedibrews.free-online.co.uk/
Founded: 1978
Staff: 3
Activities: manufacturing astrophotographic developers * supplying specialist films
City Reference Coordinates: 000°34'00"W 51°20'00"N

Stafford and District Astronomical Society (SDAS)
6 Elm Walk
Penkridge ST19 5NL
Telephone: (0)1785-712065
Founded: 1982
Membership: 20
Activities: lectures * observing * visits to astronomical places
City Reference Coordinates: 002°07'00"W 52°48'00"N (Stafford)

Stanley Thornes Ltd.
Ellenborough House
Wellington Street
Cheltenham GL50 1YW
Telephone: (0)1242-228888
Telefax: (0)1242-221914
Electronic Mail: cservice@thornes.co.uk
WWW: http://www.thornes.co.uk/
Founded: 1972
• Publisher
City Reference Coordinates: 002°04'00"W 51°54'00"N

Starlight Xpress
Briar House
Foxley Green Farm
Ascot Road
Holyport SL6 3LA
Telephone: (0)1628-777126
Telefax: (0)1628-580411
Electronic Mail: tplatt@starlight.win-uk.net (Terry Platt)
              mhattey@fdeltd.win-uk.net (Michael Hattey)
WWW: http://www.ibmpcug.co.uk/~starlite
       http://www.demon.co.uk/astronomer/sxadd/
Founded: 1991
Staff: 6
Activities: designing and constructing astronomical CCD cameras
City Reference Coordinates: 000°44'00"W 51°32'00"N (Maidenhead)

STARLINK Project
• See "Council for the Central Laboratory of the Research Councils (CCLRC), Rutherford Appleton Laboratory (RAL), STARLINK Project"

Stratford-upon-Avon Astronomical Society (SONAAS)
c/o Graham Fraser (Secretary)
2 Fining Court
Leamington Spa CV32 5FG
or :
c/o Robin Swinbourne (Secretary)
18 Old Milverton
Leamington Spa CV32 6SA
Telephone: (0)1926-316065
Telefax: (0)1926-316065
Electronic Mail: 100255.66@compuserve.com
                 ssmith@kidsons.compulink.co.uk
WWW: http://www.kidsons.co.uk/kidsons/sonaas
       http://www.compulink.co.uk/~kidsons
       http://www.cix.co.uk/~kidsons/
Founded: 1993
Membership: 50
Activities: monthly star parties * lectures * conventions
Periodicals: "Pegasus", "Equuleus"
City Reference Coordinates: 001°31'00"W 52°18'00"N (Leamington Spa)
                            001°41'00"W 52°12'00"N (Stratford-upon-Avon)

Surrey Satellite Technology Ltd. (SSTL)
c/o Surrey Space Centre
University of Surrey
Guilford GU2 7XH
Telephone: (0)1483-879278
Telefax: (0)1483-879503
Electronic Mail: m.sweeting@ee.surrey.ac.uk (M.N. Sweeting, Managing Director/CEO)
WWW: http://www.sstl.co.uk/
Founded: 1985
Staff: 50
Activities: spacecraft and ground equipment * designing, manufacturing and testing * launch services * short course, training and technology transfer packages
City Reference Coordinates: 000°35'00"W 51°14'00"N

Sutton Astronomical Society (MSAS) (Mansfield and_)
● See "Mansfield and Sutton Astronomical Society (MSAS)"

Swansea Astronomical Society
c/o Maurice Convey (Secretary)
132 Eaton Crescent
Uplands
Swansea SA1 4QR
Wales
Telephone: (0)1792-466814
Electronic Mail: sjw@astrabio.demon.co.uk (S.J. Wainwright, Webmaster)
WWW: http://www.swan.ac.uk/astra/starpage.htm
Founded: 1948
City Reference Coordinates: 003°57'00"W 51°38'00"N

Tavistock Astronomical Society
c/o Science Department
Kelly College
Tavistock PL19 0HZ
WWW: http://ourworld.compuserve.com/homepages/Mike_Loader/tavastro.htm
City Reference Coordinates: 004°08'00"W 50°33'00"N

Taylor & Francis Ltd.
1 Gunpowder Square
London EC4A 3DE
Telephone: (0)171-5830490
Telefax: (0)1715830581
WWW: http://www.tandf.co.uk/
Founded: 1798
● Publisher
City Reference Coordinates: 000°10'00"W 51°30'00"N

Telescope Technologies Ltd. (TTL)
1 Morpeth Wharf
Birkenhead CH41 1NQ
Telephone: (0)151-6503100
Telefax: (0)151-6503113
Electronic Mail: info@gnat.com
WWW: http://www.ngat.com/
Founded: 1995
Staff: 27
Activities: manufacturing telescopes and associated equipment
City Reference Coordinates: 003°02'00"W 53°24'00"N

The Astronomer
16 Westminster Close
Basingstoke RG22 4PP
Telephone: (0)1256-471074
Telefax: (0)1256-471074
Electronic Mail: guy@tahq.demon.co.uk
                        gmh@astrophysics.starlink.rutherford.ac.uk
WWW: http://www.demon.co.uk/astronomer
Founded: 1964
Membership: 400
Activities: observing projects * rapid publication of results * e-mail news service
Periodicals: (12) "The Astronomer"; "Discovery Circulars", "E-Mail Circulars"
Awards: (1) "Best Contributor to Magazine", "Best Contributor to Cover Photograph"
City Reference Coordinates: 001°05'00"W 51°15'00"N

The Astronomy Bookshop
2 Pell Hill Cottages
Wadhurst TN5 6DS
Telephone: (0)1892-783652
Electronic Mail: perfect@pobox.com
WWW: http://www.pobox.com/~perfect/astron.htm
Founded: 1992
Staff: 1
Activities: publishing and bookselling via the Internet
City Reference Coordinates: 000°21'00"E 51°01'00"N

The Astronomy Roadshow
167 Shakespeare Road
Gillingham ME7 5QB
Telephone: (0)1634-853741
                 (0)973-335544

Electronic Mail: astroroadshow@lineone.net
Founded: 1993
Staff: 2
Activities: operating mobile planetarium
City Reference Coordinates: 000°33'00"E 51°24'00"N

The Computer Journal
c/o Peter Hammersley
Middlesex Polytechnic
The Burroughs
Hendon
London NW4 4BT
Telephone: (0)181-2021487
              (0)181-3681299 x4219
Telefax: (0)181-2021539
Electronic Mail: p.hammersley@cluster.middlesex.ac.uk
                     editor@computer-journal.com
WWW: http://www.oup.co.uk/oup/smj/journals/ed/titles/computer_journal/
Founded: 1958
Staff: 2
● Journal (6) (ISSN 0010-4620)
City Reference Coordinates: 000°10'00"W 51°30'00"N

The Island Planetarium
Fort Victoria Country Park
Westhill Lane
Norton
Yarmouth PO41 0RR
Isle of Wight
Telephone: (0)1983-761555
Electronic Mail: paule@astrofvp.demon.co.uk (Paul England)
WWW: http://www.astrofvp.demon.co.uk/
Founded: 1992
City Reference Coordinates: 001°29'00"W 50°42'00"N

The Skeptic
P.O. Box 475
Manchester M60 2TH
Telephone: (0)7020-935370
Telefax: (0)7020-935372
Electronic Mail: edit@skeptic.org.uk
WWW: http://www.skeptic.org.uk/
Founded: 1987
● Journal (6) (ISSN 0959-5228, circ.: 600)
City Reference Coordinates: 002°15'00"W 53°30'00"N

Thornes Ltd. (Stanley_)
● See "Stanley Thornes Ltd."

Torbay Astronomical Society (TAS)
c/o Ian Walsh (President)
St. Brelades
Lummaton Cross
Torquay TQ2 8ET
Telephone: (0)1803-323600
Electronic Mail: tas@halien.net
WWW: http://www.halien.com/TAS/
Founded: 1956
City Reference Coordinates: 003°30'00"W 50°28'00"N

True Technology Ltd.
Woodpecker Cottage
Red Lane
Aldermaston RG7 4PA
Telephone: (0)1189-700777
Telefax: (0)1189-701661
Electronic Mail: truetech@dircon.co.uk
WWW: http://www.users.dircon.co.uk/~truetech
Founded: 1993
City Reference Coordinates: 001°09'00"W 51°23'00"N

UK Gemini Support Group (UKGSP)
● See "Oxford University, UK Gemini Support Group (UKGSP)"

UMI Ltd.
The Old Hospital
Ardingly Road
Cuckfield RH17 5JR
Telephone: (0)1444-445000
Telefax: (0)1444-445050
Electronic Mail: umi@umi.uk.com
WWW: http://www.umi.com/
Founded: 1938
Staff: 15
Activities: on-demand copies of dissertations, theses and out-of-print books * electronic databases * journal sin microform
● Publisher
City Reference Coordinates: 000°09'00"W 51°00'00"N

United Kingdom Industrial Space Committee (UKISC)
c/o A.G. Hicks (Secretary General)
P.O. Box 14
Wisbech PE13 1JZ
Telephone: (0)1945-64975
Telefax: (0)1945-61988
Electronic Mail: icks.ukisc@btinternet.com
WWW: http://www.ukspace.com/
Founded: 1975
Membership: 40
Activities: representing the collective interests of the UK space industry
Periodicals: (1) "Annual Report"
City Reference Coordinates: 000°10'00"E 52°40'00"N

United Kingdom Students for the Exploration and Development of Space (UKSEDS)
c/o Space Education Trust
Royal Aeronautical Society
4 Hamilton Place
London W1V 0BQ
Telephone: (0)1795-521784
Telefax: (0)1795-520880
Electronic Mail: info@uk.seds.org
WWW: http://www.uk.seds.org/
Founded: 1988
Periodicals: "Ecliptic"
City Reference Coordinates: 000°10'00"W 51°30'00"N

University College London (UCL), Department of Physics and Astronomy
Gower Street
London WC1E 6BT
Telephone: (0)171-3877050
Telefax: (0)171-3807145
Electronic Mail: <userid>@star.ucl.ac.uk
WWW: http://www.phys.ucl.ac.uk/
Founded: 1828
Staff: 30
Activities: education * research in UV, optical and IR astronomy * space sciences * upper atmosphere * planetary geology
Coordinates: 000°14'24"W 51°36'00"N H75m
City Reference Coordinates: 000°10'00"W 51°30'00"N

University College London (UCL), Department of Physics and Astronomy, Planetary Image Centre
33-35 Daws Lane
London NW7 4SD
Telephone: (0)181-9590421
Telefax: (0)171-3807145
Electronic Mail: <userid>@ps.ucl.ac.uk
WWW: http://www.ps.ucl.ac.uk/
        http://www.phys.ucl.ac.uk/
Founded: 1980
Activities: planetary geology * archive of planetary images
City Reference Coordinates: 000°10'00"W 51°30'00"N

University College London (UCL), Department of Physics and Astronomy, University of London Observatory (ULO)
Mill Hill Park
London NW7 2QS
Telephone: (0)181-9590421
Telefax: (0)181-9064161
Electronic Mail: <userid>@ulo.ucl.ac.uk
WWW: http://www.ulo.ucl.ac.uk/
        http://www.phys.ucl.ac.uk/

Founded: 1928
Staff: 12
Activities: interstellar spectroscopy * geological analysis of planetary surfaces * CP stars * education
Periodicals: "Communications of University of London Observatory" (ISSN 0458-2128)
Coordinates: 000°14'27"W 51°36'46"N H81m
City Reference Coordinates: 000°10'00"W 51°30'00"N

University College London (UCL), Mullard Space Science Laboratory (MSSL)
Holmbury
Saint Mary
Dorking RH5 6NT
Telephone: (0)1306-70292
            (0)1483-274111
Telefax: (0)1306-70201
       (0)1483-278312
Electronic Mail: <userid>@mssl.ucl.ac.uk
WWW: http://mssla3.mssl.ucl.ac.uk/
       http://www.mssl.ucl.ac.uk/
City Reference Coordinates: 000°20'00"W 51°14'00"N

University Microfilms International (UMI), Godstone
● See "UMI Ltd."

University of Bath, Astronomy Society
(AstroSoc)
c/o Union General Office
Bath University Students' Union
University of Bath
Bath BA2 7AY
Electronic Mail: su4as@bath.ac.uk
WWW: http://www.bath.ac.uk/~su4as/home.html
City Reference Coordinates: 002°22'00"W 51°23'00"N

University of Birmingham, School of Physics and Space Research, Astrophysics and Space Research Group
Edgbaston Park Road
Birmingham B15 2TT
Telephone: (0)121-4146453
Telefax: (0)121-4143722
Electronic Mail: <userid>@star.sr.bham.ac.uk
               bhvad::<userid>.starlink
               bison@bison.ph.bham.ac.uk (Birmingham Solar Oscillations Network - BiSON)
WWW: http://www.sr.bham.ac.uk/
       http://www.birmingham.ac.uk/physics/
       http://www.bham.ac.uk/physics/
       http://bison.ph.bham.ac.uk/ (BiSON)
Staff: 40
Activities: gamma-ray astronomy * X-ray astronomy * solar physics * space technology
Observatories: 1 (Wast Hills Observatory)
City Reference Coordinates: 001°50'00"W 52°30'00"N

University of Bradford, Faculty of Engineering, Engineering in Astronomy (EIA) Group
Bradford BD7 1DP
Telephone: (0)1274-234024
            (0)1274-234070
Telefax: (0)1274-236600
Electronic Mail: <userid>@bradford.ac.uk
WWW: http://www.telescope.org/
       http://www.eia.brad.ac.uk/eia.html
       http://www.eia.brad.ac.uk/rti/ (Bradford Robotic Telescope)
Founded: 1990
Staff: 12
City Reference Coordinates: 001°45'00"W 53°48'00"N

University of Bristol, H.H. Wills Physics Laboratory
Tyndall Avenue
Bristol BS8 1TL
Telefax: (0)117-9255624
Electronic Mail: <userid>@bristol.ac.uk
WWW: http://www.star.bris.ac.uk/
       http://www.bris.ac.uk/
Founded: 1876 (University)
Staff: 40
Activities: astrophysics * particle physics * polymer physics * microstructural physics * theoretical physics * health physics * liquid physics * ...

City Reference Coordinates: 002°35'00"W 51°27'00"N

University of Cambridge, Department of Applied Mathematics and Theoretical Physics (DAMTP)
Silver Street
Cambridge CB3 9EW
Telephone: (0)1223-337900
Telefax: (0)1223-337918
Electronic Mail: enquiries@damtp.cam.ac.uk
WWW: http://www.damtp.cam.ac.uk/
Founded: 1959
Staff: 50
City Reference Coordinates: 000°08'00"E 52°13'00"N

University of Cambridge, Institute of Astronomy (IoA)
The Observatories
Madingley Road
Cambridge CB3 0HA
Telephone: (0)1223-337548
Telefax: (0)1223-337523
Electronic Mail: <userid>@ast.cam.ac.uk
WWW: http://www.ast.cam.ac.uk/
         http://www.ast.cam.ac.uk/IOA/IOA.html
         http://www-xray.ast.cam.ac.uk/ (X-Ray Astronomy Group)
Founded: 1820
Staff: 70
Activities: stellar radial velocities * stellar spectroscopy * faint objects * cosmology * QSOs * AGNs * clusters * distant galaxies * galaxy * stars * compact objects * helioseismology * instrumentation
Periodicals: (1) "Annual Report"
Coordinates: 000°05'41"W 52°12'52"N H22m
City Reference Coordinates: 000°08'00"E 52°13'00"N

University of Cambridge, Isaac Newton Institute for Mathematical Sciences
● See "Isaac Newton Institute for Mathematical Sciences"

University of Central Lancashire, Centre for Astrophysics
Preston PR1 2HE
Telephone: (0)1772-892000
Telefax: (0)1772-892903
Electronic Mail: <userid>@uclan.ac.uk
WWW: http://sa1.star.uclan.ac.uk/
Founded: 1926
Staff: 10
Activities: interstellar dust * star formation * flare stars * solar physics * STP * close binaries * white dwarfs * active galaxies * observational cosmology * multiwaveband astronomy
Coordinates: 002°42'12"W 53°46'30"N H34m (Jeremiah Horrocks Observatory)
             002°35'30"W 53°48'06"N H63m (Alston Observatory)
City Reference Coordinates: 002°42'00"W 53°46'00"N

University of Durham, Department of Physics
South Road
Durham DH1 3LE
Telephone: (0)191-3742000
Telefax: (0)191-3743749
Electronic Mail: <userid>@durham.ac.uk
               physics.office@durham.ac.uk
WWW: http://www.dur.ac.uk/
         http://www.dur.ac.uk/~dph0www/
Staff: 30
Activities: observing * theoretical cosmology * cosmic rays * gamma rays * extragalactic astronomy * polarimetry * ISM * galactic structure * instrumentation * image sharpening * historical astronomy
City Reference Coordinates: 001°34'00"W 54°47'00"N

University of Edinburgh, Department of Mathematics and Statistics
King's Buildings
Edinburgh EH9 3JZ
Scotland
Telephone: (0)131-6505035
Telefax: (0)131-6506553
Electronic Mail: <userid>@ed.ac.uk
               maths@ed.ac.uk
WWW: http://www.maths.ed.ac.uk/
Staff: 1
Activities: stellar dynamics * globular star clusters * three-body problem * n-body simulations * dynamics of binary stars
City Reference Coordinates: 003°13'00"W 55°57'00"N

University of Edinburgh, Institute for Astronomy
Royal Observatory
Blackford Hill
Edinburgh EH9 3HJ
Scotland
Telephone: (0)131-6688356 (E. Gibson, Administrative Assistant)
            (0)131-6688100 (Switchboard)
Telefax: (0)131-6688416
Electronic Mail: <userid>@roe.ac.uk
WWW: http://www.roe.ac.uk/ifa/
Staff: 19
Activities: education * research * mobile planetarium
City Reference Coordinates: 003°13'00"W 55°57'00"N

University of Glasgow, Department of Aerospace Engineering
James Watt Building
Glasgow G12 8QQ
Scotland
Telephone: (0)141-3398855
Telefax: (0)141-3305560
Electronic Mail: <userid>@aero.gla.ac.uk
WWW: http://www.aero.gla.ac.uk/
Founded: 1952
Staff: 12
Activities: aeronautics and space technology * orbital dynamics * spacecraft guidance and control * mission analysis * solar sails * subcontractor for NASA, ESA and BNSC
City Reference Coordinates: 004°15'00"W 55°53'00"N

University of Glasgow, Department of Physics and Astronomy, Astronomy and Astrophysics Group
Glasgow G12 8QQ
Scotland
Telephone: (0)141-3305182
Telefax: (0)141-3305183
Electronic Mail: <userid>@astro.gla.ac.uk
WWW: http://www.astro.gla.ac.uk/
        http://www.physics.gla.ac.uk/Home.html
Founded: 1760
Staff: 7
Activities: solar flares * accreting binaries * jets * plasma astrophysics * cosmology * gravitational waves * statistical astronomy * solar and stellar spectropolarimetry
City Reference Coordinates: 004°15'00"W 55°53'00"N

University of Glasgow, Observatory
Acre Road
Glasgow G20 0TL
Scotland
Telephone: (0)141-9465213
Telefax: (0)141-3305183
Electronic Mail: <userid>@astro.gla.ac.uk
Founded: 1967
Staff: 3
Activities: education * stellar photometry and polarimetry * developing small optical instrumentation * planetarium
Coordinates: 004°18'20"W 55°54'08"N H53m
            004°24'13"W 55°56'21"N H150m
City Reference Coordinates: 004°15'00"W 55°53'00"N

University of Hertfordshire, Department of Physical Sciences, Division of Physics and Astronomy
College Lane
Hatfield AL10 9AB
Telephone: (0)1707-284602
Telefax: (0)1707-284644
Electronic Mail: phyqinfo@herts.ac.uk
                <userid>@star.herts.ac.uk
WWW: http://www.herts.ac.uk/natsci/Physics/PhysicsHome.html
Founded: 1968
Staff: 25
Activities: education * optical and IR astronomy
City Reference Coordinates: 000°13'00"W 51°46'00"N

University of Hertfordshire, Observatory
Bayfordbury
Near Hertford SG13 8LD
Telephone: (0)1707-285560
Telefax: (0)1992-503498
WWW: http://www.herts.ac.uk/
        http://www.herts.ac.uk/astro-ub

http://www.herts.ac.uk/natsci/Physics/Observatory.html
Founded: 1968
Staff: 20
Activities: education * spectroscopy * image processing * polarimetry * active galaxies * protostars
Coordinates: 000°05'34"W 51°46'28"N H100m
City Reference Coordinates: 000°05'00"W 51°48'00"N (Hertford)

University of Keele, Astronomical Society
c/o Department of Physics
Keele ST5 5BG
Telephone: (0)1782-583308
Telefax: (0)1782-583342
Electronic Mail: ae@astro.keele.ac.uk
WWW: http://www.astro.keele.ac.uk/astrosoc/home.html
Founded: 1960
Membership: 110
Activities: talks by leading astronomers * films * visits * observing
Periodicals: "Newsletter", "Sidereal Times"
City Reference Coordinates: 002°17'00"W 53°00'00"N

University of Keele, Department of Physics
Keele ST5 5BG
Telephone: (0)1782-583342
Telefax: (0)1782-711093
Electronic Mail: <userid>@astro.keele.ac.uk
WWW: http://www.keele.ac.uk/depts/ph/phhome.html
        http://www.astro.keele.ac.uk/ (Astrophysics Group)
Founded: 1952
Staff: 6
Activities: interacting binaries * cataclysmic variable stars * circumstellar dust
City Reference Coordinates: 002°17'00"W 53°00'00"N

University of Kent, Electronic Engineering Laboratory, Radio Astronomy Group
Canterbury CT2 7NT
Telephone: (0)1227-764000
Telefax: (0)1227-456084
Electronic Mail: <userid>@ukc.ac.uk
            star@star.ukc.ac.uk
WWW: http://www-star.ukc.ac.uk/
      http://stork.ukc.ac.uk/electronics/
Founded: 1969
Staff: 7
Activities: molecular-line astronomy * SIS receiver development
City Reference Coordinates: 001°05'00"E 51°17'00"N

University of Kent, Unit for Space Sciences and Astrophysics
Physics Laboratory
Canterbury CT2 7NR
Telephone: (0)1227-764000 x3788
          (0)1227-459616 (direct)
Telefax: (0)1227-762616
Electronic Mail: <userid>@ukc.ac.uk
WWW: http://wwwspace.ukc.ac.uk/
Founded: 1985
Staff: 25
Activities: comets (IR, visual) * space instrumentation * cosmic dust * minor planets * hypervelocity impacts * space debris * planetary surfaces
City Reference Coordinates: 001°05'00"E 51°17'00"N

University of Leeds, Department of Applied Mathematical Studies
Leeds LS2 9JT
Telephone: (0)113-2335110
Telefax: (0)113-2429925
               <userid>@amsta.leeds.ac.uk
WWW: http://www.amsta.leeds.ac.uk/applied/
Staff: 20
Activities: research * education * applied mathematics * fluid dynamics and computation (including astrophysics)
City Reference Coordinates: 001°35'00"W 53°50'00"N

University of Leeds, Department of Physics and Astronomy
E.C. Stoner Building
Woodhouse Lane
Leeds LS2 9JT
Telephone: (0)113-2333861
Telefax: (0)113-2336017

         (0)113-2333900
Electronic Mail: info@physics.leeds.ac.uk
               <userid>@ast.leeds.ac.uk
               <userid>@sun.leeds.ac.uk
WWW: http://www.leeds.ac.uk/physics
      http://ast.leeds.ac.uk/ (Astronomy Group)
      http://ast.leeds.ac.uk/haverah/hav-home.html (High Energy Cosmic Ray Group)
Founded: 1996 (1876 as "Department of Physics")
Activities: ultra-high-energy gamma-ray astronomy * globular clusters * molecular clouds
Coordinates: 001°35'00"W 53°50'00"N
City Reference Coordinates: 001°35'00"W 53°50'00"N

University of Leicester, Astronomy Society
● "Leicester University Astronomy Society"

University of Leicester, Department of Physics and Astronomy, Astronomy Group
University Road
Leicester LE1 7RH
Telephone: (0)116-2522073
Telefax: (0)116-2522070
Electronic Mail: <userid>@star.le.ac.uk
WWW: http://www.star.le.ac.uk/astron/
Founded: 1965
Staff: 6
Activities: QSOs * AGNs * IR astronomy * accretion in close binary systems * brown dwarfs
Coordinates: 001°06'00"W 52°37'00"N
City Reference Coordinates: 001°06'00"W 52°37'00"N

University of Leicester, Department of Physics and Astronomy, Radio and Space Plasma Physics Group
University Road
Leicester LE1 7RH
Telephone: (0)116-2523563
Telefax: (0)116-2523555
Electronic Mail: <userid>@ion.le.ac.uk
WWW: http://ion.le.ac.uk/
Staff: 35
City Reference Coordinates: 001°05'00"W 52°38'00"N

University of Leicester, Department of Physics and Astronomy, X-Ray Astronomy Group
University Road
Leicester LE1 7RH
Telephone: (0)116-2523494
Telefax: (0)116-2523311
Electronic Mail: <userid>@star.le.ac.uk
WWW: http://www.star.le.ac.uk/
      http://ledas-www.star.le.ac.uk (Leicester Database and Archive Service - LEDAS)
Staff: 80
Activities: X-ray astrophysics * space instrumentation * data centre for ROSAT and GINGA * XMM Survey Science Centre
City Reference Coordinates: 001°05'00"W 52°38'00"N

University of Liverpool, Department of Applied Mathematics and Theoretical Physics
Senate House
Abercromby Square
P.O. Box 147
Liverpool L69 3BX
Telephone: (0)151-7094012
Telefax: (0)151-7086502
Electronic Mail: <userid>@liverpool.ac.uk
WWW: http://www.liv.ac.uk/
Activities: celestial mechanics
● See also "Liverpool John Moores University"
City Reference Coordinates: 002°55'00"W 53°25'00"N

University of Liverpool, Department of Physics
P.O. Box 147
Liverpool L69 3BX
Telephone: (0)151-7943370
Telefax: (0)151-7943348
Electronic Mail: <userid>@ns.ph.liv.ac.uk
WWW: http://www.ph.liv.ac.uk/
Founded: 1881
Staff: 30
Activities: nuclear astrophysics * nuclear structure * particle physics * solid-state physics * magnetism * surface science
● See also "Liverpool John Moores University"

City Reference Coordinates: 002°55'00"W 53°25'00"N

University of London, Goldsmith's College, Department of Mathematical and Computer Sciences

New Cross
London SE14 6NW
Telephone: (0)181-6927171
Electronic Mail: <userid>@gold.lon.ac.uk
WWW: http://www-maths.gold.ac.uk/
Founded: 1905
Staff: 4
Activities: solar-system dynamics * three-body problem * statistics of minor-planet and comet distributions * early star formation * galactic dark clouds
City Reference Coordinates: 000°10'00"W 51°30'00"N

University of London, Queen Mary and Westfield College (QMWC), Astronomy Unit

Mile End Road
London E1 4NS
Telephone: (0)171-9755454
Telefax: (0)181-9819587
Electronic Mail: astronomy@qmw.ac.uk
WWW: http://www.maths.qmw.ac.uk/Astronomy/
        http://www-star.qmw.ac.uk/
        http://www.maths.qmw.ac.uk/~www/astro/home.html
Founded: 1983
Staff: 30
Activities: theoretical astronomy * observing * data analysis * space experiments * solar physics * magnetospheric physics * solar system * stars * galaxies * cosmology * gravitation * celestial and galactic dynamics * star formation
City Reference Coordinates: 000°10'00"W 51°30'00"N

University of London, Queen Mary and Westfield College (QMWC), Astrophysics Group

Mile End Road
London E1 4NS
Telephone: (0)171-9755555
Telefax: (0)181-9755500
Electronic Mail: <userid>@qmw.ac.uk
WWW: http://www-star.qmw.ac.uk/AstroGroup.html
Founded: 1890
Staff: 23
Activities: molecular clouds * star formation * starburst * IRAS * ISO * sub-mm detector and receiver systems * active galaxies * starburst galaxies * planetary atmospheres * IR and sub-mm astronomy
City Reference Coordinates: 000°10'00"W 51°30'00"N

University of London Observatory (ULO)

● See "University College London (UCL), Department of Physics and Astronomy, University of London Observatory (ULO)"

University of Manchester, Department of Physics and Astronomy

Oxford Road
Manchester M13 9PL
Telephone: (0)161-2754100
Telefax: (0)161-2754297
Electronic Mail: <userid>@ast.man.ac.uk
WWW: http://www.ph.man.ac.uk/physics/
        http://www.ast.man.ac.uk/ (Astronomy Group)
        http://www.jb.man.ac.uk/ (Jodrell Bank Obs. - see separate entry)
Staff: 9
Activities: galactic structure * galactic encounters * QSOs * active galaxies * SNR * novae * HII regions * HH objects * bipolar nebulae * pulsars
Periodicals: "Astronomical Contributions from the University of Manchester, Series III"
City Reference Coordinates: 002°15'00"W 53°30'00"N

University of Manchester, Department of Physics and Astronomy, Jodrell Bank Observatory

Macclesfield SK11 9DL
Telephone: (0)1477-571321
Telefax: (0)1477-571618
Electronic Mail: <userid>@jb.man.ac.uk
WWW: http://www.jb.man.ac.uk/
        http://www.jb.man.ac.uk/merlin/ (MERLIN)
Activities: radio-astronomy research * planetarium (see separate entry)
Coordinates: 002°18'26"W 53°14'10"N H78m (Jodrell Bank)
            002°32'03"W 53°09'22"N H47m (Darnhall)
            002°08'35"W 52°06'01"N (Defford)
            002°59'45"W 52°47'24"N (Knockin)
            002°59'46"W 53°06'45"N (Wardle)
            002°26'38"W 53°17'18"N (Pickmere)
            000°02'20"E 52°09'59"N (Cambridge)

● Konwn previously as "Nuffield Radio Astronomy Laboratories (NRAL)"
City Reference Coordinates: 002°07'00"W 53°16'00"N

University of Manchester Institute of Science and Technology (UMIST), Department of Physics, Astrophysics Group
P.O. Box 88
Manchester M60 1QD
Telephone: (0)161-2003677
Telefax: (0)161-2003669
Electronic Mail: <userid>@ast.ma.umist.ac.uk
WWW: http://saturn.ma.umist.ac.uk:8000/
Founded: 1983
Staff: 3
Activities: interstellar chemistry and physics * interstellar and interplanetary dust * circumstellar chemistry * evolution of interstellar clouds * star formation * chemical data * sub-mm observing * nucleation of dust particles
City Reference Coordinates: 002°15'00"W 53°30'00"N

University of Newcastle-upon-Tyne, Department of Physics
6 Kensington Terrace
Newcastle-upon-Tyne NE1 7RU
Telephone: (0)191-2227411
Telefax: (0)191-2227361
WWW: http://www.phys.ncl.ac.uk/
Founded: 1871
Staff: 16
Activities: quantum gravity * cosmology * laboratory astrophysics
Coordinates: 001°48'01"W 54°59'19"N H50m (Close House Observatory)
Awards: (1/2) "Robinson Prize"
City Reference Coordinates: 001°35'00"W 54°59'00"N

University of Nottingham, Department of Physics
University Park
Nottingham NG7 2RD
Telephone: (0)115-9515164
Telefax: (0)115-9515180
Electronic Mail: <userid>@nottingham.ac.uk
WWW: http://www.ccc.nottingham.ac.uk/
City Reference Coordinates: 001°01'00"W 52°58'00"N

University of Oxford, Department of Physics, Astrophysics
Nuclear and Astrophysics Laboratory
Keble Road
Oxford OX1 3RH
Telephone: (0)1865-273303
Telefax: (0)1865-273390
Electronic Mail: sec@astro.ox.ac.uk
                        <userid>@astro.ox.ac.uk
WWW: http://www-astro.physics.ox.ac.uk/
          http://www.physics.ox.ac.uk/
Founded: 1960
Staff: 12
Activities: observational and theoretical cosmology * galactic dynamics * high-energy astrophysics * stellar winds * chromospheres * coronae * stellar and laboratory spectroscopy * instrumentation
City Reference Coordinates: 001°15'00"W 51°46'00"N

University of Oxford, Department of Physics, Atmospheric, Oceanic and Planetary Physics
Clarendon Laboratory
Parks Road
Oxford OX1 3PU
Telephone: (0)1865-272933
Telefax: (0)1865-272923
Electronic Mail: <userid>@atm.ox.ac.uk
WWW: http://www-atm.atm.ox.ac.uk/
          http://www-atm.atm.ox.ac.uk/aopp/
Founded: 1920
Staff: 50
Activities: atmospheric, oceanic and planetary physics * planetatry and Earth observation * ocean modelling * education
City Reference Coordinates: 001°15'00"W 51°46'00"N

University of Oxford, Department of Physics, Theoretical Physics
1 Keble Road
Oxford OX1 3NP
Telephone: (0)1865-273999
Telefax: (0)1865-273947
Electronic Mail: <userid>@astro.ox.ac.uk

<userid>@ermine.ox.ac.uk
WWW: http://www.physics.ox.ac.uk/Theory.html
Activities: structure and evolution of galaxies * stellar dynamics * UV and X-ray spectroscopy * stellar chromospheres, coronae and winds
City Reference Coordinates: 001°15'00"W 51°46'00"N

University of Portsmouth, School of Computer Science and Mathematics, Relativity and Cosmology Group
Mercantile House
Hampshire Terrace
Portsmouth PO1 2EG
Telephone: (0)1705-843108
Telefax: (0)1705 843106
Electronic Mail: <userid>@sms.port.ac.uk
WWW: http://euclid.sms.port.ac.uk/cosmos/
Founded: 1994
Staff: 4
Activities: cosmology * relativity
City Reference Coordinates: 001°05'00"W 50°48'00"N

University of Saint Andrews, Astronomy Society
Saint Andrews KY16 9SS
Scotland
Electronic Mail: astrosoc@st-and.ac.uk
WWW: http://www.st-and.ac.uk/~www_sa/socs/astrosoc/
Founded: 1972
Membership: 60
City Reference Coordinates: 002°48'00"W 56°20'00"N

University of Saint Andrews, Mathematical and Computational Sciences Department, Solar Theory Group
Saint Andrews KY16 9SS
Scotland
Telephone: (0)1334-463716
Telefax: (0)1334-463748
Electronic Mail: <userid>@dcs.st-and.ac.uk
WWW: http://www-solar.dcs.st-and.ac.uk/
Founded: 1972 (University: 1410)
Staff: 30
Activities: research on theory of solar magnetic fields
City Reference Coordinates: 002°48'00"W 56°20'00"N

University of Saint Andrews, School of Physics and Astronomy
North Haugh
Saint Andrews KY16 9SS
Scotland
Telephone: (0)1334-463100
Telefax: (0)1334-463104
Electronic Mail: <userid>@star.st-and.ac.uk
physics@st-and.ac.uk
WWW: http://www.st-and.ac.uk/~www_pa/
http://star-www.st-and.ac.uk/astronomy/Welcome.html
Founded: 1939 (University: 1410)
Staff: 6
Activities: stellar astrophysics * accretion in AGN and binaries * extra-solar planets * undergraduate and graduate education
Coordinates: 002°48'54"W 56°20'12"N H30m
City Reference Coordinates: 002°48'00"W 56°20'00"N

University of Sheffield, Department of Physics and Astronomy
Hicks Building
Hounsfield Road
Sheffield S3 7RH
Telephone: 114-2223519
Telefax: 114-2728079
Electronic Mail: star@sheffield.ac.uk
WWW: http://www.shef.ac.uk/~phys/
http://www.shef.ac.uk/~phys/research/astro/ (Astronomy Group)
http://www.shef.ac.uk/~phys/research/pa/ (Particle Astrophysics)
Founded: 1883 (as "Department of Physics")
City Reference Coordinates: 001°30'00"W 53°23'00"N

University of Southampton, Physics and Astronomy Department
Highfield
Southampton SO17 1BJ
Telephone: (0)1703-592093

Telefax: (0)1703-593910
Electronic Mail: <userid>@astro.soton.ac.uk
WWW: http://www.astro.soton.ac.uk/
Founded: 1952
Staff: 6
Activities: gamma and hard X-ray astronomy * telescope development * observational astrophysics
City Reference Coordinates: 001°25'00"W 50°55'00"N

University of Sussex, Astronomy Centre
c/o School of Chemistry, Physics and Environmental Science
Falmer
Brighton BN1 9QJ
Telephone: (0)1273-606755
Telefax: (0)1273-678097
Electronic Mail: <userid>@star.cpes.susx.ac.uk
           <userid>@sussex.ac.uk
WWW: http://star-www.cpes.susx.ac.uk/
Founded: 1966
Staff: 20
Activities: stellar structure * stellar hydrodynamics * plasma astrophysics * cosmical magnetism * nucleosynthesis * star formation * galaxy formation, clustering and evolution * general relativity * cosmology * astro-particle physics * pulsars * cataclysmic binaries * surface imaging of stars
City Reference Coordinates: 000°08'00"W 50°50'00"N

University of Sussex, Science Policy Research Unit (SPRU)
Mantell Building
Brighton BN1 9RF
Telephone: (0)1273-686758
Telefax: (0)1273-685865
WWW: http://www.susx.ac.uk/spru/
Founded: 1966
Staff: 97
Activities: research * education
City Reference Coordinates: 000°08'00"W 50°50'00"N

University of Sussex, Space Science Centre
School of Engineering
Falmer
Brighton BN1 9QT
Telephone: (0)1273-678421
Telefax: (0)1273-678399
Electronic Mail: m.p.gough@sussex.ac.uk (Michael Paul Gough, Reader)
WWW: http://www.sussex.ac.uk/engg/research/space/homepage.html
Founded: 1970
Staff: 6
Activities: space physics * space instrumentation * particle correlation * wave-particle interactions * intelligent instrumention * fault tolerance * content-based data retrieval * data compression
City Reference Coordinates: 000°08'00"W 50°50'00"N

University of Teesside, School of Computing and Mathematics
Middlesbrough
Cleveland TS1 3BA
Telephone: (0)1642-342625
Telefax: (0)1642-230527
Electronic Mail: <userid>@tees.ac.uk
           scm@tees.ac.uk
WWW: http://wheelie.tees.ac.uk/
Activities: origin of solar system * numerical methods for dynamical astronomy
City Reference Coordinates: 001°14'00"W 54°35'00"N (Teesside)

University of Wales, Aberystwyth, Department of Physics
Ceredigion SY23 3BZ
Telephone: (0)1970-622813
Telefax: (0)1970-622826
Electronic Mail: <userid>@aber.ac.uk
WWW: http://www.aber.ac.uk/~dphwww/
Founded: 1884
Activities: solar-terrestrial physics * atmospheric physics * radio and space physics
City Reference Coordinates: 004°05'00"W 52°25'00"N

University of Wales, College of Cardiff (UWCC), Physics and Astronomy Department
P.O. Box 913
Cardiff CF2 3YB
Wales
or :

North Building
5 The Parade
Cardiff
Wales
Telephone: (0)1222-874458
Telefax: (0)1222-874056
Electronic Mail: <userid>@astro.cf.ac.uk
                    cardif::<userid>.starlink
WWW: http://www.astro.cf.ac.uk/
           http://www.cm.cf.ac.uk/
Founded: 1988
Staff: 18
Activities: galaxy observing * ST research * gravitational waves * general relativity * star formation theory * cosmic abundances * telescope design * galaxy simulations * cosmic magnetic fields
City Reference Coordinates: 003°13'00"W 51°13'00"N

University of Warwick, Physics Department, Space and Astrophysics Group
Coventry CV4 7AL
Telephone: (0)1203-524916
Telefax: (0)1203-692016
Electronic Mail: <userid>@warwick.ac.uk
WWW: http://www.astro.warwick.ac.uk/
City Reference Coordinates: 001°30'00"W 52°25'00"N

UoSAT Spacecraft Engineering Research Group
• See "University of Surrey, UoSAT Spacecraft Engineering Research Group"

Variorum
• See now "Ashgate Publishing Ltd."

Wadhurst Astronomical Society (WAS)
c/o Uplands Community College
Wadhurst TN5 6BA
WWW: http://users.aol.com/wadastro/
Founded: 1997
Membership: 40
Activities: meetings * observing
City Reference Coordinates: 000°21'00"E 51°04'00"N

Walsall Astronomical Society
c/o Alan Ledbury (Secretary)
71 Tame Street East
Walsall WS1 3LB
Electronic Mail: g.ledbury@cableinet.co.uk (Alan Ledbury, Secretary)
WWW: http://wkweb5.cableinet.co.uk/G.Ledbury/WASINDEX.HTM
        http://wemfas.future.easyspace.com/walsallas.html
Activities: meetings * observing
Founded: 1990
City Reference Coordinates: 001°58'00"W 52°35'00"N

Webb Society
c/o Don Miles
96 Marmion Road
Southsea PO5 2BB
Telephone: (0)1705-591146
Telefax: (0)1705-862466
Electronic Mail: rwa@ast.cam.ac.uk (R.W. Argyle, President)
WWW: http://www.webbsociety.org/
Founded: 1967
Membership: 450
Activities: education * publishing * annual meeting
Sections: Nebulae & Clusters * Southern Sky * Double Stars * Galaxies
Periodicals: (4) "Webb Society Quarterly Journal" (ISSN 0043-1680); (2) "Observing Section Reports" (ISSN 0953-8100), "The Deep-Sky Observer" (ISSN 0967-6139)
Awards: (1) "Webb Society Award"; "Webb Society Graphic Award", "Webb Society Technical Award"
City Reference Coordinates: 001°05'00"W 50°46'00"N

Welsh Border Astronomers (WBA)
c/o Pete Williamson
The Observatory
Top Street
Whittington SY11 4DR
Telephone: (0)1691-662076
Electronic Mail: pjw@wba.org.uk (Pete Williamson)
WWW: http://www.wba.org.uk/

Founded: 1996
Membership: 30
City Reference Coordinates: 003°00'00"W 52°52'00"N

Wessex Astronomical Society (WAS)
c/o Leslie A. Fry (Secretary)
Flat 7
816 Christchurch Road
Boscombe BH7 6DF
WWW: http://ourworld.compuserve.com/homepages/dstrange/was.htm
Founded: 1948
City Reference Coordinates: 001°50'00"W 50°44'00"N

West Midlands Federation of Astronomical Societies (WeMFAS)
c/o Andy Salmon
Olympus Mons
13 Jacmar Crescent
Smethwick B67 7LF
Telephone: (0)121-5654845
Electronic Mail: andy_salmon@compuserve.com
WWW: http://ourworld.compuserve.com/homepages/Michael_Bryce/WeMFAS.htm
Founded: 1995
Membership: 6
Activities: lectures * star parties * plnetarium shows * trips * acting as coordinating body for local space and astronomy societies in the Midlands
Periodicals: "Newsletter"
City Reference Coordinates: 001°58'00"W 52°30'00"N

West of London Astronomical Society (WOLAS)
c/o D. Radbourne (Secretary)
28 Tavistock Road
Edgware HA8 6DA
WWW: http://ourworld.compuserve.com/homepages/howard_bg/wolas.htm
Founded: 1967
Membership: 170
Activities: monthly meetings * observing
City Reference Coordinates: 000°16'00"W 51°36'00"N

West Yorkshire Astronomical Society (WYAS)
c/o Rosse Observatory
Carleton Grange
Carleton Road
Pontefract WF8 3RJ
or :
c/o Ken Willoughby
11 Hardistry Drive
Pontefract WF8 4BU
Electronic Mail: wyas@ndl.co.uk
WWW: http://staff.ndl.co.uk/terry/wyas/
Founded: 1973
Membership: 70
Activities: education * observing (Sun, planets, deep sky) * photography * computing
Periodicals: (12) "Phobos"
Coordinates: 001°17'14"W 53°40'34"N (Rosse Observatory)
Awards: (1) "Ralph Emerson Memorial Award", "Joynes Award", "Photographic Award"
City Reference Coordinates: 001°18'00"W 53°42'00"N

Wiley and Sons Ltd. (John_)
● See "John Wiley and Sons Ltd."

William Herschel Museum
● See "William Herschel Society"

William Herschel Society
19 New King Street
Bath BA1 2BL
Telephone: (0)1225-311342
Electronic Mail: comp@glam.ac.uk (Francis Ring, Chairman)
WWW: http://www.bath-preservation-trust.org.uk/museums/herschel/
Founded: 1976 (Museum: 1981)
Activities: managing William Herschel Museum * promoting work of William Herschel
Periodicals: (4) "Quarterly Bulletin"
City Reference Coordinates: 002°22'00"W 51°23'00"N

Wills Physics Laboratory (H.H._)
● See "University of Bristol, H.H. Wills Physics Laboratory"

Wolverhampton Astronomical Society
c/o Michael J. Bryce (Secretary)
16 Yellowhammer Court
Kidderminster DY10 4RR
Telephone: (0)1562-863545
Telefax: (0)1562-639329
Electronic Mail: michael_bryce@hotmail.com
WWW: http://wemfas.future.easyspace.com/was/was.html
Founded: 1951
Membership: 50
Activities: lectures * discussions * meetings twice a month
Periodicals: (6) "Lyra"
City Reference Coordinates: 002°14'00"W 52°23'00"N (Kidderminster)
                                                   002°08'00"W 52°36'00"N (Wolverhampton)

Worth Hill Observatory (WHO)
Seacombe House
Worth Matravers
Swanage BH19 3LF
Telephone: (0)1929-439210
Electronic Mail: 100614.3525@compuserve.com (David Granham Strange)
WWW: http://ourworld.compuserve.com/homepages/dstrange/
Founded: 1976
Staff: 1
Activities: private observatory * comets * SNs * dwarf novae * open to school groups and public observing with the Wessex
Astronomical Society (see separate entry)
Coordinates: 002°02'00"W 50°32'00"N H120m
City Reference Coordinates: 001°58'00"W 50°37'00"N

Worthing Astronomical Society (WAS)
15 Newham Lane
Steyning BN44 3LR
Telephone: (0)1903-814090
Electronic Mail: was@nquinn.demon.co.uk
WWW: http://www.nquinn.demon.co.uk/was/
Founded: 1965
Membership: 90
Activities: amateur astronomy * observing
Periodicals: (12) "WAS News"
Coordinates: 000°25'44"W 50°48'59"N H8m
City Reference Coordinates: 000°20'00"W 50°53'00"N (Steyning)
                                                   000°23'00"W 50°48'00"N (Worthing)

W.W. Norton and Co. Ltd.
10 Coptic Street
London WC1A 1PU
Telephone: (0)171-3231579
Telefax: (0)171-4364553
Founded: 1980
● Publisher
City Reference Coordinates: 000°10'00"W 51°30'00"N

Wycombe Astronomical Society (WAS)
3 Queens Court
Queens Road
High Wycombe HP13 6BA
Electronic Mail: paul@wolf359.demon.co.uk (Paul S. Scott, Chairman)
WWW: http://wycombeastro.org.uk/
Founded: 1981
Periodicals: (2) "Cygnus"
City Reference Coordinates: 000°46'00"W 51°38'00"N

York Astronomical Society (YAS)
c/o Martin Whipp (Secretary)
5 Whitham Drive
Huntington YO3 9YD
Telephone: (0)1904-763845
Electronic Mail: simonh@argonet.co.uk (Secretary)
WWW: http://www.argonet.co.uk/users/simonh/
Founded: 1972
Membership: 25

Activities: observing * talks *education * popularization
Periodicals: (4) "Algol"; "Skynotes"
City Reference Coordinates: 001°04'00"W 54°01'00"N (Huntington)
                                         001°05'00"W 53°58'00"N (York)

York Observatory
Museum Gardens
York YO1 7FR
Telephone: (0)1904-629745
Telefax: (0)1904-651221
Founded: 1831
Staff: 1
Activities: education * exhibitions * public observing
Coordinates: 001°05'12"W 53°57'40"N
City Reference Coordinates: 001°05'00"W 53°58'00"N